Second Edition

BIOACTIVE COMPOUNDS from NATURAL SOURCES

Natural Products as Lead Compounds in Drug Discovery

Second Edition

BIOACTIVE COMPOUNDS from NATURAL SOURCES

Natural Products as Lead Compounds in Drug Discovery

Edited by
Corrado Tringali

CRC Press
Taylor & Francis Group
Boca Raton London New York

CRC Press is an imprint of the
Taylor & Francis Group, an **informa** business

CRC Press
Taylor & Francis Group
6000 Broken Sound Parkway NW, Suite 300
Boca Raton, FL 33487-2742

First issued in paperback 2021

ISBN 13: 978-1-03-223581-3 (pbk)
ISBN 13: 978-1-4398-2229-6 (hbk)

Library of Congress Cataloging-in-Publication Data

Bioactive compounds from natural sources : natural products as lead compounds in drug discovery / editor, Corrado Tringali. -- 2nd ed.
    p. cm.
Includes bibliographical references and index.
ISBN 978-1-4398-2229-6 (hardback)
  1. Pharmacognosy. 2. Natural products. I. Tringali, Corrado, 1950- II. Title.

RS160.B55 2011
615.3'21--dc23
                                         2011023881

Visit the Taylor & Francis Web site at
http://www.taylorandfrancis.com

and the CRC Press Web site at
http://www.crcpress.com

Contents

Preface

The first edition of *Bioactive Compounds from Natural Sources* was published in the year 2000, in a period of renewed attention to biologically active compounds of natural origin. This followed two decades of declining interest in natural products, partly due to the expectations of the pharmaceutical industry toward new technologies such as combinatorial chemistry and high-throughput screening. The situation changed just at the beginning of the twenty-first century as a consequence of the observed reduction in the number of new molecular entities approved for drug marketing during the late twentieth century coupled with the increasing awareness of the unrivaled chemical diversity originated by nature through millions of years of evolution. This led to a new drive toward the study of bioactive compounds, both directly obtained from natural sources or derived/inspired by those obtained from natural products. Thus, natural products are again under the spotlight, in particular for their possible pharmacological applications, and this is the main reason for the publication of this second edition, largely focusing on natural products as lead compounds in drug discovery. This book may be considered as complementary to the first edition, which was devoted to the "isolation, characterization, and biological properties" of natural compounds, with the aim of connecting "lead discovery" to "lead optimization" in the field of bioactive natural products. This second edition is actually a completely new book, containing surveys of selected recent advances in an interdisciplinary area covering chemistry of natural products, medicinal chemistry, biochemistry, and other related topics. Some of the authors of the first edition have also contributed to this second edition with entirely new chapters, but the majority of chapters have been compiled by new authors who are among the most reputed scientists in the field.

Chapter 1 is an original contribution and introduces the reader to strategies and methods in the search for bioactive natural products; the following two chapters report on various natural sources of bioactive compounds such as aquatic cyanobacteria, filamentous fungi, and tropical plants (Chapter 2), or on the tremendous potentiality of metabolic engineering of natural product biosynthesis (Chapter 3). Two chapters are devoted to the contribution of emerging or developing technologies to the study of bioactive natural compounds, namely, computational methods (Chapter 4) and circular dichroism (Chapter 5). Some of the subsequent chapters highlight the potential of natural or natural-derived compounds for specific therapeutic applications: viral diseases (Chapter 6), regulation of hypoxia-inducible factors (Chapter 7), and modulation of angiogenesis in antitumor or wound-healing activity (Chapter 8). Further chapters focus on selected examples of natural product families and related synthetic analogues, namely, polyphenols (Chapters 9 and 10) and campthotecins (Chapter 11). Chapter 12 deals with natural inhibitors of cancer cell multidrug resistance. The search for bioactive compounds from locally available sources, combined with modern targets of pharmacological research, is discussed in Chapters 13 and 14, respectively, devoted to calmodulin inhibitors from Mexican plants and fungi and neuroprotective compounds from Chinese traditional medicine.

Chapter 15 deals with the antimalarial agents based on natural products, whereas Chapter 16 is devoted to endophytic microorganisms, an emerging source of new bioactive compounds. The importance of marine organisms as sources of peculiar natural products is highlighted in Chapter 17, a survey of the variety of products isolated from Australian marine organisms.

We hope that this book may prove valuable for graduate and PhD students of medicinal chemistry who are keen to explore the potential of bioactive natural products. Of course, the book also seeks to provide useful information for scientists working in various research fields where natural products may have a primary role.

Corrado Tringali
Department of Chemical Sciences
University of Catania
Catania, Italy

Editor

Professor Corrado Tringali was born in Catania, Italy, on July 25, 1950. He received his high school diploma in scientific studies in 1968 in Catania and graduated in chemistry in 1973 (110/110 cum laude) at the University of Catania. After a period as research fellow, he was appointed assistant professor of organic chemistry at the University of Catania, Faculty of Mathematical, Physical and Natural Sciences in 1977. In 1985, he became an associate professor of organic chemistry in the same faculty, and in 2001, he was finally made full professor of organic chemistry. At present, he is responsible for teaching courses in organic chemistry, natural products chemistry, and physical methods in organic chemistry and serves as the coordinator of the international Doctorate of Research in Chemical Science Program at the University of Catania (Cycles XXII–XXIV). He is also chairman of the Research Commission of the Faculty of Mathematical, Physical and Natural Sciences at the University of Catania.

Professor Tringali has trained in bioassay-guided purification methods at the laboratory of Professor K. Hostettmann (École de Pharmacie, Lausanne University, Switzerland, 1986) and in modern NMR techniques at the laboratory of Professor E. Breitmaier (Institute für Organische Chemie und Biochemie, Bonn University, Germany, 1987). His research interests include (a) isolation, structural determination, and/or NMR study of bioactive natural products from marine algae, fungi, and medicinal and food plants; and (b) chemical and enzymatic synthesis of analogues of bioactive natural products. He is currently involved in collaborative projects with various research teams and individuals in Italy and abroad, including Professor A. Oskarsson (University of Uppsala, Uppsala, Sweden), Professor N. Latruffe (Université de Bourgogne, France), Professor A. Di Pietro (Université de Lyon, France), Dr. G. Srinivas (Sree Chitra Tirunal Institute for Medical Sciences and Technology, Kerala, India), and Dr. H. Lin (National University of Singapore, Singapore) among others. Professor Tringali has authored or coauthored many original papers and contributions in international journals or books, including chapters for the following books: *Modern Methods of Plant Analysis*, vol. XII (1991); *Current Topics in Phytochemistry*, vol. 4 (2000); *Studies in Natural Products Chemistry*, vol. 28 (2003); *Targets in Heterocyclic Systems* (2007); and *Olives and Olive Oil in Health and Disease Prevention* (2010). He is regularly consulted as a referee by *Chemistry*; the *European Journal of Organic Chemistry*; *Tetrahedron Letters*; the *European Journal of Medicinal Chemistry*; *ChemMedChem*; the *Journal of Natural Products*; *Steroids*; the *Journal of Agricultural and Food Chemistry*; *Phytochemistry*, and other international journals and financial grant commissions.

Corrado Tringali enjoys photography and traveling, and his photos have been presented in various exhibitions and travel reportages.

Contributors

Cecilia Alsmark
Department of Medicinal Chemistry
Uppsala University
Uppsala, Sweden

Anders Backlund
Department of Medicinal Chemistry
Uppsala University
Uppsala, Sweden

Nina Berova
Department of Chemistry
Columbia University
New York, New York

Lars Bohlin
Department of Medicinal Chemistry
Uppsala University
Uppsala, Sweden

Ib Christian Bygbjerg
Department of International Health,
    Immunology, and Microbiology
University of Copenhagen
Copenhagen, Denmark

Robert J. Capon
Institute for Molecular Bioscience
The University of Queensland
St Lucia, Queensland, Australia

S. Brøgger Christensen
Department of Medicinal Chemistry
University of Copenhagen
Copenhagen, Denmark

Gordon M. Cragg
Division of Cancer Treatment and
    Diagnosis
National Cancer Institute at Frederick
Frederick, Maryland

Dominique Delmas
Laboratory of Biochemistry
    of Metabolism and Nutrition
University of Burgundy
Dijon, France

Attilio Di Pietro
Institut de Biologie et Chimie des
    Protéines
CNRS–Université Lyon 1
Lyon, France

George Ellestad
Department of Chemistry
Columbia University
New York, New York

Ernesto Fattorusso
Dipartimento di Chimica delle Sostanze
    Naturali
Università di Napoli Federico II
Napoli, Italy

Mario Figueroa
Departamento de Farmacia
Universidad Nacional Autónoma
    de México
México City, México

Gabriele Fontana
Research and Development
Indena Spa
Milano, Italy

Martín González-Andrade
Departamento de Farmacia
Universidad Nacional Autónoma
    de México
México City, México

Ulf Göransson
Department of Medicinal Chemistry
Uppsala University
Uppsala, Sweden

Nobuyuki Harada
Department of Chemistry
Columbia University
New York, New York

Ken-ichi Kimura
The United Graduate School of
    Agricultural Sciences
Iwate University
Iwate, Japan

Yoshiyuki Kimura
Department of Basic Medical Research
Ehime University Graduate School
    of Medicine
Ehime, Japan

A. Douglas Kinghorn
College of Pharmacy
The Ohio State University
Columbus, Ohio

Mattias Klum
Department of Medicinal Chemistry
Uppsala University
Uppsala, Sweden

Virginia Lanzotti
Dipartimento di Scienze degli Alimenti
Università di Napoli Federico II
Napoli, Italy

Norbert Latruffe
Laboratory of Biochemistry
    of Metabolism and Nutrition
University of Burgundy
Dijon, France

Gérard Lizard
Laboratory of Biochemistry
    of Metabolism and Nutrition
University of Burgundy
Dijon, France

Rachel Mata
Departamento de Farmacia
Universidad Nacional Autónoma
    de México
México City, México

Lucio Merlini
Dipartimento di Scienze Molecolari
    Agroalimentari
Università di Milano
Milano, Italy

Philippe Meunier
Institute of Molecular Chemistry of
    Burgundy–CNRS
University of Burgundy
Dijon, France

Dale G. Nagle
Department of Pharmacognosy
University of Mississippi
University, Mississippi

Koji Nakanishi
Department of Chemistry
Columbia University
New York, New York

David J. Newman
Division of Cancer Treatment and
    Diagnosis
National Cancer Institute at Frederick
Frederick, Maryland

Nicholas H. Oberlies
Department of Chemistry
    and Biochemistry
University of North Carolina
    at Greensboro
Greensboro, North Carolina

Jimmy Orjala
Department of Medicinal Chemistry
    and Pharmacognosy
University of Illinois at Chicago
Chicago, Illinois

Cedric J. Pearce
Mycosynthetix, Inc.
Hillsborough, North Carolina

Kangjian Qiao
Department of Chemical and
    Biomolecular Engineering
University of California, Los Angeles
Los Angeles, California

Isabel Rivero-Cruz
Departamento de Farmacia
Universidad Nacional Autónoma
    de México
México City, México

Judith M. Rollinger
Institute of Pharmacy/Pharmacognosy
    and Center for Molecular
    Biosciences Innsbruck
University of Innsbruck
Innsbruck, Austria

Yoshihito Shiono
Department of Food, Life, and
    Environmental Sciences
Yamagata University
Yamagata, Japan

Carmela Spatafora
Dipartimento di Scienze Chimiche
Università di Catania
Catania, Italy

Steven M. Swanson
Department of Medicinal Chemistry
    and Pharmacognosy
University of Illinois at Chicago
Chicago, Illinois

Orazio Taglialatela-Scafati
Dipartimento di Chimica delle Sostanze
    Naturali
Università di Napoli Federico II
Napoli, Italy

Xi Can Tang
Shanghai Institute of Materia Medica
Chinese Academy of Sciences
Shanghai, People's Republic of China

Yi Tang
Department of Chemical and
    Biomolecular Engineering
University of California, Los Angeles
Los Angeles, California

Corrado Tringali
Dipartimento di Scienze Chimiche
Università di Catania
Catania, Italy

Dominique Vervandier-Fasseur
Institute of Molecular Chemistry of
    Burgundy–CNRS
University of Burgundy
Dijon, France

Christina Wedén
Department of Medicinal Chemistry
Uppsala University
Uppsala, Sweden

Gerhard Wolber
Institute of Pharmacy, Pharmaceutical
    Chemistry
Freie Universität Berlin
Berlin, Germany

Xinkai Xie
Department of Chemical and
    Biomolecular Engineering
University of California, Los Angeles
Los Angeles, California

Hai Yan Zhang
Shanghai Institute of Materia Medica
Chinese Academy of Sciences
Shanghai, People's Republic of China

Yu-Dong Zhou
Department of Pharmacognosy
University of Mississippi
University, Mississippi

1 Strategies and Methods for a Sustainable Search for Bioactive Compounds*

Lars Bohlin, Cecilia Alsmark, Ulf Göransson, Mattias Klum, Christina Wedén, and Anders Backlund

CONTENTS

---

* Dedicated to Professor Emeritus Finn Sandberg on his 90th birthday.

1.1 INTRODUCTION

Mankind has, since ancient times, used nature in the search for food and bioactive organisms for use as poisons or to treat different diseases. The base for this are photosynthetic organisms on land and in the sea that have the capacity to, together with water and sunlight, capture carbon dioxide and use their enzymatic machinery to produce organic compounds. Change in the global climate has been predicted to affect the future society in different ways. Destruction of rain forests and pollution of the oceans are serious threats against the ecosystem and the access and sustainable use of natural resources both as food and future discovery of bioactive compounds. The most important source for such compounds has been medicinal plants. Today, their existence is threatened due to destruction of their natural habitat, but also because of the pressure on living space for specific ethnic groups and their traditional knowledge of how to use nature in their environment. However, a global awareness of biodiversity and the rights for specific ethnic groups have resulted in several international conventions with the purpose of protecting their natural resources but also to create new possibilities for the development of knowledge-based bioeconomy and ecosystem services in developing countries. It is an increased global interest to develop renewable natural products for industrial production and at the same time meet the demand for conserving biodiversity.

In this chapter, we want to emphasize the connection between biology and chemistry, discuss limitations in today's research, and suggest some strategies and methods for a sustainable search for bioactive natural products to secure molecules with natural origin as potential leads in drug development and also to reveal potential new targets in pathological organisms in nature.

1.2 BIOACTIVE NATURAL PRODUCTS

The enzymatic machinery of organisms in nature has developed under evolutionary pressure, resulting in a range of different biosynthetic pathways producing primary and, in particular, secondary metabolites with a large variety of basic skeletons and functional groups. The reason for this evolutionary trend is probably due to competition and coevolution between different organisms not only for nutrients and living space but also for communication, defense, synergism, and predation. In the search for bioactive compounds emphasis has, since the beginning of the nineteenth century, been to isolate pure components from the complex mixture of a bioactive extract. A typical example of this is the isolation of morphine by Sertürner, one of the first cases of bioassay-guided isolation of a pure bioactive natural product, more widely recognized a few years later (Sertürner 1806; Sertürner 1817a,b). For a long time after, natural product research focused on alkaloids, probably due to their often profound biological activity and relative ease with which they could be isolated. Other secondary metabolites, e.g., flavonoids and monoterpenes responsible, respectively, for color and scent, were considered to be waste products generally lacking important pharmacological activities and only occasionally explored for their supposed phylogenetic information (Harborne 1973, 1977). Today the scene has changed and many of these compound classes have received strong attention and have shown a diverse set of pharmacological activities. Despite this, we continue to neglect in research and literature some substance classes that, in many general schemes for natural products isolation, are recommended to be removed before any testing of biological activity, e.g., fatty acids and tannins. The usual argument, logical as it may seem, is that they disturb the outcome of some in vitro bioassays. However, some important discoveries are most probably missed and literally thrown away in this way. Furthermore, in natural products research, in general, and in pharmaceutical research, in particular, much focus has been on small molecules with molecular weight below 1000 that are soluble in hydrophilic solvents. The reason for this is presumably to a large extent purely historical, but is this appropriate? This question is addressed later in this chapter by Göransson who specializes in the field of study concerned with larger cyclic polypeptides, i.e., cyclotides. These polypeptides with amino acids in the range of 28–37 are produced in cocktails in a similar fashion to that of conotoxins known from the marine *Conus* snails. What could be the reason for some organisms to produce such complex mixtures of similar compounds? Both these classes of peptides, cyclotides and conotoxins, were discovered based on detailed clinical and ecological observation, respectively, and opened up for the understanding of new types of chemical structures. This (Göransson et al. 1999; Gray 1981) and other recent research also shows that it is still possible to discover new substance classes in nature such as quassidines exemplified by quassidine A (**1**) (Jiao et al. 2010), paltolides with paltolide A (**2**) (Plaza et al. 2010), and pycnanthulignenes with pycnanthulignene A (**3**) (Nono et al. 2010).

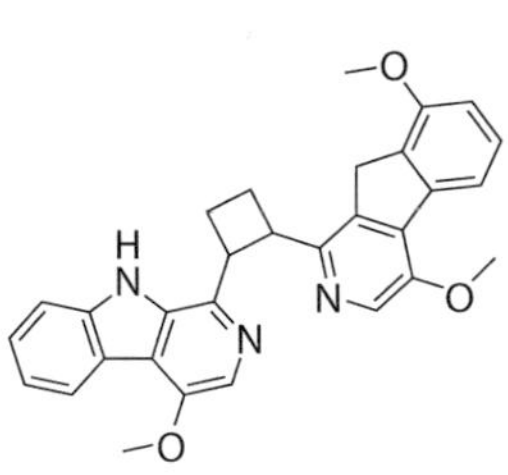

1 - Quassidine A

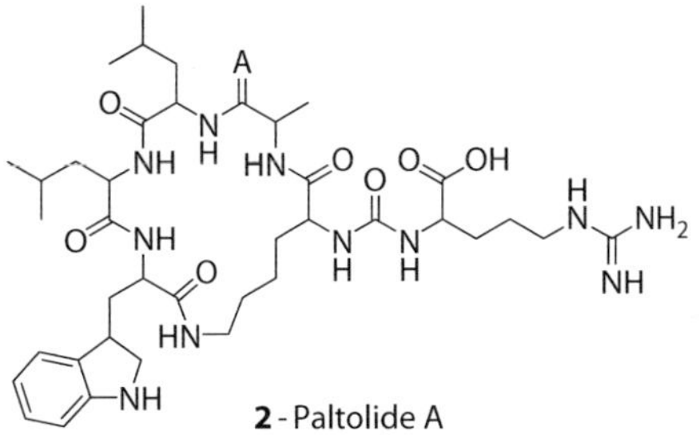

2 - Paltolide A

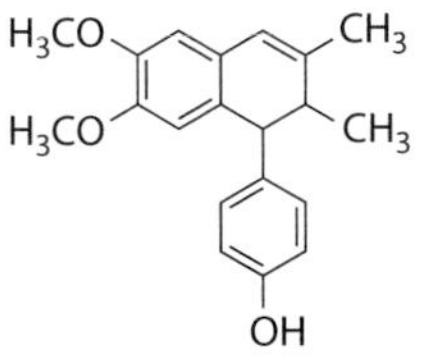

3 - Pycnanthulignene A

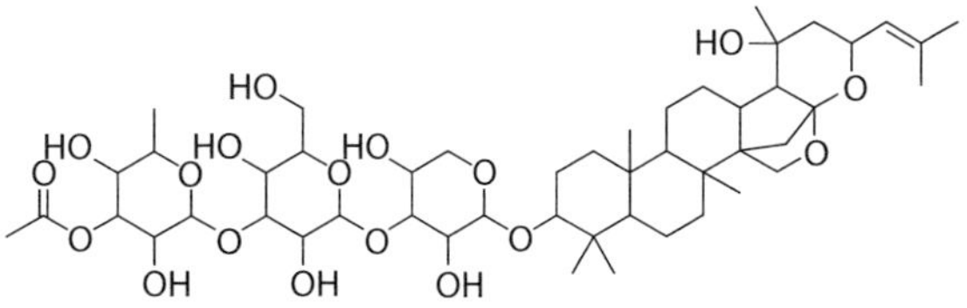

4 - Zizyphoiside

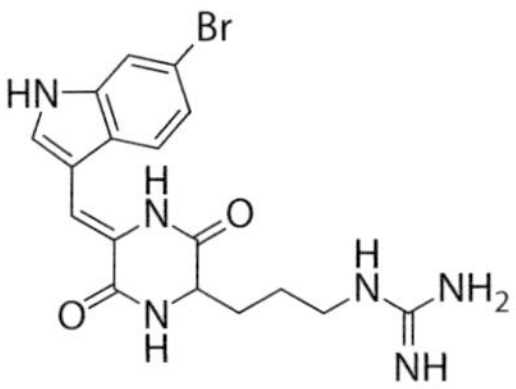

5 - Barettin

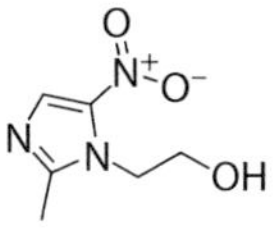

6 - Metronidazole

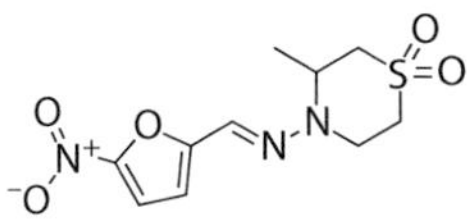

7 - Nifurtimox

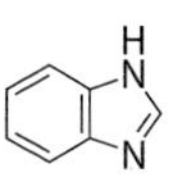

8 - Benzimidazole

Other very interesting questions are, more specifically, why are the secondary metabolites found initially produced, where in the organism, and by which organism? With the advancement of genomic research, new possibilities have opened up to study symbiotic bacteria, which in many cases has been shown to be the real producer of bioactive substances (Piel 2009). However, a very limited number of microorganisms in terrestrial and marine organisms have proven possible to produce in larger amounts in cultivation, which has influenced studies of the metagenome of unculturable microorganisms in environmental samples (Brady et al. 2009).

In addition, in a series of cases, it has been reported that endosymbionts, small and often highly specialized organisms (most often fungi or bacteria that live inside other organisms and sometimes inside their cells), are involved in biosynthesis of molecules earlier believed to be produced by the plant (Zhang et al. 2006). It is obvious that we are far from having enough knowledge about the biology of organisms in the field of natural product research. Furthermore, we do not only have groups of organisms that have not been explored, but the vast majority of species on Earth are not yet known to science! Wedén shows exciting possibilities in the bioprospecting of fungi later in the chapter, an often neglected group of organisms that are the evolutionary "sister group" of the animals. Fungi do not only pose difficult pharmaceutical problems due to their close evolutionary relation to man, they also exhibit a flabbergasting chemical creativity. Exploring this large yet unstudied wealth, the question leaps to mind if it is still possible to discover not only new substance classes but also new targets among organisms in nature? One of the authors, Alsmark, has developed modern bioinformatic tools and methods to address this question focused on potential new targets in pathogenic eukaryote parasites.

Traditionally, ethnopharmacological, ecological, or toxicological observations in nature have been the starting point for many research projects, as discussed later in this chapter. However, are some of these projects started due more to tradition or accessibility of organisms to study? How do we then select the organisms that we want to study? It is probably very important to have different starting points, but do we also need a theoretical platform and strategy for the selection of organisms? Backlund has taken this approach in his studies and discusses the interaction between biological diversity and chemical diversity using modern tools and chemography and phylogeny. Using a rigid theoretical framework for the selection and evaluation also eventually provides the basis for predictions of, e.g., biologic activity or chemical compound occurrences.

Despite all these advances in the natural sciences, it has also become of fundamental importance to contemplate the future limitations and possibilities in a globalized world where both terrestrial and marine organisms are threatened by extinction. Klum, with his broad experience of observation of different ecosystems, gives a view on the balance between the endangered biodiversity and the sustainable use and development of natural products in a future knowledge-based bioeconomy.

1.3 CHEMODIVERSITY AND THE IMPORTANCE OF DIFFERENCE

Contrary to the widespread misconception that modern drugs originate by a sort of "divine intervention" in a laboratory for organic synthesis, most drugs are based on the chemistry found in nature. This is also true for novel drugs or "new chemical entities" registered by FDA, a pattern that becomes even clearer considering their development during the last 25 years (Cragg and Newman 2009; Newman and Cragg 2007).

There are several possible explanations to this trend. One, as discussed already by Larsson and coworkers (Larsson et al. 2007), being that natural products have already been pre-validated and fine-tuned for biological systems by evolutionary forces during millions of years. Another could be that natural products biosynthesized from complex enzymatic pathways give us access to parts of chemical space to which traditional organic synthesis will not be able to take us.

In modern drug development, at major pharmaceutical companies we usually find a compound collection at the core of the research facilities. Such collections typically contain $1–2 \times 10^6$ compounds, which is close to the verge of what can be effectively managed. As we have knowledge of and at least hypothetical access to far more than 10×10^6 compounds and a hypothetical 1×10^{60} possible small drug-like carbon-based compounds, it is obvious that such collections require an intense curation to maintain a broad representation of the chemodiversity known.

1.3.1 ChemGPS-NP

One way to test and illustrate the chemodiversity of large compound sets can be, e.g., by global mapping of physical-chemical properties using ChemGPS-NP. In Figure 1.1, we can see two sample sets of compounds; 20,434 natural products (red) from the ZINC-NP database (zinc.docking.org) and 57,627 synthetic compounds (orange) from the often referred Maybridge's Screening Collection (www.maybridge.com).

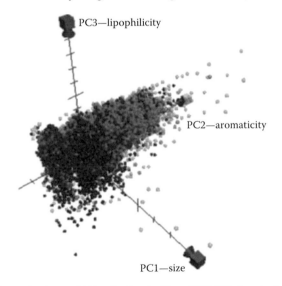

FIGURE 1.1   (See color insert.) Distribution in ChemGPS-NP chemical property space by natural products from ZINC-NP (20,434 compounds, red) and synthetic compounds from Maybridge Screening Collection (57,627 compounds, orange). The red axis (PC1) corresponds mainly to size parameters, the yellow axis (PC2) to aromaticity-related properties and the green axis (PC3) to lipophilicity. For more details on the descriptor loadings and influence in the different principal components please refer to Larsson et al. (2007).

From these two, comparably small, datasets it is already obvious that natural products and synthetic compounds exhibit fundamental differences in their physical-chemical properties. Several studies have addressed the nature of these differences (Backlund 2010). Not surprising, one of the most important features is the larger degree of complexity among the natural compounds. While it would be difficult to imagine the complexity of, e.g., palytoxin, most chemists deliberately strive to construct as "uncomplex" and clear-cut molecules as possible while retaining their biological activity and specificity. There are several obvious reasons to this mode of operation, one being that it will be easier to provide the substance, another that molecular docking and binding experiments and estimates will be much easier to perform.

The chemical diversity known and yet encountered in nature is, however woefully large, still only a minute fraction of all possible small carbon-based molecules estimated to at least 10^{60} (Bohacek et al. 1996). In this perspective, it may prove helpful to consider the concepts of *known, possible*, and *tangible* molecules as explored and discussed by, e.g., Oprea and coworkers (Hann and Oprea 2004).

As mentioned briefly earlier, natural products are biosynthesized by complex enzymatic machineries themselves results of evolutionary forces. These compounds as well as their machineries have been honed over millennia and tuned to a specific biological activity. Many natural products are large compared with the suggested limit of 500 g/mol for orally available drugs defined in Lipinski's "rule of five" (Ro5) (Lipinski et al. 1997). Despite this, a large proportion of these compounds appear to have a low "Lipinski Alert Index" (LAI, a parameter estimated from the complete set of Ro5) and hence suitable "drug-like" properties. Scrutinizing the parameters of natural products dataset indicates that the reason for the observed anomaly is that despite their larger size, natural products often have more biologically suitable lipophilicity-related properties than their man-made counterparts, thereby favoring their bioavailability.

1.4   CONCEPT OF BIODIVERSITY

Evident from surveys of known natural products and their related mother organisms is that there exists a correlation between available chemodiversity and biodiversity—the diversity of living organisms. Unfortunately, in many instances, there is a widespread belief that biological diversity is synonymous with the number of species. While a large number of species indeed do vouch for a measurable diversity, it is obvious for most people that there is a much larger biological diversity between the three species dandelion (*Taraxacum vulgare*), horse (*Equus caballus*), and golden chanterell (*Cantharellus cibarius*) than there is between the estimated 3200 species of the single genus *Hieracium* (hawksweeds). Even though the hawksweeds represent a larger number of species, the genetic differences between these are several orders of magnitude smaller than those between the horse and the dandelion—with the chanterell falling between these. Thus the proper method of estimating biological diversity is not found in the so popular "number of species" but instead invokes application of phylogenetic measures such as evolutionary (or more simplified genetic) distances. As we transverse the proverbial "Tree of Life" we can notice that the rate at which evolution proceeds varies greatly, such that even two different groups within the same genus can show different patterns. This was convincingly demonstrated already in 2001 by Richardson and coworkers (Richardson et al. 2001) who studied the plant genus *Phylica* found on the South African mainland and several islands in the southern Atlantic and Indian oceans. From a thorough phylogenetic analysis, they showed that the two main groups of the genus exhibited such differences in mutation rates that they would influence their estimate of the two groups' age, respectively. If such differences are not taken into account, we will not be able to measure "true" biological diversity, but rather the diversity of "characters as human happens to notice." This is in part connected also to recent discussions on so-called cryptic biodiversity that we must use our powers of observation to see beyond what is immediately apparent.

1.4.1 CHEMICAL AND BIOLOGICAL DIVERSITY

That biological and chemical diversity are related might appear intuitive and simple, but has nevertheless been challenged by some authors. If, however, chemical diversity is defined as a multidimensional volume defined by differences in physical-chemical property space of the compounds, it follows by necessity that adding additional compounds can never decrease the volume—only maintain or expand it further. One example where there is a very strong relation between biological and chemical diversity is that of the sesquiterpene lactones. In the series of graphs (Figure 1.2a through m), it is demonstrated step by step how expansion of the biological groups included results in an expansion of the multidimensional chemical space volume claimed. In this particular case, it is also shown that in some instances may the chemical diversity encountered be highly unevenly distributed over the phylogenetic tree.

With this understanding, it is important to remember that even between the three obviously different exemplar species mentioned earlier, the dandelion, horse, and chanterell, we only find a small portion of the global biological diversity represented. Hence, when discussing preservation of biodiversity, it is increasingly important to remember that this relates not only to the well-known organisms that we can see, but also to ensure the preservation of the entire ecosystem with all their components from bacteria to mammals, so-called *cryptic biodiversity* (Esteban and Finlay 2010).

1.5 SUSTAINABLE USE OF NATURAL RESOURCES

At present, under pressure from overexploitation and global climatic changes, species are becoming extinct at a rate surpassing that of the great Triassic extinction (Cuarón 1993; Jin et al. 2000). IUCN estimates that numerous species of larger animals and plants are disappearing every day, at an estimated rate of 0.1%–0.01% per year, and that there is right now among the species we know another 17,291 that are estimated to be critically endangered or at the verge of extinction (IUCN 2009) (Figure 1.3).

Among, probably, the most important threats against terrestrial ecosystems today we find major deforestation in Southeast Asian, Amazonian, and central African rain forests. All of these processes will eventually, in a serious way, reduce the Earth's biodiversity and, hence, the available chemodiversity. One of the worse aspects of these trends is not only that species become extinct, but also that they do so before we even have a chance to know that they existed. In the worst-case scenario, this invokes

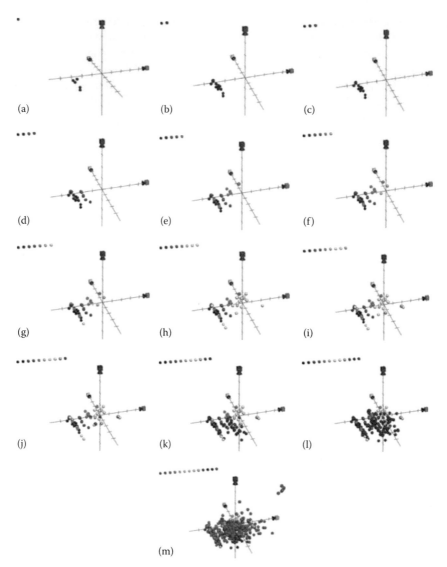

FIGURE 1.2 **(See color insert.)** In this figure, we can follow the expansion of sesqui-terpene lactones in ChemGPS-NP chemical property space as a function of the number of major evolutionary groups added with: a = fungi, b = liverworts, c = conifers, d = basal magnolioids, e = monocotyledons, f = rosids, g = lamids, h = apioids, i = Asteraceae—subfamily Mutisioideae, j = Asteraceae—Gochnatioideae, k = Asteraceae—Carduoideae, l = Asteraceae—Chicorioideae, and m = Asteraceae—Asteroideae.

not only the loss of a living entity that may have been refined and honed through evolutionary processes for millions of years, but also that this species might harbor valuable potentials for mankind. One striking example to this problem can be seen in the genus *Papaver* (poppies) where there are circa 50 species worldwide. Although they all produce alkaloids, and some produce alkaloids of the morphine-type, there

(a)

(b)

FIGURE 1.3   (a) The tropical rain forest of Southeast Asia is still among the most diverse biomes known on Earth. Parallel to the still largely unappreciated biological diversity, this is also a source for a likewise unexplored chemical diversity. (b) **(See color insert.)** In the mosaic of the rain forest, we find not only the sky-scraping trees and other green plants, but also ponds and streams harboring thousands of species, all uniquely adapted to their ecological niche. (Photos by Mattias Klum.)

is only one of all these species (and of *all* organisms yet studied on Earth) that is known to perform the last step in the biosynthesis and form morphine—which has been used for thousands of years and is still one of the most valuable drugs known to mankind. If we continue to destroy natural habitats and allow species to go extinct before they are known, we will never be able to know what we might have missed. Unfortunately the problem grows even bigger when taking into account the often-marginalized problem that while destroying biological diversity, and thus losing chemical diversity, we also lose cultural diversity. As the tropical forests of Borneo and the Amazonas shrink, so do the possibilities for indigenous cultures to retain

FIGURE 1.4 Despite their global importance to mankind, the tropical rain forest is today under siege in many areas. In this view, we can see how the wide plains in the background has already been exploited by plantations of oil palm, *Elaeis guineensis*, and how the plantation area is enlarged, extending into the hills. Such plantations transform the biodiversity-rich rain forest into a barren, yet palm oil–productive, monoculture. The use of palm oil in various alimentary processes is still increasing, despite several recent alarming reports of health risks (e.g., Leong et al. 2009; de Moraes Mizurini et al. 2010). (Photo by Mattias Klum.)

their mode of life, and with this their knowledge of the possible uses of plants, fungi, and animals in their environment (Figure 1.4).

While these issues, global processes and senseless destruction for short-term economic benefits, remain a major problem, it is important to not become an eco-fundamentalist. There are many people that are depending on forestry and agriculture for their income, and we must find a balance between preserving and using nature—today, usually referred to as a "sustainable use." In the widely accepted definition of sustainable development from the report of the Brundtland commission, *Our Common Future*, published in 1987 we find it very clearly stated that sustainability requires a holistic approach to problems, ranging from conservation biology to politics—and not excluding natural products chemistry. Issues with particular interest in our field relate to the CITES and the legal and intellectual rights that protect and occasionally prevent, e.g., the collecting and sampling of material in the field. Despite the annoyance a researcher might experience upon these restrictions, it becomes increasingly important to adhere to the applicable rules and regulations. By doing so we take our part in paving the way for a development of a bioeconomy or so-called ecosystem services, where natural resources are valued in another perspective than the few euros paid for the material to a teak table or the last few rhinoceros horns. As has already been shown in several places, the crucial point in nature conservation and development of sustainable use is when the people living in, or by, a rain forest realize that the forest is more worth standing there. And this is so because tourists and researchers pay for going there to see and study—just because it has not yet been cut down and replaced with palm oil plantations. Hence, to support this in the long run beneficial development, we have to take the responsibility of trying not only to solve our own scientific problems but at the same time contributing toward addressing other issues as well. How can we expect to gain respect among these groups of people if we do not value their assets—and engage our research also in fields relevant to them.

1.6   ETHNOPHARMACOLOGICAL APPROACHES

Medicinal plants used in traditional medicine have been during long time the starting point for many research projects. Most of the developed countries have forgotten or neglected their traditions concerning the use of medicinal plants. For many developing countries, the situation is the opposite where the only option is to use natural resources to treat their diseases. The knowledge is usually by tradition orally transferred from generation to generation, which is the case, e.g., in West Samoa (Cox 1993). However, the traditional medicine in developing countries is not homogeneous. Some healers act as magicians and believe that the cause of the disease can be cured by soothing the spirits by using decoctions of plants together with other more physical exercises. Other healers that act as herbal doctors believe that some organisms in nature contain properties that can be used to treat different diseases (Samuelsson 1987). It is obvious but not always the case in research projects based on ethnopharmacology that there must be a respectful relationship between the healer and the scientist coming from a developed country. These types of projects have been performed in different ways. Many questions that do not get enough attention are the diagnoses of the disease, how the plant is extracted, which dose, the form of administration, and the period of treatment. A trained clinical pharmacologist who observes the use of a medicinal plant could probably answer these questions and observe if the treatment has any effect on the particular disease or results in adverse effects. If possible, it would be of great value if these observations could be followed up by a placebo-controlled clinical study in the same country (Figure 1.5).

1.6.1   BIOLOGICAL TESTING

To discover bioactive molecules in natural sources, biological test methods (bioassays) must be used, which serve different purposes. As a first step, a bioassay is used to detect the presence of bioactivity in, e.g., plant tissue or an extract of a microbial broth, plant, or marine organism. For the isolation of a pure bioactive compound,

FIGURE 1.5 **(See color insert.)** In the rain forest, there still live people with an immense knowledge of possible uses of the plants and animals around them. In this picture we see chief Tebaran demonstrating two of the more than 60 species with specific uses that he collected within a radius of 2 m in a randomly selected spot in the Danum valley of Borneo. (Photo by Mattias Klum.)

the fractionation process must be guided by test methods that can detect the bio-activity that is related to the observed effect of the extract. Hyphenated methods, where chromatography is connected to bioassay using microliter plate in vitro, can identify specific bioactive peaks in a chromatogram, but larger amounts are needed for the evaluation of the preliminary biological characterization of a pure isolated substance. For an in-depth understanding of chemically well-defined bioactive molecules, an initial biological screening based on only one test method is not enough. A secondary bioassay can be useful to determine which hits from the initial screening are important for further studies and also to understand their mode of action. In the first screening, it is important to have the capacity to handle a large number of samples and relatively fast answers, while in a secondary or even tertiary assay, a clear strategy using specific experimental protocols and pharmacological tools and time for analyses is needed. Vollmar and Dirsch (Leikert et al. 2002) have shown that such a strategy can give results that can be published in well-known scientific journals not usually containing scientific results based on natural product research. In an earlier publication by Claeson and Bohlin (1997), different types of bioassays were discussed: single-target specific assays and multi-target functional assays. The first category has a high degree of specificity, resulting in a specific mode of action, e.g., receptor affinity or enzyme inhibition. The multi-target func-tional assay includes biological testing on whole animals, isolated organs or intact cells where the observed activity can be related to several possible levels of interaction. The selection of a bioassay for a specific study, e.g., of medicinal plant, should be based on the aim of the research project where the single target assay has the capacity to identify molecules with novel chemical structure but with known biological activity. To find compounds that act on unknown biological targets, multi-target functional bioassays are needed. Even if such a natural product is known, the unique structure–activity relationship of biological novelty of the product is very important and possible to protect in a patent procedure (Claeson and Bohlin 1997).

For successful natural product discovery programs, much more attention must be given to secure a high quality of the test material from collected organisms, which can be achieved by using new bioinformatic models described earlier in this chapter. The same approach can also be used for an in-depth understanding of the interpretation of bioassay results and further selection of organisms and targets. Further attention is also needed to develop strategies for efficient combination of bioassays oriented for a better understanding of the reaction of the material being tested. Very important for the development of possible treatment of diseases in developing countries using the local resources is access to relevant test models. Recently, UNIDO took the initiative to conduct an expert group meeting with the aim to consolidate the advances in and need for such bioscreening techniques that resulted in a practical manual (Bohlin and Huss 2006).

1.6.2 Different Strategies

A few examples of our own ethnopharmacological studies are given here, which represent different approaches from different ethnic groups around the world.

An observation of the use of arrow poison in Central Africa started a project with the aim to understand the botanical, chemical, and pharmacological aspects of the use of the *Erythrophleum* and *Strychnos* species. This led to several inventories and collections of different species and a careful screening of extracts using in vivo test models, e.g., Hippocratic screening in rats and mice. Muscle relaxant active extracts were further separated and tested on relevant organ models, e.g., rat diaphragm muscle for bioassay-guided isolation of pure indole alkaloids with muscle relaxant properties (Bohlin 1978). Further structure–activity studies on the cell level were performed with synthesized strychnine analogs and their mode of action (Eckernäs et al. 1980).

This type of screening project was later, in our research, changed to more specific medicinal plants, which had an ethnopharmacological history combined with clinical observations.

Extract from the fern *Polypodium decumanum* in Honduras called *Calaguala* was reported to be used orally against psoriasis, both traditionally and in capsules sold in pharmacies. As clinical studies of this extract were also published, some controlled double-blind studies where *Calaguala*-treated patients with psoriasis showed improvement; this gave a strong motivation to start chemical and pharmacological studies on this plant. The results are described in several publications. However, it is important in this context to note the difficulty in observing the effect of *Calaguala* treatment in man due to lack of an in vivo psoriatic model. Instead, we focused on a presumed immunological reason for psoriasis and used a skin grafting model in mice where the rejection time of skin transplant was compared between *Calaguala* extract and cyclosporin (Tuominen et al. 1991). In a further bioassay-guided isolation of pure compounds, the effect on the exocytosis of elastase from human neuthrophiles and the influence on the inflammatory mediators leucotriene B4 and platelet-activating factor were measured in vitro (Vasänge et al. 1997a, b; Vasänge-Tuominen et al. 1994; Tuominen et al. 1992).

The plant *Ipomoëa pes-caprae*, called *Pak bung ta lae* in Thai traditional medicine, is used to treat different degrees of skin damages caused by contact with poisonous jellyfishes. A lipohilic extract was produced from the leaves and a cream (IPA) was tested on patients after contact with jellyfishes. The treatment resulted in significantly less pain, inflammation, and even scar formation. In our further research, two in vivo rat models were used to evaluate the effect of the plant extract. A significant anti-inflammatory activity was observed both in carrageenan-induced rat paw edema and arachidonic acid–induced mouse ear edema. Further isolation was then followed using both in vivo models but also in vitro observing the inhibition of the enzyme cyclooxygenase involved in production of the inflammatory mediator prostaglandin resulting in a number of bioactive compounds (Pongprayoon et al. 1991).

To get access to the true traditional knowledge in different ethnic groups, it is necessary to have a deep respect for the traditions and to have the confidence from the local healers. In West Samoa we collaborated with ethnobotanist Paul Cox who can not only speak their language and as a botanist could identify the plants used for different diseases, but also is highly respected by the local people. The inner bark of the rain forest tree *Alphitonia zizyphoides* is important for treatment of different inflammatory disorders. Both muscle relaxant activity and anti-inflammatory activity were shown in our screening and resulted in both known and novel chemical compounds with different biosynthetic origin, such as zizyphoiside A (**4**) (Dunstan et al. 1998). However, in a screening of extracts and fractions on different cells, a novel effect was discovered by one of the fractions, namely, an enhancement of plating efficiency of lymphocytes and bone marrow cells in culture (Dunstan et al. 1994).

A last example from our own culture shows the importance of old literature. In the middle of the eighteenth century, Linnaeus wrote about the use of the rhizome of *Menyanthes trifoliata*, for treatment of nephritis and arthritis. Some years ago we were informed about the use of a decoction of the rhizome of this plant for treatment of severe glomerulonephritis. In a pharmacological study, we could show that a decoction of fresh rhizomes had a significant inhibitory effect on induced ischemia in a renal inflammation rat model (Tunon et al. 1994). The common use of this plant has, for a long time, been as a tonicum prepared from the leaves and not the rhizome. Thus the combination of information from old literature together with personal experience confirmed the old observation using a modern pharmacological model in vivo.

This type of research could be seen as a classical natural product research resulting in a number of bioactive molecules with both known and novel chemical structures. Do we really need this type of research today? Should we continue with screening of larger and larger number of extracts? Is it important to develop high-speed automated hyphenated methods combining isolation, detection, and bioassay with the purpose to identify more and more molecules?

In the following section, a new class of molecules are presented that were discovered based on clinical observations of the use of a medicinal plant.

1.7 FROM TRADITIONAL MEDICINE TO NOVEL BIOTECHNOLOGY

Exploring traditional medicine is a proven concept for the discovery of new drugs and new chemical entities, but can leads from ethnopharmacology play a role also in today's pharmaceutical and medical research for which biotechnology and peptide- and protein-based drugs become more and more important? Undeniably, natural products research has been biased toward low molecular weight, and relatively hydrophobic, compounds (Sticher 2008) by tradition in the pharmacognosy laboratory and the isolation procedures used therein. For example, in most extraction protocols apolar or medium polar organic solvents are used (e.g., petroleum ether, chloroform, ethylacetate, and methanol) that favor compound classes such as terpenes and flavonoids. Although many proteinaceous compounds likely have been overlooked by these traditions, such unfavorable conditions for protein and peptide research have also led to the discovery of some unique peptide structures, which resemble small molecules in being thermally, biologically, and chemically stable as well as being exceptionally hydrophobic compounds. Here, we will describe one such protein family, the cyclotides, which is an extraordinary example of the virtually untapped source of structural and biological diversity that plant peptides and proteins represent. Hence, in the following paragraphs, we will demonstrate that plants can provide unique protein structural motifs with potent biological activities and that may serve as templates for protein engineering. Furthermore, the targeted isolation of cyclotides, and other plant proteins, will be discussed.

The cyclotide family of plant peptides is defined by a unique structural motif, the circular cystine knot (CCK) (Craik et al. 1999), in which the circular peptide backbone is interlocked by three internal disulfide bonds (Göransson and Craik 2003; Rosengren et al. 2003; Wang et al. 2009). As shown in Figure 1.6, two of these disulfide bonds form a ring together with the interconnecting peptide backbone; this ring is then penetrated by the third disulfide bond. Together, these features render these peptides their extreme stability and give them a hydrophobic surface. In one aspect, these peptides can be regarded as proteins turned inside out: Hydrophobic amino acid side chains are pushed to the molecular surface as disulfide bonds occupy the interior of the cyclotide molecule.

The discovery of this protein family is a classic example of pharmacognostic research, in which the Norwegian physician Lorents Gran played a leading role. During a Red Cross session in Congo (Zaire) in the 1960s, he observed that native women of the Lulua tribe used a herbal decoction to accelerate childbirth (Gran 1973a). The preparation was sipped as a tea, but also applied directly at the birth canal that induced very strong contractions, which in some cases led to cervical spasms that made acute Caesarean sections necessary (Gran 1973a). Although the ingredients of the decoction were kept a secret by the native women during that first session, the

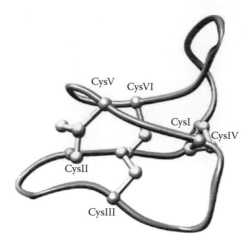

FIGURE 1.6 The seamless amide backbone and the three disulfide bonds formed by six conserved cysteine residues constitute the cyclic cystine knot. The "knot" arises from the disulfide bond CysIII–CysVI that penetrates the ring formed by the peptide backbone and the CysI–CysIV and CysII–CysV bonds. Around this stable framework, plants appear to have built their own combinatorial libraries of around 30 amino acids. The diversity in these libraries is based on the sequence variations in the loops, i.e., the sequences between the cysteines.

identity of the plant was revealed as *Oldenlandia affinis* DC. (Rubiaceae), or *Kalata-kalata* in the native language, when Gran returned a few years later. It is interesting to note that *Oldenlandia affinis* had been reported for the same use, to facilitate childbirth, from the Central African Republic, a few years earlier (Sandberg 1965). Intrigued by these experiences, Gran brought the herb back to Norway for chemical and pharmacological analyses, which showed that the plant contained a suite of peptides with potent uterotonic activity (Gran 1973b). The major peptide component, named kalata B1, was isolated but its full sequence and its cyclic structure was not determined until some 20 years later when its three-dimensional structure was also described (Saether et al. 1995). That peptide is the archetypical cyclotide.

In the early 1990s, four additional cyclic peptides were isolated: circulin A and B from *Chassalia parvifolia* Schum. (Rubiaceae), (Gustafson et al. 1994) cyclopsychotride A from *Psychotria longipes* Muell. Arg. (Rubiaceae), (Witherup et al. 1994) and violapeptide I from *Viola arvensis* Murr (Schoepke 1993). Triggered by the opportunity to work in a niche that was, and still is, underdeveloped, namely, plant proteins and aqueous extracts, we developed a fractionation protocol in our laboratory to provide an extract enriched with peptides and proteins between 10 and 50 amino acids. In that protocol (Claeson et al. 1998), shown in Figure 1.7, we merged dereplication steps commonly used in the isolation of phytochemicals, with isolation procedures mostly used in biochemistry (and thus for the isolation of proteins and peptides from animal sources). Efforts were made to decrease the amount of highly abundant plant substances that could interfere with isolation and bioassay procedures. For example, ethanol (50% v/v) was included to maximize protein solubility and minimize solubility of polysaccharides, and tannins were removed by filtration through polyamide. It was applied to circa 200 plant species to provide a standardized extract

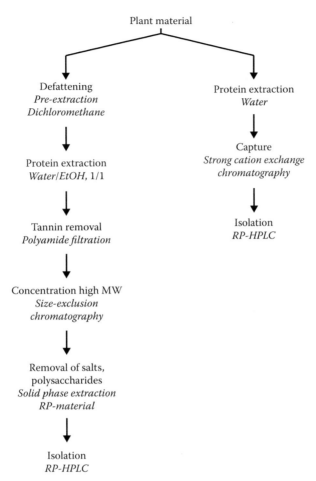

FIGURE 1.7   The isolation protocol described in Claeson et al. (1998) was designed for indiscriminate isolation of plant polypeptides in the range of 10–50 amino acids. Key features were the dereplication of highly lipophilic compounds (defattening), the solvent for the main extraction and the inclusion of size exclusion chromatography on Sephadex G-10 to separate out compounds with molecular weights lower than ~700. Today, we have developed protocols directly aimed at the isolation of cyclotides, and toward subfamilies of cyclotides. The protocol to the left illustrates a strategy for isolation of bracelet cyclotides. That cyclotide subfamily is characterized by a higher number of positively charged amino acids (in this case lysines), which is exploited in the ion exchange capture step.

for chemical and biological assays for protein discovery and activity studies. A direct result of this screening for plant proteins was the isolation of a suite of cyclotides from *Viola arvensis*, the varv peptides A-H (Göransson et al. 1999). This was the first indication that cyclotides were used as nature's own combinatorial peptide library.

To date, approximately 200 cyclotides have been reported from three plant families (besides Violaceae and Rubiaceae, peptides containing the CCK motif has been found in Cucurbitaceae); but recent findings suggest that there might be >9000 cyclotides in the Violaceae plant family alone. The proteins fall into two major sub-families referred to as "bracelets" and "Möbius."

1.7.1 STRUCTURE AND PROTEIN ENGINEERING

It is now clear that this multitude of peptides can carry many different biological activities. In addition to the effects mentioned earlier, cyclotides have been reported to have insecticidal, (Jennings et al. 2001) nematocidal, (Colgrave et al. 2008) anti-microbial, (Tam et al. 1999) anti-fouling, (Göransson et al. 2004) trypsin inhibitory, (Hernandez et al. 2000) and cytotoxic activities (Herrmann et al. 2008; Lindholm et al. 2002). Although the precise mechanisms of actions are not yet known, it seems probable that many of these effects are connected to the membrane disruptive effect of cyclotides that was recently described (Svangård et al. 2007). Cyclotides cause leakage in phospholipid model membranes, and the interaction has been described on a molecular level by NMR studies (Shenkarev et al. 2006). These studies show that the hydrophobic part of the cyclotide molecule is inserted into the membrane. In this context, it is interesting to note that the two major cyclotide subfamilies are ori-entated differently in the membrane (Wang et al. 2009), which seems to contribute to the differences in potency, in particular for their cytotoxic and hemolytic activity, that can be observed between subfamilies.

Today, the structure–activity relationships of these proteins are explored through studies of native, chemically modified, and synthetic cyclotides, with the ultimate aim to exploit them in drug design and development. For example, using a strategy for chemical modification of functional side chains, the key functional and structural role of a conserved glutamic acid residue was identified; an alanine scan has identi-fied a cluster of hydrophilic residues to be connected to the insecticidal and hemo-lytic effects (Simonsen et al. 2008). This increase in knowledge of how structure and sequence are connected to activity in combination with advances in cyclotide syn-thesis and folding (Daly et al. 1999; Gunasekera et al. 2006; Leta Aboye et al. 2008) has led to the development of the CCK motif as a first-rate template for protein engi-neering. The idea is to utilize the stable cyclotide framework as a bearer for grafted, pharmacologically active, sequence epitopes. Such sequences are then inserted into one of the variable loops, i.e., the sequences between cysteines, to be exposed on the molecular surface. Recently, two examples of grafting bioactivities onto the cyclotide framework have been published showing proof of concept: an antagonist of vascular endothelial growth factor (VEGF) has been designed to inhibit angiogen-esis (Gunasekera et al. 2008) and inhibitors of the enzymes beta tryptase and human leukocyte elastase (Thongyoo et al. 2009).

The fact that cyclotides, as most other proteins and peptides, are gene products opens a window of opportunity for production of cyclotides and cyclotide mutants using biotechnology instead of synthetic chemistry. Although some key information still is missing to describe the complete cyclotide biosynthesis (e.g., the ring closure mechanism and chromosomal arrangement), a plant cell culture of *Oldenlandia affi-nis* has been established that expresses large quantities of cyclotides (Dörnenburg 2009). Although mutated, i.e., grafted, cyclotide has not yet been engineered into the plant, the necessary technology to express mutated proteins is in place. The opportunity to utilize plants as an expression system (i.e., "molecular farming") gives a facile route to practically endless variants, quantities, and applications of cyclotides.

1.7.2 Making Ends Meet

Despite that water is the most commonly used extraction solvent in traditional medicine, in the form of teas or decoctions (Samuelsson et al. 1985), it remains a fact that only a few pharmacognosy laboratories focus on those macromolecular compounds that can be found in aqueous extracts. As discussed earlier, one reason for scarcity of plant protein research is based on traditions, but another reason is that proteins and peptides for a long time were considered unsuitable as drug compounds: they were simply thought to be too large, too unstable, too scarce, costly, and notoriously difficult to administrate. Moreover, what characters could a protein from a natural extract possibly have—originating from a plant, marine organism, microorganism, or insect—to give it a pharmacological potential (O'Keefe 2001)? Today the tide has turned, and the potency and selectivity of peptide- and protein-based drugs have led to the availability of more than 200 peptide-based drugs on the market in 2009 (Vlieghe et al. 2009), some of which can be classified as "natural products." The prime example is conotoxin MVIIa, which is marketed as ziconitide/Prialt® with the indication of severe chronic pain.

With the technology that is available today, protein and peptide research has also become more accessible. For example, many aspects of protein structural characterization has become routine, advances in genetic analyses and bioinformatics offer endless opportunities, and efficient methods for peptide synthesis and folding have been developed, as well as routine systems for protein expression. However, as in the case of the cyclotides, the potential increases exponentially when making ends meet, and when our knowledge about plant chemistry is applied together with the new biotechnology.

1.8   MARINE ECOSYSTEMS

Although most often discussed in conjuncture with deforestation of tropical woodland, the processes of global warming and ecosystem destruction discussed earlier are today equally obvious in virtually all of planet Earth's ecosystems. This includes many of the often unique habitats found in limnic and marine environments. The tropical coral reefs, which form some of the most species-rich and biologically most diverse ecosystems known, have been under threat from human activities since the 1950s. Initially, the foci was put on bio-piracy, collecting or harvesting rare shells and fish for retail, followed by ruthless dynamite fishing ventures, introduction of sea trawling in coastal areas and from the turn of the last century a growing understanding that increasing temperatures of the oceans and the bleaching of coral associated algae.

In total circa 70% of Earth is covered by oceans where the living conditions for the organisms are affected by physical conditions such as lack of sunlight, high salinity, and extreme pressure—all differences from features met by organisms on land. Despite, or maybe because of, these harsh conditions, the sea hosts an incredible biological diversity, and marine organisms are represented in more phyla than

(a)

(b)

FIGURE 1.8 (a) In the oceans covering almost two-thirds of the Earth's surface, still only a minute portion has been explored. Among organisms that have recently been studied for their contents of natural products with potential health benefits are several animals such as sponges, soft and hard corals, and several species of fish. (b) **(See color insert.)** In addition to these, numerous species of bacteria, fungi, and various groups of plants including red and brown algae have contributed to our knowledge of chemical diversity. (Photos by Mattias Klum.)

those from any other ecosystem on Earth (May 1994). As on land, the flora and fauna of the sea are continuously confronted and challenged by predators, parasites, and microorganisms. Furthermore, biofouling organisms such as biofilm-forming bacteria, fungi, epiphytic algae, and invertebrates rapidly attempt to colonize the surface of macroorganisms (Figure 1.8).

To withstand these threats, efficient defense mechanisms have been developed where, in many cases, secondary metabolites play a key role for survival of the organisms. Recent research in marine chemical ecology has resulted in isolation of several such bioactive secondary metabolites. These findings lay the foundation for a better understanding of interactions between different marine organisms like herbivores feeding on algae, behavior of settlement of biofouling organisms, prey–predator relationships, and competitive behavior (Paul and Ritson-Williams 2008). Many molecules discovered from marine organisms have shown novel chemical skeletons in many cases with incorporation of halogens, not found in terrestrial organisms. Most secondary metabolites isolated originate from organisms collected in tropical and temperate waters. Even if cold waters have not so far been studied to that extent, the chemical diversity is probably much less than in warm waters. However, in-depth studies of a specific organism with ecological relevance in in vivo bioassay models could probably be helpful in understanding the role of the produced secondary metabolites and the role of these metabolites in the interaction with other organisms in the surrounding environment.

1.8.1 MARINE COLD WATER SPONGE REEFS

On the continental shelf of the North Atlantic, in the Norwegian fjords and in the Swedish northern west-coast (the Koster fjord) the cold-water marine sponge *Geodia barretti* is living in very old reef-like fields of sponges. These reefs are considered to be important hotspots of biodiversity but also very vulnerable to, e.g., the impact made while trawling for fish (Klitgaard 1995). The observation that this sponge had almost no on-growth of fouling organisms on its surface raised scientific questions about biosynthesis of potential molecules for chemical defense. In a first screening attempt to identify bioactivity, the sponge was extracted with different solvents with varying polarity. Lipophilic and hydrophilic extracts were tested for bioactivity in vivo (behavioral study in mice) and in vitro (effect on guinea-pig ileum), resulting in significant activity in comparison with positive reference compounds (Andersson et al. 1983). Furthermore, the extracts showed antibacterial and antiviral activity. However, these tests were not directly relevant to answer if the sponge produced this bioactivity for protection against predators or as a chemical defense against fouling organisms. Therefore, an ecologically relevant test model, where cyprid larvae of the barnacle *Balanus improvisus*, was used to evaluate if the extract could effect the mobility or settlement of the larvae. In a concentration-response way the settlement of the cyprids was inhibited by the sponge extract in a micromolar (μM) range. The isolation procedure guided by this assay resulted in three novel molecules, so-called barettins—of which the first was barettin (5)—which all showed a reversible inhibition of larvae settlement.

However, so far it was shown that an extract from the sponge *Geodia barretti* contained peptides affecting the cyprids, but are these substances released from the sponge in the sea? A remotely operated vehicle was used for in situ sampling of water from the surface of the sponge, which was pumped through a reversed phase chromatography column. An in-line mass spectrometer could then detect barettins and prove that the compounds are released into the ambient water. To gain further understanding of the structure–activity relationship of the antifouling activity, barretin analogs were synthesized and tested. Comparative studies revealed that the native compound isolated from the sponge exhibited the highest activity in the assay used (Sjögren 2006). Another minor dibrominated compound was isolated later with even higher activity (Hedner 2008).

In search for the target of these molecules produced by the sponge, and also to evaluate a potential pharmacological lead structure, a test of barretin affinity to human serotonin (5HT) receptors using human embryonic kidney cells was performed.

A selective interaction was found in concentrations close to endogenous serotonin levels in humans (Hedner 2007). Further studies have also shown that the produced barettins act in synergy (Sjögren 2011). Several scientific questions in this project have been answered but many more appear and are waiting to be solved, such as the true producer of the barettins and the ecotoxicological effect of barettins on other marine organisms.

This research project shows that also marine organisms from cold water can result in novel knowledge when studied in depth, knowledge that could be used to understand the importance of protection of specific ecological features in nature such as the rain forests as well as ancient sponge reefs.

1.9 BIOPROSPECTING OF FUNGI

Pharmacognosy has its traditional foundation in plant science. Fungi, although historically thought to be primitive plants and still being considered as a part of botany, form a whole new base of enormous potential when it comes to finding new substances in nature. The fungal kingdom comprises an estimated total number of 1.1 million species, making it five times as species rich as the plant kingdom (Mueller 2007). Despite being evolutionarily more closely related to animals than plants, fungi are immobile and lack several characteristic features, such as a gastric cavity or gastrointestinal system and a nervous system, of animals. On the other hand, they do have a very large surface area compared to their surrounding environment, which together with an array of excreted enzymes provides efficient substitutes for nutrient uptake. Fungal cells, as plant cells, are furnished with a cell wall, but instead of the cellulose in plants, the fungal cell wall consists primarily of chitin. Chitin is also found, e.g., in the exoskeleton of insects and will protect the fungal cell while producing enzymes designed to disintegrate the cellulose cell wall of plants—a primary target of many fungi. These features, and the subsequent damage they may inflict on plants, such as huge trees or wide fields, is one reason to dwell more in the field of mycology in search for bioactive natural products. Another reason is that fungi, also from a biochemical perspective, are of great interest as they have the ability to produce, e.g., unusual amino acids including the D-enantiomers, resulting in otherwise anomalous compounds. A well-known pharmacological application of this ability is the use of cyclosporine, an anomalous compound first derived from the soil fungus *Cordyceps subsessilis*, better known under the scientific name *Tolypocladium inflatum*, initially given to the asexual stages of this same species. Cyclosporine is, since its approval in the United States in 1983, an essential immunosuppressant agent during organ transplants (Traber et al. 1989). As a group of organisms, fungi are ecologically, biologically, and chemically extremely diverse organisms of a magnitude that is yet to be discovered. Exploring the diversity of fungi—of which it is estimated that only a few percent of the species are yet known to science—may most likely lead to new classes of bioactive natural products.

1.9.1 ROLE OF BACTERIA IN FUNGAL FRUIT BODIES

In many higher phyla, bacterial interactions are well-known phenomena. Bacteria inside many marine organisms are responsible for large parts of their peculiar chemistry, and symbiotic bacteria have, in some cases, shown to be the real producers of bioactive substances (Piel 2009). In animals, endophytic bacteria are functionally essential, exemplified by the gastrointestinal systems of, e.g., *Bos taurus* (cows) and *Homo sapiens* (humans). In plants, examples are found in the rhizobia of the genus *Alnus* (alder), i.e., nodes on the root of the tree by which the tree accumulates nitrogen through N-fixing bacteria. Such interactions, of ecological benefit for both organisms, are also found among fungi where there are fungal species taking both cyanobacteria and unicellular plants as "room mates" as in lichens, but also fungi that form mycorrhizae with higher plants providing them with a vastly improved surface area for nutrient uptake and efficient chemical defense in exchange for mechanical protection and energy-rich nutrients. As fascinating as they seem, the most interesting perspective on such interactions may actually be how they are controlled or balanced. Fungi are nonphotosynthetic, but in the case of, e.g., lichens and mycorrhizae, the fungus has found a way to control and take advantage of the photosynthesis of algae or cyanobacteria and plants, respectively. Also fruit bodies of macrofungi may contain high numbers of bacteria. For example, in chanterelles (*Cantharellus cibarius*) millions of bacteria have been shown to coexist without harming the mycelial structures of the fruit body (Rangel-Castro et al. 2002). While bacteria are known to be present in fruit bodies of several fungal species, the dynamics and role of this symbiosis or coexistence is poorly understood.

In contrast, in the project discussed by Alsmark given in the following section, we have a parasite-host-driven evolution between protists and their mammalian hosts. There is adaptive evolution of parasitic eukaryotes, which finds an evolutionary edge by preying on bacteria and their genes.

1.9.2 UNDERGROUND FUNGAL BIOLOGICAL DIVERSITY

The ability to use DNA analyses to aid identification of fungal species has greatly increased our knowledge of the variation in species richness and ecology of fungi. Inventories to determine the presence, and abundance, of fungal species are most often made by recording the occurrence of fruiting specimens. This method is far from optimal as it will only give a momentary picture, which has proven to reflect only part of the total abundance of species (Dahlberg et al. 2000; Tedersoo et al. 2003). By sampling mycorrhizal roots, species with inconspicuous or overlooked fruit bodies, or species that may never fruit, will also be detected. While morphological mycorrhizal identification previously has been proven a difficult and tedious work (Agerer 1991), identification using DNA sequencing instead enables a fast and more accurate identification (Iotti and Zambonelli 2006; Koljalg et al. 2005; Tedersoo et al. 2003). In this project, we focus on hypogeous macrofungi (truffles), which are often overlooked in conventional inventories (due to their hypogeous mode of life) as shown by the great number of specimens found when using a trained dog instead (Wedén 2004). Better distribution and ecological data for these fungi will not only enable better programs for an appropriate sustainable environmental management to benefit these environments, but also provide a better basis to start to entangle complex interactions between the fungi and their mycorrhizal host (Figure 1.9).

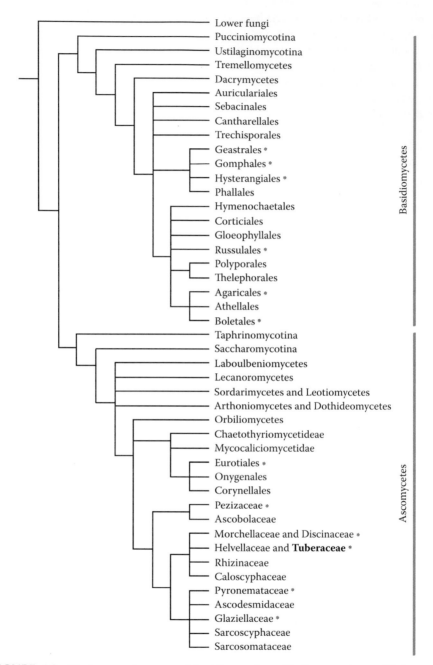

FIGURE 1.9 Phylogeny of asco- and basidiomycetes with focus on taxa with hypogeous fruiting bodies, truffles, based on a combination of several published studies (Celio et al. 2006; Hibbet et al. 2007; Læssøe and Hansen 2007). Taxa marked with * indicate evolutionary groups that have developed truffles. The most well-known truffle-producing family, Tuberaceae, comprising, e.g. the edible Alba and Burgundy truffles, is further emphasized with bold lettering. (Figure by Christina Wedén and Anders Backlund.)

Increased basic knowledge and a strong coupling between academic research and industry have lead to the initiation of substantial truffle cultivation (Wedén et al. 2009). Since the start in 1999, the truffle industry has grown to become an important and viable branch on the island of Gotland. In modern Swedish forestry there has been little incentive for new plantation of oak trees. The possibility to grow and collect truffles play an important role in increasing rural economy and thereby the incentives for planting oak and hazel. Among nature conservationists, the beneficial effects of replantation and management of open oak and hazel meadows have been repeatedly pointed out. The exceptional biological diversity in such environments, with large numbers of endangered species from many phyla including insects, birds, vascular plants, and fungi, is well known (Gärdenfors 1994). Collecting truffles in the wild also helps keeping these species-rich biotopes well managed, favoring high nature conservation values for the future. The use of trained dogs ensures minimum impact during collection of truffles and sustainable truffle production. Today, these forest meadows are diminishing, as a result of the change-over in modern agriculture toward increased cereal farming, posing a threat also to its many associated species. Investigating species richness in natural truffle habitats and understanding the dynamics of mycorrhizal associations will form an important foundation in the work to reestablish these threatened biotopes. Truffle cultivation can thus contribute to a stable bioeconomy, both supporting the documented high biodiversity associated with *Quercus* spp. (Gärdenfors 1994) and contributing to the local economy of rural areas. Truffle orchards may eventually become a new core of sustainable, high-diversity biotopes worth protecting.

1.10 PREDICTION OF TARGETS USING BIOINFORMATIC TOOLS

Entamoeba histolytica, *Trypanosoma cruzi*, *Trypanosoma brucei*, *Leishmania major*, and *Trichomonas vaginalis* are eukaryotic microorganisms and pathogens of man. These increasingly difficult-to-treat parasites affect many millions of people and animals yearly in the developed and developing world but are still often neglected in research, national health programs, and the pharmaceutical industry. Recent reports about failed treatment due to emerging resistant strains in all of the above-mentioned pathogens (Legros et al. 2002; Lossick 1990; Nozaki et al. 1996; Samarawickrema et al. 1997) highlights the urgent need for new drug targets. Furthermore, because these pathogens are also eukaryotes, they have few clear-cut metabolic and genetic differences to mammals making the search for drugs without severe side effects not only urgent but also exceedingly difficult.

Entamoeba histolytica is the third leading cause of lethal parasitic infection in humans, exceeded only by malaria and schistosomiasis. Up to 50 million people yearly are affected by symptomatic infections and *Entamoeba histolytica* is estimated to be responsible for 100,000 deaths every year (Stanley 2003). The symptoms range from chronic, mild diarrhea to aggressive dysentery, and although most infections occur in the digestive tract, other tissues, such as the liver and skin, may be invaded. Reports of failed drug treatment and differences in drug susceptibilities among *Entamoeba histolytica* strains probably herald the development of drug resistance in this parasite.

Trichomonas vaginalis is the most common, nonviral, sexually transmitted parasite in humans of the developed and developing world, infecting 200 million people
each year (Petrin et al. 1998). The implications of infection are particularly serious
for women. *Trichomonas vaginalis* infection is linked to heightened HIV susceptibility, cervical cancer, birth complications, and sterility. Trichomoniasis is currently
treated with metronidazole **(6)** or derivatives, but resistance is increasing and there
is a need for new therapeutics.

Trypanosomatids (*Trypanosoma* and *Leishmania*) are flagellated protozoan organisms of the order Kinetoplastida. The order is characterized by one flagellum and a
single mitochondrion containing the kinetoplast, a characteristic DNA-containing
organelle. The species of this parasite colonize vertebrates, invertebrates, and plants.
Trypanosoma bruceii is causing African trypanosomiasis, also known as sleeping sickness, and the cattle wasting disease nagana, in sub-Saharan Africa. For untreated cases
of trypanosomiasis, death is certain and it is estimated that 70,000 people die because
of lack of diagnosis and treatment of *Trypanosoma bruceii* infection yearly (Barrett
2006). Geographically limited to Central and South America, the parasite that causes
Chagas disease, *Trypanosoma cruzi*, is the world's leading cause of heart diseases. It
is transferred by blood transfusion or by the bite of bugs of the genus *Rhodinus* and
Triatoma. Chagas disease is sometimes called the "kiss of death" since the Triatome
bugs usually bite around the lips. It is estimated that 8–11 million people are infected
with Chagas disease and around 15,000 deaths are reported yearly (Barrett et al. 2003).
The *Leishmania* parasite is an intracellular pathogen of the immune system and it is
transferred to its human host from *Phlebotumus* sandflies. Each year 1.5–2 million
people are diagnosed with leishmaniasis worldwide. Symptoms range from partial or
total destruction of the mucous membranes of the nose, mouth, and throat cavities and
surrounding tissues to irregular bouts of fever, substantial weight loss, swelling of the
spleen and liver, and anemia. The disabling and degrading outcomes of leishmaniasis
can result in victims being humiliated and cast out from society. If left untreated, the
fatality rate in developing countries is close to 100%.

The need for new drugs to treat infection by eukaryotic parasites is becoming
increasingly evident as new reports of resistant strains continually emerge. At present, metronidazole is the drug of choice for *Entamoeba histolytica* and *Trichomonas
vaginalis*. However, it has numerous side effects such as nausea, vomiting, diarrhea,
hypersensitivity, and encephalopathy (Cudmore et al. 2004). Metronidazole has also
been reported to be carcinogenic in rodents (Bendesky et al. 2002) and is known to
pass the placental barrier in humans and is therefore ruled out in treatment for pregnant women. In addition, strains resistant to metronidazole are being increasingly
reported. *Trypanosoma bruceii* infections are predominantly treated with melarsoprol despite the fact that this drug exposes around 20% of patients to serious risk
of arsenic encephalopathy and almost a thousand people die each year following
treatment with melarsoprol (Enanga et al. 2002). Eflornithine is a far less danger-
ous treatment for sleeping sickness, but because of its high costs it is not available
to most patients. Also, eflornithine is efficient only for one out of three infectious
strains of *Trypanosoma bruceii*. A number of plant extracts are traditionally used
to treat Chagas disease, e.g., *Sangre de Drago*—dragon blood—a plant extract from

Croton roborensis, which is sold by herbal vendors throughout Bolivia. *Sangre de Drago* is effective against the symptoms of heart disease but any parasiticide properties are yet to be scientifically verified. Nifurtimox (7) and benzimidazole (8) are currently used to treat *Trypanosoma cruzi* infection although both drugs give negative side effects so severe that some patients are forced to discontinue treatment.

1.10.1 LATERAL GENE TRANSFERS AS TARGETS FOR NEW DRUGS

Recent research has presented compelling evidence that lateral gene transfer (LGT) has played an important role in the evolution of microbial prokaryotes and eukaryotes (Doolittle et al. 2003; Keeling and Palmer 2008). LGT—the transfer of genetic material between distinct evolutionary lineages—contrasts with the transmission of mutations and traits from parents to offspring, also known as vertical inheritance. Point mutations are quantitatively the major mode of evolution but they generally exert their effects very slowly. In contrast, LGT offers the opportunity of evolution by acquisition whereby radical new capabilities may be acquired quickly. The genes transferred include pathogenicity islands as well as drug resistance genes that increase the virulence of medically important organisms. They also include genes that encode proteins that facilitate colonization or allow the acquisition of nutrients from the host, an attractive opportunity to scavenge ready-made metabolites in a limited environment.

LGTs in pathogenic eukaryotes are attractive candidates for drug targets since their prokaryotic origin make them less likely to have a vulnerable homolog in mammalian genomes, and, hence, the drugs may have fewer side effects on patients. Better understanding of the metabolic impact of LGT in eukaryotes will guide the screen for potential drug targets. In collaboration with TIGR and Sanger Institutes, we have made genome-wide tree-based screens for LGT in the genomes of *Entamoeba histolytica* (Clark et al. 2007; Loftus et al. 2005), the trypanosomatides (Berriman et al. 2005; El-Sayed et al. 2005), and *Trichomonas vaginalis* (Carlton et al. 2007). In order to achieve an effective but reliable screen of these large datasets, we combined rapid screening methods (such as homology searches and distance phylogeny) for LGT followed by a more detailed Bayesian phylogenetic analysis of genes that pass the primary screen. All Bayesian trees were manually inspected and all cases where the tree topology shows one of our chosen parasites clustered with prokaryote sequences separated from any other eukaryote by at least one well-supported node were considered as an LGT in that species for the gene analyzed. The conservative selection thresholds singled out recent LGTs that probably only represent a subset of the complete transferome in our selected pathogens. The analyses showed that many of the metabolic differences between these parasites and man are due to LGT in the parasite genomes. Many of the LGTs detected lack a homologue in mammalian genomes, e.g., tagatose-6-phosphate kinase, which is active in galactose metabolism in *Entamoeba histolytica*, but not in human. Other LGTs, inferred by phylogeny as bacterial like, are likely to be structurally different to the ancestral eukarytotic homologue, e.g., isovaleryl-CoA dehydrogenase in the trypanosoma.

1.10.2   LGT IN *TRICHOMONAS VAGINALIS*

In *Trichomonas vaginalis*, a total of 24,159 trees were retrieved, and 650 of these were selected from the primary screen as representing potential LGT. The 650 candidate genes were then processed through the secondary screen to make better trees with a Bayesian approach. Among the 650 candidates we identified 76 genes where the most straightforward interpretation of trees indicated that lateral gene transfer from a prokaryote to an ancestor of *Trichomonas vaginalis* had occurred. We also identified 77 genes where *Trichomonas* and diverse prokaryotes contained a particular gene, but we were unable to find a convincing homologue among other eukaryotes. The remaining 497 cases remained unresolved, but do not show vertical inheritance. Similar results were achieved from the trypanosomatides and *Entamoeba histolytica*. In summary, the LGT analyses of these pathogenic protists have generated around 300 potential targets for drugs. The putative LGTs are integrated into diverse metabolic pathways, including nucleotide and amino acid metabolism, synthesis of lipophosphoglycan, salvage pathways, biosynthesis of cofactors and cell wall components, and many more. Thus, in the broadest sense, LGT must be affecting the fitness of the recipient organism. These results indicate strongly that recent gene transfers have fundamentally shaped the content of eukaryotic genomes. Furthermore, these LGTs have produced numerous pathways and genes that are present in individual parasitic protozoan species but that are not present on the mammalian genome, thus providing a large and variable pool of potential drug targets across protozoan diversity.

1.11   CONCLUSIONS AND FUTURE PERSPECTIVES

Several recent reviews have been written about new discoveries of bioactive compounds with a natural origin (Cragg et al. 2009 and references therein). Some of these compounds have shown remarkable chemical structures and biological activity never seen before. These novel chemical entities have opened up new fields of basic science and have also led to new applications as new drugs, precursors for semisynthesis, models for synthesis or pharmacological tools. The pharmaceutical industry has in recent time decreased their interest in natural product discovery programs due to problems with accessibility, complexity of chemical structures with many chiral centers, time-consuming isolation procedures, and intellectual property rights and bioconventions.

However, discovery of new chemical structures of natural origin has an increased importance for inspiration of the new combinatorial chemistry where new methodology can be used for total synthesis of natural product like compounds resulting in greater diversity but with less complexity. Engineering of the biosynthesis through genetic manipulations gives combinatorial biosynthesis new possibilities.

An increased understanding of natural products in mixtures and evaluation of synergy and further well-performed clinical trials would be very important for a greater acceptance and respect for the pharmacological potential of herbal remedies.

From these perspectives, it becomes obvious that a multidisciplinary approach is necessary in order to obtain a set of information and material as complete as possible. At present, such endeavors might appear "unfocused" and we must therefore strive for a wider acceptance of a broader approach. In the present situation when pollution, deforestation, and destruction of the marine environment is a first-degree threat against the Earth's biodiversity, we must take the uttermost advantage of every chance to discover, describe, and communicate the riches of nature—and this, we believe, is no task for a lone wolf. To overcome environmental denialists, the only way forward is through information, education, and understanding and this is exactly where science unfortunately often fails. To get our message through and make it trustworthy, we are in dire need also of communication experts.

ACKNOWLEDGMENTS

All present and former collaborators are gratefully acknowledged as well as the economic support from the Swedish research councils VR and FORMAS, VINNOVA, and Swedish foundation for strategic research.

REFERENCES

Agerer, R. 1991. Characterization of ectomycorrhiza. In *Methods in Microbiology*, eds. J. R. Norris, D. J. Read, and A. K. Varma. London, U.K.: Academic.

Andersson, L., G. Lidgren, L. Bohlin, L. Magni, S. Ögren, and L. Afzelius. 1983. Studies of Swedish marine organisms. I. Screening of biological activity. *Acta Pharm Suec* 20:401–414.

Backlund, A. 2010. Topical chemical space in relation to biological space. In *Comprehensive Natural Products II Chemistry and Biology* Vol. 3, eds. L. Mander and H. W. Lui. Oxford, U.K.: Elsevier.

Barrett, M. P. 2006. The rise and fall of sleeping sickness. *Lancet* 367:1377–1378.

Barrett, M. P., R. J. Burchmore, A. Stich, J. O. Lazzari, A. C. Frasch, J. J. Cazzulo, and S. Krishna. 2003. The trypanosomiases. *Lancet* 362:1469–1480.

Bendesky, A., D. Menendez, and P. Ostrosky-Wegman. 2002. Is metronidazole carcinogenic? *Mutat Res* 511:133–144.

Berriman, M., E. Ghedin, C. Hertz-Fowler, G. Blandin, H. Renauld, D. C. Bartholomeu, N. J. Lennard, E. Caler, N. E. Hamlin, B. Haas, U. Bohme, L. Hannick, M. A. Aslett, J. Shallom, L. Marcello, L. Hou, B. Wickstead, U. C. Alsmark, C. Arrowsmith, R. J. Atkin, A. J. Barron, F. Bringaud, K. Brooks, M. Carrington, I. Cherevach, T. J. Chillingworth, C. Churcher, L. N. Clark, C. H. Corton, A. Cronin, R. M. Davies, J. Doggett, A. Djikeng, T. Feldblyum, M. C. Field, A. Fraser, I. Goodhead, Z. Hance, D. Harper, B. R. Harris, H. Hauser, J. Hostetler, A. Ivens, K. Jagels, D. Johnson, J. Johnson, K. Jones, A. X. Kerhornou, H. Koo, N. Larke, S. Landfear, C. Larkin, V. Leech, A. Line, A. Lord, A. Macleod, P. J. Mooney, S. Moule, D. M. Martin, G. W. Morgan, K. Mungall, H. Norbertczak, D. Ormond, G. Pai, C. S. Peacock, J. Peterson, M. A. Quail, E. Rabbinowitsch, M. A. Rajandream, C. Reitter, S. L. Salzberg, M. Sanders, S. Schobel, S. Sharp, M. Simmonds, A. J. Simpson, L. Tallon, C. M. Turner, A. Tait, A. R. Tivey, S. Van Aken, D. Walker, D. Wanless, S. Wang, B. White, O. White, S. Whitehead, J. Woodward, J. Wortman, M. D. Adams, T. M. Embley, K. Gull, E. Ullu, J. D. Barry, A. H. Fairlamb, F. Opperdoes, B. G. Barrell, J. E. Donelson, N. Hall, C. M. Fraser, S. E. Melville, and N. M. El-Sayed. 2005. The genome of the African trypanosome *Trypanosoma brucei*. *Science* 309:416–422.

Bohacek, R. S., C. McMartin, and W. C. Guida. 1996. The art and practice of structure-based drug design: A molecular modeling perspective. *Med Res Rev* 16:3–50.

Bohlin, L. 1978. Some strychnos alkaloids: Their occurrence, structure and biological activity. [Thesis]. Uppsala: Acta universitatis upsaliensis. Abstracts of Uppsala dissertations from the Faculty of Pharmacy, 32.

Bohlin, L. and U. Huss. 2006. Evaluation of anti-inflammatory activity of natural products based on enzyme inhibitors. *ICS, UNIDO* biological screening of plant constituents-training manual.

Brady, S. F., L. Simmons, J. H. Kim, and E. W. Schmidt. 2009. Metagenomic approaches to natural products from free-living and symbiotic organisms. *Nat Prod Rep* 26:1488–1503.

Carlton, J. M., R. P. Hirt, J. C. Silva, A. L. Delcher, M. Schatz, Q. Zhao, J. R. Wortman, S. L. Bidwell, U. C. Alsmark, S. Besteiro, T. Sicheritz-Pontén, C. J. Noel, J. B. Dacks, P. G. Foster, C. Simillion, Y. Van de Peer, D. Miranda-Saavedra, G. J. Barton, G. D. Westrop, S. Muller, D. Dessi, P. L. Fiori, Q. Ren, I. Paulsen, H. Zhang, F. D. Bastida Corcuera, A. Simoes-Barbosa, M. T. Brown, R. D. Hayes, M. Mukherjee, C. Y. Okumura, R. Schneider, A. J. Smith, S. Vanacova, M. Villalvazo, B. J. Haas, M. Pertea, T. V. Feldblyum, T. R. Utterback, C. L. Shu, K. Osoegawa, P. J. de Jong, I. Hrdy, L. Horvathova, Z. Zubacova, P. Dolezal, S. B. Malik, J. M. Logsdon, Jr., K. Henze, A. Gupta, C. C. Wang, R. L. Dunne, J. A. Upcroft, P. Upcroft, O. White, S. L. Salzberg, P. Tang, C. H. Chiu, Y. S. Lee, T. M. Embley, G. H. Coombs, J. C. Mottram, J. Tachezy, C. M. Fraser-Liggett, and P. J. Johnson. 2007. Draft genome sequence of the sexually transmitted pathogen *Trichomonas vaginalis. Science* 315:207–212.

Celio, G. J., M. Padamsee, B. T. Dentinger, R. Bauer, and D. J. McLaughlin. 2006. Assembling the fungal tree of life: Constructing the structural and biochemical database. *Mycologia* 98:850–859.

Claeson, P. and L. Bohlin. 1997. Some aspects of bioassay methods in natural-product research aimed at drug lead discovery. *Trends Biotechnol* 15:245–248.

Claeson, P., U. Göransson, S. Johansson, T. Luijendijk, and L. Bohlin. 1998. Fractionation protocol for the isolation of polypeptides from plant biomass. *J Nat Prod* 61:77–81.

Clark, C. G., U. C. Alsmark, M. Tazreiter, Y. Saito-Nakano, V. Ali, S. Marion, C. Weber, C. Mukherjee, I. Bruchhaus, E. Tannich, M. Leippe, T. Sicheritz-Pontén, P. G. Foster, J. Samuelson, C. J. Noel, R. P. Hirt, T. M. Embley, C. A. Gilchrist, B. J. Mann, U. Singh, J. P. Ackers, S. Bhattacharya, A. Bhattacharya, A. Lohia, N. Guillen, M. Duchene, T. Nozaki, and N. Hall. 2007. Structure and content of the *Entamoeba histolytica* genome. *Adv Parasitol* 65:51–190.

Colgrave, M. L., A. C. Kotze, D. C. Ireland, C. K. Wang, and D. J. Craik. 2008. The anthelmintic activity of the cyclotides: Natural variants with enhanced activity. *ChemBioChem* 9:1939–1945.

Cox, P. A. 1993. Saving the ethnopharmacological heritage of Samoa. *J Ethnopharmacol* 38:181–188.

Cragg, G. M., P. G. Grothaus, and D. J. Newman. 2009. Impact of natural products on developing new anti-cancer agents. *Chem Rev* 109:3012–3043.

Cragg, G. M. and Newman, D. J. 2009. Nature: A vital source of leads for anticancer drug development. *Phytochem Rev* 8:313–331.

Craik, D. J., N. L. Daly, T. Bond, and C. Waine. 1999. Plant cyclotides: A unique family of cyclic and knotted proteins that defines the cyclic cystine knot structural motif. *J Mol Biol* 294:1327–1336.

Cuarón, A. D. 1993. Extinction rate estimates. *Cuarón* 118.

Cudmore, S. L., K. L. Delgaty, S. F. Hayward-McClelland, D. P. Petrin, and G. E. Garber. 2004. Treatment of infections caused by metronidazole-resistant *Trichomonas vaginalis. Clin Microbiol Rev* 17:783–793.

Dahlberg, A., H. Croneborg, and T. Hallingbäck. 2000. Mykorrhizasvampar: var femte art är rödlistad. *Svensk Bot Tidskr* 94:286–292.

Daly, N. L., S. Love, P. F. Alewood, and D. J. Craik. 1999. Chemical synthesis and folding pathways of large cyclic polypeptides: studies of the cystine knot polypeptide kalata B1. *Biochemistry* 38:10606–10614.

de Moraes Mizurini, D., I. da Costa Maia, F. Lucia de Carvalho Sardinha, R. de Queiroz Monteiro, S. Ortiz-Costa, and M. das Graças Tavares do Carmo. 2011. Venous thrombosis risk: Effects of palm oil and hydrogenated fat diet in rats. *Nutrition* 27:233.

Doolittle, W. F., Y. Boucher, C. L. Nesbo, C. J. Douady, J. O. Andersson, and A. J. Roger. 2003. How big is the iceberg of which organellar genes in nuclear genomes are but the tip? *Philos Trans R Soc Lond B Biol Sci* 358:39–57; discussion 57–58.

Dornenburg, H. 2009. Progress in kalata peptide production via plant cell bioprocessing. *Biotechnol J* 4:632–645.

Dunstan, C. A., J. Andersson, L. Bohlin, P. A. Cox, and K. O. Grönvik. 1994. A plant extract which enhances the plating efficiency of lymphoid cell lines and enhances the survival of normal lymphoid cells in vitro. *Cytotechnology* 14:27–38.

Dunstan, C., B. Liu, C. J. Welch, P. Perera, and L. Bohlin. 1998. Alphitol, a phenolic substance from *Alphitonia zizyphoides* which inhibits prostaglandin biosynthesis in vitro. *Phytochemistry* 48:495–497.

Eckernäs, S. A., L. Bohlin, and L. Sahlström. 1980. Inhibition of brain synaptosomal uptake of choline by some 2-carboxamidostrychnine derivatives with muscle-relaxant properties. *Acta Pharmacol Toxicol Copenh* 47:81–83.

El-Sayed, N. M., P. J. Myler, G. Blandin, M. Berriman, J. Crabtree, G. Aggarwal, E. Caler, H. Renauld, E. A. Worthey, C. Hertz-Fowler, E. Ghedin, C. Peacock, D. C. Bartholomeu, B. J. Haas, A. N. Tran, J. R. Wortman, U. C. Alsmark, S. Angiuoli, A. Anupama, J. Badger, F. Bringaud, E. Cadag, J. M. Carlton, G. C. Cerqueira, T. Creasy, A. L. Delcher, A. Djikeng, T. M. Embley, C. Hauser, A. C. Ivens, S. K. Kummerfeld, J. B. Pereira-Leal, D. Nilsson, J. Peterson, S. L. Salzberg, J. Shallom, J. C. Silva, J. Sundaram, S. Westenberger, O. White, S. E. Melville, J. E. Donelson, B. Andersson, K. D. Stuart, and N. Hall. 2005. Comparative genomics of trypanosomatid parasitic protozoa. *Science* 309:404–409.

Enanga, B., R. J. Burchmore, M. L. Stewart, and M. P. Barrett. 2002. Sleeping sickness and the brain. *Cell Mol Life Sci* 59:845–858.

Esteban, G. F. and B. J. Finlay. 2010. Conservation work is incomplete without cryptic biodiversity. *Nature* 463:293.

Gärdenfors, U. 1994. Eken—utnyttjad av tusentals organismer. In *Ekfrämjandet 50 år.* (ed.) U. Olsson. Ronneby, Sweden: Ekfrämjandet och skogsvårdsstyrelsen Ronneby.

Göransson, U. and D. J. Craik. 2003. Disulfide mapping of the cyclotide kalata B1: Chemical proof of the cyclic cystine knot motif. *J Biol Chem* 278:48188–48196.

Göransson, U., T. Luijendijk, S. Johansson, L. Bohlin, and P. Claeson. 1999. Seven novel macrocyclic polypeptides from *Viola arvensis. J Nat Prod* 62:283–286.

Göransson, U., M. Sjögren, E. Svangård, P. Claeson, and L. Bohlin. 2004. Reversible antifouling effect of the cyclotide cycloviolacin O2 against barnacles. *J Nat Prod* 67:1287–1290.

Gran, L. 1973a. Oxytocic principles of *Oldenlandia affinis. Lloydia* 36:174–178.

Gran, L. 1973b. On the effect of a polypeptide isolated from "Kalata-kalata" (*Oldenlandia affinis* DC.) on the oestrogen dominated uterus. *Acta Pharmacol Toxicol* 33:400–408.

Gray, W. R., A. Luque, B. M. Olivera, J. Barrett, and L. J. Cruz. 1981. Peptide toxins from *Conus geographus* venom. *J Biol Chem* 256:4734–4740.

Gunasekera, S., N. L. Daly, M. A. Anderson, and D. J. Craik. 2006. Chemical synthesis and biosynthesis of the cyclotide family of circular proteins. *IUBMB Life* 58:515–524.

Gunasekera, S., F. M. Foley, R. J. Clark, L. Sando, L. J. Fabri, D. J. Craik, and N. L. Daly. 2008. Engineering stabilized vascular endothelial growth factor-A antagonists: Synthesis, structural characterization, and bioactivity of grafted analogues of cyclotides. *J Med Chem* 51:7697–7704.

Gustafson, K. R., R. C. Sowder, II, L. E. Henderson, I. C. Parsons, Y. Kashman, J. H. Cardellina, II, J. B. McMahon, R. W. Buckheit, L. K. Pannell, and M. R. Boyd. 1994. Circulins A and B: Novel HIV-inhibitory macrocyclic peptides from the tropical tree *Chassalia parvifolia*. *J Am Chem Soc* 116:9337–9338.

Hann, M. M. and T. I. Oprea. 2004. Pursuing the leadlikeness concept in pharmaceutical research. *Curr Opin Chem Bio* 8:255–263.

Harborne, J. B. 1973. A chemotaxonomic survey of flavonoids and simple phenols in leaves of the Ericaceae. *J Linn Soc Bot* 66:37–54.

Harborne, J. B. 1977. Flavonoids and the evolution of the Angiosperms. *Biochem Syst Ecol* 5:7–22.

Hedner, E. 2007. Bioactive compounds in the chemical defence of marine sponges: Structure-activity relationships and pharmacological targets. [Thesis]. Uppsala: Acta Universitatis Upsaliensis. Digital comprehensive summaries of Uppsala dissertations from the Faculty of Pharmacy, 63.

Hedner, E., M. Sjögren, S. Hodzic, R. Andersson, U. Göransson, P. R. Jonsson, and L. Bohlin. 2008. Antifouling activity of a dibrominated cyclopeptide from the marine sponge *Geodia barretti*. *J Nat Prod* 71:330–333.

Hernandez, J. F., J. Gagnon, L. Chiche, T. M. Nguyen, J. P. Andrieu, A. Heitz, T. T. Hong, T. T. C. Pham, and D. L. Nguyen. 2000. Squash trypsin inhibitors from *Momordica cochinchinensis* exhibit an atypical macrocyclic structure. *Biochemistry* 39:5722–5730.

Herrmann, A., R. Burman, J. S. Mylne, G. Karlsson, J. Gullbo, D. J. Craik, R. J. Clark, and U. Göransson. 2008. The alpine violet, *Viola biflora*, is a rich source of cyclotides with potent cytotoxicity. *Phytochemistry* 69:939–952.

Hibbett, D. S., M. Binder, J. F. Bischoff, M. Blackwell, P. F. Cannon, O. E. Eriksson, S. Huhndorf, T. James, P. M. Kirk, R. Lucking, H. Thorsten Lumbsch, F. Lutzoni, P. B. Matheny, D. J. McLaughlin, M. J. Powell, S. Redhead, C. L. Schoch, J. W. Spatafora, J. A. Stalpers, R. Vilgalys, M. C. Aime, A. Aptroot, R. Bauer, D. Begerow, G. L. Benny, L. A. Castlebury, P. W. Crous, Y. C. Dai, W. Gams, D. M. Geiser, G. W. Griffith, C. Gueidan, D. L. Hawksworth, G. Hestmark, K. Hosaka, R. A. Humber, K. D. Hyde, J. E. Ironside, U. Koljalg, C. P. Kurtzman, K. H. Larsson, R. Lichtwardt, J. Longcore, J. Miadlikowska, A. Miller, J. M. Moncalvo, S. Mozley-Standridge, F. Oberwinkler, E. Parmasto, V. Reeb, J. D. Rogers, C. Roux, L. Ryvarden, J. P. Sampaio, A. Schussler, J. Sugiyama, R. G. Thorn, L. Tibell, W. A. Untereiner, C. Walker, Z. Wang, A. Weir, M. Weiss, M. M. White, K. Winka, Y. J. Yao, and N. Zhang. 2007. A higher-level phylogenetic classification of the fungi. *Mycol Res* 111(Pt 5):509–547.

Iotti, M. and A. Zambonelli. 2006. A quick and precise technique for identifying ectomycorrhizas by PCR. *Mycol Res* 110(Pt 1):60–65.

IUCN. The IUCN red list of threatened species 2009. Available from http://www.iucnredlist.org//, accessed on February 26, 2009.

Jennings, C., J. West, C. Waine, D. J. Craik, and M. Anderson. 2001. Biosynthesis and insecticidal properties of plant cyclotides: The cyclic knotted proteins from *Oldenlandia affinis*. *Proc Natl Acad Sci USA* 98:10614–10619.

Jiao, W. H., H. Gao, C. Y. Li, F. Zhao, R. W. Jiang, Y. Wang, G. X. Zhou, and X. S. Yao. 2010. Quassidines A-D, bis-beta-carboline alkaloids from the stems of *Picrasma quassioides*. *J Nat Prod*. 73 (e-pub ahead of publication):167–171.

Jin, Y. G., Y. Wang, W. Wang, Q. H. Shang, C. Q. Cao, and D. H. Erwin. 2000. Pattern of marine mass extinction near the Permian-Triassic boundary in South China. *Science* 289:432–436.

Keeling, P. J. and J. D. Palmer. 2008. Horizontal gene transfer in eukaryotic evolution. *Nat Rev Genet* 9:605–618.

Klitgaard, A. B. 1995. The fauna associated with outer shelf and upper slope sponges (Porifera, Demospongiae) at the Faroe Islands, Northeastern Atlantic. *Sarsia* 80:1–22.

Koljalg, U., K. H. Larsson, K. Abarenkov, R. H. Nilsson, I. J. Alexander, U. Eberhardt, S. Erland, K. Hoiland, R. Kjoller, E. Larsson, T. Pennanen, R. Sen, A. F. Taylor, L. Tedersoo, T. Vralstad, and B. M. Ursing. 2005. UNITE: A database providing web-based methods for the molecular identification of ectomycorrhizal fungi. *New Phytol* 166:1063–1068.

Laessoe, T. and K. Hansen. 2007. Truffle trouble: What happened to the Tuberales? *Mycol Res* 111:1075–1099.

Larsson, J., J. Gottfries, S. Muresan, and A. Backlund. 2007. ChemGPS-NP: Tuned for navigation in biologically relevant chemical space. *J Nat Prod* 70:789–794.

Legros, D., G. Ollivier, M. Gastellu-Etchegorry, C. Paquet, C. Burri, J. Jannin, and P. Buscher. 2002. Treatment of human African trypanosomiasis—Present situation and needs for research and development. *Lancet Infect Dis* 2:437–440.

Leikert, J. F., T. R. Rathel, P. Wohlfart, V. Cheynier, A. M. Vollmar, and V. M. Dirsch. 2002. Red wine polyphenols enhance endothelial nitric oxide synthase expression and subsequent nitric oxide release from endothelial cells. *Circulation* 106:1614–1617.

Leong, X. F., M. N. Najib, S. Das, M. R. Mustafa, and K. Jaarin. September 2, 2009. Intake of repeatedly heated palm oil causes elevation in blood pressure with impaired vasorelaxation in rats. *Tohoku J Exp Med* 19(1):71–78.

Leta Aboye, T., R. J. Clark, D. J. Craik, and U. Göransson. 2008. Ultra-stable peptide scaffolds for protein engineering-synthesis and folding of the circular cystine knotted cyclotide cycloviolacin O_2. *ChemBioChem* 9:103–113.

Lindholm, P., U. Göransson, S. Johansson, P. Claeson, J. Gullbo, R. Larsson, L. Bohlin, and A. Backlund. 2002. Cyclotides: A novel type of cytotoxic agents. *Mol Cancer Ther* 1:365–369.

Lipinski, C. A., F. Lombardo, B. W. Dominy, and P. J. Feeney. 1997. Experimental and computational approaches to estimate solubility and permeability in drug discovery and development settings. *Adv Drug Deliv Rev* 23:3–25.

Loftus, B., I. Anderson, R. Davies, U. C. Alsmark, J. Samuelson, P. Amedeo, P. Roncaglia, M. Berriman, R. P. Hirt, B. J. Mann, T. Nozaki, B. Suh, M. Pop, M. Duchene, J. Ackers, E. Tannich, M. Leippe, M. Hofer, I. Bruchhaus, U. Willhoeft, A. Bhattacharya, T. Chillingworth, C. Churcher, Z. Hance, B. Harris, D. Harris, K. Jagels, S. Moule, K. Mungall, D. Ormond, R. Squares, S. Whitehead, M. A. Quail, E. Rabbinowitsch, H. Norbertczak, C. Price, Z. Wang, N. Guillen, C. Gilchrist, S. E. Stroup, S. Bhattacharya, A. Lohia, P. G. Foster, T. Sicheritz-Ponten, C. Weber, U. Singh, C. Mukherjee, N. M. El-Sayed, W. A. Petri, Jr., C. G. Clark, T. M. Embley, B. Barrell, C. M. Fraser, and N. Hall. 2005. The genome of the protist parasite *Entamoeba histolytica. Nature* 433:865–868.

Lossick, J. G. 1990. Treatment of sexually transmitted vaginosis/vaginitis. *Rev Infect Dis* 12(Suppl 6):S665–S681.

May, R. M. 1994. Biological diversity: differences between land and sea. *Philo Trans: Biol Sci* 343:105–111.

Mueller, G. M., J. P. Schmit, P. R. Leacock, B. Buyck, J. Cifuentes, D. E. Desjardin, R. E. Halling, K. Hjortstam, T. Iturriaga, K.-H. Larsson, J. Lodge, T. W. May, D. Minter, M. Rajchenberg, S. A. Redhead, L. Ryvarden, J. M. Trappe, R. Watling, and Q. Wu. 2007. Global diversity and distribution of macrofungi. *Biodivers Conserv* 16:37–48.

Newman, D. J. and G. M. Cragg. 2007. Natural products as sources of new drugs over the last 25 years. *J Nat Prod* 70:461–477.

Nono, E. C., P. Mkounga, V. Kuete, K. Marat, P. G. Hultin, and A. E. Nkengfack. 2010. Pycnanthulignenes A-D, antimicrobial cyclolignene derivatives from the roots of *Pycnanthus angolensis. J Nat Prod* 73:213–216.

Nozaki, T., J. C. Engel, and J. A. Dvorak. 1996. Cellular and molecular biological analyses of nifurtimox resistance in *Trypanosoma cruzi. Am J Trop Med Hyg* 55:111–117.

O'Keefe, B. R. 2001. Biologically active proteins from natural product extracts. *J Nat Prod* 64:1373–1381.

Paul, V. J. and R. Ritson-Williams. 2008. Marine chemical ecology. *Nat Prod Rep* 25:662–695.

Petrin, D., K. Delgaty, R. Bhatt, and G. Garber. 1998. Clinical and microbiological aspects of *Trichomonas vaginalis*. *Clin Microbiol Rev* 11:300–317.

Piel, J. 2009. Metabolites from symbiotic bacteria. *Nat Prod Rep* 26:338–362.

Plaza, A., J. L. Keffer, J. R. Lloyd, P. L. Colin, and C. A. Bewley. 2010. Paltolides A-C, anabaenopeptin-type peptides from the Palau sponge *Theonella swinhoei*. *J Nat Prod* 73:485–488.

Pongprayoon, U., P. Bäckström, U. Jacobsson, M. Lindström, and L. Bohlin. 1991. Compounds inhibiting prostaglandin synthesis isolated from *Ipomoea pes-capræ*. *Planta Med* 57:515–518.

Rangel-Castro, J. I., J. J. Levenfors, and E. Danell. 2002. Physiological and genetic characterization of fluorescent *Pseudomonas* associated with *Cantharellus cibarius*. *Can J Microbiol* 48:739–748.

Richardson, J. E., F. M. Weitz, M. F. Fay, Q. C. B. Cronk, H. P. Linder, G. Reeves, and M. W. Chase. 2001. Rapid and recent origin of species richness in the Cape flora of South Africa. *Nature* 412:181–183.

Rosengren, K. J., N. L. Daly, M. R. Plan, C. Waine, and D. J. Craik. 2003. Twists, knots, and rings in proteins. Structural definition of the cyclotide framework. *J Biol Chem* 278:8606–8616.

Saether, O., D. J. Craik, I. D. Campbell, K. Sletten, J. Juul, and D. G. Norman. 1995. Elucidation of the primary and three-dimensional structure of the uterotonic polypeptide kalata B1. *Biochemistry* 34:4147–4158.

Samarawickrema, N. A., D. M. Brown, J. A. Upcroft, N. Thammapalerd, and P. Upcroft. 1997. Involvement of superoxide dismutase and pyruvate: Ferredoxin oxidoreductase in mechanisms of metronidazole resistance in *Entamoeba histolytica*. *J Antimicrob Chemother* 40:833–840.

Samuelsson, G. 1987. Plants used in traditional medicine as sources of drugs. *Bull Chem Soc Ethiop* 1:47–54.

Samuelsson, G., G. Kyerematen, and M. H. Farah. 1985. Preliminary chemical characterization of pharmacologically active compounds in aqueous plant extracts. *J Ethnopharm* 14:193–201.

Sandberg, F. 1965. Étude sur les plantes médicinales et toxiques d'Afrique équatoriale. *Cahièrs de la Maboké* 3:5–31.

Schoepke, T., M. I. Hasan Agha, R. Kraft, A. Otto, and K. Hiller. 1993. Compounds with hemolytic activity from *Viola tricolor* and *Viola arvensis*. *Sci. Pharm.* 61:145–153.

Sertürner, F. W. 1806. Apotheker und Chemisten. *Trommsdorff's Journal der Pharmazie für Ärzte* 14:47–93.

Sertürner, F. W. 1817a. Über das Morphium, eine neue salzfahige Grundlage, und die Mekonsäure, als Hauptbestandtheile des Opiums. *Gilbert's Annalen der Physik* 55:56–89.

Sertürner, F. W. 1817b. De la morphine et de l'acide meconique, consideres comme parties essentielles de l'opium. *Annales de Chimie et de Physique* 5:21–42.

Shenkarev, Z. O., K. D. Nadezhdin, V. A. Sobol, A. G. Sobol, L. Skjeldal, and A. S. Arseniev. 2006. Conformation and mode of membrane interaction in cyclotides. Spatial structure of kalata B1 bound to a dodecylphosphocholine micelle. *Febs J* 273:2658–2672.

Simonsen, S. M., L. Sando, K. J. Rosengren, C. K. Wang, M. L. Colgrave, N. L. Daly, and D. J. Craik. 2008. Alanine scanning mutagenesis of the prototypic cyclotide reveals a cluster of residues essential for bioactivity. *J Biol Chem* 283:9805–9813.

Sjögren, M. 2006. Bioactive compounds from the marine sponge *Geodia barretti*: Characterization, antifouling activity and molecular targets. [Thesis]. Uppsala: Acta universitatis upsaliensis. Digital comprehensive summaries of Uppsala dissertations from the Faculty of Pharmacy, 32.

Sjögren, M., P. R. Jonsson, M. Dahlström, T. Lundälv, R. Burman, U. Göransson, and L. Bohlin. 2011. Two brominated cyclic dipeptides released by the coldwater marine sponge *Geodia barretti* act in synergy as chemical defense. *J Nat Prod* 74:449–454.

Stanley, S. L., Jr. 2003. Amoebiasis. *Lancet* 361:1025–1034.

Sticher, O. 2008. Natural product isolation. *Nat Prod Rep* 25:517–554.

Svangård, E., R. Burman, S. Gunasekera, H. Lövborg, J. Gullbo, and U. Göransson. 2007. Mechanism of action of cytotoxic cyclotides: Cycloviolacin O_2 disrupts lipid membranes. *J Nat Prod* 70:643–647.

Tam, J. P., Y. A. Lu, J. L. Yang, and K. W. Chiu. 1999. An unusual structural motif of antimicrobial peptides containing end-to-end macrocycle and cystine-knot disulfides. *Proc Natl Acad Sci USA* 96:8913–8918.

Tedersoo, L., U. Koljalg, N. Hallenebrg, and K. H. Larsson. 2003. Fine scale distribution of ectomycorrhizal fungi and roots across substrate layers including coarse woody debris in a mixed forest. *New Phytologist* 159:153–165.

Thongyoo, P., C. Bonomelli, R. J. Leatherbarrow, and E. W. Tate. 2009. Potent inhibitors of beta-tryptase and human leukocyte elastase based on the MCoTI-II scaffold. *J Med Chem* 52:6197–6200.

Traber, R., H. Hofmann, and H. Kobel. 1989. Cyclosporins-new analogues by precursor directed biosynthesis. *J Antibiot* 42:591–597.

Tunon, H., L. Bohlin, and G. Öjteg. 1994. The effect of *Menyanthes trifoliata* L. on acute renal failure might be due to PAF-inhibition. *Phytomedicine* 1:39–45.

Tuominen, M., L. Bohlin, L. O. Lindbom, and W. Rolfsen. 1991. Enhancing effect of Calaguala on the prevention on rejection of skin transplants in mice. *Phytother Res* 5:234–236.

Tuominen, M., L. Bohlin, and W. Rolfsen. 1992. Effects of Calaguala and an active principle, adenosine, on platelet activating factor. *Planta Med* 58:306–310.

Vasänge, M., B. Liu, C. J. Welch, W. Rolfsen, and L. Bohlin. 1997a. The flavonoid constituents of two *Polypodium* species (Calaguala) and their effect on the elastase release in human neutrophils. *Planta Med* 63:511–517.

Vasänge, M., W. Rolfsen, and L. Bohlin. 1997b. A sulphonoglycolipid from the fern *Polypodium decumanum* and its effect on the platelet activating-factor receptor in human neutrophils. *J Pharm Pharmacol* 49:562–566.

Vasänge-Tuominen, M., P. Perera-Ivarsson, J. Shen, L. Bohlin, and W. Rolfsen. 1994. The fern *Polypodium decumanum*, used in the treatment of psoriasis, and its fatty acid constituents as inhibitors of leukotriene B4 formation. *Prostaglandins Leukot Essent Fatty Acids* 50:279–284.

Vlieghe, P., V. Lisowski, J. Martinez, and M. Khrestchatisky. 2009. Synthetic therapeutic peptides: science and market. *Drug Discov Today* 15:40–56.

Wang, C. K., M. L. Colgrave, D. C. Ireland, Q. Kaas, and D. J. Craik. 2009. Despite a conserved cystine knot motif, different cyclotides have different membrane binding modes. *Biophys J* 97:1471–1481.

Wang, C. K., S. H. Hu, J. L. Martin, T. Sjögren, J. Hajdu, L. Bohlin, P. Claeson, U. Göransson, K. J. Rosengren, J. Tang, N. H. Tan, and D. J. Craik. 2009. Combined x-ray and NMR analysis of the stability of the cyclotide cystine knot fold that underpins its insecticidal activity and potential use as a drug scaffold. *J Biol Chem* 284:10672–10683.

Wedén, C. 2004. Black truffles of Sweden: Systematics, population studies, ecology and cultivation of tuber aestivum syn. T. uncinatum. [Thesis]. Uppsala: Acta Universitatis Upsaliensis. Comprehensive summaries of Uppsala dissertations from the Faculty of Science and Technology, 1043.

Wedén, C., L. Pettersson, and E. Danell. 2009. Truffle cultivation in Sweden: results from *Quercus robur* and *Corylus avellana* field trials on the island of Gotland. *Scandinavian J Forest Res* 24:37–53.

Witherup, K. M., M. J. Bogusky, P. S. Anderson, H. Ramjit, R. W. Ransom, T. Wood, and M. Sardana. 1994. Cyclopsychotride A, a biologically active, 31-residue cyclic peptide isolated from *Psychotria longipes. J Nat Prod* 57:1619–1625.

Zhang, H. W., Y. C. Song, and R. X. Tan. 2006. Biology and chemistry of endophytes. *Nat Prod Rep* 23:753–771.

2 Discovery of Potential Anticancer Agents from Aquatic Cyanobacteria, Filamentous Fungi, and Tropical Plants

Jimmy Orjala, Nicholas H. Oberlies,
Cedric J. Pearce, Steven M. Swanson,
and A. Douglas Kinghorn

CONTENTS

2.1   INTRODUCTION

Cancer leads to one in four deaths in the United States, and is also a major public health hazard in countries all over the world. Despite improvements in the survival rates in several major types of cancer, in the United States for the year 2009, over 560,000 deaths and nearly 1,480,000 new cancer cases were projected to occur (Jemal et al., 2009). In addition to cancer prevention and early detection, one of the strategies for improving the dire consequences of the scourge of cancer is improved treatment options inclusive of chemotherapy (Jemal et al., 2009). Natural products have played an integral role in cancer chemotherapy, not only in terms of providing new drugs and new drug leads for synthetic modification, but in also affording substances to probe cellular and molecular mechanisms of action germane to cancer inhibition (Cragg et al., 2005, 2009). In a now widely cited analysis, it was established that out of 81 nonbiological small-molecule anticancer drugs introduced into therapy in North America, western Europe, and Japan over a 25 year period from 1981 to 2006, 11.1% of these were unmodified natural products and 30.9% natural product semisynthetic derivatives, with only 22.2% classified as being totally synthetic compounds produced from an initial lead by random screening (Newman and Cragg, 2007). Several compounds of terrestrial microbe and higher plant origin are now approved as drugs in cancer chemotherapy (Cragg et al., 2005), and recently the first such marine-derived compound, ecteinascidin 743, has been approved in certain European countries for patients with advanced soft tissue sarcoma as an orphan drug (Bailly, 2009; Sashidhara et al., 2009). Judging by the relatively large number of natural products in clinical trials as potential new oncolytic agents (Butler, 2008, 2010; Harvey, 2008; Sashidhara et al., 2009), it can be expected that new natural product-derived drugs will be approved to treat cancer in the next few years, but from even more diverse types of organisms than has been the case to date.

The natural world offers the researcher interested in anticancer drug discovery a very large number of potential organisms on which to work, with the earth's biodiversity divisible into six kingdoms (Eubacteria, Archaea, Protoctista, Plantae, Fungi, and Animalia) (Tan et al., 2007). The authors of this chapter have a joint interest in the discovery of potential new cancer chemotherapeutic agents from three of the above-named kingdoms, namely, cyanobacteria of aquatic origin, filamentous fungi, and higher plants of tropical origin. A preliminary account of our current joint work has appeared in the literature, inclusive of the organizational aspects of our project (Kinghorn et al., 2009). In this chapter, the potential of cyanobacteria, fungi, and higher plants to afford new anticancer drugs will be described in turn, with specific information provided on how these types of organisms may be accessed and handled in the laboratory. The success of any drug discovery program depends on the effective use of an array of in vitro and in vivo bioassays, and this aspect of the process also will be discussed, prior to some concluding remarks. The authors of this chapter are currently collaborating on a multidisciplinary "program project" funded by the U.S. National Cancer Institute, directed toward the discovery of new natural product anticancer agents from diverse organisms. Aspects of our joint work in this endeavor are mentioned in Sections 2.2 through 2.6.

2.2   CYANOBACTERIA AS SOURCES OF POTENTIAL ANTICANCER AGENTS

Cyanobacteria (also known as blue-green algae) are Gram-negative bacteria and represent the only group of prokaryotes that are able to perform oxygenic photosynthesis similar to plants. Fossil evidence indicates that cyanobacteria have populated the earth for around 3.5 billion years and ancient cyanobacteria are believed to have been instrumental in the creation of the Earth's oxygen-rich atmosphere (De Marais, 2000; Schopf, 2000). In addition, chloroplasts, the organelles responsible for photosynthesis in algae and plants, are thought to have evolved from endosymbiosis with cyanobacteria (Douglas, 1994). Cyanobacteria are distributed worldwide and can be found most in terrestrial, marine, brackish, and freshwater environments. Cyanobacteria are also found in extreme environments such as hot springs where they live in temperatures up to 70°C, arid desert soils, frigid lakes in Antarctica, the Arctic tundra, and areas of high salinity.

2.2.1   BIOACTIVE COMPOUNDS FROM CYANOBACTERIA

Several strains of cyanobacteria are known to produce toxins associated with toxic water blooms and represent a public health hazard due to their presence in water reservoirs for drinking water supplies (for recent reviews see Shaw and Lam, 2007; van Apeldoorn et al., 2007; Welker, 2008). The cyanotoxins are divided into two categories: hepatotoxins (e.g., microcystins, nodularins, and cylindrospermopsin) and neurotoxins (e.g., anatoxin-a and saxitoxin). Chemical and biological investigations into these toxic constituents have indicated that cyanobacteria produce a diverse array of secondary metabolites with unique chemical structures and biological activity profiles (Figure 2.1). For example, the microcystins and nodularins were found to be potent inhibitors of eukaryotic protein serine/threonine phosphatases 1 and 2A. Microcystin-LR (**1**), frequently present in cyanobacterial blooms in rivers and lakes, is currently used as a molecular probe in pharmacological studies.

Cyanobacteria are considered a promising source for new pharmaceutical lead compounds and a large number of chemically diverse metabolites have been obtained from cyanobacteria. Screens for new antineoplastic agents from cyanobacteria yielded a hit rate of approximately 7%, which is comparable to that found for other microorganisms such as Actinobacteria (Carmichael, 1992). The rate of rediscovery of known bioactive compounds was significantly lower among the cyanobacteria than from other microorganisms. Both marine and non-marine cyanobacteria have proven to be a rich source of diverse new metabolites (for recent reviews, see: Gerwick et al., 2001; Singh et al., 2005; Sielaff et al., 2006; Welker and von Döhren, 2006; Tan, 2007). Several potential anticancer agents of cyanobacterial origin have been discovered. Three of these are highlighted here: the cryptophycins, dolastatin 10, and dolastatin 15.

The pioneering work of Moore and Patterson in Hawaii led to the discovery of cryptophycin (**2**) as an antimitotic agent from the terrestrial cyanobacterium *Nostoc* sp. (GSV 224) (Golakoti et al., 1994). Cryptophycin was originally isolated from another terrestrial *Nostoc* sp. (ATCC 53789) by a group from Merck as an

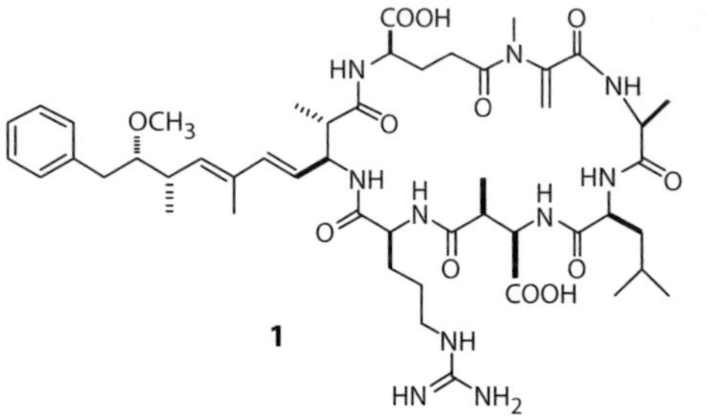

1

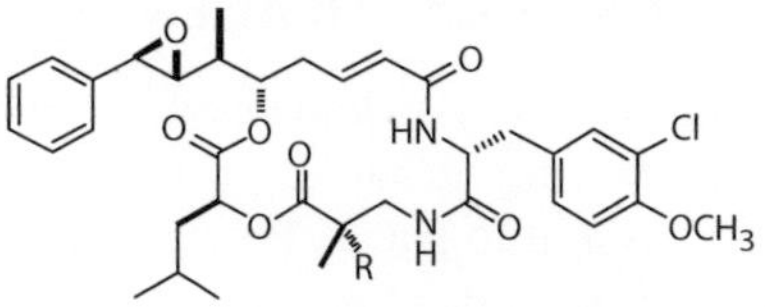

2 R = H
3 R = CH_3

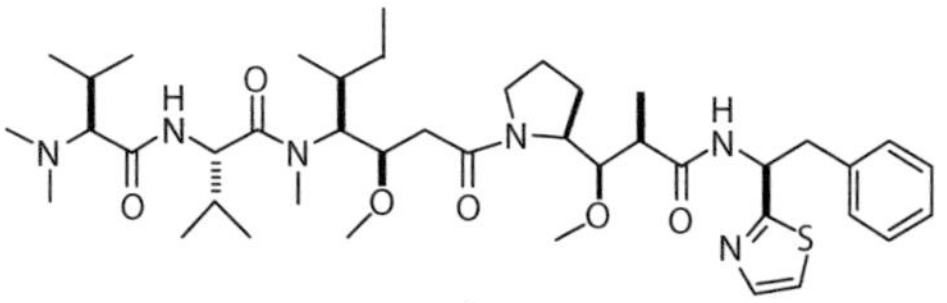

4

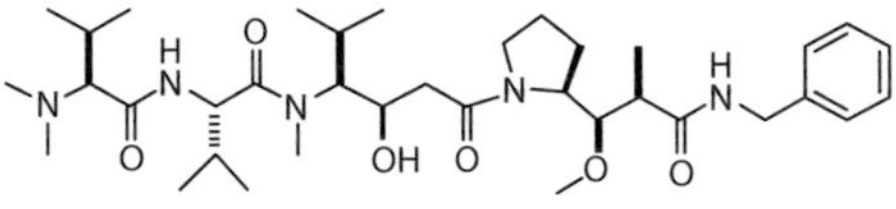

5

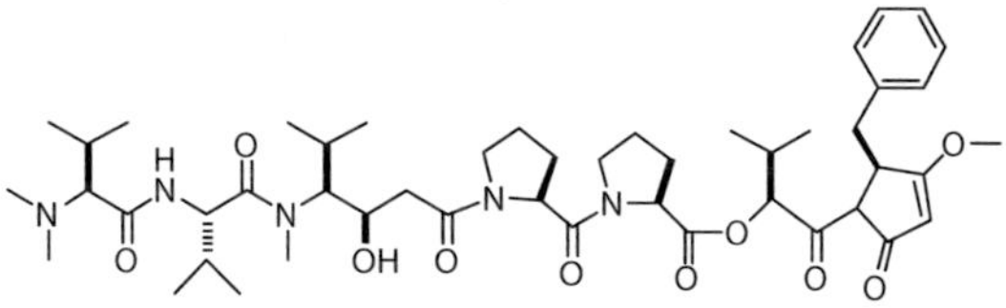

6

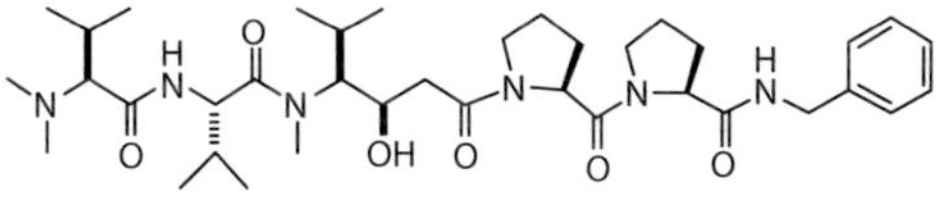

7

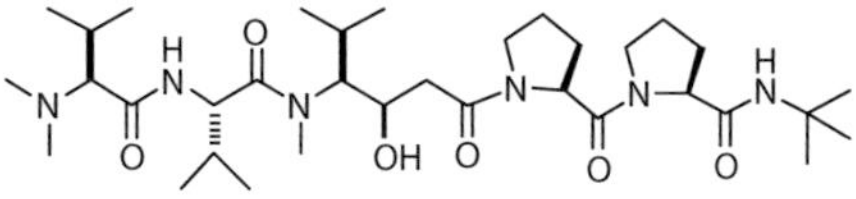

8

FIGURE 2.1 Structures of bioactive compounds from cyanobacteria.

antifungal agent (Schwartz et al., 1990). Cryptophycin is a depsipeptide with tubulin-destabilizing activity. Synthetic efforts led to the development of cryptophycin 52 (**3**), a synthetic analogue that was advanced into clinical trials (Shih and Teicher, 2001). Cryptophycin 52 was discontinued during phase II clinical trials due to toxic side effects (Newman and Cragg, 2004). Potential second-generation analogues of cryptophycin with significantly increased potency have recently been reported (Liang et al., 2005).

Dolastatin 10 (**4**), a linear peptide, was originally isolated from the sea hare *Dolabella auricularia* in the early 1970s by Pettit's group (Pettit et al., 1987). The structure elucidation took almost 15 years due to the low concentration in the sea hare (1 mg for 100 kg; Pettit et al., 1987). It was long speculated that the cyanobacterial diet was the true source of dolastatin 10 for *D. auricularia*. This hypothesis was confirmed as dolastatin 10 was found in the marine cyanobacterium *Symploca* sp. (Luesch et al., 2001). Dolastatin 10 is a potent inhibitor of tubulin polymerization and entered clinical trials in the 1990s using synthetically produced material. Dolastatin 10 was discontinued during the phase II evaluation due to lack of significant antitumor activity (Newman and Cragg, 2004). A synthetic derivative of dolastatin 10, TZT-1027 (**5**), recently completed a phase I study for the treatment of non-small cell lung cancer (Horti et al., 2008).

Dolastatin 15 (**6**), a linear depsipeptide, was also obtained from the sea hare *D. auricularia* at low concentration (Pettit et al., 1989a). Again, a cyanobacterial diet is the probable source of dolastatin 15 as several related depsipeptides have been found in cyanobacteria. Dolastatin 15 inhibits microtubule assembly and induces apoptosis. However, the low yield and poor water solubility prevented it from entering clinical trials (Newman and Cragg, 2004). Two synthetic derivatives of dolastatin 15, cematodin (**7**) and ILX-651 (**8**), have progressed to clinical trials. Cematodin is a stable and water-soluble second-generation dolastatin 15 analogue (De Arruda et al., 1995). It was discontinued during phase II evaluations due to lack of objective responses (Newman and Cragg, 2004). ILX-651 is an orally active third-generation dolastatin 15 analogue that is currently in phase II clinical trials (Simmons et al., 2005; Ray et al., 2007).

2.2.2 PROCESSING OF CYANOBACTERIA FOR ANTICANCER DRUG DISCOVERY

Studies of bioactive metabolites from cyanobacteria have been performed with both field-collected material and laboratory-grown cultures. Field-collected material (water blooms or tuft-forming marine cyanobacteria) often contains an assemblage of multiple organisms, which can raise a concern regarding the true origin of an isolated natural product. This also makes the recollection and re-isolation of a particular natural product difficult. In addition, field collections are only feasible for cyanobacteria that naturally have dense growth. Laboratory-grown cultures use unialgal strains and allow for the investigation of species that may not grow to sufficient density in the wild. Culture in the laboratory also affords the ability to easily obtain additional biomass for the re-isolation of any natural products of interest. For these reasons, we have chosen to focus on laboratory-grown cyanobacteria centered on a culture collection of freshwater and terrestrial cyanobacteria obtained from the Upper Midwest and Great Lakes region. This region provides a large variety of freshwater microenvironments, which are home to a diverse set of cyanobacteria. Aquatic cyanobacteria are being collected and evaluated as part of our current program project (Kinghorn et al., 2009).

Multiple techniques have been devised to obtain unialgal strains from material collected in the field. Two commonly used techniques are isolation by micropipette and streak plate (Andersen and Kawachi, 2005). Isolation by micropipette uses this device to isolate and wash cyanobacterial cells under a dissecting microscope. Once the cells are washed sufficiently, they are transferred to a tube containing a sterile medium and incubated at 20°C with an 18/6 h light/dark cycle. We employ three different media for the isolation of freshwater and terrestrial cyanobacteria: Z, BG-12, and BG-12$_0$ (Falch et al., 1995; Chlipala et al., 2009). Isolation by streak plate is commonly used for microbiological isolation work and has been successfully adapted to isolation of microalgae. Each collection is typically streaked onto a plate with medium and cycloheximide. The plates are incubated at 20°C with an 18/6 h light/dark cycle. Once algal colonies are visible (approximately 2–4 weeks), they are isolated and washed using a micropipette, as detailed above. Unialgal strains are then grown in inorganic medium at 22°C. Light is provided via fluorescent lamps with an 18/6 h light/dark cycle and a mean luminance at shelf level of 1.93 klx. Initially, a 150 mL of culture in a 250 mL Erlenmeyer flask is cultured for approximately 2 weeks. Once there is sufficient growth, this biomass is used to inoculate two 2.8 L Fernbach flasks with 2 L of medium each (total volume: 4 L). These cultures are aerated with sterile-filtered air and allowed to grow for 6–8 weeks. The harvested material is freeze-dried and extracted using methanol-dichloromethane (1:1). The resulting extract is dried in vacuo and a small portion is used to create a library stock solution at 10 mg/mL in DMSO. The extract library is stored at −80°C and used to supply extracts for biological evaluation. The remaining extract is fractionated using a solid-phase extraction (SPE) column packed with Diaion HP-20SS and a step gradient of water and 2-propanol to give eight fractions. The fractions are dried in vacuo and dissolved at 10 mg/mL in DMSO. This fraction library is also stored at −80°C.

The slow growth rate and low biomass yield is a challenge when studying bioactive metabolites from cultured cyanobacteria: It takes almost 2 months to obtain 2×2 L of cyanobacterial culture and the average extract amount obtained is 120 mg. We use microanalytical techniques to overcome the limitations of low sample mass, thereby accelerating our investigation of cultured cyanobacteria. The procedure used in laboratory is outlined in Figure 2.2. Depending on assay configuration, either the extract library or fraction library is used for the initial screening. For extracts showing activity in a given assay, the corresponding SPE fractions are immediately evaluated. Active SPE fractions are further separated by semi-preparative high-performance liquid chromatography (HPLC) using C_{18} or C_8 phase-bonded silica gel columns (Onyx™, Phenomenex, Torrance, CA). The column material and the gradient

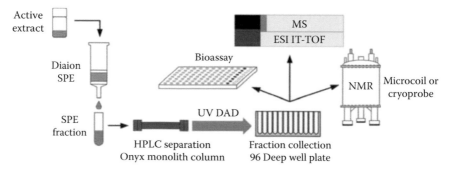

FIGURE 2.2   Processing of active cyanobacterial extracts.

employed are optimized to match the polarity of the SPE fraction. HPLC fractions are collected directly in a 96 deep-well microtiter plate and dried in vacuo. The HPLC fractions are re-dissolved and aliquots are transferred to a clean 96-well plate for biological evaluation. Additional aliquots from the HPLC fractions are evaluated using mass spectrometry (MS) and NMR. Accurate mass measurements are obtained using an electrospray ionization-ion trap-time-of-flight spectrometer equipped with a 96-well plate injector. Proton NMR spectra of the HPLC fractions are obtained using microcoil NMR (Lin et al., 2008) or cryoprobe NMR spectrometers. The activity data of HPLC fractions is plotted as a function of fraction number. This bioactivity chromatogram is then compared with UV, MS, and NMR data to identify the active constituent(s). Initially, spectroscopic data are used in conjunction with in-house as well as publicly available databases (SciFinder, Beilstein, MarinLit) to determine if the metabolite(s) have been previously described. This process is called "dereplication." If no previously reported metabolites are found, complete structure determination of the active principle(s) follows. This may initially include further purification steps, if deemed necessary, or HPLC fractionation of the remaining SPE fraction sample to obtain sufficient material to obtain two-dimensional (2D) NMR spectra. If needed, additional biomass is obtained by inoculating additional Fernbach flasks. The increased NMR sensitivity obtained by microcoil and cryoprobe NMR affords the discovery of natural products at the nanomole scale (Molinski, 2009), and we are able to perform many of our studies with the initial cyanobacterial growth volume of 4 L. The quantity of pure active natural product obtained using our methods is low, but sufficient for structure determination as well as to measure the level of biological activity against selected targets. Follow-up studies require re-isolation; however, this effort is spent only for novel compounds with a promising activity profile.

2.3   FILAMENTOUS FUNGI AS SOURCES OF POTENTIAL ANTICANCER AGENTS

Plants, microorganisms, and animals have all yielded novel compounds displaying potent anticancer activity (Kinghorn, 2008; Cragg et al., 2009; Pearce et al., 2010). As a source, vascular plants may be more well known, having led to many life-saving antineoplastic agents, such as paclitaxel and camptothecin (Oberlies and Kroll, 2004). However, compounds derived from microbial sources have also had an impact on anticancer drug discovery. Recently, the Bristol-Myers Squibb derivative of the myxobacterial product, epothilone B (Ixabepilone®), was approved for clinical use against chemotherapy-resistant advanced breast cancer (Pronzato, 2008), and epothilone derivatives are being investigated for additional cancer applications. Several other bacterial products are in various stages of clinical trials for cancer, including the HSP-90-binding geldanamycin derivatives, tanespimycin (Erlichman, 2009) and alvespimycin (Bailly, 2009). Just a few weeks before finalizing this chapter, the cyclic depsipeptide, romidepsin (aka FK228 and FR901228), which inhibits histone deacetylase (HDAC), was approved by the FDA for cutaneous T-cell lymphoma (CTCL) (Glaser, 2007; Piekarz et al., 2009). In short, secondary metabolites from microorganisms continue to have a profound impact on anticancer drug discovery (Bailly, 2009).

2.3.1 WHY FUNGI?

Fungi have been the source of many important medicinal compounds (Pearce et al., 2010), and chemically diverse fungal metabolites have demonstrated potent in vitro and in vivo anticancer activities. Fungi are found widely in nature, and although 1.5 million individual fungal species are estimated in the world, only about 80,000 have been described in the literature (Hawksworth and Rossman, 1997). Taken together with new genome data reinforcing the biosynthetic potential of these organisms (Misiek and Hoffmeister, 2007), fungi are a highly promising source for new medicines, including those for treating cancer. There is a fungal component in our current collaborative natural product anticancer drug program project (Kinghorn et al., 2009). Below, we include a brief account of active metabolites found from filamentous fungi garnered from the literature. We posit that fungi are an under-explored resource for novel chemistries. Moreover, with the trend to destruction of natural ecosystems throughout the world, we hope that we will be able to investigate some of these organisms before they disappear forever.

Fungi generate a vast chemical diversity, largely from a variety of simple chemical starting materials, including organic acids, sugars, amino acids, terpenes, and bases such as purines and pyrimidines, via pathways such as polyketide synthases (PKS) and non-ribosomal peptide synthetases (NRPS) (Turgeon et al., 2008). This chemical diversity has translated into a wealth of pharmacological activities, including those associated with the following endpoints: antibacterial (penicillin and cephalosporin), antifungal (pneumocandin and strobilurin), antiviral (3-O-methylviridicatin and emmyguyacins), antiparasitic (apicidin), immunomodulators (cyclosporin and mycophenolic acid), enzyme inhibition (statins), receptor binding (lysergic acid derivatives), gene regulation (e.g., caffeic acid phenethyl ester that specifically inhibits DNA binding of NF-κB), and especially anticancer (discussed later) (Keller et al., 2005).

One of the more challenging issues for those engaged in research on bioactive compounds from fungi is the apparent lack of relationship between morphological aspects of a fungus and its secondary metabolites. Specifically, knowledge of fungal taxonomy does not necessarily predict the types of compounds that are produced. Other fungal researchers have expressed similar frustrations (e.g., Nielsen and Smedsgaard, 2003), and this is in stark contrast to plant-based drug discovery, where general structural classes can often (but not always) be predicted based on the taxonomy. Moreover, once compounds are discovered, it is not possible to predict a priori the pharmacological activity of the vast majority of new chemical entities with any certainty, although this limitation is probably true for all forms of drug discovery.

2.3.2 GENOMIC CONSIDERATIONS OF CHEMICAL DIVERSITY

Gene sequencing, which is being conducted on an increasing number of fungi (McAlpine, 2009), has demonstrated that there are multiple so-called "silent" biosynthetic pathways. Thus, there is genetic information present within the fungal genome that encodes, for example, for polyketide synthesis that has not been accounted for at the secondary metabolite level. Unlocking these pathways could lead to the discovery of many more complex polyketides as well as other metabolites (i.e., those derived from NRPS). Gene sequences for as many as 30 or more novel products per fungus have been detected (Pel et al., 2007). Taken together with the prediction that there are approximately 1.5 million fungal species (Hawksworth and Rossman, 1997) and using the median figure of 15 possible products per organism, there is a potential for in excess of 20 million metabolites. Approximately 7000 fungal metabolites have been reported to date (Berdy, 2005), and, thus, there remains more than 99% of possible fungal metabolites yet to be discovered (for recent reviews see: Keller et al., 2005; Misiek and Hoffmeister, 2007). Of the many PKS and NRPS identified through sequence analysis, in some cases the products of these enzymes are known, but in other cases they are unknown; currently, there are insufficient data available regarding the biosynthetic rules to enable the prediction of the structure of PKS products of fungi, although this seems to be rapidly changing (Crawford et al., 2008, 2009). In a parallel manner, as our understanding of the rules for biological activity becomes more refined, their associated biological activities and potential as anticancer compounds, or as medicines for other applications, may become more predictable as well.

2.3.3 Medicinal Compounds from Fungi

Biologically active compounds abound in the fungi, illustrating the promise of this rich source of pharmaceuticals (Pearce, 1997; Pearce et al., 2010). A number of major drugs of the twentieth century have been derived from fungal metabolites, dating from the discovery of penicillin G (Fleming, 1946) and cephalosporin C (Newton and Abraham, 1955), antibiotics isolated from *Penicillium chrysogenum* and *Cephalosporium acremonium*, respectively. The immunosuppressant peptide metabolite, cyclosporin A (Dreyfus et al., 1976), used following organ transplant, was isolated from two fungi imperfecti identified as *Cylindrocarpon lucidum* and *Tolypocladium inflatum* Gams, later reclassified as *Tolypocladium nivea* and then to *Beauvaria nivea*. Cyclosporin A is produced by a NRPS, which is synthesized from one of the largest gene sequences known, at 45.8 kb (Weber et al., 1994). The statin cholesterol-lowering agents, a large class of HMG-CoA reductase inhibitors, were discovered from fungi; lovastatin (see Vagelos, 1991 for a detailed review) was discovered from *Aspergillus terreus*, and compactin (Brown et al., 1976) was discovered from *Penicillium citrinum* and *Penicillium brevicompactum*. Derivatization of these compounds has led to the discovery of a number of compounds used clinically, including pravastatin, which is prepared using microbial hydroxylation of compactin (Buss et al., 2003). This group includes the most successful drug in the history of medicine as measured by sales, Lipitor®, which had revenues in excess of 12 billion dollars in 2007 (Herper, 2008). Recently, a new class of antifungal agents that inhibit glucan biosynthesis has been discovered; these are the lipopeptide echinocandins (Tkacz, 1992) found from a number of sources including *Aspergillus nidulans*. One of these lipopeptides, caspofungin acetate, was developed for clinical use and introduced recently by Merck for *Candida* infections (Rybowicz and Gurk-Turner, 2002). In short, the list of compounds discovered from fungi that have advanced into drugs spans the development of the modern pharmacopeia, from agents that were key life savers during World War II (i.e., the penicillins) to compounds just now finding their way to the market (e.g., caspofungin acetate).

2.3.4 FUNGAL METABOLITES AS ANTICANCER LEAD COMPOUNDS

A number of fungal products have been investigated for anticancer activity (Figure 2.3), some of which are currently undergoing clinical trials and two of which, polysaccharide K and Lentinan, are approved to treat cancer patients in Japan. These compounds illustrate a diversity of chemical structures and mechanisms of action, where these are known. Most of the compounds mentioned were discovered 30 years or more ago, and then shown subsequently to have potential anticancer activity. Again, this illustrates the potential for new discoveries to lead to new disease treatments.

Brefeldin A (**9**) is a macrocyclic lactone produced by *Penicillium brefeldianum*, *P. decumbens*, and *P. cyaneum* and shown subsequently to possess a variety of biological activities, including antifungal, antiviral, and anticancer effects. Brefeldin A is biosynthesized via a PKS mechanism (Hutchinson et al., 1981, 1983). A number of derivatives have been prepared through synthetic chemistry (Zhu et al., 1997; Anadu et al., 2006; Gao et al., 2008), and it can be imagined that through genetic engineering of the PKS, other new brefeldins could be produced as well. In either case new analogues could serve to address some of brefeldin A's limitations, which include poor solubility and pharmacokinetics and neurotoxicity in some animal models (Kikuchi et al., 2003; Anadu et al., 2006). Brefeldin was shown to disrupt intracellular protein transport by dissociation of the endoplasmic reticulum (Perkel et al., 1989; Satiat-Jeunemaitre and Hawes, 1993). This compound also has selectivity for

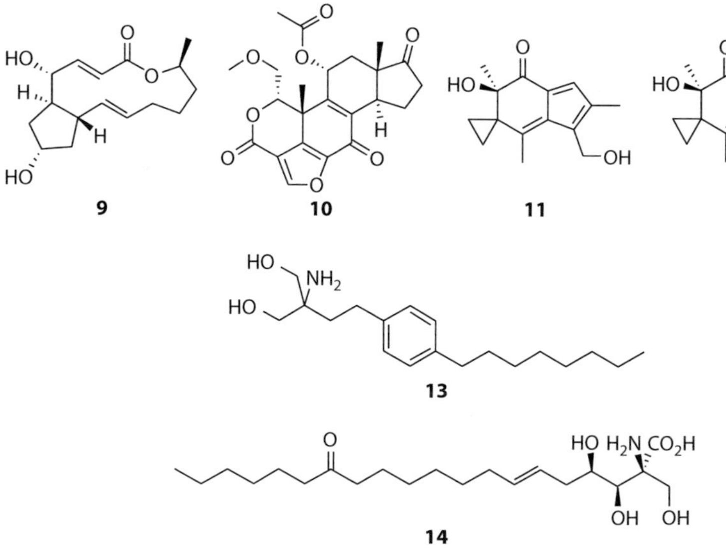

FIGURE 2.3 Structures of bioactive compounds from fungi.

certain cell types in the 60 human tumor cell line panel (McCloud et al., 1995), which is used by the U.S. National Cancer Institute (NCI) to detect potential new chemotherapeutic agents, thus raising more interest in this class of compounds.

Wortmannin (**10**) is a polycyclic lactone containing compound produced by *Talaromyces wortmanni*, which was classified initially as *Penicillium wortmanni* (MacMillan et al., 1972). It was described originally as an antifungal antibiotic, and then later found to have anticancer activity and shown subsequently to be a potent inhibitor of phosphatidylinositol-3-kinase (Powis et al., 1995; Schultz et al., 1995).

Illudins were reported initially from *Clitocybe illudens* (McMorris and Anchel, 1963). Irofulven (**11**) was prepared from illudin S (**12**), and the medicinal chemistry of this and related compounds has been disclosed (McMorris et al., 2007). The illudins, especially irofulven, have demonstrated potent anticancer properties; and, as of 2006, irofulven was in phase II clinical trials against several different solid tumors (Paci et al., 2006). Irofulven is thought to work through a mechanism involving macromolecule adduct formation, possibly via enzymatic activation, producing an unstable intermediate that is subject to nucleophilic attack, S-phase arrest, and induction of apoptosis (Van Laar et al., 2004; Paci et al., 2006).

One of the more promising groups of fungal metabolites are the polysaccharide constituents extracted from the fruiting bodies of fungi. In Japan, extracts of β-glucans from shitake mushroom (*Lentinula edodes*) have been approved for use clinically and are marketed as Lentinan, although this is not available in the United States currently. Lentinan and related products such as polysaccharide K from *Coriolus versicolor* have been shown to be immunostimulants (Sullivan et al., 2006) and are used in conjunction with traditional anticancer therapies. The fungal polysaccharides have been reviewed elsewhere (Wasser, 2002; Zhang et al., 2007).

An exciting new addition to this traditional list of fungal metabolites and their derivatives is fingolimod (**13**), an analog of myriocin (**14**), which was first isolated from *Myriococcum albomyces*, a thermophilic fungus, and later obtained from the fruiting bodies of *Isaria sinclairii* (Bagli et al., 1973). Fingolimod (also known as FTY720) has been in clinical trials for multiple sclerosis and results from the phase II studies have shown it to be a promising agent for remitting-relapsing MS (Rammohan and Shoemaker, 2010). Fingolimod is a sphingosine 1-phosphate receptor antagonist. It has been investigated for anticancer activity as well. For example, recent reports showed it to be a potential antimetastatic lead (Chua et al., 2005, 2009), and it has been shown to be active against breast and colon cancer cell lines (Nagaoka et al., 2008).

2.3.5 Selection of Fungi for Anticancer Screening

A number of observations help direct the selection of fungi from our library for inclusion in our collaborative screening program project. The most significant is the capacity of the organism for producing secondary metabolites. We have employed previous observations regarding the production of biological activity as determined by a combination of medicinal and agricultural assays, for example, antibacterial, gene regulation, enzyme inhibition, pesticide production; these data are stored in the Mycosynthetix database, which spans more than 25 years of research into the current fungal library. Other groups also employ pre-qualification of fungi as productive by using HPLC analysis to demonstrate the presence of multiple metabolites in organic solvent extracts (A.D. Buss, MerLion Pharmaceuticals, Singapore, personal communication, 2009). Using this approach, we have observed as much as a 10-fold increase in the rate of active cultures compared to the use of randomly selected fungi. Solid-phase media were used to grow fungi for our current anticancer screening program. We have characterized 5 novel biologically active metabolites from the initial batch of 200 fungi investigated. Additional advantages of employing fungi as a source for new medicinal agents include our ability to culture these organisms, that is, they are renewable without having to recollect active leads. Optimization of fermentation processes, taken together with strain development, often enables the production of enough material for preclinical development and early stage clinical trials. By making adjustments to the fermentation conditions, we can often produce compounds of slightly different structural types, providing materials for preliminary structure–activity relationship analyses. For the current program project, we are continuing to investigate the extensive fungal world for novel anticancer compounds, both as potential medicinal leads and as probes for new mechanisms of action.

2.4 HIGHER PLANTS AS SOURCES OF POTENTIAL ANTICANCER AGENTS

Over the last nearly 200 years, higher or vascular plants (also known as tracheo-phytes) have afforded numerous drugs and lead compounds of relevance in treating many different types of disease (Jones et al., 2006; Pan et al., 2009). Thus, in addition to terrestrial microbes, higher plants have been one of the two most prolific sources of natural product-derived anticancer agents (Kinghorn, 2008). Thus far, there are four major structural categories of plant-derived anticancer agents utilized in conventional drug therapy, which were introduced in the following order, and with the approved compounds included in parentheses: Vinca (*Catharanthus*) alkaloids (vinblastine, vincristine, vinorelbine); epipodophyllotoxin derivatives (etoposide, teniposide, etoposide phosphate); taxane diterpenoids (paclitaxel, docetaxel, Abraxane); and the camptothecin alkaloid derivatives (topotecan, irinotecan, belotecan) (Cragg et al., 2005, 2009; Saklani and Kutty, 2008). Since these standard oncology agents have been covered in detail recently by other authors (Cragg et al., 2005, 2009; Cragg and Newman, 2005, Itokawa et al., 2008), it is not intended to discuss them further in this chapter.

2.4.1 NEW ANTICANCER PLANT-DERIVED LEAD COMPOUNDS

Currently, in addition to the approved compounds mentioned above, there are various plant-derived small organic molecules that are being advanced toward marketing approval as new cancer chemotherapeutic agents, with diverse mechanisms of action (Butler, 2008, 2010; Saklani and Kutty, 2008) (Figure 2.4). For example, there are several analogues based on the lead compounds camptothecin, 10-deacylbaccatin III (an analogue of paclitaxel), and vinblastine that are currently in advanced clinical trials, including cositecan (**15**), larotaxel (**16**), and vinflunine (**17**), respectively.

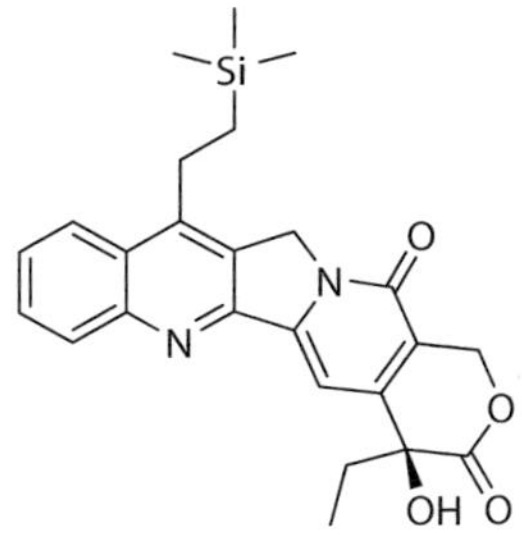

FIGURE 2.4 Structures of biologically active agents from plants.

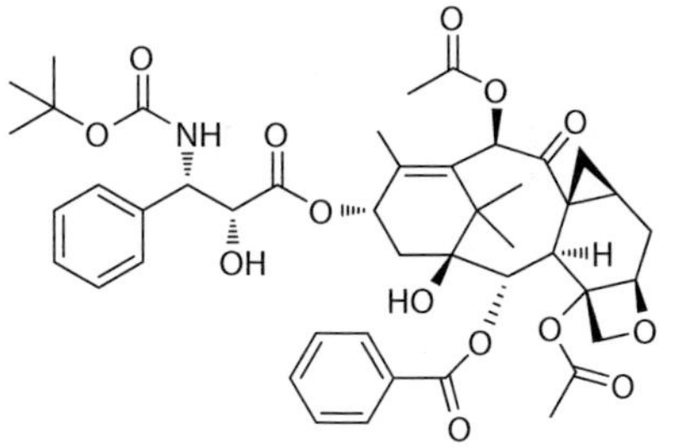

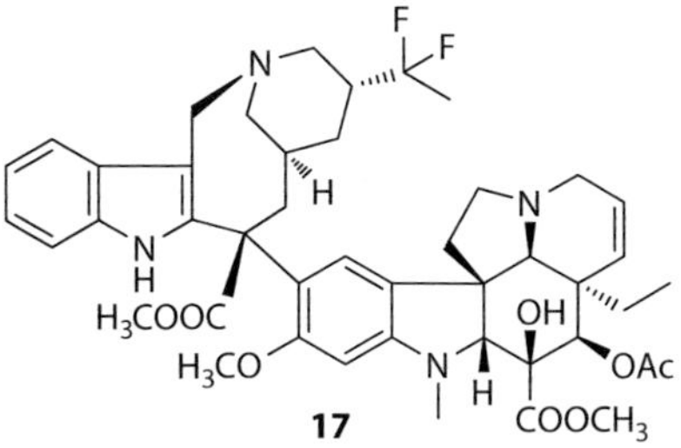

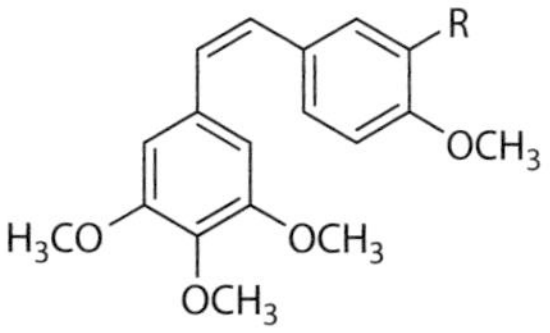

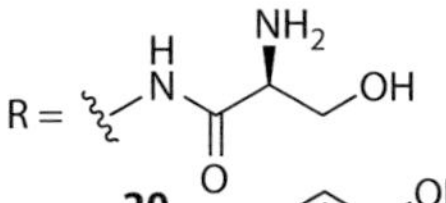

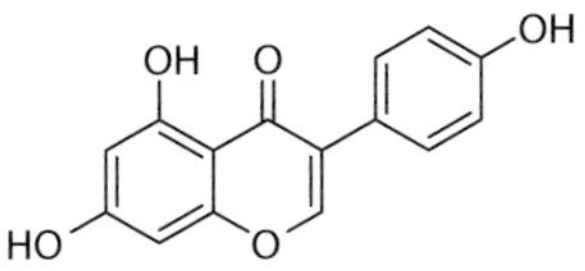

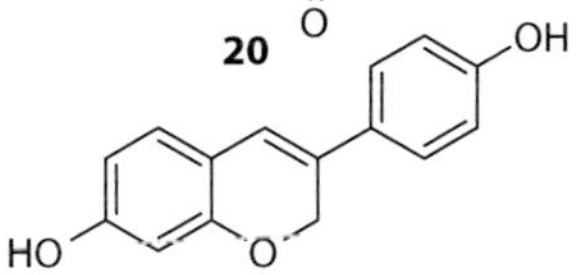

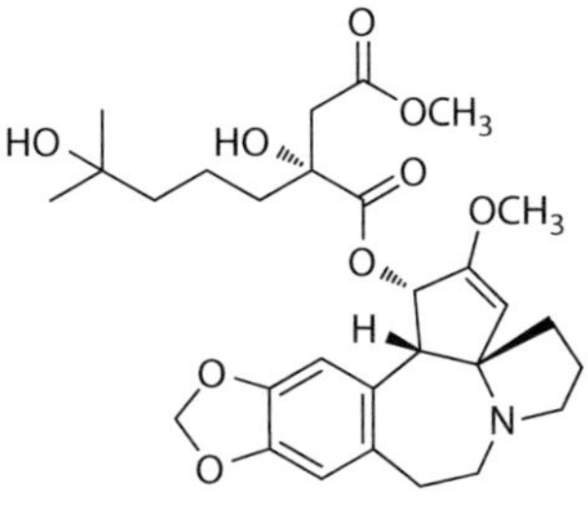

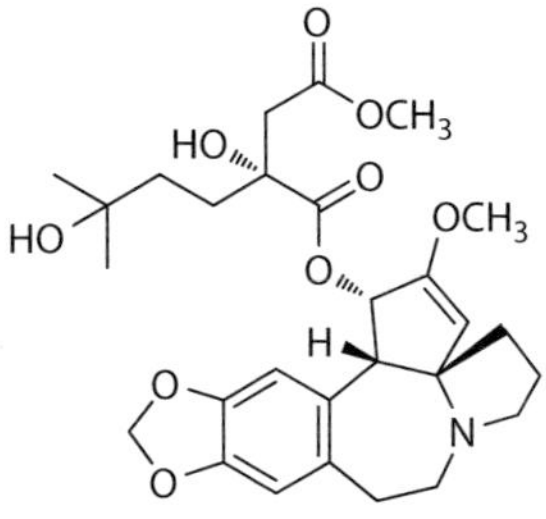

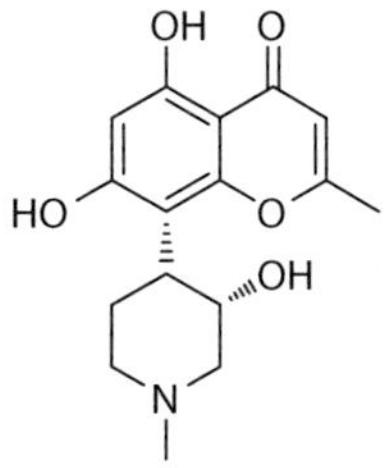

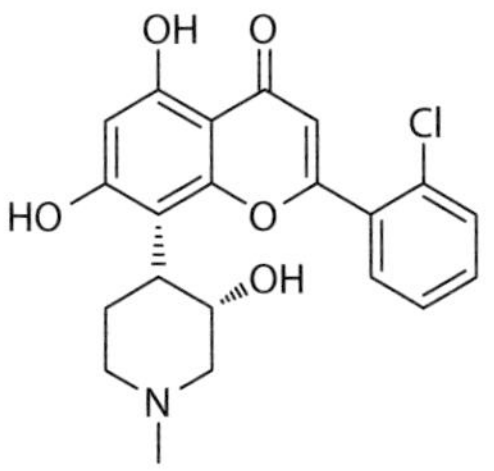

Cositecan (**15**) is a topoisomerase I inhibitor and a highly lipophilic (trimethylsilyl)ethyl camptothecin derivative with high δ-lactone stability, which promotes improved pharmacokinetic properties (Versace, 2003). This substance is now in phase III clinical trials for refractory advanced ovarian cancer (Butler, 2010). In turn, larotaxel (**16**), a microtubule-stabilizing semisynthetic taxane, has been developed in Europe as an agent with decreased multidrug resistance sensitivity and the ability to cross the blood–brain barrier (Carlson, 2008). Phase III trials of this compound for the treatment of advanced breast cancer and for refractory pancreatic cancer are either completed or ongoing (Butler, 2010). Vinflunine (**17**), a microtubule inhibitor, is based structurally on the lead compound vinblastine, and a close analog of vinorelbine, is also in an advanced state of development (Carlson, 2008). In a recent phase III clinical report on compound **17**, it was found that this compound has the potential for use against transitional cell carcinoma of the urothelial tract (Bellmunt et al., 2009). There has been a Medicines Authorization Application filed with the European Medicines Agency on behalf of this semisynthetic bisindole alkaloid as an anticancer agent (Butler, 2010).

However, a high proportion of the new anticancer leads from plants that are now in clinical trials are based on carbon skeletons unrelated to the four classes of plant-derived compounds currently used in treating cancer patients. For example, Pettit and colleagues at Arizona State University isolated the first of a number of structurally related benzenoids from the southern African tree, *Combretum caffrum* Kuntze (Combreaceae), nearly 30 years ago (Pettit et al., 1982). A major lead compound in this series is combretastatin A-4 (**18**), a potent tubulin inhibitor and a *cis*-stilbenoid (Pettit et al., 1989b). The development of the more water-soluble phosphate derivative of compound **19** (CA4P) toward clinical trials as a vascular disrupting (or targeting) agent has been reviewed thoroughly (Pinney et al., 2005; Carlson, 2008). This compound, known as fosbretabulin, is currently in various advanced clinical trials for the treatment of cancers such as anaplastic thyroid, head and neck, and platinum-resistant ovarian (Butler, 2010). Ombrabulin (**20**), a synthetic combretastatin derivative, is a further vascular disrupting agent currently being evaluated in advanced clinical trials for patients with advanced stage tissue sarcoma (Butler, 2010). A second example of a structurally different plant chemotype of interest for cancer drug development is the isoflavonoid, genistein (**21**), found commonly in the soybean [*Glycine max* Merr.; Leguminosae]. Phenoxodiol (**22**) is a synthetic analog of compound **21**, and is a multiple signal transduction inhibitor that results in degranulation of various antiapoptotic proteins (Choueiri et al., 2006). Although phenoxodiol is produced by synthesis, and several synthetic schemes have been proposed (e.g., Liepa, 1981), it also occurs as a natural product that was isolated from *Lespedeza homoloba* Nakai (Leguminosae), and designated as haginin E (Miyase et al., 1999). Phenoxodiol is currently in a phase III clinical trial in combination with carboplatin for the treatment of ovarian cancer, and it is being evaluated also for its potential beneficial effects in cervical and prostate cancer (Butler, 2010).

Two examples of promising plant-derived antineoplastic agents from species used in traditional medicine will be briefly described, namely, homoharringtonine and flavopiridol. The first of these, homoharringtonine (**23**), was reported initially by a U.S. group as an antileukemic constituent of *Cephalotaxus harringtonia*

(Knight ex J. Forbes) K. Koch (Cephalotaxaceae) (Powell et al., 1970, 1972). This plant is indigenous to eastern Asia, and the bark is used in Chinese traditional medicine for a variety of conditions (Itokawa et al., 2005). A mixture of homoharringtonine and a related alkaloid, harringtonine (**24**), has been used in mainland China to treat acute myeloid leukemia for over 30 years (Kantarjian et al., 2001). Homoharringtonine has been found to be a protein synthesis inhibitor (Baaske and Heinstein, 1977), and blocks the elongation phase of translation by preventing substrate binding at the acceptor site on the 60S ribosome subunit (Itokawa et al., 2005). There have been extensive clinical trials conducted on compound **23**, and these have been reviewed (Itokawa et al., 2005). Currently, homoharringtonine (also known as omacetaxine mepesuccinate) is in advanced clinical trials for the treatment of patients with refractory chronic myeloid leukemia (Butler, 2010). Rohitukine (**25**) is a chromone alkaloid that was isolated by an Indian group from *Dysoxylum binectariferum* Hiern (Meliaceae), and found to exhibit anti-inflammatory and immunomodulatory activity (Naik et al., 1988). The structure and relative configuration of this alkaloid was determined earlier as a constituent of *Amoora rohituka* (Roxb.) Wight & Arn. (Meliaceae) (Harmon et al., 1979). *Dysoxylum binectariferum* is described in the Ayurvedic system of traditional medicine in south Asia (Aggarwal et al., 2006). Flavopiridol (**26**) (also called alvocidib) is an analog of rohitukine with selectivity for cyclin-dependent kinases and the compound was prepared originally by scientists at Hoechst India in Mumbai (Newman and Cragg, 2005). Flavopiridol has been subjected to extensive in vitro and in vivo testing (e.g., Arguello et al., 1998), and the clinical efficacy was improved as a result of noting a large difference in protein binding of this substance between human plasma and bovine serum used in cell culture (Byrd et al., 2007). Flavopiridol has been subjected to several advanced clinical trials, particularly in reference to chronic lymphocytic leukemia and other hematological malignancies (Butler, 2010; Christian et al., 2009).

2.4.2 COLLECTION AND PROCESSING OF CANDIDATE PLANTS FOR ANTICANCER SCREENING

The portion of our current federally funded collaborative program project dealing with the discovery of anticancer agents from tropical plants (Kinghorn et al., 2009) has followed on from an earlier National Collaborative Drug Discovery Group's project (1990–2005) in which the entire focus was on plants (Kinghorn et al., 1999, 2003). There are several reasons why tropical plant species specifically have been chosen compared with their temperate counterparts. First, it is generally recognized that tropical plants sustain great biodiversity, and it may be rationalized that they will afford more chemical diversity in their secondary metabolites than plants from temperate countries (Kinghorn et al., 2009). Second, we have been able to collect from countries that are regarded as containing "biodiversity hotspots," of which many such regions are under major threat of habitat loss due to a variety of factors (Myers et al., 2000). Third, we have been able to develop suitable plant collection agreements, inclusive of intellectual property aspects, with a number of tropical countries, most notably with Indonesia (Soejarto et al., 2002; Kinghorn et al., 2003). The development of a plant collection agreement with a source country requires a detailed awareness of the United Nations Convention on Biological Diversity (CBD). Two of the main stipulations of the CBD are the need for the conservation of biodiversity, and that each source country has a sovereign right of ownership of its own genetic resources (Soejarto et al., 2002; Cordell, 2010).

In an effort to streamline plant collection activities, our group has carried out a correlative ANOVA variance analysis of plant collection profiles (country of origin, plant part, and taxonomy), with the resultant cytotoxic activity determined with a small panel of cancer cell lines, for over 2500 plants over a 15 year period of study (Balunas et al., 2006). Several interesting results emerged from this study. First, it was found that collections made in Ecuador of the six countries where a substantial number of species were obtained, resulted in the most potent cytotoxic activity overall. Second, below-ground plant organs (e.g., roots and rhizomes) tended on average to have greater cytotoxicity than aerial plant parts (e.g., bark, leaves, and twigs). Finally, of 145 families included in the collection, species in the Clusiaceae, Elaeocarpaceae, Meliaceae, and Rubiaceae were the most cytotoxic on average, with those in the genera *Aglaia*, *Casearia*, *Exostema*, *Mallotus*, and *Trichosanthes* being of particular promise, and worthy of further investigation (Balunas et al., 2006).

We endeavor to collect between 0.3 and 1.0 kg of each dried plant specimen for initial processing and testing. A previously agreed-upon number of primary plant samples is obtained during each plant collection trip to a given source country, which is planned and coordinated for our ongoing program project by Dr. D. Doel Soejarto of the University of Illinois at Chicago, Illinois, who is trained as a plant taxonomist. One of the advantages of working with macroorganisms like higher plants as opposed to cyanobacteria or filamentous fungi is that their taxonomy is generally well established, which enables plants in rarer genera that are not so highly investigated phytochemically to be targeted for collection. Once plants are dried, inventoried, and imported to the United States after agricultural inspection, they are then powdered and stored. A small aliquot of each powdered sample (ca. 20–30 g) is subjected to a standard extraction scheme developed for our project by late Dr. Monroe E. Wall, of Research Triangle Institute, which reduces the presence of plant polyphenols in each resultant organic-soluble extract (Wall et al., 1996). Selected extracts are then subjected to a panel of cell-based and mechanism-based bioassays (see Section 2.5 for examples of biological assays used), with selected plants then subjected to activity-guided fractionation (Kinghorn et al., 2009).

2.5 BIOASSAYS IN SUPPORT OF THE DISCOVERY OF NATURAL PRODUCT CANCER CHEMOTHERAPEUTIC AGENTS

Our approach to the evaluation of natural products for anticancer activity has been based on models developed at the NCI and pharmaceutical companies, such as Bristol-Myers Squibb, with whom we collaborate on our current program project supported by the NCI. Our strategy is to begin with the simplest system, an isolated enzyme target challenged with a mixture of natural products in the form of an extract of a cyanobacterium, fungus, or plant. Extracts displaying activity are fractionated, reevaluated against the molecular target, and the most active fractions are subsequently tested in cell-based assays. Ultimately, pure compounds are isolated from the fractions and the most promising agents are studied in murine assays for anticancer activity. In the following paragraphs, we describe some of our bioassays and how they are used in concert for the discovery of natural anticancer agents that work by inhibiting the proteasome.

The 26S proteasome is a large complex of proteins that is responsible for degradation of proteins within the cytoplasm and complements the activity of lysosomes, which digest membrane-bound proteins and proteins taken up from the outside of the cell (Ciechanover, 1994). The 26S proteasome consists of a 20S cylindrical catalytic core capped at each end by 19S subunits that restrict entry into the 20S core to only those proteins marked for degradation by ubiquitin tags (Navon and Ciechanover, 2009). The ubiquitin-proteasome system plays a pivotal role in the control of cellular pathways that govern proliferation, apoptosis, and differentiation in normal and cancer cells (Pickart, 2004). It is believed that dysregulation of the degradation of key proteins regulating these pathways can render cancer cells particularly vulnerable to destruction. Accordingly, proteasome inhibitors are now recognized as potential anticancer agents. Bortezomib (Velcade®) is the first proteasome inhibitor approved by the U.S. Food and Drug Administration and is indicated for the treatment of multiple myeloma (Adams, 2003).

To assess the capacity of natural products to inhibit proteasome activity, we employ the commercially available 20S Proteasome Assay Kit (BIOMOL International, Plymouth Meeting, PA 19462) following the manufacturer's protocol (Su et al., 2004). In brief, a 96-well plate containing a test buffer with the proteasome 20S enzyme is incubated with the test substance for 10 min. A substrate (Suc-LLVY-AMC) is then added and the fluorescence is measured every minute during a 15 min time course. Substances found active in the proteasome assay are then evaluated in intact human cancer cells, in which we measure the accumulation of ubiquitinated proteins in cells as a function of treatment with putative proteasome inhibitors or the positive control bortezomib. Specifically, we evaluate p53 and p21 accumulation since these proteins are well known to be involved in cancer and are tightly regulated by the action of the proteasome. An estimated 50% of human tumors have mutations in p53, and many other tumors are associated with alterations in the positive (ARF) or negative (MDM2) regulators of p53 (Momand et al., 2000; Sherr and Weber, 2000; Soussi, 2000). Wild-type p53 promotes the expression of MDM2, a ubiquitin ligase E3 (Honda et al., 1997), which in turn, ubiquitinates p53 targeting it for rapid degradation by the proteasome (Maki et al., 1996; Haupt et al., 1997; Kubbutat et al., 1997). Gene expression of p21 is induced by p53 (El-Deiry et al., 1993) and its protein half-life is regulated by the proteasome (Bloom et al., 2003). Therefore, p21 protein formation and degradation are heavily influenced by proteasome activity (Zhu et al., 2007). Cells are incubated in the absence or presence of 100 nM bortezomib or test substance and the expression of p53 and p21 is measured in total cell extracts by immunoblotting.

Once it is established that a substance is capable of inhibiting the 20S catalytic subunit of the proteasome in an isolated system and can induce the accumulation of proteasome substrates in intact cells, the test material is subjected to analyses in vivo. The choice of human cancer cell lines employed for the in vitro and in vivo assays is an important consideration. We have chosen lines that are sensitive to an established proteasome inhibitor, bortezomib, and are known to be tumorigenic when grafted subcutaneously in immunodeficient mice (Alley et al., 2004). These lines include MCF-7, which is a human breast carcinoma used as one of the "gatekeeper" lines of the screening program run by the Developmental Therapeutics Program of the NCI (Shoemaker, 2006). We also employ MD-MBA-435, which was originally thought to be a human breast cancer line, but was subsequently determined to be of melanoma derivation (Rae et al., 2004). Lastly, we use the human colon cancer line designated HT29. In addition to being well established and thoroughly characterized, each of these lines propagates well when cultured in hollow fibers, a requisite property for use in the hollow fiber assay.

The hollow fiber assay was developed at the NCI by Hollingshead and associates (Hollingshead et al., 1995) as a means of prioritizing leads from data generated in the NCI's 60-cell line panel for cancer cell toxicity. In this assay, human cancer cells are propagated in hollow fibers containing pores that have a molecular weight exclusion of about 500 kDa, which is too small to allow the cells to escape, but big enough for large proteins to pass through. The cell-filled fibers are implanted into immunocompromised hosts, which are then exposed to four daily injections of test substance. After treatments are concluded, the fibers are retrieved and cell number assessed by a MTT [3-(4,5-dimethylthiazol-2-yl)-2,5-diphenyltetrazolium bromide] assay. The hollow fiber assay bridges in vitro cytotoxicity studies and traditional xenograft experiments combining the speed and relative simplicity of the culture experiments within the context of an animal. We have recently reviewed the technology and methodology of the assay (Mi et al., 2009) and the NCI Developmental Therapeutics Program Web site has a detailed experimental protocol (NCI Biological Testing Branch).

Cells are maintained in RPMI-1640 medium supplemented with fetal bovine serum (FBS, 5% vol/vol) and 2 mM glutamine at 37°C in a 5% CO_2 atmosphere. Adherent cells in late log-phase growth are released by brief digestion with trypsin, washed and suspended in medium supplemented with FBS at the seeding density predetermined as optimal for the line (Hollingshead et al., 1999; Mi et al., 2009). The cells are then carefully injected into sterile, conditioned (Hollingshead et al., 1995) polyvinylidene fluoride hollow fibers. The fibers are then heat sealed at 2 cm intervals and cut in the middle of the seals to generate fibers for the study. Prior to implantation, the fibers are cultured overnight and on the following day, a set of fibers from each cell line is evaluated for viable cell mass by MTT assay (Alley et al., 1991). Another set of fibers remains in culture to confirm sterility. The remaining fibers are transplanted (day 0) into the intraperitoneal (IP) or subcutaneous (SC) space of immunodeficient male and female NCr nu/nu mice. Each mouse can carry up to six fibers distributed evenly between IP and SC locations. The test compound is administered in four daily IP injections on days 3, 4, 5, and 6 followed by fiber retrieval on day 7. During agent administration, each mouse is weighed daily and carefully monitored for toxicity, which is objectively determined as a 20% or greater loss of body weight or subjectively judged by lethargic behavior, scruffy coat, or hunched posture.

On day 7 of the experiment, all mice are sacrificed, and the fibers are retrieved. Necropsies are performed on each mouse to assess and record gross toxicity to major organs. The fibers are then placed into six-well plates, with each well containing culture medium and allowed to equilibrate for 30 min at 37°C. The viable cell mass contained within each hollow fiber is determined using an MTT dye conversion method (Alley et al., 1991). The percent net growth for each cell line in each treatment group is calculated by subtracting the day 0 absorbance from the day 7 absorbance and dividing this difference by the net growth in the day 7 vehicle-treated controls minus the day 0 values. A 50% or greater reduction in net cell growth in the treated samples compared to the vehicle control samples is considered a positive result.

2.6 CONCLUSIONS

In a recent detailed analysis of natural products being developed as drugs, of a total of 225 compounds, 108 substances from plants, 25 from bacteria, 7 from fungi, 24 from animals, and 61 semi-synthetics were either in the preclinical, phase I–III, or prereg-istration development phases, respectively. Furthermore, 86 (38%) of these leads were being developed for cancer treatment (Harvey, 2008). Therefore, based on these data, it seems reasonable to target as diverse an array of organisms as possible in order to discover new anticancer agents from natural sources. According to a recent review by Li and Vederas, although natural products do not dovetail with the current model of the pharmaceutical industry in terms of drug discovery, they do offer an almost unlimited array of new lead structures for future consideration (Li and Vederas, 2009).

Although the groups of organisms described in this chapter are by no means comprehensive in terms of the overall number of taxa that it would be possible to study, they each offer particular advantages in terms of the prospects of discovering new natural product lead compounds of potential interest in cancer chemotherapy. Cyanobacteria (blue-green algae) of marine origin have been extremely valuable in the past as sources of bioactive compounds. However, the study of their aquatic freshwater counterparts in this regard is still not well developed, and appears very promising in future. Fungi are a proven source for medicinal agents and continue to provide new therapies and novel leads. Surprisingly only a fraction of the fungi thought to exist have been investigated; in addition, the latest genome analysis has revealed that in many, possibly all, fungi there are so-called silent biosynthetic pathways, the products of which have yet to be reported. As part of the research collaboration mentioned in this chapter, filamentous fungi have been investigated for potential anticancer agents and offer great promise for the elucidation of new chemical entities. Higher plants have traditionally been a mainstay for the sourcing of new anticancer agents. While many of the new leads from plants that are now in advanced development for oncological clinical therapy are based on the camptoth-ecin and paclitaxel chemotypes, new compounds based on other structural types are also coming to the fore.

ACKNOWLEDGMENTS

Laboratory studies referred to in this chapter in connection with the current pro-gram project (5P01 CA125066; Principal Investigator, A.D. Kinghorn) are being supported by the U.S. National Cancer Institute and National Institutes of Health, Bethesda, Maryland. We are grateful for the committed efforts of all participants in this multidisciplinary effort.

REFERENCES

Adams, J. 2003. Potential for proteasome inhibition in the treatment of cancer. *Drug Discov. Today* 8:307–315.

Aggarwal, B. B., H. Ichikawa, P. Garodia, P. Weerasinghe, G. Sethi, I. D. Bhatt, M. K. Pandey, S. Shishodia, and M. G. Nair. 2006. From traditional medicine to modern medicine: Identification of therapeutic targets for suppression of inflammation and cancer. *Expert Opin. Ther. Targets* 10:87–118.

Alley, M. C., M. G. Hollingshead, D. J. Dykes, and W. R. Waud. 2004. Human tumor xenograft models in NCI drug development. In *Anticancer Drug Development Guide*, eds. B. A. Teicher and P. A. Andrews, pp. 125–152. Totowa, NJ: Humana Press.

Alley, M. C., C. M. Pacula-Cox, M. L. Hursey, L. R. Rubinstein, and M. R. Boyd. 1991. Morphometric and colorimetric analyses of human tumor cell line growth and drug sensitivity in soft agar culture. *Cancer Res.* 51:1247–1256.

Anadu, N. O., V. J. Davisson, and M. Cushman. 2006. Synthesis and anticancer activity of brefeldin A ester derivatives. *J. Med. Chem.* 49:3897–3905.

Andersen, R. A. and M. Kawachi. 2005. Traditional microalgae isolation techniques. In *Algal Culturing Techniques*, ed. R. A. Andersen, pp. 83–100. San Diego, CA: Academic Press.

Arguello, F., M. Alexander, J. A. Sterry, G. Tudor, E. M. Smith, N. T. Kalavar, J. F. Greene, W. Koss, Jr., C. D. Morgan, S. F. Stinson, T. J. Siford, W. G. Alvord, R. L. Klabansky, and E. A. Sausville. 1998. Flavopiridol induces apoptosis of normal lymphoid cells, causes immunosuppression, and has potent antitumor activity in vivo against human leukemia and lymphoma xenografts. *Blood* 91:2482–2490.

Baaske, D. M. and P. Heinstein. 1977. Cytotoxicity and cell cycle specificity of homoharringtonine. *Antimicrob. Agents Chemother.* 12:298–300.

Bagli, J. F., D. Kluepfel, and M. St. Jacque. 1973. Elucidation of structure and stereochemistry of myriocin. A novel antifungal antibiotic. *J. Org. Chem.* 38:1253–1260.

Bailly, C. 2009. Ready for a comeback of natural products in oncology. *Biochem. Pharmacol.* 77:1447–1457.

Balunas, M. J., W. P. Jones, Y.-W. Chin, Q. Mi, N. R. Farnsworth, D. D. Soejarto, G. A. Cordell, S. M. Swanson, J. M. Pezzuto, H.-B. Chai, and A. D. Kinghorn. 2006. Relationships between inhibitory activity against a cancer cell line panel, profiles of plants collected, and compound classes isolated in an anticancer drug discovery project. *Chem. Biodivers.* 3:897–915.

Bellmunt, J., C. Théodore, T. Demkov, B. Komyakov, L. Sengelov, G. Daugaard, A. Caty, J. Carles, A. Jagiello-Gruszfeld, O. Karyakin, F. M. Delgado, P. Hurteloup, N. Morsli, Y. Salhi, S. Culine, and H. von der Maase. 2009. Phase III trial of vinflunine plus best supportive care compared with best supportive care alone after a platinum-containing regimen in patients with advanced transitional cell carcinoma of the urothelial tract. *J. Clin. Oncol.* 27:4454–4461.

Berdy, J. 2005. Bioactive microbial metabolites. A personal view. *J. Antibiot. (Tokyo)* 58:1–26.

Bloom, J., V. Amador, F. Bartolini, G. DeMartino, and M. Pagano. 2003. Proteasome-mediated degradation of p21 via N-terminal ubiquitinylation. *Cell* 115:71–82.

Brown, A. G., T. C. Smale, T. J. King, R. Hasenkamp, and R. H. Thompson. 1976. Crystal and molecular structure of compactin, a new antifungal metabolite from *Penicillium brevicompactum*. *J. Chem. Soc., Perkin Trans.* 1:1165–1170.

Buss, A. D., B. Cox, and R. D. Waigh. 2003. Natural products as leads for new pharmaceuticals. In *Burger's Medicinal Chemistry and Drug Discovery*, 6th edn., Vol. 1, Drug discovery, ed. D. J. Abraham, pp. 847–900. Hoboken, NJ: Wiley-Interscience.

Butler, M. S. 2008. Natural products to drugs: Natural product-derived compounds in clinical trials. *Nat. Prod. Rep.* 25:475–516.

Butler, M. S. 2010. A snapshot of natural product-derived compounds in late stage clinical development at the end of 2008. In *Natural Product Chemistry for Drug Discovery*, eds. A. D. Buss and M. S. Butler, pp. 321–354. Cambridge, U.K.: Royal Society of Chemistry.

Byrd, J. C., T. S. Lin, J. T. Dalton, D. Wu, M. A. Phelps, B. Fischer, M. Moran, K. A. Blum, B. Rovin, M. Brooker-McEldowney, S. Broering, L. J. Schaaf, A. J. Johnson, D. M. Lucas, N. A. Heerema, G. Lozanski, D. C. Young, J.-R. Suarez, A. D. Colevas, and M. R. Grever. 2007. Flavopiridol administered using a pharmacologically derived schedule is associated with marked clinical efficacy in refractory, genetically high-risk chronic lymphocytic leukemia. *Blood 109*:399–404.

Carlson, R. O. 2008. New tubulin targeting agents currently in clinical development. *Expert Opin. Invest. Drugs 17*:707–722.

Carmichael, W. W. 1992. Cyanobacteria secondary metabolites—The cyanotoxins. *J. Appl. Bacteriol. 72*:445–459.

Chlipala, G., S. Mo, E. J. Carcache de Blanco, A. Ito, S. Bazarek, and J. Orjala. 2009. Investigation of antimicrobial and protease-inhibitory activity from cultured cyanobacteria. *Pharm. Biol. 47*:53–60.

Choueiri, T. K., R. Wesolowski, and T. M. Mekjail. 2006. Phenoxodiol: Isoflavone analog with antineoplastic activity. *Curr. Oncol. Rep. 8*:104–107.

Christian, B. A., M. R. Grever, J. C. Byrd, and T. S. Lin. 2009. Flavopiridol in chronic lymphocytic leukemia: A concise review. *Clin. Lymph. Myel. 9*(Suppl. 3):S179–S185.

Chua, C. W., Y. T. Chiu, H. F. Yuen, K. W. Chan, K. Man, X. Wang, M. T. Ling, and Y. C. Wong. 2009. Suppression of androgen-independent prostate cancer cell aggressiveness by FTY720: Validating Runx2 as a potential antimetastatic drug screening platform. *Clin. Cancer Res. 15*:4322–4335.

Chua, C. W., D. T. Lee, M. T. Ling, C. Zhou, K. Man, J. Ho, F. L. Chan, X. Wang, and Y. C. Wong. 2005. FTY720, a fungus metabolite, inhibits in vivo growth of androgen-independent prostate cancer. *Int. J. Cancer 117*:1039–1048.

Ciechanover, A. 1994. The ubiquitin-proteasome proteolytic pathway. *Cell 79*:13–21.

Cordell, G. A. 2010. The convention on biological diversity and its impact on natural product research. In *Natural Product Chemistry for Drug Discovery*, eds. A. D. Buss and M. S. Butler, pp. 81–139. Cambridge, U.K.: Royal Society of Chemistry.

Cragg, G. M., P. M. Grothaus, and D. J. Newman. 2009. Impact on natural products on developing new anti-cancer agents. *Chem. Rev. 109*:3012–3043.

Cragg, G. M., D. G. I. Kingston, and D. J. Newman. 2005. *Anticancer Agents from Natural Products*. Boca Raton, FL: CRC/Taylor & Francis.

Cragg, G. M. and D. J. Newman. 2005. Plants as a source of anticancer agents. *J. Ethnopharmacol. 100*:72–79.

Crawford, J. M., T. P. Korman, J. W. Labonte, A. L. Vagstad, E. A. Hill, O. Kamari-Bidkorpeh, S. C. Tsai, and C. A. Townsend. 2009. Structural basis for biosynthetic programming of fungal aromatic polyketide cyclization. *Nature 461*:1139–1143.

Crawford, J. M., P. M. Thomas, J. R. Scheerer, A. L. Vagstad, N. L. Kelleher, and C. A. Townsend. 2008. Deconstruction of iterative multidomain polyketide synthase function. *Science 320*:243–246.

De Arruda, M., C. A. Cocchiaro, C. M. Nelson, C. M. Grinnell, B. Janssen, A. Haupt, and T. Barlozzari. 1995. LU103793 (NSC D-669356): A synthetic peptide that interacts with microtubules and inhibits mitosis. *Cancer Res. 55*:3085–3092.

De Marais, D. J. 2000. Evolution. When did photosynthesis emerge on earth? *Science 289*:1703–1705.

Douglas, S. E. 1994. Chloroplast origins and evolution. In *The Molecular Biology of Cyanobacteria*, ed. D. A. Bryant, pp. 91–118. Dordrecht, the Netherlands: Kluwer Academic Publishers.

Dreyfus, M., E. Harri, H. Hofmann, W. Pache, and H. Tscherter. 1976. Cyclosporine A and C new metabolites from *Trichoderma polysporum*. *Eur. J. Appl. Microbiol.* 3:125–133.

El-Deiry, W. S., T. Tokino, V. E. Velculescu, D. B. Levy, R. Parsons, J. M. Trent, D. Lin, W. E. Mercer, K. W. Kinzler, and B. Vogelstein. 1993. *Waf1*, a potential mediator of p53 tumor suppression. *Cell* 75:817–825.

Erlichman, C. 2009. Tanespimycin: The opportunities and challenges of targeting heat shock protein 90. *Expert Opin. Investig. Drugs* 18:861–868.

Falch, B. S., G. M. Koenig, A. D. Wright, O. Sticher, C. K. Angerhofer, J. M. Pezzuto, and H. Bachmann. 1995. Biological activities of cyanobacteria: Evaluation of extracts and pure compounds. *Planta Med.* 61:321–328.

Fleming, A. 1946. *Penicillin, Its Practical Application*, 380pp. London, U.K.: Butterworth & Co., Ltd.

Gao, J., Y. X. Huang, and Y. K. Wu. 2008. Enantioselective total synthesis of 13-*O*-brefeldin A. *Tetrahedron* 64:11105–11109.

Gerwick, W., L. T. Tan, and N. Sitachitta. 2001. Nitrogen-containing metabolites from marine cyanobacteria. In *The Alkaloids*, Vol. 57, ed. G. A. Cordell, pp. 75–184. San Diego, CA: Academic Press.

Glaser, K. B. 2007. HDAC inhibitors: Clinical update and mechanism-based potential. *Biochem. Pharmacol.* 74:659–671.

Golakoti, T., I. Ohtani, D. J. Patterson, R. E. Moore, T. H. Corbett, F. A. Valerlote, and L. Demchik. 1994. Total structures of cryptophycins, potent antitumor depsipeptides from the blue-green alga *Nostoc* sp. strain GSV 224. *J. Am. Chem. Soc.* 116:4729–4737.

Harmon, A. D., U. Weiss, and J. V. Silverton. 1979. The structure of rohitukine, the main alkaloid of *Amoora rohituka* (syn. *Aphanamixis polystachya*) (Meliaceae). *Tetrahedron Lett.* 721–724.

Harvey, A. L. 2008. Natural products in drug discovery. *Drug Discov. Today* 13:894–901.

Haupt, Y., R. Maya, A. Kazaz, and M. Oren. 1997. MDM2 promotes the rapid degradation of p53. *Nature* 387:296–299.

Hawksworth, D. L. and A. Y. Rossman. 1997. Where are all the undescribed fungi? *Phytopathology* 87:888–891.

Herper, M. 2008. Pfizer wins longer life for Lipitor. Forbes.com. June 18, 2008.

Hollingshead, M. G., M. C. Alley, R. F. Camalier, B. J. Abbott, J. G. Mayo, L. Malspeis, and M. R. Grever. 1995. In vivo cultivation of tumor cells in hollow fibers. *Life Sci.* 57:131–141.

Hollingshead, M. G., J. Plowman, M. C. Alley, J. G. Mayo, and E. A. Sausville. 1999. The hollow fiber assay. In *Relevance of Tumor Models for Anticancer Drug Development*, eds. H. H. Fiebig and A. M. Burger, pp. 109–120. Basel, Switzerland: Karger AG.

Honda, R., H. Tanaka, and H. Yasuda. 1997. Oncoprotein mdm2 is a ubiquitin ligase E3 for tumor suppressor p53. *FEBS Lett.* 420:25–27.

Horti, J., E. Juhasz, Z. Monostori, K. Maeda, S. Eckhardt, and I. Bodrogi. 2008. Phase I study of TZT-1027, a novel synthetic dolastatin 10 derivative, for the treatment of patients with non-small cell lung cancer. *Cancer Chemother. Pharmacol.* 62:173–180.

Hutchinson, C. R., I. Kurobane, C. T. Mabuni, R. W. Kumola, A. G. McInnes, and J. A. Walter. 1981. Biosynthesis of macrolide antibiotics. 3. Regiochemistry of isotopic hydrogen labeling of brefeldin A by acetate. *J. Am. Chem. Soc.* 103:2474–2477.

Hutchinson, C. R., S. W. Li, A. G. McInnes, and J. A. Walter. 1983. Comparative biochemistry of fatty-acid and macrolide antibiotic (brefeldin-A)—Formation in *Penicillium brefeldianum*. *Tetrahedron* 39:3507–3513.

Itokawa, H., S. L. Morris-Natschke, T. Akiyama, and K.-H. Lee. 2008. Plant-derived natural product research aimed at new drug discovery. *J. Nat. Med.* 62:263–280.

Itokawa, H., X. Wang, and K.-H. Lee. 2005. Homoharringtonine and related compounds. In *Anticancer Agents from Natural Products*, eds. G. M. Cragg, D. G. I. Kingston, and D. J. Newman, pp. 47–70. Boca Raton, FL: CRC/Taylor & Francis.

Jemal, S., R. Siegel, E. Ward, Y. Hao, J. Xu, and M. J. Thun. 2009. Cancer statistics, 2009. *CA Cancer J. Clin. 59*:225–249.

Jones, W. P., Y.-W. Chin, and A. D. Kinghorn. 2006. The role of pharmacognosy in modern medicine and pharmacy. *Curr. Drug Targets 7*:247–264.

Kantarjian, H. M., M. Talpaz, V. Santini, A. Mungo, B. Cheson, and S. M. O'Brien. 2001. Homoharringtonine: History, current research, and future direction. *Cancer 92*:1591–1605.

Keller, N. P., G. Turner, and J. W. Bennett. 2005. Fungal secondary metabolism—From biochemistry to genomics. *Nat. Rev. Microbiol. 3*:937–947.

Kikuchi, S., K. Shinpo, S. Tsuji, I. Yabe, M. Niino, and K. Tashiro. 2003. Brefeldin A-induced neurotoxicity in cultured spinal cord neurons. *J. Neurosci. Res. 71*:591–599.

Kinghorn, A. D. 2008. Drug discovery from natural products. In *Foye's Principles of Medicinal Chemistry*, 6th edn., eds. T. Lemke and D.A. Williams, pp. 12–25. Baltimore, MD: Wolters Kluwer/Lippincott Williams & Wilkins.

Kinghorn, A. D., E. J. Carcache-Blanco, H.-B. Chai, J. Orjala, N. R. Farnsworth, D. D. Soejarto, N. H. Oberlies, M. C. Wani, D. J. Kroll, C. J. Pearce, S. M. Swanson, R. A. Kramer, W. C. Rose, C. R. Fairchild, G. D. Vite, S. Emanuel, D. Jarjoura, and F. O. Cope. 2009. Discovery of anticancer agents of diverse natural origin. *Pure Appl. Chem. 81*:1051–1063.

Kinghorn, A. D., N. R. Farnsworth, D. D. Soejarto, G. A. Cordell, J. M. Pezzuto, G. O. Udeani, M. C. Wani, M. E. Wall, H. A. Navarro, R. A. Kramer, A. T. Menendez, C. R. Fairchild, K. E. Lane, S. Forenza, D. M. Vyas, K. S. Lam, and Y.-Z. Shu. 1999. Novel strategies for the discovery of plant-derived anticancer agents. *Pure Appl. Chem. 71*:1611–1618.

Kinghorn, A. D., N. R. Farnsworth, D. D. Soejarto, G. A. Cordell, S. M. Swanson, J. M. Pezzuto, M. C. Wani, M. E. Wall, N. C. Oberlies, D. J. Kroll, R. A. Kramer, W. C. Rose, G. D. Vite, C. R. Fairchild, R. W. Peterson, and R. Wild. 2003. Novel strategies for the discovery of plant-derived anticancer agents. *Pharm. Biol. 41*(Suppl.):53–67.

Kubbutat, M. H., S. N. Jones, and K. H. Vousden. 1997. Regulation of p53 stability by MDM2. *Nature 387*:299–303.

Li, J. W.-H. and J. C. Vederas. 2009. Drug discovery and natural products: End of an era or an endless frontier? *Science 325*:161–165.

Liang, J., R. E. Moore, E. D. Moher, J. E. Munroe, R. S. Alawar, D. A. Hay, D. L. Varie, T. Y. Zhang, J. A. Aikins, M. J. Martinelli, C. Shih, J. E. Ray, L. L. Gibson, V. Vasudevan, L. Polin, K. White, J. Kushner, C. Simpson, S. Pugh, and T. H. Corbett. 2005. Cryptophycins-309, 249 and other cryptophycin analogs: Preclinical efficacy studies with mouse and human tumors. *Invest. New Drugs 23*:213–224.

Liepa, A. J. 1981. A synthesis of hydroxylated isoflavylium salts and their reduction products. *Aust. J. Chem. 34*:2647–2655.

Lin, Y., S. Schiavo, J. Orjala, P. Vouros, and R. Kautz. 2008. Microscale LC-MS-NMR platform applied to the identification of active cyanobacterial metabolites. *Anal. Chem. 80*:8045–8054.

Luesch, H., R. E. Moore, V. J. Paul, S. L. Mooberry, and T. H. Corbett. 2001. Isolation of dolastatin 10 from the marine cyanobacterium *Symploca* species VP642 and total stereochemistry and biological evaluation of its analogue symplostatin 1. *J. Nat. Prod. 64*:907–910.

MacMillan, J., A. E. Vanstone, and S. K. Yeboah. 1972. Fungal products. Part III. Structure of wortmannin and some hydrolysis products. *J. Chem. Soc., Perkin Trans. 1*:2898–2903.

Maki, C. G., J. M. Huibregtse, and P. M. Howley. 1996. In vivo ubiquitination and proteasome-mediated degradation of p53(1). *Cancer Res.* 56:2649–2654.

McAlpine, J. B. 2009. Advances in the understanding and use of the genomic base of micro-bial secondary metabolite biosynthesis for the discovery of new natural products. *J. Nat. Prod.* 72:566–572.

McCloud, T. G., M. P. Burns, F. D. Majadly, G. M. Muschik, D. A. Miller, K. K. Poole, J. M. Roach, J. T. Ross, and W. B. Lebherz, III. 1995. Production of brefeldin-A. *J. Ind. Microbiol.* 15:5–9.

McMorris, T. C. and M. Anchel. 1963. Structures of basidiomycete metabolites illudin S and illudin M. *J. Am. Chem. Soc.* 85:831–852.

McMorris, T. C., R. Chimmani, M. Gurram, M. D. Staake, and M. J. Kelner. 2007. Synthesis and antitumor activity of amine analogs of irofulven. *Bioorg. Med. Chem. Lett.* 17:6770–6772.

Mi, Q., J. M. Pezzuto, N. R. Farnsworth, M. C. Wani, A. D. Kinghorn, and S. M. Swanson. 2009. Use of the in vivo hollow fiber assay in natural products anticancer drug discov-ery. *J. Nat. Prod.* 72:573–580.

Misiek, M. and D. Hoffmeister. 2007. Fungal genetics, genomics, and secondary metabolites in pharmaceutical sciences. *Planta Med.* 73:103–115.

Miyase, T., M. Sano, H. Nakai, M. Muraoka, M. Nakazawa, M. Suzuki, K. Yoshino, Y. Nishihara, and J. Tanai. 1999. Antioxidants from *Lespedeza homoloba*. (I). *Phytochemistry* 52:303–310.

Molinski, T. F. 2009. Nanomole-scale natural products discovery. *Curr. Opin. Drug Discov. Dev.* 12:197–206.

Momand, J., H. H. Wu, and G. Dasgupta. 2000. MDM2–Master regulator of the p53 tumor suppressor protein. *Gene* 242:15–29.

Myers, N., R. A. Mittermeier, G. A. B. Da Fonseca, and J. Kent. 2000. Biodiversity hotspots for conservation priorities. *Nature (London)* 403:853–858.

Nagaoka, Y., K. Otsuki, T. Fujita, and S. Uesato. 2008. Effects of phosphorylation of immuno-modulatory agent FTY720 (fingolimod) on antiproliferative activity against breast and colon cancer cells. *Biol. Pharm. Bull.* 31:1177–1181.

Naik, R. G., S. L. Kattige, S. V. Bhat, B. Alreja, N. J. De Souza, and R. H. Rupp. 1988. An antiinflammatory and immunomodulatory piperidinylbenzopyranone from *Dysoxylum binectariferum*: Isolation, structure and total synthesis. *Tetrahedron* 44:2081–2086.

Navon, A. and A. Ciechanover. 2009. The 26S proteasome: From basic mechanisms to drug targeting. *J. Biol. Chem.* 284:33713–33718.

NCI Biological Testing Branch. Hollow fiber assay protocols for tumor cell lines. http://dtp.nci.nih.gov/branches/btb/pdf/cancer_protocol_hollow_fiber.pdf

Newman, D. J. and G. M. Cragg. 2004. Marine natural products and related compounds in clinical and advanced preclinical trials. *J. Nat. Prod.* 67:1216–1238.

Newman, D. J. and G. M. Cragg. 2005. Developments and future trends in anticancer natural products drug discovery. In *Anticancer Agents from Natural Products*, eds. G. M. Cragg, D. G. I. Kingston, and D. J. Newman, pp. 553–571. Boca Raton, FL: CRC/Taylor & Francis.

Newman, D. J. and G. M. Cragg. 2007. Natural products as sources of new drugs over the last 25 years. *J. Nat. Prod.* 70:461–477.

Newton, G. G. F. and E. P. Abraham. 1955. Cephalosporin-C, a new antibiotic containing sulphur and *d*-alpha-aminoadipic acid. *Nature* 175:548.

Nielsen, K. F. and J. Smedsgaard. 2003. Fungal metabolite screening: Database of 474 myco-toxins and fungal metabolites for dereplication by standardised liquid chromatography-UV-mass spectrometry methodology. *J. Chromatogr. A* 1002:111–136.

Oberlies, N. H. and D. J. Kroll. 2004. Camptothecin and taxol: Historic achievements in natu-ral products research. *J. Nat. Prod.* 67:129–135.

Paci, A., K. Rezai, A. Deroussent, D. De Valeriola, M. Re, S. Weill, E. Cvitkovic, C. Kahatt, A. Shah, S. Waters, G. Weems, G. Vassal, and F. Lokiec. 2006. Pharmacokinetics, metabolism, and routes of excretion of intravenous irofulven in patients with advanced solid tumors. *Drug Metab. Dispos. 34*:1918–1926.

Pan, L., E. J. Carcache de Blanco, and A. D. Kinghorn. 2009. Plant-derived natural products as leads for drug discovery. In *Plant-Derived Natural Products: Synthesis, Function, and Application*, eds. A. E. Osbourn and V. Lanzotti, pp. 546–567. New York: Springer.

Pearce, C. 1997. Biologically active fungal metabolites. *Adv. Appl. Microbiol. 44*:1–80.

Pearce, C., P. Eckard, I. Gruen-Wollny, and F. G. Hanske. 2010. Microorganisms: Their role in the discovery and development of medicines. In *Natural Product Chemistry for Drug Discovery*, eds. A. D. Buss and M. S. Butler, pp. 215–244. Cambridge, U.K.: The Royal Society of Chemistry.

Pel, H. J., J. H. de Winde, D. B. Archer, P. S. Dyer, G. Hofmann, P. J. Schaap, G. Turner, R. P. de Vries, R. Albang, K. Albermann, M. R. Andersen, J. D. Bendtsen, J. A. E. Benen, M. van den Berg, S. Breestraat, M. X. Caddick, R. Contreras, M. Cornell, P. M. Coutinho, E. G. J. Danchin, A. J. M. Debets, P. Dekker, P. W. M. van Dijck, A. van Dijk, L. Dijkhuizen, A. J. M. Driessen, C. d'Enfert, S. Geysens, C. Goosen, G. S. P. Groot, P. W. J. de Groot, T. Guillemette, B. Henrissat, M. Herweijer, J. P. T. W. van den Hombergh, C. A. M. van den Hondel, R. T. M. van der Heijden, R. M. van der Kaaij, F. M. Klis, H. J. Kools, C. P. Kubicek, P. A. van Kuyk, J. Lauber, X. Lu, M. J. E. C. van der Maarel, R. Meulenberg, H. Menke, M. A. Mortimer, J. Nielsen, S. G. Oliver, M. Olsthoorn, K. Pal, N. N. M. E. van Peij, A. F. J. Ram, U. Rinas, J. A. Roubos, C. M. Sagt, M. Schmoll, J. Sun, D. Ussery, J. Varga, W. Vervecken, P. J. J. van de Vondervoort, H. Wedler, H. A. B. Woesten, A.-P. Zeng, A. J. J. van Ooyen, J. Visser, and H. Stam. 2007. Genome sequencing and analysis of the versatile cell factory *Aspergillus niger* CBS 513.88. *Nat. Biotechnol. 25*:221–231.

Perkel, V. S., Y. Miura, and J. A. Magner. 1989. Brefeldin-A inhibits oligosaccharide processing of glycoproteins in mouse hypothyroid pituitary tissue at several subcellular sites. *Proc. Soc. Exp. Biol. Med. 190*:286–293.

Pettit, G. R., G. M. Cragg, D. L. Herald, J. M. Schmidt, and P. Lohavanijaya. 1982. Antineoplastic agents. 84. Isolation and structure of combretastatin. *Can. J. Chem. 60*:1374–1376.

Pettit, G. R., Y. Kamano, C. Dufresne, R. L. Cerny, C. L. Herald, and J. M. Schmidt. 1989a. Isolation and structure of the cytostatic linear depsipeptide dolastatin 15. *J. Org. Chem. 54*:6005–6006.

Pettit, G. R., Y. Kamano, C. L. Herald, A. A. Tuinman, F. E. Boettner, H. Kizu, J. M. Schmidt, L. Baczynskyj, K. B. Tomer, and R. J. Bontems. 1987. The isolation and structure of a remarkable marine animal antineoplastic constituent: Dolastatin 10. *J. Am. Chem. Soc. 109*:6883–6885.

Pettit, G. R., S. B. Singh, E. Hamel, C. M. Lin, D. S. Alberts, and D. Garcia-Kendall. 1989b. Isolation and structure of the strong cell growth and tubulin inhibitor combretastatin A-4. *Experientia. 45*:209–211.

Pickart, C. M. 2004. Back to the future with ubiquitin. *Cell 116*:181–190.

Piekarz, R. L., R. Frye, M. Turner, J. J. Wright, S. L. Allen, M. H. Kirschbaum, J. Zain, H. M. Prince, J. P. Leonard, L. J. Geskin, C. Reeder, D. Joske, W. D. Figg, E. R. Gardner, S. M. Steinberg, E. S. Jaffe, M. Stetler-Stevenson, S. Lade, A. T. Fojo, and S. E. Bates. 2009. Phase II multi-institutional trial of the histone deacetylase inhibitor romidepsin as monotherapy for patients with cutaneous T-cell lymphoma. *J. Clin. Oncol. 27*:5410–5417.

Pinney, K. G., C. Jelinek, K. Edvardsen, D. J. Chaplin, and G. R. Pettit. 2005. The discovery and development of the combretastatins. In *Anticancer Agents from Natural Products*, eds. G. M. Cragg, D. G. I. Kingston, and D. J. Newman, pp. 23–46. Boca Raton, FL: CRC/Taylor & Francis.

Powell, R. G., D. Weisleder, and C. R. Smith. 1970. Structures of harringtonine, isoharringtonine, and homoharringtonine. *Tetrahedron Lett.* *11*:815–818.

Powell, R. G., D. Weisleder, and C. R. Smith. 1972. Antitumor alkaloids from *Cephalotaxus harringtonia*: Structure and activity. *J. Pharm. Sci.* *61*:1227–1230.

Powis, G., M. Berggren, A. Gallegos, T. Frew, S. Hill, A. Kozikowski, R. Bonjouklian, L. Zalkow, R. Abraham, C. Ashendel, R. Shultz, and R. Merriman. 1995. Advances with phospholipid signalling as a target for anticancer drug development. *Acta Biochim. Pol.* *42*:395–403.

Pronzato, P. 2008. New therapeutic options for chemotherapy-resistant metastatic breast cancer. *Drugs 68*:139–146.

Rae, J. M., S. J. Ramus, M. Waltham, J. E. Armes, I. G. Campbell, R. Clarke, R. J. Barndt, M. D. Johnson, and E. W. Thompson. 2004. Common origins of MDA-MB-435 cells from various sources with those shown to have melanoma properties. *Clin. Exp. Metastasis 21*:543–552.

Rammohan, K. W. and J. Shoemaker. 2010. Emerging multiple sclerosis oral therapies. *Neurology 74*:S47–S53.

Ray, A., T. Okouneva, T. Manna, H. P. Miller, S. Schmid, L. Arthaud, R. Luduena, M. A. Jordan, and L. Wilson. 2007. Mechanism of action of the microtubule-targeted antimitotic depsipeptide tasidotin (formerly ILX651) and its major metabolite tasidotin C-carboxylate. *Cancer Res.* *67*:3767–3776.

Rybowicz, J. and C. Gurk-Turner. 2002. Caspofungin: The first agent available in the echinocandin class of antifungals. *Proc. Baylor Univ. Med. Cent.* *15*:97–99.

Saklani, A. and S. K. Kutty. 2008. Plant-derived compounds in clinical trials. *Drug Discov. Today 13*:161–171.

Sashidhara, K. V., K. N. White, and P. Crews. 2009. A selective account of effective paradigms and significant outcomes in the discovery of inspirational marine natural products. *J. Nat. Prod.* *72*:588–602.

Satiat-Jeunemaitre, B. and C. Hawes. 1993. The distribution of secretory products in plant cells is affected by brefeldin-A. *Cell Biol. Int.* *17*:183–193.

Schopf, J. W. 2000. The fossil record: Tracing the roots of the cyanobacterial lineage. In: *The Ecology of Cyanobacteria. Their Diversity in Time and Space*, eds. B. A. Whitton and M. Potts, pp. 13–35. Dordrecht, the Netherlands: Kluwer Academic Publishers.

Schultz, R. M., R. L. Merriman, S. L. Andis, R. Bonjouklian, G. B. Grindey, P. G. Rutherford, A. Gallegos, K. Massey, and G. Powis. 1995. In vitro and in vivo antitumor activity of the phosphatidylinositol-3-kinase inhibitor, wortmannin. *Anticancer Res.* *15*:1135–1139.

Schwartz, R. E., C. F. Hirsch, D. F. Sesin, J. E. Flor, M. Chartrain, R. E. Fromtling, G. H. Harris, M. J. Salvatore, J. M. Liesch, and K. Yudin. 1990. Pharmaceuticals from cultured algae. *J. Ind. Microbiol.* *5*:113–124.

Shaw, G. and P. K. S. Lam. 2007. Health aspects of freshwater cyanobacterial toxins. *Water Sci. Technol. Water Supply 7*:193–203.

Sherr, C. J. and J. D. Weber. 2000. The Arf/p53 pathway. *Curr. Opin. Genet. Dev. 10*:94–99.

Shih, C. and B. A. Teicher. 2001. Cryptophycins: A novel class of potent antimitotic antitumor depsipeptides. *Curr. Pharm. Des.* *7*:1259–1276.

Shoemaker, R. H. 2006. The NCI 60 human tumour cell line anticancer drug screen. *Nat. Rev. Cancer 6*:813–823.

Sielaff, H., G. Christiansen, and R. Schwecke. 2006. Natural products from cyanobacteria: Exploiting a new source for drug discovery. *IDrugs 9*:119–127.

Simmons, T. L., E. Andrianasolo, K. McPhail, P. Flatt, and W. H. Gerwick. 2005. Marine natural products as anticancer drugs. *Mol. Cancer Ther.* *4*:333–342.

Singh, S., B. N. Kate, and U. C. Banerjee. 2005. Bioactive compounds from cyanobacteria and microalgae: An overview. *Crit. Rev. Biotechnol.* *25*:73–95.

Soejarto, D. D., J. A. Tarzien Sorensen, C. Gyllenhaal, G. A. Cordell, N. R. Farnsworth, H. H. S. Fong, A. D. Kinghorn, and J.M. Pezzuto. The evolution of the University of Illinois' policy on benefit-sharing in research on natural products. 2002. In *Ethnobiology and Biocultural Diversity*, eds. J. R. Stepp, F. S. Wyndham, and R. Zarger, pp. 21–30. Athens, GA: University of Georgia Press.

Soussi, T. 2000. The p53 tumor suppressor gene: From molecular biology to clinical investigation. *Ann. N. Y. Acad. Sci. 910*:121–137; discussion 37–39.

Su, B. N., B. Y. Hwang, H. Chai, E. J. Carcache-Blanco, L. B. Kardono, J. J. Afriastini, S. Riswan, R. Wild, N. Laing, N. R. Farnsworth, G. A. Cordell, S. M. Swanson, and A. D. Kinghorn. 2004. Activity-guided fractionation of the leaves of *Ormosia sumatrana* using a proteasome inhibition assay. *J. Nat. Prod. 67*:1911–1914.

Sullivan, R., J. E. Smith, and N. J. Rowan. 2006. Medicinal mushrooms and cancer therapy—Translating a traditional practice into Western medicine. *Perspect. Biol. Med. 49*:159–170.

Tan, L. T. 2007. Bioactive natural products from marine cyanobacteria for drug discovery. *Phytochemistry 68*:954–979.

Tan, G., C. Gyllenhaal, and D. D. Soejarto. 2007. Biodiversity as a source of anticancer drugs. *Curr. Drug Targets 7*:265–277.

Tkacz, J. S. 1992. *Emerging Targets in Antibacterial and Antifungal Chemotherapy*, eds. N. H. Georgopapadakou and J. A. Sutcliffe, 606pp. New York: Chapman & Hall.

Turgeon, B. G., S. Oide, and K. Bushley. 2008. Creating and screening *Cochliobolus heterostrophus* non-ribosomal peptide synthetase mutants. *Mycol. Res. 112*:200–206.

Vagelos, P. R. 1991. Are prescription drug prices high? *Science 252*:1080–1084.

van Apeldoorn, M. E., H. P. van Egmond, G. J. A. Speijers, and G. J. I. Bakker. 2007. Toxins of cyanobacteria. *Mol. Nutr. Food Res. 51*:7–60.

Van Laar, E. S., S. Roth, S. Weitman, J. R. MacDonald, and S. J. Waters. 2004. Activity of irofulven against human pancreatic carcinoma cell lines in vitro and in vivo. *Anticancer Res. 24*:59–65.

Versace, R. W. 2003. The silatecans, a novel class of lipophilic camptothecins. *Expert Opin. Ther. Patents 13*:751–760.

Wall, M. E., M. C. Wani, D. M. Brown, F. Fullas, J. B. Oswald, F. F. Josephson, N. M. Thornton, J. M. Pezzuto, C. W. W. Beecher, N. R. Farnsworth, G. A. Cordell, and A. D. Kinghorn. 1996. Effect of tannins on screening of plant extracts for enzyme inhibitory activity and techniques for their removal. *Phytomedicine 3*:281–285.

Wasser, S. P. 2002. Medicinal mushrooms as a source of antitumor and immunomodulating polysaccharides. *Appl. Microbiol. Biotechnol. 60*:258–274.

Weber, G., K. Schorgendorfer, E. Schneider-Scherzer, and E. Leitner. 1994. The peptide synthetase catalyzing cyclosporine production in *Tolypocladium niveum* is encoded by a giant 45.8-kilobase open reading frame. *Curr. Genet. 26*:120–125.

Welker, M. 2008. *Cyanobacterial hepatotoxins*: Chemistry, biosynthesis, and occurrence. In *Seafood and Freshwater Toxins*, 2nd edn., ed. L. M. Botana, pp. 825–843. Boca Raton, FL: CRC Press.

Welker, M. and H. von Döhren. 2006. Cyanobacterial peptides—Nature's own combinatorial biosynthesis. *FEMS Microbiol. Rev. 30*:530–563.

Zhang, M., S. W. Cui, P. C. K. Cheung, and Q. Wang. 2007. Polysaccharides from mushrooms: A review on their isolation process, structural characteristics and antitumor activity. *Trends Food Sci. Technol. 18*:4–19.

Zhu, J. W., H. Hori, H. Nojiri, T. Tsukuda, and Z. Taira. 1997. Synthesis and activity of brefeldin A analogs as inducers of cancer cell differentiation and apoptosis. *Bioorg. Med. Chem. Lett. 7*:139–144.

Zhu, Q., G. Wani, J. Yao, S. Patnaik, Q. E. Wang, M. A. El-Mahdy, M. Praetorius-Ibba, and A. A. Wani. 2007. The ubiquitin-proteasome system regulates p53-mediated transcription at p21(Waf1) promoter. *Oncogene 26*:4199–4208.

3 Metabolic Engineering of Natural Product Biosynthesis

Xinkai Xie, Kangjian Qiao, and Yi Tang

CONTENTS

3.1 INTRODUCTION

Natural products represent a prolific resource for developing pharmaceutical compounds. Over 60% of the approved drugs and pre-New Drug Application candidates in the anticancer and anti-infective fields during the period 1984–1995 are natural products or derived from natural products (Cragg et al. 1997). At present, natural products remain as important leads for drug development. More than 30 natural product-derived drugs were launched in the United States, Europe, and Japan between 1998 and 2007 (Butler 2005, 2008). Obtaining large and sustainable quantities of natural products, drugs are therefore critical not only for performing clinical trials, but also for supplying large therapeutic demand. Chemical synthesis, while powerful as a method of obtaining complex compounds from simple precursors, is difficult to implement and economically impractical on a large scale for multistep reactions required for many of the clinically relevant natural products. Therefore, fermentation remains an essential method of cultivating the producing organisms and harvesting desired natural products. One avenue that has showed promise in boosting the production titer and obtaining analogues of natural product therapeutics through fermentation is metabolic engineering.

Metabolic engineering can be defined as redirection of an organism's metabolic flux, whether primary or secondary, through alteration of the host's genetic material. Many successful examples of metabolic engineering of primary metabolism toward the production of simple chemicals, fuels, and foreign metabolites have been demonstrated in the last 20 years. In the metabolic engineering of secondary metabolism, natural products or natural product-derived drugs have become important targets (Khosla and Keasling 2003). Since most natural products are assembled via building blocks of primary metabolism, such as acetate, amino acids, and sugars, engineering of secondary metabolism often includes rewiring of primary metabolism for precursor supply. Metabolic engineering of natural product pathways can be coarsely separated into two categories: (1) increasing the yield of target compound; and (2) obtaining analogues of known natural products through alteration of the biosynthetic enzymes (Zhang and Tang 2008). In recent years, different strategies have been developed and employed to realize these objectives for a number of natural product families. Some of these strategies include the following: (1) pathway reconstitution in heterologous hosts, often those that are genetically simple to manipulate and can be cultured to high density; (2) overexpression of rate-limiting enzymes in a given pathway or improving the activity of the rate-limiting enzyme through protein engineering and directed evolution; (3) elimination of competing metabolic pathways, including those that consume key precursors or those that produce byproducts that are difficult to remove during purification; and (4) metabolic flux rebalancing to direct primary metabolism toward dedicated, key precursor metabolites. These strategies have been coordinately used to achieve the desired goal of industrial pharmaceutical production.

In this chapter, we focus on the recent advances in the field of metabolic engineering toward drug development. The natural products included in this chapter are either directly used as pharmaceuticals or served as precursor compounds in drug discovery. All the pharmaceuticals shown in this chapter have been, or are currently being, subjected to metabolic engineering efforts. This chapter has been divided into subtopics based on the families of natural products. The main families of compounds presented are polyketides, nonribosomal peptides, and isoprenoids, all of which have important and successful examples of metabolic engineering.

3.2 POLYKETIDES

Polyketides are secondary metabolites synthesized by bacteria, fungi, and plants. They are a family of structurally diverse natural products widely used as pharmaceuticals. The diverse polyketide therapeutics include antibiotics (erythromycin, tetracycline, and rifamycins), immunosuppressants (rapamycin and tacrolimus), anticancer agents (daunorubicin and doxorubicin), antifungals (amphotericin and griseofulvin), and cholesterol-lowering agents (lovastatin and compactin) (Figure 3.1).

Polyketides are synthesized by a group of enzymes called polyketide synthases (PKSs). PKSs operate in a similar fashion to fatty acid synthases (FASs) in building a carbon backbone from simple carbon building blocks, such as acetate. Both PKSs and FASs are composed of catalytic domains that work coordinately to condense units of activated acyl-CoAs. The polyketide synthases perform successive decarboxylative Claisen condensations between the growing polyketide chain attached to the ketosynthase (KS) domain and the extender unit attached to acyl-carrier protein (ACP) domain to synthesize the polyketide backbone (Figure 3.2). Following elongation of the polyketide backbone, other accessory tailoring domains such as β-keto-reductase (KR), dehydrase (DH), and enoyl-reductase (ER) selectively modify the growing polyketide chain. These tailoring domains can be used in different combinations to produce backbone β-carbons of different oxidation states. All polyketide ACP domains must first be post-translationally modified at the active site serine within the DSL motif by phosphopantetheine to become the *holo* form. This modification of the ACP domain is catalyzed by a phosphopantetheinyl transferase (PPTase), either

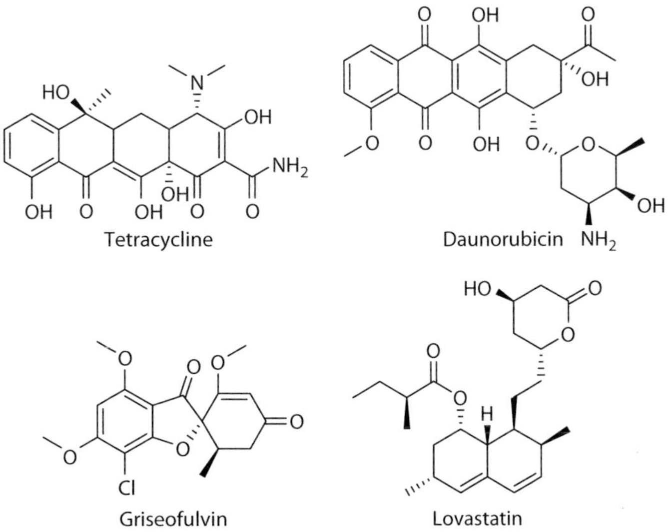

FIGURE 3.1 Examples of polyketides.

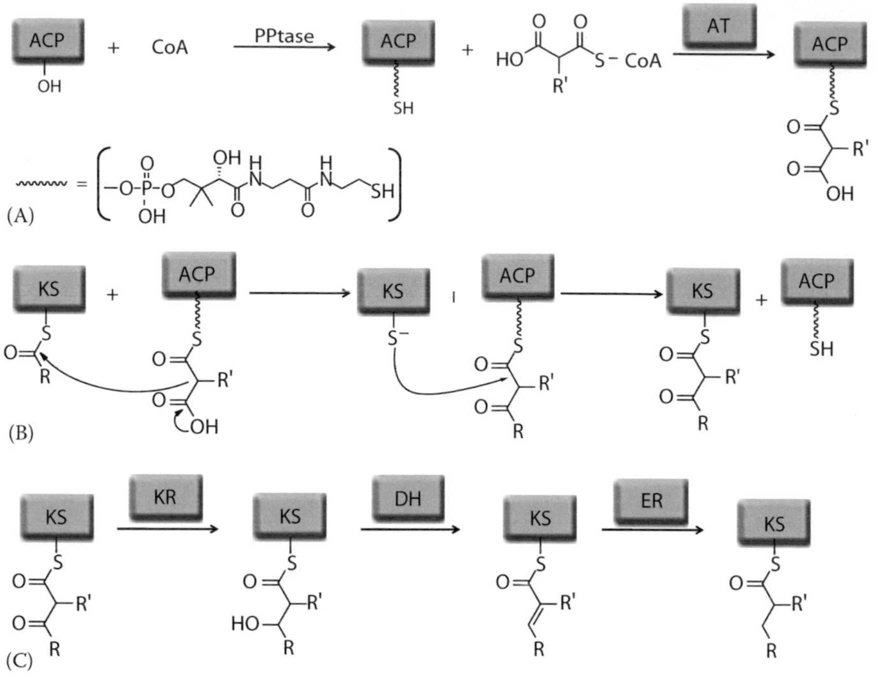

FIGURE 3.2 Polyketide biosynthesis. (A) Phosphopantetheinyl transfer modification of apo-ACP is catalyzed by a PPTase to form the holo-ACP. (B) The extension of the polyketide chain is catalyzed by successive decarboxylative Claisen condensations. (C) The growing polyketide chain can be selectively modified by KR, DH, and ER to a reduced polyketide backbone.

dedicated to the gene cluster or shared with FAS or other PKS clusters in the host. The AT domain selectively activates the extender unit and transfers it to the ACP domain for chain extension. Common extender units include malonyl-CoA, methylmalonyl-CoA, and ethylmalonyl-CoA (Chan et al. 2009). Extensive coverage of the mechanisms of polyketide biosynthesis is available in a number of reviews (Khosla et al. 1999; McDaniel et al. 2005; Hill 2006; Hertweck et al. 2007).

3.2.1 ERYTHROMYCIN AND ITS BIOSYNTHETIC PATHWAY

Erythromycin is naturally biosynthesized in the Gram-positive bacterium *Saccharopolyspora erythrea* (*S. erythrea*), and is commercially administered as a broad-spectrum macrolide antibiotic. The mechanism of action of erythromycin and its derivatives is through its reversible binding to the 50S ribosomal subunit of bacteria. *S. erythrea* itself is unaffected by erythromycin because the producer strain is self-resistant via methylation of an adenine base (A2058) in its 23S RNA, which interferes with erythromycin binding (Walsh 2003). Once produced, the macrolide antibiotic is then exported from the cell by an ATP-binding cassette-type protein.

The biosynthesis of erythromycin has been thoroughly studied since the early 1990s after the discovery of the gene cluster (Donadio et al. 1991). The pathway has been subjected to numerous combinatorial biosynthesis and metabolic engineering efforts to yield erythromycin analogues. The polyketide core of erythromycin, 6-deoxyerthronolide B (6-dEB), is synthesized by the type I modular PKS 6-deoxyerthronolide B synthases (DEBSs): DEBS1, DEBS2, and DEBS3, which are encoded by *eryAI*, *eryAII*, and *eryAIII* (Cortes et al. 1990; Donadio et al. 1991) (Figure 3.3). In general, type I PKSs may be classified as either modular or iterative. In the modular type I PKSs found in bacteria, such as DEBs, each catalytic domain is used only once from the N-terminus to C-terminus of the enzymatic assembly line. The modular nature of biosynthetic logic and colinearity of the assembly are therefore ideal for rationally modifying the final structure of the polyketide through domain insertion, inactivation, deletion, or scrambling. Biosynthesis 6-dEB by DEBS is initiated by loading of propionate starter unit onto the ACP domain of the loading didomain, followed by chain transfer to the KS domain of the first module to prime polyketide biosynthesis. The growing chain is then elongated through successive Claisen-like condensation with six methylmalonyl-CoA units by the six modules. After each decarboxylative condensation step, the β-keto position of elongated polyketide can be selectively modified by the built-in combinations of KR/DH/ER domains. Collectively, at the end of the sixth and final module, a linear polyketide precursor that incorporates all the stereocenters of 6-dEB is synthesized. The linear chain can then be macrocyclized and released from the assembly line by the terminal TE domain to form 6-dEB. Once offloaded, 6-dEB is subjected to downstream tailoring modifications such as oxidation and glycosylation with deoxysugars to yield the bioactive natural products erythromycin A (EA) and its derivatives erythromycin B, C, D (EB, EC, and ED).

3.2.2   Metabolic Engineering of *Saccharopolyspora erythrea* for the Production of Erythromycin and Its Derivatives

The first analogue of erythromycin produced by pathway engineering is through the inactivation of the KR domain in module 5 (KR5) of the wild type *S. erythrea*. The resultant polyketide product was shown to be 5,6-dideoxy-3α-mycarosyl-5-oxoerythronolide B **1** (Figure 3.4) (Donadio et al. 1991). The 813-bp in-frame deletion of the targeted KR gene was facilitated by homologous recombination, followed by the selection of correct mutant strains. Similar modifications on the erythromycin biosynthetic pathway genes were performed subsequently to produce numerous analogues directly from *S. erythrea*. For example, Donadio et al. mutated the active site ER domain of module 4 (ER4) to produce $\Delta^{6,7}$-anhydroerythromycin C **2** (Donadio et al. 1993). The point mutation was favored over domain deletion because it minimized the disturbance to the folding of the mutated PKS. Stassi et al. swapped the AT domain from DEBS module 4, which selects methylmalonyl-CoA, with module 5 AT domain from the niddamycin PKS, which selects ethylmalonyl-CoA as extender unit (Stassi et al. 1998). Upon exogenous addition of ethylmalonate in the culture, *S. erythrea* produced small amounts of 6-desmethyl-6-ethyl-EA **3** along with erythromycin A. The low yield of the targeted analogous was attributed to the low intracellular concentration of the ethylmalonyl-CoA extender unit. To address this limitation, crotonyl-CoA reductase was then introduced in the host and resulted in the biosynthesis of compound **3** as the predominant product.

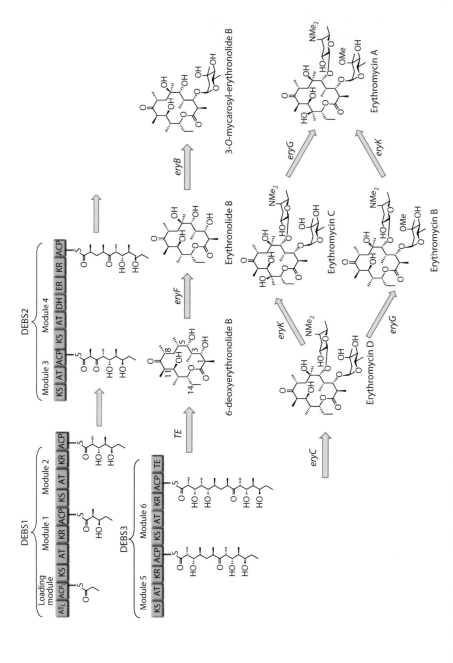

FIGURE 3.3 Erythromycin biosynthetic pathway.

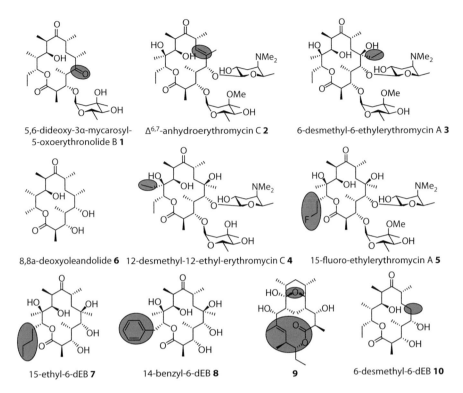

5,6-dideoxy-3α-mycarosyl-
5-oxoerythronolide B **1**

Δ^{6,7}-anhydroerythromycin C **2**

6-desmethyl-6-ethylerythromycin A **3**

8,8a-deoxyoleandolide **6**   12-desmethyl-12-ethyl-erythromycin C **4**   15-fluoro-ethylerythromycin A **5**

15-ethyl-6-dEB **7**   14-benzyl-6-dEB **8**   **9**   6-desmethyl-6-dEB **10**

FIGURE 3.4   Erythromycin analogues. The modified parts are highlighted.

Replacing the loading didomain of DEBS1 with that from the avermectin modular PKS from *Streptomyces avermitilis* led to the production of several erythromycin derivatives that incorporated propionate, butyrate, and pentanoate acyl groups as starter units (Marsden et al. 1998). The success of this strategy was based on the substrate promiscuity of the avermectin loading didomain, which accepts more than 40 different carboxylic acids as start units. When the KS domain from module 1 of DEBS was inactivated by mutating the active site cysteine to alanine (KS1°), no erythromycin can be produced (Jacobsen et al. 1998). Precursor feeding using the diketide-SNAC, (2*S*, 3*R*)-2-ethyl-3-hydroxy-pentanoyl-*N*-acethylcysteamine ((2*S*, 3*R*)-2-ethyl-3- hydroxy-pentanoyl-SNAC), produced 12-desmethyl-12-ethyl-EC **4**. Therefore, precursor-directed biosynthesis using the mutant DEBSs represents a convenient avenue to derivatize the starter unit of the macrolide. However, a follow-up study found that the diketide SNAC was degraded by the host by unknown pathways, which significantly decreases the efficiency of incorporating the precursor (Frykman et al. 2001). Locating and eliminating the corresponded enzyme responsible for acyl-SNAC degradation can therefore improve the overall efficiency of this approach. Based on the precursor-directed biosynthesis approach, Ward et al. deleted the loading didomain and module 1 from DEBS1 and the domain deletion afforded a DEBS mutant that is more efficient at polyketide production when supplied with acyl-SNAC starter units (Ward et al. 2007). Addition of (2*S*, 3*R*)-5-fluoro-3-hydroxyl-2-methylpentanoyl-SNAC to *S. erythrea* expressing this mutant produced 15-fluoro-Erythromycin A **5** at a 10-fold higher yield compared to the KS1° mutant strain.

The production of erythromycin analogues by feeding 15-methyl-6-dEB to the KS1° strain was also optimized (Carreras et al. 2002). It was found that high dissolved oxygen levels, neutral pH, and specific aglycon feed concentration (15-methyl-6-dEB/dry cell weight ~0.06) were preferred for higher production of erythromycin analogues and higher ratio of EA analogue to other analogues (EB, EC, ED). Since EA is biologically much more potent than EB, EC, and ED, it is desirable to convert EB, EC, and ED to EA completely. The ratio of EA to other analogues can also be improved by overexpression of *eryK* (Desai et al. 2004). When extra copies of *eryK* was introduced to the KS1° strain and fed with 15-fluoro-6-dEB, the new strain produced 1.3 g/L compound **5** with 75%–80% molar yield on fed precursor, while the KS1° strain produced only 0.9 g/L compound **5** with 50%–55% molar yield. Introduction of both *eryK* and *eryG* genes into *S. erythrea* not only almost completely eliminated undesired EB, EC, and ED, but also slightly increased EA production by 25% (Chen et al. 2008). When extra copies of the methylmalonyl-CoA mutase operon was introduced into the *S. erythrea* wild type strain, endogenous methylmalonyl-CoA concentration was increased and in return 50% more erythromycin was produced (Reeves et al. 2007).

Replacing the entire DEBS genes (*eryAI*, *eryAII*, and *eryAIII*) in the wild type *S. erythrea* strain with those from an industrially-used overproducing strain demonstrated that the overproduction was primarily due to genotypical differences in non-PKS genes (Rodriguez et al. 2003). The industrial overproduction strain was found to express the erythromycin genes several days longer than the wild type. DNA microarrays indicated that the accumulated mutations via classical mutagenesis affected the timing and rate of erythromycin synthesis, which may result from alterations to the regulatory genes involved in erythromycin biosynthesis and precursor biosynthesis (Lum et al. 2004). Subsequently, the involved regulatory protein was found to be an ortholog of BldD, a key regulator of development in *Streptomyces coelicolor* (*S. coelicolor*) (Chng et al. 2008). The regulator binds to all five promoter regions in the *ery* gene cluster as well as to its own promoter. The removal of BldD in *S. erythrea* decreased erythromycin production by sevenfold in liquid cultures. The function of BldD was further confirmed by higher titer expression of BldD in the industrially optimized strain. To further increase the high levels of erythromycin (~4 g/L) produced in the industrial strain, a bacterial hemoglobin (VHB) gene originally isolated from *Vitreoscilla sp.* was integrated into the overproduction strain (Minas et al. 1998). VHB alleviates the defects of hypoxic conditions through promoting oxygen delivery (Wei and Chen 2008). The resultant recombinant strain *S. erythrea::vhb* produced >7 g/L erythromycin, a 70% improvement from the original industrial strain.

3.2.3 Streptomyces as Heterologous Hosts for the Production of 6-dEB, Erythromycin and Their Derivatives

Streptomyces coelicolor and *Streptomyces lividans* are widely used as heterologous hosts for producing polyketides due to the availability of well-established genetic protocols. The workhorse mutant *S. coelicolor* CH999 was created by deleting of the entire actinorhodin (*act*) gene cluster (McDaniel et al. 1993). The deletion of this PKS removed endogenous *S. coelicolor* PKS and afforded a host organism with a clean genetic/chemical/visual background. CH999 was the first host engineered for the heterologous expression of the entire DEBS gene cluster (Kao et al. 1994). The genes encoding DEBS1, DEBS2, and DEBS3 were cloned into a *Streptomyces-E. coli* shuttle vector to give pCK7. Upon transformation of CH999 with pCK7, the resultant strain produced 6-dEB and 8,8a-deoxyoleandolide **6** (primed by acetyl-CoA instead of propionyl-CoA). The heterologous host/vector pair proved to be straightforward to use and manipulate, as in the subsequent years, numerous 6-dEB analogues were biosynthesized via genetic engineering. To showcase the power of a heterologous expression system, a combinatorial library of more than 50 6-dEB analogues was generated by swapping AT and β-carbon processing domains (KR, DH, and ER) between DEBS and the rapamycin PKS (RAPS) (McDaniel et al. 1999). Single, double, and triple modifications of 6-dEB were obtained from single, double, and triple domain swapping experiments. The ability to generate large libraries of macrolide analogues opens the door for screening improved compounds as antibiotics.

One method to improve 6-dEB production in *S. coelicolor* is to increase the intracellular methylmalonyl-CoA concentration. To accomplish this, *matB* and *matC*, which encode a malonyl-CoA synthase that convert methylmalonate to methylmalonyl-CoA and a putative dicarboxylate transport protein, respectively, were overexpressed in the CH999 strain harboring the DEBS genes (Lombo et al. 2001). The addition of methylmalonate to the culture produced large amount of 6 dEB, resulting in a 300% enhancement of macrolactone titers compared to the original strain CH999/pCK7.

To equip CH999/pCK7 for precursor-directed biosynthesis, Jacobsen et al. introduced the KS1° mutant of DEBS1 genes to CH999 (Jacobsen et al. 1997). The resultant transformant CH999/pJRJ2 lost the ability to produce 6-dEB as expected. Exogenous feeding of designed synthetic molecules (2*S*, 3*R*)-2-methyl-3-hydroxyl-pentanoyl-SNAC, (2*S*, 3*R*)-2-methyl-3-hydroxyl-heptanoyl-SNAC, (2*S*, 3*R*)-2-methyl-3-hydroxyl-4-benzyl-butyryl-SNAC, and (2*E*, 4*S*, 5*R*)-2,4-dimethyl-5-hydroxyl-2-heptenoyl-SNAC) produced large amounts of 6-dEB and its analogues 15-ethyl-6-dEB (55 mg/L) **7**, 14-benzyl-6-dEB **8** (22 mg/L), and compound **9** (25 mg/L), respectively. 15-ethyl-6-dEB and 14-benzyl-6-dEB were further converted to erythromycin D derivatives by feeding these aglycons to a *S. erythrea* mutant (A34) that was blocked in 6-dEB production. Compound **9** is a 16-membered macrolactone, and was isolated in the hemiketal form (Kinoshita et al. 2001). The substrate (2*E*, 4*S*, 5*R*)-4-dimethyl-5-hydroxyl-2-heptenoyl-SANC, which lacks a methyl group at the C-2 position, did not yield any macrolactone products. During optimization of the production of 6-dEB analogues from CH999/pJRJ2, it was found that productivity depended strongly on the feed concentration of the diketide starter unit (Leaf et al. 2000). It was determined that the best time to feed the starter unit was 2 days after inoculation. Furthermore, lower pH (5.5) and implementation of glucose feeding in shake flasks were found to greatly enhance the productivity of 6-dEB to ~1.5 g/L. To further improve the production of 6-dEB analogues, this strain was first subjected to classical random mutagenesis followed by selection of higher producer mutants (Desai et al. 2004). The best mutant was then cured of the previous plasmid and transformed with pJRJ2 for 6-dEB analogue production. Hydrophobic resin XAD-16HP was added to the fermentor to absorb the product coupled with slow release of the diketide precursor. A final production of 1.3 g/L of 15-methyl-6-dEB was achieved. Lastly, to address the process expense of using *N*-acetyl cysteamine (SNAC), several alternative thioesters were discovered (Murli et al. 2005). One of these, methyl thioglycolate, was verified to be an efficient thioester carrier for 6-dEB analogue biosynthesis.

3.2.4 *Escherichia coli* as Heterologous Hosts for the Production of 6-dEB, Erythromycin, and Their Derivatives

E. coli remains one of the most powerful organisms used for recombinant protein expression and heterologous chemical biosynthesis. Many biologically active compounds have been engineered and produced from *E. coli*, including polyketides (Zhang et al. 2008), nonribosomal peptides (Watanabe et al. 2006), terpenoids (Chang et al. 2007), and alkaloids (Minami et al. 2008). The advantages of using *E. coli* as a heterologous expression host come from its faster growth characteristics, abundant genetic tools, and better understanding of its primary metabolic pathways compared to other hosts.

One of the key deficiencies in *E. coli*-based expression of functional PKSs is the lack of the ability to post-translationally modify the ACP domains by a compatible PPTase. This was addressed by overexpression of a broadly specific PPTase, *sfp*, from *Bacillus subtilis* in *E. coli* BL21(DE3) (Quadri et al. 1998). To eliminate the need to express Sfp episomally, the *sfp* gene was incorporated into the genome of *E. coli* BL21(DE3) under the T7 promoter to yield the *E. coli* BAP1 strain (Pfeifer et al. 2001). To produce 6-dDEBS in *E. coli*, additional metabolic engineering efforts are required to elevate the low endogenous concentration of propionyl-CoA (~5.3 μM) (Bennett et al. 2009), as well as to install a suitable pathway for the accumulation of the extender unit (2*S*)-methylmalonyl-CoA. The first objective was accomplished by deleting the propionate catabolism pathway genes *prpRBCD* (while at the same time integrating *sfp* gene). Overexpression of *prpE* converted the exogenously supplied propionate to propionyl-CoA (Pfeifer et al. 2001). The second objective was addressed by overexpressing the propionyl-CoA carboxylase gene from *S. coelicolor* and an endogenous biotin ligase gene, *birA*, which can enhance the activity of propionyl-CoA carboxylase. When all three DEBS genes were introduced into the engineered *E. coli* strain (BAP1/pBP130/pBP144) and induced with isopropyl β-D-1-thiogalactopyranoside (IPTG), substantial amounts of 6-dEB were produced. During recombinant expression, temperature must be maintained at ~22°C to prevent aggregation of DEBS proteins and loss of activity. Using this host, biosynthesis of 6-dEB analogues 13-benzyl-6-dEB was first demonstrated after exchanging the loading didomain between DEBS and RAPS.

High-cell-density fed-batch cultivation was developed to increase the yield of 6-dEB production using the engineered *E. coli* strain harboring the DEBS genes (Pfeifer et al. 2002). F1 minimal media was selected and gave the best yield of 6-dEB up to 100 mg/L after 6 days' fermentation. Co-expression of an accessory thioesterase (TEII, encoded by *ery-ORF5*) further improved the yield of 6-dEB to ~180 mg/L, which is ~200 times higher than the shake flask titer. Different methylmalonyl-CoA biosynthetic pathways were used to compare the efficiency of 6-dEB production

(Murli et al. 2003). When malonyl/methylmalonyl-CoA ligase (MatB) was expressed in *E. coli* and methylmalonate was fed to the culture, 90% of all acyl-CoAs were detected to be methylmalonyl-CoA. Surprisingly, methylmalonyl-CoA generated by this method was not utilized by DEBS to form 6-dEB. The reason for this is still unknown. On the other hand, the *Propionibacteria shermanii* methylmalonyl-CoA mutase/epimerase pathway can generate methylmalonyl-CoA up to 30% of all acyl-CoAs, the same amount as PCC pathway. However, the titer of 6-dEB was fivefold higher for the strain containing the PCC pathway compared to the *E. coli* strain containing the mutase pathway. Both PCC and mutase pathways were then integrated into the *E. coli* genome to create K207-3. The authors also used more compatible origins of replication on different plasmids containing DEBS genes to generate K207-3/pKOS207-129/pBP130, which improved the stability of the plasmids. The engineered new strain produced ~280 mg/L of 6-dEB in shake flash experiment (Lau et al. 2004). During further optimization of *E. coli* fermentation conditions for 6-dEB production, it was noted that the accumulation of excess ammonia greatly impacts the productivity of the cells. After careful control and removal of ammonia, the F1-minimal medium-based high-cell-density fed-batch fermentation produced 1.1 g/L 6-dEB. Subsequently, to avoid the instability of the multi-plasmids system, the PCC pathway genes and DEBS genes were all integrated into BAP1 genome (Wang and Pfeifer 2008). However, the overall production of 6-dEB was decreased probably due to lower transcriptional levels of the single-copy DEBS genes.

Kennedy et al. overexpressed an acetoacetyl-CoA:acetyl-CoA transferase, AtoAD, in *E. coli* to synthesize butyryl-CoA from supplemented butyrate (Kennedy et al. 2003). Since the endogenous propionyl-CoA concentration is low in wild type *E. coli* and the loading didomain of DEBS has relaxed specificity toward butyryl-CoA, the strain harboring the DEBS genes utilized butyryl-CoA as the starter unit and synthesized 15-methyl-6-dEB directly. To synthesize erythromycin in *E. coli*, another 17 heterologous genes were introduced to the 6-dEB-producing strain, including the mycarose operon that converts glucose-1-phosphate to TDP-L-mycarose, the desosamine operon that converts TDP-4-keto-6-deoxyglucose to TDP-D-desosamine, and ErmE and resistance genes that confer *E. coli* resistance to macrolide antibiotics (Peiru et al. 2005). Both EC and ED were produced in this engineered host, which demonstrated the metabolic dexterity of *E. coli* toward the production of complicated natural products.

3.2.5 OTHER POLYKETIDES

Similar strategies used in metabolic engineering of erythromycin can be applied to obtain derivatives and improve yields of other type I modular PKSs. However, for those polyketides originated from fungal type I iterative PKSs (megasynthase and iterative in nature), such as lovastatin (Kennedy et al. 1999) and compactin (Abe et al. 2002), the biosynthetic programming rules were much more complicated (Ma et al. 2009). Strategies like domain inactivation and swapping are not readily applicable. For example, improving the titers of lovastatin mainly focused on strain optimization by traditional chemical/physical mutagenesis coupled with screening method (Kumar et al. 2000), overexpression of regulator genes such as *lovE* and *laeA* (Bok and Keller 2004), elimination of competing metabolic pathways

(Couch and Gaucher 2004), and optimization of the fermentation conditions (Novak et al. 1997; Manzoni and Rollini 2002). For those polyketides synthesized from type II PKSs (dissociate and iterative in nature), such as tetracycline and doxo-rubicin, similar strategies including random mutagenesis and overexpression of reg-ulatory genes have been used to improve the yields of target compounds.

3.3   NONRIBOSOMAL PEPTIDES

Nonribosomal peptides represent a large family of secondary metabolites from a variety of organisms. A vast majority of these natural products exhibit important biological activities, such as antibiotics penicillin (Smith et al. 1990) and vanco-mycin (Hubbard and Walsh 2003; Oberthur et al. 2005), siderophore enterobactin (Gehring et al. 1997; Gehring et al. 1998), and immunosuppressive agent cyclospo-rine A (Weber et al. 1994) (Figure 3.5).

Nonribosomal peptides are biosynthesized by a group of megasynthetases called nonribosomal peptide synthetases (NRPSs) (Schwarzer et al. 2003). Typical NRPSs

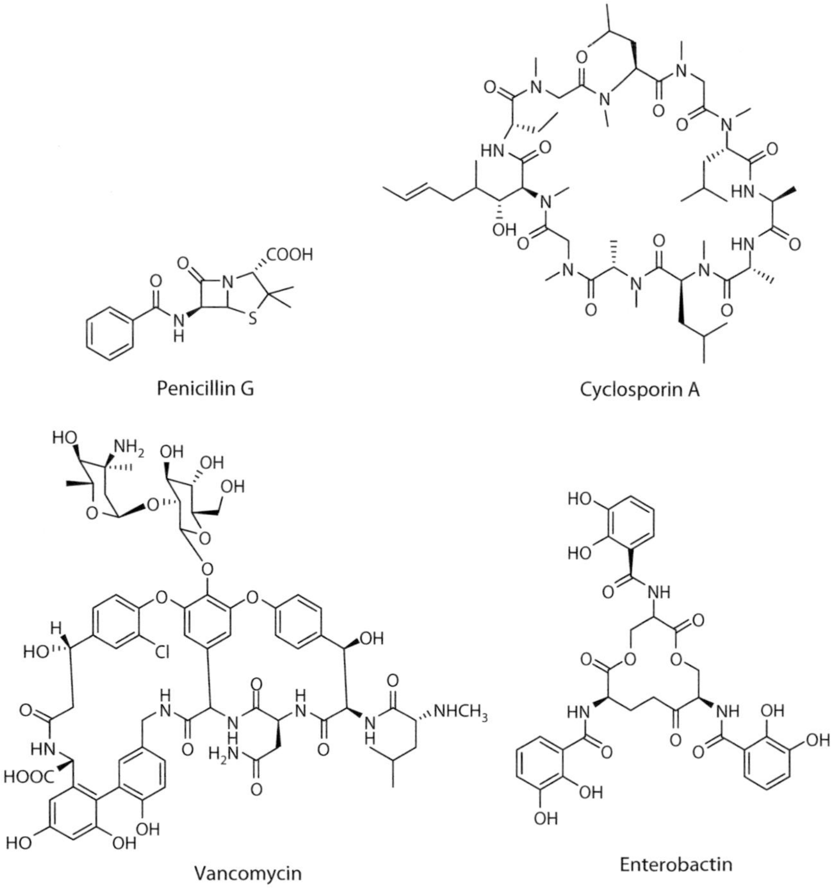

Penicillin G

Cyclosporin A

Vancomycin

Enterobactin

FIGURE 3.5   Examples of nonribosomal peptides.

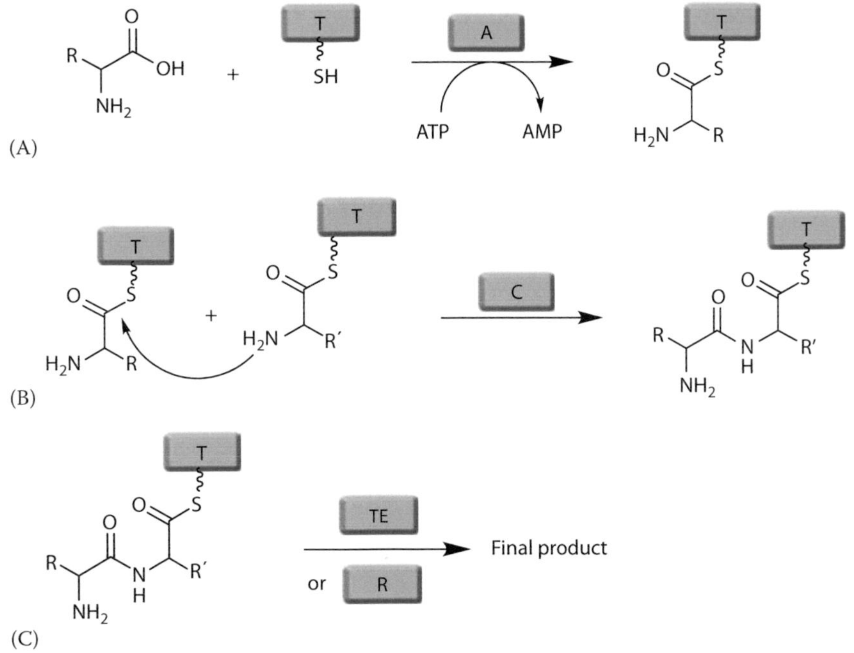

FIGURE 3.6 Biosynthesis of nonribosomal peptides. (A) Recognition, activation and loading of amino acid. (B) Condensation between the downstream and upstream aminoacyl-S-T. (C) Product releasing from thiolation domain.

share the similar programming logic with PKSs but utilize a more diverse set of building blocks—amino acids (von Dohren et al. 1999). Each NPRS is composed of a series of modules, whereby each module harbors unique domain organization and catalyzes one round of addition of a specific amino acid (Marahiel et al. 1997). A classical NRPS module is organized as C-A-T. Adenylation (A) domain activates one amino acid to aminoacyl-O-AMP by consumption of ATP and then tethers it to the phosphopantetheine arm of thiolation (T) domain, while condensation (C) domain catalyzes the formation of amide bond between two amino acids (Figure 3.6). Some tailoring domains are also involved to amplify the diversity of NRPs. For example, oxidase (Ox) domain in epothilone biosynthesis is used to transform dihydrothiazole into thiozole. Epimerase (E) domain alters the stereochemistry of α positions of aminoacyl groups. The releasing domain, either a thioesterase (TE) or a reduction (R) domain, is responsible for releasing the nascent aminoacyl chain from the T domain of the last module. Three different releasing mechanisms are present (Figure 3.7): first, a linear peptide can be formed via direct hydrolysis by TE; second, macrocyclization-catalyzed TE domains can afford peptidyl macrolactones or macrolactams; and lastly, an R domain can catalyze release of products on many PKS/ NRPS hybrid assembly lines via either reduction or Dieckmann cyclization (Sims and Schmidt 2008) (Figure 3.7). Detailed reviews of NPRSs are available in numerous review papers (Schwarzer et al. 2003; Fischbach and Walsh 2006).

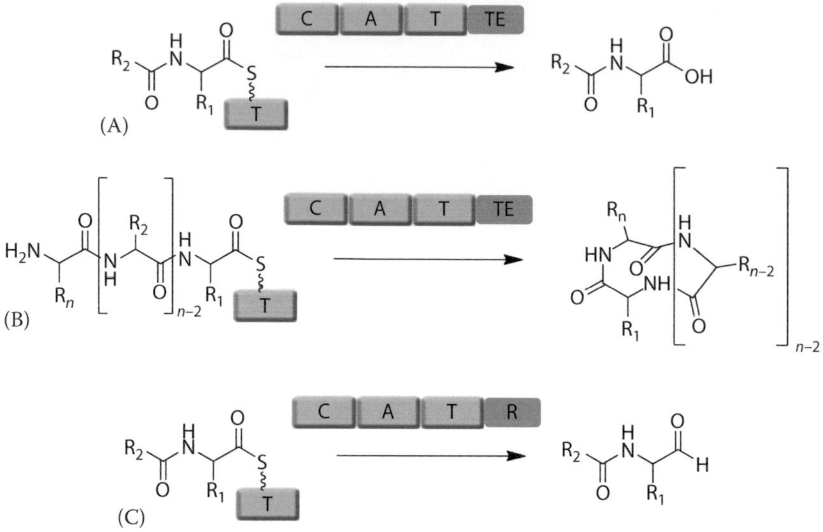

FIGURE 3.7 Different releasing mechanisms of NRPSs. (A) TE-catalyzed hydrolytic release (B) TE-mediated intra-molecular macrolactamization. (C) R-directed reductive release.

3.3.1 DAPTOMYCIN

Daptomycin is a cyclic lipopeptide that was approved by U.S. Food and Drug Administration in 2003 for treating skin infections caused by Gram-positive pathogens (Kirkpatrick et al. 2003). It was initially discovered by Eli Lilly in 1987 in a program of searching soil samples for novel antibiotics producing strains (Debono et al. 1987). Daptomycin and its close relatives, in the presence of Ca^{2+}, can inhibit peptidoglycan biosynthesis and can therefore starve the Gram-positive bacteria to death (Eliopoulos et al. 1985; Baltz et al. 2005). To date, no known resistance toward daptomycin has been developed in pathogenic bacteria, which makes this family of compounds highly effective and powerful (Lamp et al. 1992; Sauermann et al. 2008). Structurally, daptomycin is composed of two portions: NRP core and a tethered fatty acyl chain (Figure 3.8). The NRP core contains 13 amino acids, including 7 natural amino acid residues (Trp_1, Asp_3, Thr_4, Gly_5, Asp_7, Asp_9, Gly_{10}), 3 D-amino acids (D-Asn_2, D-Ala_8, D-Ser_{11}), and 3 non-proteinogenic amino acids (ornithine$_6$, (2S, 3R)-3-methylglutamic acid$_{12}$, kynurenine$_{13}$). The decanoyl moiety is linked to the N-terminus of the first residue tryptophan. Daptomycin is produced via fermentation from *Streptomyces roseosporus*, which normally produces a set of cyclic peptides known as A21978C. Each member of A21978C contains the same cyclic peptide core and is acylated with different fatty-acyl chains at the first tryptophan residue. Daptomycin can be produced from *Streptomyces roseosporus* upon feeding of the culture with decanoic acid.

Because of the powerful activity of daptomycin, it is of significant interest to study and engineer the biosynthetic pathway genes of daptomycin to generate more daptomycin derivatives that have improved biological activities. To accomplish this, metabolic engineering strategies were applied: (1) to elucidate and understand daptomycin biosynthetic pathway, (2) to produce daptomycin in the natural host *Streptomyces*

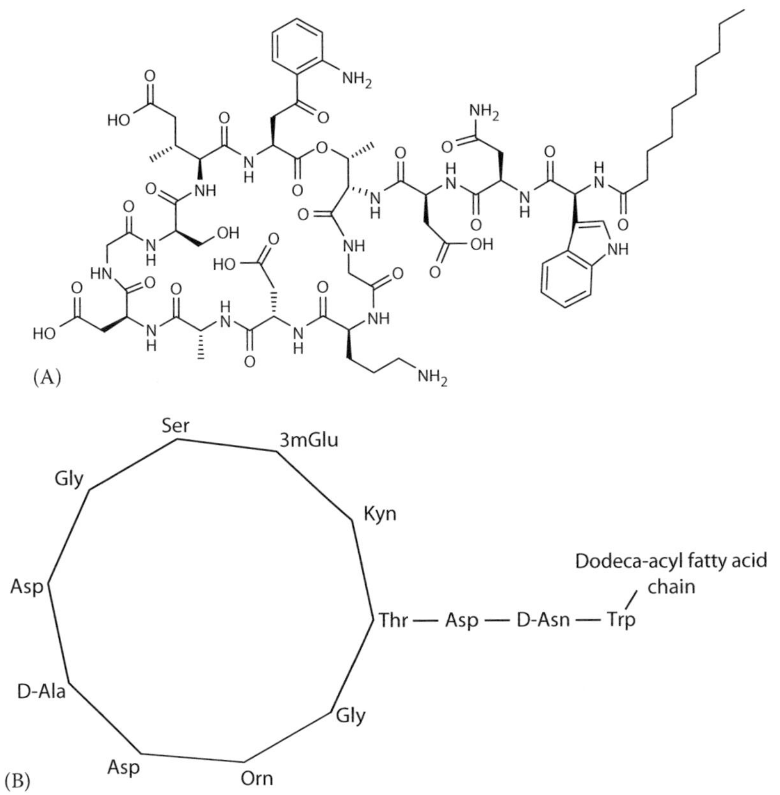

FIGURE 3.8 (A) Molecular structure and (B) structure map of daptomycin.

roseosporus, (3) to heterologously produce daptomycin in *Streptomyces lividans*, and (4) to generate daptomycin analogues. Most of the advances in our understanding of daptomycin biosynthesis have been published from the Baltz group from Cubist Pharmaceuticals. The same group has also published numerous reviews on this subject (Baltz et al. 2005; Baltz 2008).

3.3.2 ELUCIDATION OF DAPTOMYCIN BIOSYNTHETIC PATHWAY

Transposon mutagenesis revealed that the daptomycin gene cluster is located at one end of the chromosome of *S. roseosporus* (Mchenney et al. 1998) (Figure 3.9). Part of the genes were cloned, sequenced, and confirmed to encode NRPS. Later on, the entire gene cluster of daptomycin was cloned in pStreptoBAC V and sequenced. Functions of the *dpt* genes were assigned through bioinformatics study and heterologous expression: the NRPS for daptomycin biosynthesis is composed of three giant multi-modular subunits, DptA, DptBC, and DptD, encoded by *dpt*A, *dpt*BC, and *dpt*D genes, respectively. *dpt*E and *dpt*F genes lie upstream of NRPS and encode acyl-CoA ligase and acyl carrier protein, respectively. *dpt*I and *dpt*J lie downstream of NRPS and encode methyltransferase and 2,3-dioxygenase, respectively. Biosynthesis of daptomycin is consistent with the colinearity rule of NRPS: five-module DptA

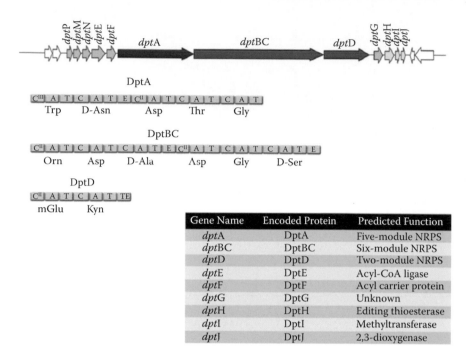

Gene Name	Encoded Protein	Predicted Function
*dpt*A	DptA	Five-module NRPS
*dpt*BC	DptBC	Six-module NRPS
*dpt*D	DptD	Two-module NRPS
*dpt*E	DptE	Acyl-CoA ligase
*dpt*F	DptF	Acyl carrier protein
*dpt*G	DptG	Unknown
*dpt*H	DptH	Editing thioesterase
*dpt*I	DptI	Methyltransferase
*dpt*J	DptJ	2,3-dioxygenase

FIGURE 3.9 Daptomycin gene cluster and domain organizations.

catalyzes the binding, activation, and incorporation of the first five amino acids. The first module of DptA crosstalks with DptE and DptF to transfer the fatty-acyl chain to the N-terminus of Trp_1. DptBC continues the assembly process to incorporate six amino acids and DptD adds three more amino acids. The nascent peptide chain is released by the TE domain of DptD and is macrolactonized between the side chain hydroxyl group of Thr_4 and kynurenine$_{13}$ to form the final product daptomycin.

In order to verify the proposed functions of the above genes, several genetic studies have been performed. The native host, *S. roseosporus*, was proven to be a suitable host for molecular genetic manipulation since it is relatively nonrestricting for heterologous DNA (McHenney and Baltz 1996). In the sequencing study of *dpt* gene cluster, a Tn5099 insertion in *dpt*BC abolished the production of daptomycin completely, indicating that NRPS subunit *dpt*BC gene is essential for the biosynthesis of daptomycin (Mchenney et al. 1998). A gene knockout system with pRHB538 plasmid containing *rps*L gene encoding streptomycin resistance marker was developed. A series of chromosomal deletions of single or multiple genes in *S. roseosporus* were conducted by using double-crossover homologous recombination (Hosted and Baltz 1997; Miao et al. 2006; Nguyen et al. 2006) (Table 3.1). Other deletion mutants were constructed by incorporating genes into the knockout strain KN100 (Δ*dpt*BCD) and KN125 (Δ*dpt*BCDGHIJ) via conjugation with *E. coli* and ΦC31 *attB* site-specific integration (Nguyen et al. 2006). Disruptions of one or more NRPS genes led to loss of A21978C biosynthesis, which showed that all NRPS genes are essential for biosynthesis of daptomycin. Both DptG and DptH are necessary for the maximal production of A21978C complex. DptH appears to be an editing thioesterase, while

TABLE 3.1
dpt **Gene Knockout Strains of *S. roseosporus***

Strains (*S. roseosporus*)	Relevant Features	Reference
UA343	A21978C wide-type producer	Miao et al. (2006)
MM140	UA343::pRHB157	Mchenney et al. (1998)
UA117	UA343 rpsL7(SmR)	Miao et al. (2006)
UA378	UA117 ΔdptD::ermE	Miao et al. (2006)
UA474	UA117 ΔdptAD::ermE	Coeffet-Le Gal et al. (2006)
UA431	UA117 ΔdptEFABCDGHIJ	Nguyen et al. (2006)
KN100	UA117 ΔdptBCD	Nguyen et al. (2006)
KN125	UA117 ΔdptBCDGHIJ	Nguyen et al. (2006)

the function of DptG is still unknown (Baltz 2008). Since the deletion of *dpt*I led to the substitution of 3-methylglutamine with glutamine in the lipopeptide products, DptI is proposed to be a methyltransferase that catalyzes the formation of 3-methylglutamine (Mahlert et al. 2007). However, deletion of *dpt*J causes merely 50% loss of A27918C yield, indicating that perhaps another functional dioxygenase may partially complement the role of DptJ.

3.3.3 METABOLIC ENGINEERING OF *STREPTOMYCES ROSEOSPORUS* FOR THE PRODUCTION OF DAPTOMYCIN

Naturally, *S. roseosporus* produces lipopeptide complex A21978C with the same tridecapeptide portion but with different fatty acyl chains. These natural pendants are mainly branched-chain C11, C12, and C13 fatty acids. Daptomycin is a single lipopeptide compound with a straight-chain C10 lipid side chain (decanoate), and was initially developed from A21978C through a semisynthetic method, including chemical deacylation and reacylation with decanoate. Later, a group of penicillin-active deacylases from *Actinoplanes*, especially *Actinoplanes utahensis* NRRL12052, were demonstrated to be effective in removing the lipid chains from A21978C complex (Boeck et al. 1988). In 2000, Kreuzman and coworkers cloned the deacylase gene from *A. utahensis* into *S. lividans* and engineered a strain capable of producing deacylated lipopeptide core (Kreuzman et al. 2000).

Despite the enzymatic deacylation step, the production of daptomycin using semisynthetic approaches involves numerous steps with low yield. Therefore, a precursor-directed fermentation method was designed by Huber and his coworkers to synthesize daptomycin biosynthetically (Huber et al. 1988; Julia et al. 2006). By supplying decanoic acid to *S. roseosporus*, daptomycin was produced instead of A21978C complex with a yield 100 mg/L under the optimum conditions. In the precursor feeding process, a special formulation with a 1:1 mixture of decanoic acid and methyl oleate was used to prevent the decanoic acid from forming waxy solid at cultivation temperature of 30°C. However, high concentration of decanoic acid lead to complete lysis of *S. roseosporus* cells. Therefore, the supplementation of decanoic acid needs to be carefully controlled.

3.3.4 Heterologous Production of Daptomycin in *Streptomyces lividans*

Just as *S. coelicolor*, *S. lividans* has long been recognized as a "fermentation friendly" heterologous host. It is also genetically closely related to *S. roseosporus*. To produce daptomycin heterologously, the entire 128 kb *dpt* gene cluster was integrated into *S. lividans* genome via *E. coli*-mediated conjugation (Miao et al. 2005). However, the recombinant strain produced A21978C with a significantly lower yield than those produced by *S. roseosporus*, in which yields range from 150 mg/L to 1 g/L. In addition, the complex background of host metabolites from *S. lividans*, which includes actinorhodin and a series of metabolites belonging to CDA complex (another family of cyclic lipopeptides), further complicated the production of daptomycin. To address this issue, Penn and coworkers knocked out the genes encoding actinorhodin cluster in *S. lividans*, which greatly improved the titer of A21978C (Julia et al. 2006). Furthermore, by adjusting phosphate concentration in a defined fermentation medium, additional improvement in A21978C titer from 20 to 55 mg/L was observed. The yield of daptomycin was predicted to be further increased by deleting the CDA-related genes, since both biosynthetic pathways compete for some of the amino acid precursors (Baltz 2008).

3.3.5 Metabolic Engineering of *Streptomyces roseosporus* to Produce Novel Daptomycin Derivatives

There is no known literature example of the total synthesis of daptomycin-like compounds. In addition, previous medicinal chemical studies on daptomycin were limited to modifications of the lipid pendant of daptomycin (Debono et al. 1988). Therefore, given the complexity and limitation of the chemical synthesis of daptomycin analogues, metabolic engineering strategies turned out to be an ideal approach to generate novel daptomycin analogues.

Based on previous studies, it is believed that the daptomycin biosynthetic genes were transcribed into a single message RNA, which impeded setting up a robust system for metabolic engineering to combinatorially synthesize new lipopeptide antibiotics. By utilizing two types of integration sites, ΦC31 and IS117 *attB*, successful ectopic expressions of daptomycin NRPS genes were achieved. In addition, the installation of a strong constitutive ermE promoter allowed strong expression of the genes in *S. roseosporus*. Aided by these discoveries, engineered *dpt* NRPS genes or ancillary genes from different lipopeptide clusters were inserted into the *dpt* knockout strains UA431 (ΔdptEFABCDGHIJ), KN100, and KN125 at the two integration sites. The genetically engineered *S. roseosporus* strains were capable of producing a number of new derivatives of daptomycin (Table 3.2).

The first generation of daptomycin analogues were generated by swapping the *dpt* NRPS genes with its counterparts from different lipopeptide gene clusters. A54145 (*lpt*) from *S. fradiae* is a complex lipopeptide with 10-member ring cyclic peptide core, while CDA (*cdaPS3*) from *S. coelicolor* is a series of lipopeptides of which the structures were only partially characterized. Both gene clusters were cloned and analyzed and the relationships between product structures and biosynthetic genes were elucidated to some extent. Although only 53%–57% similarity was shared among

TABLE 3.2

Combinatorial Biosynthesis and Antibacterial Study of Daptomycin Analogues

Compound No.	Position 8	Position 11	Position 12	Position 13	Yield (mg/L)	MIC (μg/mL)	Reference
Daptomycin	D-Ala	D-Ser	3mGlu	Kyn	250	0.5–1	Coeffet-Le Gal et al. (2006)
11	D-Ala	D-Ser	3mGlu	**Ile/Val**	94	4	Coeffet-Le Gal et al. (2006)
12	D-Ala	D-Ser	3mGlu	**Trp**	168	1	Miao et al. (2006)
13	**D-Ser**	D-Ser	3mGlu	Kyn	200	1	Miao et al. (2006)
14	D-Ala	D-Ser	3mGlu	**Asn**	84	128	Nguyen et al. (2006)
15	D-Ala	**D-Ala**	3mGlu	Kyn	200	0.5–1	Nguyen et al. (2006)
16	D-Ala	**D-Asn**	3mGlu	Kyn	200	1	Nguyen et al. (2006); Gu et al. (2007)
17	**D-Asn**	D-Ser	3mGlu	Kyn	N/A	120	Nguyen et al. (2006)
18	D-Ala	D-Ser	**Glu**	Kyn	N/A	8	Nguyen et al. (2006)
19	**D-Asn**	D-Ser	**Glu**	Kyn	N/A	128	Nguyen et al. (2006)
20	D-Ala	**D-Asn**	**Glu**	Kyn	150	32	Nguyen et al. (2006)
21	**D-Asn**	D-Ser	3mGlu	**Ile**	N/A	16	Nguyen et al. (2006)

Note: Changes in amino acid residues of daptomycin are indicated in bold.

the last NRPS module DptD, LptD, and CdaPS3, a highly conserved interpeptide docking sequence facilitated the construction and functional crosstalk of the hybrid assembly lines. By inserting the *lpt*D or *cda*PS3 genes at the IS117 site of UA378 under the control of the ermE promoter, new lipopeptides **11** or **12** were produced at good yields, respectively. Overexpression of dptA via integration at the ΦC31 site under the control of the ermE promoter further increased the yields of **11** and **12**.

The modular nature and the colinearity of NRPSs allow engineered biosynthesis of novel lipopeptide antibiotics via module or domain swapping (Schwarzer et al. 2003). Module swapping has proven to be an effective way to generate new derivatives of NRP (Stachelhaus et al. 1995; Doekel and Marahiel 2000). Similar strategy was applied to the engineering of the daptomycin pathway. For example, swapping between the DptD module responsible for incorporation of $3mGlu_{12}$ and the LptD module responsible for incorporation of Ile_{13}, or with the CdaPS3 module responsible for inserting Trp_{13}, led to the synthesis of compounds **11** and **12** at titers that were higher than those achieved via whole subunit exchange. To introduce structural diversity into the daptomycin molecule at position 13, an asparagine-specific module from LptC replaced the last module in DptD, which activated kynurenine$_{13}$. The resulting recombinant strain produced the expected analogue of daptomycin in which the last amino acid is occupied by asparagine (Table 3.2). Replacements of the module that activates, epimerizes, and incorporates D-serine$_{11}$ or D-Alanine$_8$ with each other in DptBC resulted in amino acid substitutions at positions 8 and 11. These recombinant strains produced targeted daptomycin analogues in titers up to 200 mg/L. Meanwhile, exchange between the module responsible for inserting D-Serine$_{11}$ with a module responsible for inserting D-asparagine from LptC successfully produced analogue **14** (Nguyen et al. 2006; Gu et al. 2007).

Finally, by combining the aforementioned strategies of subunit exchange, module exchange, and ancillary gene manipulation, an additional seven new daptomycin derivatives **15–21** were produced. As shown in Table 3.2, at least five of these compounds were confirmed to be as active as daptomycin in antibacterial assays. Therefore, metabolic engineering approaches were successfully employed in the biosynthesis of daptomycin-related lipopeptides that have not been obtained via total synthesis or semisynthesis.

3.4   ISOPRENOIDS

Isoprenoids (terpenes) are a large and diverse class of natural products produced by nearly all species. There are approximately 50,000 known isoprenoids. All of them are derived from five carbon isoprene building blocks. According to the numbers of isoprene unit, isoprenoids are classified as hemiterpenoids (C5), monoterpenoids (C10), sesquiterpenoids (C15), diterpenoids (C20), etc. Isoprenoids play a number of essential biological roles, including membrane fluidity, protein prenylation, electron transport, and cellular development (McCaskill and Croteau 1998). In addition, many isoprenoids produced by microorganisms and plants have been shown to possess powerful biological properties. Two of the best known isoprenoid pharmaceuticals from the metabolic engineering perspectives are the antimalarial agent artemisinin and the anticancer agent paclitaxel.

The biosynthetic pathways of many isoprenoids are still under investigation, whereas the biosynthesis of the universal building blocks, isopentenyl diphosphate (IPP) and dimethylallyl diphosphate (DMAPP) are well studied (Figure 3.10). They are synthesized via the mevalonate (MVA) pathway, primarily found in eukaryotes and Archaea; and the 1-deoxyxylulose-5-phosphate (DXP) pathway primarily found in eubacteria and plastid-bearing eukaryotes (Eisenreich et al. 1998; Lange et al. 2000). In plant, both pathways exist: the MVA enzymes operate in the cytosol and transform acetyl-CoA to IPP in six steps, whereas the DXP enzymes operate in the plastid and transform glyceraldehydes-3-phosphate and pyruvic acid in seven steps to both IPP and DMAPP with a ratio of 5:1. An IPP isomerase (IDI) is present to balance IPP and DMAPP. Downstream prenyltransferases convert IPP and DMAPP to different isoprenoid precursors. Geranyl diphosphate (GPPS) converts one IPP and one DMAPP to geranyl diphosphate (GPP); farnesyl diphosphate synthase (FPPS) converts two IPPs and one DMAPP to FPP; geranylgeranyl diphosphate synthase (GGPPS) converts three IPPs and one DMAPP to GGPP, etc. (Figure 3.10).

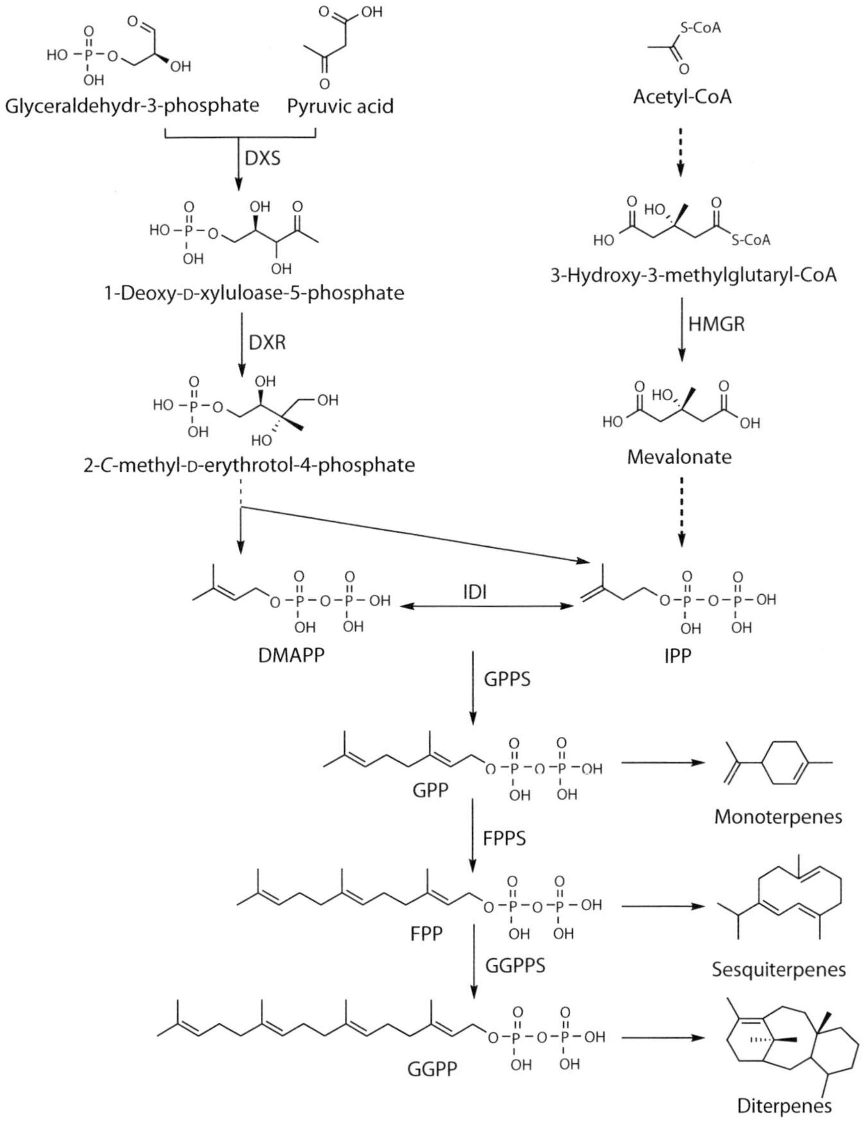

FIGURE 3.10 Isoprenoid biosynthetic pathway. DXP pathway on the left mostly exists in prokaryotes; MVA pathway on the right mostly exists in eukaryote; plants have both pathways.

These linear precursors are subsequently cyclized by diverse terpene cyclases into the large number of structures in the terpene family.

3.4.1 ARTEMISININ AND ITS BIOSYNTHETIC PATHWAY

Artemisinin is a sesquiterpene lactone endoperoxide obtained from *Artemisia annua* L. It is the active component of the traditional Chinese herb qing hao, Chinese name of *Artemisia annua* L (Klayman 1985). The artemisinin derivatives including

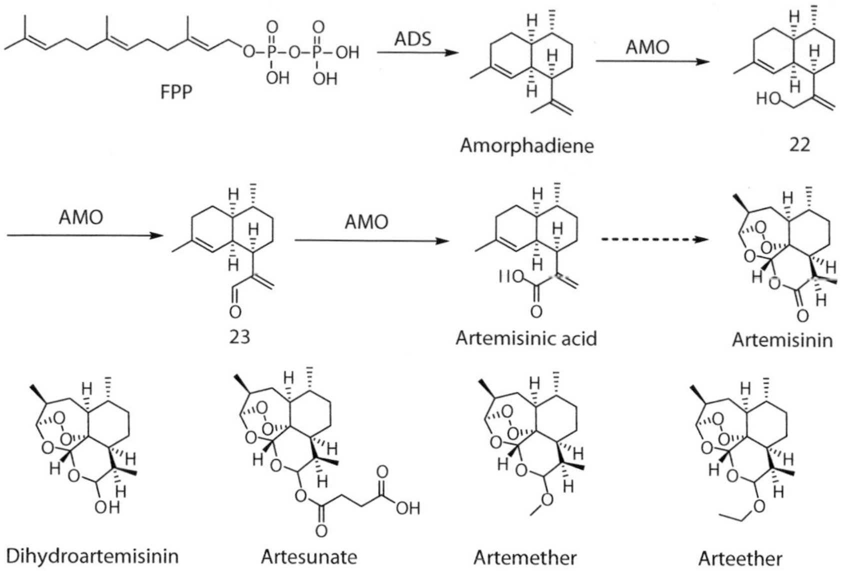

FIGURE 3.11 Artemisinin biosynthetic pathway from FPP to artemisinic acid and artemisinin derivatives for ATC.

dihydroartemisinin, artesunate, artemether, and arteether, all of which are components of artemisinin combination therapies (ATC), are used for treatment of malaria (Figure 3.11). Malaria is one of the most infectious diseases around the world and is caused by protozoan parasites. There are about 3–5 million cases of malaria each year, which kills more than 1 million people worldwide (Snow et al. 2005). As a consequence, the demand for artemisinin is very significant. An estimate of 114 ton of artemisinin is required yearly, considering the requirement of 0.5 g of artemether per dose (Haynes 2006). The high demand places great pressures on the supply of artemisinin. Many advances have been made to improve the production of artemisinin in its original plant host as well as in microbial hosts.

The intermediates and enzymes involved in artemisinin biosynthesis in *Artemisia annua* L have been identified and characterized (Figure 3.11). The biosynthesis of FPP from the mevalonate pathway and the DXP pathway is common in isoprenoid biosynthesis. The next step involves an amorpha-4,11-diene synthase (ADS), which converts FPP to amorphadiene (Mercke et al. 2000; Wallaart et al. 2001; Martin et al. 2003). Amorphadiene is then further tailored by consecutive oxidation to dihydroartemisinic acid by a cytochrome P450 oxygenase (CYP71AV1/CPR) (Bertea et al. 2005; Ro et al. 2006). The conversion from dihydroartemisinic acid to artemisinin may not be controlled by enzymes since it undergoes spontaneous autoxidation to artemisinin in aprotic solvents exposed to light, in air as well as in the presence of chlorophyll sensitizer (Wallaart et al. 1999). The total synthesis of artemisinin is not commercially viable at present. However, semisynthesis from artemisinic acid to artemisinin is a feasible practical route (Roth and Acton 1989, 1991). Metabolic engineering for the production of artemisinic acid has therefore been intensively investigated, together with the production of artemisinin itself and its analogues. We will describe the production of artemisinin and its analogues based on different hosts since different strategies have been utilized in different species.

3.4.2 METABOLIC ENGINEERING OF PLANTS FOR ARTEMISININ INTERMEDIATES AND DERIVATIVES PRODUCTION

Before the genes responsible for artemisinin biosynthesis were cloned, methods to increase artemisinin production were limited to overexpression of pathway genes for FPP biosynthesis and other regulator genes. *Agrobacterium*-mediated transformation and homologous recombination were commonly used to genetically modify the genes in plants. This process is highly time intensive in plants. It can take up to several months or years to obtain a single transgenic plant. The transformation of *A. annua* L was established in the mid-1990s using *A. tumefaciens* and *A. rhizogenes* (Cai et al. 1995; Jaziri et al. 1995). A cDNA encoding FPPS from *Gossypium arboreum* was transformed into *A. annua* L (Chen et al. 2000). The target gene was placed under the CaMV 35S promoter, which is a strong constitutive promoter and gives high levels of gene expression in plants. The expression of the target gene was confirmed by PCR, southern and northern blot. The transgenic plants regenerated from shoot lines produced $8 \sim 10$ mg/g DW (dry weight) of artemisinin, which is $2 \sim 3$ times higher than the yield of the wild type *A. annua* L. When FPPS from the original plant *A. annu* L was overexpressed after transforming into a high producing (6 mg/g DW) *A. annu* L (Han et al. 2006), the transgenic plant produced ~ 9 mg/g DW, a 34% improvement in artemisinin production. These results reflect that FPPSs from different plants have no significant difference in terms of artemisinin overproduction, which is reasonable since FPPS is only involved in the biosynthesis of the FPP precursor. The same strategy was applied to transfer an isopentenyl transferase gene (*ipt*) responsible for the rate-limiting step in cytokinin biosynthesis. The transformation of *ipt* into *A. annua* L via *A. tumefaciens* that showed overexpression of the *ipt* gene increased cytokinin concentrations by two- to threefold (Sa et al. 2001). In return it improved chlorophyll content by 20%–60% and resulted in elevated artemisinin production by 30%–70% compared to the control.

The first pathway-specific gene in artemisinin biosynthesis is ADS, which converts FPP to amorpha-4,11-diene (amorphadiene). It was cloned from the *A. annua* L genomic DNA and the enzyme was functionally expressed in *E. coli* (Chang et al. 2000; Wallaart et al. 2001). Overexpression of ADS in tobacco, which is a nonamorphadiene-producing plant, accumulated amorphadiene ranging from 0.2 to 0.7 ng/g FW (fresh weight) (Wallaart et al. 2001). Later studies found that the biosynthesis of isoprenoids was compartmentalized in plants. Monoterpenes and diterpenes are produced in the plastid through the DXP pathway, while sesquiterpenes and triterpenes are produced in the cytosol through the MVA pathway. To take advantage of the compartmentalization, coexpression of ADS and FPPS genes in plastid of tobacco produced amorphadiene at a level of 25 µg/g FW (Wu et al. 2006). This was a 25,000-fold improvement compared to transformants harboring the ADS gene in the cytosol where the FPPS gene already exists. The redirection of the flux from the

DXP pathway to amorphadiene through expression of FPPS in the plastid greatly improved sesquiterpene production.

Optimization of the metabolic flux was also accomplished outside of the organism. Different fermentation conditions such as shake flask, mist reactor, and bubble column reactor were studied using the hairy roots of *A. annu* L (Souret et al. 2003). It was found that oxygen levels, light, root packing density, and fermentor conditions affected artemisinin gene expression and production. This initial study provides the groundwork for artemisinin production from *A. annu* L using high cell density fermentation.

3.4.3 METABOLIC ENGINEERING OF *ESCHERICHIA COLI* FOR THE PRODUCTION OF ARTEMISININ INTERMEDIATES

Engineering the artemisinin pathway in a microbial host for high-level production is an attractive alternative because the overall yield of artemisinin in both original and engineered plants is too low. The benefits of using *E. coli* as a heterologous expression host have been discussed earlier. *E. coli* is a prokaryote and uses the DXP pathway to produce IPP and DMAPP, both of which are essential for the prenylation of tRNAs and cell wall biosynthesis. IspA, an endogenous FPPS from *E. coli*, is present for FPP biosynthesis. To obtain high-level expression of ADS in *E. coli*, the ADS gene was subjected to codon optimization for bacterial expression (Martin et al. 2003). Expressing the synthetic ADS gene in *E. coli* DH10B (designated as E1) offered a maximum concentration of amorphadiene of 0.086 μg caryophyllene equivalent/mL (CEM) when cultured in LB medium. Coexpression of DXS, IDI from *Haematococcus pluvialis*, and IspA in E1 (designated as E2) increased FPP concentration and increased amorphadiene yield to 0.313 μg CEM.

To further improve amorphadiene yield, the MVA pathway from *Saccharomyces cerevisiae* was integrated to the genome of E2 (designated as E3) (Martin et al. 2003). The high flux through the MVA pathway greatly increased amorphadiene concentrations to a maximum of 3.1 μg CEM. It has to be mentioned that the accumulation of IPP inhibited the growth rate of *E. coli* while FPP had little inhibition effect. Converting them to amorphadiene can therefore minimize the negative effects on *E. coli* growth. To increase compound production, glycerol was added to LB medium as an additional carbon source. The optical density of E3 increased by more than threefold and yielded 24 μg CEM, an almost 300-fold improvement in amorphadiene total production compared to E1 when glycerol was added to the media. This excellent work created a solid foundation for isoprenoids production in *E. coli*. When the E3 strain was cultured in terrific broth (TB) supplemented with 1% (v/v) glycerol as an additional carbon source, 0.48 mg/mL amorphadiene was produced, another 27-fold increase from 24 μg CEM.

With the introduction of the *S. cerevisiae* MVA pathway into *E. coli*, the accumulation of pathway intermediate 3-hydroxy-3-methyl-glutaryl-coenzyme A (HMG-CoA) caused flux imbalance and inhibited cell growth (Pitera et al. 2007). This inhibition could be partially counteracted by the addition of palmitic acid (16:0) and oleic acid (*cis*-Δ^9-18:1) to the medium based on studies that accumulated HMG-CoA impeded *E. coli* type II fatty acid synthesis (Kizer et al. 2008). To overcome the metabolic flux imbalance, yeast genes for HMG-CoA synthase and HMG-CoA reductase were replaced with more active, functionally equivalent genes from *Staphylococcus aureus*, which resulted in doubling the production of amorphadiene from *E. coli* (Tsuruta et al. 2009). Amorphadiene production was further improved by optimizing nitrogen and carbon delivery in a fed-batch fermentation process, which yielded 90 g/L dry cell weight and an average titer of 27.4 g/L amorphadiene. An over 10-million-fold improvement in the level of amorphadiene production compared to E1 showcases the great potential of metabolic engineering as a means for producing plant natural products in microorganisms.

One of the most difficult problems for *E. coli* to synthesize active isoprenoids is the functional expression of cytochrome P450s, which modify nascent terpenes into their corresponding active forms. The native oxygenase CYP71AV1 (AMO) was first discovered and expressed in *S. cerevisiae* (Ro et al. 2006). However, the expression of native AMO led to no-detectable recombinant cytochrome P450 in either in vivo or in vitro studies performed with *E. coli* (Chang et al. 2007). To aid the recombinant expression of a functional cytochrome P450 in *E. coli*, the codons of the AMO gene were optimized. In addition, the predicted transmembrane sequence of the AMO gene was replaced with the N-terminal sequences of other P450s, a common strategy used to aid functional heterologous expression of P450s in *E. coli*. One hybrid, A13AMO where the N-terminal was from CYP52A13 of *Candida tropicalis*, was functionally expressed. For the P450 to be fully active, a redox partner was required. Both A13AMO and the reductase ctCPR from *C. tropicalis* were co-transformed to E3 with a resultant production of 0.45 mg/L of compound **22** (alcohol congener of artemisinic acid). The replacement of ctCPR with aaCPR from the native host *A. annua* L produced 5.6 mg/L compound **22**, a 12-fold improvement. By switching the expression vector from pETDUET-1 to pCWori, which was designed for functional expression of P450s (Barnes et al. 1991), fully oxidized artemisinic acid was detected together with compounds **22** and **23** (aldehyde congener of artemisinic acid). When the parent *E. coli* strain was switched from DH10B to DH1, more than 1000-fold increase in oxygenase activity was detected. The total production of compounds **22**, **23** and artemisinic acid reached 553 mg/L when a two-phase fermentation method was applied. When the organic layer was removed, artemisinic acid was the dominant product with a yield of 105 mg/L.

3.4.4 METABOLIC ENGINEERING OF *SACCHAROMYCES CEREVISIAE* FOR THE PRODUCTION OF ARTEMISININ INTERMEDIATES

The yeast strain *S. cerevisiae* is another ideal host for heterologous expression of enzymes and production of natural products. It has the same benefits as *E. coli* in terms of relatively rapid growth, highly developed genetic tools, advanced fermentation science, and well-studied metabolic pathways. Moreover, it possesses functions that *E. coli* does not have, such as ability to carry out post-translational modifications. To engineer *S. cerevisiae* for the production of amorphadiene, the ADS gene was placed under the control of a *GAL1* promoter and was subsequently introduced into *S. cerevisiae* to give EPY201, which produced 4.4 mg/L amorphadiene (Ro et al. 2006). Overexpression of a truncated rate-limiting step HMG-CoA reductase (HMGR) (EPY208) improved amorphadiene production by 5.5-fold to

22 mg/L. Downregulation of FPP to sterol biosynthesis using a methionine-repressible promoter yielded EPY225 which produced 45 mg/L amorphadiene. All of these modifications were made by chromosome integration to ensure genetic stability of *S. cerevisiae*. Overexpression of *upc2-1*, a semi-dominant mutant allele (EPY213), enhanced the activity of *UPC2*, which regulates the biosynthesis of sterols in *S. cerevisiae* and produced amorphadiene up to 105 mg/L. Integration of a second copy of the HMGR gene (EPY219) increased amorphadiene production by another ~50% to 149 mg/L. Finally, overexpression of FPPS (EPY224) resulted in a small increase in amorphadiene production to 153 mg/L. In order to increase the supply of acetyl-CoA for IPP biosynthesis, an acetaldehyde dehydrogenase gene together with a *Salmonella enterica* acetyl-CoA synthetase gene was integrated into EPY224 to create strain PDB108 (Shiba et al. 2007). The introduction of acetaldehyde dehydrogenase and acetyl-CoA synthase greatly increased acetyl-CoA concentration through the pyruvate dehydrogenase bypass which converts pyruvate into acetyl-CoA. The resultant PDB108 strain increased amorphadiene production by another 22%.

To create a strain that can produce artemisinic acid, the cytochrome P450 gene (AMO) responsible for oxidation was isolated from *A. annua* L using degenerate primers highly specific to P450 subfamilies in *Asteraceae*. When AMO and the redox partner aaCPR were integrated into EPY224, 32 mg/L artemisinic acid together with a small amount of compounds **22** and **23** were produced in shake flask cultivation. In a 1 L aerated bioreactor, 115 mg/L artemisinic acid was produced (Ro et al. 2006). Most of the artemisinic acid was excreted by yeast cells and remained on the outer cell surface after protonation in acidic culture conditions. This made purification of artemisinic acid extremely easy by simply washing the cell pellet with alkaline buffer.

With the development of high-yielding artemisinic acid strain in both *E. coli* and *S. cerevisiae*, a 30%–60% reduction in the cost of ATC was projected by switching the plant extracted artemisinin to the microbial production of artemisinic acid (Hale et al. 2007). In addition to the lower manufacturing cost, it reduces the pollution by using much less organic solvent as well as alleviates pressure on agricultural land for planting *A. annua* L.

3.4.5 OTHER ISOPRENOIDS

Metabolic engineering has been less successful on other biologically important isoprenoids such as the anticancer agents paclitaxel (Taxol) and eleutherobin—the feeding deterrent and growth disruptant of insects azadirachtin—largely due to the limited elucidation of corresponding biosynthetic pathways from plants. Taxol is one of the most intensive studied subjects and partial pathways were reconstituted in both *E. coli* and *S. cerevisiae*. An important intermediate of Taxol, taxidiene, was successfully produced from *S. cerevisiae* (DeJong et al. 2006) and the yield was optimized to 8.7 mg/L, a 40-fold improvement (Engels et al. 2008). Strategies were similar to those used in artemisinic acid production including codon optimization, expression of a truncated version of HMG-CoA reductase isoenzyme 1, a mutant regulatory protein UPC2-1, and a more efficient GGPPS from *Sulfolobus acidocaldarius* (Engels et al. 2008). The main bottlenecks for full reconstitution of isoprenoid biosynthetic pathways in microorganisms are uncovering and functional expression of corresponding cytochrome P450 oxygenases. With the expansion of biotechnology and bioinformatic tools, full pathway reconstitution and high-level production of more complex isoprenoids in microorganisms can be realized in the near future.

REFERENCES

Abe, Y., T. Suzuki, C. Ono et al. 2002. Molecular cloning and characterization of an ML-236B (compactin) biosynthetic gene cluster in *Penicillium citrinum. Mol. Genet. Genomics* 267: 636–646.

Baltz, R. H. 2008. Biosynthesis and genetic engineering of lipopeptide antibiotics related to daptomycin. *Curr. Top. Med. Chem.* 8: 618–638.

Baltz, R. H., V. Miao, and S. K. Wrigley 2005. Natural products to drugs: Daptomycin and related lipopeptide antibiotics. *Nat. Prod. Rep.* 22: 717–741.

Barnes, H. J., M. P. Arlotto, and M. R. Waterman 1991. Expression and enzymatic-activity of recombinant cytochrome-P450 17-alpha-hydroxylase in *Escherichia-Coli. Proc. Natl. Acad. Sci. USA* 88: 5597–5601.

Bennett, B. D., E. H. Kimball, M. Gao et al. 2009. Absolute metabolite concentrations and implied enzyme active site occupancy in *Escherichia coli. Nat. Chem. Biol.* 5: 593–599.

Bertea, C. M., J. R. Freije, H. van der Woude et al. 2005. Identification of intermediates and enzymes involved in the early steps of artemisinin biosynthesis in *Artemisia annua. Planta Med.* 71: 40–47.

Boeck, L. D., D. S. Fukuda, B. J. Abbott et al. 1988. Deacylation of A21978c, an acidic lipopeptide antibiotic complex, by actinoplanes-utahensis. *J. Antibiot.* 41: 1085–1092.

Bok, J. W. and N. P. Keller 2004. LaeA, a regulator of secondary metabolism in *Aspergillus* spp. *Eukaryot. Cell* 3: 527–535.

Butler, M. S. 2005. Natural products to drugs: Natural product derived compounds in clinical trials. *Nat. Prod. Rep.* 22: 162–195.

Butler, M. S. 2008. Natural products to drugs: Natural product-derived compounds in clinical trials. *Nat. Prod. Rep.* 25: 475–516.

Cai, G., G. Li, and H. Ye 1995. Hairy root culture of *Artemisia annua* L. by Ri plasmid transformation and biosynthesis of artemisinin. *Chin. J. Biotechnol.* 11: 227–235.

Carreras, C., S. Frykman, S. Ou et al. 2002. *Saccharopolyspora erythraea*-catalyzed bioconversion of 6-deoxyerythronolide B analogs for production of novel erythromycins. *J. Biotechnol.* 92: 217–228.

Chan, Y. A., A. M. Podevels, B. M. Kevany et al. 2009. Biosynthesis of polyketide synthase extender units. *Nat. Prod. Rep.* 26: 90–114.

Chang, M. C. Y., R. A. Eachus, W. Trieu et al. 2007. Engineering *Escherichia coli* for production of functionalized terpenoids using plant P450s. *Nat. Chem. Biol.* 3: 274–277.

Chang, Y. J., S. H. Song, S. H. Park et al. 2000. Amorpha-4,11-diene synthase of *Artemisia annua*: cDNA isolation and bacterial expression of a terpene synthase involved in artemisinin biosynthesis. *Arch. Biochem. Biophys.* 383: 178–184.

Chen, Y., W. Deng, J. Wu et al. 2008. Genetic modulation of the overexpression of tailoring genes eryK and eryG leading to the improvement of erythromycin A purity and production in *Saccharopolyspora erythraea* fermentation. *Appl. Environ. Microbiol.* 74: 1820–1828.

Chen, D., H. Ye, and G. Li 2000. Expression of a chimeric farnesyl diphosphate synthase gene in *Artemisia annua* L. transgenic plants via *Agrobacterium tumefaciens*-mediated transformation. *Plant Sci.* 155: 179–185.

Chng, C., A. M. Lum, J. A. Vroom et al. 2008. A key developmental regulator controls the synthesis of the antibiotic erythromycin in *Saccharopolyspora erythraea. Proc. Natl. Acad. Sci. USA* 105: 11346–11351.

Coeffet-Le Gal, M. F., L. Thurston, P. Rich et al. 2006. Complementation of daptomycin dptA and dptD deletion mutations in trans and production of hybrid lipopeptide antibiotics. *Microbiology* 152: 2993–3001.

Cortes, J., S. F. Haydock, G. A. Roberts et al. 1990. An unusually large multifunctional polypeptide in the erythromycin-producing polyketide synthase of *Saccharopolyspora erythraea*. *Nature* 348: 176–178.

Couch, R. D. and G. M. Gaucher 2004. Rational elimination of *Aspergillus terreus* sulochrin production. *J. Biotechnol.* 108: 171–178.

Cragg, G. M., D. J. Newman, and K. M. Snader 1997. Natural products in drug discovery and development. *J. Nat. Prod.* 60: 52–60.

Debono, M., B. J. Abbott, R. M. Molloy et al. 1988. Enzymatic and chemical modifications of lipopeptide antibiotic A21978c—The synthesis and evaluation of daptomycin (Ly146032). *J. Antibiot.* 41: 1093–1105.

Debono, M., M. Barnhart, C. B. Carrell et al. 1987. A21978c, a complex of new acidic peptide antibiotics—Isolation, chemistry, and mass-spectral structure elucidation. *J. Antibiot.* 40: 761–777.

DeJong, J. M., Y. L. Liu, A. P. Bollon et al. 2006. Genetic engineering of taxol biosynthetic genes in *Saccharomyces cerevisiae*. *Biotechnol. Bioeng.* 93: 212–224.

Desai, R. P., T. Leaf, Z. Hu et al. 2004. Combining classical, genetic, and process strategies for improved precursor-directed production of 6-deoxyerythronolide B analogues. *Biotechnol. Prog.* 20: 38–43.

Desai, R. P., E. Rodriguez, J. L. Galazzo et al. 2004. Improved bioconversion of 15-fluoro-6-deoxyerythronolide B to 15-fluoro-erythromycin A by overexpression of the eryK Gene in *Saccharopolyspora erythraea*. *Biotechnol. Prog.* 20: 1660–1665.

Doekel, S. and M. A. Marahiel 2000. Dipeptide formation on engineered hybrid peptide synthetases. *Chem. Biol.* 7: 373–384.

Donadio, S., J. B. McAlpine, P. J. Sheldon et al. 1993. An erythromycin analog produced by reprogramming of polyketide synthesis. *Proc. Natl. Acad. Sci. USA* 90: 7119–7123.

Donadio, S., M. J. Staver, J. B. McAlpine et al. 1991. Modular organization of genes required for complex polyketide biosynthesis. *Science* 252: 675–679.

Eisenreich, W., M. Schwarz, A. Cartayrade et al. 1998. The deoxyxylulose phosphate pathway of terpenoid biosynthesis in plants and microorganisms. *Chem. Biol.* 5: R221–R233.

Eliopoulos, G. M., C. Thauvin, B. Gerson et al. 1985. In vitro activity and mechanism of action of A21978C1, a novel cyclic lipopeptide antibiotic. *Antimicrob. Agents Chemother.* 27: 357–362.

Engels, B., P. Dahm, and S. Jennewein 2008. Metabolic engineering of taxadiene biosynthesis in yeast as a first step towards Taxol (Paclitaxel) production. *Metab. Eng.* 10: 201–206.

Fischbach, M. A. and C. T. Walsh 2006. Assembly-line enzymology for polyketide and nonribosomal peptide antibiotics: Logic, machinery, and mechanisms. *Chem. Rev.* 106: 3468–3496.

Frykman, S., T. Leaf, C. Carreras et al. 2001. Precursor-directed production of erythromycin analogs by *Saccharopolyspora erythraea*. *Biotechnol. Bioeng.* 76: 303–310.

Gehring, A. M., K. A. Bradley, and C. T. Walsh 1997. Enterobactin biosynthesis in *Escherichia coli*: Isochorismate lyase (EntB) is a bifunctional enzyme that is phosphopantetheinylated by EntD and then acylated by EntE using ATP and 2,3-dihydroxybenzoate. *Biochemistry* 36: 8495–8503.

Gehring, A. M., I. Mori, and C. T. Walsh 1998. Reconstitution and characterization of the *Escherichia coli* enterobactin synthetase from EntB, EntE, and EntF. *Biochemistry* 37: 2648–2659.

Gu, J. Q., K. T. Nguyen, C. Gandhi et al. 2007. Structural characterization of daptomycin analogues A21978C(1–3)(D-Asn(11)) produced by a recombinant *Streptomyces roseosporus* strain. *J. Nat. Prod.* 70: 233–240.

Hale, V., J. D. Keasling, N. Renninger et al. 2007. Microbially derived artemisinin: A biotechnology solution to the global problem of access to affordable antimalarial drugs. *Am. J. Trop. Med. Hyg.* 77: 198–202.

Han, J. L., B. Y. Liu, H. C. Ye et al. 2006. Effects of overexpression of the endogenous farnesyl diphosphate synthase on the artemisinin content in *Artemisia annua* L. *J. Integr. Plant Biol.* 48: 482–487.

Haynes, R. K. 2006. From artemisinin to new artemisinin antimalarials: Biosynthesis, extraction, old and new derivatives, stereochemistry and medicinal chemistry requirements. *Curr. Top. Med. Chem.* 6: 509–537.

Hertweck, C., A. Luzhetskyy, Y. Rebets et al. 2007. Type II polyketide synthases: Gaining a deeper insight into enzymatic teamwork. *Nat. Prod. Rep.* 24: 162–190.

Hill, A. M. 2006. The biosynthesis, molecular genetics and enzymology of the polyketide-derived metabolites. *Nat. Prod. Rep.* 23: 256–320.

Hosted, T. J. and R. H. Baltz 1997. Use of rysL for dominance selection and gene replacement in *Streptomyces roseosporus*. *J. Bacteriol.* 179: 180–186.

Hubbard, B. K. and C. T. Walsh 2003. Vancomycin assembly: Nature's way. *Angew. Chem. Int. Ed.* 42: 730–765.

Huber, F. M., R. L. Pieper, and A. J. Tietz 1988. The formation of daptomycin by supplying decanoic acid to *Streptomyces roseosporus* cultures producing the antibiotic complex A21978c. *J. Biotechnol.* 7: 283–292.

Jacobsen, J. R., C. R. Hutchinson, D. E. Cane et al. 1997. Precursor-directed biosynthesis of erythromycin analogs by an engineered polyketide synthase. *Science* 277: 367–369.

Jacobsen, J. R., A. T. Keatinge-Clay, D. E. Cane et al. 1998. Precursor-directed biosynthesis of 12-ethyl erythromycin. *Bioorg. Med. Chem.* 6: 1171–1177.

Jaziri, M., K. Shimomura, K. Yoshimatsu et al. 1995. Establishment of normal and transformed root cultures of *Artemisia annua* L for artemisinin production. *J. Plant Physiol.* 145: 175–177.

Julia, P., L. Xiang, A. Whiting et al. 2006. Heterologous production of daptomycin in *Streptomyces lividans*. *J. Ind. Microbiol. Biotechnol.* 33: 121–128.

Kao, C. M., L. Katz, and C. Khosla 1994. Engineered biosynthesis of a complete macrolactone in a heterologous host. *Science* 265: 509–512.

Kennedy, J., K. Auclair, S. G. Kendrew et al. 1999. Modulation of polyketide synthase activity by accessory proteins during lovastatin biosynthesis. *Science* 284: 1368–1372.

Kennedy, J., S. Murli, and J. T. Kealey 2003. 6-deoxyerythronolide B analogue production in *Escherichia coli* through metabolic pathway engineering. *Biochemistry* 42: 14342–14348.

Khosla, C., R. S. Gokhale, J. R. Jacobsen et al. 1999. Tolerance and specificity of polyketide synthases. *Annu. Rev. Biochem.* 68: 219–253.

Khosla, C. and J. D. Keasling 2003. Metabolic engineering for drug discovery and development. *Nat. Rev. Drug Discov.* 2: 1019–1025.

Kinoshita, K., P. G. Williard, C. Khosla et al. 2001. Precursor-directed biosynthesis of 16-membered macrolides by the erythromycin polyketide synthase. *J. Am. Chem. Soc.* 123: 2495–2502.

Kirkpatrick, P., A. Raja, J. LaBonte et al. 2003. Daptomycin. *Nat. Rev. Drug. Discov.* 2: 943–944.

Kizer, L., D. J. Pitera, B. F. Pfleger et al. 2008. Application of functional genomics to pathway optimization for increased isoprenoid production. *Appl. Environ. Microbiol.* 74: 3229–3241.

Klayman, D. L. 1985. Qinghaosu (artemisinin): An antimalarial drug from China. *Science* 228: 1049–1055.

Kreuzman, A. J., R. L. Hodges, J. R. Swartling et al. 2000. Membrane-associated echinocandin B deacylase of *Actinoplanes utahensis*: Purification, characterization, heterologous cloning and enzymatic deacylation reaction. *J. Ind. Microbiol. Biotechnol.* 24: 173–180.

Kumar, M. S., P. M. Kumar, H. M. Sarnaik et al. 2000. A rapid technique for screening of lovastatin-producing strains of *Aspergillus terreus* by agar plug and *Neurospora crassa* bioassay. *J. Microbiol. Methods* 40: 99–104.

Lamp, K. C., M. J. Rybak, E. M. Bailey et al. 1992. In vitro pharmacodynamic effects of concentration, pH, and growth phase on serum bactericidal activities of daptomycin and vancomycin. *Antimicrob. Agents Chemother.* 36: 2709–2714.

Lange, B. M., T. Rujan, W. Martin et al. 2000. Isoprenoid biosynthesis: The evolution of two ancient and distinct pathways across genomes. *Proc. Natl. Acad. Sci. USA* 97: 13172–13177.

Lau, J., C. Tran, P. Licari et al. 2004. Development of a high cell-density fed-batch biopro-cess for the heterologous production of 6-deoxyerythronolide B in *Escherichia coli. J. Biotechnol.* 110: 95–103.

Leaf, T., L. Cadapan, C. Carreras et al. 2000. Precursor-directed biosynthesis of 6-deoxyerythronolide B analogs in *Streptomyces coelicolor*: Understanding precursor effects. *Biotechnol. Progress* 16: 553–556.

Lombo, F., B. Pfeifer, T. Leaf et al. 2001. Enhancing the atom economy of polyketide bio-synthetic processes through metabolic engineering. *Biotechnol. Progress* 17: 612–617.

Lum, A. M., J. Q. Huang, C. R. Hutchinson et al. 2004. Reverse engineering of industrial pharmaceutical-producing actinomycete strains using DNA microarrays. *Metab. Eng.* 6: 186–196.

Ma, S. M., J. W. Li, J. W. Choi et al. 2009. Complete reconstitution of a highly reducing itera-tive polyketide synthase. *Science* 326: 589–592.

Mahlert, C., F. Kopp, J. Thirlway et al. 2007. Stereospecific enzymatic transformation of alpha-ketoglutarate to (2S,3R)-3-methyl glutamate during acidic lipopeptide biosynthe-sis. *J. Am. Chem. Soc.* 129: 12011–12018.

Manzoni, M. and N. Rollini 2002. Biosynthesis and biotechnological production of statins by filamentous fungi and application of these cholesterol-lowering drugs. *Appl. Microbiol. Biotechnol.* 58: 555–564.

Marahiel, M. A., T. Stachelhaus, and H. D. Mootz 1997. Modular peptide synthetases involved in nonribosomal peptide synthesis. *Chem. Rev.* 97: 2651–2673.

Marsden, A. F. A., B. Wilkinson, J. Cortes et al. 1998. Engineering broader specificity into an antibiotic-producing polyketide synthase. *Science* 279: 199–202.

Martin, V. J., D. J. Pitera, S. T. Withers et al. 2003. Engineering a mevalonate pathway in *Escherichia coli* for production of terpenoids. *Nat. Biotechnol.* 21: 796–802.

McCaskill, D. and R. Croteau 1998. Some caveats for bioengineering terpenoid metabolism in plants. *Trends Biotechnol.* 16: 349–355.

McDaniel, R., S. Ebert-Khosla, D. A. Hopwood et al. 1993. Engineered biosynthesis of novel polyketides. *Science* 262: 1546–1550.

McDaniel, R., A. Thamchaipenet, C. Gustafsson et al. 1999. Multiple genetic modifications of the erythromycin polyketide synthase to produce a library of novel "unnatural" natural products. *Proc. Natl. Acad. Sci. USA* 96: 1846–1851.

McDaniel, R., M. Welch, and C. R. Hutchinson 2005. Genetic approaches to polyketide anti-biotics. 1. *Chem. Rev.* 105: 543–558.

McHenney, M. A. and R. H. Baltz 1996. Gene transfer and transposition mutagenesis in *Streptomyces roseosporus*: Mapping of insertions that influence daptomycin or pigment production. *Microbiology* 142(Pt 9): 2363–2373.

Mchenney, M. A., T. J. Hosted, B. S. Dehoff et al. 1998. Molecular cloning and physical mapping of the daptomycin gene cluster from *Streptomyces roseosporus. J. Bacteriol.* 180: 143–151.

Mercke, P., M. Bengtsson, H. J. Bouwmeester et al. 2000. Molecular cloning, expression, and characterization of amorpha-4,11-diene synthase, a key enzyme of artemisinin biosyn-thesis in *Artemisia annua* L. *Arch. Biochem. Biophys.* 381: 173–180.

Miao, V., M. F. Coeffet-LeGal, P. Brian et al. 2005. Daptomycin biosynthesis in *Streptomyces roseosporus*: Cloning and analysis of the gene cluster and revision of peptide stereochemistry. *Microbiology* 151: 1507–1523.

Miao, V., M. F. Coeffet-Le Gal, K. Nguyen et al. 2006. Genetic engineering in *Streptomyces roseosporus* to produce hybrid lipopeptide antibiotics. *Chem. Biol.* 13: 269–276.

Minami, H., J. S. Kim, N. Ikezawa et al. 2008. Microbial production of plant benzylisoquinoline alkaloids. *Proc. Natl. Acad. Sci. USA* 105: 7393–7398.

Minas, W., P. Brunker, P. T. Kallio et al. 1998. Improved erythromycin production in a genetically engineered industrial strain of *Saccharopolyspora erythraea*. *Biotechnol. Prog.* 14: 561–566.

Murli, S., J. Kennedy, L. C. Dayem et al. 2003. Metabolic engineering of *Escherichia coli* for improved 6-deoxyerythronolide B production. *J. Ind. Microbiol. Biotechnol.* 30: 500–509.

Murli, S., K. S. MacMillan, Z. Hu et al. 2005. Chemobiosynthesis of novel 6-deoxyerythronolide B analogues by mutation of the loading module of 6-deoxyerythronolide B synthase 1. *Appl. Environ. Microbiol.* 71: 4503–4509.

Nguyen, K. T., D. Kau, J. Q. Gu et al. 2006. A glutamic acid 3-methyltransferase encoded by an accessory gene locus important for daptomycin biosynthesis in *Streptomyces roseosporus*. *Mol. Microbiol.* 61: 1294–1307.

Nguyen, K. T., D. Ritz, J. Q. Gu et al. 2006. Combinatorial biosynthesis of novel antibiotics related to daptomycin. *Proc. Natl. Acad. Sci. USA* 103: 17462–17467.

Novak, N., S. Gerdin, and M. Berovic 1997. Increased lovastatin formation by *Aspergillus terreus* using repeated fed-batch process. *Biotechnol. Lett.* 19: 947–948.

Oberthur, M., C. Leimkuhler, R. G. Kruger et al. 2005. A systematic investigation of the synthetic utility of glycopeptide glycosyltransferases. *J. Am. Chem. Soc.* 127: 10747–10752.

Peiru, S., H. G. Menzella, E. Rodriguez et al. 2005. Production of the potent antibacterial polyketide erythromycin C in *Escherichia coli*. *Appl. Environ. Microbiol.* 71: 2539–2547.

Pfeifer, B. A., S. J. Admiraal, H. Gramajo et al. 2001. Biosynthesis of complex polyketides in a metabolically engineered strain of *E. coli*. *Science* 291: 1790–1792.

Pfeifer, B., Z. Hu, P. Licari et al. 2002. Process and metabolic strategies for improved production of *Escherichia coli*-derived 6-deoxyerythronolide B. *Appl. Environ. Microbiol.* 68: 3287–3292.

Pitera, D. J., C. J. Paddon, J. D. Newman et al. 2007. Balancing a heterologous mevalonate pathway for improved isoprenoid production in *Escherichia coli*. *Metab. Eng.* 9: 193–207.

Quadri, L. E., P. H. Weinreb, M. Lei et al. 1998. Characterization of Sfp, a *Bacillus subtilis* phosphopantetheinyl transferase for peptidyl carrier protein domains in peptide synthetases. *Biochemistry* 37: 1585–1595.

Reeves, A. R., I. A. Brikun, W. H. Cernota et al. 2007. Engineering of the methylmalonyl-CoA metabolite node of *Saccharopolyspora erythraea* for increased erythromycin production. *Metab. Eng.* 9: 293–303.

Ro, D. K., E. M. Paradise, M. Ouellet et al. 2006. Production of the antimalarial drug precursor artemisinic acid in engineered yeast. *Nature* 440: 940–943.

Rodriguez, E., Z. Hu, S. Ou et al. 2003. Rapid engineering of polyketide overproduction by gene transfer to industrially optimized strains. *J. Ind. Microbiol. Biotechnol.* 30: 480–488.

Roth, R. J. and N. Acton 1989. A simple conversion of artemisinic acid into artemisinin. *J. Nat. Prod.* 52: 1183–1185.

Roth, R. J. and N. Acton 1991. A facile semisynthesis of the antimalarial drug qinghaosu. *J. Chem. Educ.* 68: 612–613.

Sa, G., M. Mi, Y. He-chun et al. 2001. Effects of ipt gene expression on the physiological and chemical characteristics of *Artemisia annua* L. *Plant Sci.* 160: 691–698.

Sauermann, R., M. Rothenburger, W. Graninger et al. 2008. Daptomycin: A review 4 years after first approval. *Pharmacology* 81: 79–91.

Schwarzer, D., R. Finking, and M. A. Marahiel 2003. Nonribosomal peptides: From genes to products. *Nat. Prod. Rep.* 20: 275–287.

Shiba, Y., E. M. Paradise, J. Kirby et al. 2007. Engineering of the pyruvate dehydrogenase bypass in *Saccharomyces cerevisiae* for high-level production of isoprenoids. *Metab. Eng.* 9: 160–168.

Sims, J. W. and E. W. Schmidt 2008. Thioesterase-like role for fungal PKS-NRPS hybrid reductive domains. *J. Am. Chem. Soc.* 130: 11149–11155.

Smith, D. J., M. K. Burnham, J. Edwards et al. 1990. Cloning and heterologous expression of the penicillin biosynthetic gene cluster from *Penicillum chrysogenum*. *Biotechnology (New York)* 8: 39–41.

Snow, R. W., C. A. Guerra, A. M. Noor et al. 2005. The global distribution of clinical episodes of *Plasmodium falciparum* malaria. *Nature* 434: 214–217.

Souret, F. F., Y. Kim, B. E. Wyslouzil et al. 2003. Scale-up of *Artemisia annua* L. hairy root cultures produces complex patterns of terpenoid gene expression. *Biotechnol. Bioeng.* 83: 653–667.

Stachelhaus, T., A. Schneider, and M. A. Marahiel 1995. Rational design of peptide antibiotics by targeted replacement of bacterial and fungal domains. *Science* 269: 69–72.

Stassi, D. L., S. J. Kakavas, K. A. Reynolds et al. 1998. Ethyl-substituted erythromycin derivatives produced by directed metabolic engineering. *Proc. Natl. Acad. Sci. USA* 95: 7305–7309.

Tsuruta, H., C. J. Paddon, D. Eng et al. 2009. High-level production of amorpha-4,11-diene, a precursor of the antimalarial agent artemisinin, in *Escherichia coli*. *PLoS One* 4: e4489.

von Dohren, H., R. Dieckmann, and M. Pavela-Vrancic 1999. The nonribosomal code. *Chem. Biol.* 6: R273–R279.

Wallaart, T. E., H. J. Bouwmeester, J. Hille et al. 2001. Amorpha-4,11-diene synthase: Cloning and functional expression of a key enzyme in the biosynthetic pathway of the novel antimalarial drug artemisinin. *Planta* 212: 460–465.

Wallaart, T. E., W. van Uden, H. G. Lubberink et al. 1999. Isolation and identification of dihydroartemisinic acid from *Artemisia annua* and its possible role in the biosynthesis of artemisinin. *J. Nat. Prod.* 62: 430–433.

Walsh, C. 2003. *Antibiotics: Actions, Origins, Resistance*. ASM Press: Washington, DC, pp. 96–99.

Wang, Y. and B. A. Pfeifer 2008. 6-Deoxyerythronolide B production through chromosomal localization of the deoxyerythronolide B synthase genes in *E. coli*. *Metab. Eng.* 10: 33–38.

Ward, S. L., R. P. Desai, Z. H. Hu et al. 2007. Precursor-directed biosynthesis of 6-deoxyerythronolide B analogues is improved by removal of the initial catalytic sites of the polyketide synthase. *J. Ind. Microbiol. Biotechnol.* 34: 9–15.

Watanabe, K., K. Hotta, A. P. Praseuth et al. 2006. Total biosynthesis of antitumor nonribosomal peptides in *Escherichia coli*. *Nat. Chem. Biol.* 2: 423–428.

Weber, G., K. Schorgendorfer, E. Schneiderscherer et al. 1994. The peptide synthetase catalyzing cyclosporine production in *Tolypocladium niveum* is encoded by a giant 45.8-kilobase open reading frame. *Curr. Genet.* 26: 120–125.

Wei, X. X. and G. Q. Chen 2008. Applications of the VHb gene vgb for improved microbial fermentation processes. *Methods Enzymol.* 436: 273–287.

Wu, S., M. Schalk, A. Clark et al. 2006. Redirection of cytosolic or plastidic isoprenoid precursors elevates terpene production in plants. *Nat. Biotechnol.* 24: 1441–1447.

Zhang, W., Y. Li, and Y. Tang 2008. Engineered biosynthesis of bacterial aromatic polyketides in *Escherichia coli*. *Proc. Natl. Acad. Sci. USA* 105: 20683–20688.

Zhang, W. and Y. Tang 2008. Combinatorial biosynthesis of natural products. *J. Med. Chem.* 51: 2629–2633.

4 Computational Approaches for the Discovery of Natural Lead Structures

Judith M. Rollinger and Gerhard Wolber

CONTENTS

4.1 INTRODUCTION

Drug discovery and development are time- and resource-consuming processes. Thousands of chemical entities must be evaluated to find a hit, and nevertheless, the outcome usually remains an unpredictable challenge (Drews 2003). Statistics reveal natural sources as the most promising pool for drug candidates or drug leads (Newman and Cragg 2007). Natural compounds derive from the phenomenon of biodiversity as response to the richness in a variety of organisms in the ecosphere (Hadacek 2002). The evolution of these so-called secondary metabolites has resulted in diverse, more or less complex natural chemicals able to interact with a variety of targets from other organisms and the environment. They can therefore be considered as "privileged structures" (Koch and Waldmann 2004). The biosynthesis and selection of these chemical entities enhance the organisms' survival and competitiveness (Hadacek 2002; Waterman 1992).

From the myriad of natural chemical scaffolds resulting from nature's combinatorial chemistry efforts, some 230,000 natural products (*Dictionary of Natural Products on DVD 2007*) (Buckingham 2007) have been identified until now, which is just the tip of the iceberg. In recent years, several studies have explored structural differences between natural products, drug substances, and other chemicals (Feher and Schmidt 2003; Ganesan 2008; Grabowski and Schneider 2007; Henkel et al. 1999; Stahura et al. 2000) concluding that natural products interrogate a different area of chemical space than synthetics. Because of their immensely high chemical and biological diversity, it has been suggested to feed structural libraries with natural product scaffolds for panning biologically relevant chemical space (Breinbauer et al. 2002; Ertl et al. 2008; Koch and Waldmann 2004). The innately bioactive nature of natural compounds and the possibility to fish in an infinite pool of sophisticated secondary metabolites on the other hand render these chemical entities the most profitable and efficient source of new drug leads. These circumstances emerged as a basis of a renewed interest in exploitation of natural products as starting point for drug discovery and development (Butler 2004; Harvey 2008; Koehn and Carter 2005; Li and Vederas 2009).

Natural product preparations, mainly herbal remedies, have historically been the major source of pharmaceutical agents. Even today, about 70% of the world's population relies upon medicinal plants for its primary pharmaceutical care (McChesney et al. 2007). These natural products can be summarized as being either an extract consisting of a complex mixture of different metabolites or single chemical entities derived from a natural matrix.

Natural product research is a demanding task in the sense of supply of natural material, analytics, reproducibility, isolation procedure, structure elucidation, chemical complexity, etc. As far as their bioactivity is concerned, pharmacologists dealing with natural materials usually have to struggle with phenomenological effects of extracts that should be tracked down on a molecular level for a proper assignment of interactions according to western medical practice. Accordingly, in the pharmaceutical industry, drug discovery efforts mainly focus on developing and marketing single chemical entities, which are clearly defined from a physical and chemical perspective.

Here we will provide an overview of the basic computational technologies for drug discovery and advances of *in silico* concepts in natural product research, and demonstrate some recent studies where these techniques helped in natural lead finding and rationalization.

4.2   HOW TO ACCESS NATURAL PRODUCTS' BIOACTIVITY

Since the first isolation of a single chemical entity from natural matrices, that is, morphine from opium, the latex of *Papaver somniferum*, by Friedrich Wilhelm Sertürner in 1804 (Müller-Jahncke and Friedrich 1996), it is one of the pharmacognosists' goals to enrich the known chemical space by the isolation of natural compounds and establishing their structures. This is not practiced as an end in itself; however, it provides interesting novel templates for synthetic chemists and serves as chemotaxonomic tools to support phylogenetic relationships. Importantly, these new scaffolds provide a glance into structural requirements for an evolutionary trimmed bioactivity, albeit the target is usually unknown in the first place. Clues for an idea about the bioactivity of a novel natural compound may be provided from the interacting ecosphere itself, for example, microorganisms or herbivores that can be repelled successfully, or animals are observed with respect to their exposure to natural materials and their subconscious, instinctive, or conscious application to treat and cure impairments (Robles et al. 1995). One of the most successful strategies used to discover nature's bioactivity is the exploration of curing or preventive agents traditionally employed or observed by man, called ethnopharmacology (Bruhn and Holmstedt 1981). Approximately, 10,000 of the world's plants have documented medicinal use; however, only about 150–200 plant materials are utilized in western medicine. All these biorational approaches can help tremendously in the search for any bioactivity, and there is still enough potential for many more important findings of pharmaceutical interest (McChesney et al. 2007).

The advancement in bioassay technology and high throughput screening has perhaps been one of the strongest factors for the reemerging role of natural products in drug discovery and development over the last decades (Littleton et al. 2005; Potterat and Hamburger 2006). Highly automated, miniaturized, selective, and very specific bioassays are available to not only screen compound libraries of synthetic or natural origin, but also handle complex natural product preparations. The challenge however remains in the analysis and evaluation of the obtained results: Where is the activity threshold for extracts, and can it be expanded? What about low-concentrated, but highly active ingredients—do they drown in the crowd? What about assay disturbing metabolites, for example, saponins, tannins, colored or fluorescent ingredients (depending on the assay)? What about weakly active natural compounds that only exert a certain activity in their combination (additive and synergistic effects)? All these concerns arise when dealing with multicomponent mixtures like natural preparations, and they complicate the outcome of a bioassay screen.

Once a biological activity has been demonstrated as being worthy of detailed investigation, sophisticated chromatographic separation and structure elucidation technologies facilitate the isolation and identification of the active constituents. Nevertheless, an iterative assaying of fractions, subfractions, and isolates are required, unless the separation procedure is hyphenated with a bioassay online detection, which enables the direct assignment of a detected activity to a recorded chromatographic peak. Ideally, these hyphenated techniques are additionally coupled with high-resolution mass spectrometry and/or powerful nuclear magnetic resonance (NMR) methodologies. With these techniques in hand and a careful evaluation of the obtained results, an early dereplication is enabled. Thus, a time and cost consuming isolation of metabolites with already known activity or unexciting scaffolds can be skipped in an early phase.

All the above-mentioned strategies for natural lead finding are heavily oriented toward experiments and highly sophisticated equipment, and thus consume time and resources. With the advent of computational techniques, there is an ever growing effort to apply the predictive *in silico* power to the combined chemical and biological space in order to streamline drug discovery, design, development, and optimization. In the last two decades, computational techniques have rapidly gained in popularity, implementation, and appreciation. They have evolved aiming at analyzing, understanding, and overall predicting the interaction of a compound with respect to a specific biological target.

4.3   PROBLEMS IN LEAD FINDING

Methods providing an accurate prediction of which target a chemical entity may show an interaction are in high demand, especially in natural product research. A random screening has not shown to be target oriented, while selections based on criteria mentioned earlier, such as traditional use or ecological observations can substantially increase the likelihood of finding bioactive natural products.

The understanding of biological and physiological pathways and targets of human-pathological relevance coming out of the human genome project is continuously growing. The increasing molecular knowledge of druggable targets together with the advances that have been made in biotechnology and bioassaying are indispensable tools for the discovery and development of more specific and selective drugs. All the gathered molecular and structural information is not only essential for a target-oriented therapy approach; it is also a crucial requirement to establish working hypotheses in medicinal chemistry and chemoinformatics to streamline the search for novel and better drugs.

Seeking a small compound for exactly one target that has to be blocked or stimulated perfectly (best at nanomolar concentration) and exclusively (without interaction on any off-target) to abolish the focused impairment has been regarded as an ideal strategy for rational drug design according to modern western medicine. This is probably a heavily oversimplified conception.

4.4   MULTITARGET COMPOUNDS: PROBLEMS AND CHALLENGES

Many compounds and even drugs that are in clinical use or development have been found to be more promiscuous with regard to their biological targets and effects than originally anticipated (Keiser et al. 2009). Medical and pharmaceutical communities are facing a distinct rise in late stage attrition in phases 2 and 3. This can mainly be traced back to lack of efficacy on the one hand, and toxicology or missing clinical safety on the other hand. Each of these two criteria accounts for 30% of failures (Kola and Landis 2004).

In spite of all efforts with the evaluation of the postgenomic data explosion or, better to say, because of our increasing molecular insights, we realize an imperfect knowledge of disease and complex biological systems. Considering an estimated number of 3,000–10,000 druggable targets—and even more isoforms, modifications, and mutants (Meisner et al. 2004; Overington et al. 2006)—the approximately 500 currently exploited targets are not really representative enough to deem a ligand being selective and specific. There is a significant probability that an identified ligand also interacts (usually with different affinities) with a range of other macromolecules. These so-called multitarget compounds can cause harmful side effects and accordingly lead to the previously mentioned drug failures.

Paradoxically, we are often facing a lack of clinical efficacy, although a drug candidate was attested a highly potent and selective target affinity and even a quite acceptable bioavailability. It is proposed by the plausible concept of network pharmacology that a drug action at a single target might often be surpassed by partial, but multiple drug actions (Csermely et al. 2005; Hopkins 2008; Korcsmaros et al. 2007). A potent and highly selective drug affects a single gene in the cellular network, for example, by eliminating the corresponding link. In robust biological networks, redundant functions or pathways and alternative compensatory signaling routes can result in systems that are resilient against random deletion and thus lead to ineffectiveness (resistance) although dealing with ligands of high binding affinity. This inherent robustness of interaction networks has profound implications for drug discovery. On the other hand, an organism with an underlying unstable cellular network may suffer additionally by the perturbations caused by a high-affinity single-target drug (toxicity, side effects).

In any case, the multitargeted nature of compounds—and of natural products in particular—complicates our simplified view of "one-drug, one target" and challenges our search for a highly potent single-target inhibitor routinely developed in the course of a drug discovery program.

However, multitarget compounds also open up additional medical use (Hopkins 2008). Hitting more than one target might enable a compound to be applied therapeutically in several unrelated diseases. Of even more interest can be a compound's multiple and balancing interaction on biological targets within a disturbed pathway relevant to a particular disease. This has been widely recognized over recent years as being beneficial for the treatment of diseases with complex etiologies such as cancer, asthma, and psychiatric impairments (Morphy and Rankovic 2009; Roth et al. 2004). It has been suggested that a drug may have an increased efficacy for a therapeutic application by relatively weak patterns of inhibition of multiple targets as compared to a complete inhibition of a single target (Csermely et al. 2005).

4.5 HOW TO ACCESS MULTITARGET PROFILE OF NATURAL PRODUCTS

Several studies have already demonstrated the multitarget property of natural compounds (e.g., Kimura 2006; Wang et al. 2008). In natural product research, a main focus is now on determining the variety of targets that might be affected by a single chemical entity or, even more complex, to identify the varieties of targets affected by each individual constituent of a multicomponent natural preparation. In order to exploit the full therapeutic potential and minimize toxicity, it is important to identify a compound's respective target spectrum as thoroughly as possible. There are biochemical and cellular readouts as well as systematic, nonbiased methods like chemical proteomics for target profiling of a selected bioactive compound (Rix and Superti-Furga 2009).

In the search of multiple targets for individual compounds, computational strategies might give valuable hints to search for bioactivities or to scrutinize toxic interactions, and thus are primarily helpful for prioritizing experimental efforts. The available *in silico* tools and their applicability in natural product research will be discussed in the following paragraphs.

4.6 MOLECULAR MODELING

The aim of molecular modeling is to predict the biological activity of a small organic molecule using computer calculations. Such *in silico* modeling approaches use known experimental data to estimate geometry, macromolecule binding capabilities, and similar chemical properties using various rule-based and statistical computational techniques (*knowledge-based approaches*). Additionally, quantum chemistry and laws of physics can be used to simulate and understand a specific aspect of binding. The combination of both is necessary because we are still far away from completely being able to simulate complex biological systems according to the laws of physics and quantum chemistry despite the growing computational power of today's computers. The key aspect of all molecular modeling approaches is how the computer represents a chemical structure; that is, which simplifications are made for creating a computer representation and which kind of models are used.

4.6.1 MODELS FOR CHEMICAL STRUCTURES: 2D, 3D, AND OTHER REPRESENTATIONS

Traditionally, chemistry deals with 2D structures only, which has proven to be a highly efficient formalism over the years: A chemist is able to quickly recognize important structural elements and functional groups in this way and develop new analogs from known active ligands using established bioisosteric replacement rules. However, this classic way of thinking has its limits, since biochemical properties and the ability of a ligand to interact with a macromolecule depends on the 3D form of a molecule and cannot be directly derived from a 2D chemical formula. The concept of 3D molecules is much older than the use of computers in chemistry, and has already been introduced by Emil Fischer in 1890/1891, when he tried to understand how many configurations of a pentose molecule could exist. From these insights, he developed the famous key-lock metaphor, which compares the 3D complementarity between a macromolecule and a ligand to a key fitting into a lock. Linus Pauling was the first to propose the α-helix as the secondary structure of proteins: Nobel laureate James Watson describes in his book how Pauling was able to gain this insight by not drawing "the double helix" on a sheet of paper, but by building 3D models of it. Watson and Crick later followed this inspiration for their discovery of the DNA structure (Klebe 2009).

For computer-aided drug design we have to decide whether to use topological (2D) information, use a 3D characterization of a molecule, or derive properties for either of these representations. The more data that are calculated for a molecular structure, the more computational power has to be spent. More data do not necessarily mean better results, especially if the additional information is only calculated and not based on experimental observations. Concerning the molecules, especially the 2D/3D conversion step, makes a huge difference in computational speed since conformations have to be generated. The ability of widely used programs to search conformational space has been extensively examined in recent years with good results (Agrafiotis et al. 2007; Bostrom 2001; Bostrom et al. 1998; Chen and Foloppe 2008; Kirchmair et al. 2005, 2006; Perola and Charifson 2004). Therefore, many 3D approaches presample conformations once and store them in a database for subsequent virtual screening (Walters et al. 1998).

The methods described in the following sections use different molecular representations and algorithms that are optimized for specific molecular characterizations, each with different applicability domains.

4.6.2   KNOWLEDGE-BASED APPROACHES: STRUCTURE- AND LIGAND-BASED DESIGN

Knowledge-based approaches are the most frequently applied and most efficient methods in molecular modeling due to their universality. The aim of these methods is to find a representation of a molecule that reflects all available knowledge that was gained through experiments, including structural data from x-ray or NMR experiments and structure–activity relationships from biological testing. Knowledge-based methods either use experimentally determined 3D macromolecular structure information (*structure-based design*) or rely on the similarity of known active ligands (*ligand-based design*).

4.6.3   STRUCTURE-BASED DESIGN

Structure-based design relies on the availability of an experimentally determined structure of the macromolecular target under investigation. The Protein Data Bank (PDB) (Berman et al. 2000) represents the largest public repository of protein and nucleic acid structures determined by x-ray crystallography or NMR and currently (as of January 2010) comprises more than 57,000 macromolecular structures, many of them with cocrystallized ligands. One of the reasons for the increased popularity of structure-based approaches is the rapidly growing number of available coordinate files in the PDB and its illustrative nature: The result of a structure-based modeling experiment shows the 3D arrangement of a ligand bound to a macromolecule and can be interpreted in an intuitive way.

Many free and web-based computer programs are available to deal with the deposited PDB structures (Kirchmair et al. 2008). One important limit of crystal structures is that the atom coordinates are only models. Refraction patterns as derived from x-ray crystallography can only be interpreted accurately for certain areas of the macromolecule, and these inaccuracies inherent to the experimental technique are not always sufficiently understood by modelers. Another important limit is that not all macromolecules can be structurally determined and that therefore the PDB is still far away from covering all known classes. A closer look at the most prominent protein class, the enzymes, reveals that a major part (35%) of all enzyme entries in the database only cover 34 different enzyme types (Mestres 2005).

Approximately 60% of all currently known drug targets are located on the cell surface (Overington et al. 2006). However, these highly important targets, including G-protein coupled receptors (GPCRs) and ion channels, are far from being structurally well determined; too few membrane protein structures have been experimentally elucidated (White 2004).

Protein-ligand docking is the most prominent structure-based modeling technique represented by a large number of different programs and algorithms (see Section 4.6.11). Docking places a ligand flexibly into a predefined binding site of a protein and subsequently estimates which conformation of the ligand is the most plausible placement.

A different structure-based approach has been presented by Böhm with the computer program LUDI (Bohm 1992). This software searches for interaction centers in the protein and assembles potential new ligands by combining fragments from a 3D structure library. The scoring function for selecting the fragments not only depends on chemical features, but also on the number of rotational bonds available in the yet to be assembled ligand molecule.

If there is no structural information on a particular target available, homology modeling (described below) can be used to estimate a putative 3D structure of a given amino acid sequence and subsequently allow for the application of structure-based modeling. This technique constructs a model of the target protein based on a related homologous protein used as a template. The success of such a model depends on sequence similarity and the probability that important regions remain conserved between the template and the homology model.

An alternative for docking is structure-based pharmacophore modeling as implemented in the program LigandScout (Wolber and Langer 2005) described in Section 4.6.7.

4.6.4   LIGAND-BASED DESIGN

Ligand-based design is used if the structure of the target is unknown or if structure determination is impossible. Starting from a collection of molecules with known activity for a specific target, the purpose of ligand-based models is to identify molecules that are similar to already known bioactive compounds. The problem of similarity can be approached from different perspectives, such as property similarity, descriptor similarity, or structural similarity.

Many ligand-based techniques exist, including topological fingerprints, shape-based similarity, or methods that select several numeric molecular descriptors to fit statistical models. Such methods include quantitative structure–activity relationships (QSAR), neural networks, or other machine learning techniques. A review on descriptor-oriented ligand-based techniques has recently been published (Melville et al. 2009), and a brief overview about the most important techniques is provided in the following sections.

4.6.5 QUANTITATIVE STRUCTURE–ACTIVITY RELATIONSHIPS

QSAR methods statistically explore the relationship between calculated molecular properties and biological activity (Van de Waterbeemd and Gifford 2003). Calculated descriptors include scalar molecular properties like molecular weight or log P estimations, 2D descriptors characterizing chemical topology, and 3D descriptors, like molecular volume, the 3D arrangement of chemical features or functional groups, electrostatic or steric fields.

QSAR approaches are particularly used for optimizing structures after first hits have been identified (lead structure development and optimization), since the effect of the replacement of a specific moiety can be specifically interpreted. The most important prerequisite for a successful QSAR model is a solid underlying statistical validation using a training set of molecules covering a broad range of activities. Precision and statistical significance of a model are examined by applying an external test set for activity prediction. The errors between the experimentally determined activities and the predicted ones are used as a benchmark for the quality of the model.

3D-QSAR approaches are particularly useful to predict and rationalize structure–activity relationships based on the spatial arrangements of chemical properties and molecular shape. The success of these 3D approaches depends greatly on the quality of the initial 3D alignment of the compounds used for the model. Such alignments can be obtained from substructure-based overlays, pharmacophore-based approaches, as well as protein-ligand docking. Getting a plausible 3D alignment and a meaningful 3D QSAR model becomes particularly more challenging for dissimilar compounds or high binding site flexibility. A large number of QSAR tools exist: the most prominent are Comparative Molecular Field Analysis (CoMFA) (Cramer III et al. 1988) and Comparative Molecular Shape Indices Analysis (CoMSIA) (Klebe 1998). Both approaches derive statistical models that are visualized in color-coded contours around the molecule, indicating locations where electrostatic properties and spatial arrangements are favorable or unfavorable for biological activity. The GRID/GOLPE approach is another more recent QSAR technique (Baroni et al. 1993) that uses a specialized force field (see Section 4.6.10) to characterize ligands.

4.6.6   TOPOLOGICAL MOLECULAR FINGERPRINTS

Topological fingerprints consider the connection table (i.e., the 2D structure) of a molecule and ignore the respective atom coordinates. The rapid calculation and efficiency of such fingerprint methods implicate their popularity in molecular modeling. An important advantage of fingerprints is that they can be rapidly pre-computed for large databases, which offers the possibility to virtually screen millions of compounds in minutes on a single computer. Daylight fingerprints (Cragg et al. 2005) have become a de facto standard in the field of data mining and virtual screening. Another increasingly popular fingerprinting approach is Feature Trees (FTrees) representing the major building block characteristics of a compound with respect to their overall arrangement (Rarey et al. 2006; Rarey and Dixon 1998). The node-labeled tree structure is more closely related to the molecular structure than a linear representation as used in Daylight fingerprints and implies a more complex, but still efficient comparison algorithm. Hert et al. (2004) provide a more detailed overview on topological fingerprints methods and their performance during virtual screening.

A recent study (McGaughey et al. 2007) on 11 different targets showed that topological descriptors perform surprisingly well compared to 3D methods, as they are able to retrieve the largest number of actives with low computational efforts. However, the work also demonstrates that hits obtained from topological screening methods are less diverse than hits retrieved using 3D methods. Brown and Martin drew the same conclusion in 1997 (Brown and Martin 1997). However, since the diversity of lead structure candidates is of extraordinary importance for virtual screening, the higher computational effort representing molecules in 3D seems justified.

4.6.7   PHARMACOPHORE MODELING

According to the International Union of Pure and Applied Chemistry (IUPAC) definition by Camille Wermuth, a pharmacophore describes the 3D arrangement of steric and electronic features necessary to trigger or block a biological response (Wermuth et al. 1998). Pharmacophores consist of chemical features, like H-bond donor and acceptors, aromatic rings, hydrophobic groups, and positive and negative ionizable moieties. Instead of representing chemical properties of active compounds on an atomic level, chemical features encode chemical functionalities, such as the basic properties of an amine, guanidine, or amidino moieties by a universal positive ionizable feature. Pharmacophores including positive ionizable features will there-fore be able to identify different kinds of basic chemical groups that allow for scaffold hopping, which is much more important in drug development than obtaining high enrichment rates. The very few, fundamental feature types explain the wide acceptance, applicability, and robustness of this approach. One of the most important goals of molecular modeling is the capability of identifying novel scaffolds showing biological activity (McGaughey et al. 2007). Pharmacophore modeling addresses this challenge by generalizing molecular structure to chemical functions and there-fore extracting the essence of macromolecule–ligand binding.

While ligand-based pharmacophore modeling represents one of the earliest established molecular modeling approaches, structure-based pharmacophore modeling has been introduced more recently. This approach elucidates information included in structures of protein–ligand complexes in order to derive pharmacophore models. LigandScout (Wolber et al. 2006; Wolber and Kosara 2006; Wolber and Langer 2005) is the first software tool that allows for fully automated pharmacophore elucidation from protein–ligand structures. The program generates pharmacophore models based on a set of rules that automatically detects and classifies protein–ligand interactions into hydrogen bonds, charge-transfer interactions, and lipophilic areas. Automated placement of excluded volume spheres on the ligand environment occupied by the protein allows for simulating the shape of the binding site in order to reject compounds that do not fit into the binding pocket. Subsequently, these pharmacophore models can be used for rapid virtual database screening.

For ligand-based 3D pharmacophore design, the maximum common subset of chemical features aligned in 3D space is determined to create a pharmacophore model. The best possible spatial alignment of the molecule's chemical features to the ones of the model should approximately correlate with the known activity for quantitative predictions. However, even if such a correlation is not possible, qualitative models can be created from small datasets.

The pioneer program in the field of ligand-based 3D pharmacophore design is the program DISCO presented by Martin et al. (Martin et al. 1993). The programs Catalyst (now integrated into Discovery Studio), MOE (Lokhande et al. 2006) and PHASE (Dixon et al. 2006a,b) implemented similar approaches for the automated elucidation of shared features between bioactive compounds. Recent developments include the program MOGA (Cottrell et al. 2006) using multi-objective optimization techniques on the one hand, and the program LigandScout/Espresso that uses a new pattern-matching 3D alignment algorithm for improved speed and accuracy (Wolber et al. 2006; Wolber and Langer 2001) on the other hand.

A common way to computationally deduce pharmacophore models is to arrange the key interactions of active ligands in 3D space, with respect to conformational flexibility. Features common among all active compounds are considered particularly important for protein–ligand interactions. Obviously, the active molecules used for pharmacophore model development need to share the same binding mode; one pharmacophore model is able to represent only one particular binding mode. In case of diverse scaffolds of distinct binding modes, several pharmacophores can be developed and applied in a parallel way during virtual screening.

4.6.8 NEURAL NETWORKS

Artificial neural networks (ANNs) are a ligand-based method derived from knowledge acquisition and information processing methods of the human brain (Zupan and Gasteiger 2000). During the learning phase, ANNs usually adapt their structure to the external or internal information and are particularly useful to find patterns in chemical data. Kohonen maps (also known as SOMs, self-organizing maps) are among the most popular concepts of ANNs, allowing for analyzing and visualizing high-dimensional data in low-dimensional space (Rose et al. 1991; Saarinen and Kohonen 1985).

4.6.9 HOMOLOGY MODELING

If no structural information on a particular target is available, homology modeling may be employed in order to derive the putative 3D structure of a target computationally. This technique attempts to construct a model of the target protein at atomic resolution based on a related homologous protein used as a template (Vallat et al. 2008). In general, the success of homology modeling is determined by the similarity between the template and the model. Particularly difficult are regions with sequence alignment gaps (i.e., regions that are present in the target but not in the template structure) and loop regions. Homology modeling shares several aspects with protein folding predictions, a rapidly developing field in molecular modeling. The Critical Assessment of Techniques for Protein Structure Prediction (CASP) project is a biannual large-scale competition, which evaluates the current state-of-the-art of homology modeling (Lopez et al. 2007) encouraging exciting new developments. While in recent years the generation of high-quality homology models was connected to considerable manual intervention by experts, the gap between models derived by experts in combination with computational methods and automatically derived models is closing (Battey et al. 2007). The increasing amount of structural information has pushed the number of web servers offering solutions for homology modeling considerably. As these servers do have their strengths and weaknesses, meta-servers have been established that allow for comparing and combining results obtained from a collection of homology modeling servers. SWISS-MODEL is one of the most prominent available web services and is often used as a standard for comparisons (Arnold et al. 2006; Kiefer et al. 2009; Peitsch 1996; Schwede et al. 2003). Homology modeling can be used to bridge ligand- and structure-based design. However, homology models are based on certain approximations and can be inaccurate. Employing such models may be useful for understanding a binding mode in a more illustrative way, but must be regarded very critical when used for structure-based modeling, since a false macromolecular structure prediction may result in completely incorrect results.

4.6.10 MOLECULAR INTERACTION FIELDS

In 1985, Peter Goodford developed the program GRID (Goodford 1985), a force field for the analysis of molecular interactions designed for the analysis of protein-ligand binding. Since then, the force field has been steadily improved (Carosati et al. 2004; Wade et al. 1993; Wade and Goodford 1993) and today, GRID is established as a powerful, universal tool for investigating protein–ligand interaction sites. A plethora of studies have been published that successfully applied molecular interaction fields (MIFs) for modeling and screening (Ahlstroem et al. 2005; Cruciani et al. 2006;

Tintori et al. 2008) as well as a couple of GRID-based approaches, including the GRID/GOLPE approach and GBPM (GRID-based pharmacophore model) (Ortuso et al. 2006a,b).

4.6.11   PROTEIN–LIGAND DOCKING

Molecular docking aims at predicting the binding mode of a ligand to a protein. The computational process is divided into two stages: In the first step, a small organic molecule is flexibly placed into the binding site (*pose placement*). In a second step, all available poses are ranked according to different algorithms (*pose scoring*).

Several docking programs have been developed in the previous decades, most of them using different algorithmic techniques for the same problem. The most important ones are incremental ligand fragmentation and reconstruction (e.g., FlexX (Rarey et al. 1996)), volume- or shape-based algorithms (e.g., DOCK (Ewing et al. 2001)), genetic algorithms (e.g., GOLD (Jones et al. 1997)), systematic search (e.g., Glide (Halgren et al. 2004)), Monte Carlo approaches (e.g., LigandFit (Venkatachalam et al. 2003)), and surface-based molecular similarity methods (e.g., Surflex (Jain 2003)). Scoring remains an unsolved problem; especially since the entropy part of the ligand binding energy is theoretically impossible to calculate from a single geometric snapshot of a ligand-protein complex.

While pose placement in the interaction site is considered to be working reliably, scoring functions have been discussed controversially (Warren et al. 2006). As confirmed by a recent study (Hawkins et al. 2007), the predictive power in terms of ligand affinity in current docking programs is not satisfactory. Nevertheless, docking is a valuable and illustrative tool for rationalization of ligand binding and drug action and has led to several breakthroughs in drug design as discussed below.

4.6.12   MOLECULAR DYNAMICS SIMULATIONS

The main problem of a crystal structure is that it only represents a single snapshot of a dynamic biological system. Moreover, this snapshot does not represent the protein in solution (as in vivo), but in a crystallized form. Due to their experimental nature, crystal structure coordinates do not provide much information on the dynamics of a certain protein; they only provide hints about the flexibility of certain protein areas by temperature factors (B-factors) (Joosten et al. 2006; Laskowski and Swaminathan 2006). If several crystal structures of the same protein with different ligands have been determined, modelers can learn about different binding modes and the putative flexibility of the binding site from the differences of these crystals.

Molecular dynamics (MD) simulations allow for investigating movements of protein–ligand complexes for a short period of time (several nanoseconds). MD is a form of computer simulation in which atoms and molecules are allowed to interact for a period of time by approximations of the law of known physics and quantum chemistry. The foundation of MD is that statistical ensemble averages are equal to time averages of the system (ergodic hypothesis). The result of an MD simulation is a set of coordinates (trajectory) that can be analyzed in terms of protein–ligand interactions forming over time, and protein regions with higher or lower flexibility.

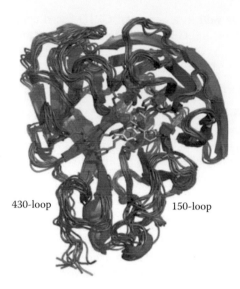

430-loop   150-loop

FIGURE 4.1   **(See color insert.)** Conformational flexibility of the binding site observed by MD simulations based on the protein structure of influenza neuraminidase (PDB entry 2hty). The figure shows 10 representative frames (gray) of a short MD simulation. The binding site is occupied by oseltamivir (Chart 4.1, **1**; orange) and katsumadain A (Chart 4.1, **2**; green; heteroatoms O, red; N, blue). (Figure provided by Dr. Johannes Kirchmair.)

Since MD computations are time-consuming, these methods are mainly used for lead optimization or rationalization rather than hit discovery.

Recently, MD simulations were applied to cope with the flexible binding site of the antiinfluenza target neuraminidase. Several publications in the field of x-ray crystallography (Russell et al. 2006) and MD simulations have previously indicated considerable flexibility of two loop regions (150- and 430-loops) that are located adjacent to the active site targeted by oseltamivir (Chart 4.1, **1**; Tamiflu®) (Amaro et al. 2007, 2009) (Figure 4.1). Small organic molecules that exploit these loop regions for ligand binding are likely to result in a significantly enhanced binding affinity. This could be demonstrated recently by docking the neuraminidase-inhibiting diarylheptanoids isolated from the seeds of *Alpinia katsumadai* in a highly probable trajectory frame obtained by MD simulations. This technique revealed as essential to render the conformational space of the target an extended binding pocket confirmation and to exploit the so far unoccupied space emerging from the relaxation of both flexible loops toward the solvents. The best scored and most active diarylheptanoid, katsumadain A (Chart 4.1, **2**), was able to inhibit the neuraminidase of human influenza virus A/PR/8/34 and of four swine influenza viruses all of subtype H1N1 in the range of 0.9–1.6 µM (Grienke et al. 2010).

4.6.13   PARALLEL VIRTUAL SCREENING: ACTIVITY PROFILING

Molecular modeling techniques can be applied to predict the activity of small organic molecules not only to a single target but also to several targets or different

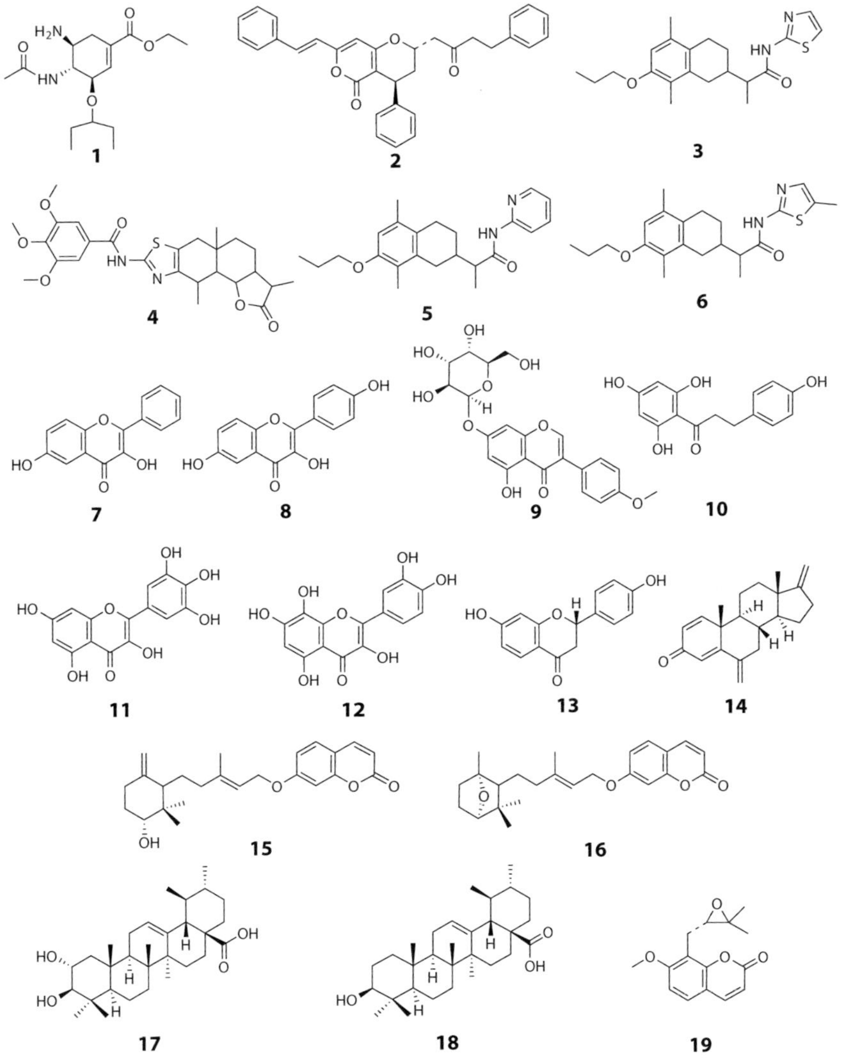

CHART 4.1   Chemical structures of compounds mentioned in the text.

protein isoforms in a single prediction step. The prerequisite for this approach is a collection of validated models for several targets. There are three main applications for activity profiling: (i) Identification of potential side effects by creating antitarget models (e.g., for the hERG potassium channel or for mechanistic cytochrome P450 inhibition), (ii) optimization of isoform specificity for compounds (e.g., a specific kinase inhibition profile is crucial for the therapeutic applicability of a compound, so models for several kinases are built and screened in parallel), and (iii) target fishing (e.g., a newly synthesized compound or isolated natural product is screened against several models to determine which biological targets it should be tested against).

Several attempts for multitarget profiling have been published in the last few years (Ekins et al. 2007). The most recent approach has been published by Keiser et al. who explored possible drug-target effects computationally by using chemical similarities between 3,665 drugs and 65,241 ligands organized into 246 targets from the MDL Drug Data Report (MDDR) database. The applied chemoinformatics methods provided tools to explore associations between drug-target effects and off-targets systematically, both to understand drug effects and to explore new opportunities for therapeutic intervention (Keiser et al. 2009).

The main challenge remaining is the determination of the applicability domain of the models used for these predictions. For example, models may be generated with a homologous series of active molecules (local model) or a structurally diverse range of molecules (global model). The modeling expert who created such a model might know how to apply similarity cut-offs and this knowledge can be lost when applied in a fully automated manner. It is important for the application of parallel screening that the results are validated critically and that the researcher performing the activity profiling experiment understands the history of the model yielding the target prediction. This requires modeling methods that are transparent and easy to understand. 3D pharmacophore modeling (Steindl et al. 2006a,b, 2007) is one of the most promising approaches for activity profiling due to the transparency of the results.

4.7 CHALLENGES FOR THE APPLICATION OF MOLECULAR MODELING TO NATURAL PRODUCT RESEARCH

4.7.1 CHALLENGES FROM CHEMOINFORMATIC POINT OF VIEW

Natural products were often viewed as being "too complex" but then have been recognized as a valuable source of interesting and diverse structures (Paterson and Anderson 2005). Wetzel et al. summarize three studies (Ertl and Schuffenhauer 2007; Feher and Schmidt 2003; Grabowski and Schneider 2007) describing the differences of natural product chemistry compared to synthetic molecules (Wetzel et al. 2007): Natural products show higher molecular weight than synthetic drugs, contain more oxygen atoms, fewer nitrogen atoms, three times more stereo centers, their degree of unsaturation is higher than in synthetic drugs and they incorporate less aromatic rings. Although natural products in general contain more rings than drugs, most of them are nonaromatic and part of single fused ring system.

The major challenge for *in silico* techniques is the increased molecular weight and the higher number of saturated bonds resulting in higher molecular flexibility. As a rule of thumb, docking and conformational sampling become slow and inaccurate when a molecule has more than seven rotatable bonds. However, these problems are not irresolvable, but have to be addressed while setting up a virtual screening run using natural products taking into account longer computational time. A too high degree of flexibility may also result in promiscuity, that is, a compound is fitted into a binding site in an implausible way, which stresses the need for visual inspection of virtual hits and selection of the most plausible virtual hits. Another putative practical problem is the higher number of stereo centers whose definitions are often neglected during database construction and therefore the programs have to enumerate all stereoisomers, which unnecessarily increase the (virtual) flexibility.

Overall, the diversity and high number of interesting natural product scaffolds—many of them containing privileged scaffolds—fully justify the additional effort for modeling and yields promising results when the limits caused by the higher flexibility of natural products are considered during the modeling process.

4.7.2   AVAILABILITY OF NATURAL PRODUCT DATABASES

Databases of natural product molecules are an important tool for natural product molecular modeling. Such databases either contain compounds that are purchasable from commercial sources or known natural products from literature. Virtual hits with known natural product sources bear the advantage over synthetic virtual hits that they can be analyzed in terms of their ethnopharmacological and pharmacognostic context and therefore provide hints to related compounds or activities.

4.7.2.1   Commercial Natural Product Databases

Several commercial suppliers of natural products and natural product derivatives offer downloadable structure-data files from their corporate websites. The largest public collection of genuine natural products is provided by Analyticon (approx. 1200 of microbial origin, 3500 of plant origin, www.analyticon-dicovery.com), smaller collections are available from BioFocus (245 compounds, http://www.biofocus.com/), GreenPharma (www.greenpharma.com, 240 compounds), SPECS (natural product collection, 429 compounds, www.specs.com), and TimTec (450 plant constituent, 190 products from bacterial, fungal, and animal sources, www.timtec.net). VitasM (www.vitasmlab.com) provides approx. 25,000 natural product derivatives.

4.7.2.2   Databases from Literature Sources

The most comprehensive database of natural products is the *Dictionary of Natural Products* (DNP) database provided by Chapman and Hall (Buckingham 2007) containing more than 214,500 compounds and is updated every 6 months with current literature data.

The DIOS database (Rollinger et al. 2008) contains 9,676 unique small molecular weight natural compounds based on information from the ethnopharmacological source *De materia medica*, witnessed by Pedanius Dioscorides (First century AD). From this collection, about 800 plants are described which were in medical use.

The Natural Product Database (NPD) (Rollinger et al. 2004) collected at the University of Innsbruck was last updated in 2009 and currently contains 146,790 unique compounds from natural sources, including constituents of animals, plants, and fungi from land-based as well as marine organism.

Traditional Chinese medicine has always been a major point of interest, which is reflected in three databases: The Chinese Herbal Medicines Database (CHMD,

10,216 compounds, (Fakhrudin et al. 2010), the Traditional Chinese Medicines Database (TCMD, 10,458 compounds, (Wang et al. 2007), and the Chinese Herbal Constituents Database (CHCD, (Ehrman et al. 2007a), 7,000 compounds from 240 Chinese herbs).

Other databases comprise the Bioactive Plant Compounds Database (BPCD, also contains biological activity information for 78 targets, 2,597 compounds, (Ehrman et al. 2007b)), and the Marine Natural Products Database (MNPD. 6,000 chemicals from more than 10,000 marine-derived materials (Lei and Zhou 2002).

4.7.3 CHALLENGES FROM PHARMACOGNOSTIC POINT OF VIEW

Once a secondary metabolite being an entry in a natural product database has been predicted by a data mining method for any bioactivity, it is necessary to decide, whether the efforts for its acquisition and experimental investigation are warrantable. The retrieval and isolation may often shape up as panning for gold without knowing if it can be discovered, and if yes, if it is really the assumed precious metal, that is a real hit.

There may be uncertainty surrounding the following issues:

- Reliance on the previously published structure elucidation of the secondary metabolite.
- Correct entry in the database: in many cases the configuration of a natural compound is not known or unambiguously solved.
- Is the aiming compound chemically and physically stable?
- Is it already known to be active on the scrutinized target? This can be seen in a positive sense as proof of concept, or in a negative sense, because its discovery is not a new finding any more.
- Is it described or predicted to be toxic? What about its bioavailability?
- Is it commercially available (which applies to very rare cases, and usually they are very expensive)?
- Can the secondary metabolite be rediscovered from the reported natural source (how many reports are available about its isolation)?
- Might it be an artifact or a native natural ingredient?
- Is it a main constituent in the reported natural material or only available in traces?
- Which is the best natural material to start with (species, organ, extract, etc.)?
- Is the natural starting material accessible and legally available for collection? Are there any conflicts with intellectual property rights or with the transfer of natural material from outside (International Cooperative Biodiversity Group (ICBG) (Soejarto et al. 2005))?
- Is the virtual hit detectable by analytical methods from the respective extract?
- Can it be isolated in an adequate amount and purity?
- Final and most important question: Is the isolated natural compound really interacting as ligand of the focused target as virtually predicted?

4.8 INTEGRATED COMPUTATIONAL APPROACHES IN NATURAL PRODUCT RESEARCH

Even a highly validated *in silico* tool is only able to give an activity prediction and enrich the pool of active candidates. This is often too vague for a natural product scientist, especially in the light of the above-mentioned compilation of uncertainties. Thus, additional selection criteria, which may help to render the bold venture of finding a natural lead structure into a successful endeavor, are in high demand. The combination of different *in silico* tools may help efficiently to decrease the number of samples to be tested (Kubinyi 2006; Walters et al. 1998).

First application scenarios of computer-assisted strategies in natural product research have already been presented in earlier reviews (Ferrara et al. 2006; Khan and Ather 2007; Rollinger 2009; Rollinger et al. 2006a,b, 2008). Therefore, in this survey we will primarily focus on applied computational strategies for natural lead discovery exemplified in some very recently published studies covering the time frame from 2007/2008 to early 2010.

Although computational approaches are always theoretical and, thus, never replace any experimental investigation, the hyphenation of diverse *in silico* tools has already been utilized successfully for lead finding from nature. These studies usually resort to already available natural compounds, mainly from vendors.

To minimize experimental efforts, Franke and coauthors applied a two-step ligand-based virtual screening protocol for the selection of putative inhibitors of the human 5-lipoxygenase (Franke et al. 2007). By using 43 known inhibitors from literature as query compounds, a similarity search based on a topological pharmacophore descriptor was performed with compound libraries from AnalytiCon (AnalytiCon Discovery GmbH, Potsdam, Germany). From the resulting 430 virtual hits, 18 candidates were selected for biological evaluation. The two most active natural compounds (Chart 4.1, **3**, and **4**), further served as query molecules to guide the second ligand-based screening cycle, which resulted in the discovery of even more potent lipoxygenase inhibitors belonging to the chemical class of tetrahydronaphthols (Chart 4.1, **5**, and **6**). They potently suppressed the leukotriene synthesis in both intact cells and a cell-free assay at low micromolar concentrations.

Structure-based pharmacophore models were used for a virtual screening focusing on β-ketoacyl-acyl carrier protein synthase III (KAS III) (Lee et al. 2009). This is an attractive target for the design of novel antimicrobial drugs, because the fatty acid biosynthesis is an essential step in bacterial cell wall formation. After experimental validation of the applied pharmacophore models, a merged in-house 3D NP database containing commercially available 865 natural compounds was virtually screened. Two flavones, 3,6-dihydroxyflavone (Chart 4.1, **7**), 3,6,4′-trihydroxyflavone (Chart 4.1, **8**), one glycosilated isoflavone (Chart 4.1, **9**), and one dihydrochalcone, phloretin (Chart 4.1, **10**), were selected for biological evaluation. Based on binding assays, all four phytochemicals were suggested to bind to *Escherichia coli* KASIII. Antibacterial activities in the range of 16–256 μM against four Gram-positive bacteria species *Staphylococcus aureus*, *Enterococcus faecalis*, methicillin-resistant *Staphylococcus aureus* (MRSA), and vancomycin-resistant *Enterococcus faecalis* could be shown for 3,6-dihydroxyflavone (Chart 4.1, **7**) and phloretin (Chart 4.1, **10**).

A database consisting of more than 8000 phytochemicals from traditional Chinese medicine was virtually prospected for novel aromatase inhibitors by Paoletta et al. (Paoletta et al. 2008). The authors used a combination of a ligand-based virtual screening (multiple decision tree) followed by a protein-based secondary *in silico* screening (docking). By such means, they effectively exploited thousands of metabolites from medicinal plants and successfully identified the flavones, myricetin (Chart 4.1, **11**), and gossypetin (Chart 4.1, **12**), as low micromolar aromatase inhibitors, and the flavanone, liquiritigenin (Chart 4.1, **13**), as being more potent (IC_{50} 0.34 μM) than the first generation aromatase inhibitor, aminoglutethimide (Chart 4.1, **14**).

The majority of natural compounds are not available at any vendors and thus have to be isolated from a natural source. This is often a time and resources consuming procedure. Thus, reasonable strategies such as the combination of an *in silico* approach and an empirical one are in high demand to reduce the risk of following a blank. These integrated workflows are labeled as *in combo* approach (Rollinger et al. 2005; Van de Waterbeemd 2005).

In Figures 4.2 and 4.3, schematic overviews of computer-assisted strategies for lead discovery from nature are given. The concept depicted in Figure 4.2 starts with the

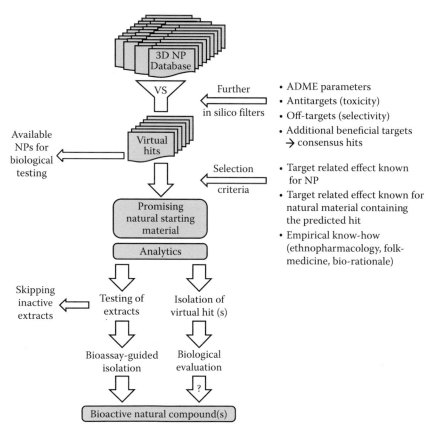

FIGURE 4.2 General workflow for a computer-assisted strategy for natural lead discovery (VS, virtual screening; NP, natural product).

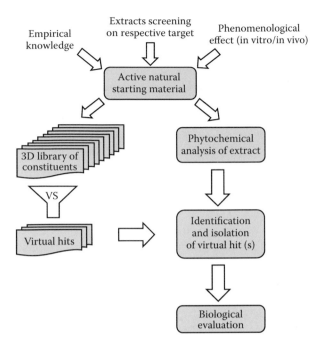

FIGURE 4.3 Computer-assisted workflow for the identification of the active metabolite/s within a natural preparation (VS, virtual screening).

computational tool for getting an idea about putative hits from nature. As soon as a sensitive data-mining tool has been developed and validated, it can be used to virtually screen a database consisting of 3D multiconformational structures of natural compounds (3D NP database). Additional virtual filtering tools for the profiling of absorption, distribution, metabolism, and excretion (ADME) parameters and toxicity (Van de Waterbeemd and Gifford 2003) might have an invaluable impact to aid a refined selection of virtual hits. Screening for related targets may help to increase the selectivity of the final virtual hits. In contrast, considering consensus hits derived from screening experiments against different targets being of synergetic and beneficial value may result in multitarget oriented candidates. In most cases, the predicted natural compounds are not commercially available; thus, they have to be isolated. Then, the selection of the natural source containing one or more of the virtual hits emerges as a crucial step. It requires a comprehensive study in literature considering all the information that can be gathered from any empirical or experimental finding, either from the individual virtual hit or the underlying natural source. This includes the hit content in the natural material, its availability, hints from traditional medicine, toxic- or target-related effects known for the virtual hit (or this chemical scaffold) or for a natural preparation including the respective hit, just to name some elements.

Once a promising natural starting material has been selected, it is advisable to analyze the extracts gained from this source in order to detect the desired compound(s). Suitable methods, for example, gas chromatography (GC) or high-performance liquid chromatography (HPLC) are combined with mass spectrometry or hyphenated with NMR. Preliminary assays with those crude extracts and fractions that contain the promising metabolite(s) may help to focus on the active ones and to skip the inactive ones. Those samples that scored well are then subjected to a bioassay-guided fractionation, conceptually ending in the isolation of the active principle(s). Alternatively, only the analyzed virtual hits are isolated from the extracts and finally subjected to biological evaluation. The latter procedure does not require an iterative testing after each chromatographic step; thus, it is more straightforward. At the same time, it is however more hazardous, since one may run the risk of (i) hunting an inactive virtual hit, and (ii) ignoring active metabolites not virtually predicted or not being an entry in the screened database (e.g., new natural compounds).

A further integrated procedure is schematized in Figure 4.3. Applying this approach, the preselection of the natural material is not guided by virtual prediction, but by previous empirical or experimental findings. Here, the virtual prediction should enable a target-oriented identification of those metabolites within a multicomponent mixture, which cause the already known biological effect of the starting material, for example, a herbal remedy. Thus, a focused 3D database is generated consisting of all the metabolites known from literature or own experiments to be included in the preparation under investigation. The resulting biased database is then subjected to a data mining experiment, for example, a virtual screening against the aiming target. Following this strategy, the putative hits may then be identified by analytical tools like liquid chromatography mass spectrometry (LC-MS) or LC-NMR in order to isolate them from the natural matrix in a target-oriented way for pharmacological testing.

In the following paragraphs, two recent examples of computational approaches integrated in classical natural lead finding strategies are presented.

The human rhinovirus (HRV) is responsible for the common cold and other upper respiratory symptoms in more than 50% of the population. Inhibitors of the HRV coat protein are promising candidates to treat these infections. In the search for natural inhibitors of this target, a previously established pharmacophore model based on a HRV coat protein crystal structure (Figure 4.4) was used for the virtual screening of the 3D NP-database DIOS. This resulted in 29 phytochemicals predicted to act as capsid binding inhibitors. For a sensible selection of the most promising natural starting material containing the predicted constituents, we consulted Pedanius Dioscorides' *Materia medica*. This was written in the first century AD and represents a great repository of botanical, medical, and pharmacological lore. The descriptions therein helped to prioritize those virtual hits, which have been reported to be constituents of asafetida. The efficacy of the applied approach could be confirmed by using a cytopathic effect inhibitory assay, in which the gum resin asafetida (derived from *Ferula assa-feotida* and related species) and its virtually predicted constituents farnesiferol B (Chart 4.1, **15**) and C (Chart 4.1, **16**) showed anti-HRV-2 effects with IC_{50} values of 2.6 and 2,5 µM, respectively (Rollinger et al. 2008). The traditionally manifested evidence for asafetida for the treatment of common cold symptoms together with the findings from the pharmacophore-based virtual screening substantially helped in the discovery of this novel class of human rhinovirus capsid binders.

In the search for the molecular mechanism underlying some anti-diabetic herbal remedies, the dichloromethane and methanol extracts of six well established

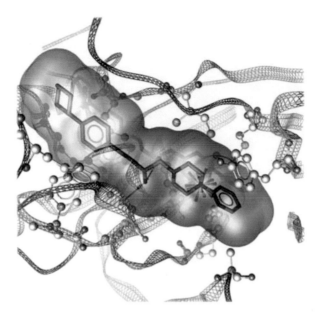

FIGURE 4.4   (See color insert.) Binding site of the human rhinovirus coat protein (PDB Code 1hrv) with cocrystallized synthetic inhibitor Sdz 35–682 (visualized with LigandScout).

medicinal plants were scrutinized for their potential to inhibit the activity of 11β-hydroxysteroid dehydrogenase 1 (11β-HSD1). This enzyme catalyzes the conversion of inactive 11-ketoglucocorticoids to active 11β-hydroxyglucocorticoids. Recent studies indicate that modulating inhibitors of 11β-HSD1 are therapeutically useful to decrease blood glucose concentration and to ameliorate metabolic abnormalities in type 2 diabetes (Atanasov and Odermatt 2007). Among the tested samples, the leaf extracts of *Eriobotrya japonica* (i.e., loquat) showed both a dose-dependent inhibition of 11β-HSD1, and a preferential inhibition of 11β-HSD1 versus 11β-HSD2 (Gumy et al. 2009). By using a previously generated and experimentally confirmed pharmacophore model (Schuster et al. 2006), the aim was to identify those secondary metabolites from *E. japonica* extracts responsible for the previously observed inhibitory activity on 11β-HSD1. By means of a pharmacophore-based virtual screening, one of the major constituents from loquat, corosolic acid (Chart 4.1, **17**), was identified and confirmed as selective inhibitor of 11β-HSD1 with an IC$_{50}$ of 0.8 μM (measured in lysates of cells expressing recombinant human 11β-HSD1) (Rollinger et al. 2010). Further active pentacyclic triterpene acids of the ursane type such as ursolic acid (Chart 4.1, **18**) were identified from the dichloromethane extract by bioassay-guided isolation. In order to further explore the properties that are mandatory for 11β-HSD1 inhibition, the most active natural compounds found in this study were flexibly aligned by pharmacophoric points using LigandScout 3.0 (Wolber et al. 2006). This resulted in a merged features pharmacophore model (Figure 4.5) revealing a high power to discriminate between active and inactive 11β-HSD1 inhibitors (Rollinger et al. 2010). The putative ligand–target interactions of four active *Eriobotrya* constituents are depicted in Figure 4.6.

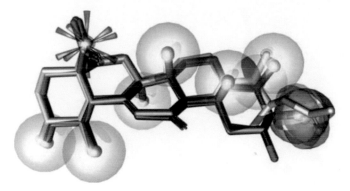

FIGURE 4.5 **(See color insert.)** Four constituents from *Eriobotrya japonica* shown from a flexible 3D overlay generated by LigandScout. Common feature points are indicated as orange mini-spheres, lipophilic areas are represented as yellow spheres, the negatively ionizable area as red star, acceptor and donor locations as red and green spheres, respectively.

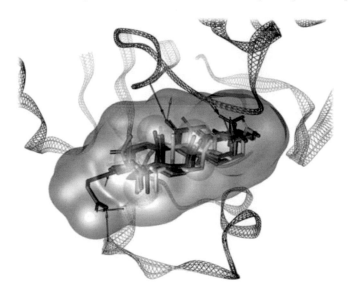

FIGURE 4.6 **(See color insert.)** 11beta-HSD1 crystal structure (PDB code 2bel, chain A) cocrystallized with carbenoxolone (blue) and possible binding modes of four constituents of *Eriobotrya japonica* (different gray shades) as predicted by structure-based molecular modeling.

4.9   VIRTUAL ACCESS TO BIOLOGICAL PROFILING AND TARGET FISHING

In recent years, several attempts have been made to predict a natural product's biological profile and to get an idea of its probable hitting biological targets, that is, target fishing.

The chemical space navigation tool ChemGPS-NP, which is based on multidimensional molecular descriptors, has been introduced by Larsson et al. for exploring a natural compound's biologically relevant chemical space (Larsson et al. 2007). As soon as the entire chemical space is populated to some extent by known active compounds, ChemGPS-NP can be used as reference system enabling characterization and comparison of molecules by evaluating the chemical space occupancy. In this way, several of the natural products were confirmed to exhibit the same activity as their drug neighbors (Rosen et al. 2009a). Recently, the ChemGPS-NPWeb portal (http://chemgps.bmc.uu.se) has been made available for the public domain to assist in compound selection and the discovery of novel lead compounds (Rosen et al. 2009b).

Another portal is provided by the Strathclyde Institute for Drug Research (SIDR). This facility enables access to the *in silico* driven Drug Discovery Portal (DDP; http://www.sidr.org/showPage.php?page=natural_products). This system enables chemists in academia to submit structures to the screening database provided they are willing and able to subject their compounds to a real screening. On the other hand, biologists can propose targets provided there are relevant biological assays available to test the predicted hits. By means of virtual screening and similarity searching, structure-activity profiles for new or existing hits can be derived. This facility accordingly aims at identifying ligands providing a setup at the chemistry–biology interface.

An interesting property prediction estimation is enabled using the computer program PASS (prediction of activity spectra for substances) implemented by Filimonov, Poroikov, and coauthors (Filimonov et al. 1999; Filimonov and Poroikov 1996). Based on the structural formula of a compound, this approach can be applied as an *in silico* tool for complex searches covering an increasing number of pharmacotherapeutic effects, mechanism of action, metabolic, and toxicity parameters (Poroikov and Filimonov 2005; Poroikov et al. 2007). The algorithm is based on the concept of "biological activity spectrum," which is an intrinsic property of a compound that reflects all its different biological activities that arise from its interaction with biological entities (Poroikov and Filimonov 2005). The applicability of the PASS approach for the prediction of biological and toxicological activities was recently demonstrated on a set of 681 cyanobacterial secondary metabolites (Devillers et al. 2007), as well as for natural cyclobutane-containing alkaloids isolated from terrestrial and marine species (Sergeiko et al. 2008).

On the basis of structural similarity to natural compounds being inhibitors of targets involved in inflammation, diabetes, and HIV, Ehrman et al. applied multiple decision trees to scrutinize multicomponent Chinese herbal remedies for active ingredients (Ehrman et al. 2007c). 8264 compounds reported from 240 herbs used in traditional Chinese medicine were subjected to Random Forest (Svetnik et al. 2003) to discriminate the phytochemicals. The distribution of potential hits collocating with known inhibitors was then evaluated by a literature search and provided evidence to support 83 predictions to pharmacological targets under investigation. A large number of natural preparations were predicted to interact with multiple targets.

Inverse docking programs have the potential to comply with the innate multitarget nature of phytochemicals. They may aid in elucidating molecular targets by virtual screening of a compound of unknown activity against a preferably high number of structurally disclosed targets. This concept of target fishing is implemented in "reverse pharmacognosy," a term coined by Philippe Bernard's group, that is, finding new biological targets by virtual or real screening of natural compounds for the identification of promising natural resources. Using their software tool Selnergy™— an inverse docking tool (Do et al. 2005)—this strategy was applied for the virtual identification of putative targets for meranzin (Chart 4.1, **19**) (Do et al. 2007). The coumarin derivative meranzin was isolated in large amounts from *Limnocitrus littoralis* and subjected for virtual profiling. Among the 400 screened proteins, the three targets COX1, COX2, and PPARγ were selected for experimental validation. The results confirmed the predictive power of the applied *in silico* approach.

A combined virtual profiling approach was performed by Huang et al. to characterize the drug-like features of a complex traditional Chinese medicinal recipe, called Xuefu Zhuyu decoction. The author used a chemical space distribution to describe potential drug-like properties, a docking protocol for target fishing and ADME prediction to derive multiple pathways and potential synergism in the curative mechanism of the investigated recipe (Huang et al. 2007). Among the 501 ingredients of the decoction derived from 11 herbs, the drug-like compounds were docked to a number of targets known to play an important role in cardiovascular disease, such as rennin, angiotensin-converting enzyme, vascular endothelial growth factor receptor (VEGFR), 3-hydroxy-3-methylglutarylcoenzyme A, and P-glycoprotein. 283 molecular ingredients were identified as virtual inhibitors of the focused targets. Intriguingly, 10 of the 11 herbs contained secondary metabolites with predicted interactions to at least two of the selected targets.

In the search for the molecular mechanism of natural compounds with known growth inhibitory activity against *Trypanosoma brucei*, Ogungbe and Setzer docked a set of phytochemicals into validated drug targets elucidated from that parasite (Ogungbe and Setzer 2009). These included trypanothione reductase, rhodesain, triosephosphate isomerase, and farnesyl diphosphate synthase. Docking calculations performed with the compound data set including iridoids, phenolics, terpenoids, alkaloids, quinones, and miscellaneous revealed that most of the top poses are phenolics and quinones. The predicted ligand–target interactions are suggested to reveal structural motifs necessary for bioactive compounds and to streamline extract purification and compound isolation schemes for target-based bioassays, which may lead to the isolation of potent antitrypanosomal agents.

Besides docking, pharmacophore-based virtual parallel screening is an excellent tool to filter for promising targets (see before). The concept is depicted in Figure 4.7. As soon as the target molecule is able to comply with all the requirements and restrictions imposed by any model, it can be assessed as rational hint. In this way, the parallel screening is not only helpful to estimate the interactions of a drug candidate with diverse antitargets or to canvass its interactions to related targets as is performed for an activity profiling. Overall, it is a computational tool for target fishing and to prioritize a few targets for experimental evaluation (Rollinger 2009).

This application scenario was recently performed with the extract of the medicinal plant *Ruta graveolens* L (Rollinger et al. 2009). Sixteen constituents were isolated and identified from the aerial parts, and consequently low energy conformers of these isolates were subjected to virtual parallel screening using 2208 pharmacophore models (model collection provided by Inte:Ligand GmbH, Vienna, Austria; (Wolber and Langer 2005)) covering 280 unique pharmacological targets. From the

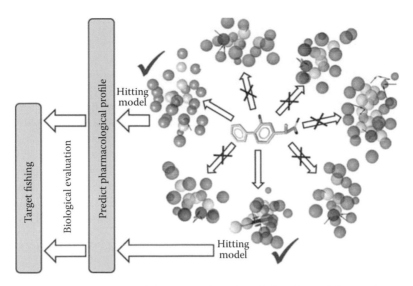

FIGURE 4.7 **(See color insert.)** Pharmacophore-based virtual parallel screening concept of a single 3D structure for the identification of potentially interacting targets.

evaluated hit lists, three pharmacological targets were selected, that is, acetylcholinesterase, the human rhinovirus coat protein, and the cannabinoid receptor 2. For a critical validation of the applied paradigm, virtual hits and nonhits were tested at the selected targets. Among the 18 predicted and biologically evaluated interactions, 14 predictions or 77.8% revealed as correct. Thus, the experimental results confirmed the high potential of the virtual parallel screening strategy as an *in silico* tool for rational target fishing and pharmacological profiling of the investigated extract.

4.10 CONCLUSIONS

Although we know about the high impact of natural compounds in drug discovery and development, plenty of secondary metabolites still await discovery. In today's research, we benefit from a wealth of available experimental structural data and a continuously increasing number of therapeutically relevant macromolecular structures. This, coupled with the growing number of isolated and elucidated compounds from nature, highlights the need to overcome the overwhelming amount of data and correlations. The use of computational techniques may prove effective in narrowing the search for the discovery of bioactive natural products, and accordingly streamline the drug discovery process.

In the last two decades, the application of predictive *in silico* tools has emerged successful in identifying chemical entities that have a high likelihood of binding to a target molecule to elicit the desired biological response. In contrast to experimental screening, for example, medium-sized or high-throughput screening, millions of compounds can be examined *in silico* for their propensity to interact with the target protein (Stockwell 2004). Intriguingly, the *in silico* techniques discussed in this review, which have so successfully been applied in medicinal chemistry, have only slowly been established in natural product research. Reasons include the previously limited availability of high quality natural product databases, the often restricted or laborious access to natural compounds for testing, and the lack of experience in molecular modeling due to the higher flexibility, chemical complexity, and multitarget nature of secondary metabolites (as discussed before). The interface between phytochemists, pharmacognosists, and pharmacologists on the one hand, and specialists in chemoinformatics on the other hand seems to be crucial for a sensible application of data mining tools in this prospering field. Computational methods and strategies on how to integrate them best in pharmacognostic workflows to access these promising hidden compounds are discussed and decorated with some recent examples from our working group and from literature. It is the authors' aim to pitch the presented computational techniques with all their tremendous possibilities and limitations to a scientific community working in the field of drug discovery from nature.

REFERENCES

Agrafiotis, D. K., A. C. Gibbs, F. Zhu, S. Izrailev, and E. Martin. 2007. Conformational sampling of bioactive molecules: A comparative study. *Journal of Chemical Information and Modeling* 47:1067–1086.

Ahlstroem, M. M., M. Ridderstroem, K. Luthman, and I. Zamora. 2005. Virtual screening and scaffold hopping based on GRID molecular interaction fields. *Journal of Chemical Information and Modeling* 45:1313–1323.

Amaro, R. E., X. Cheng, I. Ivanov, D. Xu, and J. A. McCammon. 2009. Characterizing loop dynamics and ligand recognition in human- and avian-type influenza neuraminidases via generalized born molecular dynamics and end-point free energy calculations. *Journal of the American Chemical Society* 131:4702–4709.

Amaro, R. E., D. D. L. Minh, L. S. Cheng, W. M. Lindstrom, Jr., A. J. Olson, J.-H. Lin, W. W. Li, and J. A. McCammon. 2007. Remarkable loop flexibility in avian influenza N1 and its implications for antiviral drug design. *Journal of the American Chemical Society* 129:7764–7765.

Arnold, K., L. Bordoli, J. Kopp, and T. Schwede. 2006. The SWISS-MODEL workspace: A web-based environment for protein structure homology modelling. *Bioinformatics* 22:195–201.

Atanasov, A. G. and A. Odermatt. 2007. Readjusting the glucocorticoid balance: An opportunity for modulators of 11beta -hydroxysteroid dehydrogenase type 1 activity? *Endocrine, Metabolic & Immune Disorders—Drug Targets* 7:125–140.

Baroni, M., G. Costantino, G. Cruciani, D. Riganelli, R. Valigi, and S. Clementi. 1993. Generating optimal linear pls estimations (GOLPE): An advanced chemometric tool for handling 3D-QSAR problems. *Quantitative Structure-Activity Relationships* 12:9–20.

Battey, J. N. D., J. Kopp, L. Bordoli, R. J. Read, N. D. Clarke, and T. Schwede. 2007. Automated server predictions in CASP7. *Proteins-Structure Function and Bioinformatics* 69:68–82.

Berman, H. M., J. Westbrook, Z. Feng, G. Gilliland, T. N. Bhat, H. Weissig, I. N. Shindyalov, and P. E. Bourne. 2000. The protein data bank. *Nucleic Acids Research* 28:235–242.

Böhm, H.-J. 1992. LUDI: Rule-based automatic design of new substituents for enzyme-inhibitor leads. *Journal of Computer-Aided Molecular Design* 6:593–606.

Bostrom, J. 2001. Reproducing the conformations of protein-bound ligands: A critical evaluation of several popular conformational searching tools. *Journal of Computer Aided Molecular Design* 15:1137–1152.

Bostrom, J., P.-O. Norrby, and T. Liljefors. 1998. Conformational energy penalties of protein-bound ligands. *Journal of Computer Aided Molecular Design* 12:383–396.

Breinbauer, R., I. R. Vetter, and H. Waldmann. 2002. From protein domains to drug candidates: Natural products as guiding principles in the design and synthesis of compound libraries. *Angewandte Chemie, International Edition* 41:2878–2890.

Brown, R. D. and Y. C. Martin. 1997. The information content of 2D and 3D structural descriptors relevant to ligand-receptor binding. *Journal of Chemical Information and Computational Science* 37:1–9.

Bruhn, J. G. and B. Holmstedt. 1981. Ethnopharmacology: Objectives, principles and perspectives. In *Natural Products as Medicinal Agents*, eds. J. L. Beal and E. Reinhard. Stuttgart, Germany: Hippokrates Verlag.

Buckingham, J. 2007. *Dictionary of Natural Products on DVD*. Boca Raton: Taylor & Francis Group.

Butler, M. S. 2004. The role of natural product chemistry in drug discovery. *Journal of Natural Products* 67:2141–2153.

Carosati, E., S. Sciabola, and G. Cruciani. 2004. Hydrogen bonding interactions of covalently bonded fluorine atoms: From crystallographic data to a new angular function in the GRID force field. *Journal of Medicinal Chemistry* 47:5114–5125.

Chen, I.-J. and N. Foloppe. 2008. Conformational sampling of drug like molecules with MOE and catalyst: Implications for pharmacophore modeling and virtual screening. *Journal of Chemical Information and Modeling* 48:1773–1791.

Cottrell, S. J., V. J. Gillet, and R. Taylor. 2006. Incorporating partial matches within multi-objective pharmacophore identification. *Journal of Computer-Aided Molecular Design* 20:735–749.

Cragg, G. M., D. G. I. Kingston, and D. J. Newman. 2005. Developments and future trends in anticancer natural products drug discovery. In *Anticancer Agents from Natural Products*, eds. D. J. Newman, D. G. I. Kingston, and G. M. Cragg. Boca Raton, FL: Taylor & Francis Group.

Cramer III, R. D., D. E. Patterson, and J. D. Bunce. 1988. Comparative molecular field analysis (CoMFA). 1. Effect of shape on binding of steroids to carrier proteins. *Journal of the American Chemical Society* 110:5959–5967.

Cruciani, G., Y. Aristei, R. Vianello, and M. Baroni. 2006. GRID-derived molecular interaction fields for predicting the site of metabolism in human cytochromes. In *Methods and Principles in Medicinal Chemistry*, ed. G. Cruciani. Weinheim, Germany: Wiley-VCH.

Csermely, P., V. Agoston, and S. Pongor. 2005. The efficiency of multi-target drugs: The network approach might help drug design. *Trends in Pharmacological Sciences* 26:178–182.

Devillers, J., J. C. Dore, M. Guyot, V. Poroikov, T. Gloriozova, A. Lagunin, and D. Filimonov. 2007. Prediction of biological activity profiles of cyanobacterial secondary metabolites. *SAR and QSAR in Environmental Research* 18:629–643.

Dixon, S. L., A. M. Smondyrev, E. H. Knoll, S. N. Rao, D. E. Shaw, and R. A. Friesner. 2006a. PHASE: A new design for pharmacophore perception, 3D QSAR model development, and 3D database screening: 1. Methodology and preliminary results. *Journal of Computer-Aided Molecular Design* 20:647–671.

Dixon, S. L., A. M. Smondyrev, and S. N. Rao. 2006b. Novel approach to pharmacophore modeling and 3D database searching. *Chemical Biology & Drug Design* 67:370–372.

Do, Q.-T., C. Lamy, I. Renimel, N. Sauvan, P. Andre, F. Himbert, L. Morin-Allory, and P. Bernard. 2007. Reverse pharmacognosy: Identifying biological properties for plants by means of their molecule constituents: Application to meranzin. *Planta Medica* 73:1235–1240.

Do, Q.-T., I. Renimel, P. Andre, C. Lugnier, C. D. Muller, and P. Bernard. 2005. Reverse pharmacognosy: Application of selnergy, a new tool for lead discovery. The example of e-viniferin. *Current Drug Discovery Technologies* 2:161–167.

Drews, J. 2003. Strategic trends in the drug industry. *Drug Discovery Today* 8:411–420.

Ehrman, T. M., D. J. Barlow, and P. J. Hylands. 2007a. Phytochemical databases of Chinese herbal constituents and bioactive plant compounds with known target specificities. *Journal of Chemical Information and Modeling* 47:254–263.

Ehrman, T. M., D. J. Barlow, and P. J. Hylands. 2007b. Phytochemical informatics of traditional Chinese medicine and therapeutic relevance. *Journal of Chemical Information and Modeling* 47:2316–2334.

Ehrman, T. M., D. J. Barlow, and P. J. Hylands. 2007c. Virtual screening of Chinese herbs with random forest. *Journal of Chemical Information and Modeling* 47:264–278.

Ekins, S., J. Mestres, and B. Testa. 2007. In silico pharmacology for drug discovery: Methods for virtual ligand screening and profiling. *British Journal of Pharmacology* 152:9–20.

Ertl, P., S. Roggo, and A. Schuffenhauer. 2008. Natural product-likeness score and its application for prioritization of compound libraries. *Journal of Chemical Information and Modeling* 48:68–74.

Ertl, P. and A. Schuffenhauer. 2007. Cheminformatics analysis of natural products: Lessons from nature inspiring the design of new drugs. In *Natural Products as Drugs*, eds. F. Petersen and R. Amstutz. Basel, Switzerland: Birkhaeuser Verlag.

Ewing, T. J. A., S. Makino, A. G. Skillman, and I. D. Kuntz. 2001. DOCK 4.0: Search strategies for automated molecular docking of flexible molecule databases. *Journal of Computer-Aided Molecular Design* 15:411–428.

Fakhrudin, N., A. Ladurner, A. G. Atanasov, E. H. Heiss, L. Baumgartner, P. Markt, D. Schuster, E. P. Ellmerer, G. Wolber, J. M. Rollinger, H. Stuppner, and V. M. Dirsch. 2010. Computer-aided discovery, validation and mechanistic characterisation of novel neolignan activators of PPARg. *Molecular Pharmacology* 77:559–566.

Feher, M. and J. M. Schmidt. 2003. Property distributions: Differences between drugs, natural products, and molecules from combinatorial chemistry. *Journal of Chemical Information and Computer Sciences* 43:218–227.

Ferrara, P., J. P. Priestle, E. Vangrevelinghe, and E. Jacoby. 2006. New developments and applications of docking and high-throughput docking for drug design and in silico screening. *Current Computer-Aided Drug Design* 2:83–91.

Filimonov, D. A. and V. V. Poroikov. 1996. PASS: Computerized prediction of biological activity spectra for chemical substances. In: *Bioactive Compound Design: Possibilites for Industrial Use*, eds. M. G. Ford, R. Greenwood, G. T. Brooks, and R. Franke, BIOS Scientific Publishers Inc., Oxford (U.K.), pp. 47–56.

Filimonov, D., V. Poroikov, Y. Borodina, and T. Gloriozova. 1999. Chemical similarity assessment through multilevel neighborhoods of atoms: Definition and comparison with the other descriptors. *Journal of Chemical Information and Computer Sciences* 39:666–670.

Franke, L., O. Schwarz, L. Mueller-Kuhrt, C. Hoernig, L. Fischer, S. George, Y. Tanrikulu, P. Schneider, O. Werz, D. Steinhilber, and G. Schneider. 2007. Identification of natural-product-derived inhibitors of 5-lipoxygenase activity by ligand-based virtual screening. *Journal of Medicinal Chemistry* 50:2640–2646.

Ganesan, A. 2008. The impact of natural products upon modern drug discovery. *Current Opinion in Chemical Biology* 12:306–317.

Goodford, P. J. 1985. A computational procedure for determining energetically favorable binding sites on biologically important macromolecules. *Journal of Medicinal Chemistry* 28:849–857.

Grabowski, K. and G. Schneider. 2007. Properties and architecture of drugs and natural products revisited. *Current Chemical Biology* 1:115–127.

Grienke, U., M. Schmidtke, J. Kirchmair, K. Pfarr, P. Wutzler, R. Durrwald, G. Wolber, K. R. Liedl, H. Stuppner, and J. M. Rollinger. 2010. Antiviral potential and molecular insight into neuraminidase inhibiting diarylheptanoids from *Alpinia katsumadai*. *Journal of Medicinal Chemistry* 53:778–786.

Gumy, C., C. Thurnbichler, E. M. Aubry, Z. Balazs, P. H. Pfisterer, L. Baumgartner, H. Stuppner, A. Odermatt, and J. M. Rollinger. 2009. Inhibition of 11beta-hydroxysteroid dehydrogenase type 1 by plant extracts used as traditional antidiabetic medicines. *Fitoterapia* 80:200–205.

Hadacek, F. 2002. Secondary metabolites as plant traits: Current assessment and future perspective. *Critical Reviews in Plant Sciences* 21:273–322.

Halgren, T. A., R. B. Murphy, R. A. Friesner, H. S. Beard, L. L. Frye, W. T. Pollard, and J. L. Banks. 2004. Glide: A new approach for rapid, accurate docking and scoring. 2. Enrichment factors in database screening. *Journal of Medicinal Chemistry* 47:1750–1759.

Harvey, A. L. 2008. Natural products in drug discovery. *Drug Discovery Today* 13:894–901.

Hawkins, P. C. D., A. G. Skillman, and A. Nicholls. 2007. Comparison of shape-matching and docking as virtual screening tools. *Journal of Medicinal Chemistry* 50:74–82.

Henkel, T., R. M. Brunne, H. Muller, and F. Reichel. 1999. Statistical investigation into the structural complementarity of natural products and synthetic compounds. *Angewandte Chemie, International Edition* 38:643–647.

Hert, J., P. Willett, D. J. Wilton, P. Acklin, K. Azzaoui, E. Jacoby, and A. Schuffenhauer. 2004. Comparison of topological descriptors for similarity-based virtual screening using multiple bioactive reference structures. *Organic & Biomolecular Chemistry* 2:3256–3266.

Hopkins, A. L. 2008. Network pharmacology: The next paradigm in drug discovery. *Nature Chemical Biology* 4:682–690.

Huang, Q., X. Qiao, and X. Xu. 2007. Potential synergism and inhibitors to multiple target enzymes of Xuefu Zhuyu Decoction in cardiac disease therapeutics: A computational approach. *Bioorganic & Medicinal Chemistry Letters* 17:1779–1783.

Jain, A. N. 2003. Surflex: Fully automatic flexible molecular docking using a molecular similarity-based search engine. *Journal of Medicinal Chemistry* 46:499–511.

Jones, G., P. Willett, R. C. Glen, A. R. Leach, and R. Taylor. 1997. Development and validation of a genetic algorithm for flexible docking. *Journal of Molecular Biology* 267:727–748.

Joosten, R. P., G. Chinea, G. J. Kleywegt, and G. Vriend. 2006. Protein three-dimensional structure validation. In *Comprehensive Medicinal Chemistry II*, eds. D. Triggle and J. Taylor. Amsterdam, the Netherlands: Elsevier.

Keiser, M. J., V. Setola, J. J. Irwin, C. Laggner, A. I. Abbas, S. J. Hufeisen, N. H. Jensen, M. B. Kuijer, R. C. Matos, T. B. Tran, R. Whaley, R. A. Glennon, J. Hert, K. L. H. Thomas, D. D. Edwards, B. K. Shoichet, and B. L. Roth. 2009. Predicting new molecular targets for known drugs. *Nature* 462:175–181.

Khan, M. T. H. and A. Ather. 2007. Potentials of phenolic molecules of natural origin and their derivatives as anti-HIV agents. *Biotechnology Annual Review* 13:223–264.

Kiefer, F., K. Arnold, M. Kuenzli, L. Bordoli, and T. Schwede. 2009. The SWISS-MODEL repository and associated resources. *Nucleic Acids Research* 37:D387–D392.

Kimura, I. 2006. Medical benefits of using natural compounds and their derivatives having multiple pharmacological actions. *Yakugaku Zasshi* 126:133–143.

Kirchmair, J., C. Laggner, G. Wolber, and T. Langer. 2005. Comparative analysis of protein-bound ligand conformations with respect to catalyst's conformational space subsampling algorithms. *Journal of Chemical Information and Modeling* 45:422–430.

Kirchmair, J., P. Markt, S. Distinto, D. Schuster, G. M. Spitzer, K. R. Liedl, T. Langer, and G. Wolber. 2008. The protein data bank (PDB), its related services and software tools as key components for in silico guided drug discovery. *Journal of Medicinal Chemistry* 51:7021–7040.

Kirchmair, J., G. Wolber, C. Laggner, and T. Langer. 2006. Comparative performance assessment of the conformational model generators omega and catalyst: A large-scale survey on the retrieval of protein-bound ligand conformations. *Journal of Chemical Information and Modeling* 46:1848–1861.

Klebe, G. 1998. Comparative molecular similarity indices analysis: CoMSIA. *Perspectives in Drug Discovery and Design* 12:87–104.

Klebe, G. 2009. *Wirkstoffdesign: Entwurf und Wirkung von Arzneistoffen*, 2nd edn. Heidelberg, Germany: Spektrum Akademischer Verlag.

Koch, M. A. and H. Waldmann. 2004. Natural product-derived compounds libraries and protein structure similarity as guiding principles for the discovery of drug candidates. In *Methods and Principles in Medicinal Chemistry*, eds. R. Mannhold, H. Kubinyi, and G. Folkers. Weinheim, Germany: Wiley-VHC.

Koehn, F. E. and G. T. Carter. 2005. The evolving role of natural products in drug discovery. *Nature Reviews Drug Discovery* 4:206–220.

Kola, I. and J. Landis. 2004. Opinion: Can the pharmaceutical industry reduce attrition rates? *Nature Reviews Drug Discovery* 3:711–716.

Korcsmaros, T., M. S. Szalay, C. Bode, I. A. Kovacs, and P. Csermely. 2007. How to design multi-target drugs: Target search options in cellular networks. *Expert Opinion on Drug Discovery* 2:799–808.

Kubinyi, H. 2006. Success stories of computer-aided design. In *Computer Applications in Pharmaceutical Research and Development*, eds. S. Ekins. New York: Wiley-Interscience.

Larsson, J., J. Gottfries, S. Muresan, and A. Backlund. 2007. ChemGPS-NP: Tuned for navigation in biologically relevant chemical space. *Journal of Natural Products* 70:789–794.

Laskowski, R. A. and G. J. Swaminathan. 2006. Problems of protein three-dimensional structures. In *Comprehensive Medicinal Chemistry II*, eds. D. Triggle and J. Taylor. Amsterdam, the Netherlands: Elsevier.

Lee, J.-Y., K.-W. Jeong, S. Shin, J.-U. Lee, and Y. Kim. 2009. Antimicrobial natural products as beta-ketoacyl-acyl carrier protein synthase III inhibitors. *Bioorganic & Medicinal Chemistry* 17:5408–5413.

Lei, J. and J. J. Zhou. 2002. A marine natural product database. *Journal of Chemical Information and Computer Science* 42:742–748.

Li, J. W. H. and J. C. Vederas. 2009. Drug discovery and natural products: End of an era or an endless frontier? *Science* 325:161–165.

Littleton, J., T. Rogers, and D. Falcone. 2005. Novel approaches to plant drug discovery based on high throughput pharmacological screening and genetic manipulation. *Life Sciences* 78:467–475.

Lokhande, T. N., C. L. Viswanathan, A. Joshi, and A. Juvekar. 2006. Design, synthesis and evaluation of naphthalene-2-carboxamides as reversal agents in MDR cancer. *Bioorganic & Medicinal Chemistry* 14:6022–6026.

Lopez, G., A. Rojas, M. Tress, and A. Valencia. 2007. Assessment of predictions submitted for the CASP7 function prediction category. *Proteins-Structure Function and Bioinformatics* 69:165–174.

Martin, Y. C., M. G. Bures, E. A. Danaher, J. DeLazzer, I. Lico, and P. A. Pavlik. 1993. A fast new approach to pharmacophore mapping and its application to dopaminergic and benzodiazepine agonists. *Journal of Computer-Aided Molecular Design* 7:83–102.

McChesney, J. D., S. K. Venkataraman, and J. T. Henri. 2007. Plant natural products: Back to the future or into extinction? *Phytochemistry* 68:2015–2022.

McGaughey, G. B., R. P. Sheridan, C. I. Bayly, J. C. Culberson, C. Kreatsoulas, S. Lindsley, V. Maiorov, J.-F. Truchon, and W. D. Cornell. 2007. Comparison of topological, shape, and docking methods in virtual screening. *Journal of Chemical Information and Modeling* 47:1504–1519.

Meisner, N.-C., M. Hintersteiner, V. Uhl, T. Weidemann, M. Schmied, H. Gstach, and M. Auer. 2004. The chemical hunt for the identification of drugable targets. *Current Opinion in Chemical Biology* 8:424–431.

Melville, J. L., E. K. Burke, and J. D. Hirst. 2009. Machine learning in virtual screening. *Combinatorial Chemistry & High Throughput Screening* 12:332–343.

Mestres, J. 2005. Representativity of target families in the protein data bank: Impact for family-directed structure-based drug discovery. *Drug Discovery Today* 10:1629–1637.

Morphy, R. and Z. Rankovic. 2009. Designing multiple ligands-medicinal chemistry strategies and challenges. *Current Pharmaceutical Design* 15:587–600.

Müller-Jahncke, W.-D. and C. Friedrich. 1996. *Geschichte der Arzneimitteltherapie*. Stuttgart, Germany: Dt. Apotekerverlag.

Newman, D. J. and G. M. Cragg. 2007. Natural products as sources of new drugs over the last 25 years. *Journal of Natural Products* 70:461–477.

Ogungbe, I. V. and W. N. Setzer. 2009. Comparative molecular docking of antitrypanosomal natural products into multiple *Trypanosoma brucei* drug targets. *Molecules* 14:1513–1536.

Ortuso, F., S. Alcaro, and T. Langer. 2006a. GRID-based pharmacophore models: Concept and application examples. In *Methods and Principles in Medicinal Chemistry*, eds. T. Langer and D. Hoffmann Remy. Weinheim, Germany: Wiley-VCH.

Ortuso, F., T. Langer, and S. Alcaro. 2006b. GBPM: GRID-based pharmacophore model: Concept and application studies to protein-protein recognition. *Bioinformatics* 22:1449–1455.

Overington, J. P., B. Al-Lazikani, and A. L. Hopkins. 2006. How many drug targets are there? *Nature Reviews Drug Discovery* 5:993–996.

Paoletta, S., G. B. Steventon, D. Wildeboer, T. M. Ehrman, P. J. Hylands, and D. J. Barlow. 2008. Screening of herbal constituents for aromatase inhibitory activity. *Bioorganic & Medicinal Chemistry* 16:8466–8470.

Paterson, I. and E. A. Anderson. 2005. The renaissance of natural products as drug candidates. *Science* 310:451–453.

Peitsch, M. C. 1996. SWISS-MODEL: An automated comparative protein modeling server and a model repository. *Folding & Design* 1 (Suppl.):48–49.

Perola, E. and P. S. Charifson. 2004. Conformational analysis of drug-like molecules bound to proteins: An extensive study of ligand reorganization upon binding. *Journal of Medicinal Chemistry* 47:2499–2510.

Poroikov, V. and D. Filimonov. 2005. PASS: Prediction of biological activity spectra for substances. In: *Predictive Toxicology*. ed. C. Helma, Boca Raton: Taylor & Francis, pp. 459–478.

Poroikov, V., D. Filimonov, A. Lagunin, T. Gloriozova, and A. Zakharov. 2007. PASS: Identification of probable targets and mechanisms of toxicity. *SAR and QSAR in Environmental Research* 18:101–110.

Potterat, O. and M. Hamburger. 2006. Natural products in drug discovery-concepts and approaches for tracking bioactivity. *Current Organic Chemistry* 10:899–920.

Rarey, M. and J. S. Dixon. 1998. Feature trees: A new molecular similarity measure based on tree matching. *Journal of Computer-Aided Molecular Design* 12:471–490.

Rarey, M., P. Fricker, S. Hindle, G. Metz, C. Rummey, and M. Zimmermann. 2006. Feature trees: Theory and applications from large-scale virtual screening to data analysis. In *Methods and Principles in Medicinal Chemistry*, eds. T. Langer and D. Hoffmann Remy. Weinheim, Germany: Wiley-VCH.

Rarey, M., B. Kramer, T. Lengauer, and G. Klebe. 1996. A fast flexible docking method using an incremental construction algorithm. *Journal of Molecular Biology* 261:470–489.

Rix, U. and G. Superti-Furga. 2009. Target profiling of small molecules by chemical proteomics. *Nature Chemical Biology* 5:616–624.

Robles, M., M. Aregullin, J. West, and E. Rodriguez. 1995. Recent studies on the zoopharmacognosy, pharmacology and neurotoxicology of sesquiterpene lactones. *Planta Medica* 61:199–203.

Rollinger, J. M. 2009. Accessing target information by virtual parallel screening-the impact on natural product research. *Phytochemistry Letters* 2:53–58.

Rollinger, J. M., S. Haupt, H. Stuppner, and T. Langer. 2004. Combining ethnopharmacology and virtual screening for lead structure discovery: COX-inhibitors as application example. *Journal of Chemical Information and Computer Sciences* 44:480–488.

Rollinger, J. M., D. V. Kratschmar, D. Schuster, P. H. Pfisterer, C. Gumy, E. M. Aubry, S. Brandstoetter, H. Stuppner, G. Wolber, and A. Odermatt. 2010. 11 Beta-hydroxysteroid dehydrogenase 1 inhibiting constituents from *Eriobotrya japonica* revealed by bioactivity-guided isolation and computational approaches. *Bioorganic & Medicinal Chemistry* 18:1507–1515.

Rollinger, J. M., T. Langer, and H. Stuppner. 2006a. Integrated in silico tools for exploiting the natural products' bioactivity. *Planta Medica* 72:671–678.

Rollinger, J. M., T. Langer, and H. Stuppner. 2006b. Strategies for efficient lead structure discovery from natural products. *Current Medicinal Chemistry* 13:1491–1507.

Rollinger, J. M., P. Mock, C. Zidorn, E. P. Ellmerer, T. Langer, and H. Stuppner. 2005. Application of the in combo screening approach for the discovery of non-alkaloid acetylcholinesterase inhibitors from *Cichorium intybus*. *Current Drug Discovery Technologies* 2:185–193 (Erratum: 2006, 3:89).

Rollinger, J. M., D. Schuster, B. Danzl, S. Schwaiger, P. Markt, M. Schmidtke, J. Gertsch, S. Raduner, G. Wolber, T. Langer, and H. Stuppner. 2009. In silico target fishing for rationalized ligand discovery exemplified on constituents of *Ruta graveolens*. *Planta Medica* 75:195–204.

Rollinger, J. M., T. M. Steindl, D. Schuster, J. Kirchmair, K. Anrain, E. P. Ellmerer, T. Langer, H. Stuppner, P. Wutzler, and M. Schmidtke. 2008. Structure-based virtual screening for the discovery of natural inhibitors for human rhinovirus coat protein. *Journal of Medicinal Chemistry* 51:842–851.

Rollinger, J. M., H. Stuppner, and T. Langer. 2008. Virtual screening for the discovery of bioactive natural products. *Progress in Drug Research* 65:211, 213–249.

Rose, V. S., I. F. Croall, and H. J. H. MacFie. 1991. An application of unsupervised neural network methodology (Kohonen topology-preserving mapping) to QSAR analysis. *Quantitative Structure-Activity Relationships* 10:6–15.

Rosen, J., J. Gottfries, S. Muresan, A. Backlund, and T. I. Oprea. 2009a. Novel chemical space exploration via natural products. *Journal of Medicinal Chemistry* 52:1953–1962.

Rosen, J., A. Loevgren, T. Kogej, S. Muresan, J. Gottfries, and A. Backlund. 2009b. ChemGPS-NPweb: Chemical space navigation online. *Journal of Computer-Aided Molecular Design* 23:253–259.

Roth, B. L., D. J. Sheffler, and W. K. Kroeze. 2004. Magic shotguns versus magic bullets: Selectively non-selective drugs for mood disorders and schizophrenia. *Nature Reviews Drug Discovery* 3:353–359.

Russell, R. J., L. F. Haire, D. J. Stevens, P. J. Collins, Y. P. Lin, G. M. Blackburn, A. J. Hay, S. J. Gamblin, and J. J. Skehel. 2006. The structure of H5N1 avian influenza neuraminidase suggests new opportunities for drug design. *Nature* 443:45–49.

Saarinen, J. and T. Kohonen. 1985. Self-organized formation of colour maps in a model cortex. *Perception* 14:711–719.

Schuster, D., E. M. Maurer, C. Laggner, L. G. Nashev, T. Wilckens, T. Langer, and A. Odermatt. 2006. The discovery of new 11beta -hydroxysteroid dehydrogenase type 1 inhibitors by common feature pharmacophore modeling and virtual screening. *Journal of Medicinal Chemistry* 49:3454–3466.

Schwede, T., J. Kopp, N. Guex, and M. C. Peitsch. 2003. SWISS-MODEL: An automated protein homology-modeling server. *Nucleic Acids Research* 31:3381–3385.

Sergeiko, A., V. V. Poroikov, L. O. Hanus, and V. M. Dembitsky. 2008. Cyclobutane-containing alkaloids: Origin, synthesis, and biological activities. *Open Medicinal Chemistry Journal* 2:26–37.

Soejarto, D. D., H. H. S. Fong, G. T. Tan, H. J. Zhang, C. Y. Ma, S. G. Franzblau, C. Gyllenhaal, M. C. Riley, M. R. Kadushin, J. M. Pezzuto, L. T. Xuan, N. T. Hiep, N. V. Hung, B. M. Vu, P. K. Loc, L. X. Dac, L. T. Binh, N. Q. Chien, N. V. Hai,

T. Q. Bich, N. M. Cuong, B. Southavong, K. Sydara, S. Bouamanivong, H. M. Ly, V. Thuy Tran, W. C. Rose, and G. R. Dietzman. 2005. Ethnobotany/ethnopharmacology and mass bioprospecting: Issues on intellectual property and benefit-sharing. *Journal of Ethnopharmacology* 100:15–22.

Stahura, F. L., J. W. Godden, L. Xue, and J. Bajorath. 2000. Distinguishing between natural products and synthetic molecules by descriptor Shannon entropy analysis and binary QSAR calculations. *Journal of Chemical Information and Computer Sciences* 40:1245–1252.

Steindl, T. M., D. Schuster, C. Laggner, K. Chuang, R. D. Hoffmann, and T. Langer. 2007. Parallel screening and activity profiling with HIV protease inhibitor pharmacophore models. *Journal of Chemical Information and Modeling* 47:563–571.

Steindl, T. M., D. Schuster, C. Laggner, and T. Langer. 2006a. Parallel screening: A novel concept in pharmacophore modeling and virtual screening. *Journal of Chemical Information and Modeling* 46:2146–2157.

Steindl, T. M., D. Schuster, G. Wolber, C. Laggner, and T. Langer. 2006b. High-throughput structure-based pharmacophore modelling as a basis for successful parallel virtual screening. *Journal of Computer-Aided Molecular Design* 20:703–715.

Stockwell, B. R. 2004. Exploring biology with small organic molecules. *Nature* 432:846–854.

Svetnik, V., A. Liaw, C. Tong, J. C. Culberson, R. P. Sheridan, and B. P. Feuston. 2003. Random forest: A classification and regression tool for compound classification and QSAR modeling. *Journal of Chemical Information and Computer Sciences* 43:1947–1958.

Tintori, C., V. Corradi, M. Magnani, F. Manetti, and M. Botta. 2008. Targets looking for drugs: A multistep computational protocol for the development of structure-based pharmacophores and their applications for hit discovery. *Journal of Chemical Information and Modeling* 48:2166–2179.

Vallat, B. K., J. Pillardy, and R. Elber. 2008. A template-finding algorithm and a comprehensive benchmark for homology modeling of proteins. *Proteins: Structure Function & Bioinformatics* 72:910–928.

Van de Waterbeemd, H. 2005. Which in vitro screens guide the prediction of oral absorption and volume of distribution? *Basic & Clinical Pharmacology & Toxicology* 96:162–166.

Van de Waterbeemd, H. and E. Gifford. 2003. ADMET in silico modelling: Towards prediction paradise? *Nature Reviews Drug Discovery* 2:192–204.

Venkatachalam, C. M., X. Jiang, T. Oldfield, and M. Waldman. 2003. LigandFit: A novel method for the shape-directed rapid docking of ligands to protein active sites. *Journal of Molecular Graphics and Modelling* 21:289–307.

Wade, R. C., K. J. Clark, and P. J. Goodford. 1993. Further development of hydrogen bond functions for use in determining energetically favorable binding sites on molecules of known structure. 1. Ligand probe groups with the ability to form two hydrogen bonds. *Journal of Medicinal Chemistry* 36:140–147.

Wade, R. C. and P. J. Goodford. 1993. Further development of hydrogen bond functions for use in determining energetically favorable binding sites on molecules of known structure. 2. Ligand probe groups with the ability to form more than two hydrogen bonds. *Journal of Medicinal Chemistry* 36:148–156.

Walters, W. P., M. T. Stahl, and M. A. Murcko. 1998. Virtual screening—An overview. *Drug Discovery Today* 3:160–178.

Wang, S. Q., Q. S. Du, K. Zhao, A. X. Li, D. Q. Wei, and K. C. Chou. 2007. Virtual screening for finding natural inhibitor against cathepsin-L for SARS therapy. *Amino Acids* 33:129–135.

Wang, L., G.-B. Zhou, P. Liu, J.-H. Song, Y. Liang, X.-J. Yan, F. Xu, B.-S. Wang, J.-H. Mao, Z.-X. Shen, S.-J. Chen, and Z. Chen. 2008. Dissection of mechanisms of Chinese medicinal formula Realgar-Indigo naturalis as an effective treatment for promyelocytic leukemia. *Proceedings of the National Academy of Sciences of the United States of America* 105:4826–4831.

Warren, G. L., C. W. Andrews, A.-M. Capelli, B. Clarke, J. LaLonde, M. H. Lambert, M. Lindvall, N. Nevins, S. F. Semus, S. Senger, G. Tedesco, I. D. Wall, J. M. Woolven, C. E. Peishoff, and M. S. Head. 2006. A critical assessment of docking programs and scoring functions. *Journal of Medicinal Chemistry* 49:5912–5931.

Waterman, P. G. 1992. Roles for secondary metabolites in plants. *Ciba Foundation Symposium* 171:255–275.

Wermuth, C.-G., C. R. Ganellin, P. Lindberg, and L. A. Mitscher. 1998. Glossary of terms used in medicinal chemistry (IUPAC recommendations 1997). *Annual Report of Medicinal Chemistry* 33:385–395.

Wetzel, S., A. Schuffenhauer, S. Roggo, P. Ertl, and H. Waldmann. 2007. Cheminformatic analysis of natural products and their chemical space. *Chimia* 61:355–360.

White, S. H. 2004. The progress of membrane protein structure determination. *Protein Science* 13:1948–1949.

Wolber, G., A. A. Dornhofer, and T. Langer. 2006. Efficient overlay of small organic molecules using 3D pharmacophores. *Journal of Computer-Aided Molecular Design* 20:773–788.

Wolber, G. and R. Kosara. 2006. Pharmacophores and pharmacophore searches. In *Methods and Principles in Medicinal Chemistry*, eds. T. Langer and D. Hoffmann Remy. Weinheim, Germany: Wiley-VCH.

Wolber, G. and T. Langer. 2001. Comb(i)Gen: A novel software package for the rapid generation of virtual combinatorial libraries. In: *Rational Approaches to Drug Design*, eds. H.-D. Höltje and W. Sippl. Prous Science, Barcelona, Spain, pp. 390–399.

Wolber, G. and T. Langer. 2005. LigandScout: 3-d pharmacophores derived from protein-bound ligands and their use as virtual screening filters. *Journal of Chemical Information and Modeling* 45:160–169.

Zupan, J. and J. Gasteiger. 2000. *Neural Networks for Chemists*. 2nd edn. Weinheim, Germany: Wiley-VCH.

5 Recent Advances in the Application of Electronic Circular Dichroism for Studies of Bioactive Natural Products

Nina Berova, George Ellestad,
Koji Nakanishi, and Nobuyuki Harada

CONTENTS

5.1   INTRODUCTION

Chiral molecular structures can exist in two enantiomeric forms of opposite absolute configuration (AC). Most bioactive natural products are chiral, and their biological activity is closely related to their chirality, that is, absolute configuration and conformation. Therefore, the unambiguous determination of the absolute configuration of chiral compounds is essential for the complete characterization of natural products and their interaction with biomolecular systems (Mason 1982).

Circular dichroism (CD) spectroscopy is one of the most useful methods for determining the absolute configurations of bioactive natural products and for obtaining insight into their chiral intra- and intermolecular interactions. In this chapter, the basic principles of electronic CD and applications to various bioactive natural products are reviewed. Electronic circular dichroism (ECD) is characterized by Cotton effects classified into several groups depending on the chromophoric geometry and their electronic properties. Empirical and nonempirical methods have been established that govern the relationship between the sign of the Cotton effects and absolute configuration. Among them, the CD exciton chirality method is the most useful for determining the absolute configurations of natural products in a nonempirical manner and the principles of this method are briefly explained. Another useful method for determining absolute configurations of natural products is the ab initio molecular orbital (MO) calculations of ECD and optical rotation (OR) in which the theoretically derived curves are compared with experimental ones (Autschbach 2010).

Moreover, ECD is a very sensitive and useful spectroscopic tool not only for determining absolute configuration and conformation, but also for monitoring chiral intermolecular interactions (Eliel et al. 1994, Berova et al. 2000, 2007). In the following, the basic issues of ECD spectroscopy including optical rotation and the most useful ECD methodologies for determining the absolute configurations of chiral compounds are presented together with some recent ECD applications to a variety of bioactive natural products.

This chapter will review some more recent applications by circular dichroism spectroscopy in the UV/Vis region, called electronic circular dichroism (ECD), for the determination of ACs of natural products as well as for studying the interactions of natural products with target biopolymers, for example, DNA and proteins.

5.2   MEASUREMENT OF OPTICAL ROTATION $[\alpha]_D^t$

To specify the enantiomer of a natural product, it is important to report the data of optical rotation. The specific rotation $[\alpha]_\lambda^t$ is defined as follows:

$$[\alpha]_\lambda^t = \frac{100\alpha}{(lc)} \quad \text{(solution)} \qquad (5.1)$$

$$[\alpha]_\lambda^t = \frac{\alpha}{(l\rho)} \quad \text{(liquid)} \qquad (5.2)$$

where
   λ is the wavelength of light (nm)
   t is the temperature (°C)
   α is the observed rotation angle (degree)
   l is the cell length [dm (=10 cm)]
   c is the concentration (sample weight, g/100 cm³ solution)
   ρ is the density of liquid (g/cm³)

Usually the specific rotation at Na-D line (589 nm), $[\alpha]_D^t$, is used and expressed as follows. For example,

$$[\alpha]_D^{20} + 52.8\,(c\ 0.325,\ \text{EtOH})$$

where it is a general rule that concentration c is expressed in the unit of (g solute/100 cm^3 solution).

5.3   MEASUREMENT OF CD SPECTRA

CD is caused by the difference in the absorption intensity for left- and right-circularly polarized lights, and is observed as the ellipticity angle θ. CD is generally expressed by molar ellipticity $[\theta]$ or molar CD $\Delta\varepsilon$.

$$[\theta] = \frac{\theta M}{(lc)} \tag{5.3}$$

where
  θ is the observed ellipticity angle (degree = °)
  M is the molecular weight
  l is the cell length (dm = 10 cm)
  c is the concentration (g solute/100 cm^3 solution)

In the case of organic compounds, molar CD $\Delta\varepsilon$ (dm^3 mol^{-1} cm^{-1} or M^{-1} cm^{-1}) is preferentially used, and $\Delta\varepsilon$ is the difference between the absorption coefficient ε_l for left-circularly polarized light and coefficient ε_r for right-circularly polarized light.

$$\Delta\varepsilon = \varepsilon_l - \varepsilon_r \tag{5.4}$$

Molar ellipticity $[\theta]$ is correlated to molar CD $\Delta\varepsilon$ as follows:

$$[\theta] = 3300\Delta\varepsilon \tag{5.5}$$

In practice, the next equation is useful for obtaining molar CD $\Delta\varepsilon$

$$\frac{\theta}{33} = \Delta A = \Delta\varepsilon \times c' \times l' \tag{5.6}$$

where
  ΔA is the CD absorbance
  c' is the molar concentration (mol solute/dm^3 solution)
  l' is the cell length (cm)

The intensity scale (or sensitivity) of a CD spectropolarimeter is generally expressed in millidegrees (m°)/cm, and, therefore, molar CD $\Delta\varepsilon$ is calculated as follows:

$$\text{sensitivity (m°/cm)} \times \text{CD signal (cm)}/33{,}000 = \Delta A = \Delta\varepsilon \times c' \times l' \tag{5.7}$$

Using Equation 5.6 or 5.7, $\Delta\varepsilon$ values are calculated and plotted against wavelength λ. For the CD Cotton effects, the wavelength at an extremum (λ_{ext}) and the intensity ($\Delta\varepsilon$) are described as follows. For example,

CD (EtOH) λ_{ext} 320.5 nm ($\Delta\varepsilon$ − 63.1), 295.5 (+39.7)

To calibrate the CD spectropolarimeter, the following standard samples, conditions, and CD data are suggested:

androsterone: solution (c 0.0500 g/100 cm^3 solution in 1,4-dioxane), λ_{ext} 304 nm ($\Delta\varepsilon$ + 3.39).

(−)-pantolactone: solution (c 0.0150 g/100 cm^3 solution in water), λ_{ext} 219 nm ($\Delta\varepsilon$ − 5.00).

(+)-10-camphorsulfonic acid ammonium salt: solution (c 0.0600 g/100 cm^3 solution in water, λ_{ext} 290.5 nm ($\Delta\varepsilon$ + 2.40).

5.4   CD SPECTRA AND ROTATIONAL STRENGTH

The CD spectra are characterized by positive and/or negative Cotton effects, the sign and intensity of which are governed by the rotational strength R. By using Equation 5.8, the R value is experimentally obtainable from the observed CD spectra (Harada and Nakanishi 1983, Lightner and Gurst 2000, Berova et al. 2007)

$$R = 2.296 \times 10^{-39} \int \frac{\Delta\varepsilon(\sigma)}{\sigma} \, d\sigma \text{ cgs unit} \tag{5.8}$$

where σ is the wavenumber (cm^{-1}).

The rotational strength is theoretically expressed by Equation 5.9.

$$R = \text{Im}\left\{ <0|\boldsymbol{\mu}|a> \cdot <a|\boldsymbol{M}|0> \right\} \tag{5.9}$$

where
   Im denotes the imaginary part of the terms in brackets {}
   < > denotes the integration over configurational space
   $\boldsymbol{\mu}$ and $\boldsymbol{M}$ are operators of electric and magnetic moment vectors, respectively

The dot · stands for the scalar product of two vectors, and 0 and a are wavefunctions of ground and excited states, respectively. The rotational strength R is thus equal to the imaginary part of the scalar product of electric and magnetic transition moments. If the electric and magnetic moment vectors, <0|$\boldsymbol{\mu}$|a> and <a|$\boldsymbol{M}$|0>, are parallel to each other, the rotational strength R is positive, leading to a positive Cotton effect. On the other hand, if they are antiparallel, the rotational strength R is negative and a negative CD Cotton effect is observed.

5.5   MOST USEFUL CD METHODS FOR DETERMINING ABSOLUTE CONFIGURATIONS OF NATURAL PRODUCTS

The CD methods for the absolute configurational assignment of natural products are grouped as shown in Figure 5.1.

1. *Comparison of CD*: The absolute configuration of chiral compounds can be determined by comparison of CD with that of an authentic sample or with that of a closely related compound. Since CD is generally much more sensitive than specific rotation $[\alpha]_D$, it is useful for the microgram scale determination of absolute configuration (see examples in Section 5.7.1).

2. *Inherently chiral chromophores and CD* (Eliel et al. 1994): As shown in the conjugated diene and enone (Figure 5.1a), the chromophore itself is twisted and chiral. Therefore, the $\pi-\pi^*$ and/or $n-\pi^*$ transitions generate CD Cotton effects reflecting the helicity of the chromophore. The CD intensity of this group is moderate.

3. *Exciton-coupled CD* (Harada and Nakanishi 1983, Berova and Nakanishi 2000): If a system has two chromophores exhibiting intense transitions in chiral positions (Figure 5.1b), the exciton coupling between two chromophores generates bisignate CD Cotton effects reflecting the helicity between two electric transition moments. The CD Cotton effects of this group are generally much stronger than others, and the CD generation mechanism is very clear and theoretically established as described in Section 5.6. Therefore, the CD exciton chirality method is applicable without any computer calculation, and, hence, it has been successfully applied to various natural products

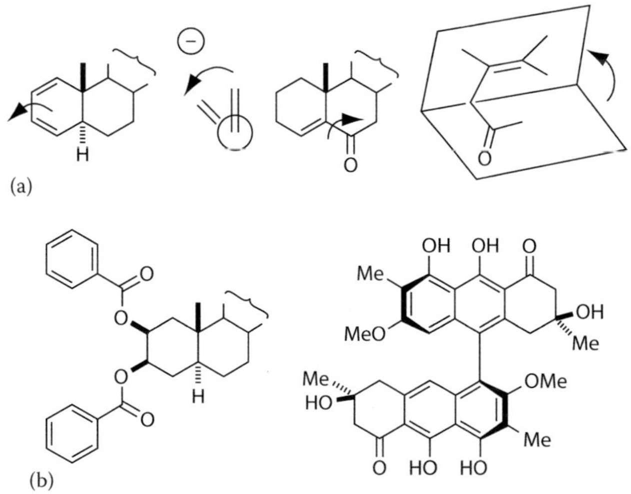

(a)

(b)

FIGURE 5.1   CD methods and Cotton effects useful for the determination of the absolute configuration of natural products. (a) Inherently chiral chromophores and helicity rule. (b) Exciton coupling systems with two or more chromophores showing intense exciton split CD.

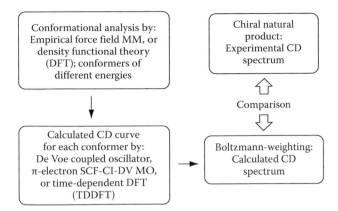

FIGURE 5.2 Procedure for the determination of absolute configuration by the theoretical calculation of CD spectra.

for determining their absolute configurations. When determining the ACs by CD spectroscopy, it is advisable to take advantage of the large Cotton effects of this group, even in the cases of theoretical calculations.

4. *Theoretical simulation of CD*: The calculation procedure is outlined in Figure 5.2. The conformational analysis of a natural product in question is carried out by the empirical force field method, for example, molecular mechanics (MM or MMFF94), and/or by the density functional theory (DFT) calculation giving several stable conformations and their relative energies.

For each conformation obtained, the CD spectral curve is calculated by several methods, such as (i) the exciton chirality CD method, (ii) DeVoe calculation, (iii) π-electron SCF-CI-DV method, and (iv) ab initio MO method as described below. In the last method, the TDDFT (time-dependent DFT) calculation is very useful for the simulation of ECD spectra of chiral compounds. Based on the Boltzmann-distribution, the population of each stable conformer is obtained, and total averaged CD curve is calculated by Boltzmann-weighting. The comparison of observed and calculated CD spectral curves leads to the determination of the absolute configuration of the natural product.

If rotational strength R defined by Equation 5.9 can be calculated by quantum mechanical theory, CD spectra can be simulated using the Gaussian curve (Stephens and Harada 2010) as shown in Equation 5.10:

$$\Delta\varepsilon(\sigma) = \left\{ \frac{1}{\left(2.296 \times 10^{-39}\sqrt{\pi}\Delta\sigma\right)} \right\} \sum_a \sigma_a R_a \exp\left[-\left\{ \frac{(\sigma - \sigma_a)}{\Delta\sigma} \right\}^2 \right] \qquad (5.10)$$

where $\Delta\sigma$ is half the bandwidth at $1/e$ peak height of the Gaussian curve and expressed in cm^{-1} units. The parameters σ_a and R_a are the excitation wavenumber and rotational strength for the transition from ground state to excited state a, respectively. So the comparison of both calculated and observed spectra leads to the determination of ACs.

For the calculation of rotational strength R, there are several theoretical methods as follows:

1. *CD exciton chirality method* (Harada and Nakanishi 1983, Berova and Nakanishi 2000): This is the simplest and most reliable CD method applicable to various natural products because the exciton-coupled CD is based on the coupled oscillator theory and the mechanism of this method has already been established as will be briefly explained in the following section. The rotational strength R and the excitation energy E are easily calculable by Equations 5.11 through 5.14 as shown in Section 5.6. Although the method is applicable without any numerical calculation, the computer-assisted CD simulation of course supports the qualitative absolute configurational assignment.

2. *DeVoe calculation* (DeVoe 1964, 1965, Superchi et al. 2004): This is a simple method based on the coupled oscillator theory, which is applicable to more complex chiral molecules composed of two or more groups.

3. *π-Electron SCF-CI-DV MO* (*Self Consistent Field–Configuration Interaction–Dipole Velocity–Molecular Orbital*) *Method* (Kemp and Mason 1966, Harada and Nakanishi 1983, Harada 1999, 2000): This is an MO method with π-electron approximation, which is applicable to chiral molecules of twisted π-electron systems. As this method treats only π-electrons, computation time is much shorter than the cases treating all electrons.

4. *Ab initio MO calculations* (Crawford 2006, Bringmann et al. 2009, Autschbach 2010). Recent years have seen great progress in the development of first-principle calculations of chiroptical properties. The development of ab initio methodologies, which include Hartree–Fock, DFT, as well as high-level correlation methods, such as coupled cluster theory, have enabled one to derive theoretical simulations of CD, optical rotation, and other chiroptical properties at very high levels of sophistication. Since these methods treat all electrons, including σ-electrons, a high-speed, large computer and long computational times are necessary, especially for calculations on conformationally flexible molecules. Some pertinent examples are shown in Section 5.7.4.

5.6 CD EXCITON CHIRALITY METHOD

The CD exciton chirality method has been successfully applied to a variety of natural products to determine their absolute configurations. This method enables one to deduce the absolute configuration of a chiral compound without any reference compound, and, therefore, it is established as a nonempirical method. The principles of the CD exciton chirality method are explained using the steroidal bis(*p*-dimethylaminobenzoate) shown below as a model compound, where the nonempirical nature of this method is easily understood (Figure 5.3) (Harada and Nakanishi 1983, Berova and Nakanishi 2000, Berova et al. 2000).

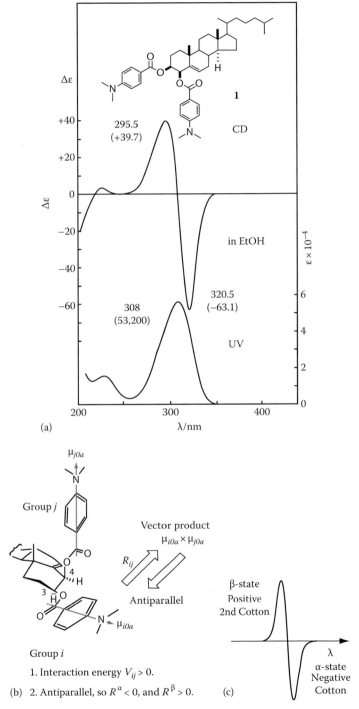

FIGURE 5.3 Application of the CD exciton chirality method to cholest-5-ene-3β,4β-diol bis(*p*-dimethylaminobenzoate) **1**: CD and UV spectra in EtOH.

As exemplified with cholest-5-ene-3β,4β-diol bis(p-dimethylaminobenzoate) **1** in Figure 5.3, when two identical chromophores (i and j) that exhibit intense UV absorption of their π–π* transition (ground state 0 → excited state a) exist in a molecule, these two chromophores interact with each other and the excited state splits into two energy levels (α and β states) (Harada and Nakanishi 1983). This phenomenon is called exciton coupling or exciton interaction. The energy E and rotational strength R of α and β states are expressed as follows:

α-State:

$$E^{\alpha} = E_a - V_{ij} \tag{5.11}$$

$$R^{\alpha} = +\left(\frac{1}{2}\right)\pi\sigma_0\, \boldsymbol{R}_{ij} \cdot (\boldsymbol{\mu}_{i0a} \times \boldsymbol{\mu}_{j0a}) \tag{5.12}$$

β-State:

$$E^{\beta} = E_a + V_{ij} \tag{5.13}$$

$$R^{\beta} = -\left(\frac{1}{2}\right)\pi\sigma_0\, \boldsymbol{R}_{ij} \cdot (\boldsymbol{\mu}_{i0a} \times \boldsymbol{\mu}_{j0a}) \tag{5.14}$$

where
 V_{ij} is the interaction energy between two electric transition moments $\boldsymbol{\mu}_{i0a}$ and $\boldsymbol{\mu}_{j0a}$
 σ_0 is the transition wavenumber
 $\boldsymbol{R}_{ij}$ is the distance vector from chromophore i to j
 × denotes the vector product

Equations 5.12 and 5.14 indicate that R^{α} is opposite in sign to R^{β}, but their absolute values are equal to each other. These equations are next applied to steroidal dibenzoate **1** in Figure 5.3 (Harada and Nakanishi 1983).

For the two electric transition moments $\boldsymbol{\mu}_{i0a}$ and $\boldsymbol{\mu}_{j0a}$ in the benzoate chromophores (Figure 5.3b), the interaction energy V_{ij} becomes positive, and, therefore, the α-state is lower in energy than the β-state. Two vectors $\boldsymbol{\mu}_{i0a}$ and $\boldsymbol{\mu}_{j0a}$ constitute a counterclockwise screw, and so the resultant vector $\boldsymbol{\mu}_{i0a} \times \boldsymbol{\mu}_{j0a}$ is antiparallel to the distance vector $\boldsymbol{R}_{ij}$. Therefore, the triple product $\boldsymbol{R}_{ij} \cdot (\boldsymbol{\mu}_{i0a} \times \boldsymbol{\mu}_{j0a})$ becomes negative, and so R^{α} is negative while R^{β} is positive. This result leads to the CD spectral pattern as shown in Figure 5.3c, where the Cotton effect at longer wavelength (named 1st Cotton effect) is negative and that at a shorter wavelength (2nd Cotton effect) is positive. These exciton-coupled CD Cotton effects with signs opposite to each other are called "bisignate Cotton effects."

The UV spectrum of **1** shows an intense π–π* absorption band (λ_{max} 308 nm, ε 53,200), which is polarized along the long axis of the chromophore (Figure 5.3a). The CD spectrum shows negative 1st and positive 2nd Cotton effects in agreement with the theoretical conclusion: 1st Cotton effect, λ_{ext} 320.5 nm, Δε − 63.1, and 2nd one λ_{ext} 295.5 nm, Δε + 39.7. The amplitude of the exciton CD is defined as

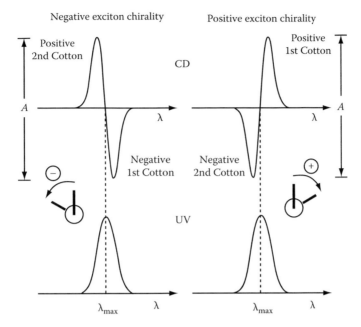

FIGURE 5.4   Typical pattern of exciton coupled CD Cotton effects and UV absorption band. In general, the CD zero-crossing point corresponds to λ_{max} of UV band.

$A = \Delta\varepsilon_1 - \Delta\varepsilon_2$, where $\Delta\varepsilon_1$ and $\Delta\varepsilon_2$ are $\Delta\varepsilon$ values of 1st and 2nd Cotton effects, respectively. In the case of dibenzoate **1**, $A = -102.8$. From these results, one can unambiguously determine the AC of the original glycol.

From the above theoretical results, the following CD exciton chirality rule is obtained:

1. *If the long axes of two interacting chromophores constitute a clockwise screw sense, the CD shows a positive 1st Cotton effect at a longer wavelength and a negative 2nd Cotton effect at a shorter wavelength (Figure 5.4).*
2. *If they make a counterclockwise screw sense, a negative 1st Cotton effect at a longer wavelength and a positive 2nd Cotton effect at a shorter wavelength are observed.*

From the mechanism of exciton coupled CD, some important features are derived.

1. The intensity of the exciton CD (*A*-value) is inversely proportional to the square of the interchromophoric distance R_{ij} provided the remaining angular part is the same (Harada and Nakanishi 1983).

$$A \propto R_{ij}^{-2} \tag{5.15}$$

2. The *A*-value of exciton split CD is the function of the dihedral angle between two transition moments. In the case of vicinal glycol dibenzoates, the sign of the exciton split Cotton effects remains unchanged from 0° to 180°.

Therefore, the qualitative definition shown in Figure 5.4 is applicable to a dibenzoate with the dihedral angle of more than 90°. The maximum A-value is around 70° (Harada and Nakanishi 1983).

3. The A-value is proportional to the square of absorption coefficient ε of the chromophore. Therefore, it is advisable to use chromophores undergoing intense $\pi-\pi^*$ transition.

4. For the exciton coupling systems with three or more chromophores, it was experimentally found that the so-called additivity rule holds. For example, for a trimer,

$$A(\text{total}) = A(1,2) + A(1,3) + A(2,3) \tag{5.16}$$

where $A(1,2)$, $A(1,3)$, and $A(2,3)$ are the A-values of component pairs.

5.7 APPLICATIONS OF ECD FOR DETERMINING THE ABSOLUTE CONFIGURATION OF SECONDARY METABOLITES FROM PLANTS, MARINE ORGANISMS, AND MICROORGANISMS

5.7.1 COMPARISON OF CD SPECTRA

5.7.1.1 Absolute Configuration of Cytotoxic Metabolites Produced by Bacteria Obtained from Marine Fish

New cytotoxic metabolites containing a chlorine atom, chaetomugilins A (**2**), B (**3**), and D (**4**) were isolated from a strain of *Chaetomium globosum*, which was originally obtained from the marine fish, *Mugil cephalus* (Figure 5.5) (Yamada et al. 2008, Yasuhide et al. 2008).

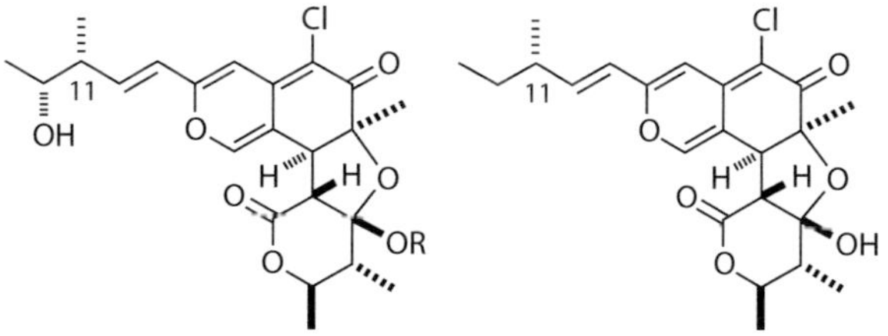

Chaetomugilin A (**2**), R = H
CD data
λ_{ext} 341 nm, $\Delta\varepsilon$ −2.0
291 nm, $\Delta\varepsilon$ +1.6
272 nm, $\Delta\varepsilon$ −3.2

Chaetomugilin B (**3**), R = Me

Absolute stereostructure
of (**3**) by x-ray:
Flack parameter = 0.1 (2)

Chaetomugilin D (**4**)
CD data
λ_{ext} 347 nm, $\Delta\varepsilon$ −1.1
291 nm, $\Delta\varepsilon$ +2.8
272 nm, $\Delta\varepsilon$ −6.4

FIGURE 5.5 Absolute stereostructures of chaetomugilins and CD spectral data. CD numerical data were taken from Yasuhide et al. (2008). (Data from Yasuhide, M. et al., *J. Antibiot.*, 61, 615, 2008.)

The relative stereostructures of **2** and **3** were determined by NMR spectroscopy including ¹H–¹H COSY, DEPT, HMQC, HMBC, and NOESY experiments, and then confirmed by x-ray crystallography of **3**. Since the compound has a chlorine atom, its absolute configuration was determined as shown by the Flack parameter = 0.1 (2). To determine the absolute configuration of hemiketal **2**, it was treated with *p*-TsOH in MeOH yielding methyl-ether **3**, and, therefore, the absolute configuration of **2** was established as shown.

The relative stereostructure of **4** was determined by NMR spectroscopy including ¹H–¹H COSY, DEPT, HMQC, HMBC, and NOESY experiments, which indicated that the relative configuration of **4**, except at C-11, is the same as that of **2**. Since the CD spectra of both compounds show similar Cotton effects as shown in Figure 5.5, they both have the same absolute configuration, except for C-11.

To determine the remaining absolute configuration at C-11, compound **4** was oxidized with chromium trioxide yielding (*S*)-2-methylbutanoic acid, which was identified by the comparison of spectral data and specific rotation with the authentic sample. The absolute stereostructure of **4** was thus determined as shown.

5.7.1.2   Rapid Detection of (*R*)-Gossypol by HPLC-Online ECD

The use of online ECD in combination with integrated, or hyphenated, high-performance liquid chromatography-UV-MS and NMR has become a useful tool in the search for new drugs from natural products from crude plant extracts. Recently, this new methodology has been used to identify the levorotatory atropisomer of gossypol, (*R*)-(–)-**5**, in enantiomeric excess of 44% [72:28 ratio between (*R*)-(–)-**5** and (*S*)-(+)-**5**], from extracts of *Thespesia danis* (Figure 5.6) (Bringman et al. 2008, Sprogoe et al. 2008, Clayden et al. 2009).

The optical purity, that is, percent ee value, was determined from the intensity of the long wavelength negative CD band at 377 nm. This is an important finding because the (*R*)-(–) atropisomer is the biologically active component for evaluation in anticancer clinical trials and for studying its high enantioselectivity in vitro antimalarial activity. Up till now, only *Gossypium barbadense* and a few other cotton varieties have produced the levorotatory isomer, but in enantiomeric excess of only 30%–35%. Thus, the use of CD combined with HPLC for the online determination of concentration and chirality of natural products is becoming an important tool for the identification of biologically active agents from natural sources. The online characterization and identification of atropisomeric gossypol by HPLC and CD coupling demonstrates the potential for the detection of single enantiomer drugs from crude plant extracts.

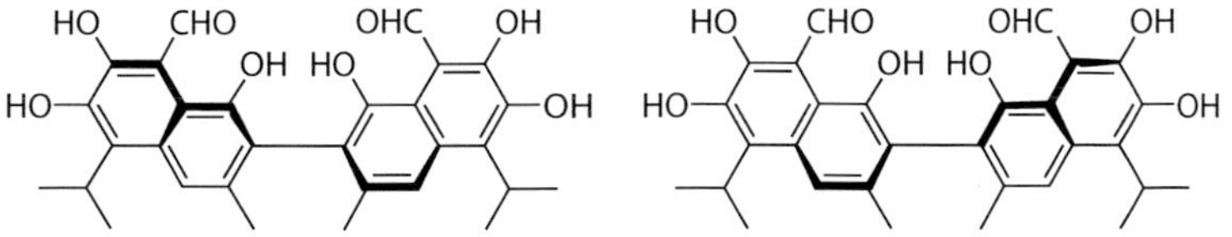

(*M*)-(*R*)-(–)-(**5**)   (*P*)-(*S*)-(+)-(**5**)

FIGURE 5.6   Absolute configurations of gossypol enantiomers (–)-(**5**) and (+)-(**5**).

5.7.2   DIENE HELICITY RULE

5.7.2.1   Absolute Configuration of Metabolites from a Marine Fungus by the Diene Helicity Rule

Novel hybrid polyketide-terpenoid metabolites, miniolutelides A (+)-(**6**) and B (−)-(**7**), were isolated from the marine-derived fungus (*Penicillium minioluteum*) (Iida et al. 2008, Ookura et al. 2008). The complex structure and relative configuration of compound **6** were determined by NMR spectroscopy and then confirmed by x-ray crystallography. To determine their absolute configurations, the CD diene helicity rule was applied as follows. The CD spectrum of **6** shows an intense positive Cotton effect around 267 nm (λ_{ext} = 267 nm, $\Delta\varepsilon$ + 18.6) due to the $\pi-\pi^*$ transition of a conjugated diene-lactone chromophore. Since the conjugated diene part has the positive helicity of + 165° in the x-ray stereostructure, the absolute configuration shown in Figure 5.7 was assigned to metabolite **6**.

Another metabolite **7** has a similar structure containing a conjugated diene–lactone chromophore, but its CD spectrum shows an intense negative Cotton effect around 267 nm (λ_{ext} = 267 nm, $\Delta\varepsilon$ − 21.3). It is interesting that despite the similar structure and absolute configuration, the CD spectrum of **7** is almost opposite in sign to that of **6**. To explain the CD data, the stable conformation of **7** was calculated by the MM2 (molecular mechanics 2) method, leading to the negative helicity of conjugated diene moiety (diene helicity −146°). The Cotton effects of opposite sign are thus explained by the conformational difference.

5.7.3   EXCITON CHIRALITY METHOD

5.7.3.1   Determination of the Absolute Stereochemistry of Spiroleptosphol from Ascomycetous Fungus

A novel γ-methylidene-spirobutanolide, spiroleptosphol (**8**) and its biosynthetic congeners were isolated from ascomycetous fungus (*Leptosheria doliolum*), and compound **8** exhibited cytotoxicity against P388 murine leukemia and HeLa human cervix carcinoma (Hashimoto et al. 2008, Murakami et al. 2008). The relative stereostructure of compound **8** was determined by NMR spectroscopy including NOE measurements.

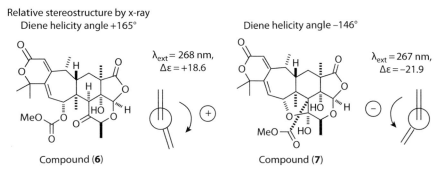

FIGURE 5.7   Absolute stereostructures of miniolutelides (A) **6** and (B) **7**, and their CD spectra.

The relative stereostructure of spiroleptosphol C (**9**), the 17-hydroxy-spiroleptosphol analog of **8**, was recently established by x-ray crystallography (Murakami et al. 2009), thus confirming the relative structure of **8** determined by NMR.

To determine its absolute configuration, compound **8** was hydrogenated giving hexahydro-derivative **10**, which was then converted to dibenzoate **11**. The CD spectrum of **11** shows typical exciton Cotton effects (λ_{ext} = 238 nm, $\Delta\varepsilon$ −18.9; λ_{ext} = 222 nm, $\Delta\varepsilon$ + 10.9) of negative chirality leading to a counterclockwise screw sense between two benzoate chromophores at the 6- and 7-positions, that is, diol moiety (Figure 5.8). The absolute configuration of **8** was thus unambiguously determined as shown.

The natural triol **8** was also converted to tribenzoate **12**, the CD spectrum of which shows intense positive 1st and negative 2nd Cotton effects (λ_{ext} = 238 nm, $\Delta\varepsilon$ + 38.4: λ_{ext} = 220 nm, $\Delta\varepsilon$ − 18.1; read from the reported figure) together with a negative one (λ_{ext} = 204 nm, $\Delta\varepsilon$ − 6.4: read from the reported figure). These CD data are interpreted as follows: the exciton chirality between the two benzoate chromophores at the 4- and 6-positions is clockwise, and that between 4- and 7-benzoates is also clockwise, while that between 4- and 6-benzoates is counterclockwise. The total exciton chirality is thus positive, and leads to positive and negative 2nd Cotton effects, confirming the absolute configuration of triol **8**.

In the CD spectrum of tribenzoate **12**, the negative, shorter wavelength exciton Cotton effect at 220 nm is overlapped with another negative band around 204 nm

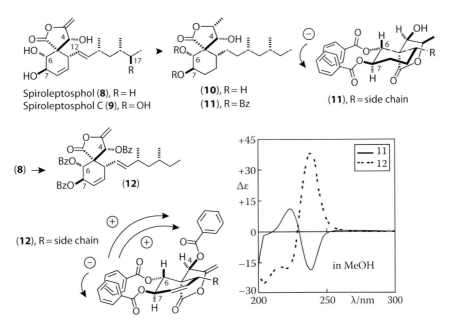

FIGURE 5.8 Determination of the absolute stereochemistry of spiroleptosphol (**8**) by the CD exciton chirality method. CD spectra. (Adapted from Murakami, T. et al., Isolation and structure of spiroleptosphol, in *Symposium Papers, 50th Symposium on the Chemistry of Natural Products*, pp. 275–280, Fukuoka, Japan, 2008. With permission. Copyright 2008 Symposium on the Chemistry of Natural Products.)

leading to an unclear pattern of the exciton coupling CD. To prevent the overlap and to obtain the unambiguous consequence of the absolute configuration, it is important to observe the typical bisignate exciton Cotton effects, and so it is advisable to use the *p*-substituted benzoate such as *p*-bromobenzoate (UV λ_{max} = 244.5 nm) or *p*-methoxybenzoate (UV λ_{max} = 257.0 nm) instead of the unsubstituted benzoate.

5.7.3.2   Absolute Configuration of α,ω-Bifunctionalized Sphingolipid Leucettamol A by Exciton-Coupled CD

This is an especially cogent example showing the power of CD exciton chirality method for determining the absolute configuration of the α,ω-bifunctionalized sphingolipid leucettamol A (**13**) using microgram amounts of material (Figure 5.9). Leucettamol A was isolated from a marine sponge *Leucetta microrhaphis* as a natural product exhibiting a variety of biological activities (antibacterial, cytotoxicity). The relative configuration of the 2-amino-3-hydroxy end groups was determined to be *erythro* by NOE enhancements in the bis-oxazolone derivative. Compound **13** was originally thought to be racemic because it did not exhibit any measureable optical rotation (Kong and Faulkner 1993). However, long-chain amino alcohols often show low optical rotations. In addition, the biosynthesis of **13** may be similar to that of sphingosine and, hence, is expected to be highly enantiospecific (condensation of alanine or serine with an activated acyl precursor followed by decarboxylation and reduction). This suggests that the 2-aminoalkanols could be optically active.

Thus, leucettamol A (**13**) was reisolated, catalytically reduced to give the perhydro derivative **14**, and then converted to *N,N',O,O'*-tetrabenzoyl derivative **15** for the ECD analysis based on the exciton coupling between the benzoate/benzamide

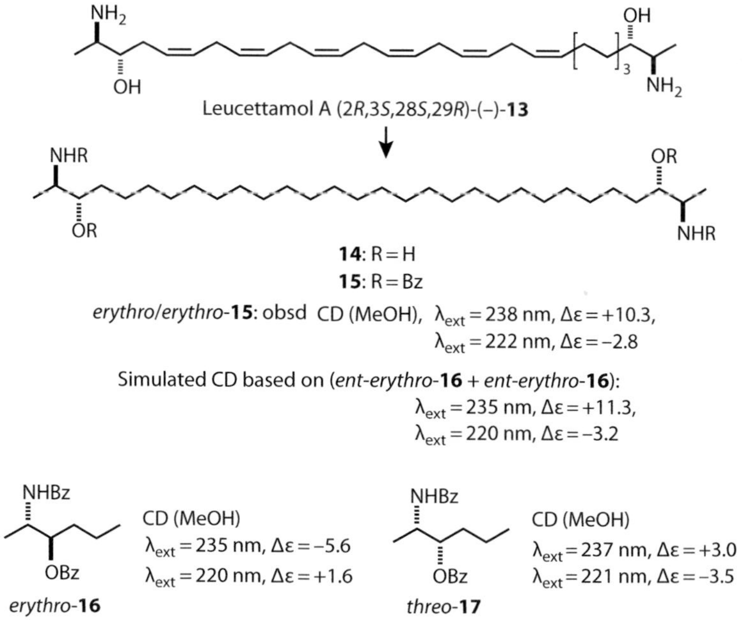

Leucettamol A (2*R*,3*S*,28*S*,29*R*)-(−)-**13**

14: R = H
15: R = Bz

erythro/erythro-**15**: obsd CD (MeOH), λ_{ext} = 238 nm, Δε = +10.3, λ_{ext} = 222 nm, Δε = −2.8

Simulated CD based on (*ent-erythro*-**16** + *ent-erythro*-**16**): λ_{ext} = 235 nm, Δε = +11.3, λ_{ext} = 220 nm, Δε = −3.2

CD (MeOH) λ_{ext} = 235 nm, Δε = −5.6 λ_{ext} = 220 nm, Δε = +1.6

erythro-**16**

CD (MeOH) λ_{ext} = 237 nm, Δε = +3.0 λ_{ext} = 221 nm, Δε = −3.5

threo-**17**

FIGURE 5.9   Absolute configuration of leucettamol A (**13**).

chromophores (Figure 5.9) (Dalisay et al. 2009). The absolute configuration of **13** was unambiguously determined by the deconvoluted exciton coupled CD as follows. Since the 2,3-*N,O*-dibenzoyl chromophores are very remote from 28,29-*N,O*-dibenzoyl chromophores, the simple additivity of ECCD is applicable. So, as a model compound, *erythro*-**16** prepared from (2*S*)-alanine was selected, which showed a negative exciton couplet (Figure 5.9) (Nicholas and Molinski 2000). The simulated ECCD curve of [*ent-erythro*-**16** + *ent-erythro*-**16**] agreed well with the observed ECCD of [*erythro/erythro*]-**15** (Figure 5.9). On the other hand, other combinations, for example, [*threo*-**17** + *threo*-**17**] or [*ent-erythro*-**16** + *threo*-**17**] led to insufficient agreement with the observed ECCD of **15**. The absolute configurations of **15** and hence **13** were unequivocally determined to be (2*R*,3*S*,28*S*,29*R*). Thus, leucettamol A is optically active and a redetermination of the specific rotation of **13** gave $[\alpha]_D = -3.8 \pm 0.1$ (*c* 4.4, MeOH) averaged over 10 measurements. The zero $[\alpha]_D$ obtained originally was under conditions likely near the detection limit.

The deconvolution ECCD method used here to determine the absolute configuration of dimeric sphingolipids readily distinguishes between both *erythro* and *threo* stereoisomers of amino alcohols.

5.7.3.3　Absolute Configuration of a New Pimarane Diterpene from a Marine Fungus by the Exciton Allylic Benzoate Method

11-Deoxydiaporthein A (**18**), a new pimarane diterpene, was isolated from a marine fungus (*Cryptosphaeria eunomia* var. *eunomia*) together with known related compounds, that is, diaporthein A, scopararane A, and diaporthein B (Yoshida et al. 2007). The relative stereostructure of compound **18** was determined by NMR spectroscopy, especially by NOESY. To determine its absolute configuration, compound **18** was converted to benzoate (**19**), the CD spectrum of which showed a positive Cotton effect ($\lambda_{ext} = 230$ nm, $\Delta\varepsilon + 5.8$) due to the interaction between benzoate and olefin chromophores (Figure 5.10). By application of the CD allylic benzoate method

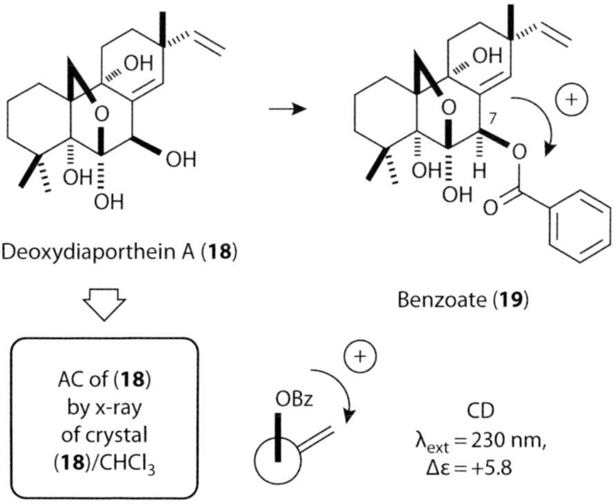

FIGURE 5.10 Determination of the absolute stereochemistry of 11-deoxydiaporthein A (**18**) by the CD allylic benzoate method and also by x-ray crystallography.

to this Cotton effect, it was concluded that the long axes of benzoate and olefin chromophores constitute a clockwise screw sense, leading to the *R* configuration at the C-7 position.

The absolute configuration determined by CD spectroscopy was confirmed by x-ray crystallography of a single crystal of compound **18** obtained by recrystallization from chloroform. It is very interesting that the crystal contained chloroform molecules as crystal solvent, and based on the strong anomalous scattering effect of chlorine atoms, the absolute configuration of 11-deoxydiaporthein A **18** was established as shown.

5.7.3.4 Absolute Configuration of an Allylic *N*-Imidazolyl-Containing Steroidal Marine Natural Product by Exciton-Coupled ECD

The antifeedant, trisulfated, steroidal alkaloid amaranzole A (**20**) was isolated from the tropical sponge *Phorbus amaranthus* at Key Largo off the coast of Florida (Figure 5.11) (Morinaka and Molinski 2008).

It contains a C24-imidazole grouping juxtaposed with a C25 double bond. The close spatial relationship between these two chromophores results in a split-CD spectrum at λ_{ext} 201 nm, $\Delta\varepsilon$ + 4.5, and λ_{ext} 189 nm, $\Delta\varepsilon$ − 6.4. A weak positive band is observed at ~256 nm that is assigned to the weakly perturbed phenol chromophore. The absolute configuration at C24 was determined to be *R* by comparison of the positive Cotton effect at ~201 nm in the CD of **20** with that of the model compound **21**. Both Cotton effects were of identical sign and similar magnitude The exciton coupled CD observed at ~201 nm (+) and 189 nm (−) appear to be solely related to the through-space interaction between the terminal olefin at ≤200 nm and the imidazole ring, but not the conjugated *p*-hydroxyphenyl group. The exciton coupling observed in the CD of **21** was absent in that of the dihydro derivative **22**, which exhibited a

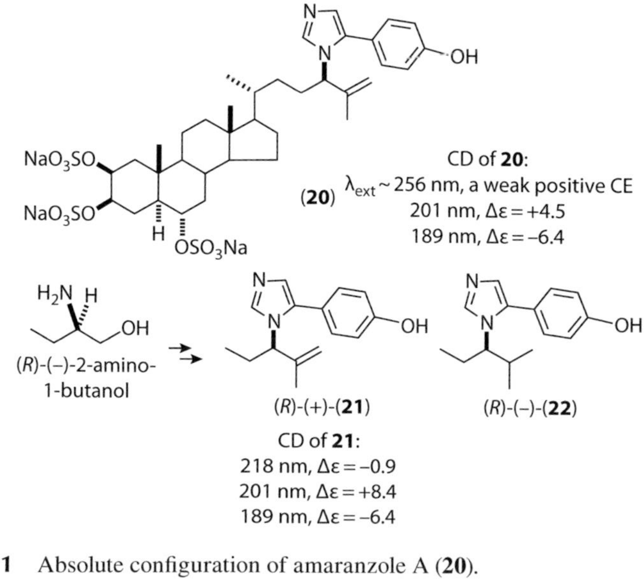

FIGURE 5.11 Absolute configuration of amaranzole A (**20**).

very weak CD arising from imidazole and phenol perturbations. The major conformation of **21** was obtained from NOE measurements and molecular mechanics calculations. The CD method developed here appears to be useful for the configurational assignments of similar acyclic *N*-allyl imidazole containing natural products.

5.7.3.5 Determination of the Absolute Configurations of Cortistatin A, an Anti-Angiogenic Steroidal Alkaloid, Based on the Exciton CD Spectrum Shown by the Natural Product Itself

Angiogenesis is the process of generating new capillary blood vessels, and their specific inhibitors are potentially promising antitumor agents. Cortistatin A (**23**), an anti-angiogenic steroidal alkaloid, was isolated together with other cortistatins B–H from the Indonesian marine sponge *Corticium simplex*.

The relative stereostructure of cortistatin A **23** was elucidated by 2D-NMR (mainly COSY and NOESY), and then confirmed by x-ray crystallography as shown in Figure 5.12. Its absolute configuration was determined by the CD exciton chirality method as follows. The π–π* transition of the conjugated diene chromophore (UV λ_{max} = ca. 234 nm) contained in the molecular skeleton couples with that of the isoquinoline chromophore (UV λ_{max} = ca. 220 nm) at the 17-position. Since the long axes of the two chromophores constitute a counterclockwise screw sense in the conformation shown in Figure 5.12, the 1st negative and 2nd positive Cotton effects are expected. The CD spectrum of cortistatin A **23** actually showed bisignate Cotton effects (λ_{ext} = 237 nm, $\Delta\varepsilon$ − 17; λ_{ext} = 217 nm, $\Delta\varepsilon$ + 35), and, therefore, its absolute configuration was determined as shown in Figure 5.12. This is a unique example in that the natural product itself shows exciton-coupled CD Cotton effects, from which the absolute configuration of the natural product was determined.

5.7.4 SIMULATION OF CD BY MO CALCULATION

5.7.4.1 Determination of the Absolute Stereochemistry of Pyranonigrin A from *Aspergillus niger* by the TDDFT Calculation of ECD

Recent progress in ab initio calculations, especially DFT and TD-DFT, enables one to calculate chiroptical spectra including electronic CD.

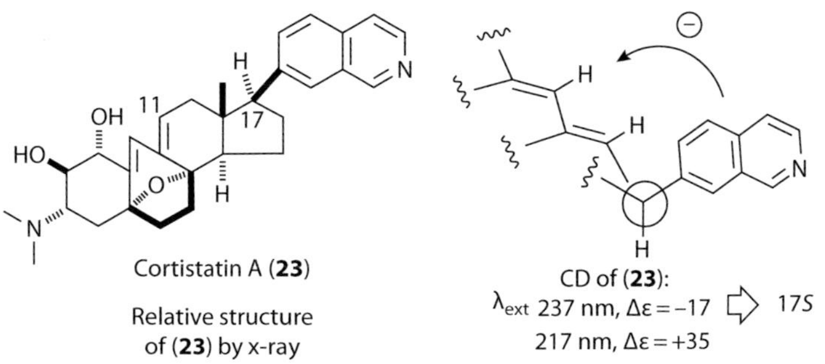

Cortistatin A (**23**)

Relative structure
of (**23**) by x-ray

CD of (**23**):
λ_{ext} 237 nm, $\Delta\varepsilon$ = −17 ⟹ 17*S*
217 nm, $\Delta\varepsilon$ = +35

FIGURE 5.12 Determination of the relative and absolute configurations of cortistatin A (**23**) by x-ray and CD exciton chirality methods.

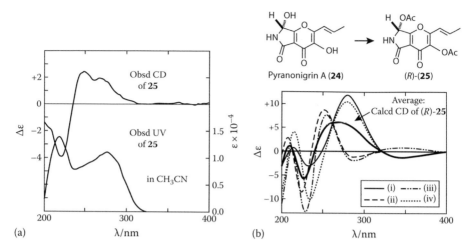

(a)   (b)

FIGURE 5.13 Determination of the absolute stereochemistry of pyranonigrin A diacetate (**25**) by the TDDFT calculation of CD spectra: (a) observed CD and UV spectra of **25** and (b) simulated CD curve of (*R*)-**25** at the B3LYP/aug-cc-pVDZ level. CD curves. (Redrawn with permission from Schlingmann, G., Taniguchi, T., He, H., Bigelis, R., Yang, H.Y., Koehn, F.E., Carter, G., and Berova, N., Reassessing the structure of pyranonigrin, *J. Nat. Prod.*, 70, 1180–1187, 2007. Copyright 2007 American Chemical Society.)

The TDDFT method was applied to pyranonigrin A diacetate (**25**) as shown in Figure 5.13 (Schlingmann et al. 2007). The conformational search for (*R*)-**25** was carried out by MMFF94 generating four stable conformers, for which CD curves were computed by TDDFT, where the *R* absolute configuration was arbitrarily selected. The averaged CD curve was next calculated by the Boltzmann-weighting as shown in Figure 5.13. Since the calculated CD curve agrees with the observed one, the *R* absolute configuration was thus assigned to natural product **24** and its diacetate **25**.

5.7.4.2 Absolute Configuration of Macropodumines B and C by Calculated, Experimental, and Solid-State ECD

Macropodumines B (**26**) and C (**27**) are remarkable pentacyclic alkaloids obtained from the leaves and fruits of *Daphniphyllum macropodum* Miq. growing in different provinces of South China. Extracts of *D. Macropodum* Miq. have been used in traditional Chinese medicine for inflammations (Figure 5.14) (Guo et al. 2009).

Based on geometry derived from an x-ray structure, the absolute configuration of (**26**) was obtained as shown (2*R*,5*S*,6*S*,18*S*) by using a solid-state CD/TDDFT approach (Figure 5.15). This method is based on comparison between a solid-state CD (KCl disc) and that obtained by calculation using the TDDFT level of theory (Figure 5.15b). The observed solid-state CD was almost identical with that recorded in acetonitrile (Figure 5.15a).

For (**27**), however, the absolute configuration (2*S*,4*R*,5*S*,6*S*,18*R*) was obtained by more traditional means whereby comparison of an experimentally determined solution CD with a calculated one based on TD-DFT theory provided the answer.

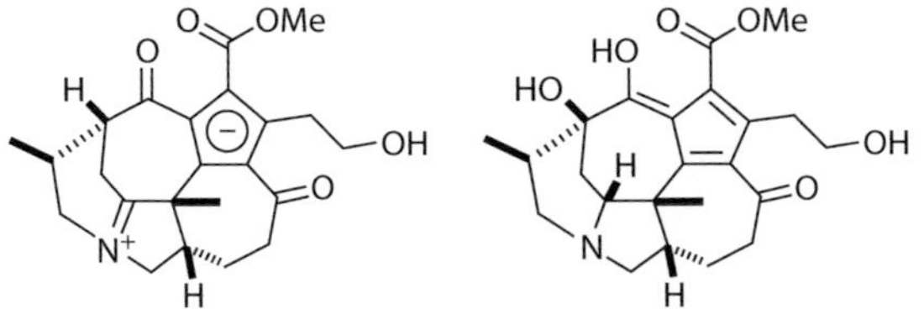

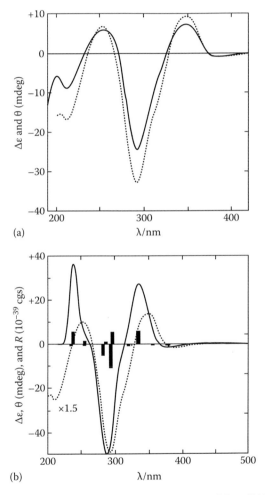

Macropodumine B (−)-(**26**)   Macropodumine C (−)-(**27**)

FIGURE 5.14   Absolute configurations of macropodumines B (**26**) and C (**27**).

(a)

(b)

FIGURE 5.15   (a) Observed CD spectra of macropodumine B (**26**): solid line, in acetonitrile; dottedline, as KCl disc. (b) Comparison of obsd. and calcd. CD spectra of (**26**): solid line, calcd CD based on the x-ray geometry by TDB3LYP/TZVP; dotted line, obsd. CD as KCl disc; vertical bars, calcd. rotational strength R in 10^{-39} cgs unit. (Adapted with permission from Guo, Y.-W., Kurtan, T., Krohn, K., Pescitelli, G., and Zhang, W.: Assignment of the absolute configuration of zwitter ionic and neutral macropodumines by means of TDDFT CD calculations. *Chirality.* 2009. 21, 561–568. Copyright Wiley-VCH Verlag GmbH & Co. KGaA.)

An advantage of the solid-state method is that it reduces the uncertainty of a conformational search along with the extensive computational effort that this requires.

5.7.4.3 Absolute Configuration of Actinophyllic Acid as Determined by Simulation of Optical Rotation and ECD

Actinophyllic acid (–)-**28**, a carboxypeptidase U inhibitor, is an alkaloid obtained from extracts of the tree *Alstonia actinophylla* (Carroll et al. 2005). Its relative configuration was determined by NMR analysis and a total synthesis was carried out by the Overman group (Figure 5.16) (Martin et al. 2008).

Separation of racemic methyl ester **29** into its enantiomers followed by optical rotation and theoretical and experimental ECD studies yielded the absolute configurations of both (+)- and (–)-enantiomers, as described below (Taniguchi et al. 2009). Chemical conversion of the methyl ester of (–)-**29** into the naturally occurring (–)-**28** provided the absolute configuration of naturally occurring actinophyllic acid **28** as shown.

Specifically, a MMFF94/Monte Carlo conformational search of low energy conformations was carried out, which provided two conformers that differ only in the orientation of the C22 ester group (Figure 5.16) and are within an energy window of 10 kcal/mol. These were further optimized by {DFT/B3LYP/6-31G(d.p.)-level} and from this latter optimization, Boltzmann population calculations revealed that these two minimal energy conformers are really only 1.4 kcal/mol different. So optical rotation properties were calculated for both conformers and then averaged based on their Boltzmann populations. Optical rotations were then calculated using TDDFT with the B3LYP functional on the methyl ester **29** with a chosen absolute configuration of (15R,16S,19S,20S,21R). This led to a predicted average $[\alpha]_D$ value of −140.6, very close to the experimentally observed value of −136 for the synthetic methyl ester (–)-**29** in methanol. The ester was then hydrolyzed to the free acid with a rotation of −29, which is in agreement with the $[\alpha]_D$ of −29 observed for the previously isolated natural (–)-**28** leading to the absolute configuration for (–)-**28** as (15R,16S,19S,20S,21R).

To confirm these assignments, ECD studies were carried out by comparing a calculated CD spectrum (from TDDFT with the B3LYP functional) with an experimentally obtained solution CD. The calculated ECD for **29** reproduced well with that obtained experimentally (Figure 5.17). Hydrolysis of the synthetic methyl ester (–)-**29** yielded the naturally occurring acid (–)-**28** and, thus, provided corroboration for its assignment. Furthermore, the absolute stereochemical assignment for (–)-**28** is consistent with the proposed biosynthetic pathway from tryptamine and (–)-secologanin.

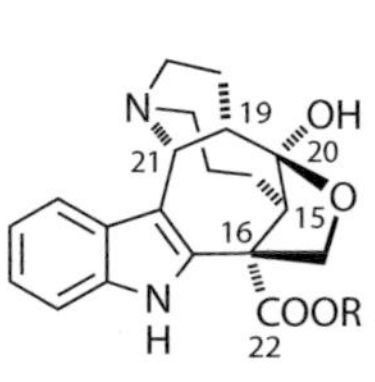

(15R,16S,19S,20S,21R)-(–)-**28**, R = H
(15R,16S,19S,20S,21R)-(–)-**29**, R = CH$_3$

FIGURE 5.16   Absolute configurations of actinophyllic acid (–)-**28** and methyl ester (–)-**29**.

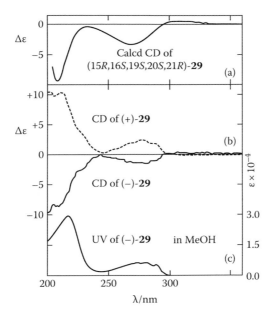

FIGURE 5.17   Comparison of calculated and observed CD spectra of actinophyllic acid methyl ester (−)-**29**: (a) CD curve of (15*R*,16*S*,19*S*,20*S*,21*R*)-**29** calculated by B3LYP/6-31G(d,p); (b) observed CD spectra of (+)- and (−)-**29** in MeOH; (c) observed UV spectrum of (−)-**29** in MeOH. (Adapted with permission from Taniguchi, T., Martin, C.L., Monde, K., Nakanishi, K., Berova, N., and Overman, L.E., Absolute configuration of actinophyllic acid as determined through chiroptical data, *J. Natl. Prod.*, 72, 430–432, 2009. Copyright 2009 American Chemical Society.)

5.7.4.4   Absolute Configuration of Brassicanal C, a Naturally Occurring Chiral Sulfinate, Determined by Optical Rotation, ECD and VCD

With the use of a model chiral sulfinate (*R*)-(+)-**30** with known absolute configuration, the absolute configuration of the cruciferous phytoalexin brassicanal C has been determined as (*S*)-(−)-**31** (Figure 5.18) (Taniguchi et al. 2008). This antimicrobial secondary metabolite is produced as a result of stresses to the plant due to invasive pathogens or UV light.

The absolute configuration was accomplished by a comparison of experimental optical rotation, ECD and VCD spectra, with those obtained by calculation using DFT. Interestingly, the average calculated [α]$_D$ of (*S*)-**31** (−443°) is much larger than that observed experimentally for **31** (−231°). The calculated value for the model compound (*R*)-(+)-**30**, whose absolute configuration was determined chemically by means of its

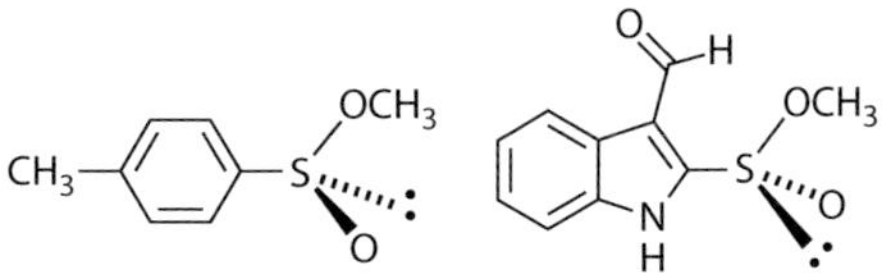

(*R*)-(+)-**30**   Brassicanal C (*S*)-(−)-**31**

FIGURE 5.18   Absolute configuration of brassicanal C (*S*)-(−)-**31**.

stereospecific conversion into a known sulfoxide (Mikolajczyk and Drabowicz 1974), was also much larger than the experimentally determined value. But the large negative values clearly indicate the *S* configuration for (–)-**31**.

Support was obtained from comparison of the calculated (TDDFT/B3LYP/aug-cc-pVDZ) with experimentally obtained ECD spectra. Both the simulated ECD curve and the experimental ECD spectrum obtained in acetonitrile show a negative, broad band at around 320 nm and a sharp positive peak at around 215 nm. Further confirmation was derived from comparison of the calculated VCD spectra with experimentally obtained VCD. The large number of distinct signals makes VCD an especially attractive method, although the milligram quantities necessary compared to microgram quantities for ECD is a detraction.

5.7.4.5 Absolute Configuration of Xylogranatins by Experimental and Calculated ECD

The xylogranatins comprise a series of novel, antifeedant limonoids isolated from the mangrove, *Xylocarpus granatuhis*. The determination of the absolute configuration of xylogranatin F (**32**) at carbons 13 and 17 provides a good example of the power of using quantum chemical CD calculated curves in comparison with experimental traces (Figure 5.19) (Wu et al. 2008).

The absolute stereochemistry at positions 3, 5, and 10 was determined by the ^{1}H NMR Mosher method and the relative configuration at carbons 13 and 17 in the δ-lactone/furan portion of the molecule determined by NOE studies.

The absolute configuration of carbons 13 and 17 was determined by a combination of experimental CD and quantum chemical CD calculations at the B3LYP/6-31G(d) level. These calculations provided six energy minima for a possible isomer (3*R*,5*S*,10*S*,13*R*,17*R*)-**32** and four for another isomer (3*R*,5*S*,10*S*,13*S*,17*S*)-**33**. The simulated CD curves for these two possibilities were obtained by using the Boltzmann statistic, and then UV corrected. It turned out that the theoretical CD curve for the isomer (3*R*,5*S*,10*S*,13*R*,17*R*)-**32** agreed well with the observed CD in the 240–400 nm region, while that of the other isomer (3*R*,5*S*,10*S*,13*S*,17*S*)-**33** was very different from the experimental curve. Therefore, the absolute configuration (3*R*,5*S*,10*S*,13*R*,17*R*) was assigned to xylogranatin F. However, the positive CD around 230 nm could not be reproduced at the B3LYP/6-31G(d) level.

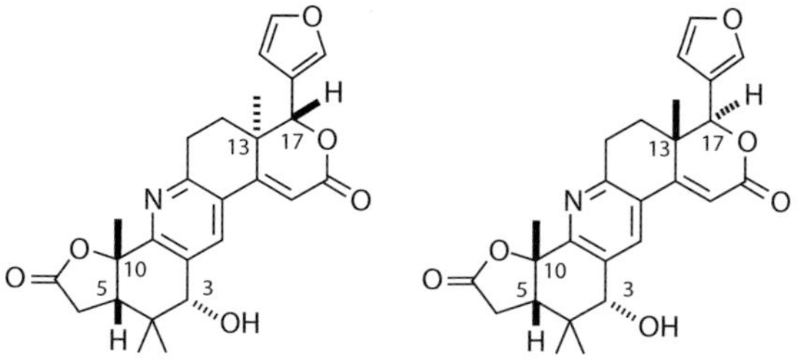

Xylogranatin F
(3*R*,5*S*,10*S*,13*R*,17*R*)-(+)-(**32**)

Another tested diastereomer
(3*R*,5*S*,10*S*,13*S*,17*S*)-(**33**)

FIGURE 5.19   Absolute configuration of xylogranatin F (+)-**32**.

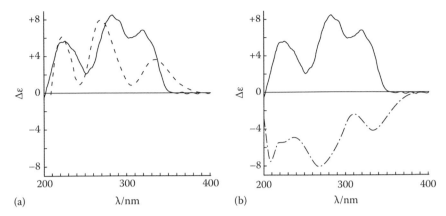

FIGURE 5.20 Comparison of observed and calculated CD spectra of xylogranatin F **32**: (a) solid line, observed CD; dotted line, calcd CD for (3*R*,5*S*,10*S*,13*R*,17*R*)-**32** by MRCI/SVP; (b) solid line, observed CD; chain line, calcd CD for (3*R*,5*S*,10*S*,13*S*,17*S*)-**33** by MRCI/SVP. (Reprinted with permission from Wu, J.W., Zhang, S., Bruhn, T., Xiao, Q., Ding, H., and Bringmann, G.: Xylogranatins F-R: Antifeedants from the Chinese mangrove, *Xylocarpus granatum*, a new biogenetic pathway to tetranortriterpenoids. *Chem. Eur. J.* 2008. 14. 1129–1144. Copyright Wiley-VCH Verlag GmbH & Co. KGaA.)

The calculation using the higher level DFT (MRCI/SVP theory) reproduced the CD spectrum well in all regions, including the positive CD at 230 nm as shown in Figure 5.20, which led to the (3*R*,5*S*,10*S*,13*R*,17*R*) absolute configuration of xylogranatin F **32**. It is thus important to try various levels of theory for obtaining a reliable determination of absolute configuration (Wu et al. 2008).

5.7.4.6 Absolute Configuration of Blennolides by Calculated and Experimental ECD

Secalonic acid is a member of the ergochrome class of fungal metabolites first isolated from *Claviceps purpurea* or ergot (Eglinton et al. 1958, Frank 1980). These potent mycotoxins have an interesting history, which dates back to medieval times where epidemics of ergot poisoning were caused by ergot alkaloids and mycotoxins due to contaminated flour by *C. purpurea*. Recently, a number of unusual chromanones along with secalonic acid B (+)-**34** were isolated from a *Blennoria* sp of fungi (Zhang et al. 2008). Among the new compounds isolated were the hemisecalonic acid, blennolide A (+)-(**35**), and rearranged products D (−)-(**36**), and E (+)-(**37**), whose structures and relative stereochemistry were determined by extensive spectroscopic analysis and confirmed by the x-ray structure of blennolide A (**35**) (Figure 5.21).

The absolute configuration of these fungal metabolites was determined by a combination of solid-state CD, ab initio calculations of theoretical CD spectra, and comparison of the calculated spectra with experimentally obtained CD spectra (Zhang et al. 2008). These metabolites exhibited strong antifungal and antibacterial activity against a number of organisms.

The atomic coordinates obtained from the x-ray structure of (+)-**35** were used directly for the TDDFT CD calculation, which was then compared with a solid-state

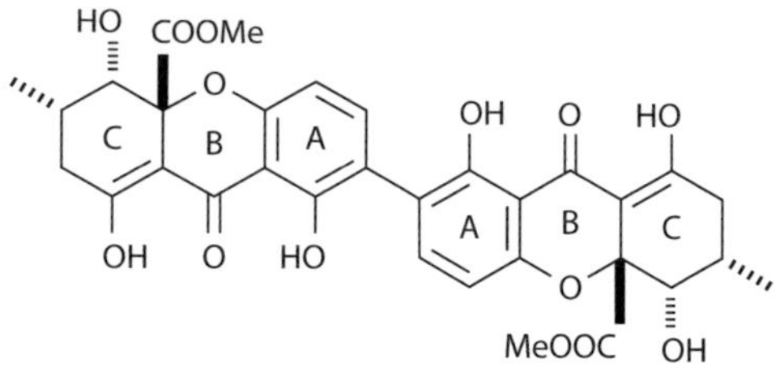

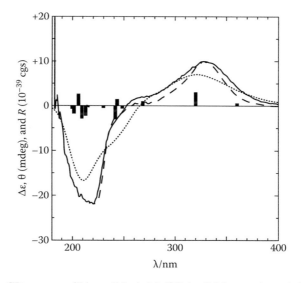

Secalonic acid B (+)-(**34**)

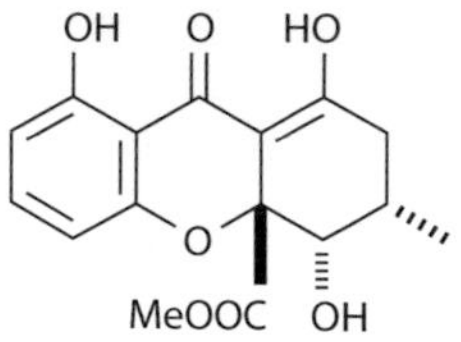

Blennolide A (+)-(**35**)

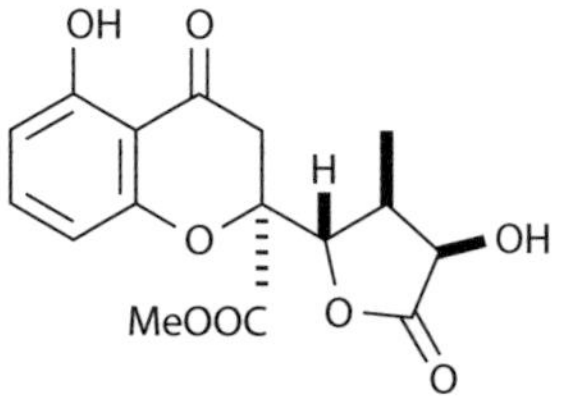

Blennolide D (–)-(**36**)

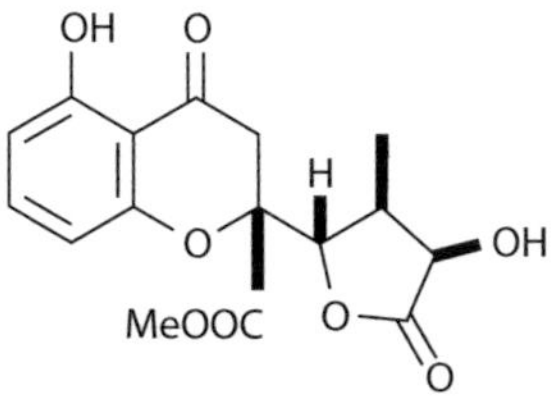

Blennolide E (+)-(**37**)

FIGURE 5.21   Absolute configurations of blennolide A (+)-(**35**) and related compounds.

FIGURE 5.22   CD spectra of blennolide A (+)-(**35**) in dichloromethane (solid line) and in the solid state as a KCl disc (dashed line) compared with the calculated TDB3LYP/TZVP CD spectrum for (5*S*,6*S*,10a*R*)-(**35**) (dotted line). Vertical bars are computed rotational strengths in 10^{-39} cgs. (Adapted with permission from Zhang, W., Krohn, K., Ullah, Z., Florke, U., Pescitelli, G., Bari, L.D., Antus, S., Kurtan, T., Rheinheimer, J., Draeger, S., and Schulz, B.: New mono- and dimeric members of the secalonic acid family: Blennolides A-G isolated from the fungus *Blennoria* sp. *Chem. Eur. J.* 2008. 14. 4913–4923. Copyright Wiley-VCH Verlag GmbH & Co. KGaA.)

CD obtained using a KCl disc and with a CD obtained from a solution of (+)-**35**. The agreement was excellent between the solid-state CD, the solution CD, and that obtained from theoretical calculations (Figure 5.22).

The absolute configurations of blennolides D (−)-**36** and E (+)-**37** were obtained by comparing the experimentally determined CD in solution with a theoretical curve calculated by the TDDFT method, since there was no solid-state structure to provide atomic coordinates as used for (+)-**35**. It should be noted that the two ^1L_a and ^1L_b transitions assigned to the chromanone chromophore between 250 and 350 nm are negative for the 2*S* configuration in **36** and positive for that in **37**, which corresponds to positive and negative helicities, respectively, of the B ring of the chromanone. The comparison of calculated and experimental CDs for blennolides D and E is not as good as for **35**, but sufficient for assigning absolute configurations with reasonable confidence.

5.8   APPLICATION OF INDUCED CD FOR STUDYING THE INTERACTIONS OF NATURAL PRODUCTS WITH DNA AND PROTEINS

5.8.1   ECD FOR INVESTIGATING THE CELLULAR UPTAKE AND TARGET OF THE ANTICANCER ANTIBIOTIC CC-1065 IN LIVE CELLS

Induced CD (ICD) has been used to study the binding of chiral or achiral molecules to biopolymers, where the monomer/biopolymer interaction leads to a strong, ICD providing a characteristic spectroscopic foot print of the binding event (Allenmark 2003). There are two different situations where ICD will be observed upon binding: (a) when the host is chiral but does not contain a UV visible chromophore, then upon binding the ICD appears within the absorption band of the achiral chromophoric guest that becomes chirally perturbed upon binding; (b) in case of an achiral chromophoric host (CD silent), binding of a chiral guest will lead to a characteristic ICD within the absorption bands of the achiral host. The CD analysis of a drug/DNA complex, however, is more difficult, in particular when both components are chiral. Thus, depending on the situation, the bound chiral drug may show new CD bands that are not observed in the free drug. And the DNA complex may also show CD changes below 300 nm reflecting DNA conformational changes upon binding.

Recently, ICD has been applied to the study of drug/DNA interactions and for drug targeting studies. In this example, ICD was used with live cells to study the DNA binding properties of some new antitumor agents based on the potent DNA alkylating agent CC-1065 (**38**) (Figure 5.23) (Tietze et al. 2009). This metabolite was isolated from fermentation broths of *Streptomyces zelensis* by Upjohn scientists (Hanka et al. 1978) and since then has been the object of a great deal of structure activity and DNA binding studies by not only the Upjohn Company but several academic groups as well. Because its preferred natural curvature is complementary to that of the DNA minor groove, it binds tightly to the DNA in a sequence specific manner with a strong preference for AT-rich regions. Upon binding, alkylation takes place via the N-3 of adenine by nucleophilic attack on the conjugated cyclopropyl group as shown in Figure 5.23.

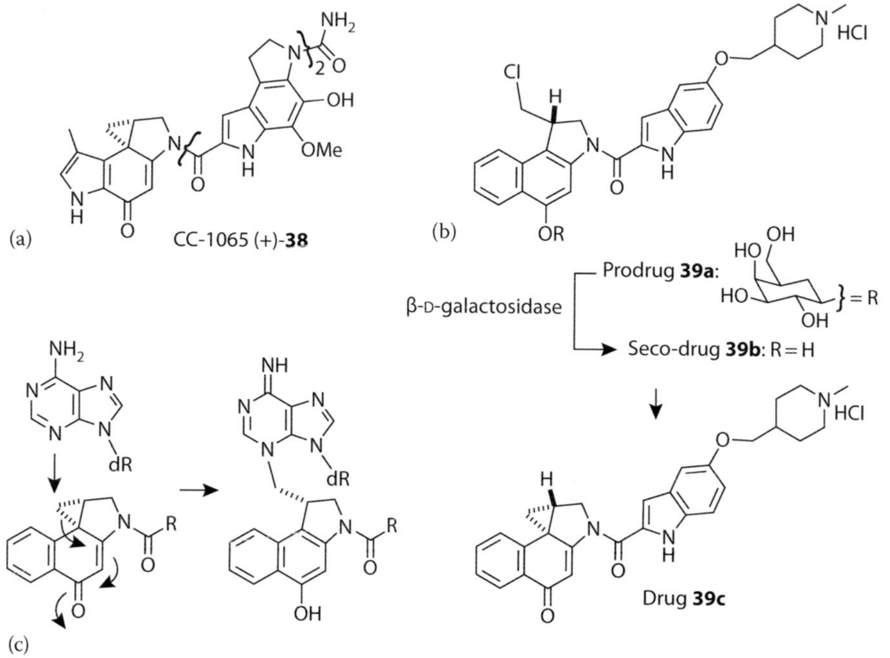

FIGURE 5.23 (a) Natural antibiotic CC-1065 (+)-**38**, (b) formation scheme of related drug **39c** from prodrug **39a** via seco-drug **39b**, and (c) DNA alkylation mechanism with drug **39c**.

Tietze et al. measured ICD of live cells to monitor the cellular uptake and DNA targeting of some CC-1065 prodrug analogs, for example, **39a**, without the need for radio-labeled drug or fluorescent dyes and to evaluate the cytotoxicity of these new agents (Tietze et al. 2009). This is based on the ICD resulting from the drug on binding to the DNA minor groove. A number of relatively nontoxic glycosylated prodrugs, on incubation with live cells, are converted to the *seco* derivatives by cellular glycosylases. Alone, these seco-drugs exhibit a weakly negative band at $\lambda_{ext} = 250\,$nm, weakly positive signals at $\lambda_{ext} = 275, 320\,$nm, and a characteristic positive 390 nm band.

On incubation of the seco-drugs with synthetic double-stranded DNA oligomers, however, a negative ICD band is observed at $\lambda_{ext} = 305\,$nm along with a positive one at $\lambda_{ext} = 335\,$nm, signals that are not observed for the oligomers alone. Furthermore, the ICD signals are significantly stronger than the CD of the seco-drugs without DNA.

Incubation of the seco-drugs with serum-free, live cell suspensions resulted in ICDs essentially identical to those observed upon drug treatment of natural DNA oligomers (minimum band at 305 nm and a positive band at 335 nm). No ICDs were observed with untreated cells. In addition, DNA binding kinetics and rates of conversion of the *seco*-analogs to the active cyclopropyl compounds were obtained, thus aiding in the prioritization of the various prodrugs for further biological evaluation.

This is the first time CD has been used to establish that cellular DNA is a target for supposed DNA damaging drugs. Clearly, this methodology has great potential for other DNA binding drugs that result in ICDs.

5.8.2   THE USE OF ECD TO SHOW THAT THE RIGHT-HANDED *P*-CONFORMER OF THE TETRAPYRROLE BILE PIGMENT BILIVERDIN IS AN ENDOGENOUS LIGAND OF HUMAN SERUM α₁-ACID GLYCOPROTEIN

Another application of ICD with biological implications has shown that the tetrapyrrole bile pigment biliverdin (BV) (**41**), a metabolite of heme (**40**), is an endogenous ligand of human serum α_1-acid glycoprotein (AAG) (Figure 5.24) (Zsila et al. 2004, Zsila and Mady 2008). AAG is an acute phase acidic protein component of human serum. Although its primary physiological function is unknown, its concentration increases in response to tissue injury, inflammation, or infection.

Two isoenergetic and enantiomeric conformers of BV, *P*-**41a** and *M*-**41b**, are in dynamic equilibrium (Figure 5.24). Measured alone in solution without chiral perturbation, they are achiral and CD silent. However, in the presence of a commercial sample of AAG (70:30 mixture of F1/S:A genetic variants), the right-handed *P*-conformer (**41a**) is preferentially bound to the protein, the complex of which exhibits an induced and strongly negative Cotton effect at ~380 nm along with a long-wavelength positive CE at ~650 nm, both Cotton effects matching in shape the UV and Vis absorption bands. ICD titration experiments were used to estimate binding constants of 10^5–10^6 M^{-1} by monitoring the negative CE and increasing molar ratios of BV/AAG. Insight into the BV binding site was determined using competition experiments with chlorpromazine, which is known to bind in the central hydrophobic β-barrel cavity of AAG.

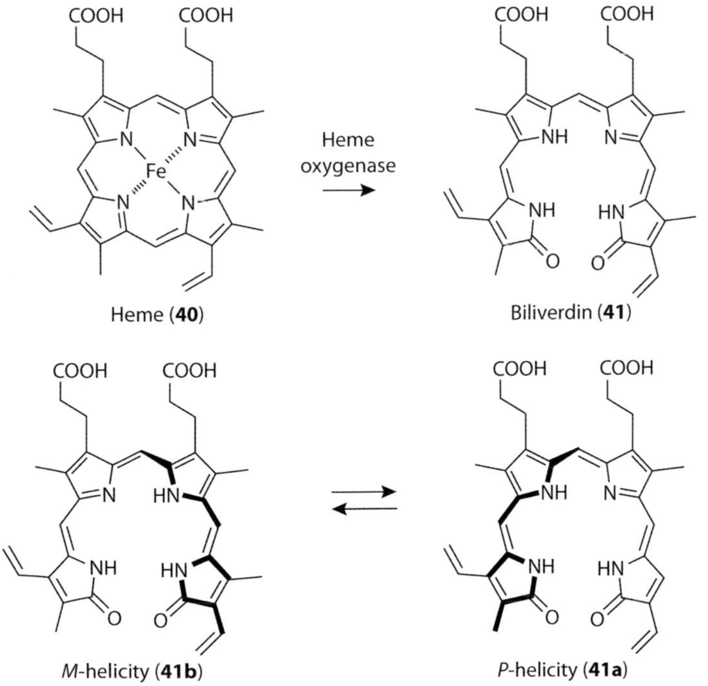

FIGURE 5.24   The dynamic equilibrium of biliverdin enantiomeric conformers (**41a**) and (**41b**) by the interaction with human serum α_1-acid glycoprotein.

Chlorpromazine was observed to displace bound BV as evidenced by the almost complete elimination of the ICD signal ~380 nm around a ratio of drug/AAG of 1.5.

5.8.3 ECD Spectroscopy Reveals Opposite Induced Chirality upon Binding of the Carotenoid Crocetin with Human and Pig Serum Albumin

Another interesting study of small molecule binding to its target biopolymer using ICD is that of crocetin and serum albumin (Zsila et al. 2002). Crocetin (**42**), the principal carotenoid of saffron, *Crocus sativus*, has been used as an anticancer agent in cancer chemotherapy and has also been put to various other medical uses in traditional and modern medicine (Figure 5.25). Thus, its interaction with serum albumin in blood, which distributes it to various tissues, is of some importance.

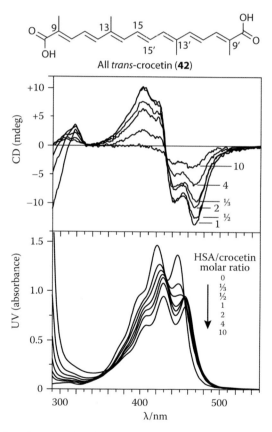

FIGURE 5.25 Induced intermolecular exciton coupling CD between all *trans*-crocetin (**42**) molecules in the **42**/HSA complex in borate buffer at pH 8.5. CD and UV spectra. (Reprinted from *Tetrahedron Asymm.*, 13, Zsila, F., Bikadi, Z., and Simonyi, M., Further insight into the molecular basis of carotenoid-albumin interactions: circular dichroism and electronic absorption study of different crocetin-albumin complexes, 273–283. Copyright 2002, with permission from Elsevier.)

An induced intermolecular exciton coupling CD between all *trans*-crocetin (**42**) molecules in the **42**/HSA complex (molar ratio 1:1) in borate buffer at pH 8.5 is left-handed with negative bands at λ_{ext} 470 nm (~ −14 mdeg) and 440 nm (~ −10), together with positive bands at λ_{ext} 430 nm (~ +8) and 410 nm (~ +10) (Figure 5.25). It is obvious that these CD spectra reflect the vibronic structure observed in the UV spectra. The left-handed exciton couplet observed with crocetin bound to HSA indicates that the long axes of the stacked crocetin molecules form a negative or counterclockwise relationship with each other. Furthermore, the couplet is red-shifted indicating that the bound crocetin molecules are located in a highly polarizable region of the protein.

Interestingly, and in contrast to the ICD observed with HSA, crocetin **42** binds to pig albumin with somewhat lower affinity and results in a positive bisignate CD couplet with positive bands at λ_{ext} 480 nm (CD ~ +8 mdeg) and 470 nm (~ +7), together with a broad negative band λ_{ext} 430 nm (~ −5) indicating a mirror image relationship to that with HSA and a positive stacking relationship between the bound crocetin molecules in the surrounding protein binding pocket. Competition binding experiments between crocetin and palmitic acid with HSA indicate a common binding site in which two carotenoid molecules bind to the fatty acid binding site.

5.9 CONCLUSION

As seen in the examples described in this chapter, CD continues to be a powerful tool for the establishment of absolute configuration of natural products of diverse chemical structures. ICD also plays an important role for studying the binding properties of natural products with target biopolymers. Thus, it is clear that recent advances in the theoretical treatment of optical activity and computational methods, along with technological developments, will further increase the importance of chiroptical methods in the structural studies of complex natural products.

REFERENCES

Allenmark S. 2003. Induced circular dichroism by chiral molecular interaction. *Chirality 15*, 409–422.

Aoki S., Watanabe Y., Sanagawa M., Setiawan A., Kotoku N., and Kobayashi M. 2006. Cortistatins A, B, C, and D, anti-angiogenic steroidal alkaloids, from the marine sponge *Corticium simplex. J. Am. Chem. Soc. 128*, 3148–3149.

Autschbach J. 2010. Computing chiroptical properties with first-principles theoretical methods: Background and illustrative examples. *Chirality* DOI:10.1002/chir.2789.

Berova N., Di Bari L., and Pescitelli G. 2007. Application of electronic circular dichroism in configurational and conformational analysis of organic compounds. *Chem. Soc. Rev. 36*, 914–931.

Berova N., Harada N., and Nakanishi K. 2000. Electronic spectroscopy: Exciton coupling, theory and applications. In *Encyclopedia of Spectroscopy and Spectrometry*. eds. Lindon J., Tranter G., and Holmes J. pp. 470–488. London, U.K.: Academic Press.

Berova N. and Nakanishi K. 2000. Exciton chirality method: Principles and application. In *Circular Dichroism: Principles and Applications*. 2nd edn. eds. Berova N., Nakanishi K., and Woody R.W. chapter 12, pp. 337–382. New York: Wiley-VCH.

Berova N., Nakanishi K., and Woody R.W. eds. 2000. *Circular Dichroism: Principles and Applications.* 2nd edn. New York: Wiley-VCH.

Bringmann G., Bruhn T., Maksimenka K., and Hemberger Y. 2009. The assignment of absolute stereostructures through quantum chemical circular dichroism calculations. *Eur. J. Org. Chem.* 2717–2727.

Bringmann G., Gulder T.A.M., Reichert M., and Gulder T. 2008. The online assignment of the absolute configuration of natural products: HPLC-CD in combination with quantum chemical CD calculations. *Chirality 20,* 628–642.

Carroll A.R., Hyde E., Smith J., Quinn R.J., Guymer G., and Forster P.I. 2005. Actinophyllic acid, a potent indole alkaloid inhibitor of the coupled enzyme assay carboxypeptidase U/hippuricase from the leaves of *Alstonia actinophylla* (Apocynaceae). *J. Org. Chem. 70,* 1096–1099.

Clayden J., Moran W.J., Edwards P.J., and LaPlante S.R. 2009. The challenge of atropisomerism in drug discovery. *Angew. Chem. Int. Ed. 48,* 6398–6401.

Crawford T.D. 2006. *Ab initio* calculation of molecular chiroptical properties. *Theor. Chem. Acc. 115,* 227–245.

Dalisay D.S., Tsukamoto S., and Molinski T. 2009. Absolute configuration of the α,ω-bifunctionalized sphingolipid leucettamol A from *Leucetta microrhaphis* by deconvoluted exciton coupled CD. *J. Natl. Prod. 72,* 353–359.

DeVoe H. 1964. Optical properties of molecular aggregates. I. Classical model of electronic absorption and refraction. *J. Chem. Phys. 41,* 393–400.

DeVoe H. 1965. Optical properties of molecular aggregates. II. Classical theory of the refraction, absorption, and optical activity of solutions and crystals. *J. Chem. Phys. 43,* 3199–3208.

Eglinton G., King F.E., Lloyd G., Loder J.W., Marshall J.R., Robertson A., and Whalley W.B. 1958. The chemistry of fungi. Part XXXV. A preliminary investigation of ergoflavin. *J. Chem. Soc.* 1833–1842.

Eliel E.L., Wilen S.H., and Mander L.N. 1994. *Stereochemistry of Organic Compounds.* chapter 13, pp. 991–1118. New York: John Wiley & Sons, Inc.

Frank B. 1980. *The Biosynthesis of the Ergochromes, in the Biosynthesis of Mycotoxins: A Study in Secondary Metabolism.* ed. Steyn P.S. pp. 157–191. New York: Academic Press.

Guo Y-W., Kurtan T., Krohn K., Pescitelli G., and Zhang W. 2009. Assignment of the absolute configuration of zwitter ionic and neutral macropodumines by means of TDDFT CD calculations. *Chirality 21,* 561–568.

Hanka L.J., Dietz A., Gerpheide S.A., Kuentzel S.I., and Martin D.G. 1978. CC-1065 (NSC-298223), a new antitumor antibiotic and production, *in vitro* biological activity, microbiological assays and taxonomy of the producing microorganism. *J. Antibiot. 31,* 1211–1217.

Harada N. 1999. Circular dichroism spectroscopy and absolute stereochemistry of biologically active compounds. In *The Biology–Chemistry Interface.* eds. Cooper R. and Snyder J.K. chapter 6, pp. 139–190. New York: Marcel Dekker, Inc.

Harada N. 2000. Circular dichroism of twisted π-electron systems: Theoretical determination of the absolute stereochemistry of natural products and chiral synthetic organic compounds. In *Circular Dichroism: Principles and Applications.* eds. Nakanishi K., Berova N., and Woody R. pp. 431–457. New York: John Wiley & Sons.

Harada N. and Nakanishi K. 1983. *Circular Dichroic Spectroscopy—Exciton Coupling in Organic Stereochemistry*, Mill Valley, CA: University Science Books and Oxford: Oxford University Press.

Hashimoto M., Tsushima T., Murakami T., Nomiya M., Takada N., and Tanaka K. 2008. Spiroleptosphol isolated from *Leptosphaeria doliolum. Bioorg. Med. Chem. Lett. 18,* 4228–4231.

Iida M., Ooi T., Kito K., Yoshida S., Kanoh K., Shizuri Y., and Kusumi T. 2008. Three new polyketide–terpenoid hybrids from *Penicillium* sp. *Org. Lett. 10,* 845–848.

Kemp C.M. and Mason S.F. 1966. The absorption and circular dichroism spectra and the absolute configuration of (+)-1-fluoro-12-methylbenzo[*c*]phenanthrene. *Tetrahedron* *22*, 629–635.

Kong F.H. and Faulkner D.J. 1993. Leucettamols A and B, two antimicrobial lipids from the calcareous sponge *Leucetta microraphis*. *J. Org. Chem.* *58*, 970–971.

Lightner D.A. and Gurst J.E. 2000. *Organic Conformational Analysis and Stereochemistry from Circular Dichroism Spectroscopy*. New York: Wiley-VCH.

Martin C.L., Overman L.E., and Rohde J.M. 2008. Total synthesis of (±)-actinophyllic acid. *J. Am. Chem. Soc.* *130*, 7568–7569.

Mason S.F. 1982. *Molecular Optical Activity and the Chiral Discrimination*. Cambridge, U.K.: Cambridge University Press.

Mikolajczyk M. and Drabowicz J. 1974. Asymmetric synthesis of sulphonic esters with sulfur atom as a sole chirality centre. *J. Chem. Soc. Chem. Comm.* 547–548.

Morinaka B.I. and Molinski T. 2008. Exciton coupling circular dichroism of an allylic *N*-imidazolyl group in amaranzole A, a marine natural product from *Phorbus amaranthus*. *Chirality 20*, 1066–1070.

Murakami T., Tsushima T., Takada N., Tanaka K., and Hashimoto M. 2008. Isolation and structure of spiroleptosphol. In *Symposium Papers, 50th Symposium on the Chemistry of Natural Products*. pp. 275–280. Fukuoka, Japan.

Murakami T., Tsushima T., Takada N., Tanaka K., Nihei K., Miura T., and Hashimoto M. 2009. Four analogues of spiroleptosphol isolated from *Leptosphaeria doliolum*. *Bioorg. Med. Chem. 17*, 492–495.

Nicholas G.M. and Molinski T.F. 2000. Enantiodivergent biosynthesis of the dimeric sphingolipid oceanapiside from the marine sponge *Oceanapia phillipensis*. Determination of remote stereochemistry. *J. Am. Chem. Soc. 122*, 4011–4019.

Ookura R., Iida M., Yuge M., Kito K., Ooi T., Kanoh K., Shizuri Y., Namikoshi M., and Kusumi T. 2008. Study on absolute stereostructures of metabolites from marine fungi. In *Symposium Papers, 50th Symposium on the Chemistry of Natural Products*. pp. 421–425. Fukuoka, Japan.

Schlingmann G., Taniguchi T., He H., Bigelis R., Yang H.Y., Koehn F.E., Carter G., and Berova N. 2007. Reassessing the structure of pyranonigrin. *J. Nat. Prod. 70*, 1180–1187.

Sprogoe K., Staerk D., Ziegler H.L., Jensen T.H., Holm-Moller S.B., and Jaroszewski J.W. 2008. Combining HPLC-PDA-MS-SPE-NMR with circular dichroism for complete natural product characterization in crude extracts: Levorotatory gossypol in *Thesespia danis*. *J. Nat. Prod. 71*, 516–519.

Stephens P.J. and Harada N. 2010. ECD Cotton effect approximated by the Gaussian curve and other methods. *Chirality 22*, 229–233.

Superchi S., Giorgio E., and Rosini C. 2004. Structural determinations by circular dichroism spectra analysis using coupled oscillator methods: An update of the applications of the DeVoe polarizability model. *Chirality 16*, 422–451.

Taniguchi T., Martin C.L., Monde K., Nakanishi K., Berova N., and Overman L.E. 2009. Absolute configuration of actinophyllic acid as determined through chiroptical data. *J. Natl. Prod. 72*, 430–432.

Taniguchi T., Monde K., Nakanishi K., and Berova N. 2008. Chiral sulfinates studied by optical rotation, ECD and VCD: The absolute configuration of a cruciferous phytoalexin brassicanal C. *Org. Biomol. Chem. 6*, 4399–4405.

Tietze L.F., Krewer B., Major F., and Schuberth I. 2009. CD-spectroscopy as a powerful tool for investigating the mode of action of unmodified drugs in live cells. *J. Am. Chem. Soc. 131*, 13031–13036.

Wu J.W., Zhang S., Bruhn T., Xiao Q., Ding H., and Bringmann G. 2008. Xylogranatins F-R: Antifeedants from the chinese mangrove, *Xylocarpus granatum*, a new biogenetic pathway to tetranortriterpenoids. *Chem. Eur. J. 14*, 1129–1144.

Yamada T., Doi M., Shigeta H., Muroga Y., Hosoe S., Numata A., and Tanaka R. 2008. Absolute stereostructures of cytotoxic metabolites, chaetomugilins A–C, produced by a *Chaetomium* species separated from a marine fish. *Tetrahedron Lett. 49*, 4192–4195.

Yasuhide M., Yamada T., Numata A., and Tanaka R. 2008. Chaetomugilins, new selectively cytotoxic metabolites, produced by a marine fish-derived chaetomium species. *J. Antibiot. 61*, 615–622.

Yoshida S., Kito K., Ooi T., Kanoh K., Shizuri Y., and Kusumi T. 2007. Four pimarane diterpenes from marine fungus: Chloroform incorporated in crystal lattice for absolute configuration analysis by X-ray. *Chem. Lett. 36*, 1386–1387.

Zhang W., Krohn K., Ullah Z., Florke U., Pescitelli G., Bari L.D., Antus S. et al. 2008. New mono- and dimeric members of the secalonic acid family: Blennolides A–G isolated from the fungus *Blennoria* sp. *Chem. Eur. J. 14*, 4913–4923.

Zsila F., Bikadi Z., Fitos I., and Simonyi M. 2004. Probing protein binding sites by circular dichroism spectroscopy. *Current Drug Discover Technologies 1*, 133–153.

Zsila F., Bikadi Z., and Simonyi M. 2002. Further insight into the molecular basis of carotenoid–albumin interactions: Circular dichroism and electronic absorption study of different crocetin–albumin complexes. *Tetrahedron Asymmetry 13*, 273–283.

Zsila F. and Mady G. 2008. Biliverdin is the endogenous ligand of human serum α_1-acid glycoprotein. *Biochem. Biophys. Res. Comm. 372*, 503–507.

6 Natural Products as Treatments for and Leads to Active Compounds in Viral Diseases*

David J. Newman and Gordon M. Cragg

CONTENTS

6.1   INTRODUCTION

By the very nature of the topic, in this chapter, we will attempt to demonstrate how natural products have led to relatively effective treatments, not "cures," for a variety of viral infections that plague *Homo sapiens.* This chapter is not intended to be exhaustive in nature as the topic could well fill a book in its own right, but will hopefully demonstrate that a large number of the agents currently in clinical use, in clinical and preclinical trials, and those still at the discovery stage, owe a significant "debt" to natural product structures, some from sources/organisms that are well known, others that are not.

Because viral infections by their very nature are parasitic, in that they have no base reproductive mechanism of their own but have to subvert the infected cells' reproductive mechanisms (either RNA or DNA depending upon the virus itself),

---

* The opinions expressed in this chapter are those of the authors, not necessarily those of the U.S. Government.

167

so that the host cell will produce virus particles that infect other cells in the "host," we will use a variety of "subdivisions" to describe the compounds in this chapter. These will be predominately the mechanism of action (MOA) with the exception of the first major subheading, modified nucleosides, as these agents have multiple activities so a single chemical class may well have derivatives that are effective against multiple viral infections via different mechanisms.

6.2 MODIFIED NUCLEOSIDES

It can be argued quite successfully and we have done so a number of times over the last 10–15 years (e.g., Newman et al. 2000) that the derivation of the nucleoside-based antiviral agents can be traced back to the early 1950s, when Bergmann and Feeney (1950, 1951) and Bergmann and Burke (1956) reported on two arabinose-compounds that they had isolated from marine sponges, spongouridine (Figure 6.1, Structure **1**) and spongothymidine (Figure 6.1, Structure **2**). For the first time, naturally occurring bioactive nucleosides were found containing sugars other than ribose or deoxyribose (thus an expansion of the basic "privileged," Evans et al. 1988, structure). These two compounds can best be thought of as the prototypes of all of the modified nucleoside and in some cases, pseudo-nucleotide analogues made by chemists that have crossed the antiviral (and antitumor) stages since those discoveries.

Prior these reports, almost any type of organic "base" with examples being benzimidazoles, benzpyrans, etc., had been linked to either ribose or deoxyribose to make "imitation nucleosides" without much success. Only simple substitutions, such as a halogen or thiol on one of the known DNA or RNA nucleoside bases, not the sugar, demonstrated any significant bioactivities. Following the demonstration of bioactive arabinose nucleosides from the marine environment, rather than continuing the routes that had been taken in the past, chemists began to substitute acyclic entities and cyclic sugars with unusual substituents for ribose/deoxyribose, leading to vast numbers of derivatives that were tested extensively as antiviral (and antitumor) agents over the next thirty plus years.

In a 1991 review, Suckling (1991) showed how such structures evolved at the (then) Wellcome laboratories, leading to the very well-known antiviral agents, acyclovir (Figure 6.1, Structure **3**), and its later prodrug derivatives, such as valaciclovir (Figure 6.1, Structure **4**) and valganciclovir (Figure 6.1, Structure **5**), together with the original anti-HIV agent AZT (Figure 6.1, Structure **6**) and incidentally to Nobel Prizes for Hitchens and Elion.

In an interesting "temporal reversal," arabinosyl–adenine (Ara-A; Figure 6.1, Structure **7**) was synthesized in 1960 as a potential antitumor agent (Lee et al. 1960) but was later found by fermentation of *Streptomyces antibioticus* NRRL3238, and isolated, together with spongouridine (Cimino et al. 1984), from the Mediterranean gorgonian *Eunicella cavolini* in 1984.

Although the FDA or its equivalent in other countries have approved a very significant number of antiviral vaccines with many more still in clinical trials, there are agents isolated directly from Nature, or small molecules based upon "modified nucleosides," that have been reported in the relatively recent literature or approved for clinical use by the relevant agencies.

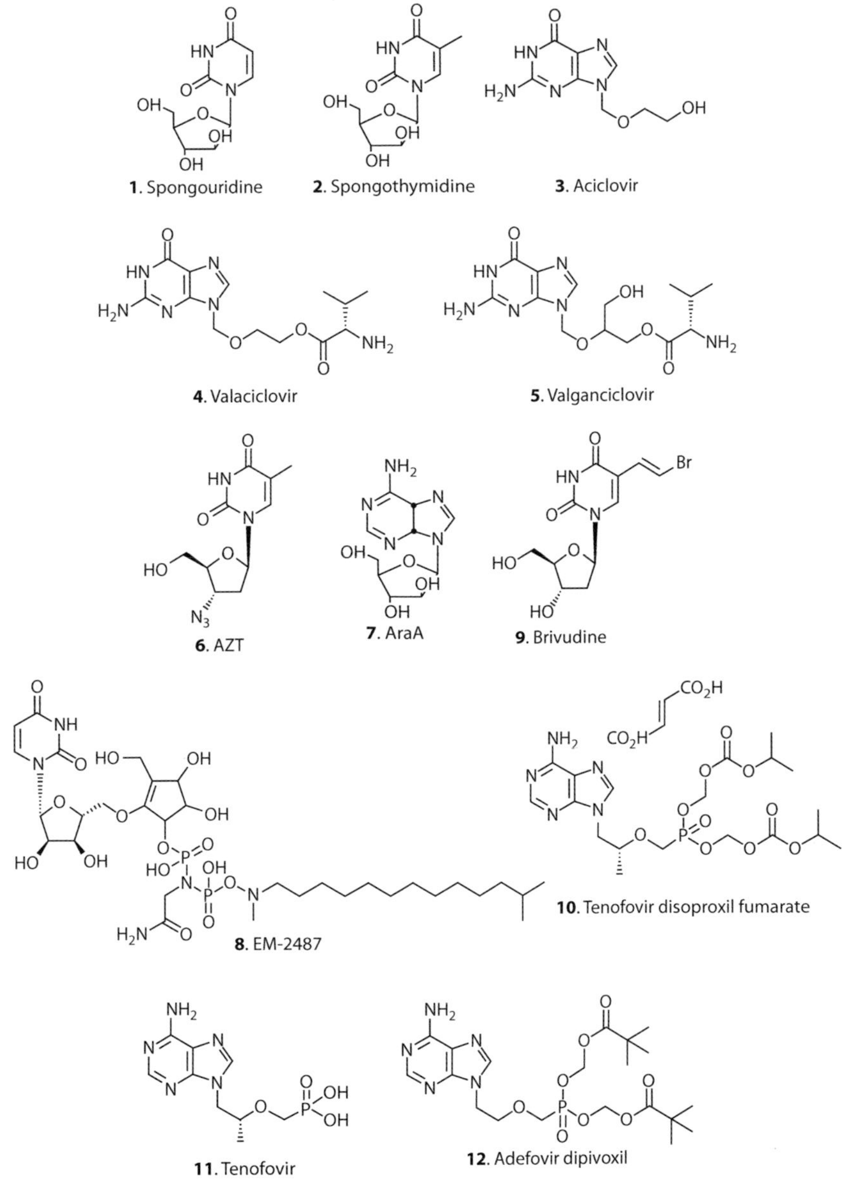

FIGURE 6.1 Modified nucleosides.

In 1999, workers at Eisai reported the isolation, structure, and preliminary properties of an unusual antiviral pseudonucleotide known as EM2487 (Figure 6.1, Structure **8**) from a terrestrial streptomycete (Takeuchi et al. 1999). In later publications, they indicated that it demonstrated inhibition of viral mRNA synthesis in OM-10.1 cells, indicating activity at the transcriptional level, and that it may well be inhibiting *tat* function (Baba et al. 1999). Very recently, a group in Japan reported that this agent inhibited porcine endogenous retrovirus (PERV), a virus that can infect certain human cells and, therefore, may be of significant import in xenotransplantation experiments, so it may yet advance further in preclinical studies (Shi et al. 2009).

Since 2000, nine such agents based upon modified nucleosides or pseudonucleo-tides have been approved for antiviral treatments covering anti-HIV, hepatitis B, herpes zoster, and cytomegalovirus (CMV). In 2001, brivudine (Figure 6.1, Structure **9**) was approved as an antiherpes drug and is currently in phase II trials for pancreatic cancer in conjunction with gemcitabine. That same year, the synthetic guanine deriv-ative valganciclovir hydrochloride (Figure 6.1, Structure **5**) was launched by Roche for the oral treatment of CMV retinitis in patients with acquired immunodeficiency syndrome (AIDS) and was later approved in 2003 for the treatment of CMV retinitis and CMV infection in transplant patients.

2001 was also the year that tenofovir disoproxil fumarate (Figure 6.1, Structure **10**), a prodrug of tenofovir (Figure 6.1, Structure **11**) was approved for treatment of HIV, subsequently being preregistered in the United States for treatment of hepatitis B; a check of the NIH clinical trials web site, as of the end of 2009, shows over 200 active trials involving one or both of these two drugs, including the use of a gel containing tenovir as a vaginal virucide in Asia. Adefovir dipivoxil (Figure 6.1, Structure **12**), an acyclic AMP analogue, was launched in 2002 as an antihepatitis B agent, though originally tested as an anti-HIV agent. The following year, 2003, emtricitabine (Figure 6.2, Structure **13**) was launched as an anti-HIV agent; and though it entered into phase III trials as an antihepatitis B agent, no details have been presented as to its current status against this virus. In 2005, Bristol–Myers Squibb launched entecavir (Figure 6.2, Structure **14**) as an antihepatitis B drug, which was followed in 2006 with the launch of telbivudine (Figure 6.2, Structure **15**) by Idenix and Novartis as an antihepatitis B drug functioning as a DNA-polymerase inhibitor. Eisai launched clevudine (Figure 6.2, Structure **16**) in Korea as a treatment for hepa-titis B in 2007, and in 2008, the older drug famciclovir (Figure 6.2, Structure **17**) was launched in Japan as an oral treatment of Herpes zoster under the code name AK-120 but the same brand name, Famvir®.

In December 2009, investigators at the Novartis Institute of Tropical Diseases in Singapore reported (Yin et al. 2009) that a variation of adenosine known by the code number NITD008 (Figure 6.2, Structure **18**), where the N-7 had been replaced by a carbon and the ribose carried an acetylene on the 2′ position, inhibited a number of flaviviruses including the four serotypes of Dengue virus, West Nile virus, Yellow fever virus, and Powassan virus. The Dengue virus was inhibited in vitro and in vivo in murine models, giving complete protection on oral dosing. However, the effect was not observed without toxicity in rat and dog models. Although this particular agent might not be "the treatment" for Dengue virus, it is at least 10 times more potent than other natural product-based agents in the cell line assays. The MOA is probably via chain termination during viral RNA synthesis as a triphosphate form of NITD008 directly inhibited the RNA-dependent RNA polymerase of Dengue virus with an IC_{50} figure of 310 nM, and no resistant clones of Dengue or West Nile viruses were observed following four months exposure to the agent at varying concentrations.

Thus, even a half-century after Bergmann's discovery of bioactive arabinose nucleosides, small molecules synthesized as a result of his discoveries are still in

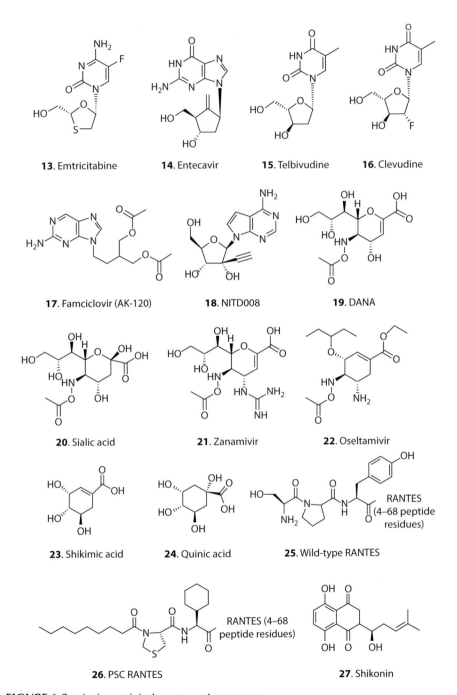

FIGURE 6.2 Active antiviral agents and precursors.

clinical use and in clinical and preclinical trials for treatment of viral diseases and also for chemotherapeutic treatments in various tumor types, though these are not covered in this chapter.

6.3 NEURAMINIDASE INHIBITORS

In 1974, the first example of a neuraminidase inhibitor (DANA; Figure 6.2, Structure **19**), a subtle variation on the very well known aminosugar, N-acetylneuraminic acid or sialic acid (Figure 6.2, Structure **20**) was reported by Meindl et al. (1974). Using this information and the later x-ray crystallographic structures of neuraminidases A and B and their complexes with sialic acid, published in 1983 by Australian investigators (Colman et al. 1983; Varghese et al. 1983), where the essential linkages to sialic acid were identified, workers at the Australian company Biota, in conjunction with the original investigators, designed the compound that finally became approved as Relenza® or zanamivir (Figure 6.2, Structure **21**) following licensing to Glaxo (now Glaxo SmithKline) in 1990 for development and marketing. Due to its insolubility, the agent has to be delivered by insufflation over a period of 5 days. Currently, the compound is not used at significant levels due to the problems of delivery, and only a synthesis from N-acetylneuraminic acid is used for the pharmaceutical preparation (Chandler et al. 1995).

Following this demonstration of the viability of the target, in 1995, Gilead Sciences patented the compound that would ultimately become known as oseltamivir (Figure 6.2, Structure **22**) or Tamiflu®, and licensed it to Roche for codevelopment of the drug. The methods leading to a viable synthesis of the phosphate salt were thoroughly described in 2004 by Abrecht et al. (2004) and the drug was launched in November of 1999. Oseltamivir is in fact a prodrug as its free acid, rather than the ethylester, was originally chosen following its initial synthesis from shikimic acid (Figure 6.2, Structure **23**) (Kim et al. 1997). Over the years since the initial synthesis, the process has been optimized by Roche using either shikimic acid or quinic acid (Figure 6.2, Structure **24**) as the starting material, with further modifications in the individual processes since then being thoroughly described in a very recent review by Magano (2009).

What is of significant import is the realization that shikimic acid, although apparently relatively abundant in China, with a quoted yield of 1 kg of shikimic acid being isolated from 30 kg of dried fruit of *Illicium verum* (the Chinese star anise), can also be obtained from a bioengineered *Escherichia coli*, which actually produced not only shikimic acid but also quinic acid with yields of 27 g L^{-1} of the former and 12 g L^{-1} of the latter on a 42 h fed-batch fermentation, thus providing a renewable resource (Draths et al. 1999). Where the significance comes in is that very recently, Shi and his group at the East China University of Science and Technology in Shanghai have reported two very robust and productive syntheses of oseltamivir. The first used a 13-step process from shikimic acid with an overall yield of 40% (Nie and Shi 2009), which was improved in a second report starting from the same material to a nine-step method with an overall yield of 47% (Nie et al. 2009).

Other natural product derived agents that inhibit various enzymatic steps in replication of the influenza virus, including the inhibition of viral RNA polymerase, inosine monophosphate dehydrogenase, and some fusion inhibitors are well described by Lagoja and De Clercq in a 2006 review (Lagoja and De Clercq 2006). To date, however, only Tamiflu and Relenza have been approved from NP-sources.

6.4   VIRAL ENTRY INHIBITORS

The well-known targets in HIV chemotherapy, such as inhibition of reverse transcriptase, the protease, or the integrase enzymes, are all involved in the reproductive processes of the virus once it enters the cell. However, if the initial entry of the virus could be inhibited, then there is the potential for novel specific agents that may have lower toxicities and hence higher therapeutic indices.

Over the last 10 or so years, the biological processes involved in the initial attachment and then the subsequent entry into the host cell(s) have been investigated from a basic molecular biology aspect. In the case of HIV, these involve at least three processes, attachment via CD4 receptor binding (Lu et al. 1995), the binding of coreceptors (Peck et al. 1997), and the fusion process (Melikyan et al. 2000).

Prior to the description of these interrelated processes, there had been a series of reports from empirical assays (usually involving survival assays of infected cells) that a variety of nonspecific soluble polyanions would block attachment to cells; a number of these agents were derived from natural products, including sterol sulfates, high and low molecular weight heparins, dextran sulfates, etc. Although there were some early clinical trials with dextran sulfate, they were not successful (Flexner et al. 1991).

In 1995, the Gallo group (Cocchi et al. 1995) identified the chemokine RANTES (Figure 6.2, Structure **25**), a 68 residue peptide, as a competitor of the binding of gp-120 (the HIV glycoprotein) to the CCR5 coreceptor and, thus, an inhibitor of HIV infection. Modification of the N-terminus of the peptide, usually limited to the first 3–5 residues, led to a series of RANTES analogues that demonstrated nanomolar activity against HIV-infection with reduced side effects and culminated in PSC-RANTES (Figure 6.2, Structure **26**). This compound is currently being evaluated as a microbicide, particularly in the form of nanoparticles using a biodegradable polylactic–polyglycolic copolymer as the carrier (Ham et al. 2009).

The reports on the small natural product compound shikonin (Figure 6.2, Structure **27**), isolated from the Chinese medicinal plant *Lithospermum erythrorhizon*, demonstrate that such molecules have multiple effects including inhibition of CCR5 at nanomolar concentrations (Chen et al. 2003). However, it is also an inhibitor of the activation of tumor necrosis factor (Staniforth et al. 2004), and imitates insulin in that it inhibits "phosphatase and tensin homolog deleted on chromosome 10" (PTEN) and other tyrosine phosphatases (Nigorikawa et al. 2006). Thus, this is an excellent example of why one should assay against entirely different enzyme/cellular activities before considering an agent as specific.

There is one fusion inhibitor in clinical use that is based upon the discovery that peptides based upon HR1 and HR2 sequences in gp41 could inhibit HIV infection (Wild et al. 1992; Jiang et al. 1993). Thus, in 2003, the U.S. FDA approved the 36 residue peptide enfuvirtide (Fuzeon®) (Figure 6.3, Structure **28**), which, though synthetic, is identical to the 127–162 residues in the C-terminal end of gp41 (or 643–678 from the N-terminus). On binding to the HR-1 region of gp41, enfuvirtide prevents

AcO$_2$C-L-Tyr-L-Thr-L-Ser-L-Leu-L-Ile-L-His-L-Ser-L-Leu-L-Ile-L-Glu-L-
Glu-L-Ser-L-Gln-L-Asn-L-Gln-L-Gln-L-Glu-L-Lys-L-Asn-L-Glu-L-Gln-L-
Glu-L-Leu-L-Leu-L-Glu-L-Leu-L-Asp-L-Lys-L-Trp-L-Ala-L-Ser-L-Leu-
L-Trp-L-Asn-L-Trp-L-Phe-NH$_2$

28. Enfuvirtide

NH$_2$-L-Thr-L-thr-L-trp-L-glup-L-alap-L-trp-
L-asp-L-arg-L-ala-L-ile-L-ala-L-glu-L-tyr-L-ala-
L-ala-L-arg-L-ile-L-glu-L-ala-L-leu-L-leu-L-arg-
L-ala-L-leu-L-gln-L-glu-L-gln-L-gln-L-glu-L-lys
-L-asn-L-glu-L-ala-L-ala-L-leu-L-arg-L-glu-L-leu-CO$_2$H

29. TRI-1144

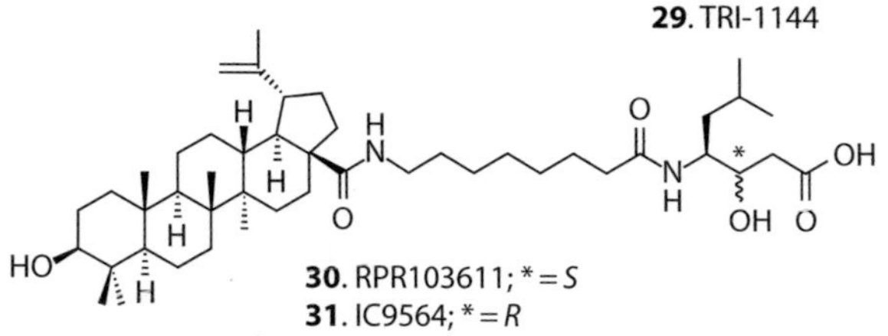

30. RPR103611; * = S
31. IC9564; * = R

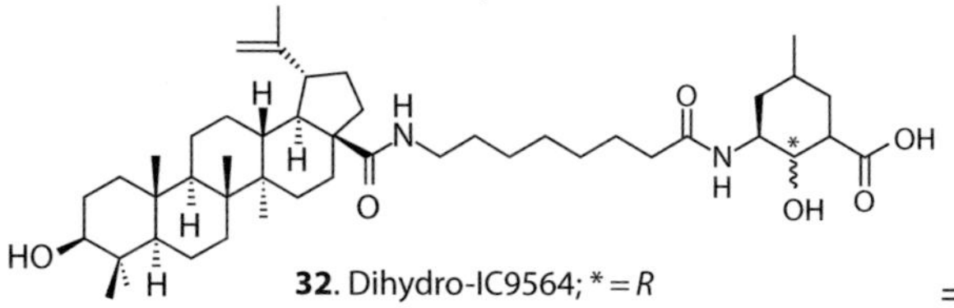

32. Dihydro-IC9564; * = R

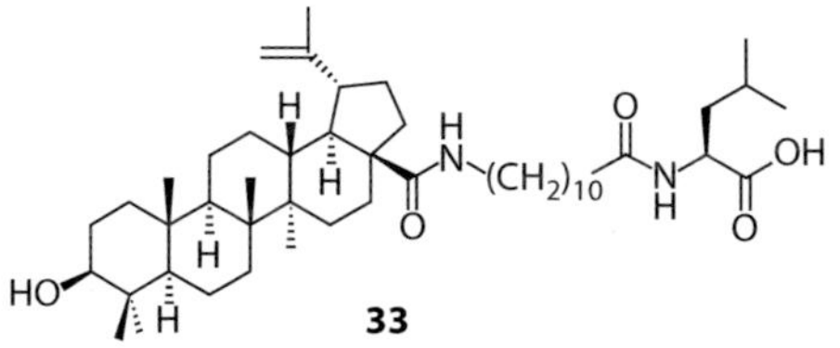

33

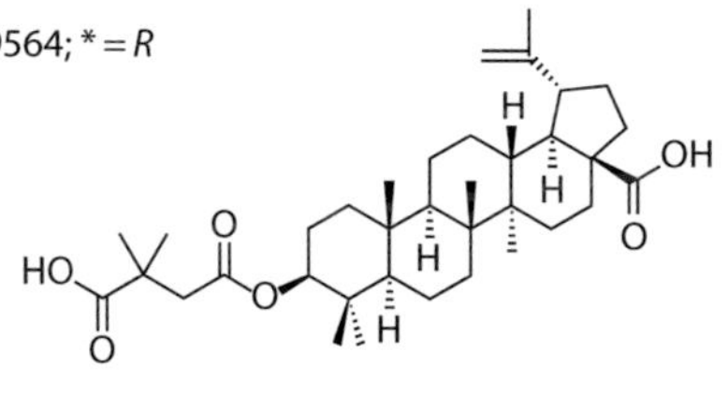

34. Bevirimat

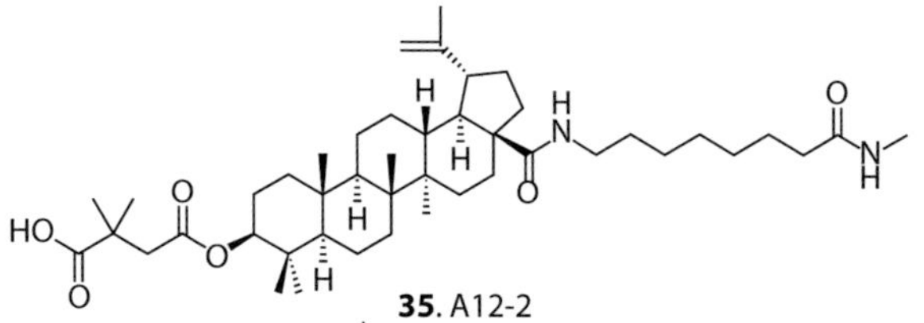

35. A12-2

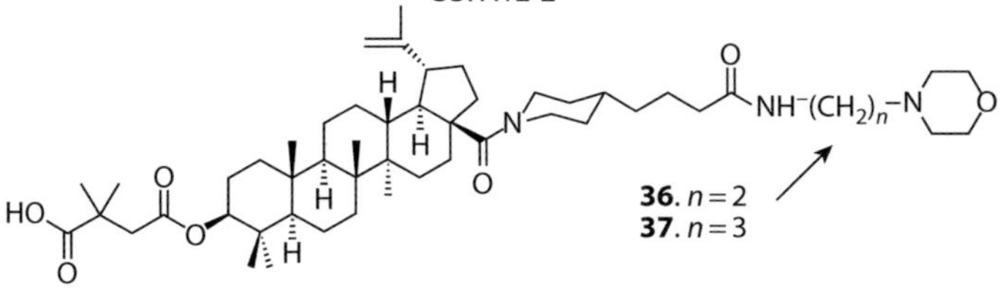

36. n = 2
37. n = 3

FIGURE 6.3   Peptidic and terpenoid antiviral agents.

formation of the six-helix arrangement necessary for further development of the infection (Weiss 2003). Recently, an interesting aspect of enfuvirtide was reported by Alexander et al. (2009) implying that attachment inhibitors such as this compound might "contribute to immune system homeostasis in individuals infected with HIV-1 that can engage CXCR4, thereby mitigating the increased risk of adverse clinical events observed in such individuals on current antiretroviral regimens."

In contrast to that report, a recent discussion of a clinical trial from Italy, where combination studies with nucleoside reverse transcriptase inhibitors (NRTIs) and/or protease inhibitors demonstrates that even using these potent agents in combination, HIV can still mutate to overcome the effects mainly by mutation in the Rev protein, one of the three regulatory proteins, *Tat, Rev,* and *Nef* (Svicher et al. 2009), though one should add that currently there are seven studies still recruiting patients under various regimens listed in the NIH "clinicaltrials.gov" Web site.

Trimeris has now entered a slightly different small peptide derived from studies on the HR1 locus into phase I clinical trials under the code number TRI-1144 (Figure 6.3, Structure **29**). Though it has not been reported other than in abstract format (Bai et al. 2007), a paper in the *Proceedings of the National Academy of Sciences* in 2007 gave a description of the rationale used to derive these peptides and should be consulted for further information (Dwyer et al. 2007), together with a recently issued patent covering the synthesis of these and other agents (Bray et al. 2009).

In addition to the compounds referred to above, a very interesting group of natural products have demonstrated activity as potential small molecule fusion inhibitors, though exactly where in the overall process they work is still under investigation. Thus, early work by Lee's group at North Carolina with betulinic acid and platanic acid demonstrated inhibition of HIV-1$_{\text{IIIB}}$ at low micromolar levels in the early 1990s (Fujioka et al. 1994). Further work by two groups demonstrated that modification of the C-28 acid side chain of betulinic acid led to RPR103611 (Figure 6.3, Structure **30**) from the De Clercq group and Rhone Poulenc Rorer (Mayaux et al. 1994; Soler et al. 1996), and its stereoisomer IC9564 (Figure 6.3, Structure **31**) from Lee's group (Sun et al. 2002) both of which could be looked at as statine derivatives. Subsequent work around IC9564 led to the dihydroderivative with an isopropyl (Figure 6.3, Structure **32**) rather than the natural isopropenyl side chain, and a chain extended version with a leucyl (Figure 6.3, Structure **33**) rather than a statine moiety as the chain terminator. Both of these compounds had activities comparable to the earlier derivatives, with an easier synthesis in the case of (Figure 6.3, Structure **33**) (Sun et al. 2002).

6.5 VIRAL MATURATION INHIBITORS

As shown above, betulinic acid derivatives modified at the C-28 site are HIV fusion inhibitors. In contrast, modification at the C-3 hydroxyl led to the phase II candidate bevirimat (Figure 6.3, Structure **34**) which originated in Lee's group, was developed initially by Panacos (Martin et al. 2008), and is now being developed by Myriad Pharmaceuticals with two clinical trials being listed as of January 2010 under the numbers NCT00511368 and NCT00967187, with an early report on human pharmacokinetics being published in early 2009 (Connor et al. 2009).

By combining both the C-3 and C-28 modifications, Lee's group together with a group at Duke University synthesized A12-2 (Figure 6.3, Structure **35**), which has both activities maintained (inhibition of both entry and maturation) in the same molecule with an IC_{50} value of 2.6 nM (Huang et al. 2006). In 2009, the same group plus investigators from Panacos Pharmaceuticals (Qian et al. 2009) reported further on molecules similar to A12-2 demonstrating retention of both activities and greater stability to pooled human liver microsomes in in vitro studies (Figure 6.3, Structures **36, 37**). They also reported extensively on the structure activity relationships (SAR) of this series of molecules in the same paper. In addition, contemporaneously with the publication of the paper referred to in the last sentence, the Lee group published an excellent review on plant-derived triterpenoids as antitumor and anti-HIV agents, which should be consulted for more of the background information on this series of compounds (Kuo et al. 2009). It will be interesting to follow the development of these agents as time progresses.

6.6 GLYCAN INTERACTIVE AGENTS

One of the first agents from nature to demonstrate significant activity against HIV infection in an in vitro setting aside from the polyanionic compounds referred to earlier was the cyanobacterial 101 aminoacid residue peptide now known as cyanovirin or CV-N. This agent was discovered by the NCI group led at that time by Boyd and was reported in detail in 1997 (Boyd et al. 1997), following its isolation and purification from *Nostoc ellipsosporum*. Later work by the NCI group and collaborators on the crystal structures and solution conformations revealed the existence of a domain-swapped dimer with two primary and two secondary carbohydrate-binding sites on opposite ends of the dimer (Bewley et al. 1998; Yang et al. 1999; Botos et al. 2002). The carbohydrate recognition sites recognize high-mannose glycans and in particular $\alpha(1,2)$-linked mannose oligomers (Bewley 2001; Bolmstedt et al. 2001).

This agent has been extensively assayed against other viruses as well as HIV and a patent has been applied for as an agent against H5N1 influenza (O'Keefe and McMahon 2008). In addition, it exhibits activity against H1N1 influenza in animal models (Smee et al. 2008), and there are recent reports that it will block herpes simplex virus I entry and cell fusion (Liu et al. 2008; Tiwari et al. 2009).

In addition to these reports, for some years, the compound has been studied for its potential to act as a vaginal virucide to help prevent HIV infection (Ndesendo et al. 2008; Buffa et al. 2009; Fischetti et al. 2009). Because of the requirement for large quantities of pure material if these studies are successful, a significant effort has been made to produce cyanovirin in plant tissues, with an example being the recent report from Drake et al. (2009) demonstrating that hydroponically grown tobacco plants may be a method of production for this and other agents, thereby addressing the major problem with all such agents, namely, the production of adequate supplies under suitable conditions and at a low price. The same group has also expanded the process to produce a fusion protein of cyanovirin linked to an HIV-neutralizing monoclonal antibody in order to extend the potential response(s) (Sexton et al. 2009).

Following on from the cyanovirin discovery, the same group of NCI scientists reported on another antiviral peptide, scytovirin, from the cyanobacterium

Scytonema varium (Bokesch et al. 2003). This particular 95 residue peptide was structurally dissimilar to cyanovirin with five intrachain disulfide bonds and found to have a pronounced affinity for α(1,2)-α(1,6)-mannose trisaccharide units (Adams et al. 2004). Further work reported from the same group demonstrated that modifications could be made to the initial peptide using deletion mutations while still maintaining reasonable anti-HIV-1 activity (Xiong et al. 2006). This work was followed by a solution NMR structural determination showing that there were novel "twists" to the base molecule with two domains having quite different carbohydrate binding affinities (McFeeters et al. 2007). The later x-ray structural analysis (Moulaei et al. 2007) did not agree in the assignment of the cysteine bridges with those determined in the NMR and sequence analyses but the peptides still had the same antiviral activities. Work is ongoing to discover the reason(s) for these discrepancies.

Although the two previous examples of entry inhibitors came from cyanobacteria, the third one, griffithsin, reported by the NCI group, was slightly higher in molecular weight and in its source was an undescribed species of the marine red alga *Griffithsia* collected in New Zealand waters as part of the NCI's marine collection program (Mori et al. 2005). Its potency against a variety of HIV strains was in the 40–600 pM range, and it targeted the terminal mannose-residues in the high mannose oligosaccharides with a total of six mannose-binding domains per homodimer. The peptide was produced using recombinant means in *E. coli* in sufficient yield in the soluble fraction, rather than as inclusion bodies, to be able to produce material for further studies. Griffithsin also existed as a homodimer in solution, with comparable activities to the naturally occurring material even after the N-terminal methionine had been cleaved (Giomarelli et al. 2006).

The x-ray structure and assignment of binding domains was performed by the same group as with cyanovirin and scytovirin and demonstrated that the molecule had a monomeric three-dimensional structure reminiscent of the jacalin lectins, but with a distinctly different dimeric structure. The investigators also reported that this peptide inhibited the SARS Coronavirus (Ziolkowska et al. 2006) and in a later paper the following year determined the binding sites when complexed with mannose by x-ray crystallography (Ziolkowska et al. 2007).

Contemporaneously with the studies on structure and mannose binding, the potential for the use of this agent as an anti-HIV virucide (microbicide) was investigated in a collaboration between the University of Washington and the NCI with a report in 2007 (Emau et al. 2007). Since the "Achilles Heel" of any naturally occurring peptide, and in particular any peptide that could be of use as an anti-HIV (or other antiviral) agent in the developing world is cost, the report by the NCI group in the early part of 2009 that griffithsin could be produced in *Nicotiana benthamiana* (tobacco plant), and that the 60+ g of material produced in one 5000 ft² enclosed greenhouse was safe and efficacious in a Macaque vaginal model, meant that a major hurdle in its early development had been overcome (O'Keefe et al. 2009; Zeitlin et al. 2009).

In addition to these agents, there are other "lectin-like" materials that have been reported and tested for their potential as antiviral agents (mainly anti-HIV). However, none are anywhere close to the development levels of the NCI-discovered agents. For further information on these other agents, one should consult the 2007 review by Balzarini (2007) and the 2008 review by Li et al. (2008).

6.7   NONNUCLEOSIDE REVERSE TRANSCRIPTASE INHIBITORS

One of the first series of natural product NNRTIs were the dipyranocoumarins known as calanolide A (Figure 6.4, Structure **38**) and B (Figure 6.4, Structure **39**) and the closely related soulattrolide (Figure 6.4, Structure **40**). The calanolides were originally reported by the NCI group led by Boyd in a paper in 1992 (Kashman et al. 1992), and over the next few years, the agents were synthesized and new sources of calanolide B were found following an inability to recover the original source tree, *Calophyllum langierum* var *austrocoriaceum*, in Sarawak (Fuller et al. 1994). In the same general time frame, workers from the then SmithKline Beecham reported (Patil et al. 1993) on the close chemical relatives, the inophyllums, which were isolated via a discovery from an intermediate host (the snail *Achatina fulica*) whose diet included leaves from the tree *Calophyllum inophyllum* Linn. These workers demonstrated that inophyllum B (Figure 6.4, Structure **41**) and its isomer P (Figure 6.4, Structure **42**) were active as NNRTIs and also inhibited HIV-1 in cell culture (Patil et al. 1993). Approximately 10 years later, a group in Mexico were able to isolate calanolides A and B and soulattrolide from the leaves of the tree *Calophyllum brasiliense* by utilizing comparative bioactivity-driven isolation techniques. These included comparison of the inhibition of HIV-reverse transcriptase with cytotoxities and high inhibition of HIV-1$_{IIIB}$ replication (Huerta-Reyes et al. 2004). Quite recently, a French group reported that inophyllums B and P could be isolated from *C. inophyllum* leaves collected in French Polynesian islands, thus demonstrating the

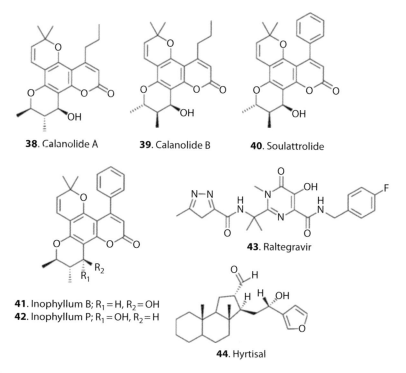

38. Calanolide A   **39**. Calanolide B   **40**. Soulattrolide

41. Inophyllum B; R$_1$ = H, R$_2$ = OH
42. Inophyllum P; R$_1$ = OH, R$_2$ = H

43. Raltegravir

44. Hyrtisal

FIGURE 6.4   NNRTI and integrase inhibitors.

widespread occurrence of these agents in *Calophyllum* species across continents and oceans (Laure et al. 2008).

The story of the discovery, syntheses by a number of groups (Sorbera et al. 1999) and subsequent preclinical development by NCI, and the early clinical trials by a joint venture between the Government of Sarawak and the small U.S. company Medichem, and its later spin off, Advanced Life Sciences, have been reported at length over the years. The authors published a short review in 2003, which should be consulted for information (Cragg and Newman 2003). The story of the establishment of the joint venture as part of the NCI's "source country commitment" involved in the licensing of any NCI patent that results from the collection program under the NCI's Letter of Collection, was extensively and favorably commented upon in a review as a "Benefit-Sharing Case Study" for the Executive Secretary of the Convention on Biodiversity (CBD) by staff of the Royal Botanic Gardens in Kew. This document is available for download from the CBD website as of the time of writing (December 2009) using the following URL <http://cbd.int/case-studies/?tab = 1> and then entering "calanolide" as the search term.

The rights to the calanolides, including all patents related to HIV treatment and syntheses, were returned to the Sarawak side of the joint venture, and currently there are two trials shown on the NCI clinical trials web site, one phase I that has completed, and another phase I that is still shown as "active" but under a nonexistent company. Hopefully in due course, further development of this agent will continue under the auspices of the Sarawak government.

6.8 INTEGRASE INHIBITORS

HIV integrase is the viral enzyme that "integrates" the viral DNA produced from the action of reverse transcriptase with the host's DNA in the nucleus. Although there are drugs that are in clinical use that will inhibit this process, none are from natural sources at this moment. However, over the years, a significant amount of work was performed at the Merck Research Laboratories in New Jersey, investigating the potential of their ~200,000 microbial extracts from over 50,000 microbial strains in a variety of assays related to integrase activities (Singh et al. 2005). Overall, 24 novel classes of natural product integrase inhibitors were identified with molecular weights ranging from 190 to 1663 Da. What was of interest was the realization later, that a α-hydroxyketo group was present in many of the compounds identified, a grouping that can bind the two Mg^{2+} ions that are at the catalytic center of the enzyme, and this group is a part of the Merck synthetic integrase inhibitor raltegravir (Figure 6.4, Structure **43**) launched in 2007.

Contemporaneously with the approval of this agent, a group in China (Du et al. 2008) demonstrated that the compound hyrtisal (Figure 6.4, Structure **44**) apparently inhibits HIV-1 integrase via a previously unrecognized binding site at the N-terminal end of the protein. This agent was previously reported (Sun et al. 2007) by the same group to be a noncompetitive inhibitor of the protein phosphatase PTP1B at about four times the concentration (~40 versus ~10 μM). Thus, there may yet be a natural product derived inhibitor of this enzyme but it will require a significant amount of chemical and pharmacological work to increase the potency of this base structure.

6.9 CONCLUSION

In this short chapter, we have attempted to demonstrate how natural product structures have led chemists and virologists to compounds that in some cases, particularly modified nucleosides, have led to drugs that are or have been in use against a variety of virus infections but mainly directed toward HIV-1. We have also demonstrated that agents of varying chemical classes, particularly peptides of around 100–140 amino acid residues isolated from marine cyanobacteria (blue-green algae) and red seaweed, together with those derived from the plant triterpenoid, betulinic acid, are significantly active as "entry inhibitors" against a variety of viruses.

The probable best case scenario for the peptidic agents is as components of vaginal virucides/microbicides, and the group that found these agents has also overcome the major hurdle when working with natural products, that of producing enough protein mass for trials by utilizing tobacco and other plants as producers, following transformation with suitable plasmids containing the necessary biosynthetic clusters/promoters. In contrast, the betulinic acid derivatives appear to be more amenable to conventional routes of administration, and it will be very informative to follow the progress of bevirimat (Figure 6.3, Structure **34**) and later derivatives in their journeys through the clinical trials programs.

We did not cover earlier agents, such as michellamine B (McMahon et al. 1995) or conocurvone (Decosterd et al. 1993) where the agents did not move out of very early preclinical trials, though in 2008, an Australian group reported the synthesis of derivatives of conocurvone and some antiviral activities (Crosby et al. 2008). Nor have we covered prostratin, perhaps the only ethnobotanical lead (albeit in a different disease setting) that is hopefully leading toward a clinical trial in antiviral therapy, though as a method of causing the efflux of masked viruses, in particular HIV, in patients undergoing Highly Active Antiretroviral Therapy (HAART) (Kulkosky et al. 2001; Brown et al. 2005).

Nor did we cover the very potent group of plant proteins known as the cyclotides, which have significant antiviral activities in addition to a multiplicity of other pharmacological activities. They are not viable drug candidates in their current form though if suitably "engineered" they may well prove to be viable candidates for a multiplicity of pharmacologic interventions in the future (Daly et al. 2009; Wang et al. 2009).

Thus, we hope that we have convinced the reader that Mother Nature's chemistry is still a viable source of structures upon which to practice drug development against viruses both old and new.

REFERENCES

Abrecht, S., P. Harington, H. Iding, M. Karpf, R. Trussardi, B. Wirtz, and U. Zutter. 2004. The synthetic development of the anti-influenza neuraminidase inhibitor oseltamivir phosphate (Tamiflu®): A challenge for synthesis and process research. *Chimia* 58:621–629.

Adams, E. W., D. M. Ratner, H. R. Bokesch, J. B. McMahon, B. R. O'Keefe, and P. H. Seeberger. 2004. Oligosaccharide and glycoprotein microarrays as tools in HIV glycobiology glycan-dependent gp120/protein interactions. *Chem. Biol.* 11:875–881.

Alexander, L., S. Zhang, B. McAuliffe, D. Connors, N. Zhou, T. Wang, M. Agler, J. Kadow, and P. F. Lin. 2009. Inhibition of envelope-mediated CD4+-T-cell depletion by human immunodeficiency virus attachment inhibitors. *Antimicrob. Agents Chemother.* 53:4726–4732.

Baba, M., M. Okamoto, and H. Takeuchi. 1999. Inhibition of human immunodeficiency virus type 1 replication in acutely and chronically infected cells by EM2487, a novel substance produced by a *Streptomyces* species. *Antimicrob. Agents Chemother.* 43:2350–2355.

Bai, X., K. L. Wilson, D. K. Davison, R. Medinas, S. A. Freel, L. Jin, S. A. Stanfield-Oakley, Z. Wang, M. L. Greenberg, and J. J. Dwyer. 2007. Structural analysis of the fusion inhibitor TRI-1144 in complex with an HR1 containing the enfuvirtide resistance mutation N43D suggests a mechanism for activity against FI-resistant virus. *Antivir. Ther.* 12:5(Suppl), Abst. S136.

Balzarini, J. 2007. Targeting the glycans of glycoproteins: A novel paradigm for antiviral therapy. *Nat. Rev. Microbiol.* 5:583–597.

Bergmann, W. and D. C. Burke. 1956. Contributions to the study of marine products. XL. The nucleosides of sponges. IV. Spongosine. *J. Org. Chem.* 21:226–228.

Bergmann, W. and R. J. Feeney. 1950. The isolation of a new thymine pentoside from sponges. *J. Am. Chem. Soc.* 72:2809–2810.

Bergmann, W. and R. J. Feeney. 1951. Contributions to the study of marine products. XXXII. The nucleosides of sponges I. *J. Org. Chem.* 16:981–987.

Bewley, C. A. 2001. Solution structure of a cyanovirin-N:Manα1–2Manα complex: Structural basis for high-affinity carbohydrate-mediated binding to gp120. *Structure* 9:931–940.

Bewley, C. A., K. R. Gustafson, M. R. Boyd, D. G. Covell, A. G. Bax, M. Clore, and A. M. Gronenborn. 1998. Solution structure of cyanovirin-N, a potent HIV-inactivating protein. *Nature Struct. Biol.* 5:571–578.

Bokesch, H. R., B. R. O'Keefe, T. C. McKee, L. K. Pannell, G. M. L. Patterson, R. S. Gardella, R. C. Sowder II et al. 2003. A potent novel anti-HIV protein from the cultured cyanobacterium *Scytonema varium. Biochemistry* 42:2578–2584.

Bolmstedt, A. J., B. R. O'Keefe, S. R. Shenoy, J. B. McMahon, and M. R. Boyd. 2001. Cyanovirin-N defines a new class of antiviral agent targeting N-linked, high-mannose glycans in an oligosaccharide-specific manner. *Mol. Pharmacol.* 59:949–954.

Botos, I., B. R. O'Keefe, S. R. Shenoy, L. K. Cartner, D. M. Ratner, P. H. Seeberger, M. R. Boyd, and A. Wlodawer. 2002. Structures of the complexes of a potent anti-HIV protein cyanovirin-N and high mannose oligosaccharides. *J. Biol. Chem.* 277:34336–34342.

Boyd, M. R., K. R. Gustafson, J. B. McMahon, R. H. Shoemaker, B. R. O'Keefe, T. Mori, R. Gulakowski et al. 1997. Discovery of cyanovirin-N, a novel human immunodeficiency virus-inactivating protein that binds virla surface envelope glycoprotein gp120: Potential applications to microbicide development. *Antimicrob. Agents Chemother.* 41:1521–1530.

Bray, B. L., B. E. Johnston, S. F. Schneider, N. A. Tvermoes, H. Zhang, and P. E. Friedrich. 2009. Novel methods of synthesis for therapeutic peptides. WO 2009042194.

Brown, H. J., W. H. McBride, J. A. Zack, and R. Sun. 2005. Prostratin and bortezomib are novel inducers of latent Kaposi's sarcoma-associated herpes virus. *Antivir. Ther.* 10:745–751.

Buffa, V., D. Stieh, N. Mamhood, Q. Hu, P. Fletcher, and R. J. Shattock. 2009. Cyanovirin-N potently inhibits human immunodeficiency virus type 1 infection in cellular and cervical explant models. *J. Gen. Virol.* 90:234–243.

Chandler, M., M. J. Bamford, R. Conroy, B. Lamont, B. Patel, V. K. Patel, I. P. Steeples et al. 1995. Synthesis of the potent influenza neuraminidase inhibitor 4-guanidino Neu5Ac2en. X-ray molecular structure of 5-acetamido-4-amino-2,6-anhydro-3,4,5-trideoxy-*D-erythro-L-gluco*-nononic acid. *J. Chem. Soc., Perkin Trans.* 1:1173–1180.

Chen, X., L. Yang, N. Zhang, J. A. Turpin, R. W. Buckheit, C. Osterling, J. J. Oppenheim, and O. M. Howard. 2003. Shikonin, a component of Chinese herbal medicine, inhibits chemokine receptor function and suppresses human immunodeficiency virus type 1. *Antimicrob. Agents Chemother.* 47:2810–2816.

Cimino, G., S. De Rosa, and S. De Stefano. 1984. Antiviral agents from a gorgonian, *Eunicella cavolini. Experientia* 40:339–340.

Cocchi, F., A. L. DeVico, A. Garzino-Demo, S. K. Ayra, R. C. Gallo, and P. Lusso. 1995. Identification of RANTES, MIP-1 alpha, and MIP-1 beta as the major HIV-suppressive factors produced by CD8+ cells. *Science* 270:1811–1815.

Colman, P. M., J. N. Varghese, and W. G. Laver. 1983. Structure of the catalytic and antigenic sites in influenza virus neuraminidase. *Nature* 303:41–44.

Connor, A., P. Evans, J. Doto, C. Ellis, and D. E. Marti. 2009. An oral human drug absorption study to assess the impact of site of delivery on the bioavailability of bevirimat. *J. Clin. Pharmacol.* 49:606–612.

Cragg, G. M. and D. J. Newman. 2003. Plants as a source of anti-cancer and anti-HIV agents. *Ann. Appl. Biol.* 143:127–133.

Crosby, I. T., M. L. Rose, M. P. Collis, P. J. de Bruyn, P. L. C. Keep, and A. D. Robertson. 2008. Antiviral agents. I. Synthesis and antiviral evaluation of trimeric naphthoquinone analogs of conocurvone. *Aust. J. Chem.* 61:768–784.

Daly, N. L., K. J. Rosengren, and D. J. Craik. 2009. Discovery, structure and biological activities of cyclotides. *Adv. Drug Del. Rev.* 61:918–930.

Decosterd, L. A., I. C. Parsons, K. R. Gustafson, J. H. Cardellina II, J. B. McMahon, G. M. Cragg, Y. Murata et al. 1993. HIV inhibitory natural products. 11. Structure, absolute stereochemistry, and synthesis of conocurvone, a potent, novel HIV-inhibitory naphthoquinone trimer from a *Conospermum* sp. *J. Am. Chem. Soc.* 115:6673–6679.

Drake, P. M. W., T. Barbi, A. Sexton, E. McGowan, J. Stadlmann, C. Navarre, M. J. Paul, and J. K.-C. Ma. 2009. Development of rhizosecretion as a production system for recombinant proteins from hydroponic cultivated tobacco. *FASEB J.* 23:3581–3589.

Draths, K. M., D. R. Knop, and J. W. Frost. 1999. Shikimic acid and quinic acid: Replacing isolation from plant sources with recombinant microbial biocatalysis. *J. Am. Chem. Soc.* 121:1603–1604.

Du, L., L. Shen, Z. Yu, J. Chen, Y. Guo, Y. Tang, X. Shen, and H. Jiang. 2008. Hyrtiosal, from the marine sponge *Hyrtios erectus*, inhibits HIV-1 integrase binding to viral DNA by a new inhibitor binding site. *Chem. Med. Chem.* 3:173–180.

Dwyer, J. J., K. L. Wilson, D. K. Davison, S. A. Freel, J. E. Seedorf, S. A. Wring, N. A. Tvermoes, T. J. Matthews, M. L. Greenberg, and M. K. Delmedico. 2007. Design of helical, oligomeric HIV-1 fusion inhibitor peptides with potent activity against enfuvirtide-resistant virus. *Proc. Natl. Acad. Sci. USA* 104:12772–12777.

Emau, P., B. Tian, B. R. O'Keefe, T. Mori, J. B. McMahon, K. E. Palmer, Y. Jiang, G. Bekele, and C. C. Tsai. 2007. Griffithsin, a potent HIV entry inhibitor, is an excellent candidate for anti-HIV microbicide. *J. Med. Primatol.* 36:244–253.

Evans, B. E., K. E. Rittle, M. G. Bock, R. M. DiPardo, R. M. Freidinger, W. L. Whitter, G. F. Lundell et al. 1988. Methods for drug discovery: Development of potent, selective, orally effective Cholecystokinin antagonists. *J. Med. Chem.* 31:2235–2246.

Fischetti, L., S. M. Barry, T. J. Hope, and R. J. Shattock. 2009. HIV-1 infection of human penile explant tissue and protection by candidate microbicides. *AIDS* 23:319–328.

Flexner, C., P. A. Barditch-Crovo, D. M. Kornhauser, H. Farzadegan, L. J. Nerhood, R. E. Chaisson, K. M. Bell, K. J. Lorentsen, C. W. Hendrix, and B. G. Petty. 1991. Pharmacokinetics, toxicity, and activity of intravenous dextran sulfate in human immunodeficiency virus infection. *Antimicrob. Agents Chemother.* 35:2544–2550.

Fujioka, T., Y. Kashiwada, R. E. Kilkuskie, L. M. Cosentino, L. M. Ballas, J. B. Jiang, W. P. Janzen, I. S. Chen, and K. H. Lee. 1994. Anti-AIDS agents, II. Betulinic acid and platanic acid as HIV principles from *Syzigium claviforum* and the anti-HIV activity of structurally-related triterpenoids. *J. Nat. Prod.* 57:243–247.

Fuller, R. W., H. R. Bokesch, K. R. Gustafson, T. C. McKee, J. H. Cardellina II, J. B. McMahon, G. M. Cragg, D. D. Soejarto, and M. R. Boyd. 1994. HIV-inhibitory coumarins from latex of the tropical rainforest tree *Calophyllum teysmannii* var *inophylloide. Bioorg. Med. Chem. Lett.* 4:1961–1964.

Giomarelli, B., K. M. Schumacher, T. E. Taylor, R. C. Sowder, J. L. Hartley, J. B. McMahon, and T. Mori. 2006. Recombinant production of anti-HIV protein, griffithsin, by auto-induction in a fermentor culture. *Protein. Expr. Purif.* 47:194–202.

Ham, A. S., M. R. Cost, A. B. Sassi, C. S. Dezzutti, and L. C. Rohan. 2009. Targeted delivery of PSC-RANTES for HIV-1 prevention using biodegradable nanoparticles. *Pharm. Res.* 26:502–511.

Huang, L., P. Ho, K. H. Lee, and C. H. Chen. 2006. Synthesis and anti-HIV activity of bi-functional betulinic acid derivatives. *Bioorg. Med. Chem.* 14:2279–2289.

Huerta-Reyes, M., M. D. C. Basualdo, F. Abe, M. Jimenez-Estrada, C. Soler, and R. Reyes-Chilpa. 2004. HIV-1 inhibitory compounds from *Calophyllum brasiliense* leaves. *Biol. Pharm. Bull.* 27:1471–1475.

Jiang, S., K. Lin, N. Strick, and A. R. Neurath. 1993. HIV-1 inhibition by a peptide. *Nature* 365:113.

Kashman, Y., K. R. Gustafson, R. W. Fuller, J. H. Cardellina II, J. B. McMahon, M. J. Currens, R. W. Buckheit Jr., S. H. Hughes, G. M. Cragg, and M. R. Boyd. 1992. The calanolides, a novel HIV-inhibitory class of coumarin derivatives from the tropical rain forest tree, *Calophyllum langierum. J. Med. Chem.* 35:2735–2743.

Kim, C. U., W. Lew, M. A. Williams, H. Liu, L. Zhang, S. Swaminathan, N. Bischofberger et al. 1997. Influenza neuraminidase inhibitors possessing a novel hydrophobic interaction in the enzyme active site: Design, synthesis, and structural analysis of carbocyclic sialic acid analogues with potent anti-influenza activity. *J. Am. Chem. Soc.* 119:681–690.

Kulkosky, J., D. M. Culnan, J. Roman, G. Dornadula, M. Schnell, M. R. Boyd, and R. J. Pomerantz. 2001. Prostratin: Activation of latent HIV-1 expression suggests a potential inductive adjuvant therapy for HAART. *Blood* 98:3006–3015.

Kuo, R.-Y., K. Qian, S. L. Morris-Natschke, and K. H. Lee. 2009. Plant-derived triterpenoids and analogues as antitumor and anti-HIV agents. *Nat. Prod. Rep.* 26:1321–1344.

Lagoja, I. M. and E. De Clercq. 2006. Anti-influenza virus agents: Synthesis and mode of action. *Med. Res. Rev.* 28:1–38.

Laure, F., P. Raharivelomanana, J.-F. Butard, J.-P. Bianchini, and E. M. Gaydou. 2008. Screening of anti-HIV-1 inophyllums by HPLC-DAD of *Calophyllum inophyllum* leaf extracts from French Polynesian Islands. *Anal. Chim. Acta* 624:147–153.

Lee, W. W., A. Benitez, L. Goodman, and B. R. Baker. 1960. Potential anticancer agents. XL. Synthesis of the beta-anomer of 9-(D-arabinofuranosyl)adenine. *J. Am. Chem. Soc.* 82:2648–2649.

Li, Y., X. Zhang, G. Chen, D. Wei, and F. Chen. 2008. Algal lectins for potential prevention of HIV transmission. *Curr. Med. Chem.* 15:1096–1104.

Liu, Z., H. Yu, Y. Yin, D. Li, and W. Zhang. 2008. Purification, renaturation and antiviral effects of recombinant cyanovirin-N on herpes simplex virus type 1. *Zhongguo Haiyang Yaowu* 27:1–5.

Lu, M., S. C. Blacklow, and P. S. Kim. 1995. A trimeric structural domain of the HIV-transmembrane glycoprotein. *Nat. Struct. Biol.* 2:1075–1082.

Magano, J. 2009. Synthetic approaches to the neuraminidase inhibitors zanamivir (Relenza) and oseltamivir phosphate (Tamiflu) for the treatment of influenza. *Chem. Rev.* 109:4398–4438.

Martin, D. E., K. Salzwedel, and G. P. Allaway. 2008. Bevirimat: A novel maturation inhibitor for the treatment of HIV-1 infection. *Antivir. Chem. Chemother.* 19:107–113.

Mayaux, J. F., A. Bousseau, R. Pauwels, T. Huet, Y. Hénin, N. Dereu, M. Evers, F. Soler, C. Poujade, and E. De Clercq. 1994. Triterpene derivatives that block entry of human immunodeficiency virus type 1 into cells. *Proc. Natl. Acad. Sci. USA* 91:3564–3568.

McFeeters, R. L., C. Xiong, B. R. O'Keefe, H. R. Bokesch, J. B. McMahon, D. M. Ratner, R. Castelli, P. H. Seeberger, and R. A. Byrd. 2007. The novel fold of scytovirin reveals a new twist for antiviral entry inhibitors. *J. Mol. Biol.* 369:451–461.

McMahon, J. B., M. J. Currens, R. J. Gulakowski, R. W. Buckheit Jr., C. Lackman-Smith, Y. F. Hallock, and M. R. Boyd. 1995. Michellamine B, a novel plant alkaloid, inhibits human immunodeficiency virus-induced cell killing by at least two distinct mechanisms. *Antimicrob. Agents Chemother.* 39:484–488.

Meindl, P., G. Bodo, P. Palese, J. Schulman, and H. Tuppy. 1974. Inhibition of neuraminidase activity by derivatives of 2-deoxy-2,3-dehydro-N-acetylneuraminic acid. *Virology* 58:457–463.

Melikyan, G. B., R. M. Markosyan, H. Hemmati, M. K. Delmedico, D. M. Lambert, and M. S. Cohen. 2000. Evidence that the transition of HIV-1 gp41 into a six-helix bundle, not the bundle configuration, induces membrane fusion. *J. Cell Biol.* 151:413–423.

Mori, T., B. R. O'Keefe, R. C. Sowder II, S. Bringans, R. Gardella, S. Berg, P. Cochran et al. 2005. Isolation and characterization of griffithsin, a novel HIV-inactivating protein, from the red alga *Griffithsia* sp. *J. Biol. Chem.* 280:9345–9353.

Moulaei, T., I. Botos, N. E. Ziolkowska, H. R. Bokesch, L. R. Krumpe, T. C. McKee, B. R. O'Keefe, Z. Dauter, and A. Wlodawer. 2007. Atomic-resolution crystal structure of the antiviral lectin scytovirin. *Protein Sci.* 16:2756–2760.

Ndesendo, V. M. K., V. Pillay, Y. E. Choonara, E. Buchmann, D. N. Bayever, and L. C. R. Meyer. 2008. A review of current intravaginal drug delivery approaches employed for the prophylaxis of HIV/AIDS and prevention of sexually transmitted infections. *AAPS Pharm. Sci. Tech.* 9:505–520.

Newman, D. J., G. M. Cragg, and K. M. Snader. 2000. The influence of natural products upon drug discovery. *Nat. Prod. Rept.* 17:215–234.

Nie, L.-D. and X.-X. Shi. 2009. A novel asymmetric synthesis of oseltamivir phosphate (Tamiflu) from (–)-shikimic acid. *Tetrahedron Asymmetr.* 20:124–129.

Nie, L.-D., X.-X. Shi, K. H. Ko, and W.-D. Lu. 2009. A short and practical synthesis of oseltamivir phosphate (Tamiflu) from (–)-shikimic acid. *J. Org. Chem.* 74:3970–3973.

Nigorikawa, K., K. Yoshikawa, T. Sasaki, E. Iida, M. Tsukamoto, H. Murakami, T. Maehama, K. Hazeki, and O. Hazeki. 2006. A naphthoquinone derivative, shikonin, has insulin-like actions by inhibiting both phosphatase and tensin homolog deleted on chromosome 10 and tyrosine phosphatases. *Mol. Pharmacol.* 70:1143–1149.

O'Keefe, B. R. and J. B. McMahon. 2008. Anti-H5N1 influenza activity of the antiviral protein cyanovirin. WO 2008022303 A2 20080221.

O'Keefe, B. R., F. Vojdani, V. Buffa, R. J. Shattock, D. C. Montefiori, J. Bakke, J. Mirsalis et al. 2009. Scaleable manufacture of HIV-1 entry inhibitor griffithsin and validation of its safety and efficacy as a topical microbicide component. *Proc. Natl. Acad. Sci. USA* 106:6099–6104.

Patil, A. D., A. J. Freyer, D. S. Eggleston, R. C. Haltiwanger, M. F. Bean, P. B. Taylor, M. J. Caranfa et al. 1993. The Inophyllums, novel inhibitors of HIV-1 reverse transcriptase isolated from the Malaysian tree, *Calophyllum inophyllum* Linn. *J. Med. Chem.* 36:4131–4138.

Peck, R. F., K. Wehrly, E. J. Platt, R. E. Atchinson, I. F. Chao, D. Kabat, B. Chesebro, and M. A. Goldsmith. 1997. Selective employment of chemokine receptors as human immunodeficiency virus type 1 coreceptors determined by individual amino acids within the envelope V3 loop. *J. Virol.* 71:7136–7139.

Qian, K., D. Yu, C.-H. Chen, L. Huang, S. L. Morris-Natschke, T. J. Nitz, K. Salzwedel, M. Reddik, G. P. Allaway, and K. H. Lee. 2009. Anti-AIDS agents. 78. Design, synthesis, metabolic stability assessment, and antiviral evaluation of novel betulinic acid derivatives as potent anti-human immunodeficiency virus (HIV) agents. *J. Med. Chem.* 52:3248–3258.

Sexton, A., S. Harman, R. J. Shattock, and J. K.-C. Ma. 2009. Design, expression, and characterization of a multivalent, combination HIV microbicide. *FASEB J.* 23:3590–3600.

Shi, M., X. Wang, M. Okamoto, S. Takao, and M. Baba. 2009. Inhibition of porcine endogenous retrovirus (PERV) replication by HIV-1 gene expression inhibitors. *Antiviral Res.* 83:201–204.

Singh, S. B., F. Pelaez, D. J. Hazuda, and R. B. Lingham. 2005. Discovery of natural product inhibitors of HIV-1 integrase at Merck. *Drugs Future* 30:277–299.

Smee, D. F., K. W. Bailey, M.-H. Wong, B. R. O'Keefe, K. R. Gustafson, V. P. Mishin, and L. V. Gubareva. 2008. Treatment of influenza A (H1N1) virus infections in mice and ferrets with cyanovirin-N. *Antivir. Res.* 80:266–271.

Soler, F., C. Poujade, M. Evers, J. C. Carry, Y. Henin, A. Bousseau, T. Huet et al. 1996. Betulinic acid derivatives: A new class of specific inhibitor of human immunodeficiency virus type 1 entry. *J. Med. Chem.* 39:1069–1083.

Sorbera, L. A., P. Leeson, and J. Castaner. 1999. Calanolide A. *Drugs Future* 24:235–245.

Staniforth, V., S. Y. Wang, L. F. Shyur, and N. S. Yang. 2004. Shikonins, phytocompounds from lithospermum erythrorhizon, inhibit the transcriptional activation of human tumor necrosis factor alpha promoter in vivo. *J. Biol. Chem.* 279:5877–5885.

Suckling, C. J. 1991. Chemical approaches to the discovery of new drugs. *Sci. Prog.* 75:323–359.

Sun, I. C., C. H. Chen, Y. Kashiwada, J. H. Wu, H. K. Wang, and K. H. Lee. 2002. Anti-AIDS agents 49. Synthesis, anti-HIV, and anti-fusion activities of IC9564 analogues based on betulinic acid. *J. Med. Chem.* 45:4271–4275.

Sun, T., Q. Wang, Z. Yu, Y. Zhang, Y. Guo, K. Chen, X. Shen, and H. Jiang. 2007. Hyrtiosal, a PTP1B inhibitor from the marine sponge *Hyrtios erectus*, shows extensive cellular effects on PI3K/AKT activation, glucose transport, and TGF-beta/Smad2 signaling. *ChemBioChem* 8:187–193.

Svicher, V., C. Alteri, R. D'Arrigo, A. Laganà, M. Trignetti, S. Lo Caputo, A. P. Callegaro et al. 2009. Treatment with the fusion inhibitor enfuvirtide influences the appearance of mutations in the human immunodeficiency virus type 1 regulatory protein rev. *Antimicrob. Agents Chemother.* 53:2816–2823.

Takeuchi, H., N. Asai, K. Tanabe, T. Kozaki, M. Fujita, T. Sakai, A. Okuda et al. 1999. EM2487, a novel anti-HIV-1 antibiotic, produced by *Streptomyces* sp. Mer-2487: Taxonomy, fermentation, biological properties, isolation and structure elucidation. *J. Anitbiot.* 52:971–982.

Tiwari, V., S. Y. Shukla, and D. Shukla. 2009. A sugar binding protein cyanovirin-N blocks herpes simplex virus type-1 entry and cell fusion. *Antivir. Res.* 84:67–75.

Varghese, J. N., W. G. Laver, and P. M. Colman. 1983. Structure of the influenza virus glycoprotein antigen neuraminidase at 2.9Å resolution. *Nature* 303:35–40.

Wang, C. K., M. L. Colgrave, D. C. Ireland, Q. Kaas, and D. J. Craik. 2009. Despite a conserved cystine knot motif, different cyclotides have different membrane binding modes. *Biophys. J.* 97:1471–1481.

Weiss, C. D. 2003. HIV gp41: Mediator of fusion and target for inhibition. *AIDS Rev.* 5:214–221.

Wild, C., T. Oas, C. McDanal, D. Bolognesi, and T. Mathews. 1992. A synthetic peptide inhibitor of human immunodeficiency virus replication: Correlation between solution structure and viral inhibition. *Proc. Natl. Acad. Sci. USA* 89:10537–10541.

Xiong, C., B. R. O'Keefe, R. A. Byrd, and J. B. McMahon. 2006. Potent anti-HIV activity of scytovirin domain 1 peptide. *Peptides* 27:1668–1675.

Yang, F., C. A. Bewley, J. M. Louis, K. R. Gustafson, M. R. Boyd, A. M. Gronenborn, G. M. Clore, and A. Wlodawer. 1999. Crystal structure of cyanovirin-N, a potent HIV-inactivating protein, shows unexpected domain swapping. *J. Mol. Biol.* 288:403–412.

Yin, Z., Y.-L. Chena, W. Schula, Q.-Y. Wanga, F. Gua, J. Duraiswamya, R. R. Kondreddia et al. 2009. An adenosine nucleoside inhibitor of dengue virus. *Proc. Natl. Acad. Sci. USA* 106:20435–20439.

Zeitlin, L., M. Pauly, and K. J. Whaley. 2009. Second-generation HIV microbicides: Continued development of griffithsin. *Proc. Natl. Acad. Sci. USA* 106:6029–6030.

Ziolkowska, N. E., B. R. O'Keefe, T. Mori, C. Zhu, B. Giomarelli, F. Vojdani, K. E. Palmer, J. B. McMahon, and A. Wlodawer. 2006. Domain-swapped structure of the potent antiviral protein griffithsin and its mode of carbohydrate binding. *Structure* 14:1127–1135.

Ziolkowska, N. E., S. R. Shenoy, B. R. O'Keefe, J. B. McMahon, K. E. Palmer, R. A. Dwek, M. R. Wormald, and A. Wlodawer. 2007. Crystallographic, thermodynamic, and molecular modeling studies of the mode of binding of oligosaccharides to the potent antiviral protein griffithsin. *Proteins* 67:661–670.

7 Natural Products as Inhibitors of Hypoxia-Inducible Factor-1

Dale G. Nagle and Yu-Dong Zhou

CONTENTS

7.1 TUMOR HYPOXIA

Rapid tumor growth outstrips the capability of existing blood vessels to supply oxygen and nutrients, and to remove metabolic waste. As a result, hypoxia (reduced oxygen tension) and acidity are signature features of the tumor microenvironment (Tatum et al., 2006; Fang et al., 2008). Clinical studies indicate that hypoxia is an important prognostic factor for a variety of cancers that range from those of the prostate to those of the brain (Tatum et al., 2006). The extent of tumor hypoxia correlates with advanced disease stages, malignant progression, poor prognosis, and treatment resistance. Hypoxic tumors are more resistant to radiation and chemotherapeutic drugs than their normoxic counterparts (Tatum et al., 2006; Moeller et al., 2007). Hypoxic tumor cells that have adapted to oxygen and nutrient deprivation are associated with a more aggressive phenotype and poor prognosis. Experimental approaches that have been explored to overcome tumor hypoxia include an assortment of methods that range from techniques to improve tumor oxygenation to efforts to develop hypoxic radiosensitizers and hypoxia-selective cytotoxins (Brown and Wilson, 2004; Tatum et al., 2006; Moeller et al., 2007). Currently, there is no clinically approved treatment method that selectively targets tumor hypoxia.

7.2 HYPOXIA-INDUCIBLE FACTORS

Hypoxia-inducible factors (HIFs) are transcription factors that mediate the important indirect effect of hypoxia—induction of genes that promote the adaptation, survival, malignant progression, and treatment resistance of hypoxic tumor cells (Semenza, 2003; Brown and Wilson, 2004; Tatum et al., 2006; Semenza, 2007). The founding member of the HIF family HIF-1 was first discovered in 1992 by Semenza and colleagues (Semenza and Wang, 1992) and has been studied extensively. Acting as a key regulator of oxygen homeostasis, HIF-1 is a heterodimer of the basic helix-loop-helix Per-ARNT-Sim (bHLH-PAS) proteins HIF-1α and HIF-1β [(also known as aryl hydrocarbon receptor nuclear translocator (ARNT)]. In general, HIF-1α protein is degraded rapidly under normoxic conditions by the 26S proteasome and is stabilized under hypoxic conditions, while HIF-1β protein is constitutively expressed. Upon hypoxic induction and activation, HIF-1 binds to the hypoxia response element (HRE) present in the promoters of target genes and activates transcription. The classical oxygen-dependent posttranslational regulation of HIF-1α protein is summarized in Figure 7.1 (Nagle and Zhou, 2010). Both the prolyl hydroxylases (PHD or HPH) that promote the degradation of HIF-1α protein and the asparaginyl hydroxylase (FIH) that inactivates HIF-1α protein require oxygen and iron (i.e., Fe^{++}) for their activities (Maxwell et al., 1999; Ivan et al., 2001; Jaakkola et al., 2001; Lando et al., 2002). Thus, HIF-1 can be activated by either physiological hypoxia or by the addition of iron chelators or transition metals (chemical hypoxia). Recent studies indicate that other oxygen-independent pathways also regulate HIF-1α degradation (Figure 7.2) (Isaacs et al., 2002; Liu et al., 2007).

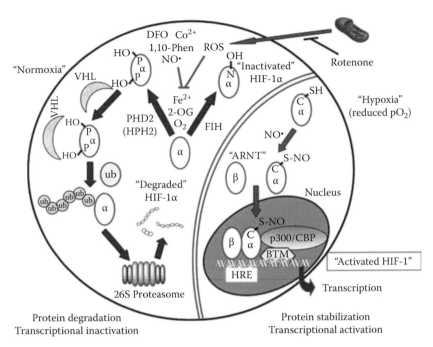

FIGURE 7.1 Hypoxic regulation of hypoxia-inducible factor-1 (HIF-1). The transcription factor HIF-1 is a heterodimer composed of an HIF-1α subunit that is regulated by cellular oxygen levels and an HIF-1β subunit (also known as ARNT) that is constitutively expressed. Under normoxic conditions, HIF-1α protein is hydroxylated at specific proline residues by Fe^{+2}/2-oxoglutarate/O_2-dependent prolyl hydroxylase enzymes (e.g., PHD2). This prolyl-hydroxylation "tags" HIF-1α protein for von Hippel-Lindau tumor suppressor protein (pVHL) E3 ubiquitin ligase-mediated polyubiquitination. The "ubiquitin-tagged" HIF-1α protein is then rapidly degraded by the 26S proteasome. Hydroxylation of an asparagine residue in the C-terminal transcriptional activation domain (CTD) contributes another level of oxygen-dependent regulation by inactivating HIF-1α protein. Like PHD2, this asparaginyl hydroxylase ["Factor Inhibiting HIF" (FIH)] is also a Fe^{+2}/2-oxoglutarate/O_2-dependent hydroxylase that modifies the asparagine residue in the CTD region of HIF-1α protein. Once hydroxylated, the interaction between HIF-1 and the co-activator CBP/p300 is disrupted and transcriptional activation is blocked. Besides hypoxic conditions, HIF-1α protein can be stabilized by addition of iron chelators, transition metals, nitric oxide radical (NO·), or inhibitors of PHDs (e.g., DMOG). Such inducing conditions inactivate the prolyl hydroxylases that tag HIF-1α protein for ubiquitination and proteasomal degradation, and suppress the asparaginyl hydroxylase that normally inactivates the transcriptional activity of HIF-1. In addition, binding between HIF-1 and CBP/p300 can be enhanced by direct nitrosylation of a sulfhydryl moiety in HIF-1α. When the level of O_2 decreased to a level below certain thresh-hold, reactive oxygen species (ROS) generated by hypoxic mitochondria inhibit PHD2 and FIH by oxidizing the Fe^{+2} in their catalytic sites. Natural products that inhibit the mitochondrial electron transport chain [e.g., rotenone (**9**)] block HIF-1 activation by suppressing the hypoxia-induced increase in ROS production by mitochondria. This promotes the PHD2-mediated degradation and FIH-facilitated inactivation of the HIF-1α subunit. (Reproduced with the permission of D.G. Nagle. © 2010 at the University of Mississippi.)

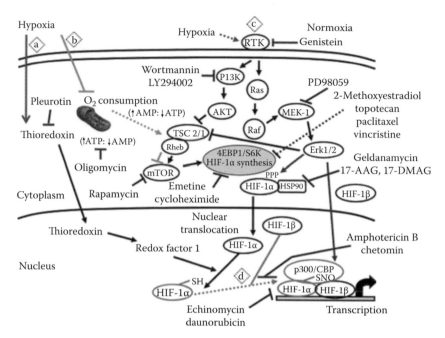

FIGURE 7.2 Natural products regulate the expression and transcriptional activation of HIF-1 under hypoxic and normoxic conditions. Under hypoxic and normoxic conditions, the transcriptional activation of HIF-1 and HIF-1α expression are tightly regulated. (a) The mushroom metabolite pleurotin (**63**) inhibits the redox protein thioredoxin. Thioredoxin is normally involved in increasing HIF-1α protein levels and maintaining a sulfhydryl moiety in the reduced state. This decreases the level of HIF-1α protein and interferes with the binding between HIF-1 and the CBP/p300 transcriptional co-activator complex under hypoxic conditions; (b) Hypoxic conditions and inhibitors of F_1F_0-ATPase (e.g., oligomycin) increase cellular AMP: ATP ratios. The elevated AMP: ATP ratio activates the TSC2-TSC1 tumor suppressor complex that blocks mTOR activation and suppresses global protein synthesis. The combined outcome of the activated PI3K/Akt and AMPK pathways (and others) led to the enhanced expression of certain proteins such as HIF-1α under hypoxic conditions. Rapamycin (**40**) and protein translation inhibitors [e.g., emetine (**43**) and cycloheximide (**29**)] inhibit HIF-1 activation by blocking the synthesis of HIF-1α protein. Antitumor agents [e.g., topotecan (**30**), 2-methoxyestradiol (**34**), paclitaxel (**35**), and vincristine (**37**)] appear to interfere with HIF-1α synthesis; (c) Upon activation of receptor tyrosine kinases (RTKs), translation of HIF-1α is regulated through PI3K/AKT and MAPK-mediated pathways. Genistein (**107**) and PD98059 (**113**) interfere with HIF-1α translation by inhibiting RTKs and MEK-1, respectively. Wortmannin (**38**), LY294002 (**39**), and other natural product–based PI3K inhibitors suppress the translation and transcriptional activation of HIF-1α. Geldanamycin (**1**) and semisynthetic geldanamycin derivatives [i.e., 17-allylamino-17-demethoxygeldanamycin (17-AAG, **2**)] interfere with the heat shock protein Hsp90. This destabilizes HIF-1α and facilitates its proteasomal degradation; (d) chetomin (**121**) and amphotericin B (**122**) directly interfere with the binding between HIF-1 and the CBP/p300 co-activator complex. Echinomycin (**114**) and anthracyclines [e.g., daunorubicin (**115**)] inhibit HIF-1 transcriptional activation by interfering with the binding of HIF-1 to the hypoxia-response elements (HREs) within the promoters of HIF-responsive genes. (Reproduced with the permission of D.G. Nagle. © 2010 at the University of Mississippi.)

Numerous preclinical and clinical studies support HIF-1 as an important molecular target for anticancer drug discovery. At the molecular level, HIF-1 activates the expression of genes that promote hypoxic adaptation and survival by (i) increasing oxygen delivery through enhancing angiogenesis, erythropoiesis, and vasodilatation; (ii) decreasing oxygen consumption through inducing genes involved in anaerobic metabolism and reducing mitochondrial oxygen consumption; and (iii) promoting survival by inducing the expression of growth factors (Semenza, 2003; Semenza, 2007). In addition, HIF-1 also increases the expression of genes that increase drug resistance, metastasis, and tumor cell dedifferentiation to assume a cancer stem cell-like phenotype (Semenza, 2003; Axelson et al., 2005; Semenza, 2007). In tumor cells, the activation of oncogenes and/or the mutation of tumor suppressor genes also lead to HIF-1 activation. Clinical studies indicate that the expression of HIF-1α protein is associated with advanced disease stage, metastasis, and a poor prognosis in cancer patients (Zhong et al., 1999; Birner et al., 2000; Bos et al., 2001; Tatum et al., 2006). In animal models, reduced tumor vascularity and suppression of tumor growth were observed when HIF-1 was inhibited by methods that vary from genetic manipulation to the use of small-molecule inhibitors (Ryan et al., 2000; Rapisarda et al., 2004; Greenberger et al., 2008). Enhanced treatment outcomes were observed when HIF-1 inhibition was combined with chemotherapy or radiation (Unruh et al., 2003; Moeller et al., 2005; Li et al., 2006; Cairns et al., 2007). At least three agents that inhibit HIF-1 have entered early phase clinical trials for cancer: EZN-2968 (an HIF-1α RNA antisense oligonucleotide antagonist), PX-478 (a compound that decreases HIF-1α gene expression), and topotecan (**30**) (NIH Clinical Trials Database, 2010).

The transcription factor HIF-2 that also mediates hypoxia-induced gene expression is composed of HIF-2α (EPAS1/HLF/HRF) and ARNT. Both HIF-1 and HIF-2 are members of the bHLH-PAS family (Ema et al., 1997; Flamme et al., 1997; Tian et al., 1997). The primary amino acid sequence of HIF-2α is 48% identical to that of HIF-1α (Tian et al., 1997). The mechanisms of hypoxia-induced HIF-2 activation are highly similar to those described for HIF-1 (Löfstedt et al., 2007). Although HIF-1 and HIF-2 can regulate the expression of a common set of target genes, studies performed in mice deficient of HIF-1A and HIF-2A genes have demonstrated that they are not redundant (Iyer et al., 1998; Ryan et al., 1998; Tian et al., 1998; Peng et al., 2000; Scortegagna et al., 2003). Recent studies suggest that the combination of inducing conditions (i.e., level of oxygen) and cell line specificity (i.e., signaling pathways, the expression of coactivators and corepressors, etc.) dictate which HIF is activated, while there is very little target gene preference between HIF-1 and HIF-2 (Löfstedt et al., 2007). Clinical studies indicated that both HIF-1α and HIF-2α proteins were induced in tumor samples in a heterogeneous pattern (Talks et al., 2000). The level of HIF-2α protein in tumor specimens has been implicated as a prognostic marker for advanced disease stages in several forms of cancer (Giatromanolaki et al., 2001; Leek et al., 2002; Yoshimura et al., 2004; Holmquist-Mengelbier et al., 2006; Koukourakis et al., 2006; Helczynska et al., 2008). Only modest progress has been made in the identification of small-molecule HIF-2 specific inhibitors.

7.3   NATURAL PRODUCTS AND ANTICANCER DRUG DISCOVERY

Natural products have been a major source of new drugs for centuries and statistics show that over 47% of approved anticancer agents are of natural origin (Newman and Cragg, 2007). The unrivaled chemical diversity of natural product–based drug discovery, empowered by functional bioassays, continues to play a key role in the discovery of chemotherapeutic agents. In addition, natural product–based drug leads are often associated with unique and dissimilar modes of action. Small-molecule HIF-1 inhibitors have served as important molecular probes to investigate the pathways that regulate HIF-1 activity (Nagle and Zhou, 2006a,b, 2009, 2010). Figure 7.2 is a schematic diagram of the pathways that regulate HIF-1 and representative small-molecule HIF-1 inhibitors. Some of these HIF inhibitors function at low nanomolar concentrations (e.g., manassantins) with a wide window between their HIF-1 inhibitory activity and cytotoxicity (Hodges et al., 2004; Hossain et al., 2005). The chemical structures and tables that present key information related to each of these inhibitors are assembled into the following six mechanistic groups: (1) agents that promote HIF-1α degradation (Table 7.1, Structures **1–25**), (2) agents that prevent the synthesis of HIF-1α protein (Table 7.2, Structures **26–61**), (3) agents that suppress HIF-1α protein accumulation (Table 7.3, Structures **62–102**), (4) multi-targeted HIF-1 inhibitors (Table 7.4, Structures **103–112**),; (5) inhibitors of HIF-1 with unique mechanisms (Table 7.5, Structures **113–124**), and (6) agents that lack confirmatory and mechanistic studies (Table 7.6, Structures **125–136**). However, it must be recognized that these mechanistic groups should be viewed as somewhat subjective and possibly overlapping in nature. For example, results that indicate that a compound suppresses HIF-1α protein accumulation may be produced by inhibiting HIF-1α protein translation and/or promoting proteasomal HIF-1α degradation. Therefore, such compounds are simply grouped into "agents that suppress HIF-1α protein accumulation," based solely upon the available experimental data. The inhibitors presented are natural products [e.g., genistein (**107**), wortmannin (**38**), rapamycin (**40**), rotenone (**9**), geldanamycin (**1**), actinomycin D (**26**), cycloheximide (**29**), etc.] or natural product–derived synthetic (or semisynthetic) compounds [e.g., PD98059 (**113**), LY294002 (**39**)].

While a number of research groups have used cell-based reporter assays to identify small-molecule HIF-1 inhibitors, others have employed approaches that range from antisense technology to chemical disruptors of protein–protein interactions. Examples of HIF-1 inhibitors identified using various approaches were provided in a recent review (Poon et al., 2009). An assortment of HIF-1 inhibitors have been discovered using cell-based reporter assays [topotecan (**30**), echinomycin (**114**), anthracyclines, and cardiac glycosides]; antisense technology (ENZ-2968 and RX0047); those that inhibit protein–protein interactions (chetomin, **121**); dominant negative strategies (dnHIF-1); and others [KRH102053 that activates PHD2, HIF oligonucleotide decoy, and intrabodies (intracellular antibodies) that inhibit HIF-1 activity]. In addition, a number of anticancer agents [e.g., taxol (**35**), doxorubicin (**118**)] and compounds/agents that target specific signaling pathways that regulate HIF activity have also been shown to inhibit HIF-1 activity (Jung et al., 2003; Mabjeesh et al., 2003; Escuin et al., 2005; Lee et al., 2009). However, the review by Aschcroft and coworkers (Poon et al., 2009) is conspicuously missing coverage of many recently reported terrestrial and marine natural products that have been identified to regulate HIF-1 activation. Two previous reviews by the authors of this chapter have described those natural products that have been discovered (in the time period up to 2005) that act as regulators of HIF-1 signaling (Nagle and Zhou, 2006a,b). In the interim period, the authors have published two highly specialized reviews that illustrate that role of marine natural products in HIF-1 inhibitor discovery (Nagle and Zhou, 2009) and that detail the impact of natural product–based HIF inhibitors as molecular probes of tumor cell biology (Nagle and Zhou, 2010), respectively. To reduce redundancy, this review highlights selected examples of natural product–based HIF-1 inhibitors that have been reported since the publication of the authors' general review on natural product–based inhibitors of HIF-1 in 2006 (Nagle and Zhou, 2006a,b). While not described in the body of text, the structures of many of these previously reported HIF inhibitors are illustrated, and their source and mechanistic details related to each of these compounds are included (Tables 7.1 through 7.6).

7.4 MITOCHONDRIAL REACTIVE OXYGEN SPECIES AND HYPOXIC ACTIVATION OF HIF-1

Traditionally, mitochondria have been viewed primarily as energy-generating organelles. Through the process of oxidative phosphorylation, mitochondria consume the majority of cellular oxygen (>90%) to produce adenosine triphosphate (ATP). Progress during the past two decades has revealed that mitochondria also function as key regulators of cellular signaling pathways (Oberst et al., 2008; Snyder and Chandel, 2009). When mammalian cells encounter a hypoxic environment ($O_2 < 5\%$), mitochondria increase reactive oxygen species (ROS) production. This seemingly paradoxical increase in ROS results from the premature transfer of electrons to oxygen by the mitochondrial electron transport chain (ETC). Mitochondrial ETC complexes I, II, and III all produce ROS. Unlike complexes I and II that may alter ROS levels in the matrix, it is believed that ETC complex III releases ROS into the intermembrane space, which are then transported into the cytosol in the form of superoxide anion radicals and/or hydrogen peroxide (Figure 7.3). The ROS enter the cytosol and inactivate the prolyl hydroxylases (PHDs) that tag HIF-1α protein for degradation, by oxidizing the PHD cofactor Fe^{2+} to Fe^{3+} through Fenton reaction. This is followed by the stabilization of HIF-1α protein and the subsequent activation of HIF-1. Mitochondrial ETC inhibitors such as rotenone (complex I inhibitor) prevent hypoxia-induced HIF-1 activation by blocking the hypoxia-induced increase in ROS production (Figures 7.1 and 7.3). Similar results are produced by the complex III inhibitors myxothiazol (**10**) and stigmatellin. The ROS produced at the Qp (Qo, outermembrane space-side) site of complex III are believed to be the mediators of hypoxic signaling, since the complex III Qi site (Q_N, matrix side) inhibitor antimycin A (**11**) gives only conflicting results when examined for its effect on hypoxia-induced HIF-1 activation. This hypoxia-mitochondrial ROS-HIF model is further

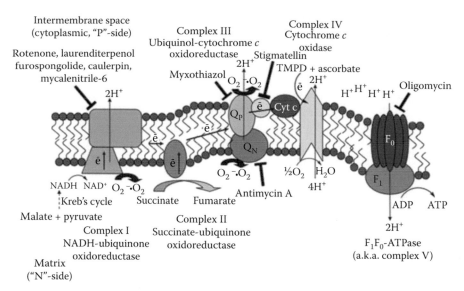

FIGURE 7.3 Natural product HIF-1 inhibitors that block the mitochondrial electron transport chain (ETC) and reduce the hypoxia-induced production of reactive oxygen species (ROS). The mitochondria of hypoxic cells release superoxide from the Qp site (Qo, outer-membrane space-side) of mitochondrial complex III (ubiquinol-cytochrome c oxidoreductase). Myxothiazole (**10**) blocks the hypoxia-induced production of ROS by directly inhibiting the Qp site of complex III. This prevents the transfer of electrons required to drive the hypoxia-induced production of reactive oxygen species at complex III. Rotenone (**9**) and the marine natural products laurenditerpenol (**15**), furospongolide (**16**), mycalenitrile-6 (**17**), and caulerpin (**19**) inhibit the mitochondrial electron transport chain (ETC) upstream of complex III by disrupting the ETC at the NADH-ubiquinone oxidoreductase (complex I) site. Superoxide anion and its metabolites (hydrogen peroxide, etc.) are believed to serve as cellular signaling molecules for the hypoxic regulation of HIF-1α stability (Figure 7.1). (Reproduced with the permission of D.G. Nagle. © 2010 at the University of Mississippi.)

supported by results obtained using genetic-based approaches that inactivated the ETC complex III subunit Rieske-Fe-S protein (Brunelle et al., 2005; Guzy et al., 2005) and cytochrome c (Mansfield et al., 2005). As anticipated, certain radical scavengers and glutathione peroxidase mimetic agents [i.e., ebselen, pyrrolidine dithiocarbamate (PDTC)] inhibited hypoxic HIF-1 activation by blocking the increase in ROS production. The growing body of evidence supports the hypothesis that mitochondrial ROS, not oxidative phosphorylation, is required for hypoxic HIF-1 activation.

7.5   AGENTS THAT PROMOTE HIF-1α DEGRADATION

Natural products and natural product–derived synthetic and semisynthetic compounds that inhibit HIF-1 activation by mechanisms that are believed to promote the degradation of HIF-1α are listed in Table 7.1 (Structures **1–25**).

TABLE 7.1

Natural Product–Based Agents That Promote HIF-1α Protein Degradation

Compound	Source	Mechanistic Data	References
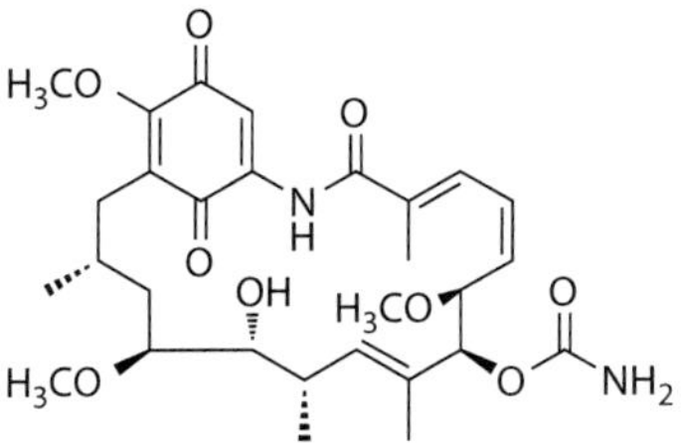 Geldanamycin **(1)**	*Streptomyces hygroscopicus*	Hsp90 inhibitor	Minet et al. (1999) Mabjeesh et al. (2002) Isaacs et al. (2002) Katschinski et al. (2002)
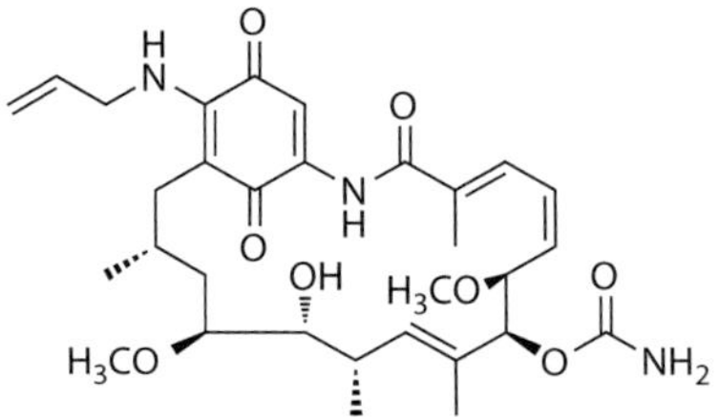 17-N-allylamino-17-demethoxygeldanamycin (17-AAG) **(2)**	Semisynthetic geldanamycin	Hsp90 inhibitor	Isaacs et al. (2002) Neckers (2006) Bisht et al. (2003)

Geldanamycin
(1)

17-N-allylamino-17-demethoxygeldanamycin (17-AAG)
(2)

(continued)

TABLE 7.1 (continued)
Natural Product–Based Agents That Promote HIF-1α Protein Degradation

Compound	Source	Mechanistic Data	References
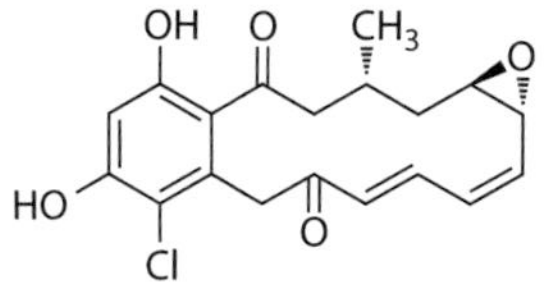 Radicicol (3)	*Humicola fuscoatra*	Hsp90 inhibitor	Hur et al. (2002)
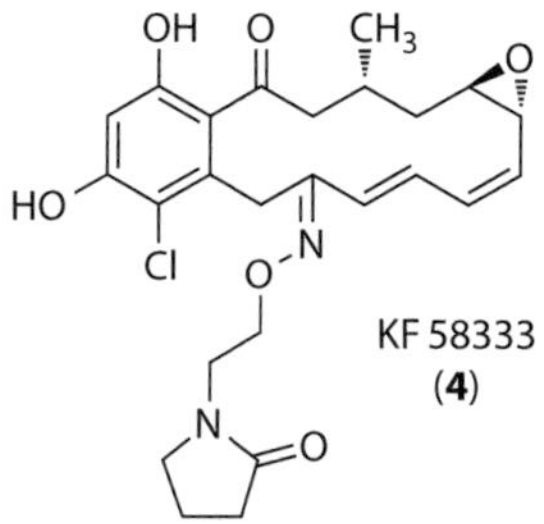 KF 58333 (4)	Semisynthetic radicicol oxime	Hsp90 inhibitor	Schulte et al. (1999)
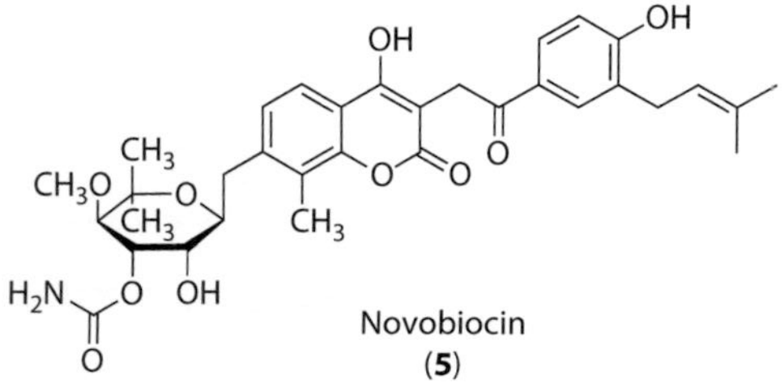 Novobiocin (5)	*Streptomyces niveus, S. speroides*	Hsp90 inhibitor	Isaacs et al. (2002) Katschinski et al. (2002)

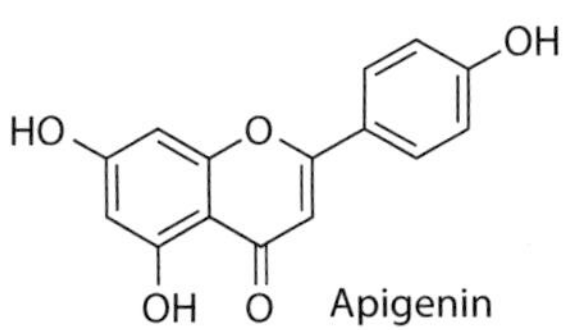 Apigenin (**6**)	Plants, fruit, vegetables	Inhibits HDM2/p53, hsp90, PI3K/AKT/p70S6K1, NFκB, HER2/neu, etc.	Osada et al. (2004) Fang et al. (2005, 2007) Liu et al. (2005) Mottet et al. (2005) Mirzoeva et al. (2008)
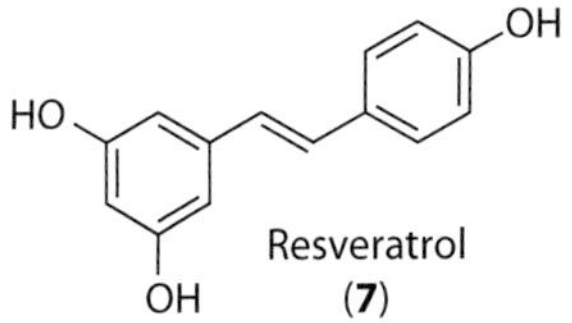 Resveratrol (**7**)	Grapes, red wine, vegetables	Inhibits IGF1-induced HIF1α expression, but only nonphysiologically relevant high concentrations	Cao et al. (2004) Zhang et al. (2005) Park et al. (2007)
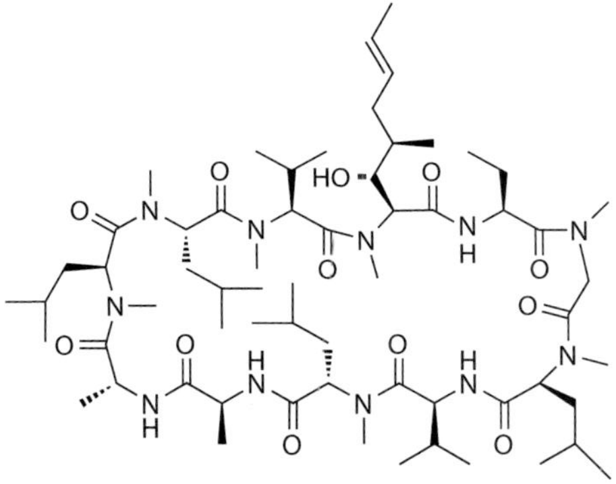 Cyclosporine A (**8**)	*Tolpocladium inflatum Gams*	Stimulates HIF prolyl hydroxylases	D'Angelo et al. (2003) Liu, YV et al. (2007b)

(continued)

TABLE 7.1 (continued)

Natural Product–Based Agents That Promote HIF-1α Protein Degradation

Compound	Source	Mechanistic Data	References
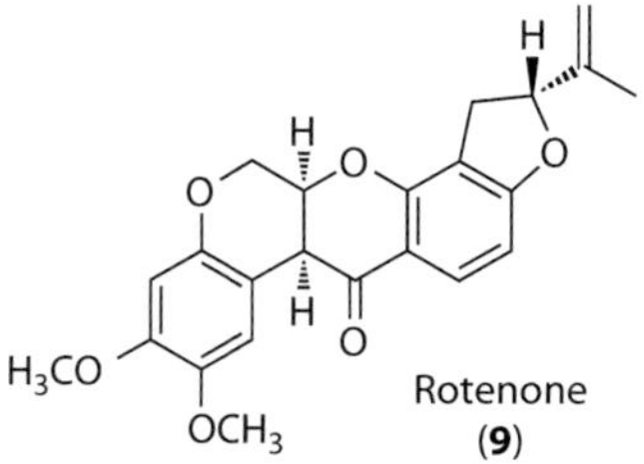 Rotenone (9)	*Lonchocarpus* spp.	Mitochondrial complex I inhibitor	Chandel et al. (1998) Agani et al. (2002) Liu, Y et al. (2009a)
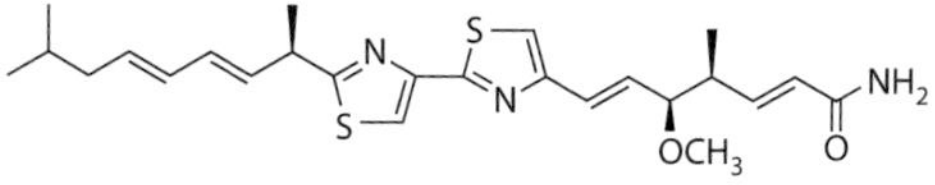 Myxothiazol (10)	*Myxococcus fulvus*	Mitochondrial complex III inhibitor	Chandel et al. (1998, 2000)
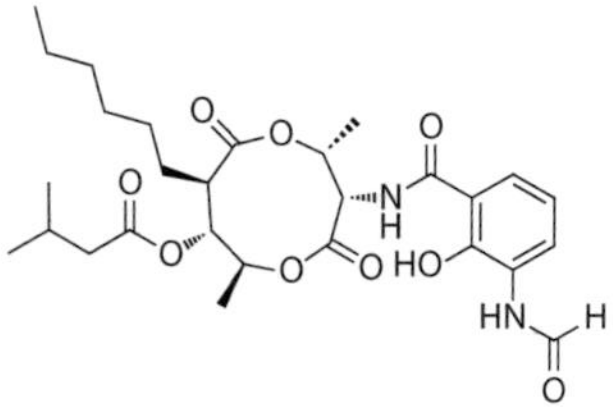 Antimycin A (11)	*Streptomyces* spp.	Mitochondrial complex III inhibitor	Hagen et al. (2003) Mateo et al. (2003) Maeda et al. (2006)

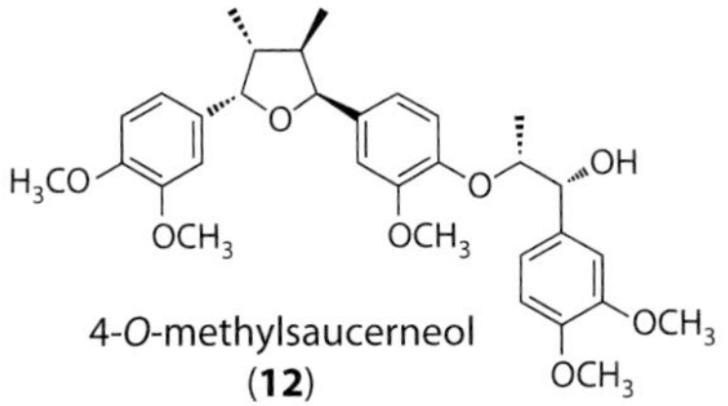

Hodges et al. (2004)
Hossain et al. (2005)

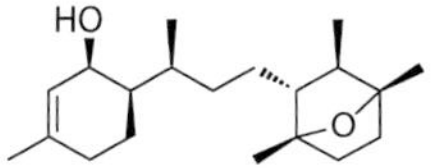

Selectively inhibits hypoxia-induced HIF-1 activation (relative to iron-chelator-induced); may inhibit mitochondrial complex I

Saururus cernuus, S. chinensis

Mohammed et al. (2004)

Mitochondrial complex I inhibitor

Laurencia intricata

(continued)

4-*O*-methylsaucerneol
(12)

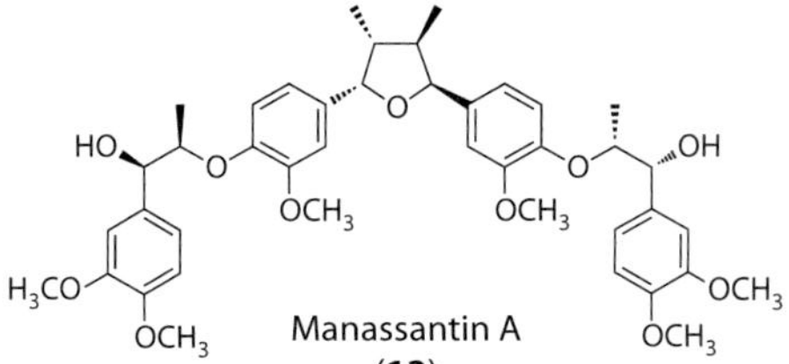

Manassantin A
(13)

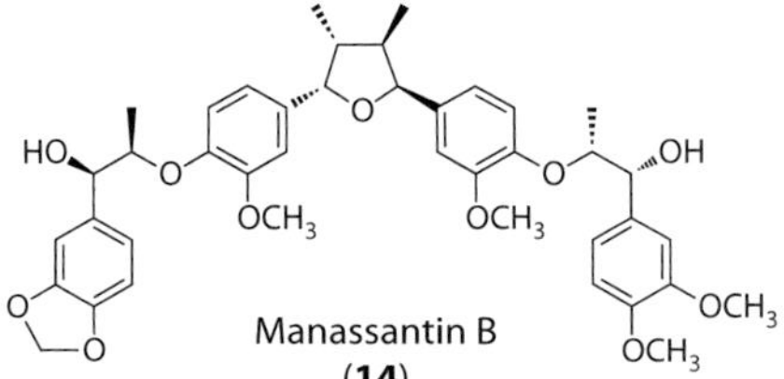

Manassantin B
(14)

laurenditerpenol
(15)

TABLE 7.1 (continued)
Natural Product–Based Agents That Promote HIF-1α Protein Degradation

Compound	Source	Mechanistic Data	References
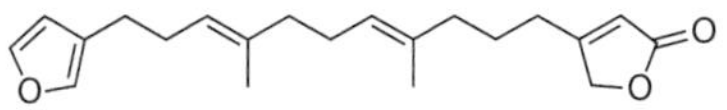 Furospongolide (**16**)	*Lendenfeldia* sp.	Mitochondrial complex I inhibitor	Liu et al. (2008)
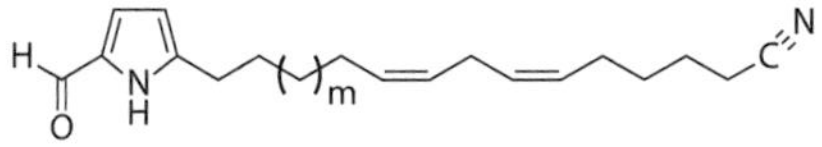 (**17**) Mycalenitrile-6 m = 9 (**18**) Mycalenitrile-7 m = 11	*Mycale* sp.	Mitochondrial complex I inhibitor	Mao et al. (2009)
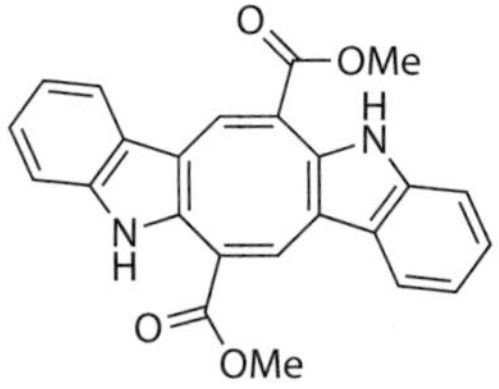 Caulerpin (**19**)	*Caulerpa* spp.	Mitochondrial complex I inhibitor	Liu et al. (2009b)
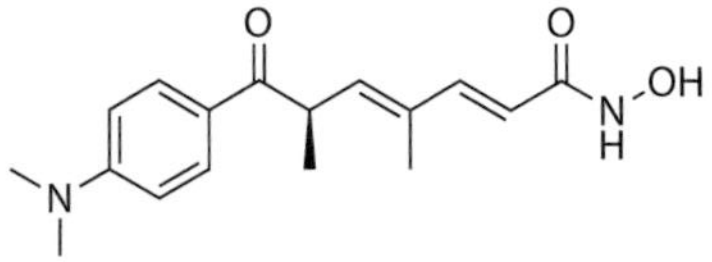 Trichostatin A (TSA) (**20**)	*Streptomyces hygroscopicus*	Histone deacetylase (HDAC) inhibitor	Kim et al. (2001) Kong et al. (2006)

Structure	Source	Activity	References
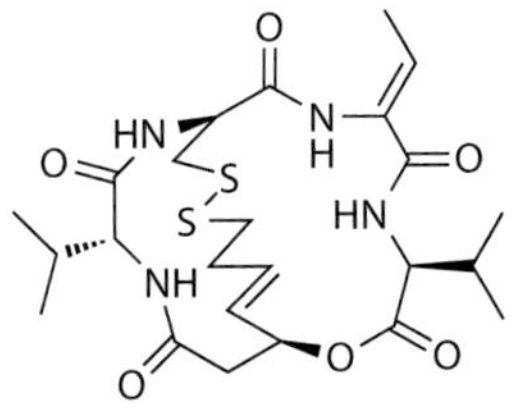 FK 228 (FR901228) (21)	*Chromobacterium violaceum*	HDAC inhibitor	Lee et al. (2003)
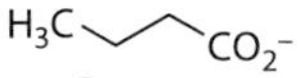 Butyrate (22)	Primary metabolite	HDAC inhibitor	Pellizzaro et al. (2002) Zgouras et al. (2003) Miki et al. (2004) Kim et al. (2007)
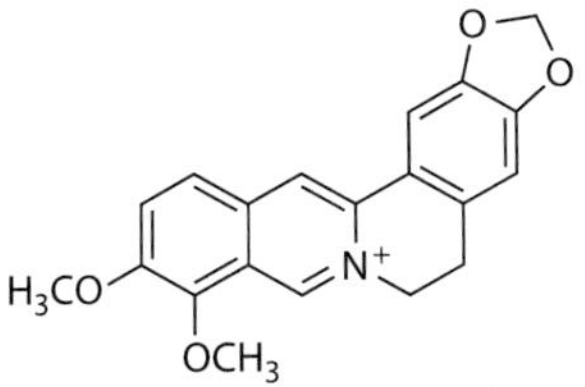 Berberine (23)	Berberidaceae, *Berberis* spp., *Hydrastis canadensis*, *Coptis chinensis*	Mitochondrial complex I inhibitor Inhibits HIF-1α Lys-532 acetylation when tested at cytotoxic concentrations	Lin et al. (2004) Turner et al. (2008)

FK 228 (FR901228)
(21)

Butyrate
(22)

Berberine
(23)

(continued)

TABLE 7.1 (continued)
Natural Product–Based Agents That Promote HIF-1α Protein Degradation

Compound	Source	Mechanistic Data	References
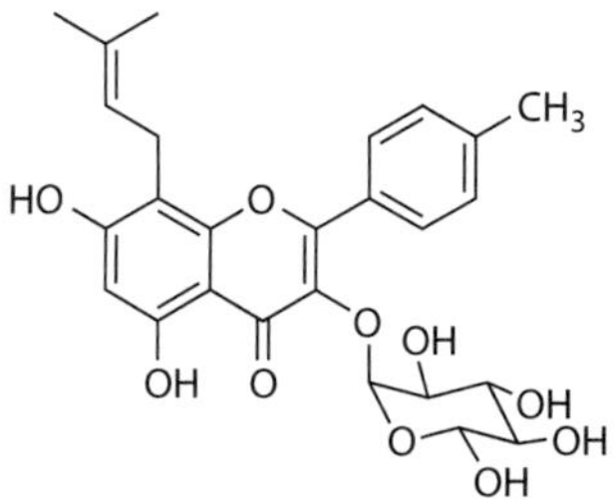 Icariside II (**24**)	*Epimedium koreanum*	May interfere with pVHL-mediated proteasome function	Choi et al. (2008)
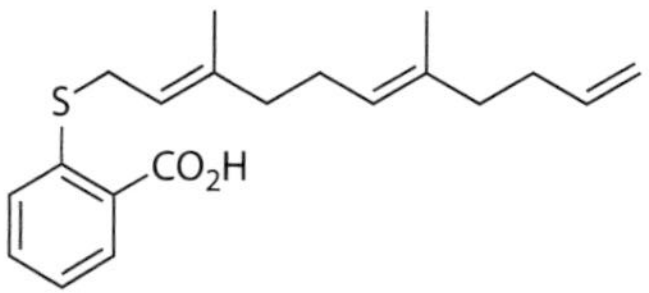 S-*trans*-farnesylthiosalicylic acid (**25**)	Synthetic prenylated salicylic acid	Ras inhibitor; inhibits PI3K/AKT/ERK	Blum et al. (2005)

7.5.1 Inhibitors of Mitochondrial ROS-Mediated HIF-1 Signaling

In the search of natural product–based HIF-1 inhibitors, research efforts in the Department of Pharmacognosy at the University of Mississippi (UM) have evaluated over 40,000 natural product–rich extracts of marine organisms and plants from the National Cancer Institute Open Repository Program. Following bioassay-guided isolation and structural elucidation, the UM program has discovered a variety of small-molecule HIF-1 inhibitors that disrupt mitochondrial function. The red algal diterpene laurenditerpenol (**15**) represented the first marine-derived mitochondrial ETC complex I inhibitor found to inhibit hypoxia-induced HIF-1 activation (Table 7.1). Laurenditerpenol inhibited hypoxia (1% O_2)-induced HIF-1 activation in a T47D breast tumor cell-based luciferase reporter assay at submicromolar concentrations (IC_{50} 0.4 µM) (Mohammed et al., 2004). The angiogenic factor vascular endothelial growth factor (VEGF) is an HIF-1 target gene that is hypoxia inducible in most cell types. Hypoxia (1% O_2) was found to increase the level of secreted VEGF protein from T47D cells by 2.6-fold and laurenditerpenol inhibited the induction by 55% at 1 µM. Mechanistic studies (in isolated mouse liver mitochondria) revealed that laurenditerpenol selectively blocks the mitochondrial ETC complex I (NADH-ubiquinone oxidoreductase, Figure 7.3) at concentrations similar to those required to inhibit hypoxia-induced HIF-1 activation in T47D cells. Supply issues hampered further mechanistic studies. Total synthesis has recently resolved the absolute configuration of laurenditerpenol and may facilitate the supply of sufficient material for more detailed biological studies (Chittiboyina et al., 2007; Jung and Im, 2009).

Sesquineolignans (e.g., 4-*O*-methylsaucerneol, **12**) and dineolignans (manassantin A, **13**; manassantin B, **14**) from the aquatic plant *Saururus cernuus* were found to act as extremely potent inhibitors of HIF-1 signaling (Table 7.1). Using the same T47D-based reporter assay, manassantins A, B, and 4-*O*-methylsaucerneol inhibited hypoxia-induced HIF-1 activation at exceptionally low concentrations (IC_{50} values 0.003, 0.003, 0.030 µM, respectively) (Hodges et al., 2004; Hossain et al., 2005). Unlike the marine algal diterpene laurenditerpenol that suppressed the mitochondrial respiration and HIF-1 activation at similar concentrations, the *S. cernuus* lignans were found only to inhibit the mitochondrial ETC at relatively high concentrations (IC_{50} values 0.320, 0.250, 0.260 µM, respectively) (Hossain et al., 2005). Thus, while disruption of the mitochondrial ETC may play a role in the mechanism of HIF-1 inhibition by *S. cernuus* lignans, the wide discrepancy between the concentrations required for each function suggest that more complex phenomena may be involved. More recent isolation and total synthesis efforts have facilitated structure–activity studies that suggest that the configurations of the side chain hydroxylated carbons in manassantins do not alter their potency (Hanessian et al., 2006; Kasper et al., 2009; Kim et al., 2009). However, the 2,3-*cis*-3,4-*trans*-4,5-*cis*-configuration of the central tetrahydrofuran moiety is critical for the potent HIF-1 inhibition observed with manassantin A (**13**) (Kasper et al., 2009).

The T47D cell-based HIF-1 assay-guided isolation and structure elucidation efforts have led to the identification of a relatively broad array of marine natural products from the U.S. National Cancer Institute's (NCI) open repository of marine invertebrate and algae extracts (Natural Products Branch, 2010) that block hypoxia-induced

HIF-1 activation (Nagle and Zhou, 2009). Some of these marine metabolites target the mitochondrial ETC (Table 7.1). The furanolipid furospongolide (**16**) was isolated from an active extract of a Saipan collection of the marine sponge *Lendenfeldia* sp. and shown to inhibit hypoxia-induced HIF-1 activation in T47D breast tumor cells (IC$_{50}$ 2.9 µM) (Liu et al., 2008). Furospongolide suppressed HIF-1 activation by inhibiting the hypoxic induction of HIF-1α protein and blocked the hypoxia-stimulated induction of secreted VEGF from T47D cells. Mechanistic studies (in T47D tumor cells, rather than isolated mitochondria) indicate that furospongolide inhibits HIF-1 activity, primarily by suppressing tumor cell respiration via the blockade of ETC complex I-mediated mitochondrial electron transfer.

As of 2009, the HIF-1 inhibitor discovery program at the University of Mississippi has evaluated more 20,000 extracts of plants and marine organisms for HIF-1 inhibitory activity. The lipid extract of a Palau collection of the marine sponge *Mycale* sp. (NCI Open Repository) inhibited hypoxia-induced HIF-1 activation in the T47D breast tumor assay system (53% inhibition at 5 µg/ml) (Mao et al., 2009). Bioassay-guided isolation yielded a series of 26 lipophilic mycalazal/mycalenitrile-type 2,5-disubstituted pyrroles. The most active compounds isolated were mycalenitrile-6 (**17**) and mycalenitrile-7 (**18**), that each inhibited hypoxia-induced HIF-1 activation with IC$_{50}$ values of 7.8 and 8.6 µM, respectively. As in the case of the sponge metabolite furospongolide, mechanistic studies revealed that the active 2,5-disubstituted pyrroles suppressed mitochondrial respiration by blocking the ETC at complex I.

The algal pigment caulerpin (**19**) was found to inhibit hypoxia-induced HIF-1 activation, to block the induction of HIF-1α protein in T47D cells, to suppress the induction of secreted VEGF, and to decrease the ability of hypoxic T47D cell-conditioned media to promote tumor angiogenesis in vitro (Liu et al., 2009b). Caulerpin was first isolated from the green algae of the genus *Caulerpa* (Aguilar-Santos, 1970), members of which (i.e., *Caulerpa taxifolia*) have been dubbed the term "killer algae." Over the last twenty years, this alga has invaded over 10,000 acres of the Mediterranean coast and has been devastating coastal ecosystems (Meinesz 1999), just as laurenditerpenol, furospongolide, and mycalenitriles, studies indicated that caulerpin selectively targeted the mitochondrial ETC at complex I to inhibit mitochondrial ROS-based hypoxic signaling.

Mateo and coworkers demonstrated that the *Streptomyces*-derived inhibitor of the mitochondrial ETC complex III inhibitor antimycin A (**11**) potently inhibited hypoxia-induced HIF-1α expression in human Tet-iNOS 293 cells (MIC 5 ng/mL) (Table 7.1) (Mateo et al., 2003) A more recent study by researchers at Showa University of Pharmaceutical Sciences in Japan has reportedly "identified" antimycin A as an inhibitor of HIF-1 signaling from bacterial culture extract screening efforts. This study also demonstrated that antimycin A inhibits HIF-1α expression and was further able to show that antimycin A potently blocks tumor angiogenesis in vivo (Maeda et al., 2006). In spite of its apparent antiangiogenic effects, previous studies suggest that the ability of antimycin A to inhibit hypoxia-induced HIF-1 activation may be less than conclusive (Chandel et al., 2000). In fact, the activity of antimycin A may depend on the nature of the cell line or the physiological conditions of the experimental model.

In general, mitochondrial ROS are required for HIF-1 activation by hypoxia, and are not essential for HIF-1 induction by chemical hypoxia (iron chelators, transition metals, DMOG, etc.) or by anoxia. As a result, mitochondrial inhibitors are selectively effective at blocking hypoxia-induced HIF-1 activation, relative to their effect on HIF-1 activation by other stimuli.

7.5.2 Other Agents That Promote HIF-1α Degradation

The flavonol glycoside icariside II (**24**) was isolated from *Epimedium koreanum* (Berberidaceae) (Choi et al., 2008). In human HOS osteosarcoma and HepG2 hepatocarcinoma cells, icariside (10 µM) suppressed hypoxia-induced accumulation of HIF-1α protein without pronounced toxicity. Icariside suppressed tumor angiogenesis and invasion in vitro and this inhibition correlated with the effects of icariside exerted on HIF-1 target genes involved in both processes. Binding studies suggested that icariside enhanced the interaction between HIF-1α and pVHL proteins under hypoxic conditions. Thus, it is most likely that icariside suppressed HIF-1 activation by facilitating the degradation of HIF-1α protein under hypoxic conditions. It was reported that icariside exhibited anti-oxidant activity and inhibited hypoxia-activated protein tyrosine kinase. It is possible that the icariside suppressed hypoxic activation of HIF-1 via multiple mechanisms.

7.6 AGENTS THAT PREVENT THE SYNTHESIS OF HIF-1α PROTEIN

Natural products and natural product–derived synthetic and semisynthetic compounds that inhibit HIF-1 activation by mechanisms that involve blocking HIF-1α protein synthesis are listed in Table 7.2 (structures **26–61**).

7.6.1 Protein Translation Inhibitors

In general, global protein synthesis can account for up to 70% of cellular ATP consumption. Many forms of stress regulate the overall mRNA translation, primarily at the initiation level. The well-characterized processes that are often affected by stress include the assembly of the eukaryotic initiation factor 4F (eIF4F) at the m^7GpppN cap structure and the formation of the eIF2-GTP-tRNA^{Met} complex. Hypoxia-induced mRNA translation responses are highly heterogeneous among individual mRNA transcripts. The translation of certain genes is stimulated by hypoxia, while overall global protein synthesis is inhibited.

Employing pharmacologic agents, Semenza and colleagues first established that the activation of HIF-1 requires functional translation. Cycloheximide (**29**), an inhibitor of general eukaryotic protein synthesis, blocked the hypoxic induction of HIF-1α protein and subsequent HIF-1 activation (Table 7.2). Since this initial discovery, an assortment of HIF-1 inhibitors has been reported to disrupt HIF-1 activation by suppressing the synthesis of HIF-1α protein under hypoxic conditions. Some of these inhibitors have been discussed in a previous review (Nagle and Zhou, 2006a).

The HIF inhibitor discovery program at the University of Mississippi discovered that the *Psychotria klugii* (Rubiaceae) metabolites klugine (**41**), isocephaeline (**42**),

TABLE 7.2
Natural Product–Based Agents That Decrease the Synthesis of the HIF-1α Protein

Compound	Source	Mechanistic Data	References
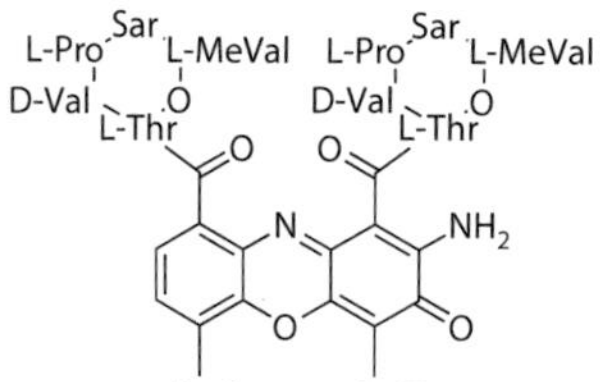 Actinomycin D (26)	*Streptomyces parvullus*	Transcription inhibitor	Wang and Semenza (1993) Page et al. (2002)
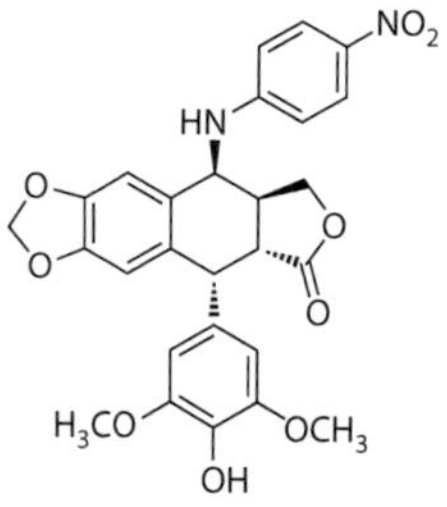 GL 331 (27)	Semisynthetic podophyllotoxin	Topoisomerase II inhibitor	Chang et al. (2003)
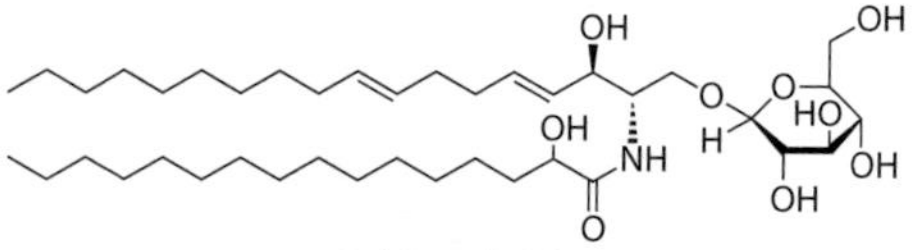 Sphingolipid (28)	*Glycine max* (Soy)		Symolon et al. (2004)

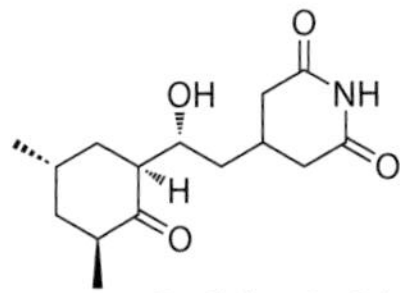

Semenza and Wang (1992)

Rapisarda et al. (2002)

Protein synthesis inhibitor

Topoisomerase I inhibitor

Streptomyces griseus

Semisynthetic camptothecin

(continued)

Cycloheximide
(29)

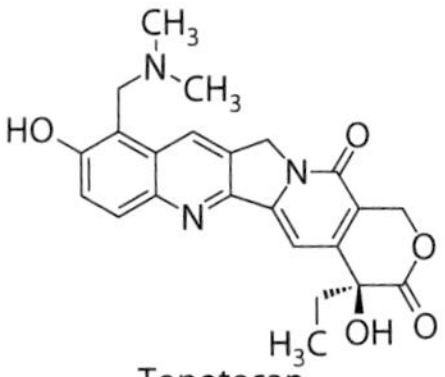

Topotecan
(30)

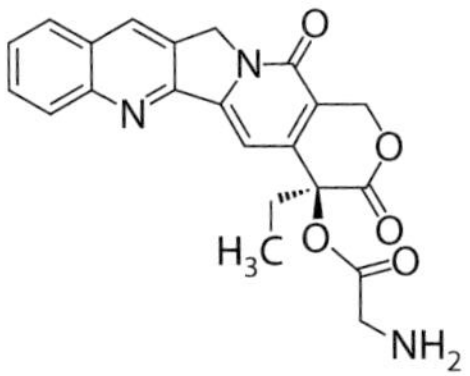

20 S-captothecin, glycine ester
(31)

TABLE 7.2 (continued)

Natural Product–Based Agents That Decrease the Synthesis of the HIF-1α Protein

Compound	Source	Mechanistic Data	References
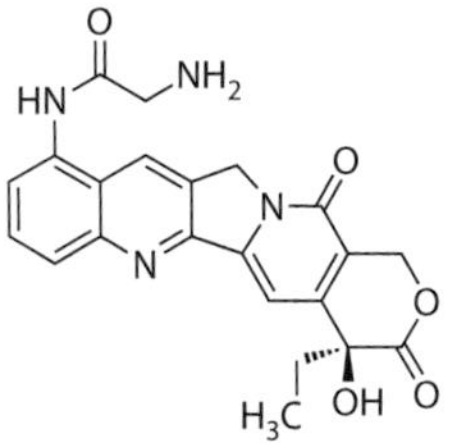 9-glycineamido-20(S)-camptothecin **(32)**			
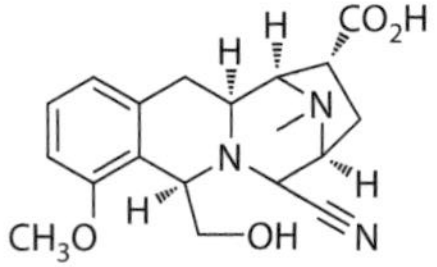 DX-52-1 **(33)**	Semisynthetic quinocarmycin analogue		Rapisarda et al. (2002)
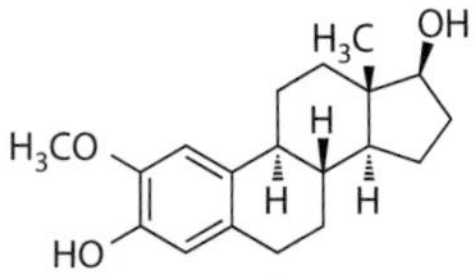 2-methoxyestradiol **(34)**	Semisynthetic estradiol	Microtubule disruptor and mitochondrial complex I inhibitor	Mabjeesh et al. (2003) Hagen et al. (2004) Escuin et al. (2005) Chua et al. (2010)

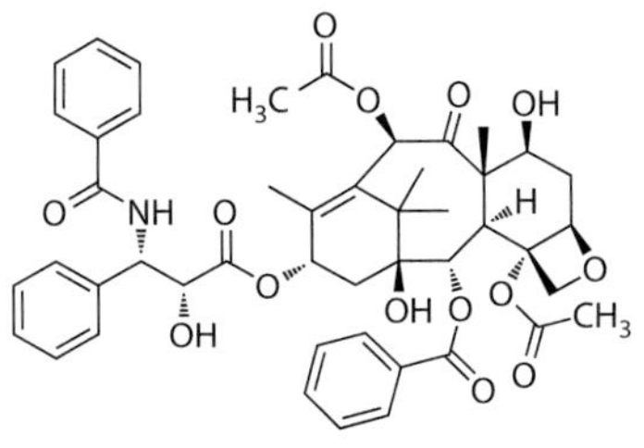

Jung et al. (2003)
Mabjeesh et al. (2003)

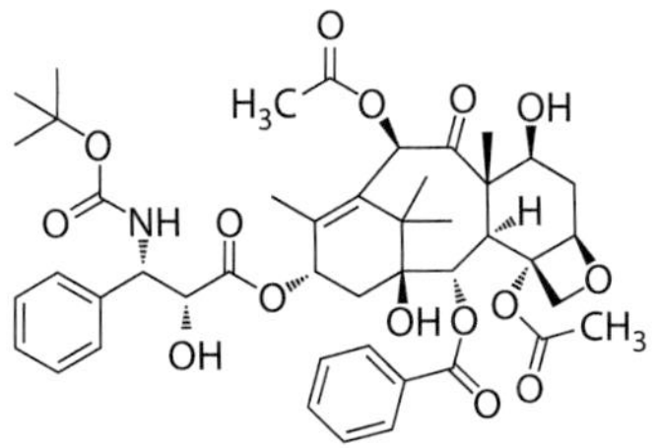

Escuin et al. (2005)

Mabjeesh et al. (2003)

(continued)

Micotubule stabilizer

Micotubule stabilizer

Microtubule disruptor

Taxus brevifolia

Semisynthetic paclitaxel
analogue

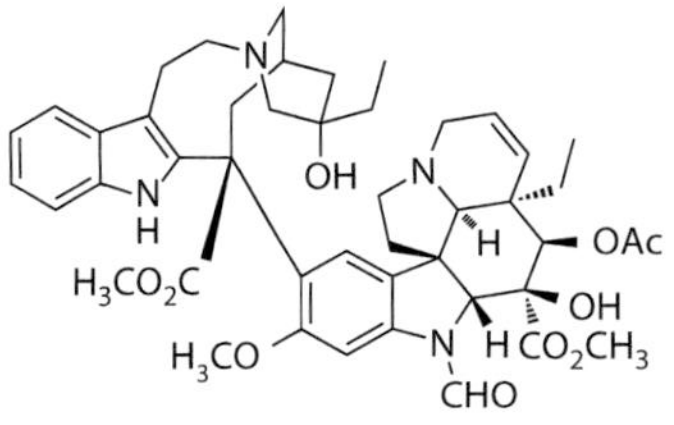

Cantharanthus roseus

Paclitaxel (taxol)
(**35**)

Docetaxel (taxotere)
(**36**)

Vincristine (**37**)

TABLE 7.2 (continued)
Natural Product–Based Agents That Decrease the Synthesis of the HIF-1α Protein

Compound	Source	Mechanistic Data	References
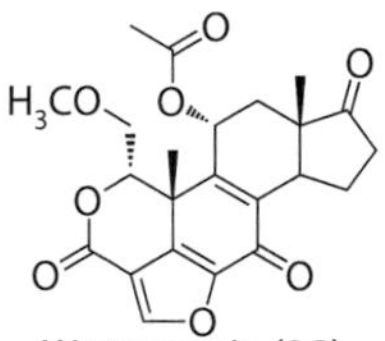 Wortmannin (**38**)	*Penicillium fumiculosum*	PI3 kinase inhibitor	Zhong et al. (2000) Jiang et al. (2001)
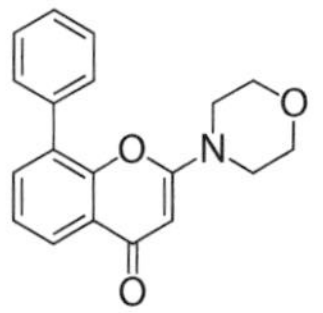 LY 294002 (**39**)	Synthetic flavonoid	PI3 kinase inhibitor	Zhong et al. (2000) Jiang et al. (2001) Pore et al. (2006)
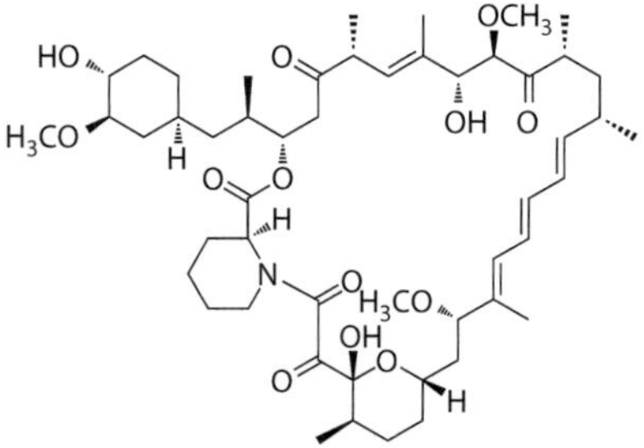 Rapamycin (sirolimus) (**40**)	*Streptomyces hygroscopicus*	mTOR inhibitor	Zhong et al. (2000) Laughner et al. (2001) Hudson et al. (2002)

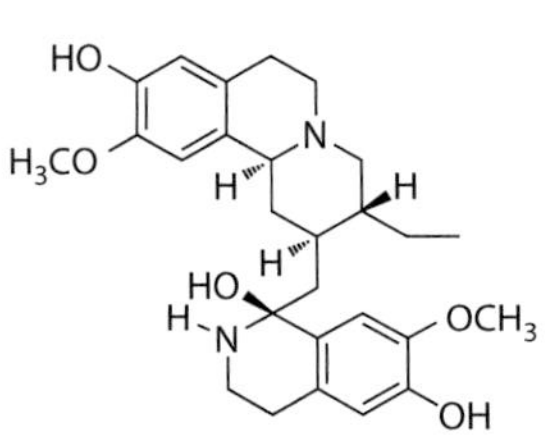

Psychotria ipecacuanha,
P. klugii

Protein translation inhibitor

Zhou et al. (2005)

(continued)

Klugine (41)

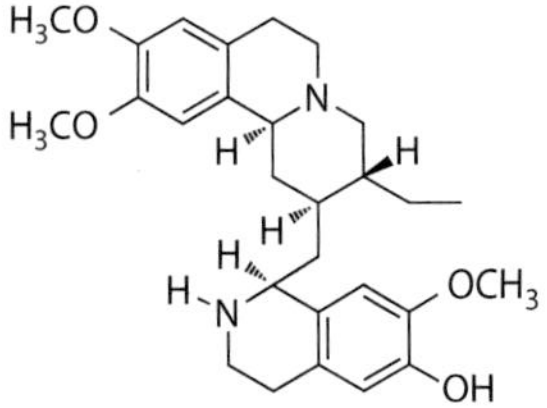

Isocephaeline (42)

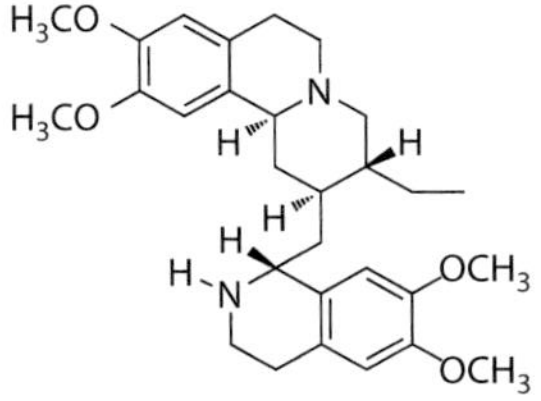

Emetine (43)

TABLE 7.2 (continued)
Natural Product–Based Agents That Decrease the Synthesis of the HIF-1α Protein

Compound	Source	Mechanistic Data	References
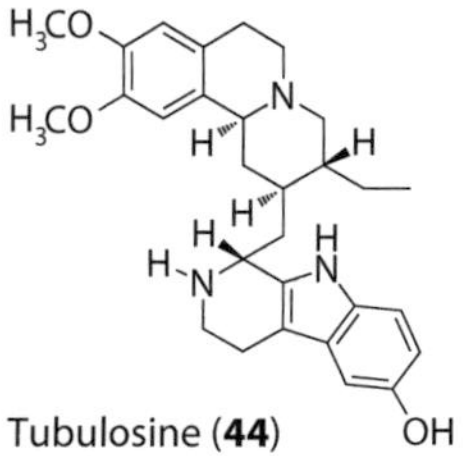 Tubulosine (**44**)	*Alangium* cf. *longiflorum*	Protein translation inhibitor	Klausmeyer et al. (2008)
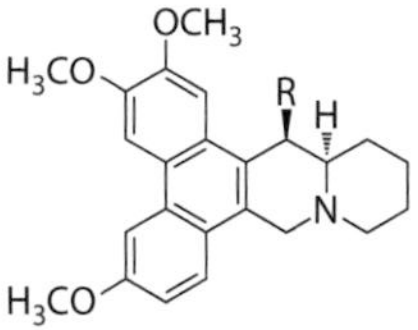 (**45**) (–)-cryptopleurine R=H (**46**) (–)-15R-hydroxycryptopleurine R=OH	*Boehmeria pannosa*	Protein translocation inhibitor	Cai et al. (2006)
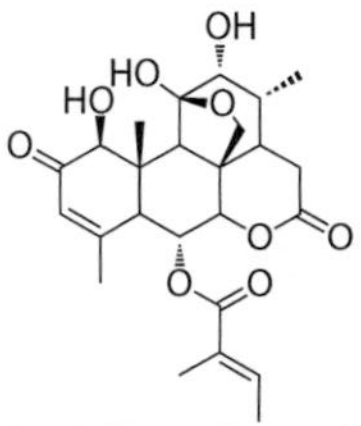 6a-tigloyoxchaparrinone (**47**)	*Ailantus altissima*	Blocks protein translation by inhibiting eIF4E phosphorylation	Jin et al. (2008)

Species	Mechanism	Reference
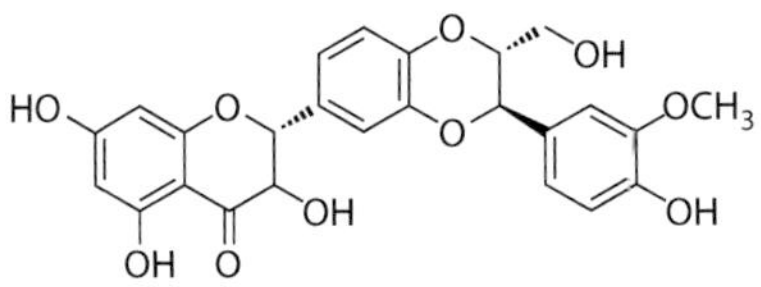 *Silybum marianum*	Extremely high concentrations inhibit mTOR/p70S6K/4E-BP1-mediated protein translation	Singh et al. (2008) Garcia-Maceira and Mateo (2009) Wang et al. (2010)
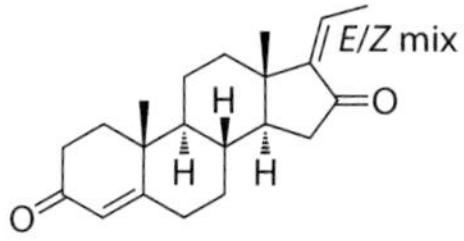 *Commiphora mukul*	Inhibits STAT-3 signaling and HIF-1α expression	Leeman-Neill et al. (2009)
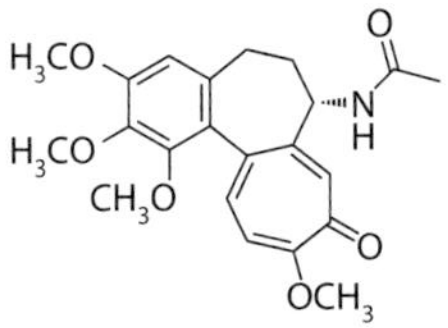 *Colchicum autumnale*	Microtubule disruptor	Escuin et al. (2005)
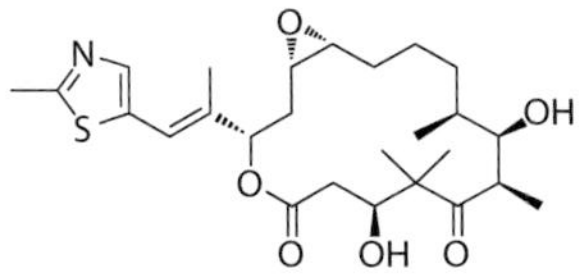 *Sorangium cellulosum*	Micotubule stabilizer	Escuin et al. (2005)

(continued)

Silibinin (**48**)

E/Z-guggulsterone (**49**), E/Z mix

Colchicine (**50**)

Epothilone A (**51**)

TABLE 7.2 (continued)

Natural Product–Based Agents That Decrease the Synthesis of the HIF-1α Protein

Compound	Source	Mechanistic Data	References
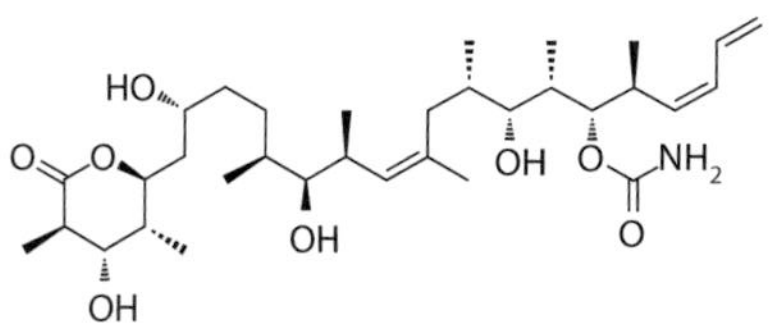 Discodermolide (**52**)	*Discodermia dissoluta*	Micotubule stabilizer	Escuin et al. (2005)
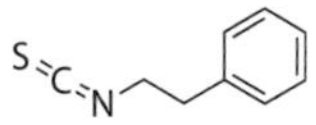 2-Phenethyl isothiocyanate (**53**)	Cruciferous vegetables— watercress, broccoli, radishes, mustard, brussel sprouts	Inhibits HIF-1α protein translation by decreasing the phosphorylation of 4E-BP1	Wang et al. (2009)
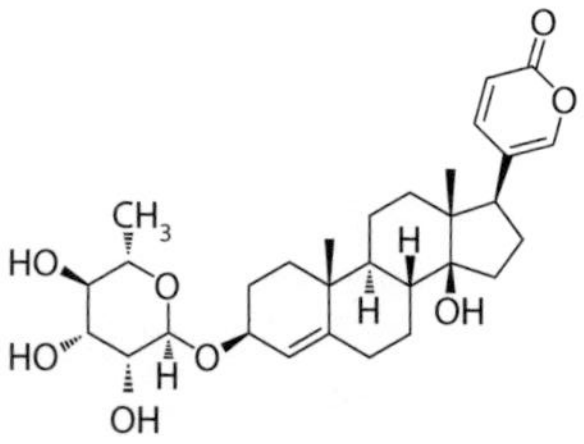 Proscillaridin A (**54**)	*Urginea* spp., *Strophanthus* spp., *Digitalis lanata*, respectively	Inhibitors of plasma membrane Na$^+$/K$^+$-ATPase	Zhang et al. (2008)

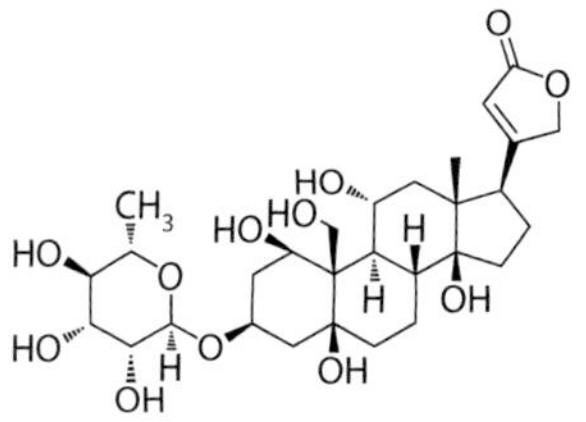

Ouabain
(55)

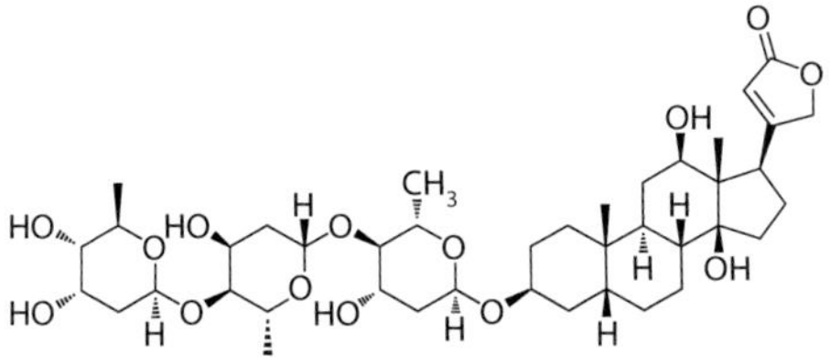

Digoxin
(56)

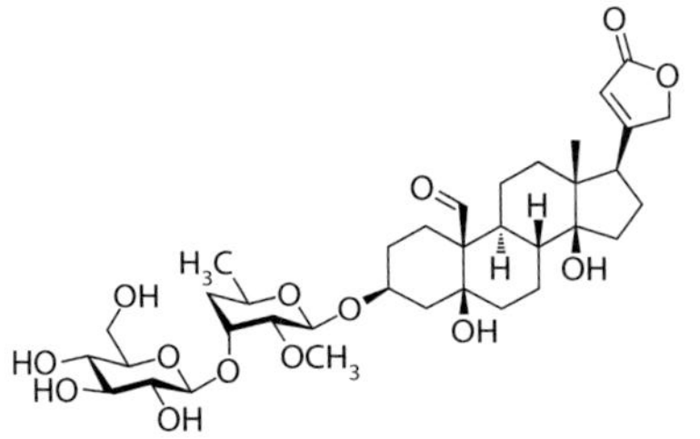

(57) Strophanthidin glycoside
(61) Strophanthidin (aglycone)

Inhibitors of plasma
membrane Na⁺/K⁺-ATPase

Crossosoma bigelovii

Klausmeyer et al. (2009)

(continued)

TABLE 7.2 (continued)
Natural Product–Based Agents That Decrease the Synthesis of the HIF-1α Protein

Compound	Source	Mechanistic Data	References

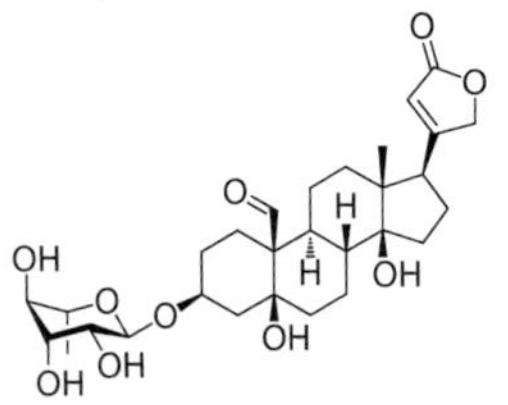

Convallatoxin
(58)

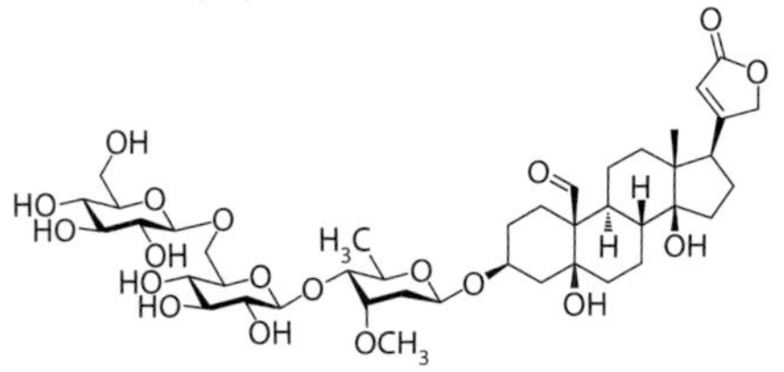

k-Strophanthoside
(59)

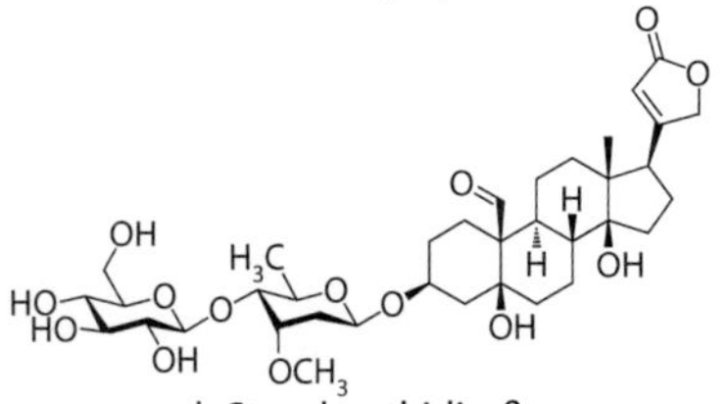

k-Strophanthidin-β
(60)

and emetine (**43**) inhibited hypoxia-induced HIF-1 activation in T47D breast tumor cells (IC$_{50}$ values 0.2, 1.1, and 0.11 μM, respectively) (Zhou et al., 2005). Terpenoid tetrahydroisoquinoline alkaloids such as emetine (also from ipecac *P. ipecacuanha*) have been shown to inhibit eukaryotic protein synthesis (Grollman, 1966). Zhou and coworkers demonstrated that *Psychotria* alkaloids inhibit both hypoxia- and iron-chelator-induced HIF-1 activation by blocking HIF-1α protein accumulation.

Just as the ipecac alkaloids klugine and emetine inhibit protein translation and HIF-1α protein expression, Klausmeyer and coworkers at the NCI have recently shown that the structurally related tubulosine alkaloids inhibit HIF-1α expression and hypoxia-induced HIF-1 activation in human U251 glioma cells (Klausmeyer et al., 2008). Tubulosines are believed to inhibit protein translation by a similar mechanism as emetine (**43**) and other related alkaloids (Gupta and Siminovitch, 1977). While IC$_{50}$ values were not published, the data would suggest that tubulosine (**44**) inhibits hypoxia-induced HIF-1 activation in U251 cells with an IC$_{50}$ of about 50 nM. Emetine appeared to be equipotent to tubulosine in the U251 stably transfected HRE/luciferase HIF-1 reporter assay, but failed to produce the same level of activity in U251 cells that were transiently transfected with a HRE/luciferase reporter. Tubulosine also suppressed HIF-1α protein levels and HIF-1 target gene expression in U251 cells with greater potency than emetine.

The phenanthroquinolizidine alkaloid cryptopleurine has been shown to inhibit protein synthesis by inhibiting translocation (Bucher and Skogerson, 1976). Researchers at the Korean Research Institute of Bioscience and Biotechnology have reported that (−)-cryptopleurine (**45**) and (−)-(15R)-hydroxycryptopleurine (**46**) from the roots of *Boehmeria pannosa* (Urticaceae) potently inhibited HIF-1 activation in human gastric adenocarcinoma AGS cancer cells (IC$_{50}$ values 8.7 and 48.1 nM, respectively) (Cai et al., 2006). As anticipated, for translation inhibitors, these alkaloids were shown to inhibit HIF-1 activation by suppressing HIF-1α protein expression. Despite these findings, the authors fail to address the role of translation inhibition as the plausible mechanism of action for these alkaloids.

Bae and coworkers at the Chungnam National University and the Korea Research Institute of Bioscience and Biotechnology established a stable human gastric adenocarcinoma MKN-45-HRE-Luc cell line–based reporter assay to monitor HIF-1 activity. Examination of a library of 1000 natural products identified the quassinoid 6α-tigloyloxychaparrinone (**47**) as an HIF-1 inhibitor (Jin et al., 2008). At low micromolar concentrations (0.3–3 μM), 6α-tigloyloxychaparrinone inhibited the induction of HIF-1α protein and the activation of HIF-1 by hypoxia (1% O$_2$). Comparable inhibitory effects were observed on chemical hypoxia (cobalt chloride)-induced HIF-1 activation. Based on an assessment of the phosphorylation and expression status of a panel of kinases and their targets, the authors concluded that 6α-tigloyloxychaparrinone inhibits HIF-1 by blocking hypoxia-induced phosphorylation of eukaryotic translation initiation factor 4E (eIF4E), a subunit of eIF4F that binds the m^{7}GpppN cap structure of mRNA. However, the observation that 6α-tigloyloxychaparrinone inhibits both hypoxia- and chemical hypoxia (cobalt chloride)-induced HIF-1 activation with comparable potency indicates that other mechanisms may be involved in addition to the blockade of hypoxia-induced eIF4E phosphorylation.

Silibinin (**48**), a flavonoid originally isolated from milk thistle (*Silybum mari-anum*), has exhibited hepatoprotective (antihepatotoxic) and antitumor properties (Li et al., 2010). A recent study reported that silibinin blocked both hypoxia- and dimethyloxaloylglycine (DMOG, an inhibitor of PHDs)-induced HIF-1α protein accumulation in human cervical cancer HeLa and hepatocarcinoma Hep3B cells at the extremely high concentration of 500 µM (Table 7.2) (Garcia-Maceira and Mateo, 2009). Based on the pathway survey results, the authors concluded that silibinin inhibited the translation of HIF-1α protein by suppressing the mTOR/p70S6K/4E-BP1 signaling pathway. However, silibinin has been reported to affect numer-ous pathways that include the activation of extracellular signal-related kinase 1/2 (ERK1/2), the inhibition of JNK1/2, p38 MAPK, Akt, and STAT signaling (Singh et al., 2009), and the induction of mitochondrial ROS (Wang et al., 2010). The com-bination of the high concentration/dose requirement and the diversity of pathways/targets known to be affected by silibinin suggest that silibinin inhibits the HIF-1 pathway in a relatively nonspecific fashion.

Originally isolated from *Streptomyces hygroscopicus*, the mTOR inhibitor rapa-mycin (**40**, sirolimus, Rapamune®) was shown to inhibit HIF-1α protein accumula-tion by at least two mechanisms: inhibition of HIF-1α protein synthesis (Laughner et al., 2001) and promotion of HIF-1α protein degradation (Hudson et al., 2002). Mechanistic studies have since defined the requirement of an mTOR signaling motif in the N-terminus of HIF-1α protein (Land and Tee, 2007). Recent studies suggest that the antitumor properties of rapamycin are mediated through its effect on HIF-mediated hypoxic signaling and VEGF-mediated angiogenesis (Wang et al., 2007; Wang et al., 2009).

7.6.2   OTHER AGENTS THAT SUPPRESS THE SYNTHESIS OF HIF-1α PROTEIN

The dietary supplement guggulsterone (**49**) has recently been shown to enhance the activity of chemotherapeutic agents on head and neck squamous cell carci-noma (HNSCC) cell lines and in xenograft models (Leeman-Neill et al., 2009). Guggulsterone (*E* and *Z* mixture, 10 µM) inhibited HIF-1α protein expression in HNSCC UM-22b and 1483 cell lines by a mechanism that may involve suppression of signal transducer and activator of transcription-3 (STAT3)-mediated signaling.

As a dynamic major cellular structural component, microtubules play an impor-tant role in cellular functions, such as division, intracellular trafficking, and signal transduction. Drugs that target microtubules constitute one class of chemothera-peutic agents for patients with advanced stage cancers. Based on the mechanisms of action, these agents are grouped as those that stabilize microtubules (i.e., Taxol® and Taxotere®) and those that destabilize microtubules (*Vinka* alkaloids). In 2003, Giannakakou and colleagues reported that the estradiol metabolite 2-methoxyestra-diol (2-ME or 2ME2, **34**) inhibited HIF-1 activity by blocking the synthesis of HIF-1α protein under normoxic, as well as hypoxic, conditions (Mabjeesh et al., 2003). The HIF-1 inhibitory activity of 2ME2 appears to be related to the disruption of micro-tubules. Other microtubule disrupting agents such as Taxol (**35**, paclitaxel) and vin-cristine (**37**) exhibited similar HIF-1 inhibitory activities. A later study suggested that 2ME2 may suppress the hypoxic induction of HIF-1α protein via the inhibition of mitochondrial ETC complex I (Hagen et al., 2004). In addition, 2ME2 did not inhibit the induction of HIF-1α protein by chemical hypoxia [N-mercaptopropionylglycine (2-oxoglutarate analogue) or DFO (iron chelator)] in human embryonic kidney (HEK) 293 cells. The related compound ENMD-1198 [2-methoxyestra-1,3,5(10),16-tetraene-3-carboxamide] was reported to suppress the hypoxic induction of HIF-1α protein in hepatocellular carcinoma HUH7 cells via a yet to be determined mechanism (Moser et al., 2008). A recent structure-activity relationship (SAR) study indicated that microtubule disruption is the primary mechanism of action for 2ME2 to suppress tumor cells (Chua et al., 2010). Since supratherapeutic concentrations of 2ME2 were required to inhibit the induction of HIF-1α protein and the activation of HIF-1, it is unlikely that 2ME2 exerts its antitumor effects via the HIF-1 pathway, at least in epithelia-derived tumor cells (e.g., MCF7, HEK293, and HCT116). This SAR study also suggested that it is unlikely that the inhibition of mitochondrial ETC complex I and the induction of cellular ROS constitute the major mechanism(s) of action for 2ME2.

Around the same time when 2ME2 was first found to be an HIF inhibitor, Neckers and colleagues reported that microtubule-depolymerizing agents [e.g., colchicine (**50**), vinblastine, and nocodazole] actually induced HIF-1α protein in certain cell lines (e.g., A549, MCF-7, Jurkat, NIH-3T3, etc.) (Jung et al., 2003). This melanoma differentiation-associated gene (MDA)-triggered HIF-induction required microtubule depolymerization and active cellular transcription. The induction of HIF-1α protein occurred in a time and concentration-dependent manner, following the activation of NFκB. A later study by Giannakakou's group challenged the MDA-inducing HIF theory (Escuin et al., 2005). In a study that examined a panel of MDAs [Taxotere (docetaxel, **36**), epothilone B (**51**), discodermolide (**52**), 2ME2 (**34**), vincristine (**37**), and colchicine (**50**)] at the concentrations that disrupt microtubule function, all of the MDAs evaluated suppressed the hypoxic induction of HIF-1α protein in human ovarian 1A9 and breast MDA-MB-231 cancer cell lines. In contrast, neither vinblastine nor colchicine induced HIF-1α protein in any of the cell lines examined (1A9, A549, MCF-7, LN229, and PC3). As observed for 2ME2, these MDAs inhibited the synthesis of HIF-1α protein. The HIF-1 inhibition occurred downstream from the disruption of microtubules. However, the difference in drug treatment times (i.e., 4 h versus 16 h) and other experimental conditions may contribute to the discrepancy between various groups when examining the effects of MDAs on the HIF pathway.

The cruciferous vegetable-derived isothiocyanates (ITCs) have exhibited both chemopreventive and anticancer activities in vitro and in vivo. One of the molecular mechanisms of action for ITCs is the depletion of intracellular stores of the major antioxidant glutathione (GSH). In human breast tumor Michigan Cancer Foundation (MCF)-7 cells, phenethyl isothiocyanate (**53**) suppressed both hypoxia (0.1% O_2) and chemical hypoxia (100 μM $CoCl_2$)-induced HIF-1 activation in a concentration-dependent manner (Wang XH et al., 2009). The hypoxic induction of both HIF-1α (0.1% O_2, 5 h) and HIF-2α (0.1% O_2, 16 h) proteins were inhibited by phenethyl isothiocyanate. Mechanistic studies suggested that phenethyl isothiocyanate inhibits the expression of HIF-1α protein by decreasing the phosphorylation of 4E-BP1. The nonphosphorylated 4E-BP1 binds to eIF4E and inhibits the translation of HIF-1α mRNA. Under normal conditions, the phosphorylation of 4E-BP1 prevents 4E-BP1 from binding eIF4E, and this enhances the translation of HIF-1α mRNA.

A human hepatocarcinoma Hep3B cell-based reporter assay was established to monitor the HIF-regulated expression of a firefly luciferase reporter under the control of HRE from the human *ENO1* gene, and the constitutive expression of a *Renilla* luciferase reporter under the control of a SV40 promoter (Zhang et al., 2008). Hypoxic exposure (1% O_2, 24 h) increased the ratio of firefly luciferase/*Renilla* luciferase by 10-fold, as an indicator of hypoxia-induced HIF activation. The cardiac glycosides proscillaridin A (**54**), ouabain (**55**), and digoxin (**56**) were shown to inhibit hypoxia-activated HIF and blocked hypoxia-induced accumulation of HIF-1α and HIF-2α proteins in Hep3B cells (Zhang et al., 2008). Further mechanistic investigations suggested that compounds such as digoxin inhibit HIF-1α protein synthesis. As a result, digoxin inhibited both hypoxia and chemical hypoxia [transition metal ($CoCl_2$), iron chelator (DFO), and PHD inhibitor (DMOG)]-induced HIF-1α protein accumulation. The HIF inhibition exerted by digoxin appeared to be independent of the mTOR and the Na^+/K^+-ATPase. Although digoxin suppressed xenograft tumor growth in severe combined immunodeficiency (SCID) mice, the concentrations of digoxin required to inhibit HIF-1 is far above the concentrations that can be tolerated by patients (Lopez-Lazaro, 2009). Therefore, while these results may appear highly intriguing, their clinical relevance is uncertain.

Bioassay-guided isolation of a *Crossosoma bigelovii* extract that inhibited HIF-1 activation yielded a new strophanthidin glycoside (**57**) (Klausmeyer et al., 2009). In a U251 human glioma cell–based reporter assay, the cardenolide and a panel of related compounds were examined for their effects on HIF-1 activation. Like other cardiac glycosides, strophanthidins inhibit Na^+/K^+-ATPase. The most potent compound convallatoxin (**58**) suppressed HIF-1 activation with an EC_{50} of 20 nM. Convallatoxin is a 3-*O*-β-L-rhamnosyl analogue of strophanthidin. The other compounds suppressed HIF-1 activation in the following order of potency: k-strophanthoside (**59**, EC_{50}: 40 nM), k-strophanthin-β (**60**, EC_{50}: 120 nM), (strophanthidin glycoside **57**, EC_{50}: 120 nM), and the aglycone strophanthidin (**61**, EC_{50}: 2 μM). However, the inhibitory effects exerted by these compounds on luciferase expression in a control cell line and on cell viability in other tumor cell lines suggested that the HIF-1 inhibitory activity might be associated with the cytotoxicity exerted by these compounds and represent an "off-target" effect.

7.7 AGENTS THAT SUPPRESS HIF-1α PROTEIN ACCUMULATION

Natural products and natural product–derived synthetic compounds that inhibit HIF-1 activation by a variety of different mechanisms that produce an overall suppression of HIF-1α accumulation are listed in Table 7.3 (Structures **62–102**).

Bioassay-guided fraction of an active extract of the sponge *Dendrilla nigra* (NCI Open Repository) yielded four new lamellarin-like phenolic pyrroles (Liu et al., 2007). The most active metabolite 7-hydroxyneolamellarin A (**67**) inhibited hypoxia- and iron-chelator [1,10-phenanthroline (10 μM)]-induced HIF-1 activation (IC_{50} values 1.9 and 3.7 μM, respectively) (Table 7.3). Hydroxyneolamellarin A also inhibited hypoxia- and iron-chelator-induced production of secreted VEGF (10 μM, T47D cells). Limited compound supply prevented further pharmacological and mechanistic studies. However, recent progress

TABLE 7.3

Natural Product–Based Inhibitors of HIF-1α Protein Accumulation

Compound	Source	Mechanistic Data	References
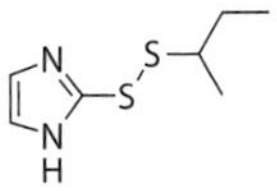 PX-12 (**62**)	Synthetic	Thioredoxin-1 inhibitor; inhibits HIF-1α at cytotoxic concentrations	Welsh et al. (2003)
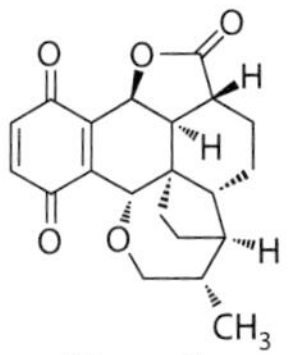 Pleurotin (**63**)	*Pleurotus griseus*	Mushroom antibiotic; inhibits thioredoxin-1 reductase; inhibits HIF-1α at cytotoxic concentrations	Welsh et al. (2003)
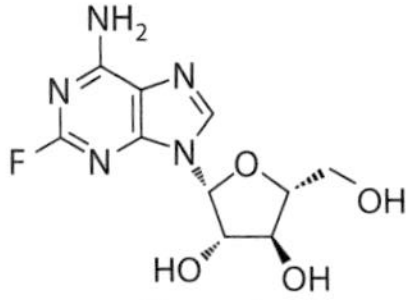 F-ara-A (**64**)	Synthetic nucleoside analog	Inhibits DNA replication and may inhibit PI3K/AKT-mediated HIF-1α regulation	Fang et al. (2004)

(continued)

TABLE 7.3 (continued)
Natural Product–Based Inhibitors of HIF-1α Protein Accumulation

Compound	Source	Mechanistic Data	References
2-Deoxy-D-ribose (65)	Thymidine metabolite	Produced only a slight decrease in HIF-1α protein	Ikeda et al. (2002)
Morphine (66)	*Papaver somniferum*	Opium alkaloid	Roy et al. (2003)
7-Hydroxyneolamellarin A (67)	*Dendrilla nigra*	Mechanism unknown	Liu, R. et al. (2007)

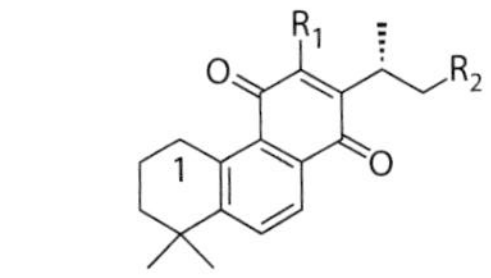

Dat et al. (2007)

Mechanism unknown

Salvia miltiorrhiza

(68) Sibiriquinone A $R_1=R_2=$ H, D^1
(69) Sibiriquinone B $R_1=R_2=$ H
(76) Neocryptotanshinone $R_1=R_2=$OH

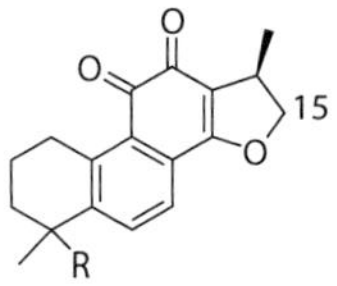

(70) Cryptotanshinone $R=CH_3$
(72) Methyl tanshinonate $R=Co_2CH_3, \Delta^{15}$
(74) Tanshinone IIA $R=CH_3, \Delta^{15}$

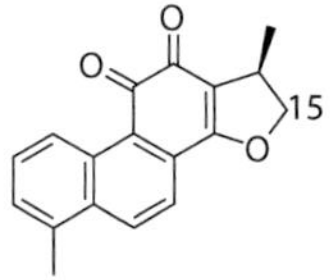

(71) 3,4-Dihydro cryptotanshinone
(73) Tanshinone I Δ^{15}

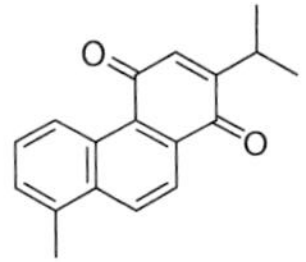

(75) Deoxytanshinquinone B

(continued)

TABLE 7.3 (continued)
Natural Product–Based Inhibitors of HIF-1α Protein Accumulation

Compound	Source	Mechanistic Data	References
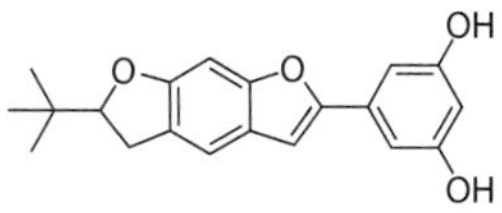 Moracin O (**77**) 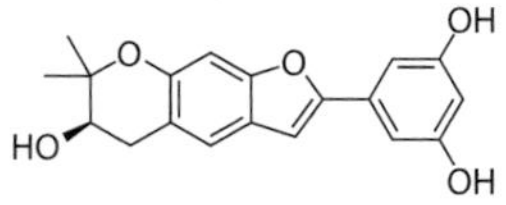 Moracin P (**78**) 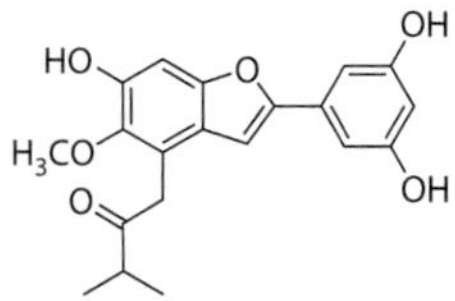 Moracin Q (**79**) 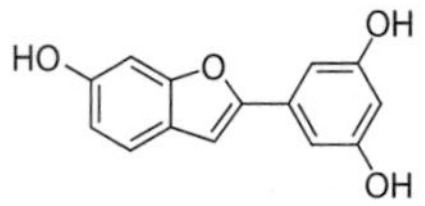 Moracin M (**80**)	*Morus bombycis*	Mechanism unknown	Dat et al. (2009)

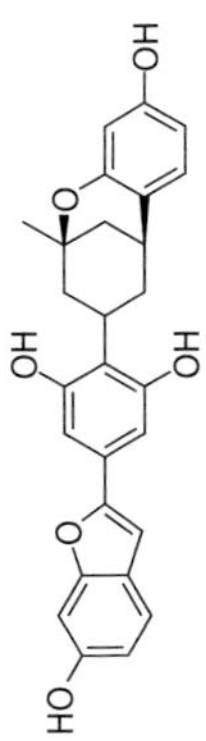

Mulberrofuran H (**81**)

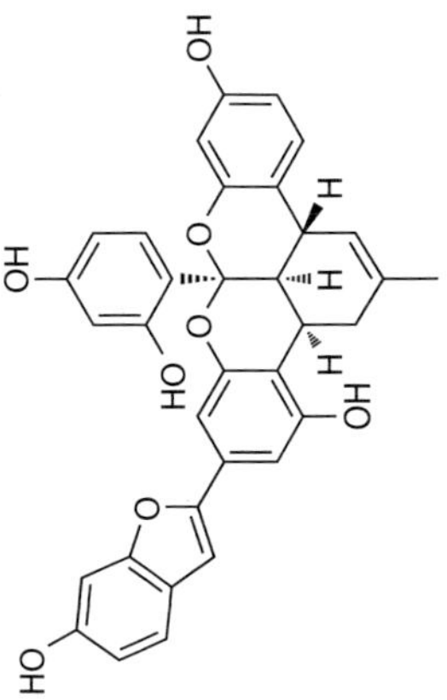

Mulberrofuran G (**82**)

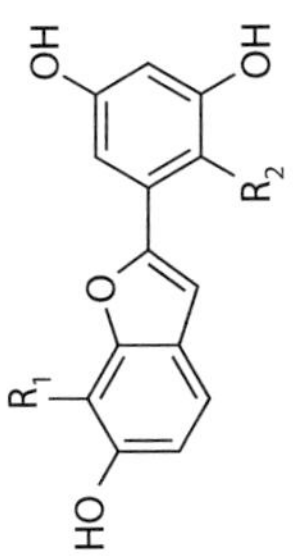

(**83**) Albafuran A $R_1 = H; R_2 = $ geranyl
(**84**) Mulberrofuran D $R_1 = $ geranyl; $R_2 = $ prenyl
(**85**) Mulberrofuran M $R_1 = H; R_2 = $ farnesyl

(continued)

TABLE 7.3 (continued)
Natural Product–Based Inhibitors of HIF-1α Protein Accumulation

Compound	Source	Mechanistic Data	References

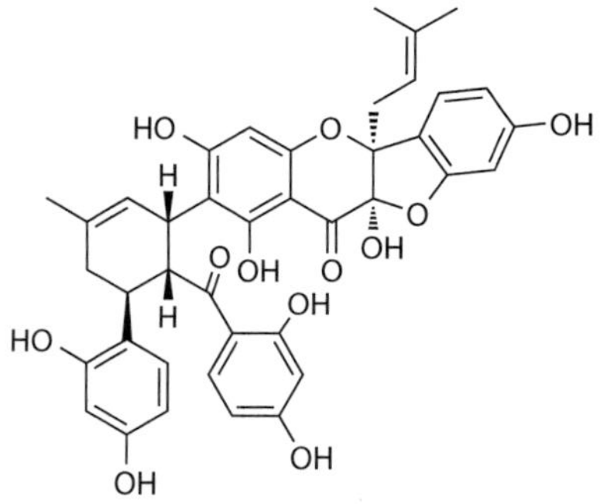

Sanggenon O (**86**)

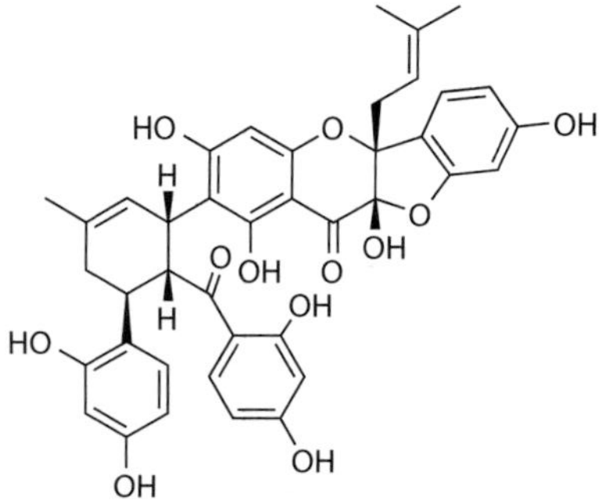

Sanggenon C (**87**)

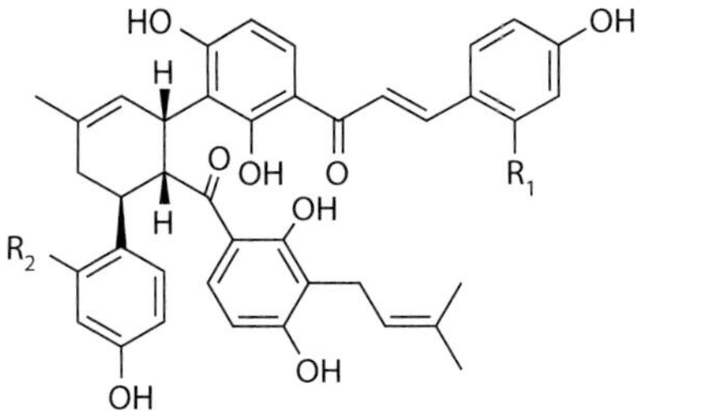

(88) Kuwanon J $R_1 = R_2 = OH$
(89) Kuwanon Q $R_1 + OH; R_2 = H$
(88) Kuwanon R $R_1 = H; R_2 = OH$
(88) Kuwanon V $R_1 = R_2 = H$

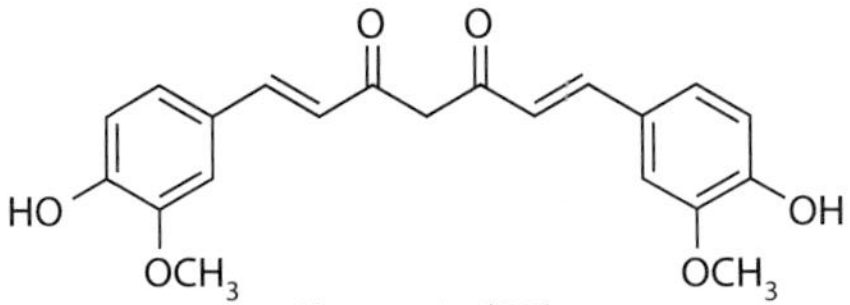

Curcumin (92)

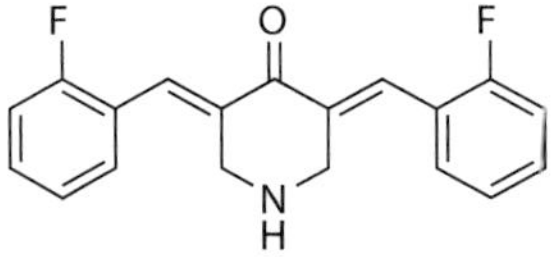

EF24 (93)

Curcuma longa, synthetic, respectively

Inhibits HIF-1A gene expression
Degrades ARNT, microtubule disruptor (EF 24)

Bae et al. (2006)
Choi et al. (2006)
Thomas et al. (2008)

(continued)

TABLE 7.3 (continued)
Natural Product–Based Inhibitors of HIF-1α Protein Accumulation

Compound	Source	Mechanistic Data	References
	Axinella sp.	Mechanism unknown	Dai et al. (2006)

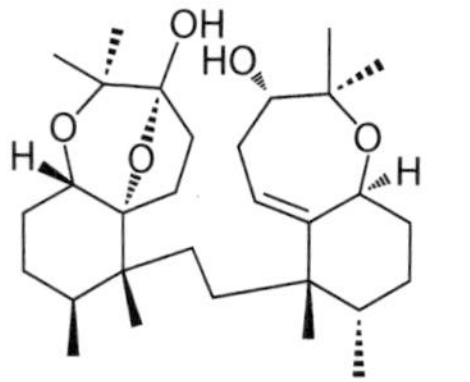

Sodwanone V
(**94**)

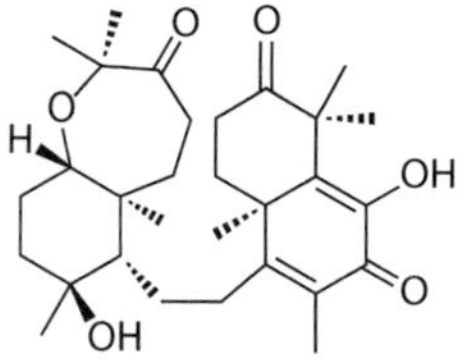

Sodwanone A
(**95**)

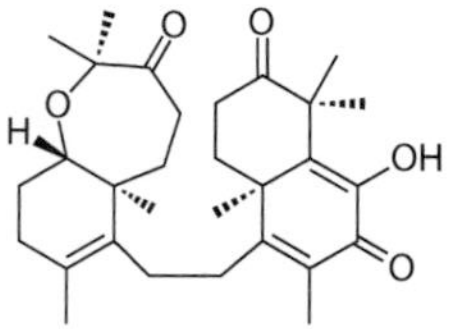

10,11-Dihydrosodwanone B
(**96**)

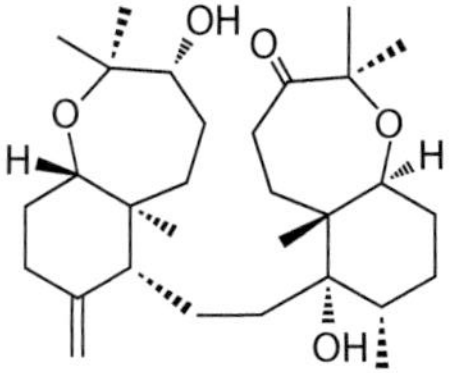

3-Epi-sodwanone K
(97)

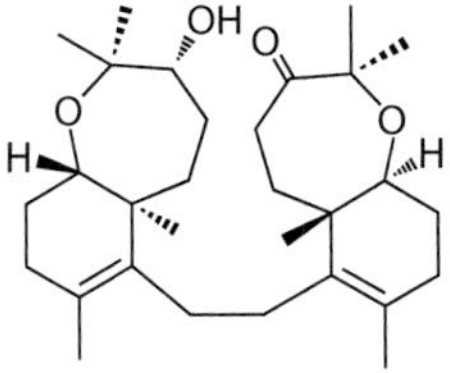

Sodwanone T
(98)

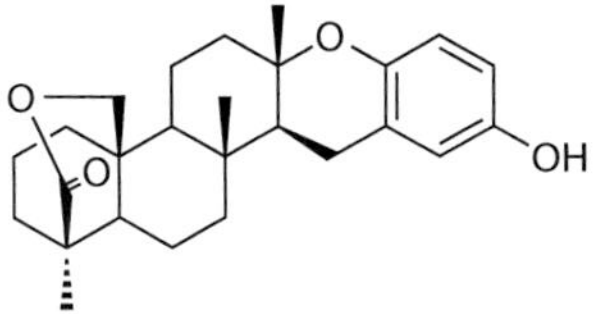

Strongylophorine 2
(99)

Petrosia
(Strongylophora)
strongylata

Mechanism unknown

Mohammed et al.
(2008)

(continued)

TABLE 7.3 (continued)
Natural Product–Based Inhibitors of HIF-1α Protein Accumulation

Compound	Source	Mechanistic Data	References
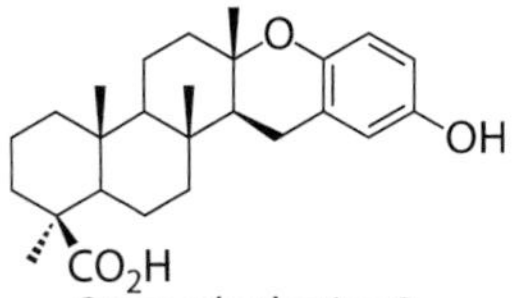 Strongylophorine 3 (**100**)			
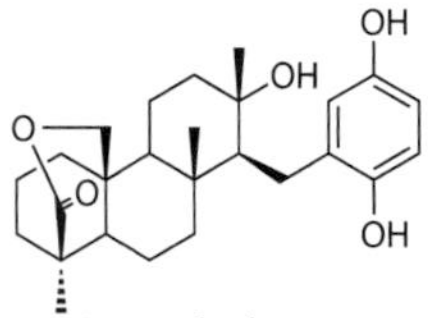 Strongylophorine 8 (**101**)			
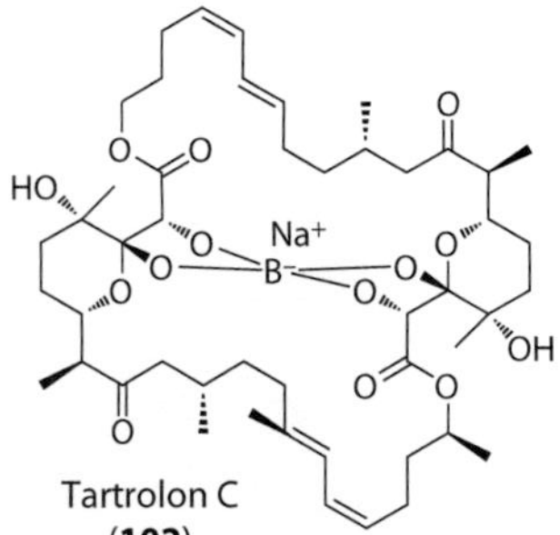 Tartrolon C (**102**)	*Streptomyces* sp.	Inhibits HIF-1 reporter activity and HIF-1α protein levels	Yamazaki et al. (2006a)

toward the total synthesis of neolamellarins may facilitate further investigation of these unusual HIF-1 inhibitors (Arafeh and Ullah, 2009).

As part of an HIF-1 inhibitor discovery effort by Lee and coworkers at the Korean Institute of Biosciences and Biotechnology, the lipid extract of Chinese (or red) sage *Salvia miltiorrhiza* (Lamiaceae) was found to inhibit hypoxia-induced HIF-1 activation in a human gastric adenocarcinoma AGS cell-based dual luciferase reporter assay system [pGL3-HRE(VEGF)/pRL-CMV] (Dat et al., 2007). Twelve abietane-type diterpenes were isolated from the active *S. miltiorrhiza* extract. Among the diterpenes isolated, the sibiriquinone A (**68**), sibiriquinone B (**69**), cryptotanshinone (**70**), and 3,4-dihydrotanshinone I (**71**) were found to be the most potent inhibitors of hypoxia-induced HIF-1 activation in the AGS cell-based reporter system (IC$_{50}$ values 0.34, 3.36, 1.58, 2.05 µM, respectively) (Table 7.3). Five other related diterpenes [methyl tanshinonate (**72**), tanshinone I (**73**), tanshinone IIA (**74**), 12-deoxytan-shinquinone B (**75**), and neocryptotanshinone (**76**)] were also found to significantly inhibit HIF-1 activation at somewhat higher concentrations. The most active compounds (**68–71**) were also found to inhibit hypoxia-induced HIF-1 activation in Hep3B human hepatocarcinoma cells at similar concentrations (IC$_{50}$ values 0.28, 3.18, 1.36, 2.29 µM, respectively). The most potent compound sibiriquinone A inhibited hypoxia-induced vascular endothelial growth factor (VEGF) mRNA expression (Hep3B cells). Both sibiriquinone A (**68**) and dihydrotanshinone I (**71**) were further shown to suppress the HIF-1α accumulation in Hep3B cells. These abietane diterpenes inhibited hypoxia-induced HIF-1 activation at concentration below their cytotoxic concentrations. However, a short-duration (24 h) cell viability assay was used that may have significantly underestimated the overall cytotoxicity of these *S. miltiorrhiza* metabolites.

Further studies by Lee and coworkers found that lipid extracts of herbal mulberry root bark preparations (Mori Cortex Radicis) and the bark of *Morus bombycis* (Moraceae) inhibited hypoxia-induced HIF-1 activation in a human hepatocarcinoma Hep3B cell-based reporter assay. Bioassay-guided isolation yielded a series of nine benzofurans and six chalcone-derived Diels-Alder adducts (Dat et al., 2009). The benzofurans moracin O (**77**, IC$_{50}$ 0.14 nM) and moracin P (**78**, IC$_{50}$ 0.65 nM) were found to be among the most potent inhibitors of HIF-1 activation discovered to date (Table 7.3). The remaining benzofurans [moracin Q (**79**), moracin M (**80**), mulberrofuran H (**81**), mulberrofuran G (**82**), albafuran A (**83**), mulberrofuran D (**84**), and mulberrofuran W (**85**)] and chalcone-derived Diels-Alder adducts [sanggenon O (**86**), sanggenon C (**87**), kuwanon J (**88**), kuwanon Q (**89**), kuwanon R (**90**), and kuwanon V (**91**)] were moderately potent HIF-1 inhibitors (IC$_{50}$ values 0.14–8.32 µM). Confirmatory studies found that these *Morus* metabolites inhibited hypoxia-induced VEGF secretion in Hep3B cells. The benzofurans (**77–79, 81, 83,** and **84**) and chalcone-derived Diels-Alder adducts (**86** and **88**) inhibited hypoxia-induced HIF-1α accumulation in Hep3B cells. However, the extremely potent activity of the two most active compounds moracin O (**77**) and moracin P (**78**) did not translate into a similarly potent inhibition of either HIF-1α accumulation or hypoxia-induced VEGF secretion in Hep3B. In order to assess whether this extremely potent activity may have been an experimental artifact, moracin O was examined for its ability to suppress luciferase expression from a control construct (pGL3-control) and on the luciferase reaction. Moracin O (**77**) inhibited the control reporter with similar potency to that observed with the pHRE reporter used to assess HIF-1 activity. This indicates that while *Morus* benzofurans and chalcone-derived Diels-Alder adducts may be moderately potent inhibitors of HIF-1 activation, caution must be used since nonspecific effects of certain moracin benzofurans may make these compounds appear as extremely potent false-positive "hits" in luciferase-based reporter assays. As in the case of the *Salvia* metabolites isolated by members of this research team, the use of a short (24 h) cell viability assay may have underestimated the potential cytotoxicity of the *Morus* metabolites.

Curcumin (**92**) has been considered the main active ingredient in the common spice, turmeric (prepared from the rhizomes of *Curcuma longa*). In human hepatocarcinoma HepG2 cells, curcumin (50 µM) blocked hypoxic induction of HIF-1α protein and this inhibitory effect was overcome by the proteasome inhibitor *N*-acetyl-Leu-Leu-norleucinal (LLnL) (Table 7.2) (Bae et al., 2006). Curcumin also suppressed HIF-1-mediated processes, such as the hypoxic induction of HIF-1 target gene VEGF and angiogenesis. It is possible that one of the mechanisms involved in the suppression of HIF-1 is that curcumin may enhance the degradation of HIF-1α protein.

In a separate study that used human hepatocarcinoma Hep3B cells, curcumin (e.g., 10 µM) suppressed hypoxia-induced HIF-1 activation and the induction of HIF-1 target genes (Choi et al., 2006). Further examination revealed that curcumin decreased the levels of HIF-1β/ARNT protein in a concentration-dependent manner under both normoxic and hypoxic conditions. Under experimental conditions (1% O_2, 16 h), curcumin did not affect the accumulation of HIF-1α protein. Similar HIF-1β/ARNT-selective inhibitory effects exerted by curcumin were observed in a panel of cell lines. Stability studies indicated that curcumin promoted proteasome-mediated degradation of ARNT in a manner that involves both oxidation and ubiquitination processes. Since curcumin is known to destabilize a number of other proteins (i.e., p53, c-Jun, CEBP, etc.), it is unlikely that the ability of curcumin to inhibit ARNT is due to an effect that is specific for the pathways involved in HIF-1 signaling. When curcumin was evaluated as a potential chemopreventive agent in clinical trials, the highest plasma concentration detected was 1.8 µM, even when it was administered at the extremely high dose of 8 g/day (~115 mg/kg/day). Giannakakou and colleagues developed the curcumin analogue EF24 (**93**) that suppressed the accumulation of HIF-1α and HIF-1β/ARNT proteins with improved potency, relative to that observed with curcumin (Thomas et al., 2008). Based on studies conducted in various cell lines where EF24 (**93**) and curcumin (**92**) were examined at concentrations that differed from one study to another, the authors concluded that both EF24 and curcumin suppressed the formation of HIF-1α protein via a proteasome-independent process that requires a functional von Hippel–Lindau (VHL) protein and that curcumin inhibited the transcription of HIF-1A. At the concentrations that inhibited the hypoxic induction of HIF-1α protein in human prostate cancer PC-3 cells (i.e., 1 µM), EF24 also induced mitotic arrest and microtubule stabilization. Given that EF24 did not stabilize polymerized tubulin in vitro, it is possible that EF24 affected an upstream pathway. Since previous studies examined the effects of curcumin on HIF-1 signaling only at concentrations that appear to be significantly above those that may be achieved in patients, the physiological relevance of experimentally high micromolar concentrations of curcumin is uncertain.

The lipid extract of a South African *Axinella* sp. sponge (NCI Open Repository) inhibited hypoxia-induced HIF-1 activation in T47D cells (Dai et al., 2006). Seven new sodwanone triterpenoids, two previously reported sodwanones, and a related yardenone-type triterpene were isolated. The most potent compound sod-wanone V (**94**) inhibited hypoxia- and iron-chelator (1,10-phenanthroline)-induced HIF-1 activation (IC_{50} 15 μM). Sodwanone V inhibited HIF-1 activation in prostate tumor PC-3 cells (IC_{50} 15 μM). Sodwanone A (**95**), 10,11-dihydrosodwanone B (**96**), 3-epi-sodwanone K (**97**), and sodwanone T (**98**) weakly inhibited hypoxia-induced HIF-1 activation (IC_{50} values 20–25 μM).

Using genetically engineered U251 human glioma cells that stably express a lucif-erase reporter gene under the control of hypoxia responsive element (Rapisarda et al., 2002), Ireland and coworkers at the University of Utah recently evaluated a library of 2376 chromatographically purified fractions derived from marine invertebrate extracts (Mohammed et al., 2008). Purified fractions from a Papua New Guinea *Petrosia* (*Strongylophora*) *strongylata* sponge extract inhibited HIF-1 activation at 1 μg/mL. Further purification resulted in the isolation of three known meroditer-penoids strongylophorines 2 (**99**), 3 (**100**), and 8 (**101**) (Salva and Faulkner, 1990) that inhibited hypoxia-induced HIF-1 activation in the U251 cell-based system (EC_{50} values 8, 13, and 6 μM, respectively) (Table 7.3). All three strongylophorines (30 μM) inhibited hypoxia-induced HIF-1α protein accumulation and suppressed VEGF expression in human glioma U251-HRE cells.

Researchers at the Numazu Biomedical Research Institute in Japan used a Chinese Hamster Ovary (CHO) cell-based high-throughput assay system (stable transformed with an HIF-1-dependent luciferase reporter (5×HRE/pGL3/VEGF/E1b) (Mabjeesh et al., 2002) to screen 5000 cultured broths of microbes for HIF-1 inhibitory activity (Yamazaki et al., 2006a). The macrodiolide tartrolone C (**102**) (Lewer et al., 2003) was isolated from the broth extract of *Streptomyces* strain 1759-27 and was found to inhibit hypoxia-induced HIF-1 activation (IC_{50} 0.17 μg/mL). Yamazaki and cowork-ers reported that tartrolone C did not display cytotoxicity at the active concentra-tions. However, no cell viability assay conditions or assay methods were reported. At relatively high concentrations ($IC_{50} > 1.0$ μg/mL), tartrolone C suppressed HIF-1α protein levels in MCF-7 breast tumor cells.

7.8 MULTI-TARGETED HIF-1 INHIBITORS

Natural products that appear to inhibit HIF-1 activation by two or more distinctly different mechanisms are listed in Table 7.4 (structures **103–112**).

7.8.1 DUAL-TARGETED AGENTS THAT INHIBIT MITOCHONDRIAL HIF-1 SIGNALING AND PROTEIN TRANSLATION

Unlike other agents that simply target the mitochondrial ETC to disrupt ROS-based hypoxic signaling and HIF-1 activation, bioassay-guided isolation of an active lipid extract of the tropical legumaceous plant *Lonchocarpus glabrescens* yielded two

TABLE 7.4

Natural Product–Based HIF-1 Inhibitors That Act through Multiple Targets

Compound	Source	Mechanistic Data	References
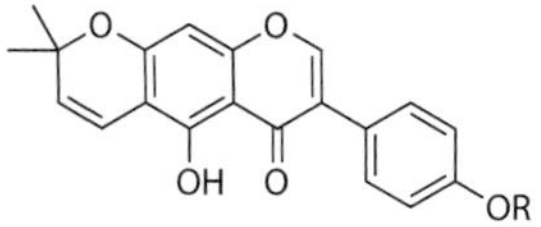 (103) Alpinum isoflavone R=H (104) 4′-O-methylalpinumisoflavone R=CH₃	Lonchocarpus glabrescens	Inhibitors of protein translation and mitochondrial complex I	Liu, Y. et al. (2009a)
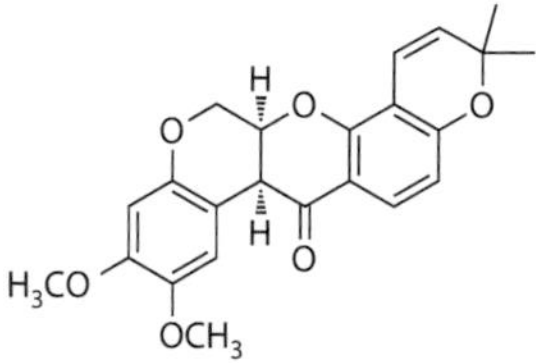 Deguelin (105)	Lorchocarpus spp.	Inhibitors of protein translation, Hsp90, and mitochondrial complex I	Oh et al. (2007) Oh et al. (2008) Fang and Casida (1999)
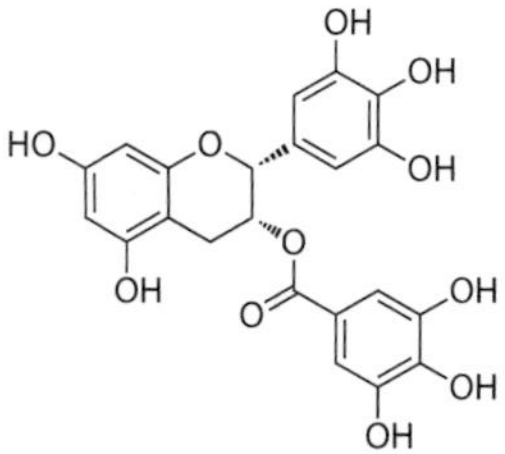 (−)-Epigallocatechin-3-gallate (EGCG) (106)	Camellia sinensis	Enhances HIF-1α degradation and inhibits protein translation	Thomas and Kim (2005) Zhang et al. (2006)

Genistein (**107**)

Isorhamnetin (**108**)

Luteolin (**109**)

Isoquercetin (quercetin 3-O-β-D-glucoside) (**110**)

Compound	Distribution	Activity	References
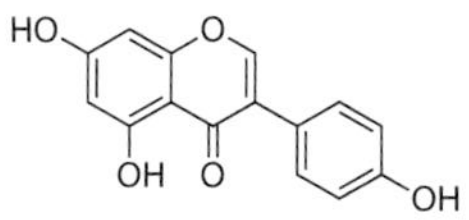 Genistein (**107**)	Broadly distributed in plants	Tyrosine kinase inhibitor	Wang et al. (1995) Hur et al. (2001)
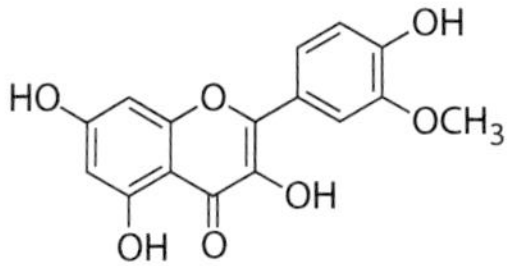 Isorhamnetin (**108**)	Broadly distributed in plants?	Tyrosine kinase and PKC inhibitor	Hasebe et al. (2003)
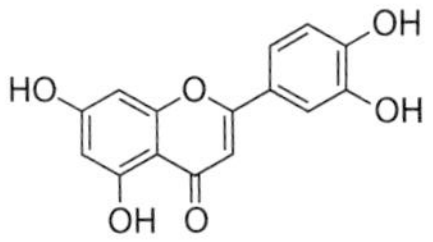 Luteolin (**109**)	Broadly distributed in plants?	Tyrosine kinase and PKC inhibitor, inhibits MAPK phosphorylation of HIF-1α	Hasebe et al. (2003), Triantafyllou et al. (2008)
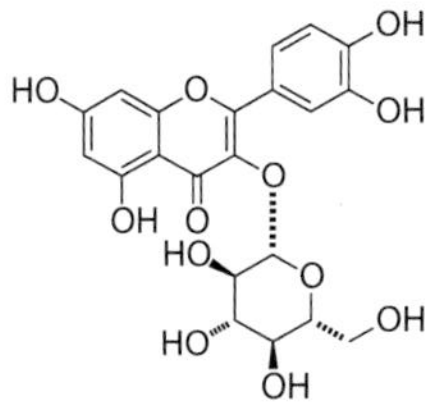 Isoquercetin (**110**)	Broadly distributed in plants?	Tyrosine and PI3 kinase and PKC inhibitor	Hasebe et al. (2003)

(continued)

TABLE 7.4 (continued)
Natural Product–Based HIF-1 Inhibitors That Act through Multiple Targets

Compound	Source	Mechanistic Data	References
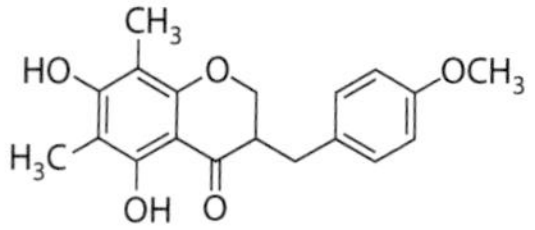 Methylophiopogonanone B (MOB) (**111**)	Broadly distributed in plants?	Tyrosine kinase inhibitor?	Hasebe et al. (2003)
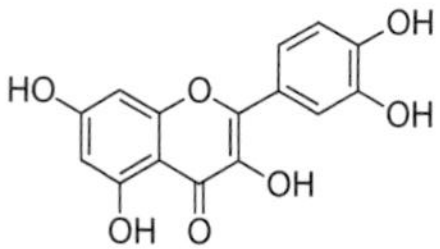 Quercetin (**112**)	Broadly distributed in plants	Tyrosine and PI3 kinase and PKC inhibitor; inhibits MAPK phosphorylation of HIF-1α	Hasebe et al. (2003), Triantafyllou et al. (2007, 2008) Jeon et al. (2007) Park et al. (2008) Lee and Lee (2008) Bach et al. (2009) Du et al. (2009) Radreau et al. (2009)

new HIF-1 inhibitors that appear to function by two distinct mechanisms (Table 7.4) (Liu et al., 2009a). Alpinumisoflavone (**103**) and 4'-*O*-methylalpinumisoflavone (**104**) inhibited hypoxia-induced HIF-1 activation (IC$_{50}$ values 5 and 0.6 μM, respectively) and were shown to inhibit the activation of HIF-1 by blocking the induction of nuclear HIF-1α protein. The compound 4'-*O*-methylalpinumisoflavone inhibited hypoxic induction of HIF-1 target genes (*CDKN1A*, *GLUT-1*, and *VEGF*), and blocked tumor angiogenesis in vitro (HUVEC Cell Tube Formation Assay). Further, 4'-*O*-methylalpinumisoflavone suppressed MDA-MB-231 breast tumor cell migration and chemotaxis. Mechanistic studies indicate that, unlike other mitochondrial inhibitors, 4'-*O*-methylalpinumisoflavone represents the first small molecule that inhibits HIF-1 activation by simultaneously blocking the mitochondrial ETC (at complex I) and disrupting protein translation in vitro.

The plant derived mitochondrial ETC complex I inhibitor rotenone (**9**) is known to suppress hypoxia-induced HIF-1 activation (Ebert et al., 1995; Agani et al., 2002). The structurally related *Lonchocarpus* rotenoid deguelin (**105**) has been shown to exert proapoptotic and antitumor properties that are mediated via potent inhibition of ETC complex I (Hail and Lotan, 2004). In collaboration with scientists at Seoul Nation University, Lee and coworkers at the University of Texas' M.D. Anderson Cancer Center found that deguelin inhibits hypoxia-induced HIF-1 activation, decreases HIF-1α expression, and suppresses VEGF levels in tumor cell lines (Oh et al., 2008). Deguelin both suppressed *de novo* HIF-1α synthesis and destabilized HIF-1α protein. The identity of deguelin as a rotenone-like ETC complex I inhibitor is not discussed, nor is the possibility that the observed destabilization of HIF-1α protein may result from well-characterized mitochondrial ROS-mediated HIF-1 signaling pathways. Lee and coworkers only describe this decrease in the HIF-1α half-life to be "…through….unknown mechanisms…" While the authors demonstrate that deguelin displays antitumor activity in vivo, a point that is not addressed is that its therapeutic potential may be limited by rotenone-like neurotoxicity. Structural homology between *Lonchocarpus* metabolites such as alpinumisoflavone (**103**) and deguelin (**105**) would tend to suggest that both sets of rotenoids might function though an identical combination of molecular mechanisms to inhibit HIF-1 activation and suppress HIF-mediated signaling in tumors. It is probable that other members of the relatively large class of plant metabolites known as rotenoids are likely to exhibit the same form of multi-targeted effect on HIF-1 signaling and may be prone to a similar toxicity profile.

7.8.2 DUAL-TARGETED AGENTS THAT ENHANCE HIF-1α DEGRADATION AND INHIBIT PROTEIN TRANSLATION

The catechins found in green tea [*Camellia sinensis* L. Ktze. (Theaceae)] have never been shown to suppress HIF-1 signaling at the submicromolar concentrations achievable though normal dietary consumption. However, significantly higher serum concentrations of these polyphenolic compounds may be achievable through the consumption of green tea extract (GTE) nutritional supplements or as part of mega-dosing cancer chemoprevention regimens. At micromolar concentrations, GTEs and green tea catechins such as (−)-epigallocatechin-3-gallate (EGCG, **106**) inhibited hypoxia-induced HIF-1 activation and the expression of HIF-1α and VEGF proteins in human cervical carcinoma (HeLa) and hepatoma (Hep2G) cells (Table 7.4) (Zhang et al., 2006). Lee and coworkers at UCLA found that relatively high concentrations of GTE and epigallocatechin-3-gallate (EGCG) (e.g., 20–100 μM) suppressed hypoxia- and serum-induced HIF-1α protein accumulation but did not affect HIF-1α mRNA levels. These studies suggested that high concentrations of GTE and EGCG may both interfere with phosphoinositide 3-kinase (PI3K)/Akt/ mTOR-mediated protein translation of HIF-1α and enhance the proteasomal degradation of HIF-1α protein. In spite of these results, these findings may be somewhat cell line–dependent and might not be universally translatable to other tumor cell models. Several other studies have observed quite different results with micromolar concentrations of green tea catechins in other cell lines. Zhou and coworkers found that the green tea catechin (–)-epicatechin-3-gallate (ECG) activated HIF-1 and enhanced the expression of HIF-1α and secreted VEGF proteins in human T47D breast tumor cells (Zhou et al., 2004). Similarly, researchers at the University of North Texas Health Science Center found that EGCG can increase HIF-1α transcription and stabilize HIF-1α in human prostate tumor PC-3 and PC-3ML cells, even under normoxic conditions (Thomas and Kim, 2005). Thomas and Kim further demonstrated that EGCG may stabilize HIF-1α protein by chelating the Fe^{++} that is required for the catalytic activity of HIF-1 prolyl hydroxylase (HPH) enzymes that hydroxylate proline residues on HIF-1α prior to its degradation by the 26S proteasome. Such findings tend to indicate that the opposing results observed by various groups may be due to a combination of the genetic diversity among tumor cell lines and specific differences in physiological (or experimental) conditions, such as the concentration of ferrous iron in the cells or media.

7.8.3 OTHER AGENTS THAT FUNCTION THROUGH MULTIPLE MECHANISMS TO DECREASE HIF-1α EXPRESSION

To investigate the structure–activity relationships of flavonoids on the induction of HIF-1α protein, the effects of representative flavonols [e.g., quercetin, (**112**)], flavones [e.g., luteolin (**109**)], flavanones, and a flavanol glycoside on the accumulation of HIF-1α protein were examined in HeLa cells (Triantafyllou et al., 2008). At extremely high concentrations (i.e., 100 μM), flavonoids induced HIF-1α protein in a concentration- and time-dependent manner. The induction of HIF-1α protein by active compounds correlated with the depletion of the intracellular labile iron pool. The observation that excess iron reversed the HIF-1α inducing effects of active flavonoids suggested that like green tea catechins, flavonoids function as iron chelators to inhibit the hydroxylases that tag HIF-1α protein for degradation. However, high concentrations (100 μM) of the same flavonoids decreased HIF-1 activation by the iron chelator desferroxamine (DFO). Further examination revealed that flavonoids inhibited HIF-1 activation by inhibiting MAPK-mediated phosphorylation of HIF-1α protein, thereby decreasing the level of nuclear HIF-1α protein by enhancing its export out of the nucleus. This was not an entirely novel observation, since some of the same flavonoids [i.e., genestine (**107**), luteolin (**109**), quercetin (**112**)] are known to function as broad-spectrum kinase inhibitors and have been previously reported to inhibit HIF-1 in a Chinese Hamster Ovary (CHO) (A4-4 clone) cell-based reporter assay (Hasebe et al., 2003). Flavonoids appear to regulate HIF-1 activation at various steps in the pathway via dissimilar mechanisms and the final outcome hinges on the balance of the activation of other associated pathways. However, the relatively weak activity of flavonoids begs the question - are these activities of any physiological relevance?

7.9 INHIBITORS OF HIF WITH UNIQUE MECHANISMS

Natural products and natural product–derived semisynthetic compounds that inhibit HIF-1 activation by unusual or unique mechanisms are listed in Table 7.5 (structures **113–124**).

7.9.1 INHIBITORS OF HIF-1/DNA-BINDING

Efforts at the NCI to discover small-molecule inhibitors of HIF-1 DNA-binding activity resulted in the identification of the polypeptide quinoxaline antibiotic echinomycin (**114**, NSC-13502, quinomycin A) (Table 7.5) (Kong et al., 2005). Echinomycin suppressed hypoxia-induced HIF-1 activation in U251 glioma cells at low nM concentrations. Echinomycin, originally isolated from *Streptomyces echinatus*, simultaneously intercalates DNA at two separate sequence-specific sites and blocks RNA synthesis and DNA replication (van Dyke and Dervan, 1984). More recent studies indicate that echinomycin lacks specificity for HIF-1 DNA-binding and similarly interferes with other transcription factors (Vlaminck et al., 2007). Moreover, low concentrations of echinomycin increase HIF-1A gene expression and increase HIF-1α protein levels under normoxic conditions. This causes echinomycin to display a dual effect, activating HIF-1 in normoxic cells and inhibiting HIF-1 activation under hypoxic conditions. The lack of specificity and potential dual activity of echinomycin is believed to render this antibiotic of no practical value for cancer therapy as an "HIF inhibitor."

In a Hep3B-c1 cell-based dual reporter assay, hypoxic exposure (1% O_2) increased the ratio of firefly luciferase to *Renilla* luciferase activities by > fivefold, indicative of the activation of HIF-1 by hypoxia. Examination of the Johns Hopkins Drug Library for HIF-1 inhibitors identified a group of anthracyclines as potent HIF inhibitors (Lee et al., 2009). Daunorubicin (**115**), epirubucin (**116**), idarubucin (**117**), and doxorubicin (**118**) all suppressed hypoxic activation of HIF-1 at submicromolar concentrations. Further mechanistic investigation revealed that these anthracyclines inhibited HIF-1 activation by disrupting the binding between HIF-1 and the HRE located on the target gene promoters. Although both daunorubicin and doxorubicin suppressed tumor progression, angiogenesis, and the mobilization of circulating angiogenic cells (CACs) in tumor-bearing mice, the HIF-1 pathway may represent one of the pathways regulated by these agents. Anthracyclines have been shown to regulate a number of other transcription factor-mediated pathways (i.e., AP1, GATA4, and sp1).

Using a Chinese Hamster Ovary (CHO) cell-based reporter assay for HIF-1 activity, extracts of cultured broths of microorganisms were examined for HIF inhibitory activities (Yamazaki et al., 2006b). The anthracyclines cinerubin (**119**) and

TABLE 7.5
Natural Product–Based Agents That Inhibit HIF-1 by Unique Mechanisms

Compound	Source	Notes	References
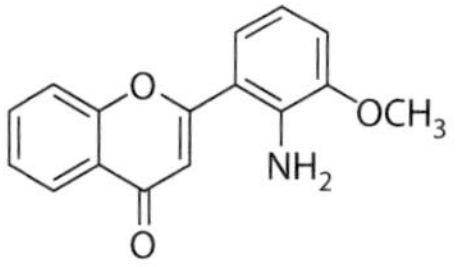 PD98059 (113)	Semisynthetic flavone	MEK1/p42/p44 selective MAP kinase (MAPK) inhibitor; inhibits HIF-1α/p300 interaction	Richard et al. (1999) Hur et al. (2001) Lee et al. (2002) Sang et al. (2003)
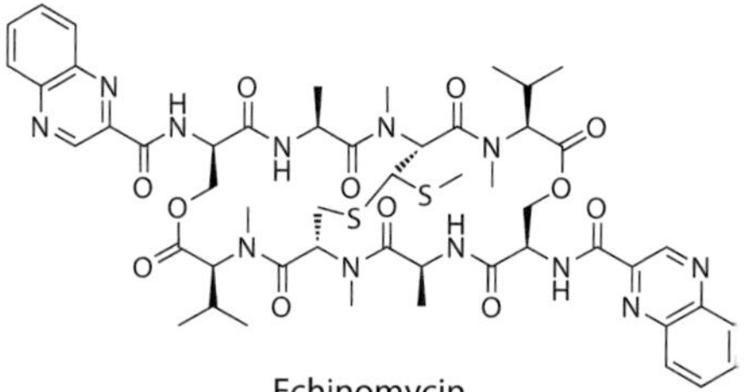 Echinomycin (114)	*Streptomyces echinatus*	Binds DNA and nonselectively interferes with binding of HIF-1 (and other transcription factors) to DNA	Kong et al. (2005) Vlaminck et al. (2007)
	Streptomyces peucetius, semisynthetic, semisynthetic, S. coeruleorubicus, respectively	Interferes with HIF-1 binding to HRE in hypoxia-responsive genes	Lee et al. (2009)

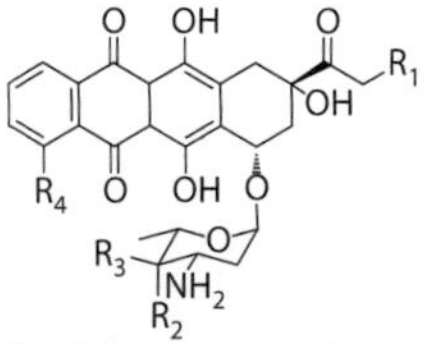

PD98059 (113)

Echinomycin (114)

(115) Daunorubicin $R_1 = H$; $R_2 = OH$; $R_3 = H$; $R_4 = OCH_3$
(116) Epirubicin $R_1 = OH$; $R_2 = H$; $R_3 = OH$; $R_4 = OCH_3$
(117) Idarubicin $R_1 = OH$; $R_2 = OH$; $R_3 = H$; $R_4 = H$
(118) Doxorubicin $R_1 = OH$; $R_2 = OH$; $R_3 = H$; $R_4 = OCH_3$

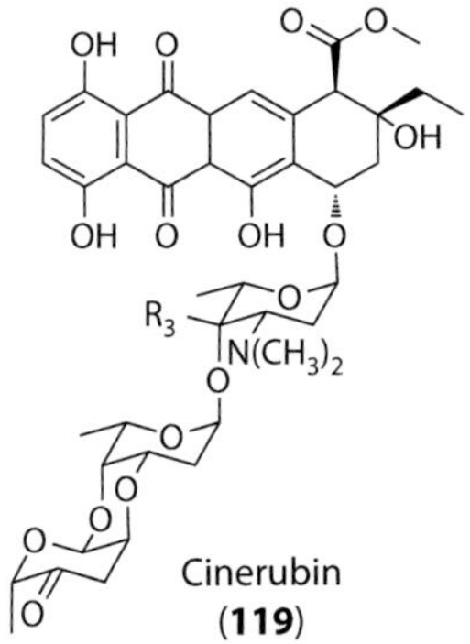

Yamazaki et al. (2006)

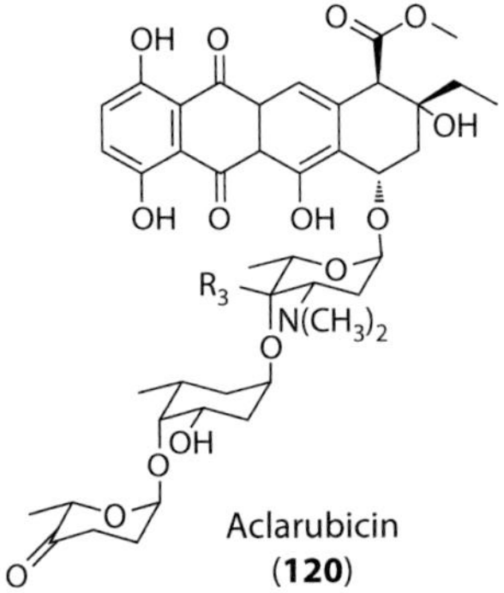

May interfere with HIF-1
binding to HRE in
hypoxia-responsive genes

Streptomyces spp.,
S. galilaeus

Cinerubin
(**119**)

Aclarubicin
(**120**)

(continued)

TABLE 7.5 (continued)
Natural Product–Based Agents That Inhibit HIF-1 by Unique Mechanisms

Compound	Source	Notes	References
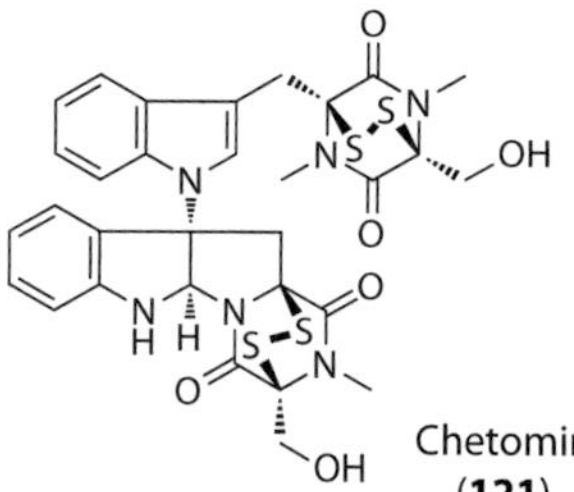 Chetomin (121)	*Chaetomium cochliodes*	Inhibits HIF-1α/p300 interaction	Kung et al. (2004) Cook et al. (2009)
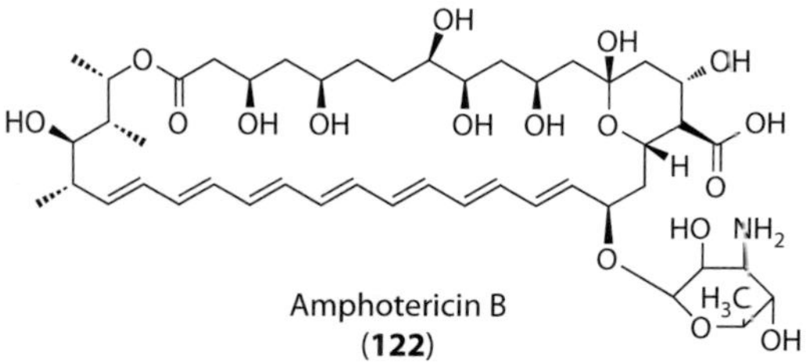 Amphotericin B (122)	*Streptomyces nodosus*	Prevents HIF-1α/p300 interaction by enhancing FIH/HIF-1α binding	Yeo et al. (2006)

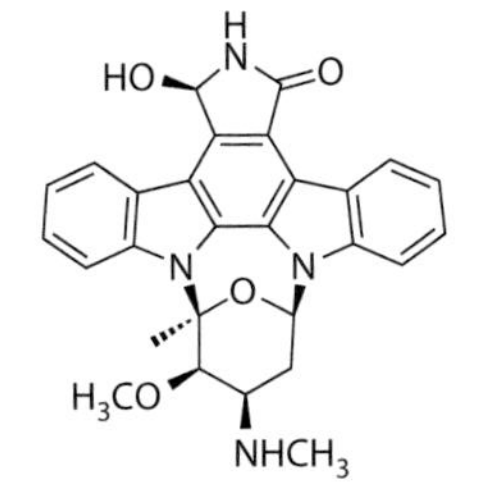

Kruger et al. (1998–1999)

Protein kinase C inhibitor

Streptomyces sp.

El Sayed et al. (2008)

Actin polymerization
inhibitor

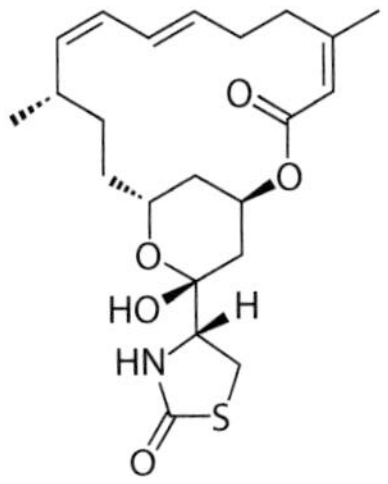

Negombata magnifica

7(*R*)-hydroxystaurosporine
(**123**)

Latrunculin A
(**124**)

aclarubicin (**120**) inhibited hypoxic activation of HIF-1 with IC_{50} values of 0.016 and 0.021 µg/mL, respectively, in the CHO cell-based reporter assay. Doxorubicin (**118**) inhibited at higher concentrations (e.g., 0.78 µg/mL) and daunorubicin had no activity at concentrations up to 3.12 µg/mL. In contrast to the results reported for other anthracyclines that appeared to indicate that they disrupt HIF-1 binding to the HRE in target genes (Lee et al., 2009), further investigation suggested that cinerubin inhibited HIF-1 by blocking the accumulation of HIF-1α protein in HepG2 cells.

7.9.2 Agents That Interfere with HIF-1/p300 Interactions

Kung and coworkers used a time-resolved fluorescence primary assay that monitored the interaction between the human HIF-1α CTAD domain (786–826) and the p300 CH1 domain (302–423) to identify small molecule disruptors of HIF-1α/p300-binding interactions required for HIF-1 transactivation (Kung et al., 2004). The fungal epidithiodiketopiperazine natural product chetomin (**121**) (isolated from *Chetomium* spp.) was identified from a library of more than 600,000 natural and synthetic compounds (Table 7.5). The epidithiodiketopiperazine (ETP) alkaloid chetomin is an antibiotic that was originally isolated by Waksman and coworkers from *C. cochliodes* in 1944 (Waksman and Bugie, 1944; Geiger et al., 1944). Chetomin was found to disrupt the tertiary structure of p300 by binding to its CH1 domain (Kung et al., 2004). Chetomin inhibited hypoxia-induced HIF-1 activation and its downstream targets in both in vitro and in vivo models. However, various factors limited its clinical potential as an antitumor agent. An examination of synthetic and natural chetomin analogs has recently resolved that only the ETP core structure is required for the ability of these alkaloids to interfere with the binding between HIF-1α and the transcriptional coactivator p300 (Cook et al., 2009). Further mechanistic studies indicate that the interaction of the ETP unit with p300 causes zinc ion ejection from the p300 complex. Support for the involvement of a zinc ejection step in the mechanism of action of these alkaloids was provided by the observation that zinc supplementation can reverse the effects of ETPs on the angiogenic factor expression and tumor cell proliferation. Dimeric ETPs have also been synthesized by Olenyuk and coworkers that exhibited significant activity in vitro (Block et al., 2009).

The antifungal *Streptomyces nodosus* metabolite amphotericin B (**122**, Fungizone®) inhibits the transcriptional activity of HIF-1 (2.5 µg/mL) in human hepatocarcinoma Hep3B and embryonic kidney HEK293 cells (Yeo et al., 2006). However, amphotericin B did not affect HIF-1α expression. The asparaginyl hydroxylase known as factor-inhibiting HIF-1 (FIH) hydroxylates Asn803 in the C-terminal transactivation domain of HIF-1α to interfere with the recruitment of the transcriptional coactivator p300. In a similar manner, mechanistic studies revealed that amphotericin B can repress the interaction of HIF-1α with p300 by enhancing FIH/HIF-1α binding. Antifungal therapy with the commonly used antifungal drug amphotericin B is known to suppress plasma levels of erythropoietin (EPO) and cause anemia in certain patients. Since EPO is a major HIF-1 target gene, suppression of HIF-1 transactivation may contribute to the anemia observed among certain patients that receive amphotericin B therapy. These results may potentially have a widespread impact on HIF-1 research, since amphotericin B is commonly used in mammalian cell culture at concentrations that may interfere with the transcriptional activation of HIF-1. It is even possible that the presence of Fungizone in cell culture systems may explain why certain groups have failed to observe inhibition of HIF-1 activation with compounds that appear to alter the expression of HIF-1α protein.

7.9.3   ACTIN INHIBITORS

Latrunculin A (**124**) is a marine macrolide that was first isolated from the Red Sea sponge *Negombata magnifica* (Podospongiidae) (Kashman et al., 1980). Latrunculin A (IC$_{50}$ 6.7 μM) and a 17-*O*-[*N*-(benzyl)-carbamate] latrunculin A derivative inhibited hypoxia-induced HIF-1 activation in T47D cells (Table 7.5) (El Sayed et al., 2008). Latrunculin A reversibly binds to actin monomers and disrupts actin polymerization and microfilament formation (Spector et al., 1989). In general, actin microfilaments are required for cell shape and organization, cellular adhesion, and cell migration (Nakaseko and Yanagida, 2001; Gachet et al., 2001). Disruption of actin polymerization by latrunculin A decreases tumor cell viability, inhibits angiogenesis, and blocks the ability of tumor cells to metastasize (El Sayed et al., 2006; Newman and Cragg, 2007). This study suggests that it is possible that the actin cytoskeleton may play a vital role in regulating HIF-1 activation.

7.10   AGENTS THAT LACK CONFIRMATORY AND MECHANISTIC STUDIES

Natural products that inhibit HIF-1 activation in initial assays but lack confirmatory studies to verify the results, and those cytotoxic natural products that are likely to inhibit HIF-1 by nonselective suppression of cellular processes are listed in Table 7.6 (structures **125–136**).

7.10.1   AGENTS THAT INHIBIT HIF-1 ACTIVATION IN REPORTER SYSTEMS BUT LACK EXPERIMENTS TO CONFIRM THE RESULTS

Cell-based reporter assays are commonly used to monitor HIF-1 activity. These reporter assays are a reliable means of rapidly evaluating the effects of potential inhibitors on the activation of a particular gene (i.e., HIF-1). However, they only model the conditions involved in actual cellular gene regulation and are susceptible to experimental interference that can result in a variety of possible false-positive and false-negative responses. For example, certain compounds that suppress cell viability may disrupt cellular signaling and appear active in a short-duration reporter assay, but do not exhibit significant cytotoxicity unless subjected to longer-term evaluation (e.g., 48 or 72 h). Standard (e.g., 48 h) cell viability assays should be performed with each potential inhibitor. Since the genetic background of any cell type determines its own unique sensitivity to cytotoxic agents, the cell viability assays must also be performed in the same cell line used for the reporter system. Compounds such as histone deacetylase (HDAC) inhibitors can differentially affect the transcription of reporter genes in cell-based assays. Fluorescent natural products and

TABLE 7.6
Natural Product–Based Agents HIF-1 Inhibitors That Lack Confirmatory, Mechanistic, or Other Studies to Distinguish HIF-1 Activity from Cytotoxicity

Compound	Source	Notes	References
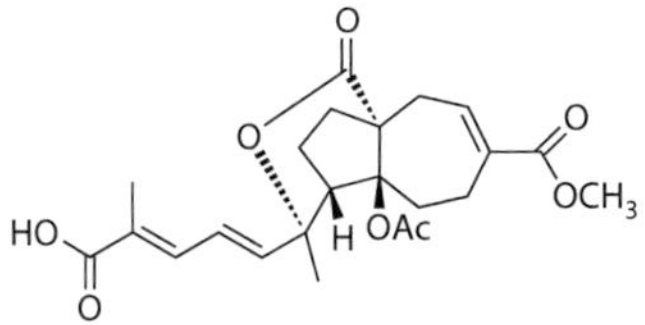 Pseudolaric acid B **(125)**	*Pseudolarix kaempferi*	Promotes HIF-1α degradation when tested above the cytotoxic concentration	Li et al. (2004)
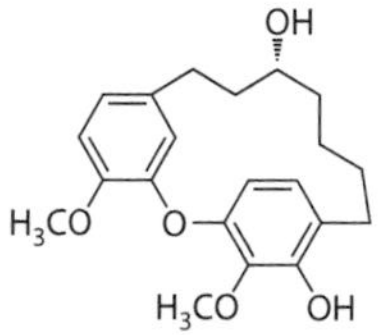 2-Oxatrycyclo[13.2.2.13,7]-eicosa-3,5,7(20),15,17,18-hexaen-10,16-diol **(126)**	*Alnus hirsuta*	Inhibits HIF-1 reporter activity No studies to confirm	Jin et al. (2007)
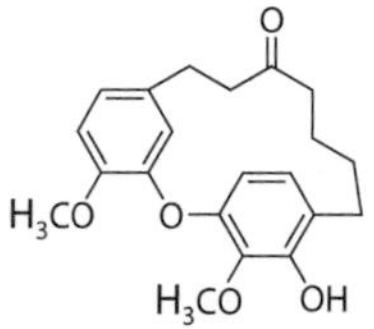 2-Oxatrycyclo[13.2.2.13,7]-eicosa-3,5,7-(20),15,17,18-hexaen-10-one **(127)**			

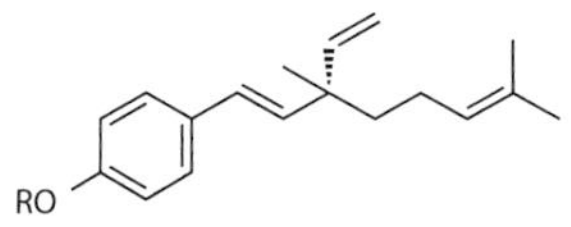

Psoralea corylifolia	Inhibits HIF-1 reporter activity No studies to confirm	Wu et al. (2008)
Crinum asiaticum var. *japonicum*	Inhibits HIF-1 reporter activity No studies to confirm	Kim et al. (2006)
Paris polyphylla	Inhibits HIF-1α mRNA levels Lack of studies to confirm	Ma et al. (2009)

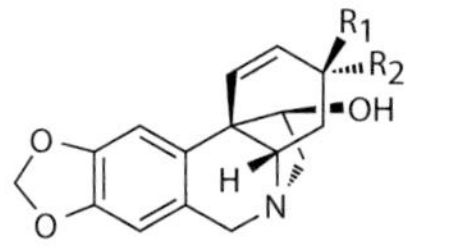

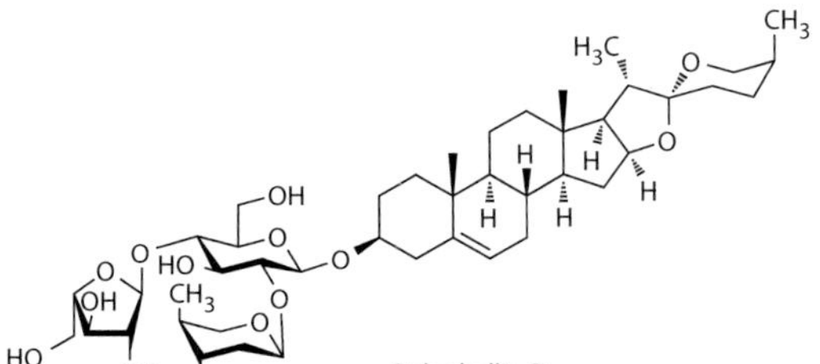

(*continued*)

(**128**) (*S*)-bakuchiol R = H
(**129**) Acetylbakuchiol R=COCH$_3$
(**130**) O-methylbakuchiol R = CH$_3$
(**131**) O-ethylbakuchiol R=CH$_2$CH$_3$

(**132**) Crinamine R$_1$ = H; R$_2$ = OCH$_3$
(**133**) 3-Epi-crinamine R$_1$ = OCH$_3$; R$_2$ = H

Polyphyllin D
(**134**)

TABLE 7.6 (continued)
Natural Product–Based Agents HIF-1 Inhibitors That Lack Confirmatory, Mechanistic, or Other Studies to Distinguish HIF-1 Activity from Cytotoxicity

Compound	Source	Notes	References
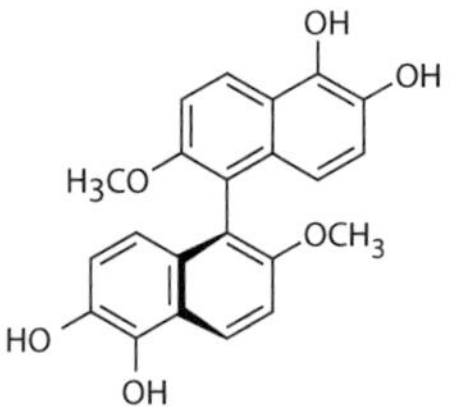 (S)-2,2'-Dimethoxy-1,1'-binaphthyl-5,5',6,6'-tetraol (135)	*Lendenfeldia* sp.	Inhibits HIF-1 reporter activity but activity may relate to cytotoxicity	Dai et al. (2007a)
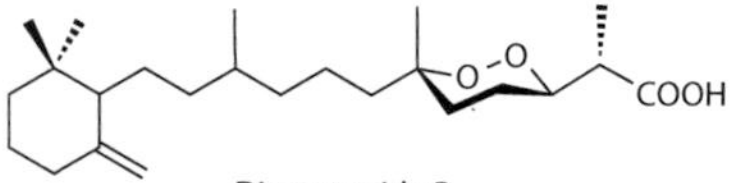 Diacarnoxide B (136)	*Diacarnus levii*	Inhibits HIF-1 reporter activity but activity may relate to hypoxia-selective cytotoxicity	Dai et al. (2007b)

those capable of quenching the fluorescence can interfere with the readout. Many substances directly block the luciferase reaction that is at the heart of the luciferase reporter assay and will display strong inhibitory activity in the assay, but will not show any effect on downstream targets. Typically, these natural products inhibit HIF-1 in reporter assays, but do not affect the expression of HIF-1 target genes or HIF-1 regulated pathways. Because of these limitations, reporter assay results must be confirmed by specific experiments that verify the initial observations on downstream HIF-1 targets.

Bae and coworkers at the Chungnam National University in Korea employed a dual luciferase reporter assay system [pGL3-HRE(VEGF)/pRL-CMV] to screen for natural products that inhibit hypoxia-induced HIF-1 activation in human gastric adenocarcinoma AGS cells (Jin et al., 2007). Two diarylheptanoids [2-oxatrycyclo[13.2.2.13,7]-eicosa-3,5,7(20),15,17,18-hexaen-10–16-diol (**126**) and 2-oxatrycyclo[13.2.2.13,7]-eicosa-3,5,7-(20),15,17,18-hexaen-10-one (**127**)] were identified from a methanol extract of the stem bark of the Manchurian Alder *Alnus hirsuta* (Betulaceae) as moderately potent inhibitors of HIF-1 activation (IC$_{50}$ values 11.2 and 12.3 µM, respectively) in the reporter assay (Table 7.6). These two triterpenes were also found to inhibit the HIF-1 reporter system due to their cytotoxicity to the AGS cells. While the diarylheptanoids displayed activity in the reporter assay, this activity was not confirmed with any additional experiments.

In collaboration with Chinese scientists, members of the Korean HIF-1 inhibitor discovery groups found that methanol extracts of *Psoralea corylifolia* (Fabaceae) seeds inhibited hypoxia-induced HIF-1 activation in AGS cells (Wu et al., 2007). Bioassay-guided isolation yielded the compound (*S*)-bakuchiol (**128**, IC$_{50}$ 6.1 µM) and two structurally novel inactive dimeric meroterpenoids (Table 7.6). A more detailed phytochemical investigation of the active extract resulted in the isolation of a series of related meroterpenes and merterpene dimers (Wu et al., 2008). Three semisynthetic derivatives of bakuchiol were prepared and tested for HIF-1 inhibitory activity along with the naturally occurring meroterpenoids. Only (*S*)-bakuchiol (**128**), acetylbakuchiol (**129**, IC$_{50}$ 5.7 µM), *O*-methylbakuchiol (**130**, IC$_{50}$ 8.7 µM), and *O*-ethylbakuchiol (**131**, IC$_{50}$ 26.3 µM) were found to significantly inhibit HIF-1 activation in AGS cells. Each of these compounds also exhibited comparable HIF-1 inhibitory activity in HeLa cells. The observation that acetylbakuchiol and bakuchiol displayed significant cytotoxicity in both AGS and HeLa cells (even in 24 h cell viability assays) suggested that the decrease in viability may have contributed to the observed inhibition of reporter activity or will most likely limit the therapeutic window for these compounds. Although only evaluated for cytotoxicity in 24 h cell viability assays, *O*-methylbakuchiol and *O*-ethylbakuchiol exhibited decreased cytotoxicity in either tumor cell line, relative to bakuchiol and acetylbakuchiol. As in the case of the diarylheptanoids from *Alnus hirsuta*, no experiments were performed to confirm or otherwise verify the activities observed in the initial HIF-1 luciferase reporter assays.

In collaboration with researchers at the NCI-Fredrick, Kim and coworkers at Chungnam National University in Korea found that the spider lily (*Crinum asiaticum* var. *japonicum*) alkaloid crinamine (**132**) and its epimer 3-epi-crinamine (**133**) inhibited hypoxia-induced HIF-1 activation in a human U251 glioblastoma cell line–based luciferase reporter assay (IC$_{50}$ values 2.7 and 5.4 µM, respectively) at concentrations found to be cytotoxic to other cell lines (Kim et al., 2006). Unspecified concentrations of crinamine failed to suppress hypoxia-induced HIF-1α expression and showed no effect in a second reporter assay for HIF-2 activation. This was taken to support the premise that crinamine is an HIF-1 specific inhibitor of hypoxic signaling. However, since the HIF-2 assay was conducted in distinctly dissimilar renal clear cell carcinoma 786-O cells, it is just as likely that cell-line specific genetic differences between the cell lines are responsible for the observed differential responses. No experiments were performed to confirm the HIF-1 luciferase reporter activities. Therefore, the HIF-1 inhibitory activity of these compounds remains to be experimentally substantiated.

7.10.2 Agents That Inhibit HIF-1α mRNA but Lack Experiments to Verify the Effects on HIF-1 Activation or HIF-1 Target Protein Expression

Since methods such as real time RT-PCR and other methods to measure cellular mRNA are subject to experimental artifacts, the results of these experiments must be experimentally verified. Preparations from the rhizome of the plant *Paris polyphylla* are used in traditional Chinese medicine and the *P. polyphylla* steroid glycoside polyphyllin D (**134**) has been shown to display cytotoxic activity in tumor cell lines. Working with scientists at the University of Missouri–Kansas City, researchers at Shandong University in China examined the effects of synthetically produced polyphyllum D on cancer cells under hypoxic conditions (Ma et al., 2009). Xiao and coworkers found that at concentrations as low as 2 μM, polyphillin D significantly inhibited hypoxic induction of HIF-1α and VEGF mRNAs in Lewis lung tumor cells (Table 7.6). No further experiments were performed to confirm the effects of polyphillin D on either HIF-1α or VEGF protein levels. Likewise, no additional studies were conducted to test whether polyphillin D can inhibit the transcriptional activity of HIF-1 or suppress VEGF-dependent angiogenesis. Consequently, it is not clear if polyphillin D should be considered an authentic HIF-1 inhibitor.

7.10.3 Cytotoxic HIF-1 Inhibitors That Suppress HIF-1 Activity at Cytotoxic Concentrations

Cytotoxic natural products such as pseudolaric acid B (**125**) (Li et al., 2004) and the marine sponge metabolites (*S*)-2,2'-dimethoxy-1,1'-binaphthyl-5,5',6,6'-tetraol (**135**) (Dai et al., 2007a) and diacarnoxide B (**136**) (Dai et al., 2007b) inhibit HIF-1 activation in tumor cell lines. However, this activity cannot be clearly distinguished from their significant cytotoxicity to the cell lines examined.

7.11 CONCLUSIONS

The search for new small-molecule inhibitors of HIF-1 signaling for use in cancer therapy has reached a nearly fevered pace in both academia and the pharmaceutical industry. Natural products have shown great promise as potential HIF-targeted therapeutic agents. In spite of this, few compounds display the sort of selectivity profile that would make them ideal drug candidates. For example, compounds that disrupt mitochondrial reactive oxygen-mediated HIF-1α stabilization are often associated with potential neurotoxicity. Likewise, many compounds that suppress HIF-1α translation are likely to exhibit cytotoxicity-associated side effects that are commonly produced by protein synthesis inhibitors. For this reason, the search for new HIF-1 inhibitors that function through novel and potentially more selective mechanisms continues at an intense rate. The broad assortment of natural product–based HIF-1 inhibitors that function through a diverse array of mechanisms highlights the overall convergence of different signaling pathways that are involved in regulating this transcription factor that is so critical to the maintenance of normal cellular oxygen homeostasis. While only a virtual handful of natural products may display the desired level of selectivity and a minimum side effect profile that will make the compounds druggable, natural products have proven to be invaluable probes of the complex cellular processes involved in HIF-1 signaling.

ACKNOWLEDGMENTS

This work was supported by the NIH-NCI CA098787 and NOAA/NIUST grant NA16RU1496.

REFERENCES

Agani, F.H., Pichiule, P., Carlos Chavez, J., and LaManna, J.C. 2002. Inhibitors of mitochondrial complex I attenuate the accumulation of hypoxia-inducible factor-1 during hypoxia in Hep3B cells. *Comp. Biochem. Physiol. A Mol. Integr. Physiol.* 132:107–109.

Aguilar-Santos, G. 1970. Caulerpin, a new red pigment from green algae of the genus *Caulerpa. J. Chem. Soc. Perkin 1* 6:842–843.

Arafeh, K.M. and Ullah, N. 2009. Synthesis of neolamellarin A, an inhibitor of hypoxia-inducible factor-1. *Nat. Prod. Commun.* 4:925–926.

Axelson, H., Fredlund, E., Ovenberger, M., Landberg, G., and Pahlman, S. 2005. Hypoxia-induced dedifferentiation of tumor cells—A mechanism behind heterogeneity and aggressiveness of solid tumors. *Semin. Cell Dev. Biol.* 16:554–563.

Bach, A., Bender-Sigel, J., Schrenk, D., Flügel, D., and Kietzmann, T. 2009. The antioxidant quercetin inhibits cellular proliferation via HIF-1-dependent induction of p21WAF. *Antioxid. Redox. Signal.* [Epub ahead of print].

Bae, M.K., Kim, S.H., Jeong, J.W., Lee, Y.M., Kim, H.S., Kim, S.R., Yun, I., Bae, S.K., and Kim, K.W. 2006. Curcumin inhibits hypoxia-induced angiogenesis via down-regulation of HIF-1. *Oncol. Rep.* 15:1557–1562.

Birner, P., Schindl, M., Obermair, A., Plank, C., Breitenecker, G., and Oberhuber, G. 2000. Overexpression of hypoxia-inducible factor 1 is a marker for an unfavorable prognosis in early-stage invasive cervical cancer. *Cancer Res.* 60:4693–4696.

Bisht, K.S., Bradbury, C.M., Mattson, D., Kaushal, A., Sowers, A., Markovina, S., Ortiz, K.L., Sieck, L.K., Isaacs, J.S., Brechbiel, M.W., Mitchell, J.B., Neckers, L.M., and Gius, D. 2003. Geldanamycin and 17-allylamino-17-demethoxygeldanamycin potentiate the *in vitro* and *in vivo* radiation response of cervical tumor cells via the heat shock protein 90-mediated intracellular signaling and cytotoxicity. *Cancer Res.* 63:8984–8995.

Block, K.M., Wang, H., Szabó, L.Z., Polaske, N.W., Henchey, L.K., Dubey, R., Kushal, S., László, C.F., Makhoul, J., Song, Z., Meuillet, E.J., and Olenyuk, B.Z. 2009. Direct inhibition of hypoxia-inducible transcription factor complex with designed dimeric epidithiodiketopiperazine. *J. Am. Chem. Soc.* 131:18078–18088.

Blum, R., Jacob-Hirsch, J., Amariglio, N., Rechavi, G., and Kloog, Y. 2005. Ras inhibition in glioblastoma down-regulates hypoxia-inducible factor-1α, causing glycolysis shutdown and cell death. *Cancer Res.* 65:999–1006.

Bos, R., Zhong, H., Hanrahan, C.F., Mommers, E.C., Semenza, G.L., Pinedo, H.M., Abeloff, M.D., Simons, J.W., van Diest, P.J., and van der Wall, E. 2001. Levels of hypoxia-inducible factor-1α during breast carcinogenesis. *J. Natl. Cancer Inst.* 93:309–314.

Brown, J.M. and Wilson, W.R. 2004. Exploiting tumour hypoxia in cancer treatment. *Nat. Rev. Cancer* 4:437–447.

Brunelle, J.K., Bell, E.L., Quesada, N.M., Vercauteren, K., Tiranti, V., Zeviani, M., Scarpulla, R.C., and Chandel, N.S. 2005. Oxygen sensing requires mitochondrial ROS but not oxidative phosphorylation. *Cell Metab.* 1:409–414.

Bucher, K. and Skogerson, L. 1976. Cryptopleurine-an inhibitor of translocation. *Biochemistry* 15:4755–4759.

Cai, X.F., Jin, X., Lee, D., Yang, Y.T., Lee, K., Hong, Y.S., Lee, J.H., and Lee, J.J. 2006. Phenanthroquinolizidine alkaloids from the roots of *Boehmeria pannosa* potently inhibit hypoxia-inducible factor-1 in AGS human gastric cancer cells. *J. Nat. Prod.* 69:1095–1097.

Cairns, R.A., Papandreou, I., Sutphin, P.D., and Denko, N.C. 2007. Metabolic targeting of hypoxia and HIF1 in solid tumors can enhance cytotoxic chemotherapy. *Proc. Natl. Acad. Sci. USA* 104:9445–9450.

Cao, Z., Fang, J., Xia, C., Shi, X., and Jiang, B.H. 2004. Trans-3,4,5′-trihydroxystibene inhibits hypoxia-inducible factor 1α and vascular endothelial growth factor expression in human ovarian cancer cells. *Clin. Cancer Res.* 10:5253–5263.

Chandel, N.S., Maltepe, E., Goldwasser, E., Mathieu, C.E., Simon, M.C., and Schumacker, P.T. 1998. Mitochondrial reactive oxygen species trigger hypoxia-induced transcription. *Proc. Natl. Acad. Sci. USA* 95:11715–11720.

Chandel, N.S., McClintock, D.S., Feliciano, C.E., Wood, T.M., Melendez, J.A., Rodriguez, A.M., and Schumacker, P.T. 2000. Reactive oxygen species generated at mitochondrial complex III stabilize hypoxia-inducible factor-1α during hypoxia: A mechanism of O_2 sensing. *J. Biol. Chem.* 275:25130–25138.

Chang, H., Shyu, K.G., Lee, C.C., Tsai, S.C., Wang, B.W., Lee, Y.H., and Lin, S. 2003. GL331 inhibits HIF-1α expression in a lung cancer model. *Biochem. Biophys. Res. Commun.* 302:95–100.

Chittiboyina, A.G., Kumar, G.M., Carvalho, P.B., Liu, Y., Zhou, Y.D., Nagle, D.G., and Avery, M.A. 2007. Total synthesis and absolute configuration of laurenditerpenol: A hypoxia inducible factor-1 activation inhibitor. *J. Med. Chem.* 50:6299–6302.

Choi, H., Chun, Y.S., Kim, S.W., Kim, M.S., and Park, J.W. 2006. Curcumin inhibits hypoxia-inducible factor-1 by degrading aryl hydrocarbon receptor nuclear translocator: A mechanism of tumor growth inhibition. *Mol. Pharmacol.* 70:1664–1671.

Choi, H.J., Eun, J.S., Kim, D.K., Li, R.H., Shin, T.Y., Park, H., Cho, N.P., and Soh, Y. 2008. Icariside II from *Epimedium koreanum* inhibits hypoxia-inducible factor-1α in human osteosarcoma cells. *Eur. J. Pharmacol.* 579:58–65.

Chua, Y.S., Chua, Y.L., and Hagen, T. 2010. Structure activity analysis of 2-methoxyestradiol analogues reveals targeting of microtubules as the major mechanism of antiproliferative and proapoptotic activity. *Mol. Cancer Ther.* 9:224–235.

Cook, K.M., Hilton, S.T., Mecinovic, J., Motherwell, W.B., Figg, W.D., and Schofield, C.J. 2009. Epidithiodiketopiperazines block the interaction between hypoxia-inducible factor-1α (HIF-1α) and p300 by a zinc ejection mechanism. *J. Biol. Chem.* 284:26831–26838.

Dai, J., Fishback, J.A., Zhou, Y.D., and Nagle, D.G. 2006. Sodwanone and yardenone triterpenes from a South African species of the marine sponge *Axinella* inhibit hypoxia-inducible factor-1 (HIF-1) activation in both breast and prostate tumor cells. *J. Nat. Prod.* 69:1715–1720.

Dai, J., Liu, Y., Zhou, Y.D., and Nagle, D.G. 2007a. Cytotoxic metabolites from an Indonesian sponge *Lendenfeldia* sp. *J. Nat. Prod.* 70:1824–1826.

Dai, J., Liu, Y., Zhou, Y.D., and Nagle, D.G. 2007b. Hypoxia-selective antitumor agents: Norsesterterpene peroxides from the marine sponge *Diacarnus levii* preferentially suppress the growth of tumor cells under hypoxic conditions. *J. Nat. Prod.* 70:130–133.

D'Angelo, G., Duplan, E., Vigne, P., and Frelin, C. 2003. Cyclosporin A prevents the hypoxic adaptation by activating hypoxia-inducible factor-1α Pro-564 hydroxylation. *J. Biol. Chem.* 278:15406–15411.

Dat, N.T., Jin, X., Lee, K., Hong, Y.S., Kim, Y.H., and Lee, J.J. 2009. Hypoxia-inducible factor-1 inhibitory benzofurans and chalcone-derived diels-alder adducts from *Morus* species. *J. Nat. Prod.* 72:39–43.

Dat, N.T., Jin, X., Lee, J.H., Lee, D., Hong, Y.S., Lee, K., Kim, Y.H., and Lee, J.J. 2007. Abietane diterpenes from *Salvia miltiorrhiza* inhibit the activation of hypoxia-inducible factor-1. *J. Nat. Prod.* 70:1093–1097.

Du, G., Lin, H., Wang, M., Zhang, S., Wu, X., Lu, L., Ji, L., and Yu, L. 2009. Quercetin greatly improved therapeutic index of doxorubicin against 4T1 breast cancer by its opposing effects on HIF-1α in tumor and normal cells. *Cancer Chemother. Pharmacol.* 65:277–287.

Ebert, B.L., Firth, J.D., and Ratcliffe, P.J. 1995. Hypoxia and mitochondrial inhibitors regulate expression of glucose transporter-1 via distinct cis-acting sequences. *J. Biol. Chem.* 270:29083–29089.

El Sayed, K.A., Khanfar, M.A., Shallal, H.M., Muralidharan, A., Awate, B., Youssef, D.T., Liu, Y., Zhou, Y.D., Nagle, D.G., and Shah, G. 2008. Latrunculin A and its C-17-*O*-carbamates inhibit prostate tumor cell invasion and HIF-1 activation in breast tumor cells. *J. Nat. Prod.* 71:396–402.

El Sayed, K.A., Youssef, D.T., and Marchetti, D. 2006. Bioactive natural and semisynthetic latrunculins. *J. Nat. Prod.* 69:219–223.

Ema, M., Taya, S., Yokotani, N., Sogawa, K., Matsuda, Y., and Fujii-Kuriyama, Y. 1997. A novel bHLH-PAS factor with close sequence similarity to hypoxia-inducible factor 1α regulates the VEGF expression and is potentially involved in lung and vascular development. *Proc. Natl. Acad. Sci. USA* 94:4273–4278.

Escuin, D., Kline, E.R., and Giannakakou, P. 2005. Both microtubule-stabilizing and microtubule-destabilizing drugs inhibit hypoxia-inducible factor-1α accumulation and activity by disrupting microtubule function. *Cancer Res.* 65:9021–9028.

Fang, J., Cao, Z., Chen, Y.C., Reed, E., and Jiang, B.H. 2004. 9-β-D-arabinofuranosyl-2-fluoroadenine inhibits expression of vascular endothelial growth factor through hypoxia-inducible factor-1 in human ovarian cancer cells. *Mol. Pharmacol.* 66:178–186.

Fang, N. and Casida, J.E. 1999. Cubé resin insecticide: Identification and biological activity of 29 rotenoid constituents. *J. Agric. Food. Chem.* 47:2130–2136.

Fang, J.S., Gillies, R.D., and Gatenby, R.A. 2008. Adaptation to hypoxia and acidosis in carcinogenesis and tumor progression. *Semin. Cancer Biol.* 18:330–337.

Fang, J., Xia, C., Cao, Z., Zheng, J.Z., Reed, E., and Jiang, B.H. 2005. Apigenin inhibits VEGF and HIF-1 expression via PI3K/AKT/p70S6K1 and HDM2/p53 pathways. *FASEB J.* 19:342–353.

Fang, J., Zhou, Q., Liu, L.Z., Xia, C., Hu, X., Shi, X., and Jiang, B.H. 2007. Apigenin inhibits tumor angiogenesis through decreasing HIF-1α and VEGF expression. *Carcinogenesis* 28:858–864.

Flamme, I., Fröhlich, T., von Reutern, M., Kappel, A., Damert, A., and Risau, W. 1997. HRF, a putative basic helix-loop-helix-PAS-domain transcription factor is closely related to hypoxia-inducible factor-1α and developmentally expressed in blood vessels. *Mech. Dev.* 63:51–60.

Gachet, Y., Tournier, S., Millar, J.B., and Hyams, J.S. 2001. A MAP kinase-dependent actin checkpoint ensures proper spindle orientation in fission yeast. *Nature* 412:352–355.

García-Maceira, P. and Mateo, J. 2009. Silibinin inhibits hypoxia-inducible factor-1α and mTOR/p70S6K/4E-BP1 signalling pathway in human cervical and hepatoma cancer cells: Implications for anticancer therapy. *Oncogene* 28:313–324.

Geiger, W.B., Conn, J.E., and Waksman, S.A. 1944. Chaetomin, a new antibiotic substance produced by *Chaetomium cochliodes*: II. Isolation and concentration. *J. Bacteriol.* 48:531–536.

Giatromanolaki, A., Koukourakis, M.I., Sivridis, E., Turley, H., Talks, K., Pezzella, F., Gatter, K.C., and Harris, A.L. 2001. Relation of hypoxia inducible factor 1α and 2α in operable non-small cell lung cancer to angiogenic/molecular profile of tumours and survival. *Br. J. Cancer* 285:881–890.

Greenberger, L.M., Horak, I.D., Filpula, D., Sapra, P., Westergaard, M., Frydenlund, H.F., Albæk, C., Schrøder, H., and Ørum, H. 2008. A RNA antagonist of hypoxia-inducible factor-1α, EZN-2968, inhibits tumor cell growth. *Mol. Cancer Ther.* 7:3598–3608.

Grollman, A.P. 1966. Structural basis for inhibition of protein synthesis by emetine and cycloheximide based on an analogy between IPECAC alkaloids and glutarimide antibiotics. *Proc. Natl. Acad. Sci. USA* 56:1867–1874.

Gupta, R.S. and Siminovitch, L. 1977. Mutants of CHO cells resistant to the protein synthesis inhibitors, cryptopleurine and tylocrebrine: Genetic and biochemical evidence for common site of action of emetine, cryptopleurine, tylocrebine, and tubulosine. *Biochemistry* 16:3209–3214.

Guzy, R.D., Hoyos, B., Robin, E., Chen, H., Liu, L., Mansfield, K.D., Simon, M.C., Hammerling, U., and Schumacker, P.T. 2005. Mitochondrial complex III is required for hypoxia-induced ROS production and cellular oxygen sensing. *Cell Metab.* 1:401–408.

Hagen, T., D'Amico, G., Quintero, M., Palacios-Callender, M., Hollis, V., Lam, F., and Moncada, S. 2004. Inhibition of mitochondrial respiration by the anticancer agent 2-methoxyestradiol. *Biochem. Biophys. Res. Commun.* 322:923–929.

Hagen, T., Taylor, C.T., Lam, F., and Moncada, S. 2003. Redistribution of intracellular oxygen in hypoxia by nitric oxide: Effect on HIF1α. *Science* 302:1975–1978.

Hail, N. Jr. and Lotan, R. 2004. Apoptosis induction by the natural product cancer chemopreventive agent deguelin is mediated through the inhibition of mitochondrial bioenergetics. *Apoptosis* 9:437–447.

Hanessian, S., Reddy, G.J., and Chahal, N. 2006. Total synthesis and stereochemical confirmation of manassantin A, B, and B₁. *Org. Lett.* 8:5477–5480.

Hasebe, Y., Egawa, K., Yamazaki, Y., Kunimoto, S., Hirai, Y., Ida, Y., and Nose, K. 2003. Specific inhibition of hypoxia-inducible factor (HIF)-1α activation and of vascular endothelial growth factor (VEGF) production by flavonoids. *Biol. Pharm. Bull.* 26:1379–1383.

Helczynska, K., Larsson, A.M., Holmquist-Mengelbier, L., Bridges, E., Fredlund, E., Borgquist, S., Landberg, G., Påhlman, S., and Jirström, K. 2008. Hypoxia-inducible factor-2α correlates to distant recurrence and poor outcome in invasive breast cancer. *Cancer Res.* 68:9212–9220.

Hodges, T.W., Hossain, C.F., Kim, Y.P., Zhou, Y.D., and Nagle, D.G. 2004. Moleculartargeted antitumor agents: The *Saururus cernuus* dineolignans manassantin B and 4-*O*-demethylmanassantin B are potent inhibitors of hypoxia-activated HIF-1. *J. Nat. Prod.* 67:767–771.

Holmquist-Mengelbier, L., Fredlund, E., Löfstedt, T., Noguera, R., Navarro, S., Nilsson, H., Pietras, A., Vallon-Christersson, J., Borg, A., Gradin, K., Poellinger, L., and Påhlman, S. 2006. Recruitment of HIF-1α and HIF-2α to common target genes is differentially regulated in neuroblastoma: HIF-2α promotes an aggressive phenotype. *Cancer Cell* 10:413–423.

Hossain, C.F., Kim, Y.P., Baerson, S.R., Zhang, L., Bruick, R.K., Mohammed, K.A., Agarwal, A.K., Nagle, D.G., and Zhou, Y.D. 2005. *Saururus cernuus* lignans—Potent small molecule inhibitors of hypoxia-inducible factor-1. *Biochem. Biophys. Res. Commun.* 333:1026–1033.

Hudson, C.C., Liu, M., Chiang, G.G., Otterness, D.M., Loomis, D.C., Kaper, F., Giaccia, A.J., and Abraham, R.T. 2002. Regulation of hypoxia-inducible factor 1α expression and function by the mammalian target of rapamycin. *Mol. Cell. Biol.* 22:7004–7014.

Hur, E., Chang, K.Y., Lee, E., Lee, S.K., and Park, H. 2001. Mitogen-activated protein kinase inhibitor PD98059 blocks the trans-activation but not the stabilization or DNA binding ability of hypoxia-inducible factor-1α. *Mol. Pharmacol.* 59:1216–1224.

Hur, E., Kim, H.H., Choi, S.M., Kim, J.H., Yim, S., Kwon, H.J., Choi, Y., Kim, D.K., Lee, M.O., and Park, H. 2002. Reduction of hypoxia-induced transcription through the repression of hypoxia-inducible factor-1α/aryl hydrocarbon receptor nuclear translocator DNA binding by the 90-kDa heat-shock protein inhibitor radicicol. *Mol. Pharmacol.* 62:975–982.

Ikeda, R., Furukawa, T., Kitazono, M., Ishitsuka, K., Okumura, H., Tani, A., Sumizawa, T., Haraguchi, M., Komatsu, M., Uchimiya, H., Ren, X.Q., Motoya, T., Yamada, K., and Akiyama, S. 2002. Molecular basis for the inhibition of hypoxia-induced apoptosis by 2-deoxy-D-ribose. *Biochem. Biophys. Res. Commun.* 291:806–812.

Isaacs, J.S., Jung, Y.J., Mimnaugh, E.G., Martinez, A., Cuttitta, F., and Neckers, L.M. 2002. Hsp90 regulates a von Hippel Lindau-independent hypoxia-inducible factor-1α-degradative pathway. *J. Biol. Chem.* 277:29936–29944.

Ivan, M., Kondo, K., Yang, H., Kim, W., Valiando, J., Ohh, M., Salic, A., Asara, J.M., Lane, W.S., and Kaelin, W.G. Jr. 2001. HIFα targeted for VHL-mediated destruction by proline hydroxylation: Implications for O_2 sensing. *Science* 292:464–468.

Iyer, N.V., Kotch, L.E., Agani, F., Leung, S.W., Laughner, E., Wenger, R.H., Gassmann, M., Gearhart, J.D., Lawler, A.M., Yu, A.Y., and Semenza, G.L. 1998. Cellular and developmental control of O_2 homeostasis by hypoxia-inducible factor 1α. *Genes Dev.* 12:149–162.

Jaakkola, P., Mole, D.R., Tian, Y.M., Wilson, M.I., Gielbert, J., Gaskell, S.J., Kriegsheim, A.V., Hebestreit, H.F., Mukherji, M., Schofield, C.J., Maxwell, P.H., Pugh, C.W., and Ratcliffe, P.J. 2001. Targeting of HIF-1α to the von Hippel-Lindau ubiquitylation complex by O_2-regulated prolyl hydroxylation. *Science* 292:468–472.

Jeon, H., Kim, H., Choi, D., Kim, D., Park, S.Y., Kim, Y.J., Kim, Y.M., and Jung, Y. 2007. Quercetin activates an angiogenic pathway, hypoxia inducible factor (HIF)-1-vascular endothelial growth factor, by inhibiting HIF-prolyl hydroxylase: A structural analysis of quercetin for inhibiting HIF-prolyl hydroxylase. *Mol. Pharmacol.* 71:1676–1684.

Jiang, B.H., Jiang, G., Zheng, J.Z., Lu, Z., Hunter, T., and Vogt, P.K. 2001. Phosphatidylinositol 3-kinase signaling controls levels of hypoxia-inducible factor 1. *Cell Growth Differ.* 12:363–369.

Jin, W.Y., Cai, X.F., Na, M.K., Lee, J.J., and Bae, K.H. 2007. Triterpenoids and diarylheptanoids from *Alnus hirsuta* inhibit HIF-1 in AGS cells. *Arch. Pharm. Res.* 30:412–418.

Jin, X., Jin, H.R., Lee, D., Lee, J.H., Kim, S.K., and Lee, J.J. 2008. A quassinoid 6α-tigloyloxychaparrinone inhibits hypoxia-inducible factor-1 pathway by inhibition of eukaryotic translation initiation factor 4E phosphorylation. *Eur. J. Pharmacol.* 592:41–47.

Jung, M.E. and Im, G.Y. 2009. Total synthesis of racemic laurenditerpenol, an HIF-1 inhibitor. *J. Org. Chem.* 74:8739–8753.

Jung, Y.J., Isaacs, J.S., Lee, S., Trepel, J., and Neckers, L. 2003. Microtubule disruption utilizes an NFκB-dependent pathway to stabilize HIF-1α protein. *J. Biol. Chem.* 278:7445–7452.

Kashman, Y., Groweiss, A., and Shmueli, U. 1980. Latrunculin, a new 2-thiazolidinone macrolide from the marine sponge *Latrunculia magnifica*. *Tetrahedron Lett.* 21:3629–3632.

Kasper, A.C., Moon, E.J., Hu, X., Park, Y., Wooten, C.M., Kim, H., Yang, W., Dewhirst, M.W., and Hong, J. 2009. Analysis of HIF-1 inhibition by manassantin A and analogues with modified tetrahydrofuran configurations. *Bioorg. Med. Chem. Lett.* 19:3783–3786.

Katschinski, D.M., Le, L., Heinrich, D., Wagner, K.F., Hofer, T., Schindler, S.G., and Wenger, R.H. 2002. Heat induction of the unphosphorylated form of hypoxia-inducible factor-1α is dependent on heat shock protein-90 activity. *J. Biol. Chem.* 277:9262–9267.

Kim, H., Kasper, A.C., Moon, E.J., Park, Y., Wooten, C.M., Dewhirst, M.W., and Hong, J. 2009. Nucleophilic addition of organozinc reagents to 2-sulfonyl cyclic ethers: Stereoselective synthesis of manassantins A and B. *Org. Lett.* 11:89–92.

Kim, S.H., Kim, K.W., and Jeong, J.W. 2007. Inhibition of hypoxia-induced angiogenesis by sodium butyrate, a histone deacetylase inhibitor, through hypoxia-inducible factor-1α suppression. *Oncol. Rep.* 17:793–797.

Kim, M.S., Kwon, H.J., Lee, Y.M., Baek, J.H., Jang, J.E., Lee, S.W., Moon, E.J., Kim, H.S., Lee, S.K., Chung, H.Y., Kim, C.W., and Kim, K.W. 2001. Histone deacetylases induce angiogenesis by negative regulation of tumor suppressor genes. *Nat. Med.* 7:437–443.

Kim, Y.H., Park, E.J., Park, M.H., Badarch, U., Woldemichael, G.M., and Beutler, J.A. 2006. Crinamine from *Crinum asiaticum* var. *japonicum* inhibits hypoxia inducible factor-1 activity but not activity of hypoxia inducible factor-2. *Biol. Pharm. Bull.* 29:2140–2142.

Klausmeyer, P., McCloud, T.G., Uranchimeg, B., Melillo, G., Scudiero, D.A., Cardellina, J.H. 2nd, and Shoemaker, R.H. 2008. Separation and SAR study of HIF-1α inhibitory tubulosines from *Alangium* cf. *longiflorum*. *Planta Med.* 74:258–263.

Klausmeyer, P., Zhou, Q., Scudiero, D.A., Uranchimeg, B., Melillo, G., Cardellina, J.H., Shoemaker, R.H., Chang, C.J., and McCloud, T.G. 2009. Cytotoxic and HIF-1α inhibitory compounds from *Crossosoma bigelovii*. *J. Nat. Prod.* 72:805–812.

Kong, X., Lin, Z., Liang, D., Fath, D., Sang, N., and Caro, J. 2006. Histone deacetylase inhibitors induce VHL and ubiquitin-independent proteasomal degradation of hypoxia-inducible factor 1α. *Mol. Cell. Biol.* 26:2019–2028.

Kong, D., Park, E.J., Stephen, A.G., Calvani, M., Cardellina, J.H., Monks, A., Fisher, R.J., Shoemaker, R.H., and Melillo, G. 2005. Echinomycin, a small-molecule inhibitor of hypoxia-inducible factor-1 DNA-binding activity. *Cancer Res.* 65:9047–9055.

Koukourakis, M.I., Bentzen, S.M., Giatromanolaki, A., Wilson, G.D., Daley, F.M., Saunders, M.I., Dische, S., Sivridis, E., and Harris, A.L. 2006. Endogenous markers of two separate hypoxia response pathways (hypoxia inducible factor 2α and carbonic anhydrase 9) are associated with radiotherapy failure in head and neck cancer patients recruited in the CHART randomized trial. *J. Clin. Oncol.* 24:727–735.

Kruger, E.A., Blagosklonny, M.V., Dixon, S.C., and Figg, W.D. 1998–1999. UCN-01, a protein kinase C inhibitor, inhibits endothelial cell proliferation and angiogenic hypoxic response. *Invas. Metas.* 18:209–218.

Kung, A.L., Zabludoff, S.D., France, D.S., Freedman, S.J., Tanner, E.A., Vieira, A., Cornell-Kennon, S., Lee, J., Wang, B., Wang, J., Memmert, K., Naegeli, H.U., Petersen, F., Eck, M.J., Bair, K.W., Wood, A.W., and Livingston, D.M. 2004. Small molecule blockade of transcriptional coactivation of the hypoxia-inducible factor pathway. *Cancer Cell* 6:33–43.

Kurebayashi, J., Otsuki, T., Kurosumi, M., Soga, S., Akinaga, S., and Sonoo, H. 2001. A radicicol derivative, KF58333, inhibits expression of hypoxia-inducible factor-1α and vascular endothelial growth factor, angiogenesis and growth of human breast cancer xenografts. *Jpn. J. Cancer Res.* 92:1342–1351.

Land, S.C. and Tee, A.R. 2007. Hypoxia-inducible factor 1α is regulated by the mammalian target of rapamycin (mTOR) via an mTOR signaling motif. *J. Biol. Chem.* 282:20534–20543.

Lando, D., Peet, D.J., Whelan, D.A., Gorman, J.J., and Whitelaw, M.L. 2002. Asparagine hydroxylation of the HIF transactivation domain a hypoxic switch. *Science* 295:858–861.

Laughner, E., Taghavi, P., Chiles, K., Mahon, P.C., and Semenza, G.L. 2001. HER2 (neu) signaling increases the rate of hypoxia-inducible factor 1α (HIF-1α) synthesis: Novel mechanism for HIF-1-mediated vascular endothelial growth factor expression. *Mol. Cell. Biol.* 21:3995–4004.

Lee, Y.M., Kim, S.H., Kim, H.S., Son, M.J., Nakajima, H., Kwon, J.H., and Kim, K.W. 2003. Inhibition of hypoxia-induced angiogenesis by FK228, a specific histone deacetylase inhibitor, via suppression of HIF-1α activity. *Biochem. Biophys. Res. Commun.* 300:241–246.

Lee, D.H. and Lee, Y.J. 2008. Quercetin suppresses hypoxia-induced accumulation of hypoxia-inducible factor-1α (HIF-1α) through inhibiting protein synthesis. *J. Cell. Biochem.* 105:546–553.

Lee, K., Qian, D.Z., Rey, S., Wei, H., Liu, J.O., and Semenza, G.L. 2009. Anthracycline chemotherapy inhibits HIF-1 transcriptional activity and tumor-induced mobilization of circulating angiogenic cells. *Proc. Natl. Acad. Sci. USA* 106:2353–2358.

Lee, E., Yim, S., Lee, S.K., and Park, H. 2002. Two transactivation domains of hypoxia-inducible factor-1α regulated by the MEK-1/p42/p44 MAPK pathway. *Mol. Cells* 14:9–15.

Leek, R.D., Talks, K.L., Pezzella, F., Turley, H., Campo, L., Brown, N.S., Bicknell, R., Taylor, M., Gatter, K.C., and Harris, A.L. 2002. Relation of hypoxia-inducible factor-2α (HIF-2α) expression in tumor-infiltrative macrophages to tumor angiogenesis and the oxidative thymidine phosphorylase pathway in human breast cancer. *Cancer Res.* 62:1326–1329.

Leeman-Neill, R.J., Wheeler, S.E., Singh, S.V., Thomas, S.M., Seethala, R.R., Neill, D.B., Panahandeh, M.C., Hahm, E.R., Joyce, S.C., Sen, M., Cai, Q., Freilino, M.L., Li, C., Johnson, D.E., and Grandis, J.R. 2009. Guggulsterone enhances head and neck cancer therapies via inhibition of signal transducer and activator of transcription-3. *Carcinogenesis* 30:1848–1856.

Lewer, P., Chapin, E.L., Graupner, P.R., Gilbert, J.R., and Peacock, C. 2003. Tartrolone C: A novel insecticidal macrodiolide produced by *Streptomyces sp.* CP1130. *J. Nat. Prod.* 66:143–145.

Li, L., Lin, X., Shoemaker, A.R., Albert, D.H., Fesik, S.W., and Shen, Y. 2006. Hypoxia-inducible factor-1 inhibition in combination with temozolomide treatment exhibits robust antitumor efficacy *in vivo*. *Clin. Cancer Res.* 12:4747–4754.

Li, M.H., Miao, Z.H., Tan, W.F., Yue, J.M., Zhang, C., Lin, L.P., Zhang, X.W., and Ding, J. 2004. *Clin. Cancer Res.* 10:8266–8274.

Li, L., Zeng, J., Gao, Y., and He, D. 2010. Targeting silibinin in the antiproliferative pathway. *Expert Opin. Investig. Drugs* 19:243–255.

Lin, S., Tsai, S.C., Lee, C.C., Wang, B.W., Liou, J.Y., and Shyu, K.G. 2004. Berberine inhibits HIF-1α expression via enhanced proteolysis. *Mol. Pharmacol.* 66:612–619.

Liu, Y.V., Baek, J.H., Zhang, H., Diez, R., Cole, R.N., and Semenza, G.L. 2007. RACK1 competes with HSP90 for binding to HIF-1α and is required for O_2-independent and HSP90 inhibitor-induced degradation of HIF-1α. *Mol. Cell* 25:207–217.

Liu, L.Z., Fang, J., Zhou, Q., Hu, X., Shi, X., and Jiang, B.H. 2005. Apigenin inhibits expression of vascular endothelial growth factor and angiogenesis in human lung cancer cells: Implication of chemoprevention of lung cancer. *Mol. Pharmacol.* 68:635–643.

Liu, Y.V., Hubbi, M.E., Pan, F., McDonald, K.R., Mansharamani, M., Cole, R.N., Liu, J.O., and Semenza, G.L. 2007b. Calcineurin promotes hypoxia-inducible factor 1α expression by dephosphorylating RACK1 and blocking RACK1 dimerization. *J. Biol. Chem.* 282:37064–37073.

Liu, Y., Liu, R., Mao, S.C., Morgan, J.B., Jekabsons, M.B., Zhou, Y.D., and Nagle, D.G. 2008. Molecular-targeted antitumor agents. 19. Furospongolide from a marine *Lendenfeldia* sp. sponge inhibits hypoxia-inducible factor-1 activation in breast tumor cells. *J. Nat. Prod.* 71:1854–1860.

Liu, R., Liu, Y., Zhou, Y.D., and Nagle, D.G. 2007. Molecular-targeted antitumor agents. 15. Neolamellarins from the marine sponge *Dendrilla nigra* inhibit hypoxia-inducible factor-1 activation and secreted vascular endothelial growth factor production in breast tumor cells. *J. Nat. Prod.* 70:1741–1745.

Liu, Y., Morgan, J.B., Coothankandaswamy, V., Liu, R., Jekabsons, M.B., Mahdi, F., Nagle, D.G., and Zhou, Y.D. 2009b. The *Caulerpa* pigment caulerpin inhibits HIF-1 activation and mitochondrial respiration. *J. Nat. Prod.* 72:2104–2109.

Liu, Y., Veena, C.K., Morgan, J.B., Mohammed, K.A., Jekabsons, M.B., Nagle, D.G., and Zhou, Y.D. 2009a. Methylalpinumisoflavone inhibits hypoxia-inducible factor-1 (HIF-1) activation by simultaneously targeting multiple pathways. *J. Biol. Chem.* 284:5859–5868.

Löfstedt, T., Fredlund, E., Holmquist-Mengelbier, L., Pietras, A., Ovenberger, M., Poellinger, L., and Påhlman, S. 2007. Hypoxia inducible factor-2α in cancer. *Cell Cycle* 6:919–926.

Lopez-Lazaro, M. 2009. Digoxin, HIF-1, and cancer. *Proc. Natl. Acad. Sci. USA* 106:E26.

Ma, D.D., Lu, H.X., Xu, L.S., and Xiao, W. 2009. Polyphyllin D exerts potent anti-tumour effects on Lewis cancer cells under hypoxic conditions. *J. Int. Med. Res.* 37:631–640.

Mabjeesh, N.J., Escuin, D., LaVallee, T.M., Pribluda, V.S., Swartz, G.M., Johnson, M.S., Willard, M.T., Zhong, H., Simons, J.W., and Giannakakou, P. 2003. 2ME2 inhibits tumor growth and angiogenesis by disrupting microtubules and dysregulating HIF. *Cancer Cell* 3:363–375.

Mabjeesh, N.J., Post, D.E., Willard, M.T., Kaur, B., van Meir, E.G., Simons, J.W., and Zhong, H. 2002. Geldanamycin induces degradation of hypoxia-inducible factor 1α protein via the proteosome pathway in prostate cancer cells. *Cancer Res.* 62:2478–2482.

Maeda, M., Hasebe, Y., Egawa, K., Shibanuma, M., and Nose, K. 2006. Inhibition of angiogenesis and HIF-1α activity by antimycin A1. *Biol. Pharm. Bull.* 29:1344–1348.

Mansfield, K.D., Guzy, R.D., Pan, Y., Young, R.M., Cash, T.P., Schumacker, P.T., and Simon, M.C. 2005. Mitochondrial dysfunction resulting from loss of cytochrome c impairs cellular oxygen sensing and hypoxic HIF-α activation. *Cell Metab.* 1:393–399.

Mao, S.C., Liu, Y., Morgan, J.B., Jekabsons, M.B., Zhou, Y.D., and Nagle, D.G. 2009. Lipophilic 2,5-disubstituted pyrroles from the marine sponge *Mycale* sp. inhibit mitochondrial respiration and HIF-1 activation. *J. Nat. Prod.* 72:1927–1936.

Mateo, J., García-Lecea, M., Cadenas, S., Hernández, C., and Moncada, S. 2003. Regulation of hypoxia-inducible factor-1α by nitric oxide through mitochondria-dependent and -independent pathways. *Biochem. J.* 376(Pt 2):537–544.

Maxwell, P.H., Wiesener, M.S., Chang, G.W., Clifford, S.C., Vaux, E.C., Cockman, M.E., Wykoff, C.C., Pugh, C.W., Maher, E.R., and Ratcliffe, P.J. 1999. The tumour suppressor protein VHL targets hypoxia-inducible factors for oxygen-dependent proteolysis. *Nature* 399:271–275.

Meinesz, A. 1999. *Killer Algae*. University of Chicago Press: Chicago, IL.

Miki, K., Unno, N., Nagata, T., Uchijima, M., Konno, H., Koide, Y., and Nakamura, S. 2004. Butyrate suppresses hypoxia-inducible factor-1 activity in intestinal epithelial cells under hypoxic conditions. *Shock* 22:446–452.

Minet, E., Mottet, D., Michel, G., Roland, I., Raes, M., Remacle, J., and Michiels, C. 1999. Hypoxia-induced activation of HIF-1: Role of HIF-1α-Hsp90 interaction. *FEBS Lett.* 460:251–256.

Mirzoeva, S., Kim, N.D., Chiu, K., Franzen, C.A., Bergan, R.C., and Pelling, J.C. 2008. Inhibition of HIF-1α and VEGF expression by the chemopreventive bioflavonoid apigenin is accompanied by Akt inhibition in human prostate carcinoma PC3-M cells. *Mol. Carcinog.* 47:686–700.

Moeller, B.J., Dreher, M.R., Rabbani, Z.N., Schroeder, T., Cao, Y., Li, C.Y., and Dewhirst, M.W. 2005. Pleiotropic effects of HIF-1 blockade on tumor radiosensitivity. *Cancer Cell* 8:99–110.

Moeller, B.J., Richardson, R.A., and Dewhirst, M.W. 2007. Hypoxia and radiotherapy: Opportunities for improved outcomes in cancer treatment. *Cancer Metastasis Rev.* 26:241–248.

Mohammed, K.A., Hossain, C.F., Zhang, L., Bruick, R.K., Zhou, Y.D., and Nagle, D.G. 2004. Laurenditerpenol, a new diterpene from the tropical marine alga *Laurencia intricata* that potently inhibits HIF-1 mediated hypoxic signaling in breast tumor cells. *J. Nat. Prod.* 67:2002–2007.

Mohammed, K.A., Jadulco, R.C., Bugni, T.S., Harper, M.K., Sturdy, M., and Ireland, C.M. 2008. Strongylophorines: Natural product inhibitors of hypoxia-inducible factor-1 transcriptional pathway. *J. Med. Chem.* 51:1402–1405.

Moser, C., Lang, S.A., Mori, A., Hellerbrand, C., Schlitt, H.J., Geissler, E.K., Fogler, W.E., and Stoeltzing, O. 2008. ENMD-1198, a novel tubulin-binding agent reduces HIF-1α and STAT3 activity in human hepatocellular carcinoma (HCC) cells, and inhibits growth and vascularization *in vivo*. *BMC Cancer* 8:206.

Mottet, D., Ruys, S.P., Demazy, C., Raes, M., and Michiels, C. 2005. Role for casein kinase 2 in the regulation of HIF-1 activity. *Int. J. Cancer* 117:764–774.

Nagle, D.G. and Zhou, Y.D. 2006a. Natural product-based inhibitors of hypoxia-inducible factor-1 (HIF-1). *Curr. Drug Targets* 7:355–369.

Nagle, D.G. and Zhou, Y.D. 2006b. Natural product-derived small molecule activators of hypoxia-inducible factor-1 (HIF-1). *Curr. Pharm. Des.* 12:2673–2688.

Nagle, D.G. and Zhou, Y.D. 2009. Marine natural products as inhibitors of hypoxic signaling in tumors. *Phytochem. Rev.* 8:415–429.

Nagle, D.G. and Zhou, Y.D. 2010. Natural products as probes of selected targets in tumor cell biology and hypoxic signaling. In *Comprehensive Natural Products Chemistry II*, Mander, L.N. ed. in chief; *Vol. 2: Structural Diversity II—Secondary Metabolite Sources, Evolution, and Selected Molecular Structures*, Moore, B.S. and Crews, P., Vol. eds., Elsevier, Oxford, U.K., Chap. 2.23, pp. 651–683.

Nakaseko, Y. and Yanagida, M. 2001. Cell biology. Cytoskeleton in the cell cycle. *Nature* 412:291–292.

Natural Products Branch 2010. *Natural Products Repository*. Developmental Therapeutics Program, National Cancer Institute, U.S. National Institutes of Health. http://dtp.nci.nih.gov//branches//npb//repository.html (accessed January 21, 2010).

Neckers, L. 2006. Using natural product inhibitors to validate Hsp90 as a molecular target in cancer. *Curr. Top. Med. Chem.* 6:1163–1171.

Newman, D.J. and Cragg, G.M. 2007. Natural products as sources of new drugs over the last 25 years. *J. Nat. Prod.* 70:461–477.

NIH Clinical Trials Database. 2010. U.S. NIH database—Clinical Trials.gov. http://www.clinicaltrials.gov//ct2//search (accessed January 22, 2010).

Oberst, A., Bender, C., and Green, D.R. 2008. Living with death: The evolution of the mitochondrial pathway of apoptosis in animals. *Cell Death Differ.* 15:1139–1146.

Oh, S.H., Woo, J.K., Ji, Q., Kang, H.J., Jeong, J.W., Kim, K.W., Hong, W.K., and Lee, H.Y. 2008. Identification of novel antiangiogenic anticancer activities of deguelin targeting hypoxia-inducible factor-1α. *Int. J. Cancer* 122:5–14.

Oh, S.H., Woo, J.K., Yazici, Y.D., Myers, J.N., Kim, W.Y., Jin, Q., Hong, S.S., Park, H.J., Suh, Y.G., Kim, K.W., Hong, W.K., and Lee, H.Y. 2007. Structural basis for depletion of heat shock protein 90 client proteins by deguelin. *J. Natl. Cancer Inst.* 99:949–961.

Osada, M., Imaoka, S., and Funae, Y. 2004. Apigenin suppresses the expression of VEGF, an important factor for angiogenesis, in endothelial cells via degradation of HIF-1α protein. *FEBS Lett.* 575:59–63.

Page, E.L., Robitaille, G.A., Pouyssegur, J., and Richard, D.E. 2002. Induction of hypoxia-inducible factor-1α by transcriptional and translational mechanisms. *J. Biol. Chem.* 277:48403–48409.

Park, S.S., Bae, I., and Lee, Y.J. 2008. Flavonoids-induced accumulation of hypoxia-inducible factor (HIF)-1α/2α is mediated through chelation of iron. *J. Cell. Biochem.* 103:1989–1998.

Park, S.Y., Jeong, K.J., Lee, J., Yoon, D.S., Choi, W.S., Kim, Y.K., Han, J.W., Kim, Y.M., Kim, B.K., and Lee, H.Y. 2007. Hypoxia enhances LPA-induced HIF-1α and VEGF expression: Their inhibition by resveratrol. *Cancer Lett.* 258:63–69.

Pellizzaro, C., Coradini, D., and Daidone, M.G. 2002. Modulation of angiogenesis-related proteins synthesis by sodium butyrate in colon cancer cell line HT29. *Carcinogenesis* 23:735–740.

Peng, J., Zhang, L., Drysdale, L., and Fong, G.H. 2000. The transcription factor EPAS-1/ hypoxia-inducible factor 2α plays an important role in vascular remodeling. *Proc. Natl. Acad. Sci. USA* 97:8386–8391.

Poon, E., Harris, A.L., and Ashcroft, M. 2009. Targeting the hypoxia-inducible factor (HIF) pathway in cancer. *Expert Rev. Mol. Med.* 11:e26, doi:10.1017/S1462399409001173.

Pore, N., Jiang, Z., Shu, H.K., Bernhard, E., Kao, G.D., and Maity, A. 2006. Akt1 activation can augment hypoxia-inducible factor-1α expression by increasing protein translation through a mammalian target of rapamycin-independent pathway. *Mol. Cancer Res.* 4:471–479.

Radreau, P., Rhodes, J.D., Mithen, R.F., Kroon, P.A., and Sanderson, J. 2009. Hypoxia-inducible factor-1 (HIF-1) pathway activation by quercetin in human lens epithelial cells. *Exp. Eye Res.* 89:995–1002.

Rapisarda, A., Hollingshead, M., Uranchimeg, B., Bonomi, C.A., Borgel, S.D., Carter, J.P., Gehrs, B., Raffeld, M., Kinders, R.J., Parchment, R., Anver, M.R., Shoemaker, R.H., and Melillo, G. 2009. Increased antitumor activity of bevacizumab in combination with hypoxia inducible factor-1 inhibition. *Mol. Cancer Ther.* 8:1867–1877.

Rapisarda, A., Uranchimeg, B., Scudiero, D.A., Selby, M., Sausville, E.A., Shoemaker, R.H., and Melillo, G. 2002. Identification of small molecule inhibitors of hypoxia-inducible factor 1 transcriptional activation pathway. *Cancer Res.* 62:4316–4324.

Rapisarda, A., Zalek, J., Hollingshead, M., Braunschweig, T., Uranchimeg, B., Bonomi, C.A., Borgel, S.D., Carter, J.P., Hewitt, S.M., Shoemaker, R.H., and Melillo, G. 2004. Schedule-dependent inhibition of hypoxia-inducible factor-1α protein accumulation, angiogenesis, and tumor growth by topotecan in U251-HRE glioblastoma xenografts. *Cancer Res.* 64:6845–6848.

Richard, D.E., Berra, E., Gothie, E., Roux, D., and Pouyssegur, J. 1999. p42/p44 mitogen-activated protein kinases phosphorylate hypoxia-inducible factor 1α (HIF-1α) and enhance the transcriptional activity of HIF-1. *J. Biol. Chem.* 274:32631–32637.

Roy, S., Balasubramanian, S., Wang, J., Chandrashekhar, Y., Charboneau, R., and Barke, R. 2003. Morphine inhibits VEGF expression in myocardial ischemia. *Surgery* 134:336–344.

Ryan, H.E., Lo, J., and Johnson, R.S. 1998. HIF-1α is required for solid tumor formation and embryonic vascularization. *EMBO J.* 17:3005–3015.

Ryan, H.E., Poloni, M., McNulty, W., Elson, D., Gassmann, M., Arbeit, J.M., and Johnson, R.S. 2000. Hypoxia-inducible factor-1α is a positive factor in solid tumor growth. *Cancer Res.* 60:4010–4015.

Salva, J. and Faulkner, D.J. 1990. Metabolites of the sponge *Strongylophora durissima* from Maricaban Island, Philippines. *J. Org. Chem.* 55:1941–1943.

Sang, N., Stiehl, D.P., Bohensky, J., Leshchinsky, I., Srinivas, V., and Caro, J. 2003. MAPK signaling up-regulates the activity of hypoxia-inducible factors by its effects on p300. *J. Biol. Chem.* 278:14013–14019.

Scortegagna, M., Ding, K., Oktay, Y., Gaur, A., Thurmond, F., Yan, L.J., Marck, B.T., Matsumoto, A.M., Shelton, J.M., Richardson, J.A., Bennett, M.J., and Garcia, J.A. 2003. Multiple organ pathology, metabolic abnormalities and impaired homeostasis of reactive oxygen species in Epas1−/− mice. *Nat. Genet.* 35:331–340.

Semenza, G.L. 2003. Targeting HIF-1 for cancer therapy. *Nat. Rev. Cancer* 3:721–732.

Semenza, G.L. 2007. Evaluation of HIF-1 inhibitors as anticancer agents. *Drug Discov. Today* 12:853–859.

Semenza, G.L. and Wang, G.L. 1992. A nuclear factor induced by hypoxia via *de novo* protein synthesis binds to the human erythropoietin gene enhancer at a site required for transcriptional activation. *Mol. Cell. Biol.* 12:5447–5454.

Singh, R.P., Gu, M., and Agarwal, R. 2008. Silibinin inhibits colorectal cancer growth by inhibiting tumor cell proliferation and angiogenesis. *Cancer Res.* 68:2043–2050.

Singh, R.P., Raina, K., Deep, G., Chan, D., and Agarwal, R. 2009. Silibinin suppresses growth of human prostate carcinoma PC-3 orthotopic xenograft via activation of extracellular signal-regulated kinase 1/2 and inhibition of signal transducers and activators of transcription signaling. *Clin. Cancer Res.* 15:613–621.

Snyder, C.M. and Chandel, N.S. 2009. Mitochondrial regulation of cell survival and death during low-oxygen conditions. *Antioxid. Redox Signal.* 11:2673–2683.

Spector, I., Shochet, N.R., Blasberger, D., and Kashman, Y. 1989. Latrunculins: Novel marine toxins that disrupt microfilament organization and affect cell growth: I. Comparison with cytochalasin D. *Cell Motil. Cytoskeleton* 13:127–144.

Symolon, H., Schmelz, E.M., Dillehay, D.L., and Merrill, A.H. Jr. 2004. Dietary soy sphingolipids suppress tumorigenesis and gene expression in 1,2-dimethylhydrazine-treated CF1 mice and $Apc^{Min/+}$ mice. *J. Nutr.* 134:1157–1161.

Talks, K.L., Turley, H., Gatter, K.C., Maxwell, P.H., Pugh, C.W., Ratcliffe, P.J., and Harris, A.L. 2000. The expression and distribution of the hypoxia-inducible factors HIF-1α and HIF-2α in normal human tissues, cancers, and tumor-associated macrophages. *Am. J. Pathol.* 157:411–421.

Tatum, J.L., Kelloff, G.J., Gillies, R.J., Arbeit, J.M., Brown, J.M, Chao, K.S., Chapman, J.D., Eckelman, W.C., Fyles, A.W., Giaccia, A.J., Hill, R.P., Koch, C.J., Krishna, M.C., Krohn, K.A., Lewis, J.S., Mason, R.P., Melillo, G., Padhani, A.R., Powis, G., Rajendran, J.G., Reba, R., Robinson, S.P., Semenza, G.L., Swartz, H.M., Vaupel, P., Yang, D., Croft, B., Hoffman, J., Liu, G., Stone, H., and Sullivan, D. 2006. Hypoxia: Importance in tumor biology, noninvasive measurement by imaging, and value of its measurement in the management of cancer therapy. *Int. J. Radiat. Biol.* 82:699–757.

Thomas, R. and Kim, M.H. 2005. Epigallocatechin gallate inhibits HIF-1α degradation in prostate cancer cells. *Biochem. Biophys. Res. Commun.* 334:543–548.

Thomas, S.L., Zhong, D., Zhou, W., Malik, S., Liotta, D., Snyder, J.P., Hamel, E., Giannakakou, P. 2008. EF24, a novel curcumin analog, disrupts the microtubule cytoskeleton and inhibits HIF-1. *Cell Cycle* 7:2409–2417.

Tian, H., Hammer, R.E., Matsumoto, A.M., Russell, D.W., and McKnight, S.L. 1998. The hypoxia-responsive transcription factor EPAS1 is essential for catecholamine homeostasis and protection against heart failure during embryonic development. *Genes Dev.* 12:3320–3324.

Tian, H., McKnight, S.L., and Russell, D.W. 1997. Endothelial PAS domain protein 1 (EPAS1), a transcription factor selectively expressed in endothelial cells. *Genes Dev.* 11:72–82.

Triantafyllou, A., Liakos, P., Tsakalof, A., Chachami, G., Paraskeva, E., Molyvdas, P.A., Georgatsou, E., Simos, G., and Bonanou, S. 2007. The flavonoid quercetin induces hypoxia-inducible factor-1α (HIF-1α) and inhibits cell proliferation by depleting intracellular iron. *Free Radic. Res.* 41:342–356.

Triantafyllou, A., Mylonis, I., Simos, G., Bonanou, S., and Tsakalof, A. 2008. Flavonoids induce HIF-1α but impair its nuclear accumulation and activity. *Free Radic. Biol. Med.* 44:657–670.

Turner, N., Li, J.Y., Gosby, A., To, S.W., Cheng, Z., Miyoshi, H., Taketo, M.M., Cooney, G.J., Kraegen, E.W., James, D.E., Hu, L.H., Li, J., and Ye, J.M. 2008. Berberine and its more biologically available derivative, dihydroberberine, inhibit mitochondrial respiratory complex I: A mechanism for the action of berberine to activate AMP-activated protein kinase and improve insulin action. *Diabetes* 57:1414–1418.

Unruh, A., Ressel, A., Mohamed, H.G., Johnson, R.S., Nadrowitz, R., Richter, E., Katschinski, D.M., and Wenger, R.H. 2003. The hypoxia-inducible factor-1α is a negative factor for tumor therapy. *Oncogene* 22:3213–3220.

van Dyke, M.M. and Dervan, P.B. 1984. Echinomycin binding sites on DNA. *Science* 225:1122–1127.

Vlaminck, B., Toffoli, S., Ghislain, B., Demazy, C., Raes, M., and Michiels, C. 2007. Dual effect of echinomycin on hypoxia-inducible factor-1 activity under normoxic and hypoxic conditions. *FEBS J.* 274:5533–5542.

Waksman, S.A. and Bugie, E. 1944. Chaetomin, a new antibiotic substance produced by *Chaetomium cochliodes*: I. Formation and properties. *J. Bacteriol.* 48:527–530.

Wang, X.H., Cavell, B.E., Syed Alwi, S.S., and Packham, G. 2009. Inhibition of hypoxia inducible factor by phenethyl isothiocyanate. *Biochem. Pharmacol.* 78:261–272.

Wang, W., Jia, W.D., Xu, G.L., Wang, Z.H., Li, J.S., Ma, J.L., Ge, Y.S., Xie, S.X., Yu, J.H. 2009. Antitumoral activity of rapamycin mediated through inhibition of HIF-1α and VEGF in hepatocellular carcinoma. *Dig. Dis. Sci.* 54:2128–2136.

Wang, G.L., Jiang, B.H., and Semenza, G.L. 1995. Effect of protein kinase and phosphatase inhibitors on expression of hypoxia-inducible factor 1. *Biochem. Biophys. Res. Commun.* 216:669–675.

Wang, H.J., Jiang, Y.Y., Wei, X.F., Huang, H., Tashiro, S., Onodera, S., and Ikejima, T. 2010. Silibinin induces protective superoxide generation in human breast cancer MCF-7 cells. *Free Radic. Res.* 44:90–100.

Wang, G.L. and Semenza, G.L. 1993. Characterization of hypoxia-inducible factor 1 and regulation of DNA binding activity by hypoxia. *J. Biol. Chem.* 268:21513–21518.

Wang, Y., Zhao, Q., Ma, S., Yang, F., Gong, Y., and Ke, C. 2007. Sirolimus inhibits human pancreatic carcinoma cell proliferation by a mechanism linked to the targeting of mTOR/ HIF-1α/VEGF signaling. *IUBMB Life* 59:717–721.

Welsh, S.J., Williams, R.R., Birmingham, A., Newman, D.J., Kirkpatrick, D.L., and Powis, G. 2003. The thioredoxin redox inhibitors 1-methylpropyl 2-imidazolyl disulfide and pleurotin inhibit hypoxia-induced factor 1α and vascular endothelial growth factor formation. *Mol. Cancer Ther.* 2:235–243.

Wu, C.Z., Cai, X.F., Dat, N.T., Hong, S.S., Han, A.R., Seo, E.K., Hwang, B.Y., Nan, J.X., Lee, D., and Lee, J.J. 2007. Bisbakuchiols A and B, novel dimeric meroterpenoids from *Psoralea corylifolia*. *Tetrahedron Lett.* 48:8861–8864.

Wu, C.Z., Hong, S.S., Cai, X.F., Dat, N.T., Nan, J.X., Hwang, B.Y., Lee, J.J., Lee, D. 2008. Hypoxia-inducible factor-1 and nuclear factor-κB inhibitory meroterpene analogues of bakuchiol, a constituent of the seeds of *Psoralea corylifolia*. *Bioorg. Med. Chem. Lett.* 18:2619–2623.

Yamazaki, Y., Hasebe, Y., Egawa, K., Nose, K., Kunimoto, S., and Ikeda, D. 2006b. Anthracyclines, small-molecule inhibitors of hypoxia-inducible factor-1α activation. *Biol. Pharm. Bull.* 29:1999–2003.

Yamazaki, Y., Someno, T., Minamiguchi, K., Kawada, M., Momose, I., Kinoshita, N., Doi, H., and Ikeda, D. 2006a. Inhibitory activity of the hypoxia-inducible factor-1 pathway by tartrolone C. *J. Antibiot. (Tokyo)* 59:693–697.

Yeo, E.J., Ryu, J.H., Cho, Y.S., Chun, Y.S., Huang, L.E., Kim, M.S., and Park, J.W. 2006. Amphotericin B blunts erythropoietin response to hypoxia by reinforcing FIH-mediated repression of HIF-1. *Blood* 107:916–923.

Yoshimura, H., Dhar, D.K., Kohno, H., Kubota, H., Fujii, T., Ueda, S., Kinugasa, S., Tachibana, M., and Nagasue, N. 2004. Prognostic impact of hypoxia-inducible factors 1α and 2α in colorectal cancer patients: Correlation with tumor angiogenesis and cyclooxygenase-2 expression. *Clin. Cancer Res.* 10:8554–8560.

Zgouras, D., Wachtershauser, A., Frings, D., and Stein, J. 2003. Butyrate impairs intestinal tumor cell-induced angiogenesis by inhibiting HIF-1α nuclear translocation. *Biochem. Biophys. Res. Commun.* 300:832–838.

Zhang, H., Qian, D.Z., Tan, Y.S., Lee, K., Gao, P., Ren, Y.R., Rey, S., Hammers, H., Chang, D., Pili, R., Dang, C.V., Liu, J.O., and Semenza, G.L. 2008. Digoxin and other cardiac glycosides inhibit HIF-1α synthesis and block tumor growth. *Proc. Natl. Acad. Sci. USA* 105:19579–19586.

Zhang, Q., Tang, X., Lu, Q.Y., Zhang, Z.F., Brown, J., and Le, A.D. 2005. Resveratrol inhibits hypoxia-induced accumulation of hypoxia-inducible factor-1α and VEGF expression in human tongue squamous cell carcinoma and hepatoma cells. *Mol. Cancer Ther.* 4:1465–1474.

Zhang, Q., Tang, X., Lu, Q., Zhang, Z., Rao, J., and Le, A.D. 2006. Green tea extract and (−)-epigallocatechin-3-gallate inhibit hypoxia- and serum-induced HIF-1α protein accumulation and VEGF expression in human cervical carcinoma and hepatoma cells. *Mol. Cancer Ther.* 5:1227–1238.

Zhong, H., Chiles, K., Feldser, D., Laughner, E., Hanrahan, C., Georgescu, M.M., Simons, J.W., and Semenza, G.L. 2000. Modulation of hypoxia-inducible factor 1α expression by the epidermal growth factor/phosphatidylinositol 3-kinase/PTEN/AKT/FRAP pathway in human prostate cancer cells: Implications for tumor angiogenesis and therapeutics. *Cancer Res.* 60:1541–1545.

Zhong, H., de Marzo, A.M., Laughner, E., Lim, M., Hilton, D.A., Zagzag, D., Buechler, P., Isaacs, W.B., Semenza, G.L., and Simons, J.W. 1999. Overexpression of hypoxia-inducible factor 1α in common human cancers and their metastases. *Cancer Res.* 59:5830–5835.

Zhou, Y.D., Kim, Y.P., Li, X.C., Baerson, S.R., Agarwal, A.K., Hodges, T.W., Ferreira, D., and Nagle, D.G. 2004. Hypoxia-inducible factor-1 activation by (−)-epicatechin gallate: Potential adverse effects of cancer chemoprevention with high-dose green tea extracts. *J. Nat. Prod.* 67:2063–2069.

Zhou, Y.D., Kim, Y.P., Mohammed, K.A., Jones, D.K., Muhammad, I., Dunbar, D.C., and Nagle, D.G. 2005. Terpenoid tetrahydroisoquinoline alkaloids emetine, klugine, and isocephaeline inhibit the activation of hypoxia-inducible factor-1 in breast tumor cells. *J. Nat. Prod.* 68:947–950.

8 Antitumor or Wound Healing Actions of Natural Products through Inhibiting or Enhancing Angiogenesis

Yoshiyuki Kimura

CONTENTS

8.1 INTRODUCTION

Cancer is the largest single cause of death in both men and women, claiming over 7 million lives each year worldwide. The surgical removal of a malignant tumor followed by radiation therapy and/or adjuvant therapy with chemotherapeutic drugs can be curative; however, the removal of certain cancers, for example, breast carcinoma, colon carcinoma, and osteogenic sarcoma, may be followed by a rapid metastasis to the lungs, liver, etc. Tumor angiogenesis involves the directional sprouting of new vessels toward a solid tumor. The stimuli that promote tumor angiogenesis may be provided directly by the tumor cells themselves or indirectly by host inflammatory cells that are attracted to the tumor site. Thus, tumor angiogenesis is important to the growth of primary as well as secondary metastatic tumors [1–3]. Tumor cells are thought to secrete angiogenic factor(s) that induce neovascularization around the tumor [4–6]. Therefore, it is necessary to develop new anticancer agents with antitumor and anti-metastatic activities but without the adverse effects, such as gastrointestinal toxicity, myelotoxicity, and immune suppression caused by cancer chemotherapeutic drugs.

Both burns and wounds initially induce a coagulative necrosis and cause a scar to form after repair. Macrophages migrate to the injured area to kill invading organisms and produce cytokines that recruit other inflammatory cells that are responsible for the diverse effects of inflammation [7,8]. Angiogenesis in the injured area is closely associated with wound healing [9]. Moreover, growth factors and cytokines are central to the healing process [10–12]. Thus, the wound-healing process is complex, involving inflammatory actions such as the migration of monocytes and production of cytokines, and growth factors and angiogenesis during re-epithelization. The antitumor and antimetastatic activities, and facilitating effect on wound-healing are closely associated with the physiological actions of vascular endothelial cell growth factor (VEGF). The inhibition of VEGF results in antitumor and antimetastatic actions, while conversely, the enhancement of VEGF results in a promotion of wound-healing. Thus, VEGF is an important key to exploring antitumor and antimetastatic substances, and substances facilitating wound-healing from medicinal plants.

8.2 ANTITUMOR AND/OR ANTIMETASTATIC ACTIONS OF VARIOUS NATURAL PRODUCTS

As shown in Table 8.1, we have already reported that various natural products inhibited tumor growth and lung metastasis by preventing angiogenesis [13–18]. Next, we will discuss the antitumor and/or antimetastatic actions of chromone, stilbenes, and anthraquinone isolated from medicinal plants, based on our own recent reports.

8.2.1 ANTITUMOR ACTIONS OF A CHROMONE DERIVATIVE, 3′-O-ACETYLHAMAUDOL*

Angelca japonica A. Gray (Hamaudo in Japanese) (Umbelliferae) is a stout perennial herb growing along the Pacific coast in the west of the Kanto region in Japan. 3′-O-acetylhamaudol (Figure 8.1) (25 or 50 mg/kg, twice daily, *po*) isolated from *A. japonica* roots inhibited tumor growth colon 26-bearing mice ($P < 0.05$) (Figure 8.2), and reduced the final tumor weight and invasion of the abdomen (Table 8.2). Body, liver, lung, spleen, and thymus weights were not significantly different among vehicle-treated colon 26-bearing control mice, and 3′-O-acetylhamaudol-treated colon 26-bearing mice. This result showed that 3′-O-acetylhamaudol has antitumor and antimetastatic actions without causing adverse reactions. Orally administered 3′-O-acetylhamaudol reduced tumor-induced angiogenesis in experiments in vivo (Figure 8.3a). Based on immunohistochemical evaluation, 3′-O-acetylhamaudol increased the number of apoptotic cells in the tumors of mice with subcutaneously

---

* From *Cancer Lett.*, 265, Kimura, Y., Sumiyoshi, M., and Baba, K., Anti-tumor actions of major component 3′-O-acetylhamaudol of *Angelica japonica* roots through dual actions, anti-angiogenesis and intestinal intraepithelial lymphocyte activation, 84–97. Copyright 2008, from Elsevier.

TABLE 8.1

Antitumor and Antimetastatic Actions of Various Natural Products

Source	Methods	Results	Time	Reference
Royal jelly	Intrasplenic Lewis lung carcinoma (LLC)-implanted C57BL mice	Royal jelly (300 and 600 mg/kg) inhibited tumor growth and liver metastasis through anti-angiogenesis	2003	[13]
Xantoangelol isolated from *Angelica keiskei*	Intrasplenic LLC-implanted C57BL mice	Xanthangelol (50 mg/kg) inhibited tumor growth and liver metastasis through anti-angiogenesis	2003	[14]
4-Hydroxyderricin isolated from *Angelica keiskei*	Intrasplenic LLC-implanted C57BL mice	4-Hyroxyderricin (50 and 100 mg/kg) inhibited tumor growth and liver metastasis through anti-angiogenesis	2004	[15]
Tuna oil	Intrasplenic LLC-implanted C57BL mice	Tuna oil (1 and 2 g/kg) inhibited liver metastasis through anti-angiogenesis	2004	[16]
Sodium pyroglutamate isolated from *Agaricus balzei*	Subcutaneously implanted LLC-implanted C57BL mice	Sodium pyroglutamate (100 and 300 mg/kg) inhibited tumor growth and lung metastasis through anti-angiogenesis and enhancement of immune function	2004	[17]
EPA and EPA derivatives	Subcutaneously implanted LLC-implanted C57BL mice	EPA and EPA derivatives (300 and 1000 mg/kg) inhibited tumor growth and lung metastasis through anti-angiogenesis	2005	[18]

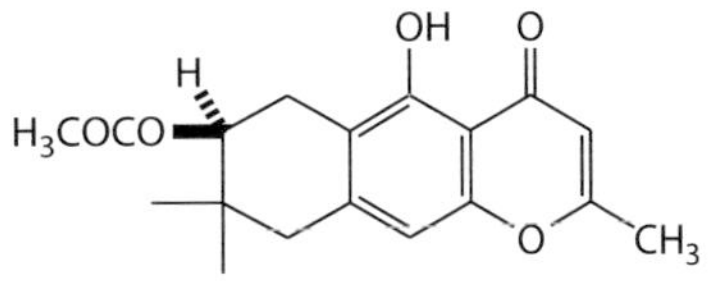

FIGURE 8.1 The structure of 3′-O-acetylhamaudol. (From *Cancer Lett.*, 265, Kimura, Y., Sumiyoshi, M., and Baba, K., Anti-tumor actions of major component 3′-O-acetylhamaudol of *Angelica japonica* roots through dual actions, anti-angiogenesis and intestinal intraepithelial lymphocyte activation, 84–97. Copyright 2008, with permission from Elsevier.)

implanted colon 26; and accordingly, reduced the expression of proliferating cell nuclear antigen (PCNA) in the tumors. The expression of CD31 (a marker of new capillaries) and hypoxia-inducible factor (HIF)-1α in the tumors was inhibited by the oral administration of 3′-O-acetylhamaudol (Tables 8.3 and 8.4). HIF-1α binds to the hypoxia response element (HRE) of target genes such as *VEGF*, and the genes for erythropoietin and glycolytic enzymes. HIF-1α bound to the HRE in the VEGF promoter acts as a major enhancer of VEGF production [20,21]. These findings suggest the antitumor and antimetastatic actions of 3′-O-acetylhamaudol to be due to the inhibition of tumor-induced angiogenesis through a reduction of VEGF production

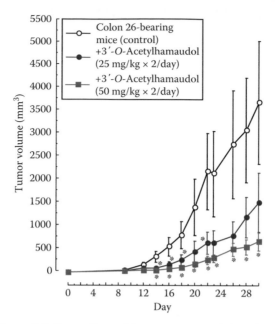

FIGURE 8.2 Effects of 3′-*O*-acetylhamudol on tumor growth in colon 26-bearing mice. Values are the mean ±SE for seven mice. *Significantly different from colon 26-bearing mice, *P* < 0.05. (From *Cancer Lett.*, 265, Kimura, Y., Sumiyoshi, M., and Baba, K., Anti-tumor actions of major component 3′-*O*-acetylhamaudol of *Angelica japonica* roots through dual actions, anti-angiogenesis and intestinal intraepithelial lymphocyte activation, 84–97. Copyright 2008, with permission from Elsevier.)

TABLE 8.2
Antitumor and Antimetastatic Actions of 3′-*O*-Hamaudol in Colon 26-Bearing Mice

	Final Tumor Weight (mg)	Tumor Invasion to Abdomen
Colon 26-bearing mice		
Control	2214.9 ± 941.3	6/7
Colon 26-bearing mice +3′-*O*-Acetylhamaudol		
25 mg/kg, twice daily	719.6 ± 261.7	3/7
50 mg/kg, twice daily	401.4 ± 124.6*	2/7

Source: *Cancer Lett.*, 265, Kimura, Y., Sumiyoshi, M., and Baba, K., Anti-tumor actions of major component 3′-*O*-acetylhamaudol of *Angelica japonica* roots through dual actions, anti-angiogenesis and intestinal intraepithelial lymphocyte activation, 84–97. Copyright (2008), with permission from Elsevier.

3′-*O*-Acetylhamaudol (25 and 50 mg/kg body weight) was administered orally twice daily for 30 days to colon 26-bearing mice. Values are the mean ± SE for seven mice.

P < 0.05, significantly different from colon 26-bearing mice (control).

via HIF-1α expression in the tumors; however, 3'-O-acetylhamaudol had no effect on VEGF production or HIF-1α expression. Therefore, its antitumor and antimetastatic actions could not be explained by the inhibition of VEGF production alone. In experiments in vitro, 3'-O-acetylhamaudol inhibited the expression of vascular endothelial growth factor receptor (VEGFR)-2 at concentrations of 50 and 100 μM, and inhibited the phosphorylation of VEGFR-2 induced by VEGF at 10–100 μM (Figure 8.3b). Furthermore, it enhanced the natural killer (NK) activity of intestinal intraepithelial lymphocytes at concentrations of 1–100 μM (Figure 8.3c). CD8+ T cell-, NK cell-, and interferon (IFN)-γ-positive cell numbers in the small intestine were reduced in colon 26-bearing mice compared with normal mice; the cell numbers were increased by the oral administration of 3'-O-acetylhamaudol (Tables 8.3 and 8.4). It is clear that interleukin (IL)-12 induces the differentiation of Th-1 cells, resulting in their production of IFN-γ, which enhances the Th-1-dominant response [22], and that

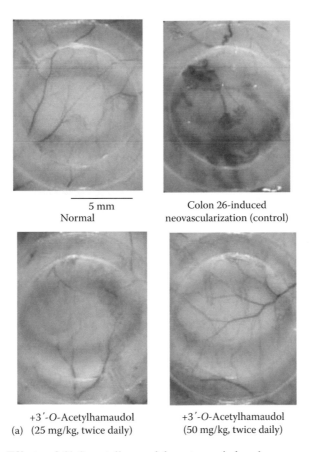

5 mm
Normal

Colon 26-induced
neovascularization (control)

+3'-O-Acetylhamaudol
(a)   (25 mg/kg, twice daily)

+3'-O-Acetylhamaudol
(50 mg/kg, twice daily)

FIGURE 8.3 Effects of 3'-O-acetylhamaudol on tumor-induced neovascularization (a), VEGFR-2 expression and VEGFR-2 phosphorylation in HUVECs (b), and NK activity in IELs (c). (a) Photographs showing inhibition of colon 26-induced neovascularization by 3'-O-acetylhamaudol in colon 26-packed chamber-bearing mice.

(continued)

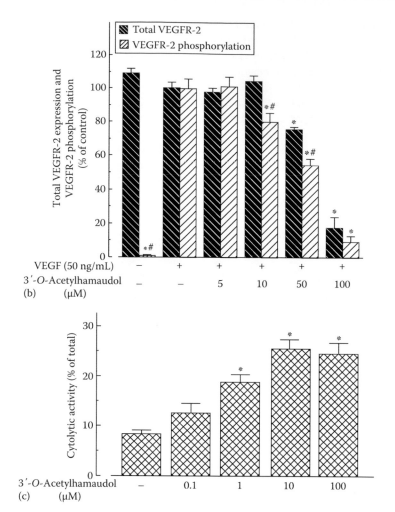

FIGURE 8.3 (continued)   (b) Effects of 3′-O-acetylhamaudol on the expression of VEGFR-2 and VEGF-induced VEGFR-2 phosphorylation in HUVECs. Values are the mean ±SE for eight experiments. *Significantly different from VEGF alone, $P < 0.05$; #Significantly different between total VEGF-2 expression and VEGFR-2 phosphorylation at the same concentration, $P < 0.05$. (c) Effects of 3′-O-acetylhamaudol on the cytotoxic activity against YAC-1 target cells of IEL. Values are the mean ±SE for four replicate experiments. *Significantly different from medium alone, $P < 0.05$. (From *Cancer Lett.*, 265, Kimura, Y., Sumiyoshi, M., and Baba, K., Anti-tumor actions of major component 3′-O-acetylhamaudol of *Angelica japonica* roots through dual actions, anti-angiogenesis and intestinal intraepithelial lymphocyte activation, 84–97. Copyright 2008, with permission from Elsevier.)

IL-12 and IFN-γ inhibit tumor growth and metastasis by activating NK cells [23,24]. This is the first report showing that 3′-O-acetylhamaudol isolated from *A. japonica* roots has antitumor and antimetastatic effects mediated through dual mechanisms, that is, anti-angiogenic actions via VEGF-induced VEGFR-2 phosphorylation and modulation of the immune system in the small intestine in tumor-bearing mice.

TABLE 8.3
Effects of 3′-O-Acetylhamaudol on Numbers of Apoptotic, PCNA-Positive and HIF-1α-Positive Cells, and CD31-Positive Areas in Tumors in Colon 26-Bearing Mice

	Positive Cell Number/Field			Positive Area/Field
	Apoptotic Cell	PCNA	HIF-1α	CD31 (μm²)
Colon 26-bearing mice				
Control	12 ± 3	576 ± 61	363 ± 55	12,010 ± 4,234
Colon 26-bearing mice +3′-O-Acetylhamaudol				
25 mg/kg, twice daily	63 ± 22	206 ± 85*	190 ± 67	921 ± 381*
50 mg/kg, twice daily	73 ± 22*	164 ± 72*	111 ± 53*	1,469 ± 667*

Source: *Cancer Lett.*, 265, Kimura, Y., Sumiyoshi, M., and Baba, K., Anti-tumor actions of major component 3′-O-acetylhamaudol of *Angelica japonica* roots through dual actions, anti-angiogenesis and intestinal intraepithelial lymphocyte activation, 84–97. Copyright (2008), from Elsevier.

3′-O-Acetylhamaudol (25 and 50 mg/kg body weight) was administered orally twice daily for 30 days to colon 26-bearing mice. Values are the mean ± SE for 7 mice.

*P < 0.05, significantly different from colon 26-bearing mice (control).

TABLE 8.4
Effects of 3′-O-Acetylhamaudol on Numbers of CD8+ T-, NK- and IFN-γ-Positive Cells in the Small Intestine of Colon 26-Bearing Mice

	Positive Cell Number/Field		
	CD8	NK	IFN-γ
Normal mice	175 ± 15*	113 ± 17*	44 ± 6*
Colon 26-bearing mice			
Control	97 ± 15	65 ± 17	29 ± 4
Colon 26-bearing mice +3′-O-Acetylhamaudol			
25 mg/kg, twice daily	251 ± 16*	189 ± 31*	44 ± 5*
50 mg/kg, twice daily	226 ± 20*	189 ± 27*	44 ± 6*

Source: *Cancer Lett.*, 265, Kimura, Y., Sumiyoshi, M., and Baba, K., Anti-tumor actions of major component 3′-O-acetylhamaudol of *Angelica japonica* roots through dual actions, anti-angiogenesis and intestinal intraepithelial lymphocyte activation, 84–97. Copyright (2008), from Elsevier.

3′-O-Acetylhamaudol (25 and 50 mg/kg body weight) was administered orally twice daily for 30 days to colon 26-bearing mice. Values are the mean ± SE for 7 mice.

*P < 0.05, significantly different from colon 26-bearing mice (control).

8.2.2 Antitumor and Antimetastatic Actions of an Anthrone-C-Glucoside, Cassialoin [25]

The antitumor effects of emodin and aloe-emodin might be due to apoptosis, reactive oxygen species (ROS)-induced DNA damage, and angiogenesis [26–36]. However, an anthrone C-glucoside, aloin, has antitumor effects by inhibiting proliferation and inducing apoptosis through cell cycle arrest in cancer cells [37–39]. The antitumor effect of aloin has yet to be demonstrated in vivo. The Thai drug "Sa mae sarn" from heartwood of *Cassia garrettiana* Craib (Leguminosae) is used as a mild cathartic in folk medicine. Hata et al. [40] isolated 10-hydroxy-10-C-D-glucosylchrysophanol-9-anthrone (cassialoin) from this plant (Figure 8.4).

Cassialoin at a dose of 10 mg/kg twice daily significantly reduced final tumor weight on day 25 compared to the tumor weight in vehicle-treated colon 26-bearing control mice, and furthermore, the number of mice with an abdominal invasion of tumors was reduced by orally administered cassialoin (2.5, 5, and 10 mg/kg, twice daily) (Table 8.5). As shown in Figure 8.5, cassialoin (5 or 10 mg/kg, twice daily)

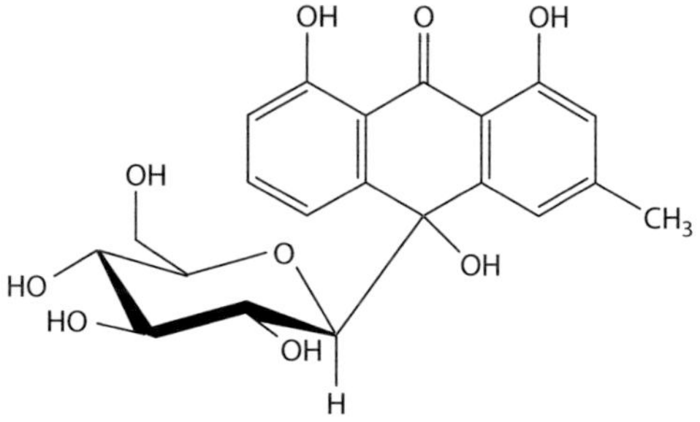

FIGURE 8.4 The structure of cassialoin. (From Kimura, Y. et al., *Cancer Sci.*, 99, 2336, 2008. Japanese Cancer Association, Blackwell-Wiley. With permission.)

TABLE 8.5
Antitumor and Antimetastatic Actions of Cassialoin in Colon 26-Bearing Mice

	Tumor Weight (mg)	Tumor Invasion to Abdomen
Colon 26-bearing mice		
Control (*n* = 16)	1619.9 ± 398.6	11/16
Colon 26-bearing mice + Cassialoin		
2.5 mg/kg, twice daily (*n* = 7)	1774.0 ± 682.1	3/7
5 mg/kg, twice daily (*n* = 7)	1315.6 ± 423.5	3/7
10 mg/kg, twice daily (*n* = 15)	364.3 ± 115.0*	2/15

Source: Kimuara, Y. et al., *Cancer Sci.*, 99, 2336, 2008.

Cassialoin (2.5, 5 and 10 mg/kg body weight) was administered orally twice daily for 25 days to colon 26-bearing mice. Values are the mean ± SE for 7–16 mice.

*$P < 0.05$, significantly different from colon 26-bearing mice (control).

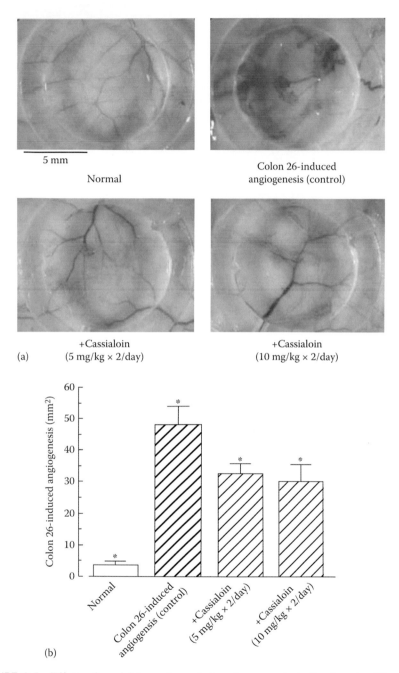

FIGURE 8.5 Effects of cassialoin on tumor-induced neovascularization in colon 26-packed chamber-bearing mice. (a) Photographs showing inhibition of colon 26-induced neovascularization by cassialoin in colon 26-packed chamber-bearing mice. (b) Values are the mean ±SE for five mice. *Significantly different from colon 26-bearing mice, $P < 0.05$. (From Kimura, Y. et al., *Cancer Sci.*, 99, 2336, 2008. Japanese Cancer Association, Blackwell-Wiley. With permission.)

significantly reduced the area of neovascularization induced by colon 26 cells compared to the control (in vivo). However, it had no effect on the formation of capillary-like tubes by human umbilical vein endothelial cells (HUVECs) in Matrigel containing VEGF (in vitro). Therefore, metabolites of cassialoin in mice may be closely associated with antitumor effects through the anti-angiogenic action. When cassialoin was administered orally to mice, it, and two metabolites were detected in the blood by HPLC; however, chrysophanol-9-anthrone, the aglycone moiety of cassialoin, was not. These two metabolites were identified as chrysophanol and aloe-emodin (Figure 8.6). Meanwhile, cassialoin and the two metabolites were detected in the stomach and small intestine, but aloe-emodin was not. The two metabolites were identified as chrysophanol-9-anthrone and chrysophanol. These findings suggest that cassialoin is metabolized to chrysophanol through chrysophanol-9-anthrone, and further, chrysophanol is metabolized to aloe-emodin by hydroxylation (Figure 8.6).

Next, we examined the mechanisms behind the antitumor and antimetastatic actions of cassialoin and its metabolites (chrysophanol, chrysophanol-9-anthrone, and aloe-emodin) in experiments in vitro. Chrysophanol-9-antherone most inhibited the VEGF-induced angiogenesis and HUVEC proliferation at concentrations of 0.5–10 μM (Figure 8.7a and b). Furthermore, it delayed the repair of HUVEC wounds in endothelial basal medium containing VEGF at concentrations of 0.5–10 μM (Figure 8.7c). It has been reported that matrix metalloproteinases (MMPs) stimulate cancer cell growth, migration, invasion, angiogenesis, and metastasis [41]. When MMPs are secreted from tumor cells, the invasion of tumor cells is facilitated by increased intravasation and extravasation through degradation of the extracellular matrix and basement membranes [42–45]. The MMP proteolytic system may also modulate tumor angiogenesis by modulating the release of a biologically active VEGF [46,47]. VEGF is one such secreted angiogenic factor and induces angiogenesis by binding to its two receptor tyrosine kinases, KDR/Flk-1 (VEGFR-2) and Flt-1 (VEGFR-1), expressed

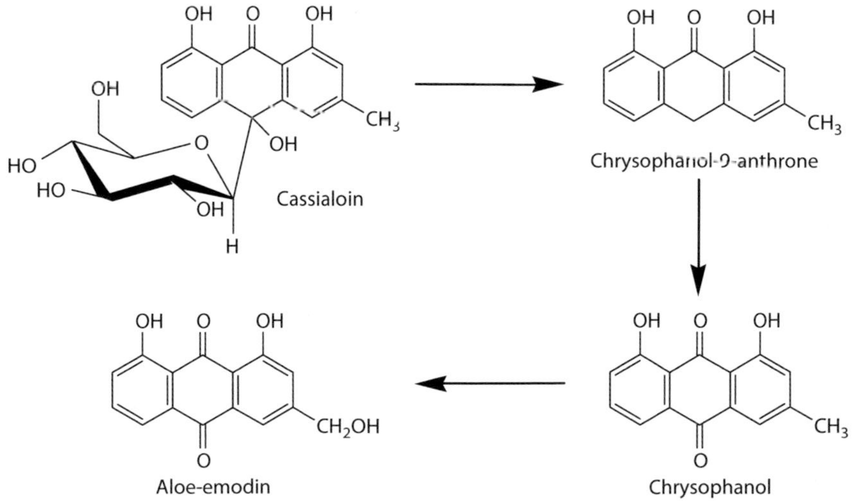

FIGURE 8.6 Metabolism of cassialoin. (From Kimura, Y. et al., *Cancer Sci.*, 99, 2336, 2008. Japanese Cancer Association, Blackwell-Wiley. With permission.)

on endothelial cells [48–51]. Cassialoin, chrysophanol, and aloe-emodin had no effect on VEGF production under hypoxic conditions (1% O_2, 5% CO_2, and 94% N_2 atmosphere) at the concentrations of 1–100 μM. Chrysophanol-9-anthrone inhibited VEGF production under hypoxic conditions at a concentration of 10 μM. Furthermore, chyrysophanol-9-anthrone inhibited MMP-9 production in colon 26 cells (Figure 8.8).

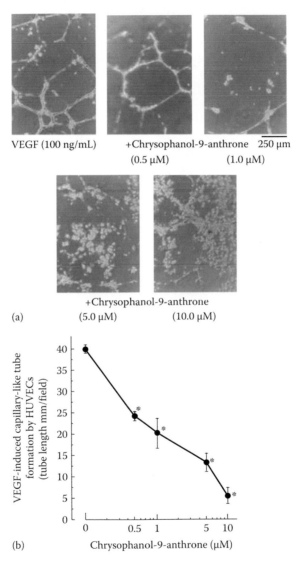

FIGURE 8.7   Effects of chrysophanol-9-anthrone on Matrigel-induced angiogenesis in HUVEC (a, b), and HUVEC migration (c). (a) Light micrographs of capillary-like tubes forming in Matrigel containing VEGF by HUVECs in the presence of chrysophanol-9-anthrone. (b) Effects of chrysophanol-9-anthrone on angiogenesis induced by Matrigel containing VEGF in HUVECs. Values are the mean ±SE for four replicates. *Significantly different from VEGF alone, $P < 0.05$.

(*continued*)

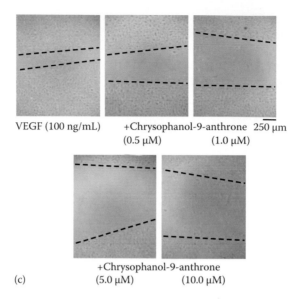

FIGURE 8.7 (continued) (c) Light micrographs of wounded HUVECs at 30 h incubation in the presence of chrysophanol-9-anthrone and VEGF. (From Kimura, Y. et al., *Cancer Sci.*, 99, 2336, 2008. Japanese Cancer Association, Blackwell-Wiley. With permission.)

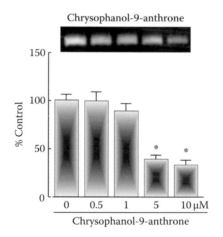

FIGURE 8.8 Effects of chrysophanol-9-anthrone on MMP-9 production in colon 26 cells. Values are the mean ±SE for four replicates. *Significantly different from medium alone, $P < 0.05$. (From Kimura, Y. et al., *Cancer Sci.*, 99, 2336, 2008. Japanese Cancer Association, Blackwell-Wiley. With permission.)

Cassialoin and chrysophanol had no effect on VEGFR-2 expression or VEGF-induced VEGFR-2 phosphorylation at 100 μM, and aloe-emodin slightly inhibited VEGFR-2 expression, and VEGF-induced VEGFR-2 phosphorylation at a concentration of 1000 μM (Figure 8.9). Chrysophanol-9-anthrone inhibited VEGFR-2 expression and VEGF-induced VEGFR-2 phosphorylation at 5 and 10 μM (Figure 8.9). Furthermore, cassialoin and aloe-emodin enhanced the Con A-induced production of IFN-γ at the concentrations

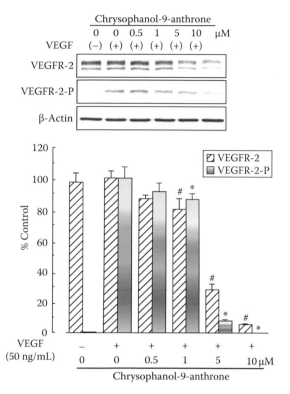

FIGURE 8.9 Effects of chrysophanol-9-anthrone on VEGFR-2 expression and VEGF-induced VEGFR-2 phosphorylation in HUVECs. Values are the mean ±SE for four replicates. #Significantly different from medium alone, $P < 0.05$. *Significantly different from VEGF stimulation, $P < 0.05$. (From Kimura, Y. et al., *Cancer Sci.*, 99, 2336, 2008. Japanese Cancer Association, Blackwell-Wiley. With permission.)

of 0.1–1 μM in splenocytes isolated from colon 26-bearing mice, but chrysophanol had no effect. Chrysophanol-9-anthrone enhanced Con A-induced IFN-γ production at 0.1 μM (Figure 8.10). None of these agents had an effect on Con A-induced IFN-γ production in splenocytes isolated from normal mice. These findings suggest the antitumor and anti-metastatic actions of orally administered cassialoin to be due to anti-angiogenesis and the inhibition of endothelial migration through the inhibition of VEGFR-2 expression and VEGFR-2 phosphorylation induced by VEGF and the inhibition of MMP-9 expression and hypoxia-induced VEGF production in tumors by the cassialoin metabolite chrysophanol-9-anthrone. Furthermore, it seems likely that another mechanism of action is the stimulation of immune function through an increase in IFN-γ expression by cassialoin and its metabolites, such as chrysophanol-9-anthrone and aloe-emodin in the spleen.

8.2.3 ANTITUMOR ACTIONS OF VARIOUS NATURAL AND SYNTHETIC STILBENES [52]

There have been a number of reports that resveratrol and resveratrol-related stilbenes inhibit tumor growth as a chemopreventive agent [53–60]. We attempted to examine

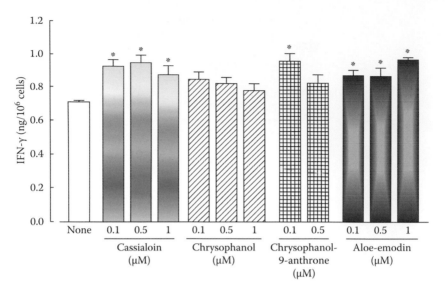

FIGURE 8.10 Effects of cassialoin, chrysophanol, chrysophanol-9-anthrone, and aloe-emodin on concanavalin A (Con A)-induced IFN-γ production in splenocytes of colon 26-bearing mice. Values are the mean ±SE for four replicates. *Significantly different from Con A stimulation, $P < 0.05$. (From Kimura, Y. et al., *Cancer Sci.*, 99, 2336, 2008. Japanese Cancer Association, Blackwell-Wiley. With permission.)

the inhibitory effects of 21 synthetic and/or natural stilbenes (Figure 8.11) on tumor MMP-9 and VEGF production as a first screening assay. We found that three hydroxystilbenes (2,3-, 3,4-, and 4,4′-dihydroxystilbenes) inhibited MMP-9 production from colon 26 cells, being most effective at concentrations of 1, 5, 10, and 25 μM (Figure 8.12). Trihydorxystilbene resveratrol and its triacetate, and tetrahydroxystilbene piceatannol, its tetraacetate, and the dimer cassiagrol A reduced MMP-9 production at 10 or 25 μM. Moreover, 2,3-dihydroxy stilbene (8 mg/kg, intraperitoneally injection) significantly inhibited tumor growth on days 9–24 and final tumor weight compared to the control, vehicle-treated colon 26-bearing mice (Figure 8.13). 3,4-Dihydroxystilbene (3 and 8 mg/kg) significantly inhibited tumor growth on days 18–24 (Figure 8.13). 4,4′-Dihydroxystilbene (3 and 8 mg/kg) significantly inhibited tumor growth on days 9–24 (Figure 8.13). To clarify the antitumor actions of these dihydroxystilbenes, we examined the effects on tumor-induced neovascularization in tumor-packed chamber-bearing mice (in vivo). The 2,3-, 3,4-, and 4,4′-dihydroxystilbenes (8 mg/kg) significantly reduced the areas of neovascularization induced by colon 26 cells compared to the control (Figure 8.14). In experiments in vitro, the 2,3- and 3,4-dihydroxystilbenes at 5, 10, 25, and 50 μM prevented capillary-like tubes forming in Matrigel containing VEGF by umbilical vein endothelial cells (HUVECs). 4,4′-Dihydroxystilbene significantly inhibited the VEGF-induced formation of capillary-like tubes at concentrations of 1, 2.5, 5, 10, 25, and 50 μM (Figure 8.15). None of the three dihydroxystilbenes had an effect on the production of VEGF under hypoxic (1% O_2, 5% CO_2, and 94% N_2 atmosphere) or normoxic conditions. Although the dihydroxystilbenes did not influence VEGFR-1 and -2 expression, or VEGF-induced VEGFR-1 phosphorylation,

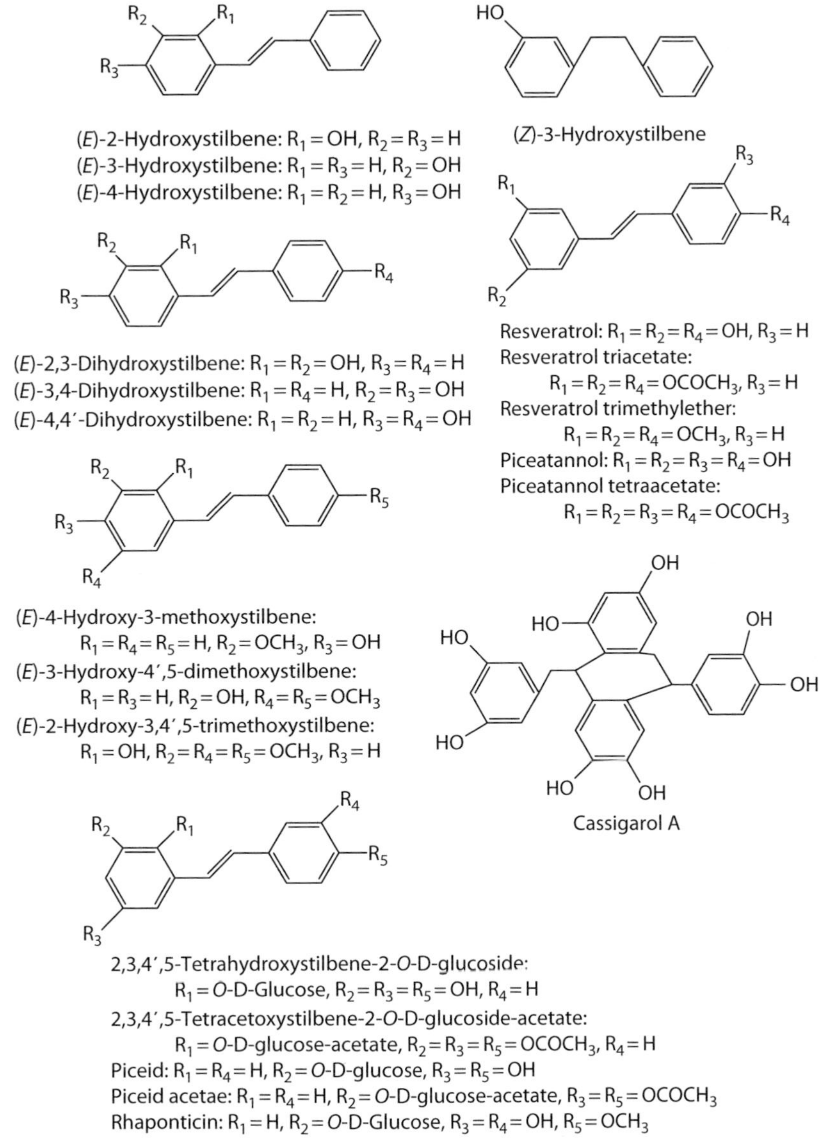

(E)-2-Hydroxystilbene: $R_1 = OH$, $R_2 = R_3 = H$
(E)-3-Hydroxystilbene: $R_1 = R_3 = H$, $R_2 = OH$
(E)-4-Hydroxystilbene: $R_1 = R_2 = H$, $R_3 = OH$

(Z)-3-Hydroxystilbene

(E)-2,3-Dihydroxystilbene: $R_1 = R_2 = OH$, $R_3 = R_4 = H$
(E)-3,4-Dihydroxystilbene: $R_1 = R_4 = H$, $R_2 = R_3 = OH$
(E)-4,4′-Dihydroxystilbene: $R_1 = R_2 = H$, $R_3 = R_4 = OH$

Resveratrol: $R_1 = R_2 = R_4 = OH$, $R_3 = H$
Resveratrol triacetate:
    $R_1 = R_2 = R_4 = OCOCH_3$, $R_3 = H$
Resveratrol trimethylether:
    $R_1 = R_2 = R_4 = OCH_3$, $R_3 = H$
Piceatannol: $R_1 = R_2 = R_3 = R_4 = OH$
Piceatannol tetraacetate:
    $R_1 = R_2 = R_3 = R_4 = OCOCH_3$

(E)-4-Hydroxy-3-methoxystilbene:
    $R_1 = R_4 = R_5 = H$, $R_2 = OCH_3$, $R_3 = OH$
(E)-3-Hydroxy-4′,5-dimethoxystilbene:
    $R_1 = R_3 = H$, $R_2 = OH$, $R_4 = R_5 = OCH_3$
(E)-2-Hydroxy-3,4′,5-trimethoxystilbene:
    $R_1 = OH$, $R_2 = R_4 = R_5 = OCH_3$, $R_3 = H$

Cassigarol A

2,3,4′,5-Tetrahydroxystilbene-2-O-D-glucoside:
    $R_1 = O$-D-Glucose, $R_2 = R_3 = R_5 = OH$, $R_4 = H$
2,3,4′,5-Tetracetoxystilbene-2-O-D-glucoside-acetate:
    $R_1 = O$-D-glucose-acetate, $R_2 = R_3 = R_5 = OCOCH_3$, $R_4 = H$
Piceid: $R_1 = R_4 = H$, $R_2 = O$-D-glucose, $R_3 = R_5 = OH$
Piceid acetae: $R_1 = R_4 = H$, $R_2 = O$-D-glucose-acetate, $R_3 = R_5 = OCOCH_3$
Rhaponticin: $R_1 = H$, $R_2 = O$-D-Glucose, $R_3 = R_4 = OH$, $R_5 = OCH_3$

FIGURE 8.11 The structure of various stilbenes. (From Kimura, Y. et al., *Cancer Sci.*, 99, 2083, 2008. Japanese Cancer Association, Blackwell-Wiley. With permission.)

they inhibited the VEGF-induced phosphorylation of VEGFR-2 at concentrations of 5–25 μM. Among the three, 4,4′-dihydroxystilbene inhibited VEGF-induced VEGFR-2 phosphorylation at the lowest concentration, 1 μM (Figure 8.16). These findings suggest that the antitumor actions of these dihydroxystilbenes (2,3-, 3,4-, and 4,4′-dihydroxystilbenes) are closely associated with their inhibition of VEGF-induced angiogenesis, through the inhibition of VEGF-induced VEGFR-2 phosphorylation in endothelial cells and MMP-9 production in colon 26 cells.

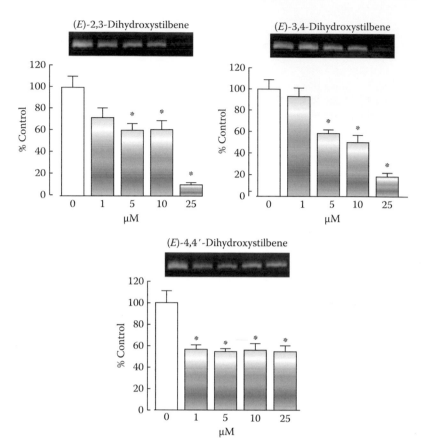

FIGURE 8.12 Effects of 2,3-, 3,4-, and 4,4′-dihydroxystilbenes on MMP-9 expression in colon 26 cells. Values are the mean ±SE for four replicates. *Significantly different from medium alone, $P < 0.05$. (From Kimura, Y. et al., *Cancer Sci.*, 99, 2083, 2008. Japanese Cancer Association, Blackwell-Wiley. With permission.)

8.3   WOUND-HEALING EFFECTS OF NATURAL PRODUCTS

Burns induce inflammatory and immune dysfunction [61–64]. In particular, the acute response to a burn results in a coordinated influx of leukocytes, such as polymorphonuclear leukocytes and macrophages, to the wound [65]. IL-1β is released from monocyte-derived macrophages during inflammation. VEGF plays a crucial role in tissue repair, as angiogenesis and increased vascular permeability are important during wound healing [66,67], and keratinocyte-derived VEGF might stimulate angiogenesis during wound healing. The inflammatory cytokine IL-1β stimulates VEGF expression in various cells including endothelial cells [68], keratinocytes [69], synovial fibroblasts [70], and colorectal carcinoma cells [71]. Thus, VEGF-induced angiogenesis in the damaged area is closely associated with the process of wound healing. Therefore, it is necessary to develop the new substances that promote angiogenesis for the treatment of skin disorders, such as bedsores and burns. Next, we will discuss the facilitating effects of ginsenoside Rb₁ and asiaticoside isolated from medicinal plants on the healing of burns.

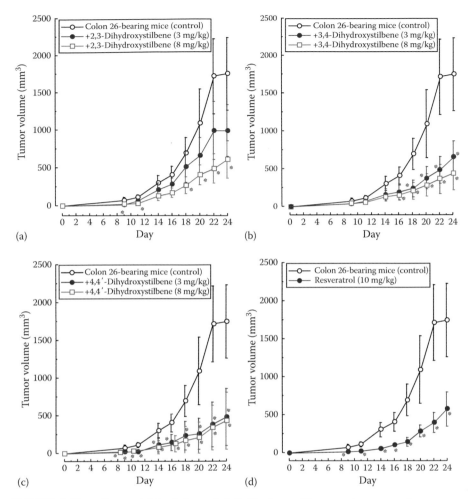

FIGURE 8.13 Effects of 2,3-, 3,4-, and 4,4′-dihydroxystilbene, and resveratrol on tumor growth in colon 26-bearing mice. Values are the mean ±SE for seven mice. *Significantly different from vehicle-treated colon 26-bearing mice, $P < 0.05$. (From Kimura, Y. et al., *Cancer Sci.*, 99, 2083, 2008. Japanese Cancer Association, Blackwell-Wiley. With permission.)

8.3.1 FACILITATING EFFECT OF GINSENOSIDE Rzb₁ ON THE HEALING OF BURNS [72]

The genus Panax, from the Greek words *pan* (all) and *akos* (healing), was named by the botanist Carl Meyer. Kanzaki et al. [73] reported that orally administered red ginseng roots stimulated the repair of intractable skin ulcers in patients with diabetes mellitus and Werner's syndrome in clinical trials, and that the local administration of ginseng saponins markedly improved wound healing in diabetic or aging rats [74]. Among the six ginsenosides tested (ginsenoside Rb_1, Rb_2, Rc, Rd, Re, and Rg_1) (Figure 8.17), ginsenoside Rb_1 enhanced wound healing the most. Then, we examined the effects of lower doses ($10^{-14}\%-10^{-8}\%$ ointment, w/w) of ginsenoside

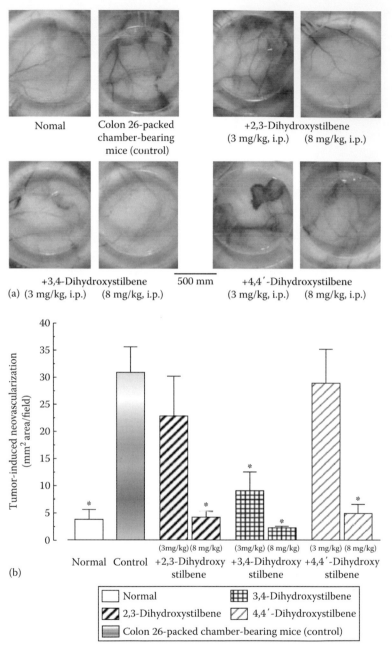

FIGURE 8.14 Effects of 2,3-, 3,4-, and 4,4′-dihydroxystilbenes on tumor-induced neovascularization in colon 26-bearing mice. (a) Photographs showing inhibition of tumor-induced neovascularization by 2,3-, 3,4-, and 4,4′-dihydoxystilbenes in colon 26-packed chamber-bearing mice. (b) Values are the mean ±SE for six mice. *Significantly different from vehicle-treated colon 26-packed chamber-bearing mice, $P < 0.05$. (From Kimura, Y. et al., *Cancer Sci.*, 99, 2083, 2008. Japanese Cancer Association, Blackwell-Wiley. With permission.)

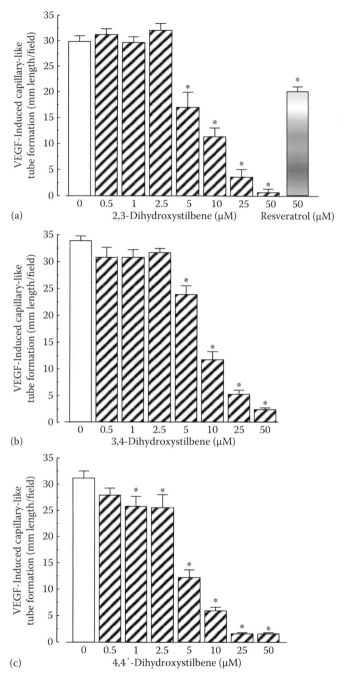

FIGURE 8.15   Effects of 2,3-, 3,4-, and 4,4′-dihydroxystilbenes on the formation capillary-like tubes by HUVECs in Matrigel containg VEGF. Values are the mean ±SE for four replicates. *Significantly different from VEGF alone, $P < 0.05$. (From Kimura, Y. et al., *Cancer Sci.*, 99, 2083, 2008. Japanese Cancer Association, Blackwell-Wiley. With permission.)

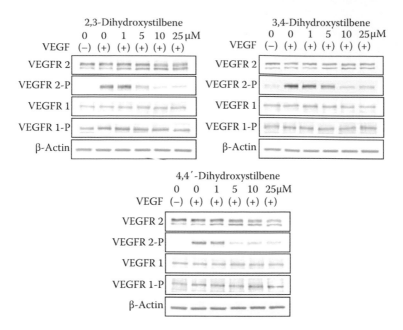

FIGURE 8.16 Effects of 2,3-, 3,4-, and 4,4′-dihydroxystilbenes on the expression of VEGFR-1 and -2, and on VEGF-induced VEGFR-1 and -2 phosphorylation in HUVECs. (From Kimura, Y. et al., *Cancer Sci.*, 99, 2083, 2008. Japanese Cancer Association, Blackwell-Wiley. With permission.)

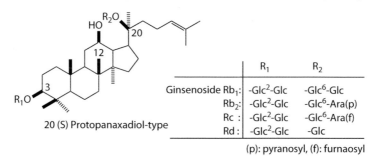

FIGURE 8.17 The structure of various ginsenosides. (From Kimura, Y., Sumiyoshi, M., Kawahira, K., and Sakanaka, M.: Effects of ginseng saponins isolated from red ginseng roots on burn wound healing in mice. *Br. J. Pharmacol.* 2006. 148. 860–870. Copyright Wiley-VCH Verlag GmbH & Co. KGaA. With permission.)

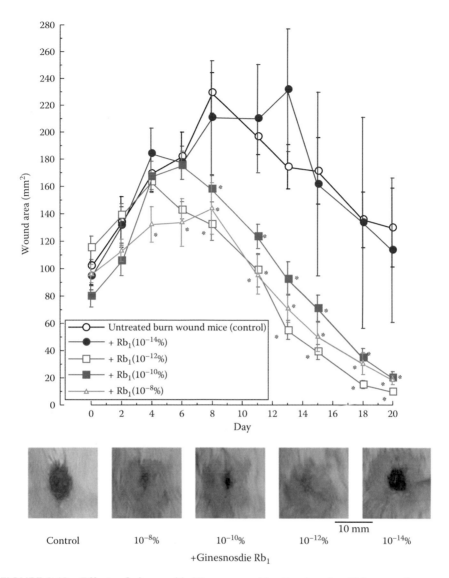

FIGURE 8.18 Effects of ginsenoside Rb$_1$ on wound healing in mice. Values are the mean ±SE for 6–12 mice. *Significantly different from vehicle-treated mice, $P < 0.05$. (From Kimura, Y., Sumiyoshi, M., Kawahira, K., and Sakanaka, M. Effects of ginseng saponins isolated from red ginseng roots on burn wound healing in mice. *Br. J. Pharmacol.* 2006. 148. 860–870. Copyright Wiley-VCH Verlag GmbH & Co. KGaA. With permission.)

Rb$_1$. The burn area in mice treated with a topical application of ginsenoside Rb$_1$ in the range of $10^{-8}\%–10^{-12}\%$ was significantly reduced on days 8–20 compared to that in vehicle-treated burn-wound control mice (Figure 8.18). To clarify the mechanism behind the facilitating effect of ginsenoside Rb$_1$ on wound healing, we examined levels of IL-1β and VEGF in exudates of the burn. The levels increased with time over 9 days. At 1 ng of ginsenoside Rb$_1$ per wounds, the level of IL-1β was increased

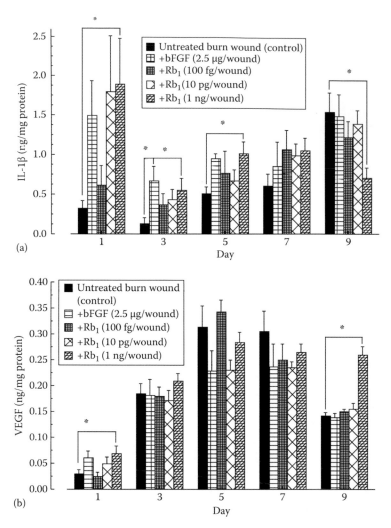

FIGURE 8.19 Effects of ginsenoside Rb₁ and bFGF on IL-b (a) and VEGF (b) production in the exudates of burns in mice. Values are the mean ±SE for six mice. *Significantly different from control, $P < 0.05$. (From Kimura, Y., Sumiyoshi, M., Kawahira, K., and Sakanaka, M. Effects of ginseng saponins isolated from red ginseng roots on burn wound healing in mice. *Br. J. Pharmacol.* 2006. 148. 860–870. Copyright Wiley-VCH Verlag GmbH & Co. KGaA. With permission.)

on days 1, 3, and 5 but significantly decreased on day 9 compared to that in vehicle-treated control mice (Figure 8.19). The topical application of bFGF (2.5 μg/wound) also increased IL-1β production on day 3. The VEGF level in the exudates from the wound increased until day 5, and then decreased. The application of ginsenoside Rb₁ increased VEGF levels on days 1 and 9 (Figure 8.19). However, that of bFGF did not affect VEGF production. The application of bFGF (2.5 mg/wound) or ginsenoside Rb₁ (100 fg, 10 pg, and 1 ng/wound) for 9 days increased the length of blood vessels by 3- to 3.5-fold and the corresponding area by 3.5- to 5.0-fold,

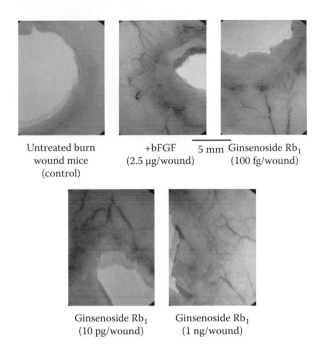

Untreated burn wound mice (control)

+bFGF (2.5 µg/wound)

5 mm Ginsenoside Rb$_1$ (100 fg/wound)

Ginsenoside Rb$_1$ (10 pg/wound)

Ginsenoside Rb$_1$ (1 ng/wound)

FIGURE 8.20 Photographs showing neovascularization from the tissue surrounding the burn and the effects of the topical application of ginsenoside Rb$_1$. (From Kimura, Y., Sumiyoshi, M., Kawahira, K., and Sakanaka, M. Effects of ginseng saponins isolated from red ginseng roots on burn wound healing in mice. *Br. J. Pharmacol.* 2006. 148. 860–870. Copyright Wiley-VCH Verlag GmbH & Co. KGaA. With permission.)

TABLE 8.6
Effect of Ginsenoside Rb$_1$ on Neovascularization from the Surrounding Burn Wound Area in Mice

	Blood Vessel Length (mm/Field)	Blood Vessel Area (mm²/Field)
Untreated burn wound		
Control	75.57 ± 24.93	10.49 ± 3.80
Burn wound + Ginsenoside Rb$_1$		
100 fg/wound	228.80 ± 38.60*	46.56 ± 15.01*
10 pg/wound	203.00 ± 17.01*	37.673 ± 5.51*
1 ng/wound	274.08 ± 37.39*	49.94 ± 4.70*
Burn wound + Basic FGF		
2.5 µg/wound	241.46 ± 28.30*	35.80 ± 5.86*

Source: Kimura, Y., Sumiyoshi, M., Kawahira, K., and Sakanaka, M.: Effects of ginseng saponins isolated from red ginseng roots on burn wound healing in mice. *Br. J. Pharmacol.* 2006. 148. 860–870. Copyright Wiley-VCH Verlag GmbH & Co. KGaA.

Values are the mean ± SE for six mice.

*Significantly different from untreated burn wound (control), $P < 0.05$.

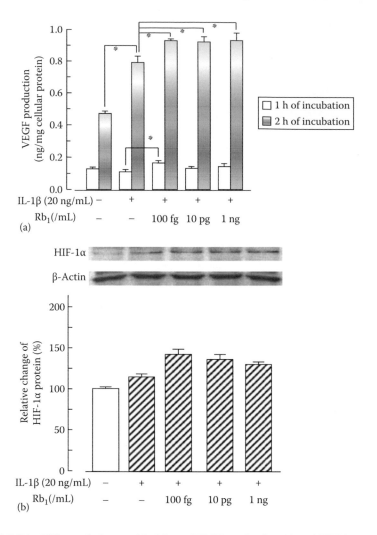

FIGURE 8.21 Effects of ginsenoside Rb₁ on VEGF production (a) and HIF-lα expression (b) with or without IL-1β in HaCaT cells. Values are the mean ±SE for six experiments. *Significantly different from medium alone, $P < 0.05$. (From Kimura, Y., Sumiyoshi, M., Kawahira, K., and Sakanaka, M. Effects of ginseng saponins isolated from red ginseng roots on burn wound healing in mice. *Br. J. Pharmacol.* 2006. 148. 860–870. Copyright Wiley-VCH Verlag GmbH & Co. KGaA. With permission.)

compared to the control (Figure 8.20 and Table 8.6). Ginsenoside Rb₁ at concentrations from 100 fg/mL to 1 ng/mL enhanced VEGF production and HIF-lα expression induced by IL-1β in the human keratinocyte cell line HaCaT (Figure 8.21). These findings suggest the enhancement of wound healing by ginsenoside Rb₁ to be due to the promotion of angiogenesis during the repair process as a result of the stimulation of VEGF production caused by the increase in HIF-1α expression in keratinocytes.

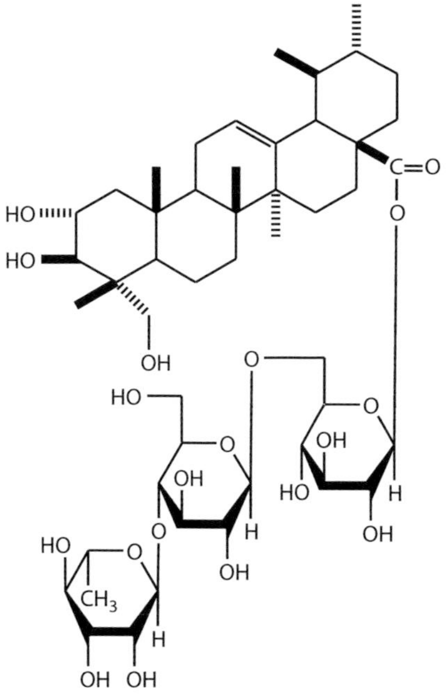

FIGURE 8.22 The structure of asiaticoside. (From *Eur. J. Pharmacol.*, 584, Kimura, Y., Sumiyoshi, M., Samukawa, K., Satake, N., and Sakanaka, M., Facilitating action of asiaticoside at low doses on burn wound repair and its mechanism, 415–423. Copyright 2008, with permission from Elsevier.)

8.3.2 FACILITATING EFFECT OF ASIATICOSIDE ON THE HEALING OF BURNS

The herb *Centella asiatica* L. (Umbelliferae) is widely cultivated as a vegetable or spice in China, Southeast Asia, India, Sri Lanka, Africa, and Oceanic countries. *C. asiatica* has also been used for the treatment of skin diseases, syphilis, rheumatism, mental illness, epilepsy, hysteria, dehydration, and leprosy in Sri Lankan and Indian Ayurvedic traditional medicine. Some reports described that asiaticoside in ointment form at doses of 0.1%–0.2% (w/w) enhanced wound repair [75–79]. In preliminary experiments, we found that the burn area was significantly reduced on days 6–8 by the topical application of asiaticoside (Figure 8.22) at 10^{-8}%–10^{-12}% (w/w) compared to that in vehicle-treated control mice (Figure 8.23). Furthermore, asiaticoside (100 ng/wound) significantly increased the IL-1β level on days 1, 3, 5, and 9, and at 1 ng/wound, also increased it on day 9 (Figure 8.24a). The topical application of asiaticoside (10 pg, 1 ng, and 100 ng/wound) increased the level of monocyte chemoattractant protein-1 (MCP-1) compared to that in vehicle-treated control mice (Figure 8.24b). VEGF levels in the exudates of the wound were also increased on days 1, 5, and 9 by the application of asiaticoside at a dose of 1 or 100 ng/wound (Figure 8.24c). It has been reported that MCP-1 is expressed at high levels in murine full-thickness dermal wounds both preceding and coinciding with maximal macrophage infiltration [80,81]. Furthermore, Low et al. [82] reported that MCP-1$^{-/-}$ mice displayed significantly delayed wound re-epithelialization, and that wound angiogenesis was also delayed, with a 48% reduction

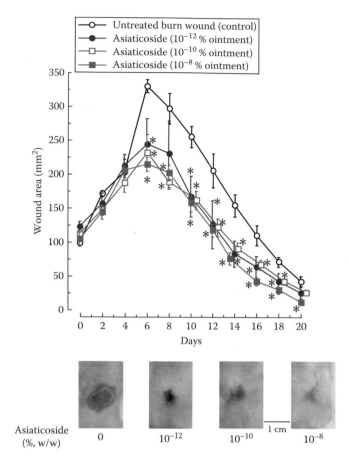

FIGURE 8.23 Effects of asiaticoside on wound healing in mice. Values are the mean ±SE for 7–11 mice. *Significantly different from control, $P < 0.05$. (From *Eur. J. Pharmacol.*, 584, Kimura, Y., Sumiyoshi, M., Samukawa, K., Satake, N., and Sakanaka, M., Facilitating action of asiaticoside at low doses on burn wound repair and its mechanism, 415–423. Copyright 2008, with permission from Elsevier.)

in capillary density at day 5 after injury. Asiaticoside increased MCP-1 production at concentrations of 10 pg, 1 ng, and 100 ng/mL in the human keratinocyte HaCaT cell line, but had no effect on VEGF production (Table 8.7). From these results, the facilitating effect of asiaticoside on wound healing cannot be explained by the direct stimulation of VEGF production in keratinocytes. IL-1β is known to be produced by the human monocyte cell line THP-1 upon stimulation by lipopolysaccharide (LPS) [83]. Asiaticoside did not enhance IL-1β production without LPS in THP-1 macrophages; however, it (10 pg, 1 ng, and 100 ng/mL) stimulated IL-1β production in the presence of MCP-1 in cultured THP-1 macrophages (Figure 8.25). Therefore, these findings suggest the enhancement of wound healing by asiaticoside to be due to the promotion of angiogenesis during the repair process as a result of the stimulation of VEGF production caused by the increase in IL-1β expression in recruited macrophages and/or through an increase in MCP-1 expression in keratinocytes and/or macrophages.

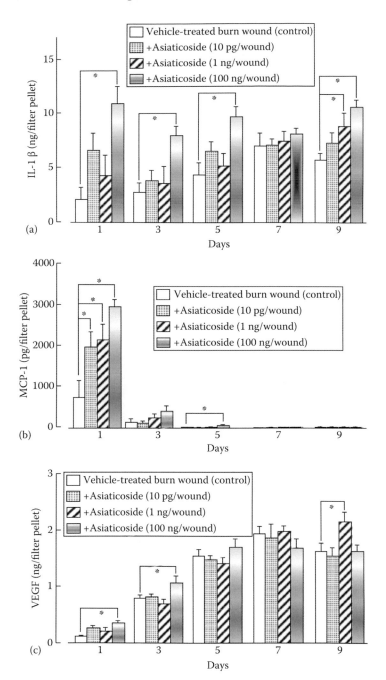

FIGURE 8.24 Effects of asiaticoside on the levels of IL-1β (a), MCP-1 (b), and VEGF (c) in the exudates of burn wound-treated mice. Values are the mean ±SE for seven mice. *Significantly different from control, $P < 0.05$. (From *Eur. J. Pharmacol.*, 584, Kimura, Y., Sumiyoshi, M., Samukawa, K., Satake, N., and Sakanaka, M., Facilitating action of asiaticoside at low doses on burn wound repair and its mechanism, 415–423. Copyright 2008, with permission from Elsevier.)

TABLE 8.7

Effects of Asiaticoside on VEGF and MCP-1 Production in the Cultured Human Keratinocyte Cell Line HaCaT (In Vitro)

	VEGF (pg/Well)	MCP-1 (ng/Well)
Medium alone	88.3 ± 8.67	2.47 ± 0.24
Medium +Asiaticoside		
10 pg/mL	99.2 ± 8.15	3.69 ± 0.35*
1 ng/mL	98.3 ± 6.25	3.84 ± 0.43*
100 ng/mL	96.4 ± 4.51	3.50 ± 0.23*

Source: *Eur. J. Pharmacol.*, 584, Kimura, Y., Sumiyoshi, M., Samukawa, K., Satake, N., and Sakanaka, M., Facilitating action of asiaticoside at low doses on burn wound repair and its mechanism, 415–423. Copyright (2008), with permission from Elsevier.

Values are the mean ± SE for six experiments.

*Significantly different from medium alone, $P < 0.05$.

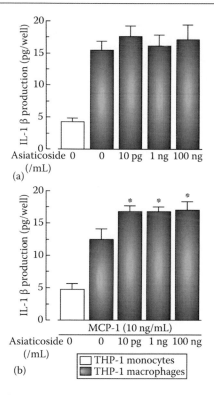

FIGURE 8.25 Effects of asiaticoside on IL-1β production in cultured THP-1 macrophages in the absence (a) or presence (b) of MCP-1. Values are the mean ±SE for four experiments. *Significantly different from THP-1 macrophages (medium alone), $P < 0.05$. (From *Eur. J. Pharmacol.*, 584, Kimura, Y., Sumiyoshi, M., Samukawa, K., Satake, N., and Sakanaka, M., Facilitating action of asiaticoside at low doses on burn wound repair and its mechanism, 415–423. Copyright 2008, with permission from Elsevier.)

8.4 CONCLUSION

The angiogenic factor VEGF is closely associated with tumor growth and tumor metastasis, but also plays an important role in the regeneration of damaged skin. Thus, VEGF and related factors closely implicated in the development of cancer, atherosclerosis, inflammation, etc., are important for the treatment of skin wounds. Among natural products, chromone, anthraquinone, chalcone, and stilbene derivatives regulated the tumor angiogenesis caused by angiogenic and related factors; conversely, saponins such as ginsenoside Rb_1 and asiaticoside enhanced the wound healing process by increasing the levels of these factors. Further study will be needed to clarify the clinical significance of these findings for the development of new agents as lead compounds.

ACKNOWLEDGMENTS

The author is grateful to the following co-workers: Prof. M. Sakanaka and Miss M. Sumiyoshi (Division of Functional Histology, Department of Functional Biomedicine, Ehime University Graduate School of Medicine), and Prof. K. Baba (Department of Pharmacognosy, Osaka University of Pharmaceutical Sciences).

REFERENCES

1. Folkman J. Angiogenesis. *Annu Rev Med* 2006; **57**: 1–18.
2. Folkman J. Antiangiogenesis in cancer therapy-endostatin and its mechanisms of action. *Exp Cell Res* 2006; **312**: 594–607.
3. Folkman J. Is angiogenesis an organizing principle in biology and medicine? *J Pediatr Surg* 2007; **42**: 1–11.
4. Folkman J, Klagsbrum M. Angiogenic factors. *Science* 1987; **235**: 442–447.
5. Klagrbrum M, D'Amore PA. Regulators of angiogenesis. *Annu Rev Physiol* 1991; **53**: 217–239.
6. Folkman J, Shing Y. Angiogenesis. *J Biol Chem* 1992; **267**: 10931–10934.
7. O'Riordain MG, Collins KH, Pilz M, Saporoschetz IB, Mannick IA, Rodricj MI. Modulation macrophage hyperactivity improves survival in a burn-sepsis model. *Arch Surg* 1992; **12**: 152–158.
8. Kataranovski M, Magic Z, Pejnovic N. Early inflammatory cytokine and acute phase protein response under the stress of thermal injury in rats. *Physiol Res* 1999; **48**: 473–482.
9. Atavilla D, Saitta A, Cucinotta D, Galeano M, Deodato B, Colonna M, Torre V et al. Inhibition of lipid peroxidation restores impaired vascular endothelial growth factor expression and stimulates wound healing and angiogenesis in the genetically diabetes mouse. *Diabetes* 2001; **50**: 667–674.
10. Brown LF, Yeo KT, Berse B, Yeo TK, Senger DR, Dvorak HF, van de Water L. Expression of vascular permeability factor (vascular endothelial growth factor) by epidermal keratinocytes during wound healing. *J Exp Med* 1992; **176**: 1375–1379.
11. Martin P. Wound healing: Aiming for perfect skin regeneration. *Science* 1997; **276**: 75–81.
12. Sen CK, Khanna S, Babior BM, Hunt TK, Ellison EC, Roy S. Oxidant-induced vascular endothelial growth factor expression in human keratinocytes and cutaneous wound healing. *J Biol Chem* 2002; **277**: 33284–33290.

13. Kimura Y, Takaku T, Okuda H. Antitumor and antimetastatic actions by royal jelly in Lewis lung carcinoma-bearing mice. *J Tradit Med* 2003; **20**: 195–200.

14. Kimura Y, Baba K. Antitumor and antimetastatic activities of *Angelica keiskei* roots, part 1: Isolation of an active substance, xanthangelol. *Int J Cancer* 2003; **106**: 429–437.

15. Kimura Y, Taniguchi M, Baba K. Antitumor and antimetastatic activities of 4-hydroxy-derricin isolated from *Angelica keiskei* roots. *Planta Med* 2004; **70**: 211–219.

16. Maeda Y, Sumiyoshi M, Kimura Y. Effects of tuna oil on tumor growth and metastasis to liver in intrasplenic Lewis lung carcinoma (LLC)-implanted mice. *J Trad Med* 2004; **21**: 215–220.

17. Kimura Y, Kido T, Takaku T, Sumiyoshi M, Baba K. Isolation of an anti-angiogenic substance from *Agaricus blazei* Murill: Its antitumor and antimetastatic actions. *Cancer Sci* 2004; **95**: 758–764.

18. Kimura Y, Sumiyoshi M. Antitumor and antimetastatic actions of eicosapentacnoic acid and ethylester and its by-products from during accelerated stability testing. *Cancer Sci* 2005; **96**: 441–450.

19. Kimura Y, Sumiyoshi M, Baba K. Anti-tumor actions of major component 3'-O-acetylhamaudol of *Angelica japonica* roots through dual actions, anti-angiogenesis and intestinal intraepithelial lymphocyte activation. *Cancer Lett* 2008; **265**: 84–97.

20. Semenza GL. Regulation of mammalian O2 homeostasis by hypoxia-inducible factor 1. *Annu Rev Cell Dev Biol* 1999; **15**: 551–578.

21. Tsuzuki Y, Fukumura D, Oosthuyse B, Koike C, Carmeliet P, Jain RK. Vascular endo-thelial growth factor (VEGF) modulation by targeting hypoxia-inducible factor-1α → Hypoxia response element → VEGF cascade differentially regulates vascular response and growth rate in tumors. *Cancer Res* 2000; **60**: 6248–6252.

22. Trinchieri G. Proinflammatory and immunoregulatory function of interleukin-12. *Int Rev Immunol* 1998; **16**: 365–396.

23. Markoviv SN, Murasko DM. Role of natural killer and T-cells in interferon induced inhibition of spontaneous metastases of the B16F10L murine melanoma. *Cancer Res* 1991; **51**: 1124–1128.

24. Lasek W, Feleszko W, Golab J, Stoklosa T, Marczak M, Dabrowska A, Malejczyk M, Jakobisiak M. Antitumor effects of the combination immunotherapy with interleukin-12 and tumor necrosis factor a in mice. *Cancer Immunol Immunother* 1997; **45**: 100–108.

25. Kimura Y, Sumiyoshi M, Taniguchi M, Baba K. Antitumor and antimetastatic actions of anthrone-C-glucoside, cassialoin isolated from *Cassia garrettiana* heartwood in colon 26-bearing mice. *Cancer Sci* 2008; **99**: 2336–2348.

26. Lee H-Z. Protein kinase C involvement in aloe-emodin- and emodin-induced apoptosis in lung carcinoma. *Br J Pharmacol* 2001; **134**: 1093–1103.

27. Lee H-Z, Hsu S-L, Liu M-C, Wu C-H. Effects and mechanisms of aloe-emodin on cell death in human lung squamous cell carcinoma. *Eur J Pharmacol* 2001; **431**: 287–295.

28. Kuo P-L, Lin T-C, Lin C-C. The antiproliferative activity of aloe-emodin is through p53-dependent and p21-dependent apoptotic pathway in human hepatoma cell lines. *Life Sci* 2002; **71**: 1879–1892.

29. Pecere T, Sarinella F, Salata C, Gatto B, Bet A, Dalla F, Vecchia FD et al. Involvement of p53 in specific anti-neuroectodermal tumor activity of aloe-emodin. *Int J Cancer* 2003; **106**: 836–847.

30. Ljubimov AV, Caballero S, Aoki AM, Pinna LA, Grant MB, Castellon R. Involvement of protein kinase CK2 in angiogenesis and retinal neovascularization. *Invest Ophthalmol Vis Sci* 2004; **45**: 4583–4591.

31. Lian L-H, Park F-J, Piao H-S, Zhao Y-Z, Sohn DH. Aloe-emodin-induced apoptosis in t-HSC/Cl-6 cells involves a mitochondria-mediated pathway. *Basic Clin Pharmacol Toxicol* 2005; **96**: 495–502.

32. Lee H-Z, Lin C-J, Yang W-H, Leung W-C, Chang S-P. Aloe-emodin induced DNA damage through generation of reactive oxygen species in human lung carcinoma cells. *Cancer Lett* 2006; **239**: 55–63.

33. Kwak H-J, Park M-J, Park C-M, Monn S-I, Yoo D-H, Lee H-C, Lee S-H et al. Emodin inhibits vascular endothelial growth factor-A-induced angiogenesis by blocking receptor-2 (KDR/Flk-1) phosphorylation. *Int J Cancer* 2006; **118**: 2711–2720.

34. Kaneshiro T, Morioka T, Inamine M, Kinjo T, Arakaki J, Chiba I, Sunagawa N, Suzui M, Yoshimi N. Anthraquinone derivative emodin inhibits tumor-associated angiogenesis through inhibition of extracellular signal-regulated kinase 1/2 phosphorylation. *Eur J Pharmacol* 2006; **553**: 46–53.

35. Lai M-Y, Hour M-J, Leung HWC, Yang W-H, Lee H-Z. Chaperones are the target in aloe-emodin-induced human lung nonsmall carcinoma H460 cell apoptosis. *Eur J Pharmacol* 2007; **573**: 1–10.

36. Lu GD, Shem H-M, Chung MCM, Ong CN. Critical role of oxidative stress and sustained JNK activation in aloe-emodin-mediated apoptotic cell death in human hepatoma cells. *Carcinogenesis* 2007; **28**: 1937–1945.

37. Esmat AY, Tomasetto C, Rio M-C. Cytotoxicity of a natural anthraquinone (aloin) against human breast cancer cell lines with and without ErbB-2-topoisomerase IIa coamplification. *Cancer Biol Ther* 2006; **5**: 97–103.

38. Nićiforovic A, Adžić M, Spasić SD, Radojćić MB. Antitumor effects of a natural anthracycline analog (aloin) involve altered activity of antioxidant enzymes in HeLaS3 cells. *Cancer Biol Ther* 2007; **6**: 1–6.

39. Buenz EJ. Aloin induces apoptosis in Jurkat cells. *Toxicol In Vitro* 2008; **22**: 422–429.

40. Hata K, Baba K, Kozawa M. Chemical studies on the heartwood of *Cassia garrettiana* Craib. I. Anthraquinones including cassialoin a new anthrone C-glycoside. *Chem Pharm Bull* 1978; **26**: 3792–3797.

41. Egeblad M, Werb Z. New functions for the matrix metalloproteinases in cancer progression. *Nat Rev Cancer* 2002; **2**: 161–174.

42. Lotta LA, Tryggvason K, Garbisa S, Hart I, Foltz CM, Shafie S. Metastatic potential correlates with enzymatic degradation of basement membrane collagen. *Nature* 1980; **284**: 67–68.

43. Macdougall JR, Matrisian LM. Contributions of tumor and stromal matrix metalloproteinases to tumor progression, invasion and metastasis. *Cancer Metastasis Rev* 1995; **14**: 351–362.

44. Di DH, Lu JS, Song CG, Li HC, Shen ZZ, Shao ZM. Over expression of aromatase protein in highly related to MMPs levels in human breast carcinomas. *J Exp Clin Cancer Res* 2005; **24**: 601–607.

45. Kim SH, Cho NH, Kim K, Lee JS, Koo BS, Kim JH, Chang JH, Choi EC. Correlations of oral tongue cancer invasion with matrix metalloproteinases (MMPs) and vascular endothelial growth factor (VEGF) expression. *J Surg Oncol* 2006; **93**: 330–337.

46. Bergers G, Brekken R, McMahon G, Vu TH, Itoh T, Tamaki K, Tanzawa K et al. Matrix metalloproteinase-9 triggers the angiogenic switch during carcinogenesis. *Nature Cell Biol* 2000; **2**: 737–744.

47. Kaliski A, Maggiorella L, Cengel KA, Mathe D, Rouffiac V, Opolon P, Lassau N, Bourhis J, Deutsch E. Angiogenesis and tumor growth inhibition by a matrix metalloproteinase inhibitor targeting radiation-induced invasion. *Mol Cancer Ther* 2005; **4**: 1717–1728.

48. Leung DW, Cachianes G, Kuang WJ, Goefel DV, Ferrara N. Vascular endothelial growth factor is a secreted angiogenic mitogen. *Science* 1989; **246**: 1306–1309.

49. Bernatchez PN, Soker S, Sirois MG. Vascular endothelia growth effect on endothelial cell proliferation, migration, and platelet-activating factor synthesis is Flk-1-dependent. *J Biol Chem* 1999; **274**: 31047–31054.

50. Davis-Smyth T, Chen H, Park J, Presta LG, Ferrara N. The second immunoglobulin-like domain of the VEGF tyrosine kinase receptor initiate a signal transduction cascade. *EMBO J* 1996; **15**: 4919–4927.

51. McMahon G. VEGF receptor signaling in tumor angiogenesis. *Oncologist* 2000; **5**: 3–10.

52. Kimura Y, Sumiyoshi M, Baba K. Antitumor activities of synthetic and natural stilbenes through antiangiogenic action. *Cancer Sci* 2008; **99**: 2083–2096.

53. Jang M, Cai L, Udeani GO, Slowing KV, Thomas CF, Beecher CWW, Fong HHS et al. Cancer chemopreventive activity of resveratrol, a natural product derived from grapes. *Science* 1997; **275**: 218–220.

54. Sun NJ, Woo SH, Cassady JM, Snapka RM. DNA polymerase and topoisomerase inhibitors from *Psoralea corylifolia*. *J Nat Prod* 1998; **61**: 362–366.

55. Clement M-V, Hirpara JL, Chawdhury S-H, Pervaiz S. Chemopreventive agent resveratrol, a natural product derived from grapes, triggers CD95 signaling-dependent apoptosis in human tumor cell. *Blood* 1998; **92**: 996–1002.

56. Sale S, Tunstall RG, Ruparelia KC, Potter GA, Steward WP, Gescher AJ. Comparison of the effects of the chemopreventive agent resveratrol and its synthetic analog trans-3,4,5,4′-tetramethoxystilbene (DMU-212) on adenoma development in the Apc (Min+) mouse and cyclooxygenase-2 in human-derived colon cancer cell. *Int J Cancer* 2005; **115**: 194–201.

57. Skinnider L, Stoessl A. The effect of the phytoalexins, lubimin, (−)-maackiain, pinosylvin, and the related compounds dehydroloroglossol and hordatine M on human lymphoblastoid cell lines. *Experientia* 1986; **42**: 568–570.

58. Lee SK, Nam KA, Hoe YH, Min HY, Kim EY, Ko H, Song S, Lee T, Kim S. Synthesis and evaluation of cytotoxicity of stilbene analogues. *Arch Pharm Res* 2003; **26**: 253–257.

59. Kim YH, Park C, Lee JO, Kim GY, Lee WH, Choi YH, Ryu CH. Induction of apoptosis by piceatannol in human leukemic U937 cells through down-regulation of Bcl-2 and activation of caspases. *Oncol Rep* 2008; **19**: 961–967.

60. Kuo PL, Hsu YL. The grape and wine constituent piceatannol inhibits proliferation of human bladder cancer cells via blocking cell cycle progression and inducing Fas/membrane bound Fas ligand-mediated apoptotic pathway. *Mol Nutr Food Res* 2008; **52**: 408–418.

61. Alecander W, Moncrief JA. Alterations of the immune response following severe thermal injury. *Arch Surg* 1966; **93**: 750–783.

62. Faunce DE, Gregory MS, Kovacs EJ. Effects of acute ethanol exposure on cellular immune responses in a murine model of thermal injury. *J Leukoc Biol* 1997; **62**: 733–740.

63. Ramzy PI, Barret JP, Herndon DN. Thermal injury. *Crit Care Clin* 1999; **15**: 333–352.

64. Leder JA, Rao LS, Freedberg IM, Simon M, Milisavljevic X, Blumenberg M. Interleukin-1 induces transcription of keratin K6 in human epidermal keratinocytes. *J Invest Dermatol* 2001; **116**: 330–338.

65. Engelhardt E, Toksoy A, Goebeler M, Debus S, Brocker EB, Gillitzer R. Chemokines IL-8, GRO alfa, MCP-1, Ip-10, and Mig are sequentially and differentially expressed during phase-specific infiltration of leukocyte subsets in human wound healing. *Am J Pathol* 1998; **153**: 1849–1860.

66. Brown LF, Yeo KT, Berse B, Yeo TK, Senger DR, Dvorak HF, van de Water L. Expression of vascular permeability factor (vascular endothelial growth factor) by epidermal keratinocytes during wound healing. *J Exp Med* 1992; **176**: 1375–1379.

67. Nissen NN, Polyverini PJ, Koch AE, Volin MV, Gamelli RL, DiPietro LA. Vascular endothelial growth factor mediates angiogenic activity during the proliferative phase of wound healing. *Am J Pathol* 1998; **152**: 1445–1452.

68. Ristimäki A, Narko K, Enholm B, Joukov V, Alitalo K. Proinflammatory cytokines regulate expression of the lymphatic endothelial mitogen vascular endothelial growth factor-C. *J Biol Chem* 1998; **273**: 8413–8418.

69. Frank S, Stallmeyer B, Kämpfer H, Kolb N, Pfeilschifter J. Nitric oxide triggers enhanced induction of vascular endothelial growth factor expression in cultured keratinocytes (HaCaT) and during cutaneous wound repair. *FASEB J* 1999; **13**: 2002–2014.

70. Ben-Av P, Crofford LJ, Wilder RL, Hla T. Induction of vascular endothelial growth factor expression in synovial fibroblasts by prostaglandin E and interleukin-1: A potential mechanism for inflammatory angiogenesis. *FEBS Lett* 1995; **372**: 83–87.

71. Konishi N, Miki C, Yoshida T, Tanaka K, Toiyama Y, Kushunoki M. Interleukin-1 receptor antagonist inhibits the expression of vascular endothelial growth factor in colorectal carcinoma. *Oncology* 2005; **68**: 138–145.

72. Kimura Y, Sumiyoshi M, Kawahira K, Sakanaka M. Effects of ginseng saponins isolated from red ginseng roots on burn wound healing in mice. *Br J Pharmacol* 2006; **148**: 860–870.

73. Kanzaki T, Morisaki N, Siina R, Saito Y. Role of transforming growth factor-b pathway in the mechanism of wound healing by saponin from Ginseng Radix rubra. *Br J Pharmacol* 1998; **125**: 255–262.

74. Morisaki N, Watanabe S, Tezuka M, Zenibayashi M, Shiina R, Koyama N, Kanzaki T, Saito Y. Mechanism of angiogenic effects of saponin from Ginseng Radix rubra in human umbilical vein endothelial cells. *Br J Pharmacol* 1995; **115**: 1188–1193.

75. Kimura Y, Sumiyoshi M, Samukawa K, Satake N, Sakanaka M. Facilitating action of asiaticoside at low doses on burn wound repair and its mechanism. *Eur J Pharmacol* 2008; **584**: 415–423.

76. Maquart FX, Bellon G, Gillery P, Wegrowski Y, Borel JP. Stimulation of collagen synthesis in fibroblast cultures by a triterpene extracted from *Centella asiatica*. *Connect Tissue Res* 1990; **24**: 107–120.

77. Maquart FX, Chastang F, Simeon A, Birembaut P, Gillery P, Wegroswski Y. Triterpenes from *Centella asiatica* stimulate extracellular matrix accumulation in rat experimental wounds. *Eur J Dermatol* 1999; **9**: 289–296.

78. Shukla A, Rasik AM, Jain GK, Shankar R, Kulshrestha DK, Dhawan BN. In vitro and in vivo wound healing activity of asiaticoside isolated from *Centella asiatica*. *J Ethnopharmacol* 1999; **65**: 1–11.

79. Macay D, Miller AL. Nutritional support for wound healing. *Altern Med Rev* 2003; **8**: 359–377.

80. DiPietro LA, Reintjes MG, Low QE, Levi B, Gamelli RL. Modulation of macrophage recruitment into wounds by monocyte chemoattractant protein-1. *Wound Repair Regen* 2001; **9**: 28–33.

81. Heinrich SA, Messingham KAN, Gergory MS, Colantoni A, Ferreira AM, DiPietro LA, Kovacs EJ. Elevated monocyte chemoattractant protein-1 levels following thermal injury precede monocyte recruitment to the wound site and are controlled, in part, by tumor necrosis factor-α. *Wound Repair Regen* 2003; **11**: 110–119.

82. Low QEH, Drugea IA, Duffner LA, Quinn DG, Cook DN, Rollins BJ, Kovacs EJ, DiPetro LA. Wound healing in MIP-1α^{-/-} and MCP-1^{-/-} mice. *Am J Pathol* 2001; **159**: 457–463.

83. Martin M, Katz J, Vogel SN, Michalek SM. Differential induction of endotoxin tolerance by lipopolysaccharides derived from *Porphyromonas gingivalis* and *Escherichia coli*. *J Immunol* 2001; **167**: 5278–5285.

9 From Natural Polyphenols to Synthetic Antitumor Agents

Carmela Spatafora and Corrado Tringali

CONTENTS

9.1 INTRODUCTION

An interesting article by Quideau (2006), also reported on the Web site of the "Groupe Polyphenols" (http://www.groupepolyphenols.com/), discusses in detail the origin and the present meaning of the term "polyphenols," at least according to the authors' opinion. About 50 years ago, this term was normally employed as an equivalent to "vegetable tannins" to indicate those vegetable substances able to convert animal skin into leather (tanning action). Later, Haslam (1998), who examined polyphenols under different points of view, proposed a definition of plant polyphenols, including specific structural characteristics common to all phenolics having a tanning property; however, this definition has subsequently been broadened in the common use to include also low-molecular-weight phenolic molecules, not necessarily water-soluble or exerting a "tanning" action. Consequently, the common feature of polyphenols has been reconfigured with regard to their biosynthetic origin, thus including phenolic metabolites biosynthetically derived through the shikimate and/or the acetate/malonate pathways. Indeed, metabolic modification of some shikimate/acetate intermediates can lead to "polyphenols" paradoxically lacking the phenolic functions (such as some lignans originated by dimerization of a phenylpropanoid unit C_6C_3, Ayers et al. 1990), or without "free" hydroxyl groups (such as some polymethoxyflavones, Walle 2007), although their carbon framework is strictly related or identical to that of the phenolic analogue. In addition, many naturally occurring derivatives such as polyphenolic glycosides, esters, and ethers are normally included in research studies on polyphenols. On the other hand, in recent times, the study of polyphenols has involved an increasing number of scientists, probably as a result of the widespread awareness that many phenolic substances possess properties of biomedical importance, potentially useful both in prevention and therapy of major degenerative diseases, such as cancer and cardiovascular/neurological pathologies (Boudet 2007). Especially worth noting is the presence of phenolic compounds in edible fruits, herbs, vegetables, as well as in foods and beverages derived from them. Thus, the present popularity of polyphenolic substances has undoubtedly broadened the original meaning of the term "polyphenols," today employed for aromatic metabolites of the shikimate/acetate biosynthetic pathways and for their derivatives or analogues, irrespective of their solubility, tanning action, molecular weight, and number of phenolic groups. The increasing interest toward polyphenols is clearly proved by Figure 9.1a, showing the number of citations for the keywords "natural polyphenols" on Scifinder Scholar in the period 1979–2009.

Among the variety of healthy properties attributed to natural polyphenols, one of the most frequently cited is their cancer chemopreventive activity; this property was originally defined as the use of natural, synthetic, or biological chemical agents to reverse, suppress, or prevent either the initial phase of carcinogenesis or the progression of neoplastic cells to cancer (Sporn et al. 1976). The hypothesis that regular assumption of dietary polyphenols may help to prevent cancer is supported by a multitude of epidemiological, in vitro and in vivo studies (among others, Surh 2003; Aggarwal and Shishodia 2006a; Nichenametla et al. 2006; Ramos 2008),

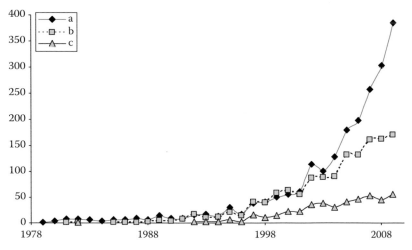

FIGURE 9.1 Number of references retrieved through Scifinder Scholar in the period 1979–2009 using the terms: (a) "natural polyphenols," (b) "antitumor polyphenols," and (c) "antitumor polyphenol analogues."

corroborated by some clinical trials (Le Marchand et al. 2000; Thomasset et al. 2006). The studies oriented to the antitumor properties of polyphenols observed in recent years are far from dwindling, as confirmed by Figure 9.1b, showing the number of citations for the keywords "antitumor polyphenols" on Scifinder Scholar in the period 1979–2009. Also noteworthy here is that a considerable number of studies are concentrated on a limited number of low-molecular-weight dietary polyphenols, for instance, the stilbenoid resveratrol (**1**) found in grape and red wine, the green tea catechins, and in particular epigallocatechin-3-gallate (EGCG, **2**), the hop chalcone xanthohumol (**3**), the soya isoflavone genistein (**4**), the yellow pigment curcumin (**5**) found in turmeric (curry), ellagic acid (**6**), a component of pomegranate, the widespread caffeic acid (**7**) and its natural derivatives chlorogenic acid (**8**), and CAPE (**9**), the latter found in honeybee propolis.

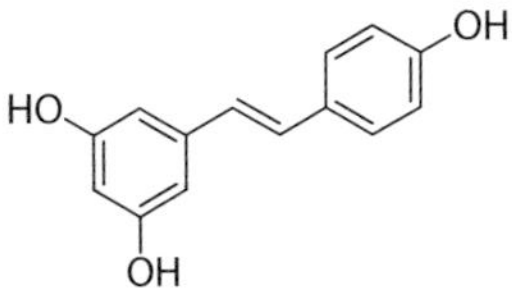

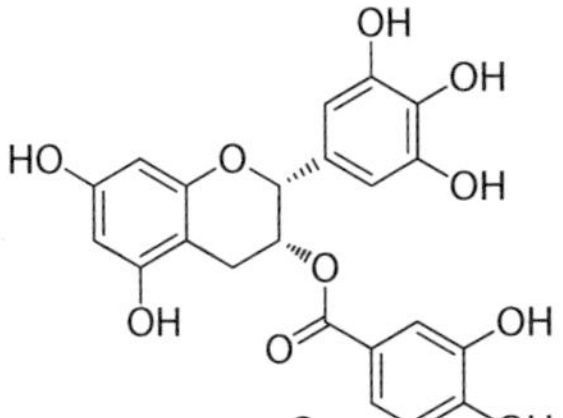

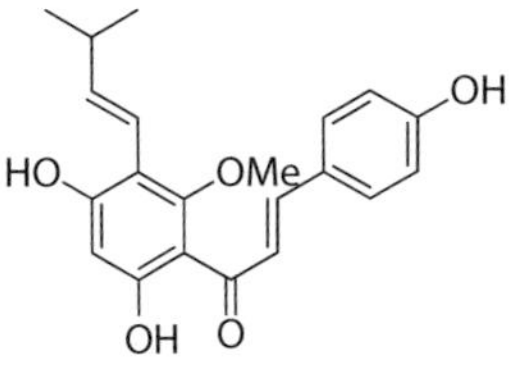

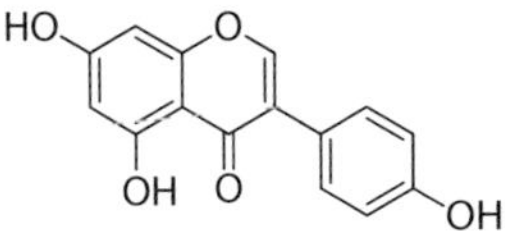

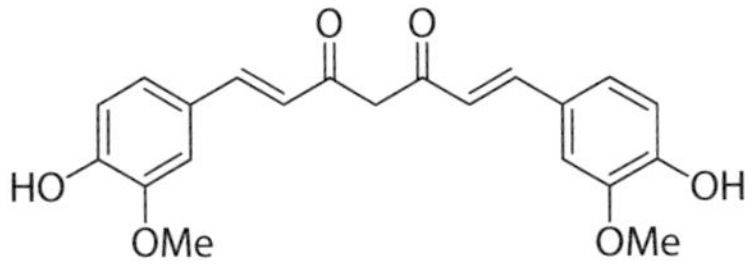

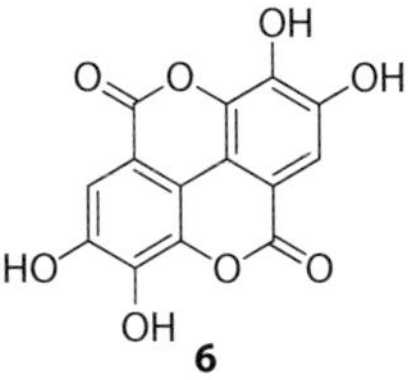

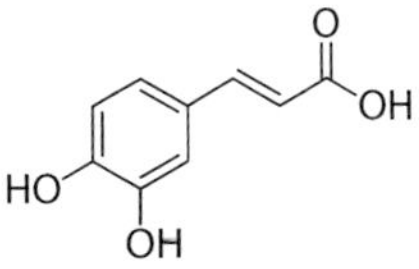

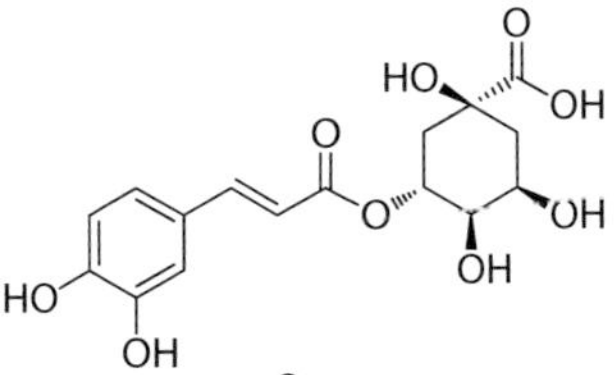

However, although many molecular mechanisms have been proposed and studied to support the preponderance of evidence indicating the chemopreventive properties of polyphenolic compounds found in fruit and vegetables (Surh 1999;

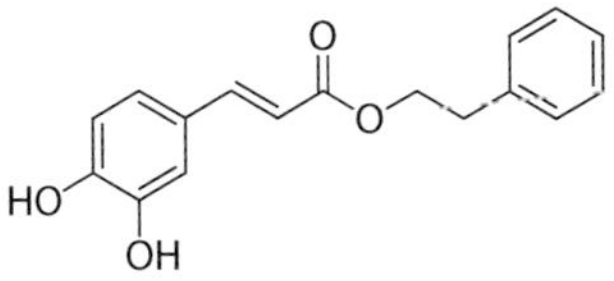

Bemis et al. 2006; Nichenametla et al. 2006), many questions still await a definitive answer: for instance, real effectiveness, bioavailability, assumption doses, and possible adverse effects such as pro-oxidant or mutagenic, synergistic or antagonistic action with other phytochemicals, and others. Although some synthetic agents with cancer chemopreventive activity have been introduced in clinical practice (raloxifene, tamoxifen, celecoxib, finasteride), the translation of chemoprevention to the clinic is not yet an established reality (William et al. 2009), and this is particularly true for natural polyphenols reputed to exert antitumor properties.

Nevertheless, in recent times, the studies aimed at the direct utilization of polyphenols as clinical chemopreventive agents have been paralleled by an increasing number of studies focused on optimization of low-molecular-weight polyphenols, considered as "lead compounds" to obtain new chemotherapeutics/ adjuvant agents. In fact, synthetic or semisynthetic analogues, possibly conserving the low or absent toxicity of the natural parent molecule, may overcome the problem of poor bioavailability and fast metabolic conversion frequently observed for natural polyphenols; the preparation of libraries of analogues may allow structure–activity relationship (SAR) studies and a better understanding of the molecular mechanisms of action of the natural polyphenols; optimized analogues may possess improved activity even through a different, more effective, mechanism of action. The trend toward preparation and evaluation of antitumor polyphenol analogues is also observed, as exemplified by Figure 9.1c, where the number of citations for the keywords "antitumor polyphenol analogues" is shown in the period 1989–2009.

On this basis, this chapter will be devoted to review selected examples from the recent literature illustrating the potential of synthetic low-molecular-weight polyphenol analogues as antitumor agents. Due to obvious space limitation, this is not an exhaustive review; a brief account of our recent studies on stilbenoid and lignan analogues is also included. Separate sections are devoted to the main families of polyphenol analogues. Many reported structures are related to dietary polyphenols, but also some analogues of polyphenols from nonedible plants are included, such as combretastatins or lignans; indeed, although these compounds are known mainly for their high cytotoxicity toward tumor cell lines, further biological activities are emerging (antiangiogenic, pro-apoptotic, MDR inhibitory, and others), thus suggesting the possible use of their analogues as adjuvants in association with current anticancer drugs.

9.2 STILBENOIDS

One of the most cited polyphenols reputed to be beneficial to health is the stilbenoid E-resveratrol (E-3,5,4′-trihydroxystilbene, **1**). Although the Z-isomer **10** is also a natural product found in *Vitis vinifera* (Katalinic et al. 2009) and other plants, the name "resveratrol" is commonly employed with reference to the isomer with E configuration of the central double bond. The popularity of resveratrol is mainly due to the so-called French paradox, namely, the inverse correlation between a high-fat diet and a low mortality risk of heart disease, observed in some southern regions of France and attributed to red wine consumption (Renaud et al. 1992).

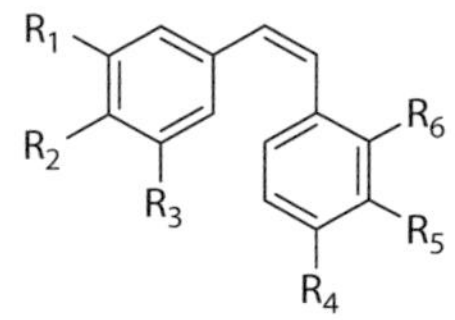

10 R$_1$=R$_3$=R$_4$=OH, R$_2$=R$_5$=R$_6$=H
14 R$_1$=R$_3$=R$_4$=OMe, R$_2$=R$_5$=R$_6$=H
15 R$_1$=R$_2$=R$_3$=R$_4$=OMe, R$_5$=OH, R$_6$=H
17 R$_1$=R$_2$=R$_3$=R$_4$=OMe, R$_5$=OPO$_3$Na$_2$, R$_6$=H
18 R$_1$=R$_2$=R$_3$=R$_4$=OMe, R$_5$=R$_6$=OPO$_3$Na$_2$
19 R$_1$=R$_2$=R$_3$=R$_4$=OMe, R$_5$=R$_6$=OH
21 R$_1$=R$_2$=R$_4$=OMe, R$_3$=R$_5$=R$_6$=OH
23 R$_1$=R$_2$=R$_3$=OMe, R$_5$=B(OH)$_2$, R$_6$=H
24 R$_1$=R$_2$=R$_3$=R$_4$=OMe, R$_5$=H, R$_6$=NH$_2$

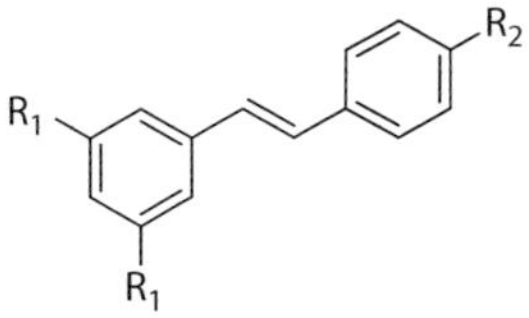

11 R$_1$=R$_2$=OCOCH$_3$
12 R$_1$=OCOCH$_2$CH$_2$CH$_3$, R$_2$=OH
13 R$_1$=R$_2$=OMe

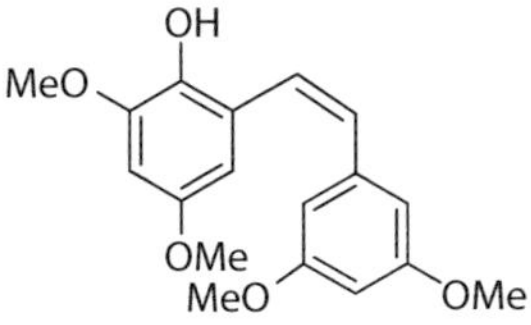

16

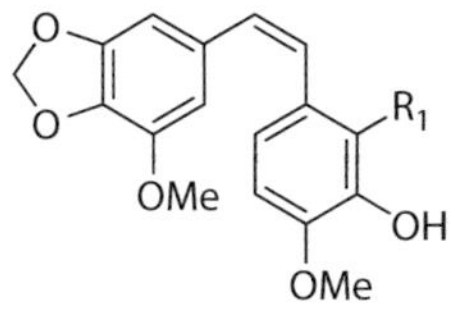

20 R=OH
22 R=H

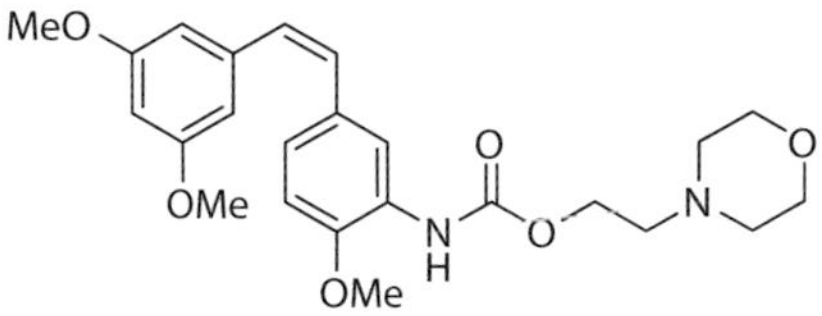

25

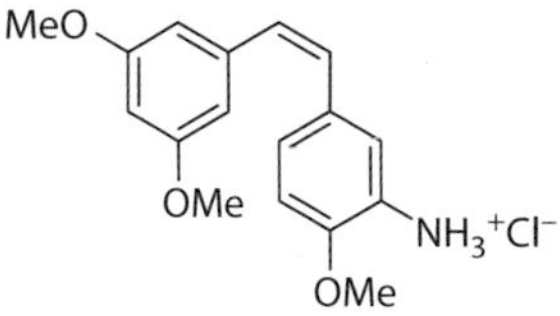

26

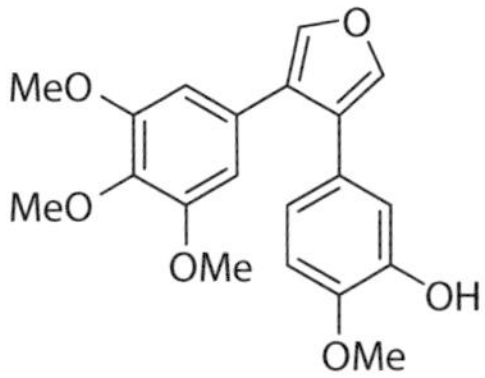

27

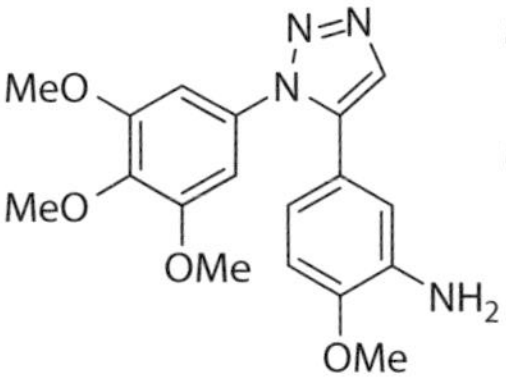

28

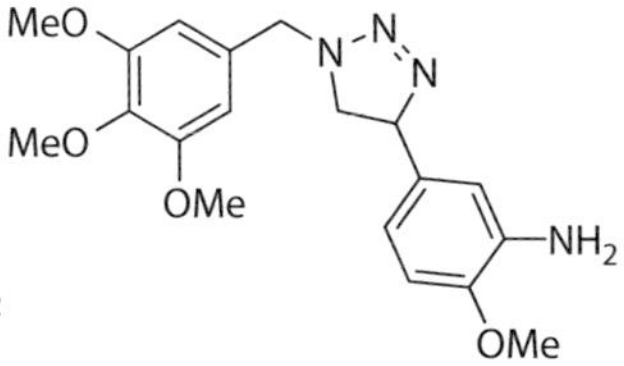

29

This stilbenoid, originally found in *Veratrum grandiflorum* (Takaoka 1940), was later obtained from the roots of *Polygonum cuspidatum* (Nonomura et al. 1963) and subsequently found in grapes and in other edible plants. A surprising variety of beneficial properties has been attributed to resveratrol, summarized in a number of reviews (Aggarwal et al. 2004; Signorelli et al. 2005; Baur et al. 2006; Burjonroppa and Fujise 2006; Delmas et al. 2006) and in a recent book (Aggarwal and Shishodia 2006b). Some aspects of the biological properties of **1** and analogues are also reviewed in Chapter 10 of this volume. Its cancer chemopreventive activity was firstly evidenced by Jang et al. (1997). Subsequent studies have shown that **1** is an inhibitor of cell survival signal transduction, an inhibitor of angiogenesis, a sensitizer to stimuli inducing apoptosis (Jazirehi et al. 2006) and that it exhibits antitumor activity also in vivo (Aggarwal et al. 2004; Baur et al. 2006). Nevertheless, the available in vivo studies indicate that **1**, though largely absorbed, has a low bioavailability, and is rapidly metabolized especially via phase II glucuronide or sulfate conjugations (Lançon et al. 2007). Thus, to overcome these serious limitations to its possible use in anticancer therapy as well as to improve its antiproliferative activity, a variety of analogues have been synthesized and submitted to bioassays toward tumor cells or more specific cancer molecular targets. We have recently reviewed the stilbenoid-based resveratrol analogues with antitumor properties (Chillemi et al. 2007), so will report here just a few examples or more recent additions. Some years ago, we obtained a series of lipophilic resveratrol analogues through either regioselective acylation/alcoholysis catalyzed by *Candida antarctica* lipase in organic solvent or other simple chemical conversions (Nicolosi et al. 2002; Cardile et al. 2005). These derivatives were submitted to antiproliferative activity bioassays against androgen nonresponsive human prostate tumor cells DU-145: most of the compounds showed either higher or comparable activity to that of **1** (GI_{50} = 24.09 μM); simple derivatives, such as 3,5,4'-tri-*O*-acetylresveratrol (**11**, G_{50} = 23.34 μM) or 3,5-di-*O*-butanoylresveratrol (**12**, G_{50} = 19.07 μM) are probably converted to resveratrol by intracellular esterases and may be, in principle, employed as lipophilic prodrugs with higher bioavailability than the natural molecule. However, the most active compound in these tests proved to be the permethylated resveratrol analogue (*E*)-3,5,4'-trimethoxystilbene (**13**, GI_{50} = 2.92 μM), previously known as a natural product (MacRae et al. 1985) and reported as a potent antiangiogenic agent (Belleri et al. 2005). In a subsequent study including its Z-isomer **14** (Cardile et al. 2007), this latter showed a potent antiproliferative activity, comparable to that of the anticancer drug vinorelbine against DU-145 and LNCaP (androgen responsive human prostate tumor) cells, and gave a GI_{50} = 0.1 μM against KB (human mouth epidermoid carcinoma) cells. This result was in perfect agreement with previous data by Schneider et al. (2003) on Caco-2 (human colon cancer) (IC_{50} = 0.25 μM) and on SW480 cells (IC_{50} = 0.23 μM) indicating **14** as a tubulin polymerization inhibitor, causing cell cycle arrest at the G2-M phase transition. Further studies in the last decade confirm that polymethoxystilbenes are a subgroup of resveratrol analogues possessing promising antitumor activity. Within this group, the majority of Z-isomers exhibit higher antiproliferative activity than their *E* analogues, in contrast with the higher activity observed for **1** with respect to that of its Z-isomer **10** (Cardile et al. 2007; Mazué et al. 2010): this suggests that a different mechanism of action may be involved for resveratrol versus its methylated analogues. In this regard, it is worth noting that there is a close structural relationship between Z-analogues of resveratrol and combretastatin A-4 (**15**), a cytotoxic stilbenoid, originally isolated by Pettit et al. (1989) from the root bark of the *Combretum caffrum* tree, known as one of the most potent inhibitors of tubulin polymerization (Tron et al. 2006; Kingston 2009). Notwithstanding the clear structural analogy, the majority of studies on resveratrol or combretastatin A-4 analogues have been carried out separately; indeed, combretastatin A-4 is

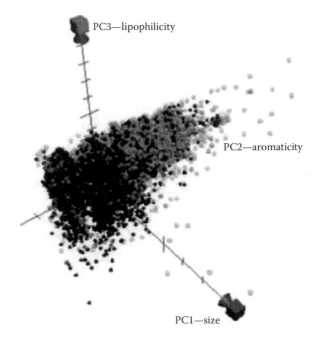

FIGURE 1.1 Distribution in ChemGPS-NP chemical property space by natural products from ZINC-NP (20,434 compounds, red) and synthetic compounds from Maybridge Screening Collection (57,627 compounds, orange). The red axis (PC1) corresponds mainly to size parameters, the yellow axis (PC2) to aromaticity-related properties and the green axis (PC3) to lipophilicity. For more details on the descriptor loadings and influence in the different principal components please refer to Larsson et al. (2007).

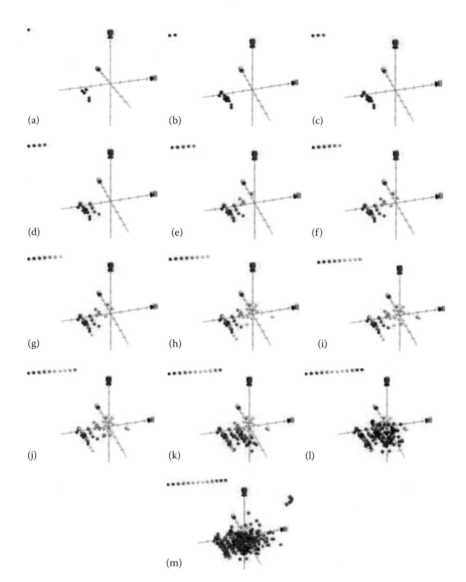

FIGURE 1.2 In this figure, we can follow the expansion of sesquiterpene lactones in ChemGPS-NP chemical property space as a function of the number of major evolutionary groups added with: a = fungi, b = liverworts, c = conifers, d = basal magnolioids, e = monocotyledons, f = rosids, g = lamids, h = apioids, i = Asteraceae—subfamily Mutisioideae, j = Asteraceae—Gochnatioideae, k = Asteraceae—Carduoideae, l = Asteraceae—Chicorioideae, and m = Asteraceae—Asteroideae.

(b)

FIGURE 1.3 (b) In the mosaic of the rain forest, we find not only the sky-scraping trees and other green plants, but also ponds and streams harboring thousands of species, all uniquely adapted to their ecological niche. (Photo by Mattias Klum.)

FIGURE 1.5 In the rain forest, there still live people with an immense knowledge of possible uses of the plants and animals around them. In this picture we see chief Tebaran demonstrating two of the more than 60 species with specific uses that he collected within a radius of 2 m in a randomly selected spot in the Danum valley of Borneo. (Photo by Mattias Klum.)

(b)

FIGURE 1.8 (b) In addition to these, numerous species of bacteria, fungi, and various groups of plants including red and brown algae have contributed to our knowledge of chemical diversity. (Photo by Mattias Klum.)

430-loop   150-loop

FIGURE 4.1 Conformational flexibility of the binding site observed by MD simulations based on the protein structure of influenza neuraminidase (PDB entry 2hty). The figure shows 10 representative frames (gray) of a short MD simulation. The binding site is occupied by oseltamivir (Chart 4.1, **1**; orange) and katsumadain A (Chart 4.1, **2**; green; heteroatoms O, red; N, blue). (Figure provided by Dr. Johannes Kirchmair.)

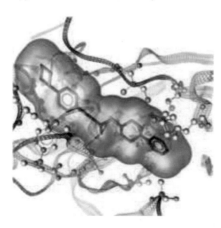

FIGURE 4.4 Binding site of the human rhinovirus coat protein (PDB Code 1hrv) with cocrystallized synthetic inhibitor Sdz 35–682 (visualized with LigandScout).

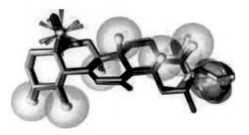

FIGURE 4.5 Four constituents from *Eriobotrya japonica* shown from a flexible 3D overlay generated by LigandScout. Common feature points are indicated as orange mini-spheres, lipophilic areas are represented as yellow spheres, the negatively ionizable area as red star, acceptor and donor locations as red and green spheres, respectively.

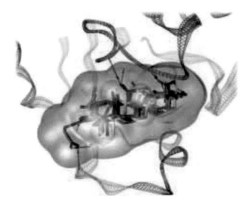

FIGURE 4.6 11beta-HSD1 crystal structure (PDB code 2bel, chain A) cocrystallized with carbenoxolone (blue) and possible binding modes of four constituents of *Eriobotrya japonica* (different gray shades) as predicted by structure-based molecular modeling.

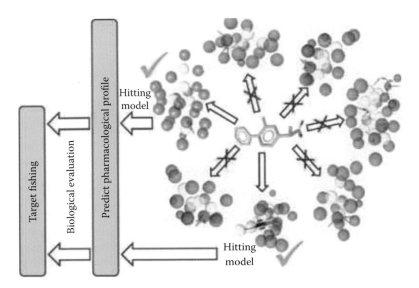

FIGURE 4.7 Pharmacophore-based virtual parallel screening concept of a single 3D structure for the identification of potentially interacting targets.

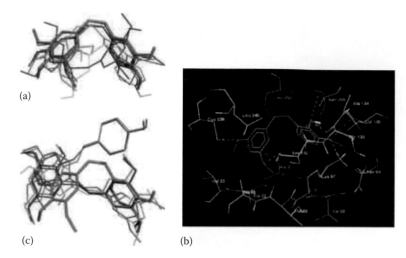

FIGURE 9.2 Computational studies and binding model of (*E*)- and (*Z*)-stilbenes isomers. For parts a and c, the binding models for all stilbenes were constructed using the combretastatin model as a template and reference ligand in the binding site. Computational docking was carried out applying the Lamarckian genetic algorithm (LGA) implemented in Auto-Dock 4.0. For fine docking, we used the following parameters: grid spacing = 0.261 Å, number of runs = 100, npts = 50-60-50 centered on combretastatin A-4, ga_num_evals = 20,000,000, ga_pop_ size = 150, and ga_num_generations = 27,000. (a) Superposition of all (*Z*)-stilbene analogues docked at the colchicine-binding site. The docked structure of combretastatin A-4, represented in orange (tube rendering), is inserted for comparison. Only polar hydrogens are represented, for clarity. Ligands are rendered as sticks with the subsequent color code: **2** yellow, **4** red, **6** green, **8** blue, **10** brown, **12** magenta, and **14** cyan. (b) Molecular docking result of compound **10** at the colchicine-binding site of tubulin. Only the amino acid residues within 4.5 Å around the inhibitor are shown for clarity. The ligand is represented with the carbon skeleton in brown with the only polar hydrogen. Dotted lines represent hydrogen bonds between ligand and receptor. The numerotation of amino acid residues is decremented by 3 units for subunit A and 56 for subunit B with respect to the original PDB file due to the strong manipulation and multiple file format conversions experienced. (c) Superposition of all (*E*)-stilbene analogues docked at the colchicine-binding site. The docked structure of combretastatin A-4, represented in orange (tube rendering), is inserted for comparison. Only polar hydrogens are represented, for clarity. Ligands are rendered as sticks with the subsequent color code: **1** yellow, **3** red, **5** green, **7** blue, **9** brown, **11** magenta, and **13** cyan. (For interpretation of the references to color in this figure legend, the reader is referred to the web version of this article.) Note the correspondence with our text: **1** = resveratrol (**1**), **2** = **10**, **3** = **13**, **4** = **14**, **5** = **37**, **10** = **16**; **14** = **38**; combretastatin A-4 = **15**; the other compounds are: **6** = (*Z*)-2-hydroxy-3,5,4′-trimethoxystilbene, **8** = (*Z*)-3,5,3′,5′-tetramethoxystilbene, **7** = (*E*)-3,5,3′,5′-tetramethoxystilbene, **9** = (*E*)-2-hydroxy-3,5,3′,5′-tetramethoxystilbene, **11** = (*E*)-3,5,3′,4′-tetramethoxystilbene, **13** = (*E*)-2-hydroxy-3,5,3′,4′-tetramethoxystilbene, **12** = (*Z*)-3,5,3′,4′-tetramethoxystilbene. (Reprinted from *Eur. J. Med. Chem.*, 45, Mazué, F., Colin, D., Gobbo, J., Wegner, M., Rescifina, A., Spatafora, C., Fasseur, D., Delmas, D., Meunier, P., Tringali, C., and Latruffe, N., Structural determinants of resveratrol for cell proliferation inhibition potency: Experimental and docking studies of new analogs, 2972–2980. Copyright 2010, with permission from Elsevier.)

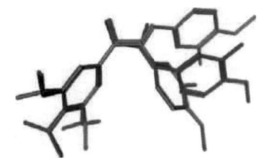

FIGURE 9.3 Superposition of **CA4** (grey), **2a** (green), and **31** (blue). Note the correspondence with our text: **CA4** = **15**; **2a** = *s-cis* chalcone, **3a** = *s-trans* chalcone. (Reprinted from *Bioorg. Med. Chem.*, 17, Ducki, S., Rennison, D., Woo, M., Kendall, A., Fournier Dit Chabert, J., McGownb, A.T., and Lawrence, N.J., Combretastatin-like chalcones as inhibitors of microtubule polymerization. Part 1: Synthesis and biological evaluation of antivascular activity, 7698–7710. Copyright 2009, with permission from Elsevier.)

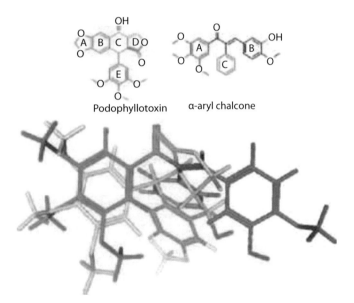

FIGURE 9.4 Pharmacophoric points between podophyllotoxin and α-aryl calchones. Note the correspondence with our text: podophyllotoxin = **59**. (Reprinted from *Bioorg. Med. Chem.*, 17, Ducki, S., Mackenzie, G., Greedy, B., Armitage, S., Fournier Dit Chabert, J., Bennett, E., Nettles, J., Snyder, J.P., and Lawrence, N.J., Combretastatin-like chalcones as inhibitors of microtubule polymerisation. Part 2: Structure-based discovery of alpha-aryl chalcones, 7711–7722, Copyright 2009, with permission from Elsevier.)

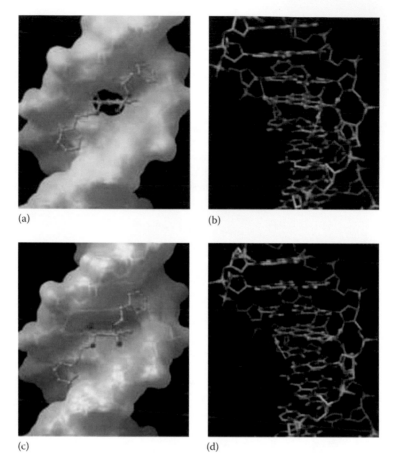

(a)

(b)

(c)

(d)

FIGURE 9.5 3D interactions of **10**-Model A (a and b) and **10**-Model B (c and d) complexes. In (a) and (c) the DNA is represented by molecular surface, and sticks and balls (colored by atom type: O, red; C, grey; polar H, sky blue; N, blue), whereas in (b) and (d) only by sticks and balls. The ligand is depicted in sticks (orange) and balls (colored as for DNA). In (a) and (c), the figure highlights the intercalation between two adjacent base pairs and the hydrogen bonds (yellow line) between ester functionality and NH$_2$ of guanine. In (b) and (d), beside the intercalation, the hydrogen bonds (yellow line) formed by OH group are shown. Note the correspondence with our text: **10 = 119**. (Di Micco, S., Daquino, C., Spatafora, C., Mazué, F., Delmas, D., Latruffe, N., Tringali, C., Riccio, R., and Bifulco, G., Structural basis for the potential antitumor activity of DNA-interacting Benzo[*kl*]xanthene lignans, *Org. Biomol. Chem.*, 9, 701–710, 2011. Copyright 2011, reprinted by permission of The Royal Society of Chemistry.)

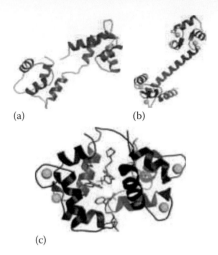

(a)   (b)

(c)

FIGURE 13.1 Structure of CaM: (a) Ca^{2+}-free form; (b) Ca^{2+}-binding; and (c) Ca^{2+}-TFP complex. (PDB entry codes: 1CFD, 1CLL and 1LIN, respectively.)

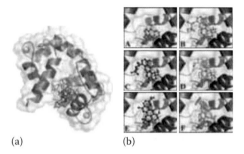

(a)   (b)

FIGURE 13.13 (a) Docking results obtained using AutoDock 4.0 inside the active site of CaM (orange): (b) (A) Docked CPZ ligand (green, sticks) into CaM appears superimposed on the cocrystallized TPF (red, lines). Top ranked binding mode of the most populated cluster of **23** (B, pale-green, sticks), **24** (C, blue, sticks), **25** (D, pale-yellow, sticks), **26** (E, gray, sticks), and **27** (F, cyano, sticks) into the binding site. Compounds **24** and **26** attach to CaM at the same position as TFP. The metal atoms (Ca^{2+}) are shown as pale-yellow color balls. Hydrogens are omitted for clarity.

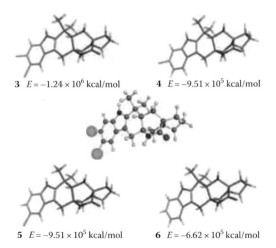

3 $E = -1.24 \times 10^{6}$ kcal/mol   **4** $E = -9.51 \times 10^{5}$ kcal/mol

5 $E = -9.51 \times 10^{5}$ kcal/mol   **6** $E = -6.62 \times 10^{5}$ kcal/mol

FIGURE 13.16 X-ray structure of **3** (center) and optimized DFT structures of **3–6** at the B3LYP/631G+ level of theory.

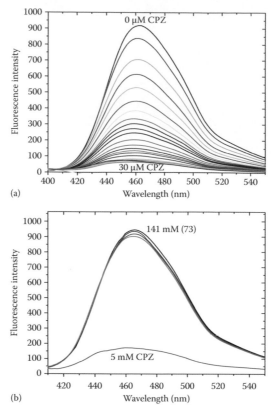

(a)

(b)

FIGURE 13.19 Titration by fluorescence of hCaM M124C-mBBr engineered hCaM with (a) CPZ and (b) compound **73**.

(a)

(b)

(c)

(d)

FIGURE 14.1 Representative plants from traditional Chinese medicine producing neuroprotective metabolites. (a) *Huperzia serrata*, (b) Caulis Sinomenii, (c) *Panax ginseng*, (d) *Apium graveolens*.

highly cytotoxic and has attracted the attention of researchers mainly as a potential anticancer therapeutic agent, whereas resveratrol is a component of edible part of plants like grapes or peanuts; it is considered a "safe molecule" and has been studied mainly for its chemopreventive properties. However, this separation is no more justified, as confirmed by the recent papers by Pettit et al. (2009a,b). On the other hand, we (Mazué et al. 2010) and other authors (Cao et al. 2008) have recently shown through molecular docking simulations that some polymethoxystilbenes are able to bind effectively to the colchicine binding site of tubulin, similarly to the binding mode of **15** (see Figure 9.2). In our study, compound **14** showed the highest antiproliferative activity toward SW480 human colorectal cell (IC$_{50}$ = 0.3 µM), but the (Z)-tetramethoxystilbene **16** (IC$_{50}$ = 7 µM) proved to have the best calculated binding affinity to the colchicine site of tubulin. Thus, Z-tetramethoxystilbenes and their hydroxylated derivatives are potentially good candidates as tubulin inhibitors but probably do not reach the tubulin pocket with the same effectiveness of **14**. It has also been noted that **15** is much more cytotoxic than expected on the basis of its activity against tubulin, and the exact mechanism of action for this compound has not yet been completely clarified.

Due to the important antiproliferative and vascular-disrupting properties of **15**, excellent reviews on combretastatins, their medicinal chemistry and synthesis of analogues have previously been reported (Hadfield et al. 2005; Pinney et al. 2005; Toze et al. 2005; Tronet et al. 2006; Chaudhary et al. 2007; Singh et al. 2009); thus, only a few exemplifying cases are reported here. Interestingly, among the variety of combretastatin A-4 analogues, combretastatin A-4 phosphate (**17**), a simple derivative with increased water solubility, is in phase III clinical development for the treatment of various forms of cancer (Kanthou et al. 2007; Banerjee et al. 2008), and the diphosphate Oxi-4503, (**18**) an hydrosoluble analogue of combretastatin A-1 (**19**) is also in phase I clinical trial (Hill et al. 2002). The stilstatins 1 (**20**) and 2 (**21**) and a series of their hydrosoluble prodrugs were tested against a panel of one murine (P388) and six human cell lines; interestingly, some prodrugs showed enhanced antineoplastic activity with respect to the parent compound combretastatin CA2 (**22**) and the authors suggest that the *o*-dihydroxy portion of **20** and **21** could give impressive anticancer activity due to the formation of an *o*-quinone, as established recently for the similar phenol system of **18** (Pettit et al. 2008). Boronic acid derivatives have moderate pH and air stability, useful properties in drug development (Kong et al. 2010): the boronic acid derivative **23**, a bioisostere of **15**, showed high tubulin polymerization inhibition (IC$_{50}$ = 1.5 µM), a [³H]-colchicine binding competitive with **15**, and potent antitumor activity toward MCF-7 human breast carcinoma cells (IC$_{50}$ = 17 nM) associated with improved solubility (Kong et al. 2005); the 2'-aminocombretastatin **24** proved to be a more potent inhibitor of tubulin polymerization than colchicine and showed IC$_{50}$ values in the range of 11–20 nM toward six tumor cell lines (Chang et al. 2006); the morpholino-carbamate derivative **25** is an hydrosoluble prodrug of **26**, the latter with tumor cell growth inhibitory activity in the nanomolar range (Simoni et al. 2009). To overcome the well-known problem of rapid *E*-isomerisation of the Z double bond in combretastatins and Z-resveratrol analogues due to light exposure or other factors, some *cis*-restricted analogues have

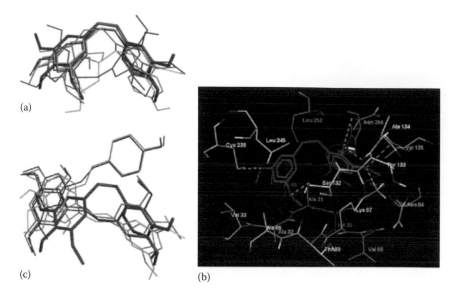

FIGURE 9.2 (See color insert.) Computational studies and binding model of (*E*)- and (*Z*)-stilbenes isomers. For parts a and c, the binding models for all stilbenes were constructed using the combretastatin model as a template and reference ligand in the binding site. Computational docking was carried out applying the Lamarckian genetic algorithm (LGA) implemented in Auto-Dock 4.0. For fine docking, we used the following parameters: grid spacing = 0.261 Å, number of runs = 100, npts = 50-60-50 centered on combretastatin A-4, ga_num_evals = 20,000,000, ga_pop_size = 150, and ga_num_generations = 27,000. (a) Superposition of all (*Z*)-stilbene analogues docked at the colchicine-binding site. The docked structure of combretastatin A-4, represented in orange (tube rendering), is inserted for comparison. Only polar hydrogens are represented, for clarity. Ligands are rendered as sticks with the subsequent color code: **2** yellow, **4** red, **6** green, **8** blue, **10** brown, **12** magenta, and **14** cyan. (b) Molecular docking result of compound **10** at the colchicine-binding site of tubulin. Only the amino acid residues within 4.5 Å around the inhibitor are shown for clarity. The ligand is represented with the carbon skeleton in brown with the only polar hydrogen. Dotted lines represent hydrogen bonds between ligand and receptor. The numerotation of amino acid residues is decremented by 3 units for subunit A and 56 for subunit B with respect to the original PDB file due to the strong manipulation and multiple file format conversions experienced. (c) Superposition of all (*E*)-stilbene analogues docked at the colchicine-binding site. The docked structure of combretastatin A-4, represented in orange (tube rendering), is inserted for comparison. Only polar hydrogens are represented, for clarity. Ligands are rendered as sticks with the subsequent color code: **1** yellow, **3** red, **5** green, **7** blue, **9** brown, **11** magenta, and **13** cyan. (For interpretation of the references to color in this figure legend, the reader is referred to the web version of this article.) Note the correspondence with our text: **1** = resveratrol (**1**), **2** = **10**, **3** = **13**, **4** = **14**, **5** = **37**, **10** = **16**; **14** = **38**; combretastatin A-4 = **15**; the other compounds are: **6** = (*Z*)-2-hydroxy-3,5,4′-trimethoxystilbene, **8** = (*Z*)-3,5,3′,5′-tetramethoxystilbene, **7** = (*E*)-3,5,3′,5′-tetramethoxystilbene, **9** = (*E*)-2-hydroxy-3,5,3′,5′-tetramethoxystilbene, **11** = (*E*)-3,5,3′,4′-tetramethoxystilbene, **13** = (*E*)-2-hydroxy-3,5,3′,4′-tetramethoxystilbene, **12** = (*Z*)-3,5,3′,4′-tetramethoxystilbene. (Reprinted from *Eur. J. Med. Chem.*, 45, Mazué, F., Colin, D., Gobbo, J., Wegner, M., Rescifina, A., Spatafora, C., Fasseur, D., Delmas, D., Meunier, P., Tringali, C., and Latruffe, N., Structural determinants of resveratrol for cell proliferation inhibition potency: Experimental and docking studies of new analogs, 2972–2980. Copyright 2010, with permission from Elsevier.)

been synthesized. Among these, we wish to cite here the combretafuran **27**, displaying $IC_{50} = 2.9\,\mu M$ on SH-SY5Y neuroblastoma cell line (Pirali et al. 2006), and two 1,2,3-triazole analogues, namely, **28**, with $IC_{50} = 0.011\,\mu M$ (Odlo et al. 2008) and **29**, with $IC_{50} = 0.38\,\mu M$ (Odlo et al. 2010), both on K562 leukemia cancer cell line; the triazole analogues were submitted to molecular modeling studies showing the binding mode to the colchicine binding site of tubulin.

The low stability of Z-analogues of resveratrol is probably one of the reasons why the research on E-stilbenoids is still highly active. For instance, the Pettit's group has recently evaluated, as inhibitors of human cancer cell lines growth and tubulin polymerization, a series of *trans* β-nitrostyrenes, exemplified by **30** and **31**, the only two active on melanoma SK-MEL-5 and renal A-498 cells ($GI_{50} < 1\,\mu g/mL$) (Pettit et al. 2009). A further contribution of the same group is the synthesis and biological evaluation of the polyhydroxylated E-stilstatin 3 (**32**) and its phosphate prodrug (**33**) that proved less active than E-resveratrol (**1**) against six human cancer cell lines. By contrast, the very simple resveratrol prodrug **34** (Aleo et al. 2008) was more active against human prostate tumor cells DU-145 than the parent compound **1**. Another simple resveratrol derivative is the 3,5-digalloylresveratrol **35**, bearing two naturally occurring galloyl pendants; the authors suggest that **35** may become useful as an adjuvant for colon cancer treatment, being able to induce apoptosis and cell cycle arrest in human HT-29 colon cancer cells (Bernhaus et al. 2009a).

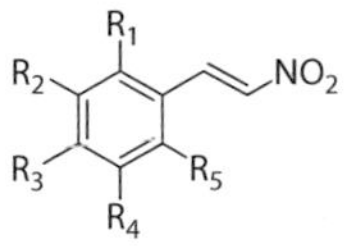

30 $R_1 = R_3 = R_5 = OMe$, $R_2 = R_4 = H$
31 $R_1 = R_2 = R_4 = R_5 = H$, $R_3 = NHCOCH_3$

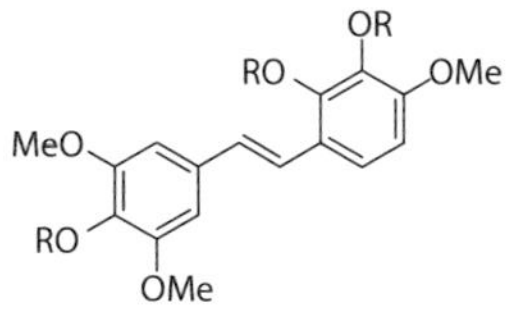

32 $R = H$
33 $R = PO_3Na_2$

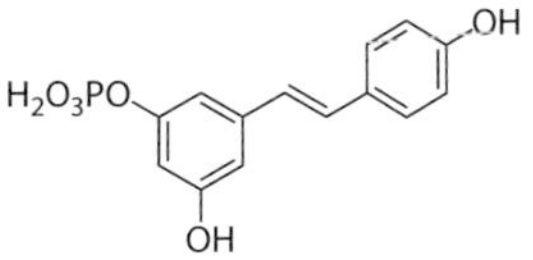

34

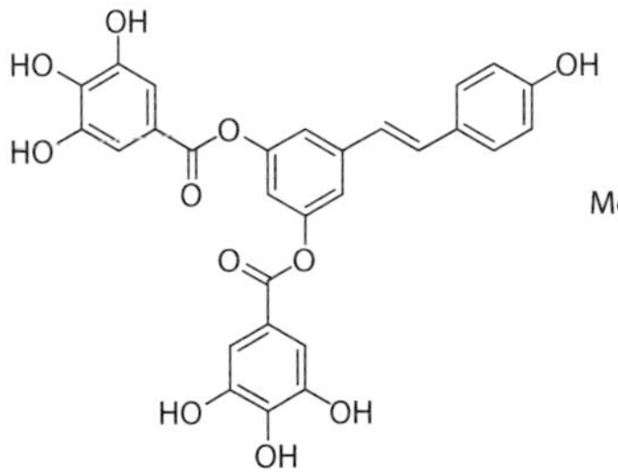

35

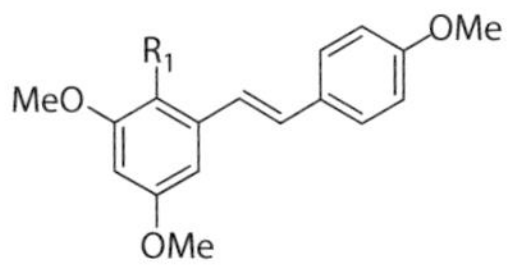

36 $R_1 = CHO$
37 $R_1 = OH$

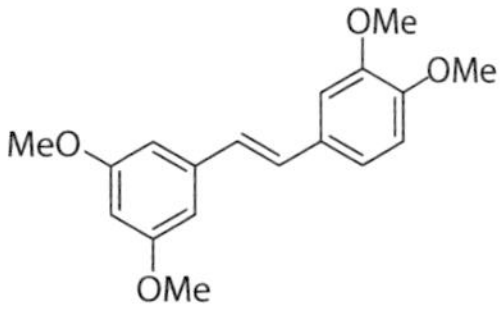

38

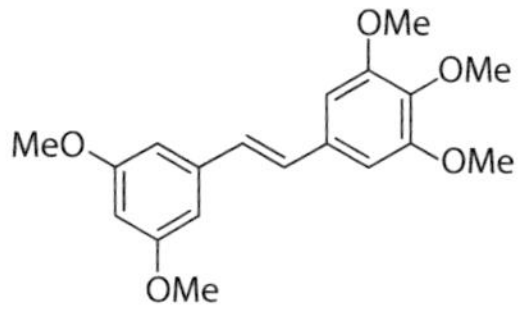

39

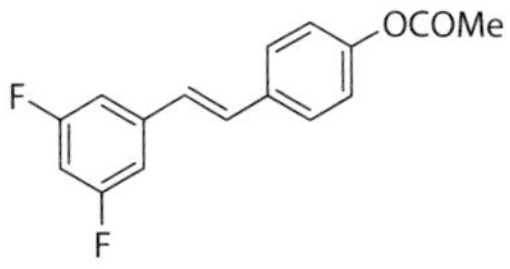

40

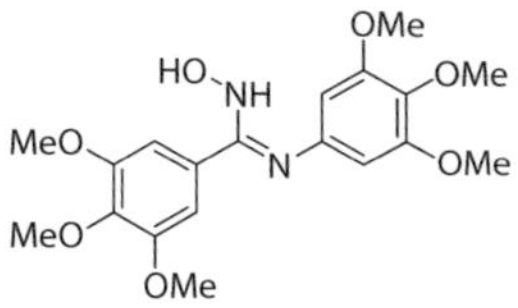

41

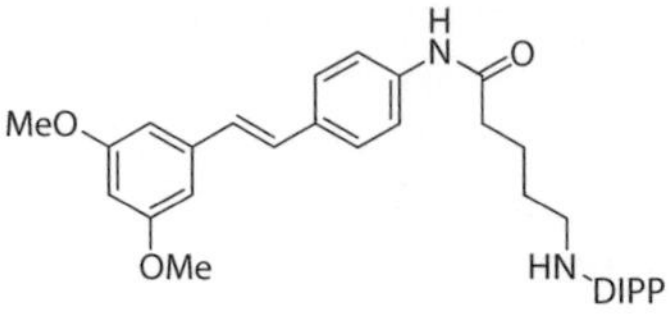

42

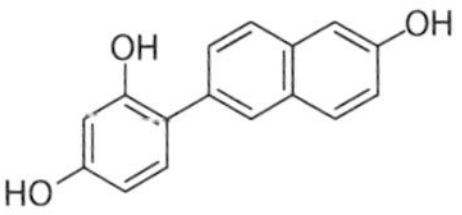

43

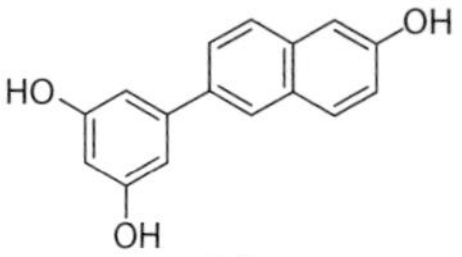

44

A series of stilbenoid derivatives modified at C-2 have been synthesized and evaluated against nasopharyngeal epidermoid tumor cell line KB: among them, (E)-2-formyl-3,5,4′-trimethoxystilbene **36** (IC_{50} = 9.2 μM) proved more active than the anticancer drug 5-fluorouracil (Huang et al. 2007). We have recently applied a mild aromatic hydroxylation method to 3,5-dimethoxystilbenes, thus obtaining 2-hydroxy derivatives; one of these, (E)-2-hydroxy-3,5,4′-trimethoxystilbene **37** proved to be a more potent antiangiogenic agent than **13**, being able to inhibit, in the range of 1–100 μM after 96 h exposure, up to the 80% of neovascularization in a porcine aortic endothelial cell in vitro model (Spatafora et al. 2009); **37** was also evaluated for its effects on swine granulosa cells where it decreased progesterone levels and inhibited VEGF output (Basini et al. 2010b). Some polymethoxystilbenes with interesting antiproliferative properties have been the subject of recent pharmacokinetic studies because of their presumed longer half-life and reduced metabolic conversion with respect to their hydroxylated analogues. For instance, the above-cited trimethoxystilbene **13** showed better oral bioavailability, longer terminal elimination half-life, greater plasma exposure and lower clearance than **1** (Lin et al. 2009). Similar pharmacokinetic results have been obtained for (E)-3,5,3′,4′-tetramethoxystilbene (**38**), previously known as a component of *Crotalaria madurensis* (Bhakuni et al. 1984) and showing antiproliferative properties toward human colon cancer cells (Saiko et al. 2008; Mazué et al. 2010), and P-gp inhibitory activity, a useful property against multidrug resistance of tumor cells (Ferreira et al. 2006); analogously, a promising pharmacokinetic profile has been observed for the (E)-3,5,3′,4′,5′-pentamethoxystilbene (**39**) (Lin et al. 2010), already known for its antitumor properties (Cushman et al. 1991; Simoni et al. 2006; Weng et al. 2009), and very recently reported as a more potent antitumor stilbenoid than **13** toward human breast carcinoma cells MCF-7 and affecting multiple cellular targets (Pan et al. 2010).

More "exotic" stilbenoids, resembling the structure of **1**, have been reported recently, among them we cite here (E)-3,5-difluoro-4′-acetoxystilbene **40**, more active than **1** against lung cancer and melanoma cells, and evaluated by NCI against a panel of 60 cell lines (Moran et al. 2009); KITC (**41**), a benzamidine derivative bearing six methoxy groups, which exerted promising antitumor activity against pancreatic cancer cells (Bernhaus et al. 2009b); the stilbenoid **42**, the most active

(IC$_{50}$ = 3.45 µM) in a series of N-phosphoryl amino acid modified resveratrol analogues evaluated for their effects against human nasopharyngeal carcinoma cell lines CNE-1 and CNE-2; HS-1793 (**43**), a resveratrol analogue including a naphthalene portion, which showed potent antitumor effects in most cancer cells tested and was able to overcome the resistance conferred to human leukemic U937 cells by the anti-apoptotic protein Bcl-2 (Jeong et al. 2009). Interestingly, **43** is very similar to **44**, a previously synthesized naphthalene analogue that was able to trigger apoptosis coupled to the induction of endogenous ceramide in MDA-MB-231 human breast cancer cells (Minutolo et al. 2005).

9.3 FLAVONOIDS

9.3.1 CHALCONES

Chalcones, characterized by an "open" flavonoid structure with only two aromatic rings, are widespread natural products and important intermediates in the biosynthesis of other flavonoids. A renewed interest in this group has been generated by recent reports on the antiproliferative activities of some natural and synthetic chalcones (Kumar et al. 2003; Go et al. 2005; Achanta et al. 2006; Modzelewska et al. 2006; Katsori et al. 2009).

The cancer chemopreventive properties of the above-cited prenylated chalcone xanthohumol (**3**), found in hops (*Humulus lupulus*) (Albini et al. 2006), were the basis to carry out a synthetic strategy to obtain **3** and a series of analogues, evaluated for their cytotoxic and antioxidant activity (Vogel et al. 2008). Some analogues, evaluated as cytotoxic agents toward HeLa cells, were more active than **3** (IC$_{50}$ = 9.4 µM); the most active was 3-hydroxyxanthohumol (**45**, IC$_{50}$ = 2.5 µM), reported as a phase 1 metabolite of **3**. Analogously to xanthohumol, the 2'-hydroxy-2,3,4',6'-tetramethoxychalcone **46**, isolated from *Caesalpinia pulcherrima* (Srinivas et al. 2003) and reported as an antitumor agent, inspired the synthesis of a series of related polyhydroxy/methoxy chalcones evaluated against a panel of 11 cancer cell lines (Boumendjel et al. 2008); from this preliminary SAR study it emerged that the methoxylation pattern is of primary importance for antitumor activity: a clear correlation between the methoxylation at 2', 4', 6' and the cell cycle arrest was observed, indicating that higher antimitotic activity is correlated to strong antiproliferative effect. The most active chalcone **47** showed IC$_{50}$ values in the range of 0.25–60 µM and was well tolerated in *in vivo* toxicity studies. The same authors reported recently, as a result of optimization of the previous leads, the chalcone analogue named JAI-51 (**48**) (Boumendjel et al. 2009), where the B ring is substituted by a N-methylindole moiety: this compound inhibited at 10 µM the proliferation of four human and one murine glioblastoma cell lines, acting as a microtubule depolymerising agent and an inhibitor of the protein efflux pumps P-gp and BCRP. Glioblastoma is a brain cancer, at present lacking in any effective chemotherapy, mainly due to the problems of blood–brain barrier (BBB) crossing and drug resistance caused by efflux pumps (see also Chapter 12). Of special interest is that **48** is able to cross the BBB and proved active in vivo on a C57BL/6 mice model bearing subcutaneous GL26 glioblastoma xenografts. A further study on polymethoxy chalcones, including also F, Cl, Br, and

NO_2 substituents at C-4', has been carried out more recently (Bandgar et al. 2010): the synthetic analogues were evaluated as possible anticancer, anti-inflammatory and antioxidant agents. Compound **49** completely inhibited five human cancer cell lines at 10 µM, proving more effective than the standards flavopiridol and gemcitabine. Bioavailability of the reported compounds was checked by an in vitro cytotoxicity study and confirmed to be nontoxic. The SAR and in silico drug relevant properties suggest that **49** and related analogues are potential candidates for future drug discovery studies.

The observation that chalcones may act as microtubule-targeting agents has stimulated a quantity of researches on new chalcone analogues, and we report here only a few examples taken from the multitude of synthetic chalcones recently reported. A series of chalcones and bis-chalcones, some of them containing boronic acid moieties, has been synthesized and evaluated for antitumor activity against human breast cancer MDA-MB-231 (estrogen receptor-negative) and MCF-7 (estrogen receptor-positive) cell lines; (Modzelewska et al. 2006). These compounds were also tested toward two normal breast epithelial cell lines, MCF-10A and MCF-12A. Among them, compounds **50** and **51** showed IC_{50} values of 3.8 and 2.0 µM, respectively, toward tumor cells, whereas the normal cell lines were four–six-fold less sensitive. Three bis-chalcones showed IC_{50} in the nanomolar range on tumor cells, and **52** was tested also on colon cancer cell lines, exhibiting a more potent inhibition of cells expressing wild-type p53 than of an isogenic cell line that was p53-null. Very recently, further boronic acid chalcones have been designed and synthesized as potential antitumor agents (Kong et al. 2010). These chalcone analogues were prepared with the aim of comparing their biological activities with those of combretastatin A-4 (**15**); in fact, the tubulin polymerization inhibitory activity of chalcone analogues are related to their structural similarities with **15**; the authors evaluated four 3,4,5-trimethoxychalcones for inhibition of [^{3}H]colchicine binding to tubulin, disruption of the microtubule network of A-10 cells, and inhibition of human MCF-7 breast cancer cells. Chalcones with the carbonyl group adjacent to the trimethoxybenzene ring proved more active than those with carbonyl adjacent to the other ring. In particular, compound **53** showed potent MCF-7 antiproliferative properties (IC_{50} = 0.9 µM), though being scarcely active as a tubulin inhibitor; when submitted to the NCI panel of human cancer cell screen, **53** proved active below 10 nM against colon cancer HCT-15, CNS cancer SBN-19, and ovarian cancer SK-OV-3. This compound was also tested as an antiangiogenic agent employing both the human umbilical vein endothelial cells (HUVEC) assay and a rat aortic ring assay, showing significant activity. Combretastatin-like chalcones, as inhibitors of microtubule polymerization, have been extensively studied recently by Ducki et al., as a continuation of their previous 5D-QSAR study (Ducki et al. 2005) of combretastatin A-4 analogues. In the first part of the more recent work (Ducki et al. 2009b), the authors report the synthesis of an array of chalcones bearing different substituents more at ring A (adjacent to the carbonyl) than at ring B. Many of these substituents are methoxy or hydroxyl groups but also Cl and F atoms, and other functionalities (-O-CH_2-O-, -O-$(CH_2)_2$-O, -NO_2, and -NMe_2) have been added to the basic skeleton of natural chalcones. In addition, chalcones bearing an alkyl group or an alkoxyl group at the α position of the double bond (adjacent to the carbonyl) or an epoxide ring instead of the latter were synthesized. Biological evaluation included

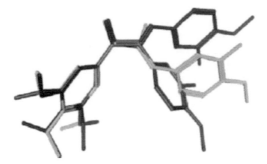

FIGURE 9.3 **(See color insert.)** Superposition of **CA4** (grey), **2a** (green), and **31** (blue). Note the correspondence with our text: **CA4** = **15**; **2a** = *s-cis* chalcone, **3a** = *s-trans* chalcone. (Reprinted from *Bioorg. Med. Chem.*, 17, Ducki, S., Rennison, D., Woo, M., Kendall, A., Fournier Dit Chabert, J., McGownb, A.T., and Lawrence, N.J., Combretastatin-like chalcones as inhibitors of microtubule polymerization. Part 1: Synthesis and biological evaluation of antivascular activity, 7698–7710. Copyright 2009, with permission from Elsevier.)

cell growth inhibition (K562 human chronic myelogenous leukemia cell line), cell cycle analysis, inhibition of tubulin polymerization, [^{3}H]colchicine competition, and HUVEC assay. The most potent chalcones, displaying cytotoxic activity in the nM range, have the same 3,4,5-trimethoxy aromatic substitution pattern of **15**. Among the series of analogues maintaining the skeleton of natural chalcones, compound **54** (IC$_{50}$ = 4.3 nM) showed the highest growth inhibitory activity. The α-alkyl and α-alkoxyl substitution causes a dramatic improvement in activity and compounds **55** (IC$_{50}$ = 0.21 nM) and **56** (IC$_{50}$ = 1.5 nM) proved much more active than **54**, and even than **15** (IC$_{50}$ = 2.0 nM). The authors conclude that the group at the α-position act as a conformational lock forcing the chalcones to adopt an *s-trans* conformation, required for higher activity (see Figure 9.3). A good correlation between cytotoxicity and tubulin binding at the colchicine site was generally observed. Several compounds also showed remarkable activity in the HUVEC shape change assay, a promising indication for possible antivascular activity in vivo. On the basis of the above summarized work, the phosphate prodrug **57** (IC$_{50}$ = 0.12 μM) and **58** (IC$_{50}$ = 8.0 μM) were prepared and are currently undergoing preclinical evaluation. The second part of this study (Ducki et al. 2009a) is focused on α-aryl chalcones, considered as a new class of antitubulin agents mimicking the well-known cytotoxic lignan podophyllotoxin (**59**). Docking studies on tubulin allowed establishing pharmacophoric points between podophyllotoxin and α-aryl chalcones (see Figure 9.4); the molecular modeling showed that the B-rings of both chalcones and combretastatin A4 did not occupy the same pocket in tubulin and belonged to different pharmacophore groups. The results were used to design the synthesis of a series of α-aryl chalcones, subsequently evaluated as leukemia K562 cell growth inhibitors. Although the most active analogue **60**, tested as *E/Z* mixture (IC$_{50}$ = 12 nM), proved less potent than the reference compounds **15** and **55**, it may be considered as a chalcone–combretastatin hybrid that could be the basis of new podophyllotoxin mimics.

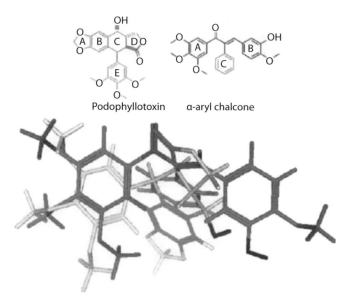

Podophyllotoxin   α-aryl chalcone

FIGURE 9.4 **(See color insert.)** Pharmacophoric points between podophyllotoxin and α-aryl calchones. Note the correspondence with our text: podophyllotoxin = **59**. (Reprinted from *Bioorg. Med. Chem.*, 17, Ducki, S., Mackenzie, G., Greedy, B., Armitage, S., Fournier Dit Chabert, J., Bennett, E., Nettles, J., Snyder, J.P., and Lawrence, N.J., Combretastatin-like chalcones as inhibitors of microtubule polymerisation. Part 2: Structure-based discovery of alpha-aryl chalcones, 7711–7722, Copyright 2009, with permission from Elsevier.)

Further, combretastatin-like chalcones have been synthesized with the aim of obtaining dual inhibitors of both inflammatory mediators such as NO and tumor cell proliferation (Yerra et al. 2009). In a series of 3,4,5-trimethoxychalcones investigated, chalcone **61** showed the highest antiproliferative activity toward HepG2 (hepatoma) and Colon 205 human cancer cell lines, with IC_{50} = 1.8 and 2.2 μM, respectively.

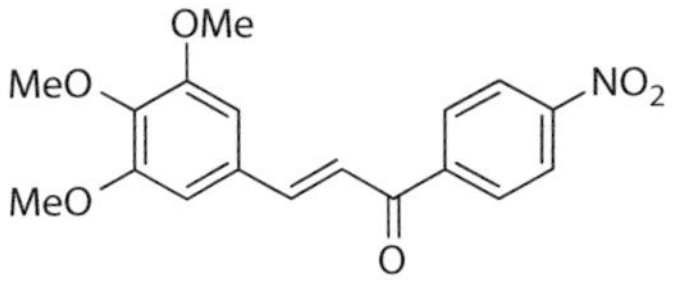

49

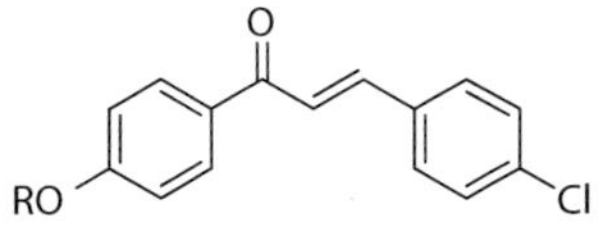

50 R=H
51 R=B(OH)₂

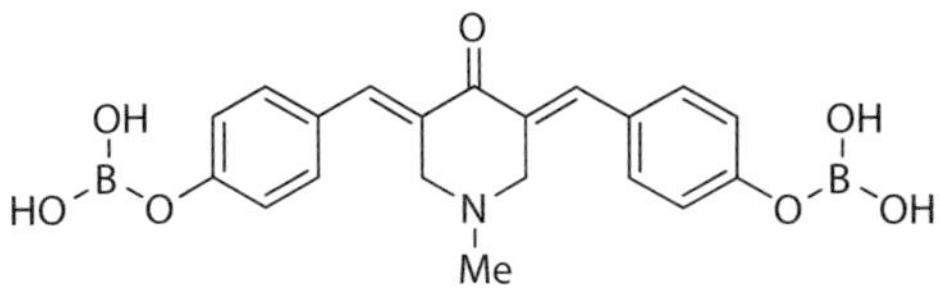

52

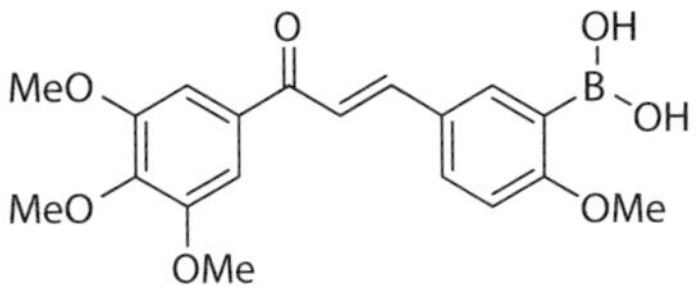

53

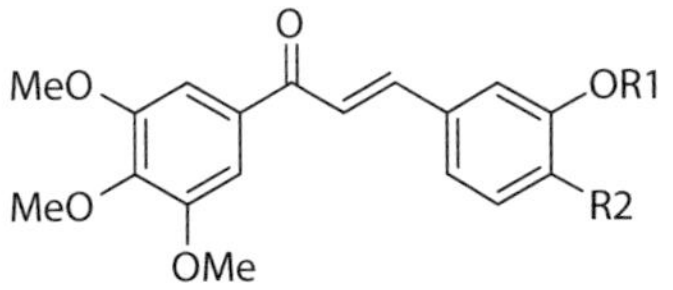

54 R₁=H, R₂=OMe
61 R₁=Me, R₂=H

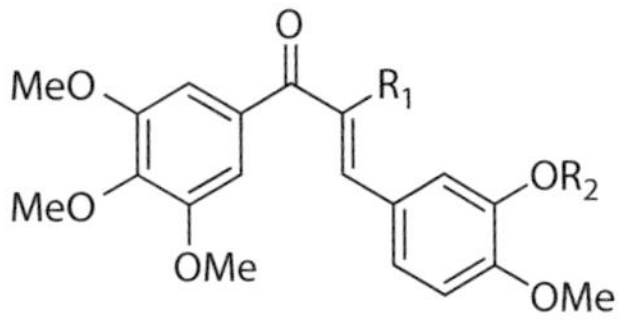

55 R₁=Me, R₂=H
56 R₁=OMe, R₂=H
57 R₁=Me, R₂=PO₃Na₂
58 R₁=OMe, R₂=PO₃Na₂

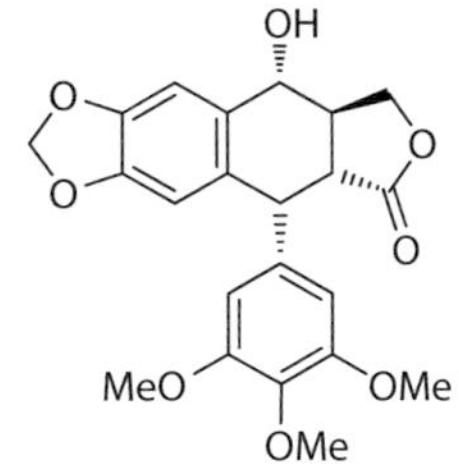

59

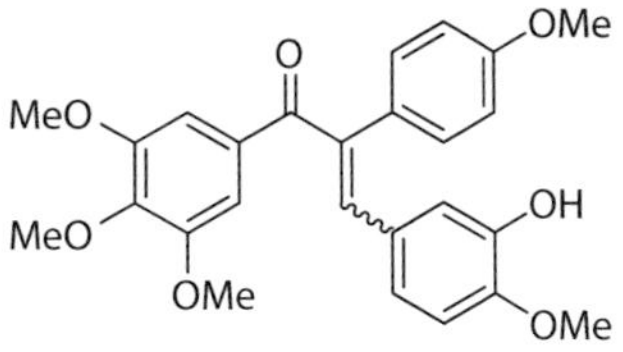

60

9.3.2 FLAVANOLS

Green tea catechins, mainly consisting of (–)-epicatechin (EC, **62**), (–)-epigallo-
catechin (EGC, **63**), (–)-epicatechin gallate (ECG, **64**), and (–)-epigallocatechin-3-
gallate (EGCG, **2**) are by far the most popular representatives of the flavanol family,
due to the large amount of studies suggesting that these phenolic constituents of
the unfermented dried leaves of *Camellia sinensis* may afford many health benefits
and in particular have cancer chemopreventive properties (Yang et al. 1999; Zaveri
2006; Huo et al. 2008). Some epidemiological studies have revealed an inverse correlation between increased green tea intake and relative risk for cancers (Imai et al.
1997; Fujiki et al. 2002). Many in vitro experiments have corroborated the antitumor potential of these polyphenols and in particular of **2**, the major component of
green tea (Nagle et al. 2006; Chen et al. 2007). Despite these promising properties,
2, as well as other catechins and polyphenols, has a low metabolic stability, resulting
in poor bioavailability; this observation prompted some researchers to synthesize
and evaluate more stable analogues of EGCG (Huo et al. 2008). The peracetylated

derivative **65** was the subject of various studies: it was demonstrated that **65** is rapidly converted to **2** by colorectal adenocarcinoma HCT116 cells, allowing reaching a 30-fold increase in intracellular concentration of the active metabolite (Lambert et al. 2006); **65** proved to have a more potent growth inhibitor than **2** toward both KYSE150 human esophageal ($IC_{50} = 10$ versus $20\,\mu M$) and HCT116 human colon cancer cells ($IC_{50} = 32$ versus $45\,\mu M$). Proteasome inhibitors have been considered as potential anticancer drugs (Almond et al. 2002) and some natural polyphenols, including EGCG, have been reported as proteasome modulators (Bonfili et al. 2008). So it is worth citing here a study where human breast cancer cells (MDA-MB-231) were treated with **65** and an enhancement of proteasome inhibition, growth suppression, and apoptosis induction was observed, compared with cells treated with natural **2** (Landis-Piwowar et al. 2007). In addition, MDA-MB-231 tumors were induced in nude mice, followed by treatment for 31 days with **65** or C4. This in vivo study showed a higher inhibition of breast tumor growth by **65** with respect to **2**, associated with increased proteasome inhibition and apoptosis induction in tumor tissues. This study was continued with the synthesis of fluorinated EGCG analogues and their acetylated derivatives, assayed as proteasome inhibitors and apoptosis-inducing agents toward leukemia Jurkat cells (Yu et al. 2008). The difluoro-benzoate analogue of EGC (**66**) proved to be a very potent inhibitor of purified 20S proteasomes (IC_{50} $0.65\,\mu M$). The acetylated analogues were employed for cell-death inducing assays on leukemia cells, and showed an order of potency similar to that of the unprotected compounds; in particular, **67** was capable of inducing 90% cell death at $50\,\mu M$ after 24 h treatment.

A further recently synthesized analogue is 3-O-(3,4,5-trimethoxybenzoyl)-(–)-epicatechin (TMECG, **68**), which showed higher stability and cellular uptake than its natural precursor, associated to an antiproliferative activity in the µM range against several cancer cell lines, especially melanoma (Sánchez-del-Campo et al. 2008). In addition, the authors demonstrated that **68** binds efficiently to human dihydrofolate reductase and downregulates folate cycle gene expression in melanoma cells. They suggest that, like other antifolate compounds, **68** could be of clinical value in cancer therapy. In a subsequent study (Sanchez-del-Campo et al. 2009), these authors made a deeper investigation of the mechanism by which **68** exerts its antiproliferative action. They report that this flavanol derivative acts as a prodrug that is selectively activated by a specific melanocyte tyrosinase; upon activation, **68** generates a stable quinone methide product that strongly inhibits dihydrofolate reductase in an irreversible manner.

Employing transesterification reactions catalyzed by a lipase PL from *Alcaligenes* sp., a series of monoester lipophilic analogues of EGCG bearing butanoyl, octanoyl, palmitoyl (**69**) groups was obtained (Matsumura et al. 2008); their in vitro antitumor activities were investigated toward colorectal carcinoma cell lines (Colon26, Caco2) and a normal cell line (rat vascular smooth muscle [RVSMC] cell) either with or without catalase. Tumor growth inhibition in vivo was also evaluated. The cytotoxicity of EGCG-monoester derivatives increased with increasing alkyl chain length, with IC_{50} values for palmitoyl analogues in the approximate range of $18–29\,\mu M$ ($93–103\,\mu M$ in RVSMC cells); **69** inhibited cell proliferation and induced apoptosis in the presence of catalase, inhibited the phosphorylation of the epidermal growth factor receptor (EGFR), and suppressed tumor growth in colorectal tumor bearing mice at a dose of 50 mg/kg.

We have already cited the increasing interest toward the antiangiogenic properties of phenolic compounds; in particular, the anti-angiogenic activity of **2** and to a lesser extent, of **62**, has been reported (Lamy et al. 2002). Thus, to improve the antiangiogenic properties of **62**, three 3-*O*-acylepicatechin derivatives (bearing C-12, C-14, and C-16 acyl chains) were synthesized and their effect on angiogenesis was examined (Matsubara et al. 2007). In addition, these lipophilic analogues of EC were evaluated for HL-60 leukemia cell growth inhibition and inhibition of DNA polymerase, a key enzyme in DNA replication. 3-*O*-Palmitoyl epicatechin (**69**) was the most potent antiangiogenic catechin analogue, resulting more active than **62** in a dose-dependent manner and statistically significant at higher than 50 μM; it was the strongest inhibitor in DNA polymerase α, β, λ with 50% inhibition doses of 2.2–15.0 μM; it showed the highest growth-inhibitory activity on HL-60 cells (IC$_{50}$ = 3.5 μM); **69** also suppressed human endothelial cell (HUVEC) tube formation on reconstituted basement membrane, suggesting that it affected not only DNA polymerase activity but also the signal transduction pathways needed for the tube formation in HUVECs.

An interesting work reports a reliable method for stereoselective functionalization of naturally occurring (+)-catechin at C4 and C8, which are the bridging positions of the flavanoid units in procyanidins, dimeric, or oligomeric polyphenols originated via a C4/C8 interflavanoid bond (Hayes et al. 2006). Gallate ester derivatives of the C4- or C8-propyl substituted catechins, respectively **70** and **71**, were prepared and submitted to cytotoxicity assays toward human cancer cell lines. The C8-propyl catechin gallate **71** proved more active (IC$_{50}$ = 31 μM) than catechin gallate or ECG against the colorectal adenocarcinoma cell line HCT116.

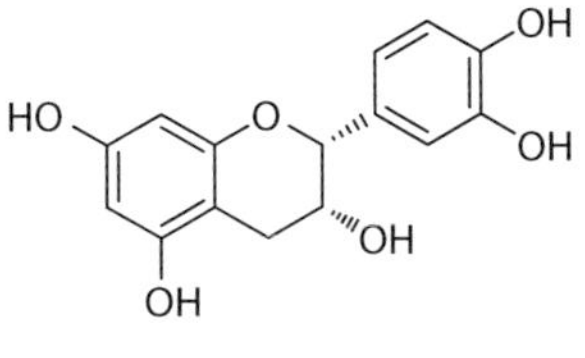

62

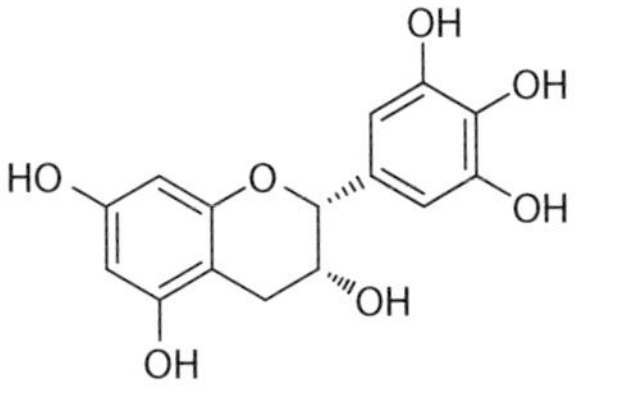

63

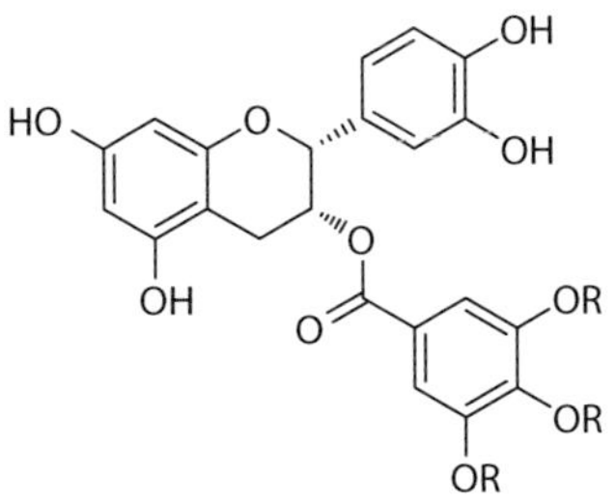

64 R = H
68 R = Me

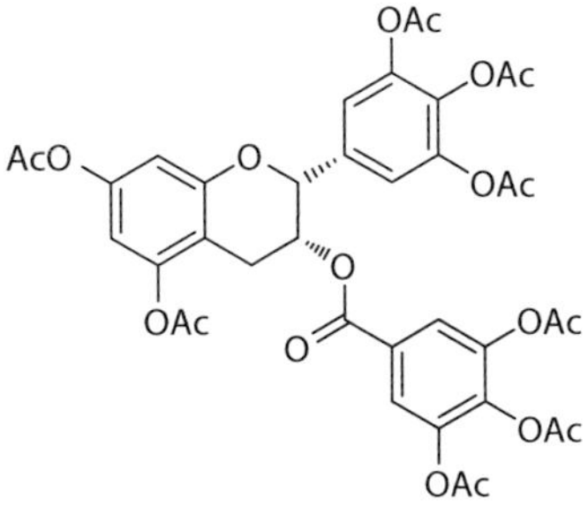

65

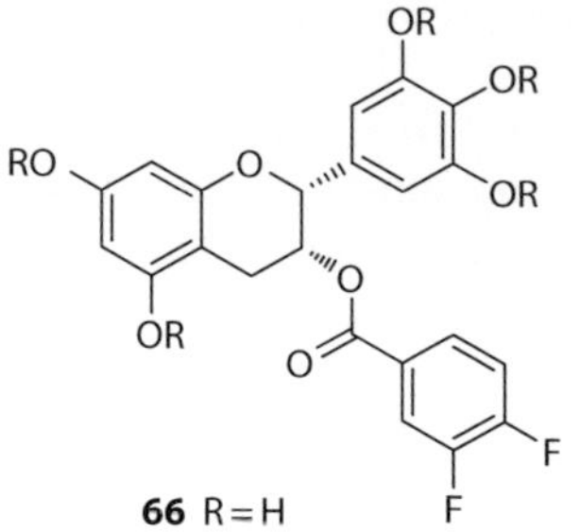

66 R = H
67 R = COCH$_3$

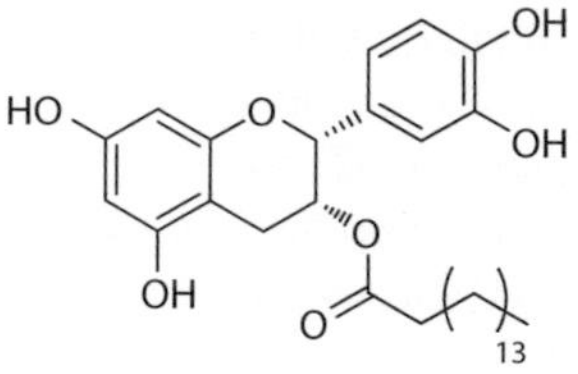

69

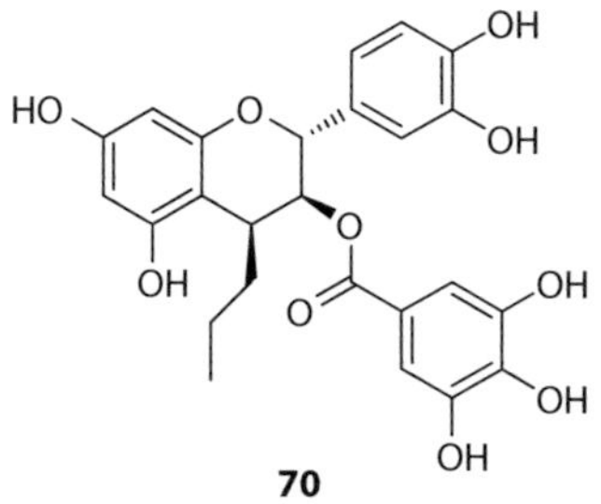

70

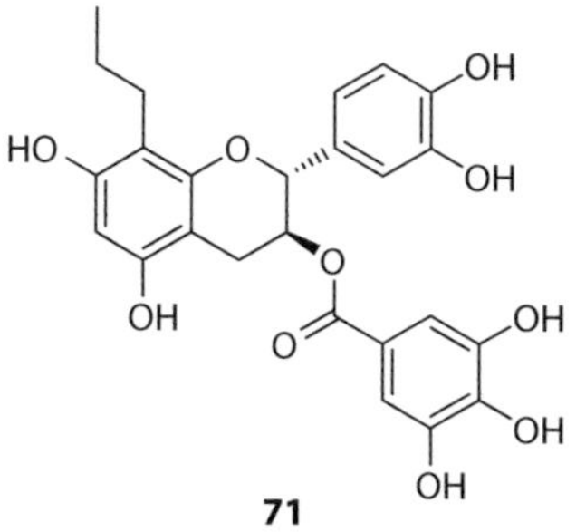

71

9.3.3   FLAVONES AND OTHER FLAVONOIDS

A group of natural flavones reputed to have significant cancer chemopreventive activity are polymethoxyflavones, which show a stronger antiproliferative activity and a higher metabolic resistance and bioavailability with respect to hydroxylated flavones (Walle 2007). In particular, the *Citrus* flavones nobiletin (**72**), tangeretin (**73**), and 5-desmethylsinensetin (**74**) proved to be the most active compounds in a screening of natural and synthetic methoxylated flavones for antiproliferative activity against six human cancer cell lines (Manthey et al. 2002). The most active synthetic compound, the limocitrin derivative **75**, had IC$_{50}$ values of 5.3, 2.6, 0.11, and 2.1 µM against the melanoma, prostate, and breast (estrogen receptor +/estrogen receptor −) cancer cell lines, respectively.

In an extensive study on the antiproliferative activity of flavonoid and caffeic acid analogues (see also Section 9.4.1), eight natural flavonoids and twenty synthetic flavones were tested against four human tumor cell lines (cervix HeLa adenocarcinoma; oropharyngeal KB carcinoma; breast MCF-7 cancer; SK-MEL-28 melanoma) and four murine cell lines (F3II and LM3, mammary adenocarcinomas; lung LP07 tumor; B16-F0 melanoma) (Cardenas et al. 2006). The natural flavone apigenin (**76**) and the synthetic 2′-nitroflavone (**77**) were among the most potent antiproliferative agents, with IC$_{50}$ values in the range of 3–22 µM. None of the natural or synthetic compounds tested affected the proliferation of normal epithelial or fibroblastic cells. Some SARs were also examined—apigenin proved significantly more active than the related flavone chrysin (**78**), thus suggesting the positive role of the OH group at C-4′. Analyses of compounds monosubstituted in B and C rings showed that a bromine atom in position 3′ and a nitro group at 4′ or 2′ yielded active compounds.

In a very recent paper (Lv et al. 2010), a series of long-chain derivatives of chrysin (**78**) were synthesized and evaluated for their antiproliferative activities against the human liver cancer cell line HT-29 and EGFR, an important kinase receptor in cancer studies. Among the 20 compounds tested, the analogues **79** and **80** showed IC_{50} lower than $10\,\mu g/mL$ toward HT-29 cells and displayed potent EGFR inhibitory activity with IC_{50} values of 0.048 and $0.035\,\mu M$ respectively, comparable to the positive control erlotinib. The result of docking studies suggested that both **79** and **80** can effectively bind the EGFR kinase and could be potential anticancer agents.

A series of 3′-amino derivatives, 5,6,7,8-tetra- or 5,7-dioxygenated on the A-ring was synthesized from either the flavone tangeretin (**73**) or the flavanone naringin (**81**), two natural *Citrus* flavonoids. These compounds were evaluated for antiproliferative activity, activation of apoptosis, and inhibition of tubulin assembly (Quintin et al. 2009). The compounds with the highest antiproliferative and pro-apoptotic activity have a common 5-hydroxy-6,7,8-trimethoxy substitution pattern on the A-ring. In particular, the tangeretin analogues **82** and **83**, the latter bearing a chlorine atom at C-3, have IC_{50} values of 0.16 and $0.14\,\mu M$, respectively, toward KB human buccal carcinoma cells. The authors suggest a contribution of the 3′-amino group to the activity, although they highlight that the natural flavonoid **84** ($IC_{50} = 0.02\,\mu M$) proved to be the most potent compound.

Recently, the widespread flavone luteolin (**85**), already known for its antioxidant and anti-inflammatory properties, has also been considered a potential anticancer agent due to its reported antitumor properties. Thus, it is worth citing here a study on the sensitizing effects on TNFα-induced apoptosis of some novel alkynyl luteolin analogues bearing various alkynyl groups at C-4′ (Cheng et al. 2008). These were firstly evaluated for their cytotoxic activity on HeLa and HepG2 tumor cells, and some of the most active compounds were tested for their effects on TNFα-induced cell death. Among the 16 compounds tested, seven showed cytotoxicity comparable to or higher than luteolin, and the analogue **86** proved to be a sensitizer on TNFα-induced apoptosis both on HeLa and HepG2 more effective than **85**.

The cyclin-dependent kinases (CDK) play a central role in cell-cycle progression and cellular proliferation and appear as novel therapeutic targets for cancer chemotherapy (Noble et al. 2004). Some flavonoids have recently emerged as possible kinase inhibitors (Ahn et al. 2007): in particular, the semisynthetic flavone flavopiridol (**87**), derived from the plant alkaloid rohitukine isolated from the stem bark of *Dysoxylum binectariferum* (Mohanakumara et al. 2010), proved active against CDK1, CDK2, CDK4, and protein–tyrosine kinase, inhibiting the proliferation of mammalian tumor cells at nM concentrations (Losiewicz et al. 1994). Flavopiridol was the first CDK inhibitor tested in clinical trials against a variety of cancers and has recently undergone a Phase II study (Lin et al. 2009). Recently, the design and synthesis of a small series of novel 8-aminoflavopiridol derivatives has been carried out (Ahn et al. 2007). These compounds were evaluated as antiproliferative agents and CDK2-Cyclin A inhibitors, but proved significantly less potent than **87**, which has IC_{50} of 7 and $2.6\,nM$ toward ovarian ID-8 and breast MCF-7 cancer cells, respectively. A molecular docking study suggested a different binding orientation inside the CDK2 binding pocket for these analogues compared to flavopiridol.

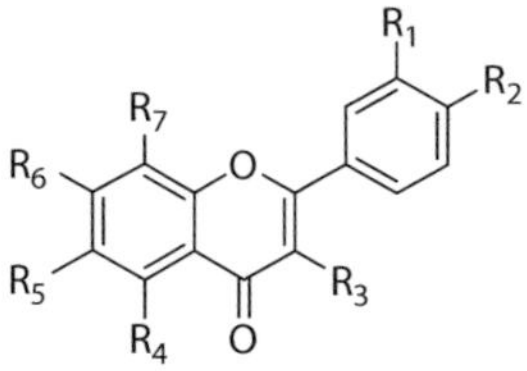

72 $R_1 = R_3 = H$, $R_2 = R_4 = R_5 = R_6 = R_7 = OMe$
73 $R_3 = H$, $R_1 = R_2 = R_4 = R_5 = R_6 = R_7 = OMe$
74 $R_3 = H$, $R_1 = R_2 = R_5 = R_6 = R_7 = OMe$, $R_4 = OH$
75 $R_1 = R_2 = R_3 = R_6 = R_7 = OMe$, $R_4 = OAc$, $R_5 = H$

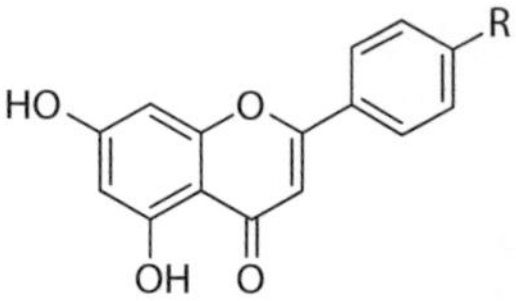

76 R = OH
78 R = H

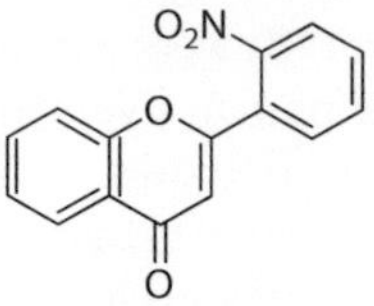

77

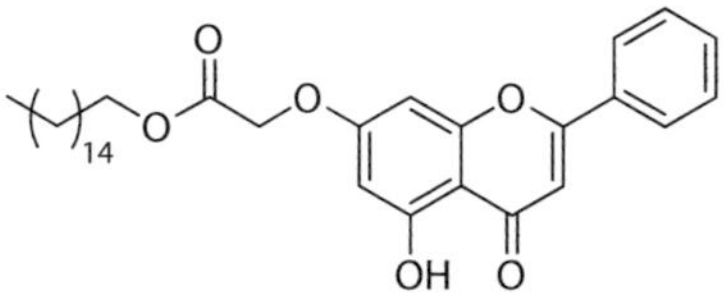

79

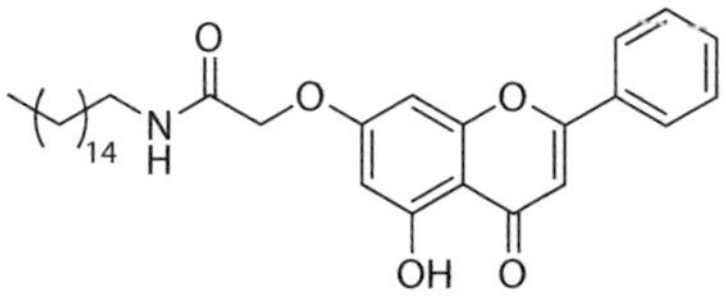

80

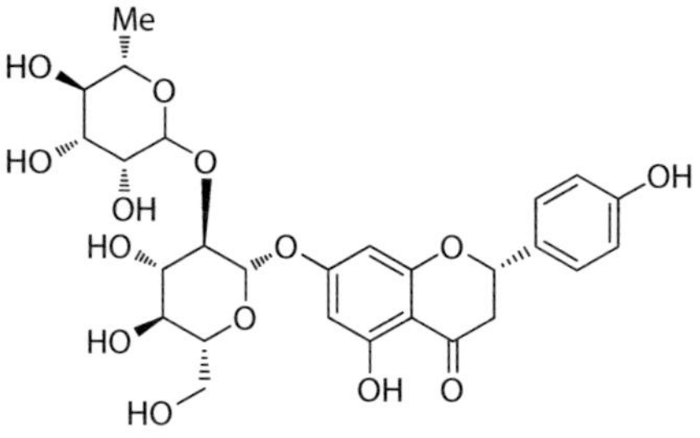

81

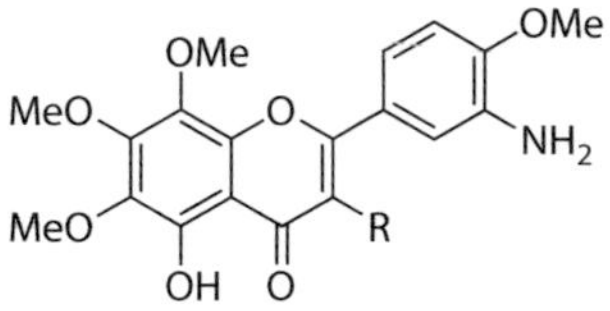

82 R = H
83 R = Cl

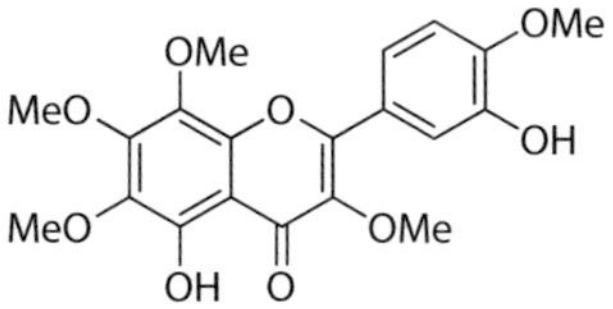

84

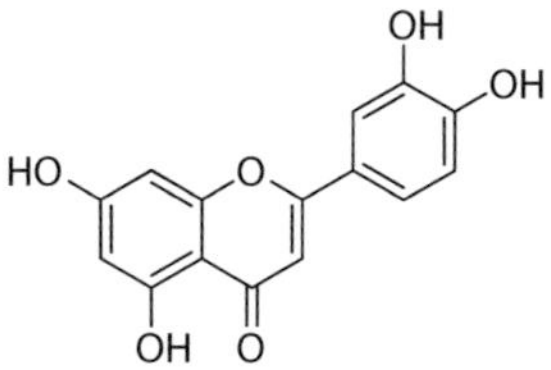

85

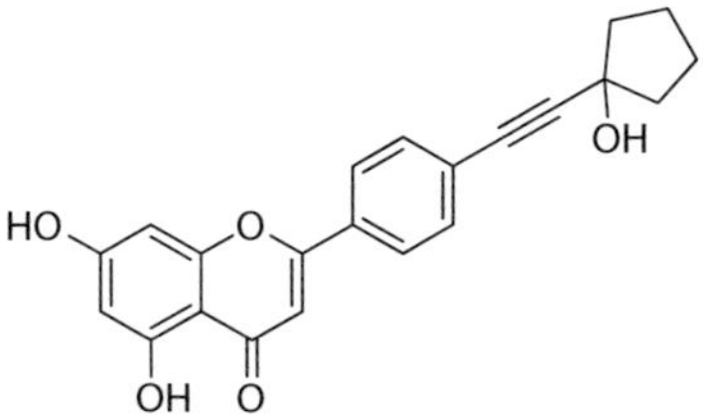

86

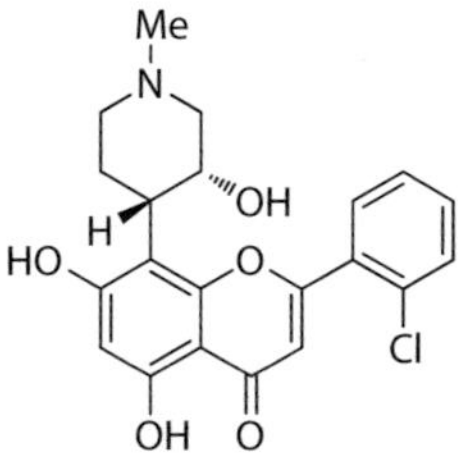

87

Genistein (**4**) is the most abundant isoflavone in soybeans and is considered a cancer chemopreventive principle responsible for relatively low incidence of hormone-dependent cancers in countries where the soy foods are widely consumed (Kim et al. 2004). In addition, **4** has been submitted to many in vitro studies indicating its moderate binding affinity to the estrogen receptor and inhibitory activity against various enzymes involved in tumor development and growth (Li et al. 2006). Nevertheless, only a limited number of research studies have addressed SAR and evaluation of genistein analogues. One recent addition to these reports was the synthesis and cytotoxicity evaluation against KB tumor cells of a series of 30 genistein analogues, bearing various substituents at the phenol groups or a SO_3H function at ring B (Li et al. 2006). It was observed that derivatives where OH substitution afforded a higher lipophilicity proved to have comparable or higher activity with respect to the positive control 5-fluorouracil (IC_{50} = 13.4 μM); in particular, the analogue **88** proved more active than the control, with IC_{50} = 8.5 μM. Also daidzein (**89**), a soy isoflavone strictly related to **4**, reported as an inhibitor of cancer cell growth at a high concentration (De Lemos 2001), has very recently been evaluated, in comparison with a series of synthetic analogues, in an extensive study aimed at investigating the mechanism underlying its anti-estrogenic effect of **89** at doses higher than 20 μM (Hacket et al. 2005). The results showed that **89** can be transformed into anti-estrogenic ligands by simple alkyl substitutions of the 7-hydroxyl hydrogen. In particular, the most effective analogue **90** reduced in vivo (xenograft mouse model) estrogen stimulated tumorigenesis in MVF-7 breast cancer cells. Molecular docking studies using the crystal structure of estrogen receptor α showed that **90** was able to bind to the anti-estrogenic conformer of ERα better than **89** and other analogues, and may be useful as a novel anti-estrogenic agent for breast cancer prevention and therapy.

The unusual prenylated isoflavone derrubone (**91**), originally isolated from *Derris robusta* (East et al. 1969), was identified as an inhibitor of the 90 kDa heat shock proteins (Hsp90), a promising chemotherapeutic target for the development of antitumor agents (Hadden et al. 2007). By means of a facile synthetic methodology, a series of 10 derrubone analogues were obtained, and their antiproliferative activity against MCF-7 (breast) and HCT-116 (colon) cancer cell lines was determined in comparison with the natural flavone. Some synthetic analogues proved more active than **91** (IC_{50} = 11.9 and 13.7 μM toward MCF-7 and HCT-116, respectively) and the 4′-methoxy analogue (**92**) showed a twofold increase in inhibitory activity (IC_{50} = 5.5 and 7.3 μM); this compound also induced the degradation of dependent client proteins in human breast cancer cells, supporting their Hsp90 inhibitory activity in vitro.

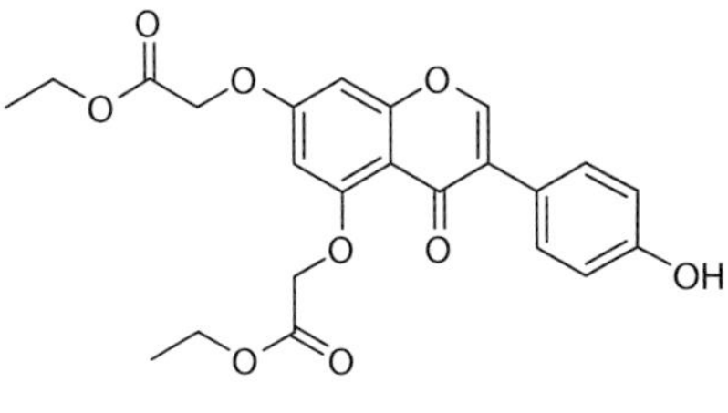

88 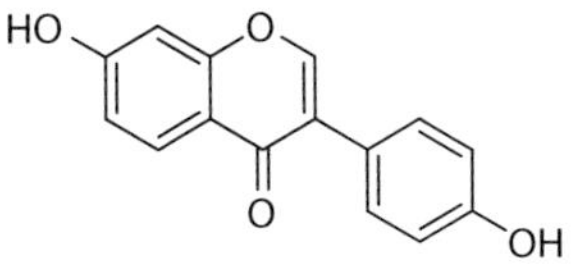   **89**

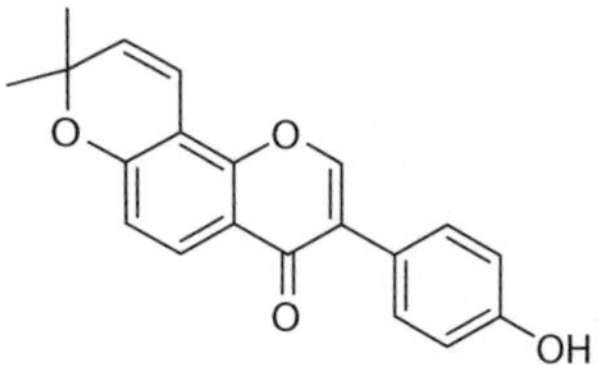

90

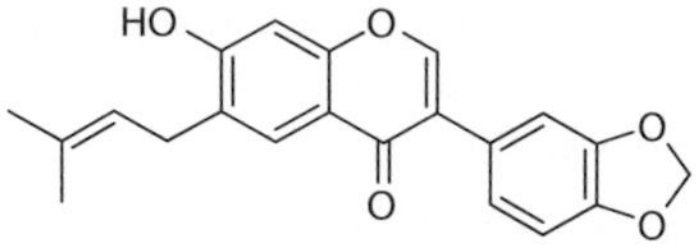

91

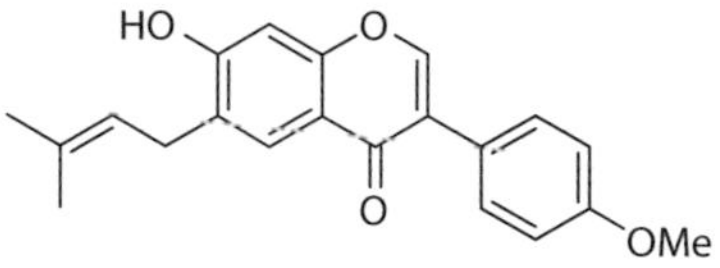

92

9.4 PHENYLPROPANOIDS AND LIGNANS

Phenylpropanoids are formed in the first steps of the shikimate biosynthetic pathway. Although many polyphenols, including flavonoids, have a phenylpropane unit in their structure, we report here only two specific groups of compounds with a C_6C_3 phenylpropane basic skeleton, namely, caffeic acid derivatives and coumarins, which have recently been cited for their antitumor properties. Lignans are frequently associated with phenylpropanoids because their biosynthetic origin is normally due to an oxidative coupling reaction between two phenylpropanoid units (Dewick 2009).

9.4.1 PHENYLPROPANOIDS

The caffeic acid (**7**) moiety is widely encountered in natural derivatives, such as chlorogenic acid (**8**) or CAPE (**9**), the caffeic acid phenethyl ester found in honeybee propolis; the latter, in particular, has been the subject of a number of studies highlighting its anticarcinogenic or antitumor properties (Lee et al. 2000; Beltran-Ramirez et al. 2008; El-Refaei et al. 2010). In addition to more recent reports, it is worth citing here a study on a small library of caffeic and dihydrocaffeic esters evaluated versus L1210 leukemia and MCF-7 breast cancer cells and submitted to QSAR study (Etzenhouser et al. 2001). The L1210 QSAR for dihydrocaffeic acid esters resembles the QSAR obtained for simple phenols and estrogenic phenols, whereas the QSAR for caffeic esters was significantly different, suggesting an important role of the olefinic double bond and lipophilicity. The *n*-octyl caffeate **93** proved more potent than CAPE as an inhibitor of cell growth versus L1210 and MCF-7 cells. Ethyl caffeate and *n*-butanoyl caffeate **94** were included in the above-cited study (Cardenas et al. 2006), mainly focused on natural and synthetic flavonoids, evaluated against various human and murine cancer cell lines (see also Section 9.3.3); **94** proved to be the most active compound, showing potent antiproliferative activity toward almost all the cell lines (IC_{50} in the range of 2–16 μM). In a further report, alkyl esters of caffeic (**7**) and ferulic (**95**) acid were synthesized and tested for inhibition of cell proliferation against MCF-7 (breast), SF-268 (CNS), NCI-H460 (lung), HCT-116 (colon), and AGS (gastric), as well as for cyclooxygenase enzymes and lipid peroxidation inhibitory activities (Jayaprakasam et al. 2006). Caffeic acid esters proved more active than ferulic acid esters in inhibiting the cell proliferation of colon and breast cancer cell lines. The most active compounds were *n*-dodecyl (**96**) and *n*-hexadecyl (**97**) caffeate, *n*-octyl (**98**) and *n*-dodecyl (**99**) ferulate with IC_{50} values in the range of 2.46–18.67 μg/mL. Lipophilicity and the number of phenolic groups appear important in defining the antiproliferative activity of these compounds. A more recent study has been carried out on three caffeic esters, namely, methyl (**100**), *n*-propyl (**101**), and *n*-octyl (**93**) caffeate, tested on human cervix adenocarcinoma cells (HeLa), in comparison with the corresponding gallate esters (Fiuza et al. 2004); experiments were also carried out in nonneoplastic fibroblasts from human embryonic lung tissue (L-132). Theoretical ab initio methods were also employed for simultaneous structural information. The ester **101** (IC_{50} = 12.0 μM) proved more active than the analogues with shorter (**100**) or longer (**93**) alkyl chain, and this was clearly related to the optimal lipophilicity for drug uptake process. A further study has been carried out on caffeic acid esters (including CAPE and analogues), evaluated as anti-HIV agents and, partly, as growth inhibitors of cancer cells, namely, hepatocellular carcinoma BEL-7404, breast adenocarcinoma MCF-7, lung adenocarcinoma A549, and gastric cancer BCG823 (Xia et al. 2008). Although CAPE (**9**) was the most active compound against BEL-7404, with IC_{50} = 5.5 μM, a number of analogues proved more potent inhibitors than CAPE or cisplatin, employed as control. In particular, phenyl caffeate (**102**) showed IC_{50} = 5.9 μM against MCF-7 cells; compound **103** had IC_{50} = 6.6 μM against BEL-7404 and the nitro-CAPE **104** showed IC_{50} = 8.7 μM against A549; this latter proved more active than cisplatin toward three cell lines (but less active toward BCG823).

Coumarins, strictly related to *o*-coumaric acid, are "old natural compounds" regarding their well-known biological properties (anticoagulant, antiviral, antimicrobial, anti-inflammatory and others); but some synthetic analogues have recently been considered as promising therapeutic agents, in particular for breast cancer, as reported in excellent new reviews (Musa et al. 2008; Riveiro et al. 2010), which make it unnecessary to treat in detail here this class of compounds, characterized mainly by anti-estrogenic activity. However, we wish to mention novobiocin (**105**), an antibiotic isolated from *Streptomyces* spp., bearing a carbamoylated sugar residue, a 3-amino-8-methyl-4,7-dihydroxycoumarin moiety (ring B), and an isopentenyl-substituted hydroxybenzoyl moiety (ring A). Recent studies evidenced that synthetic novobiocin analogues bind to the Hsp90, a promising target for cancer therapy (Riveiro et al. 2010). Late research results (Donnelly et al. 2008) showed that a number of synthetic novobiocin analogues exhibit antiproliferative activity against various cancer cell lines in the micromolar range; in particular compounds **106** and **107** have IC_{50} values of 1.1 and 1.8 μM, respectively, against androgen receptor sensitive prostate cancer cells (LnCaP).

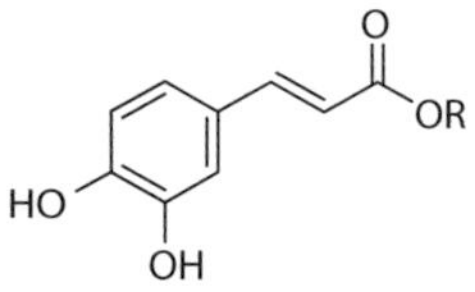

93 R = CH₂(CH₂)₆CH₃
94 R = CH₂(CH₂)₂CH₃
96 R = CH₂(CH₂)₁₀CH₃
97 R = CH₂(CH₂)₁₄CH₃
100 R = CH₃
101 R = CH₂CH₂CH₃
102 R = C₆H₅

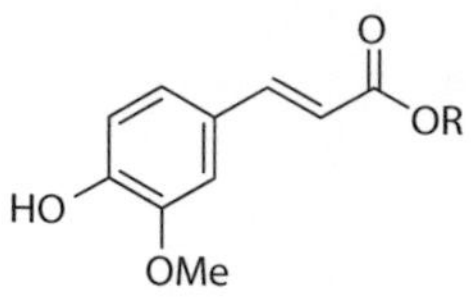

95 R = H
98 R = CH₂(CH₂)₈CH₃
99 R = CH₂(CH₂)₁₀CH₃

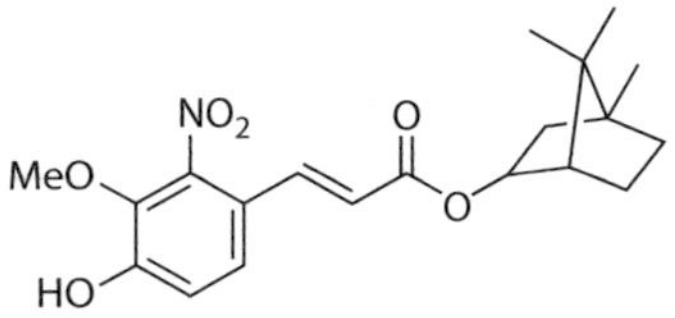

103

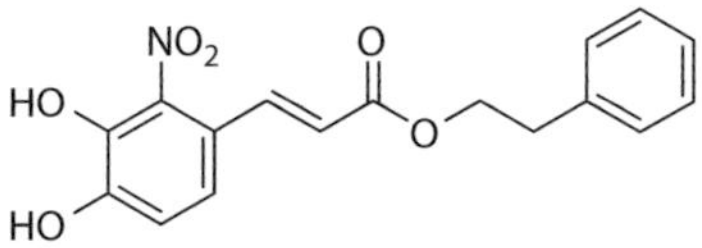

104

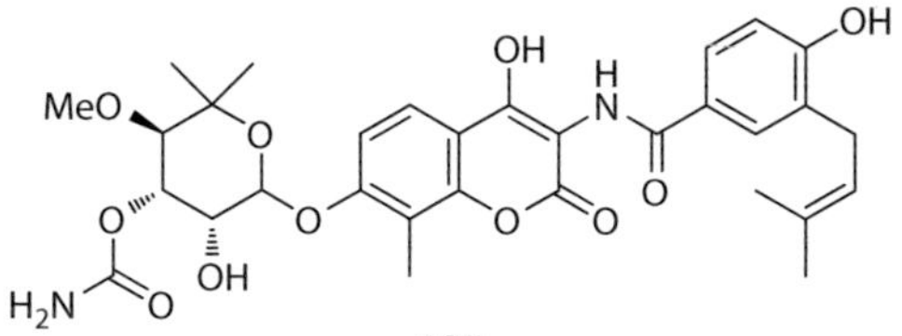

105

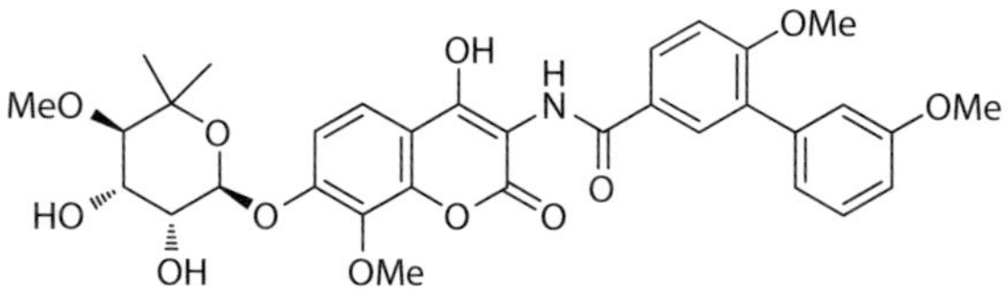

106

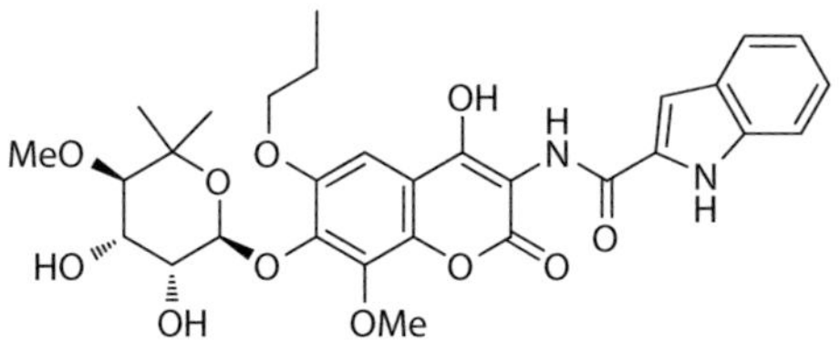

107

9.4.2 Lignans

Lignans and related compounds (neolignans, oxyneolignans, and others) are widely distributed within the plant kingdom. Their carbon skeletons are normally constituted by two phenylpropanoid (C_6C_3) units, whose β-β′ (8–8′) oxidative coupling originates the lignans both in biosynthetic pathways and in biomimetic syntheses (Spatafora et al. 2007). Lignans and other related compounds are accumulated in vascular plants, probably as chemical defense agents, and this may explain the wide variety in their reported biological activities, including cytotoxic, antimitotic, anti-leishmanial, antiangiogenic, cardiovascular, and antiviral activity (Jang et al. 1997; Apers et al. 2003). The most cited example of bioactive lignan is podophyllotoxin (**59**), the above-mentioned well-known constituent of *Podophyllum peltatum*, isolated as an antitumor principle around 1950; later work on **59** as lead compound afforded a variety of synthetic analogues, and in particular, the semisynthetic anticancer drugs etoposide (**108**), etopophos (**109**), and teniposide (**110**) (Gordaliza et al. 2000). Due to the extensive literature on podophyllotoxin analogues, which includes some excellent reviews (Bohlin et al. 1996; Imbert 1998; Lee et al. 2003; Liu et al. 2007), we cite here only a couple of recently reported unusual compounds that showed comparable cytotoxicity to that of podophyllotoxin. In a series of hybrids of podophyllotoxin with the potent tubulin synthetic inhibitor indubulin, evaluated against four human cancer cell lines including HeLa (cervix), SKOV3 (ovary), K562 (leukemia), and K562ADR (adriamycin resistant leukemia), compound **111** was the most potent and showed $IC_{50} = 0.18\,\mu M$, comparable to that of both **59** and indubulin (Yu et al. 2008). A multicomponent synthetic approach was used to obtain dihydropyrazole simplified analogues of **59** (Magedov et al. 2007). These analogues were tested against human cervical (HeLa) and breast (MCF7-AZ) adenocarcinomas and T-cell leukemia (Jurkat) cultures; among them worth mentioning here are compounds **112**, potently cytotoxic and able to induce apoptosis in Jurkat cells at $5\,\mu M$, and compound **113**, whose significant structural dissimilarity with podophyllotoxin points to the possibility of a different mechanism of action for this analogue.

According to the IUPAC recommendations, "neolignans" are phenylpropanoid dimers with a carbon linkage between two C_6C_3 units different from 8–8′. Some dihydrobenzofuran neolignans with promising biological activity have been reported in the past, among them the constituents of "dragon's blood," the blood-red latex produced by some *Croton* spp. growing in South America and employed in traditional medicine (Gupta et al. 2008). One of these neolignans, 3′,4-di-*O*-methylcedrusin (**114**), inspired the biomimetic synthesis of analogues by oxidative coupling of caffeic esters (Lemiere et al. 1995); the products were evaluated as antitumor and antileishmanial agents and the 2*R*,3*R*-enantiomer **115** separated from the racemic dimerization product of methyl caffeate, showed promising antitumor properties (average $GI_{50} = 0.3\,\mu M$; $GI_{50} < 10\,nM$ toward breast cancer cell lines) and proved by far more active than the 2*S*,3*S*-enantiomer as inhibitor of tubulin polymerization (Pieters et al. 1999). In subsequent works, **115** (or the racemate) was submitted to the chorioallantoic membrane assay (CAM) showing pronounced angiogenesis-inhibitory activity (Apers et al. 2002). Subsequently, a series of dihydrobenzofuran (and benzofuran) neolignans was obtained by dimerization of lipophilic caffeic esters: these compounds were evaluated for cytotoxic and antiprotozoal activity and some proved highly active against *Plasmodium* and *Leishmania* spp.; although no potent cytotoxic activity was observed for most of these synthetic neolignans, some showed GI_{50} values lower than $2\,\mu M$, namely, **116** (only one of enantiomers is reported), **117** (NCI-H522, non-small-cell lung), and **118** (CNS cancer cell line). More recently, (±)-**115** has been submitted to an extensive study focused on the effect of this compound on cell cycle and apoptosis (Bose et al. 2009); the results showed that (±)-**115** causes cell cycle arrest in the G2/M and induces apoptosis involving the mitochondrial controlled pathway.

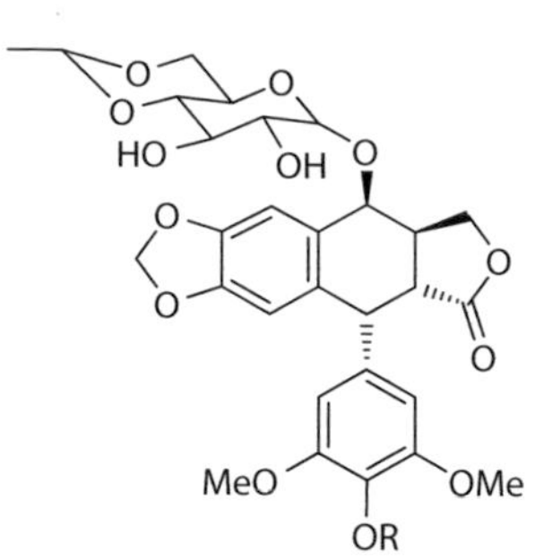

108 R=H
109 R=PO₃H₂

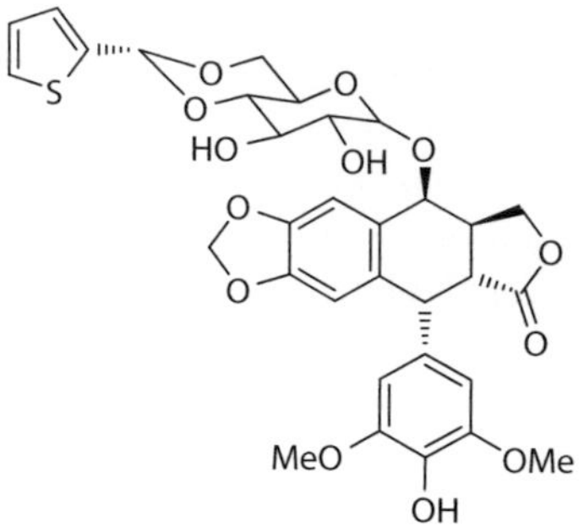

110

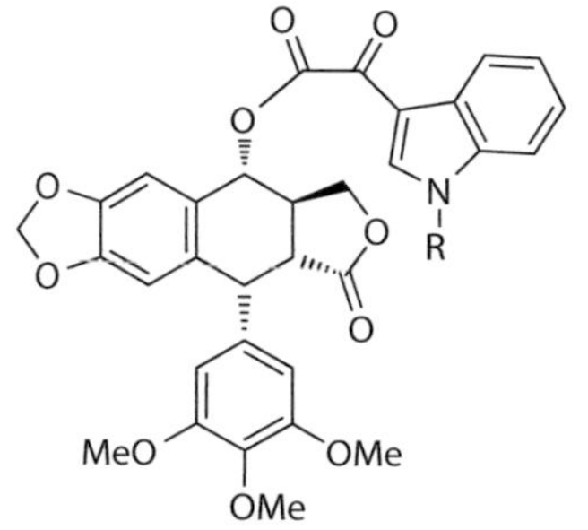

111 R=Benzo[1,3]dioxol-5-ylmethyl

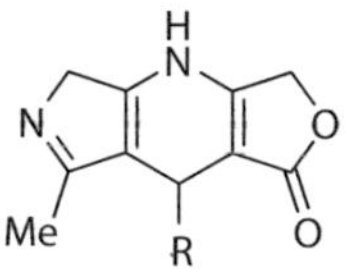

112 R=3,4,5-tri-MeO-Ph
113 R=Me

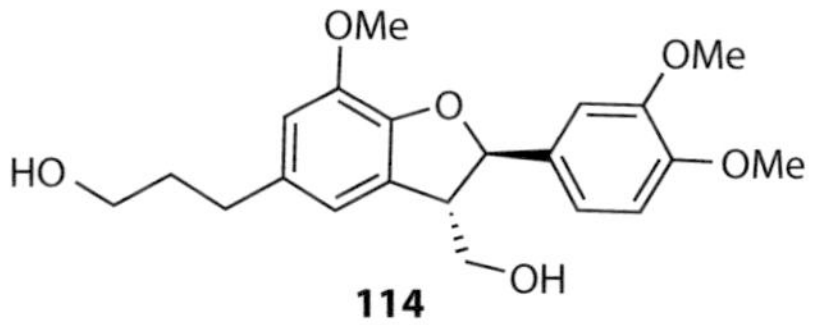

114

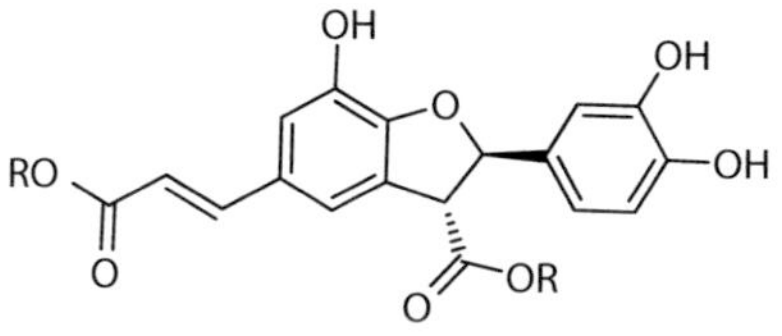

115 R=Me
116 R=t-Bu

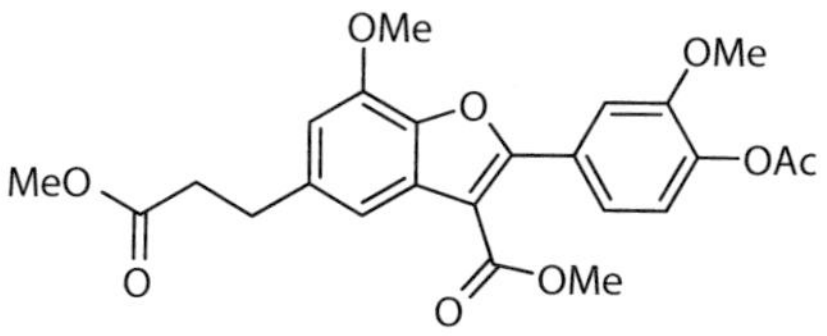

117

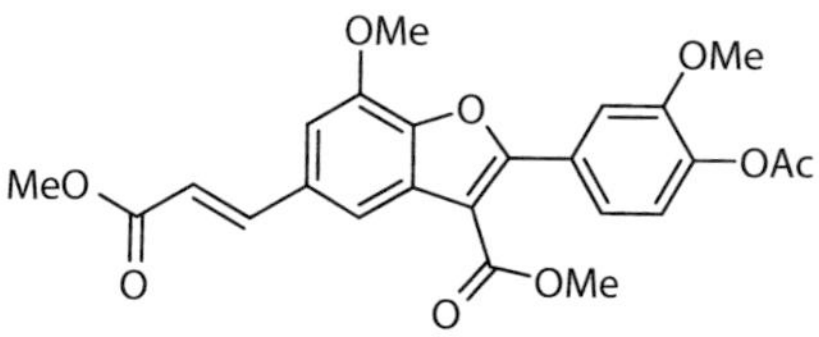

118

The above-cited antitumor properties of some lignans/neolignans prompted us to carry out biomimetic syntheses starting from CAPE (**9**), a substrate previously not employed for oxidative coupling reactions. By using Mn(OAc)₃ as an oxidative reagent, we obtained with good yield (72%) an unexpected product, the strongly fluorescent benzo[*kl*]xanthene lignan **119** (Daquino et al. 2009). This methodology was confirmed by using methyl caffeate (**100**), whose dimerization product was the lignan **120**. We also carried out a mechanistic study of this oxidative coupling reaction, thus revealing the role of the Mn^{2+} ions in the formation of these unusual products; this approach was applied to synthesize two natural benzoxanthene lignans and in particular rufescidride (**121**), originally isolated from *Cordia rufescens* (Souza da Silva et al. 2004). This is one of the few previously reported examples of benzo[*kl*]xanthene lignans, a rare group of compounds among both natural and synthetic products. Due to their unavailability, the biological properties of these lignans have not been explored, so as a first step in a study of their possible biomedical properties, we carried out an evaluation of their interaction with DNA through NMR-based approach and molecular docking, paralleled by in vitro evaluation of their antiproliferative activity toward two human cancer cell lines, namely, SW 480 (colon carcinoma) and HepG2 (hepatoblastoma) (Di Micco et al. 2011). To acquire data useful for SAR studies, the permethyl (**122** and **123**) and peracetyl (**124** and **125**) derivatives of **119** and **120** were also prepared. A significant activity was observed for compound **119** (IC_{50} = 2.57 μM and 4.76 against SW480 and HepG2, respectively), which proved more active than **120**. The activity of the peracetates was comparable to that of the parent compounds, whereas the permethylated proved inactive. From the NMR experimental results and molecular docking, it was evident that the planar chromophoric moiety of these compounds is intercalated between two DNA base pairs and the flexible chemical appendages are collocated along the grooves of nucleic acid and make contact with the external deoxyribose/backbone (see Figure 9.5). In particular, the bulky phenylethyl groups of **119** giving wider van der Waals contacts with the minor groove, improve the binding affinity, in perfect agreement with the antiproliferative activity data. Our investigation has also highlighted a crucial role of the phenolic groups in the recognition process by the biological target. Indeed, the chemical conversion of the hydroxy into methoxy functionalities prevents the hydrogen bond formation between ligands and macromolecule, lowering the affinity for the DNA, as confirmed by the bioassays. Nevertheless, acetylation does not affect the affinity for the biological target negatively, and peracetates are probably hydrolyzed to their active forms by intracellular esterases. In conclusion, these data suggest that the benzo[*kl*]xanthene lignans are suitable lead compounds for the design of DNA selective ligands with potential antitumor properties; thus, further biological evaluations of these compounds are currently in progress; a recent study on swine granulosa cells has showed promising antiangiogenic properties of compound **119** (Basini et al. 2010a).

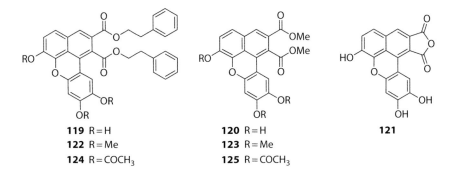

119 R=H
122 R=Me
124 R=COCH₃

120 R=H
123 R=Me
125 R=COCH₃

121

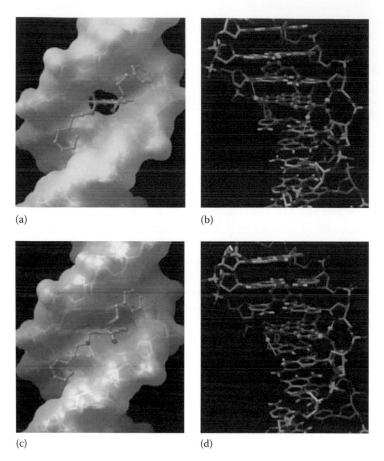

(a)   (b)

(c)   (d)

FIGURE 9.5   (See color insert.) 3D interactions of **10**-Model A (a and b) and **10**-Model B (c and d) complexes. In (a) and (c) the DNA is represented by molecular surface, and sticks and balls (colored by atom type: O, red; C, grey; polar H, sky blue; N, blue), whereas in (b) and (d) only by sticks and balls. The ligand is depicted in sticks (orange) and balls (colored as for DNA). In (a) and (c), the figure highlights the intercalation between two adjacent base pairs and the hydrogen bonds (yellow line) between ester functionality and NH_2 of guanine. In (b) and (d), beside the intercalation, the hydrogen bonds (yellow line) formed by OH group are shown. Note the correspondence with our text: **10 = 119**. (Di Micco, S., Daquino, C., Spatafora, C., Mazué, F., Delmas, D., Latruffe, N., Tringali, C., Riccio, R., and Bifulco, G., Structural basis for the potential antitumor activity of DNA-interacting Benzo[*kl*]xanthene lignans, *Org. Biomol. Chem.*, 9, 701–710, 2011. Copyright 2011, reprinted by permission of The Royal Society of Chemistry.)

9.5   CURCUMINOIDS

Curcumin (diferuloylmethane **5**) is a well-known, nontoxic natural product responsible for the color and spicy flavor of the rhizome of turmeric (*Curcuma longa*) and curry powder, the latter employed daily in South Asian and Middle Eastern cuisine. This yellow pigment has been used since ancient times as a medical treatment for a multitude of affections and has recently emerged as a possible anticancer principle, in view of its cytotoxicity toward tumor cells, nuclear factor κB (NF-κB) inhibition, antiangiogenic activity, and other promising properties, as reported in recent reviews or books (Mosley et al. 2007; Itokawa et al. 2008; Steward et al. 2008; Labbozzetta et al. 2009). These also include synthetic analogues due to the observation that curcumin, like the previously cited polyphenols, has low bioavailability and high susceptibility toward hepatic and gastrointestinal metabolism via enzyme-catalyzed conjugation and reduction. Among the most promising analogues previously reviewed, we wish to cite here dimethylcurcumin (**126**), the fluorinated derivative EF24 (**127**), and its 2-pyridine analog EF31 (**128**). In particular, **126** inhibited TNF-α-induced activation of NFκB in NFκB reporter-expressing human kidney-derived cells with an IC_{50} of 3.4 μM, as compared to 8.2 μM for **5** (Weber et al. 2006), and **127** proved potently antiproliferative and antiangiogenic against a wide range of tumor cells (Adams et al. 2004).

Some new studies have been carried out to acquire useful data for SAR analysis. A specific point of modification in the structure of curcumin is the β-diketone moiety; in fact, the keto-enol tautomerism of **5** may be an obstacle to its development as a clinically useful drug. Thus, some modified curcuminoids were prepared and evaluated as potential antiandrogenic agents; the 4-ethoxycarbonylethyl curcumin (ECECur, **129**) proved to be a potent anti-androgenic agent and exhibited selective cytotoxicity against androgen-dependent LNCaP prostate cancer cells, with IC_{50} = 1.5 μM (Lin et al. 2006a,b). Curcumin may act as a Michael acceptor, forming adducts with nucleophiles, in particular thiol groups (Mosley et al. 2007); two analogues of **127** and **128** with reduced light sensitivity and improved water solubility are their glutathione *bis*-adducts **130** and **131**, which showed essentially identical cytotoxicity (Sun et al. 2009). The effects of curcumin and some β-diketone modified analogues were evaluated by in vitro assays in the hepatocellular carcinoma HA22T/VGH cells, as well as in the MCF-7 breast cancer cell line, and in its multidrug resistant (MDR) variant MCF-7R (Simoni et al. 2008). Two analogues proved more active than **5**: **132**, with a central isoxazole ring, and the benzyl oxime derivative **133**; the latter proved to be the most potent, in particular against MCF-7R cells with IC_{50} = 9.3 μM, and inhibited constitutive NF-κB activation in the HA22T/VGH cell model. A study on the mechanism of action of the analogues *bis*-demethoxycurcumin (**134**) and curcumin diacetate (**135**) showed that both analogues, and particularly **134**, induce a p53- and p21CIP1/WAF1-independent mitotic arrest (Basile et al. 2009). This compound also induces rapid DNA double-strand breaks. As a continuation of a previous study on 1,5-diarylpentadienones with alkoxy-substituted aromatic rings (Ohori et al. 2006), a series of novel curcumin analogues were recently synthesized and screened for anticancer activity against the human colon cancer cell line HCT-116 (Yamakoshi et al. 2010). The entire data allowed establishing useful SAR on curcuminoids with 1,5-diarylpentadienone moiety and obtaining new molecular probes to target key biomolecules. The 3-oxo-1,4-pentadiene portion proved essential for eliciting cytotoxicity, and hexasubstituted compounds exhibited strong activity, with GI_{50} values in the range of 0.7–1.5 μM; symmetry is not essential for high cytotoxicity, the most potent compound being **136**.

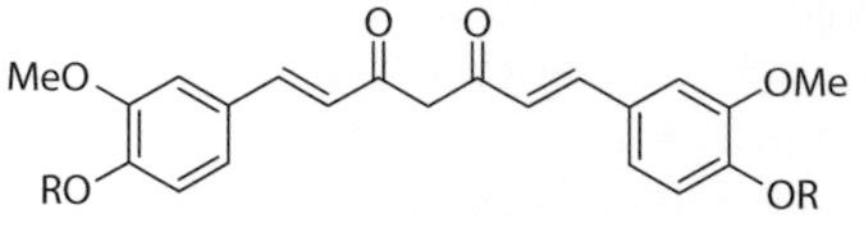

126 R = OMe
135 R = OAc

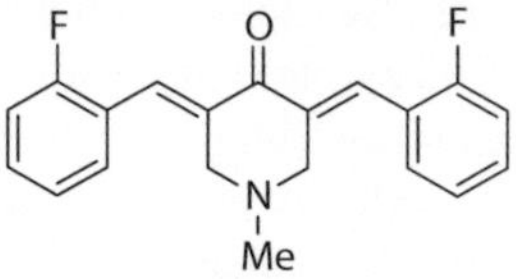

127

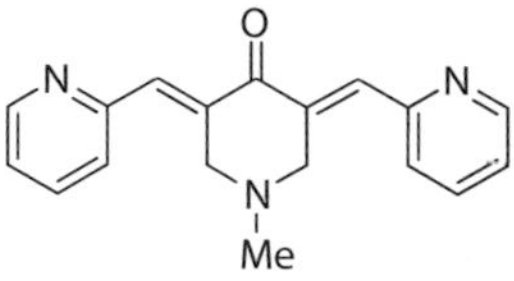

128

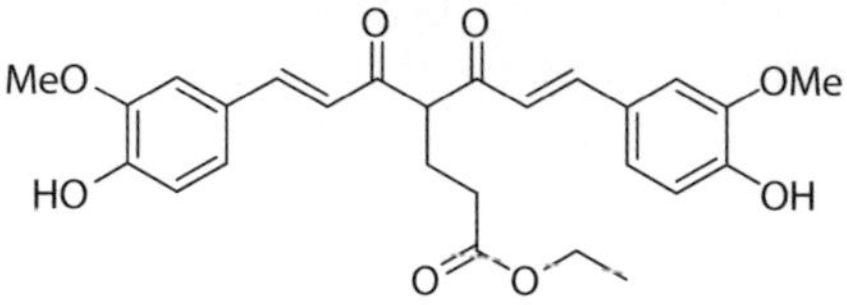

129

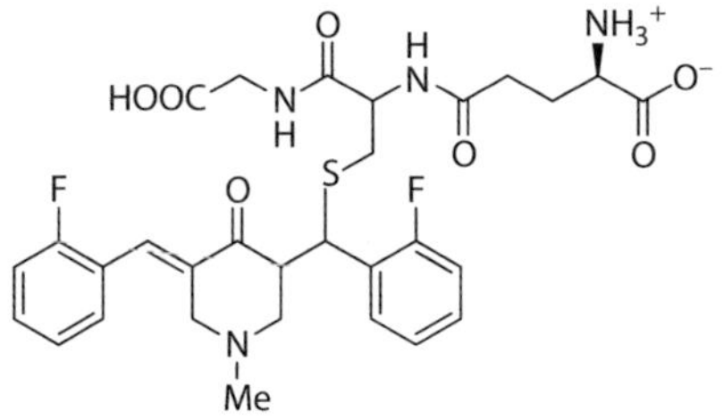

130

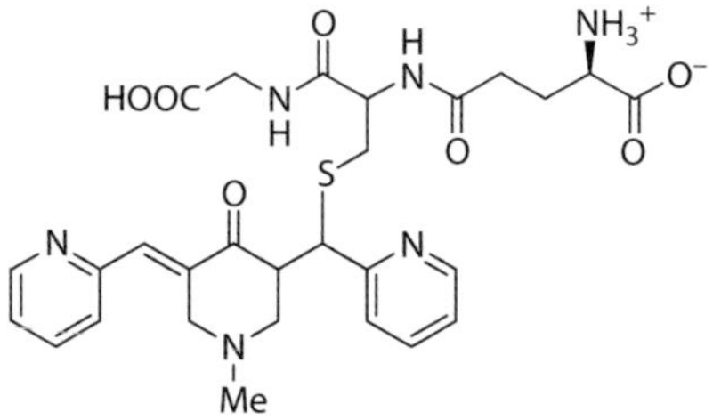

131

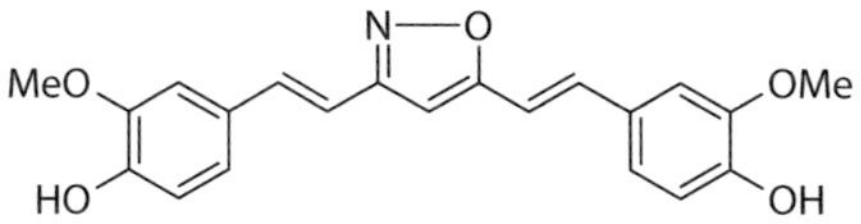

132

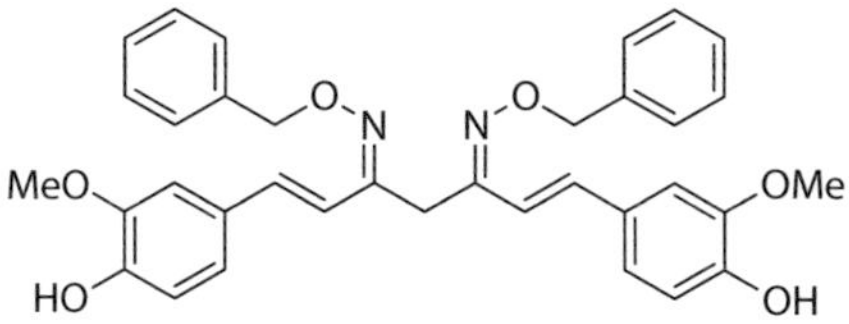

133

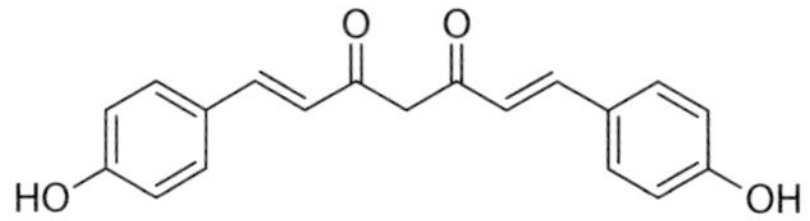

134

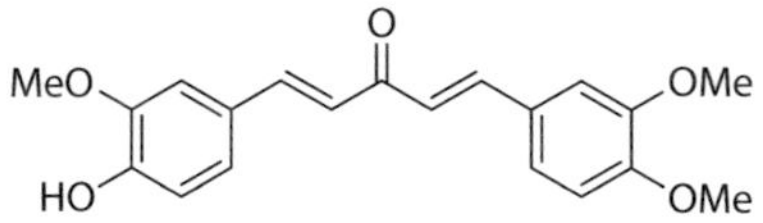

136

9.6 CONCLUSIONS

This survey of recent literature on synthetic antitumor analogues of some natural polyphenols, although far from being exhaustive (and limited-to-low-molecular-weight compounds), gives an idea of the progress in this field and suggests that some polyphenolic leads may become useful anticancer drugs in the future. Although for many studies we reported only the most potent compounds in a series, 119 synthetic analogues of natural polyphenols displaying antitumor properties are listed herewith. In a few cases, the cited compounds are already in clinical trial. In many other cases, useful SARs have been obtained, which may be the basis for future optimization. In this regard, it is worth highlighting the close structural relationships connecting some families of tubulin inhibitors, namely, analogues of resveratrol (**1**), combretastatin A-4 (**15**), and chalcones (Section 9.3.1). In turn, α-aryl chalcones appear as a new class of antitubulin agents mimicking the cytotoxic lignan podophyllotoxin (**59**). Thus, the synthesis of further hybrid molecules sharing structural moieties with the antitubulin leads reported here may be a promising target.

With regard to the low metabolic stability of polyphenols reputed to be anticarcinogenic, many interesting results have been reported and a number of analogues that are more stable to metabolic conversion and display comparable or higher antitumor activity than the parent compound, have been obtained: examples are some synthetic analogues of resveratrol (**1**, Section 9.2), EGCG (**2**, Section 9.3.2), or curcumin (**5**, Section 9.5). In some cases, analogues with higher lipophilicity showed higher activity than the parent compound, in particular, stilbenoids, flavanols, flavones, and caffeic acid derivatives.

Finally, we believe that the recently reported biomimetic synthesis of benzo[*k,l*] xanthene lignans (Section 9.4.2), such as the CAPE dimer (**119**), is also worthy of mention here. This makes a rare subgroup of lignans available, whose biological properties are almost unexplored and whose preliminary evaluation as antitumor or antiangiogenic agents seems promising.

ACKNOWLEDGMENTS

This work was supported by a grant of the Università degli Studi di Catania (Progetti di Ricerca di Ateneo, Catania, Italy) and by MIUR, Ministero dell'Università e della Ricerca (PRIN, Rome, Italy).

REFERENCES

Achanta, G., A. Modzelewska, L. Feng, S. R. Khan, and P. Huang. 2006. A boronic-chalcone derivative exhibits potent anticancer activity through inhibition of the proteasome. *Mol. Pharmacol.* 70: 426–433.

Adams, B. K., E. M. Ferstl, M. C. Davis, M. Herold, S. Kurtkaya, R. F. Camalier, M. G. Hollingshead et al. 2004. Synthesis and biological evaluation of novel curcumin analogs as anti-cancer and anti-angiogenesis agents. *Bioorg. Med. Chem.* 12: 3871–3873.

Aggarwal, B. B., A. Bhardwaj, R. S. Aggarwal, N. P. Seeram, S. Shishodia, and Y. Takada. 2004. Role of resveratrol in prevention and therapy of cancer: Preclinical and clinical studies. *Anticancer Res.* 24: 2783–2840.

Aggarwal, B. B. and S. Shishodia. 2006a. Molecular targets of dietary agents for prevention and therapy of cancer. *Biochem. Pharmacol.* 71: 1397–1421.

Aggarwal, B. B. and S. Shishodia. 2006b. *Oxidative Stress and Disease*, Vol. 20. Resveratrol in health and disease. London, U.K.: Taylor & Francis.

Ahn, Y. M., L. Vogeti, C.-J. Liu Santhapuram, K. R. Hari, J. M. White, V. Vasandani, L. A. Mitscher et al. 2007. Design, synthesis, and antiproliferative and CDK2-cyclin A inhibitory activity of novel flavopiridol analogues. *Bioorg. Med. Chem.* 15: 702–713.

Albini, A., R. Dell'Eva, R. Vene, N. Ferrari, D. R. Buhler, D. M. Noonan, and G. Fassina. 2006. Mechanisms of the antiangiogenic activity by the hop flavonoid xanthohumol: NF-κB and Akt as targets. *FASEB J.* 20: 527–529.

Aleo, D., V. Cardile, R. Chillemi, G. Granata, and S. Sciuto. 2008. Chemoenzymatic synthesis and some biological properties of *O*-phosphoryl derivatives of (*E*)-resveratrol. *Nat. Prod. Comm.* 3: 1693–1700.

Almond, J. B. and G. M. Cohen. 2002. The proteasome: A novel target for cancer chemo-therapy. *Leukemia* 16: 433–443.

Apers, S., D. Paper, J. Buergermeister, S. Baronikova, S. Van Dyck, G. Lemiere, A. Vlietinck, and L. Pieters. 2002. Antiangiogenic activity of synthetic dihydrobenzofuran lignans. *J. Nat. Prod.* 65: 718–720.

Apers, S., A. Vlietinck, and L. Pieters. 2003. Lignans and neolignans as lead compounds. *Phytochem. Rev.* 2: 201–217.

Ayers, D. C. and J. D. Loike. 1990. *Lignans. Chemical, Biological and Clinical Properties.* Cambridge, U.K.: Cambridge University Press.

Bandgar, B. P., S. S. Gawande, R. G. Bodade, J. V. Totre, and C. N. Khobragade. 2010. Synthesis and biological evaluation of simple methoxylated chalcones as anticancer, anti-inflammatory and antioxidant agents. *Bioorg. Med. Chem.* 18: 1364–1370.

Banerjee, S., Z. Wang, M. Mohammad, F. H. Sarkar, and R. M. Mohammad. 2008. Efficacy of selected natural products as therapeutic agents against cancer. *J. Nat. Prod.* 71: 492–496.

Basile, V., E. Ferrari, S. Lazzari, S. Belluti, F. Pignedoli, and C. Imbriano. 2009. Curcumin derivatives: Molecular basis of their anti-cancer activity. *Biochem. Pharmacol.* 78: 1305–1315.

Basini, G., L. Baioni, S. Bussolanti, F. Grasselli, C. Daquino, C. Spatafora, and C. Tringali. 2010a. Antiangiogenic properties of an unusual benzo[k,l]xanthene lignan derived from CAPE (Caffeic Acid Phenethyl Ester). *Invest. New Drug.* In press, DOI 10.1007/s10637-010-9550-z.

Basini, G., C. Tringali, L. Baioni, S. Bussolati, C. Spatafora, and F. Grasselli. 2010b. Biological effects on granulosa cells of hydroxylated and methylated resveratrol analogues. *Mol. Nutr. Food Res.* 54: S236–S243.

Baur, J. A. and D. A. Sinclair. 2006. Therapeutic potential of resveratrol: The *in vivo* evidence. *Nat. Rev. Drug Discov.* 5: 493–506.

Belleri, M., D. Ribatti, S. Nicoli, F. Cotelli, L. Forti, V. Vannini, L. A. Stivala, and M. Presta. 2005. Antiangiogenic and vascular-targeting activity of the microtubule-destabilizing *trans*-resveratrol derivative 3,5,4′-trimethoxystilbene. *Mol. Pharmacol.* 67: 1451–1459.

Beltran-Ramirez, O., L. Aleman-Lazarini, M. Salcido-Neyoyet, S. Hernandez-Garcia, S. Fattel-Fazenda, E. Arce-Popoca, J. Arellanes-Robledo et al. 2008. Evidence that the anticarcinogenic effect of caffeic acid phenethyl ester in the resistant hepatocyte model involves modifications of cytochrome P450. *Toxicol. Sci.* 104: 100–106.

Bemis, D. L., A. E. Katz, and R. Buttyan. 2006. Clinical trials of natural products as chemo-preventive agents for prostate cancer. *Expert Opin. Investig. Drugs* 15: 1191–1200.

Bernhaus, A., M. Fritzer-Szekeres, M. Grusch, P. Saiko, G. Krupitza, S. Venkateswarlu, G. Trimurtulu, W. Jaeger, and T. Szekeres. 2009a. Digalloylresveratrol, a new phenolic acid derivative induces apoptosis and cell cycle arrest in human HT-29 colon cancer cells. *Cancer Lett.* 274: 299–304.

Bernhaus, A., M. Ozsvar-Kozma, P. Saiko, M. Jaschke, A. Lackner, M. Grusch, Z. Horvath et al. 2009b. Antitumor effects of KITC, a new resveratrol derivative, in AsPC-1 and BxPC-3 human pancreatic carcinoma cells. *Invest. New Drugs* 27: 393–401.

Bhakuni, D. S. and R. Chaturvedi. 1984. Chemical constituents of *Crotalaria madurensis*. *J. Nat. Prod.* 47: 585–591.

Bohlin, L. and B. Rosen. 1996. Podophyllotoxin derivatives: Drug discovery and develop-ment. *Drug Discov. Today* 1: 343–351.

Bonfili, L., V. Cecarini, M. Amici, M. Cuccioloni, M. Angeletti, J. N. Keller, and A. M. Eleuteri. 2008. Natural polyphenols as proteasome modulators and their role as anti-cancer com-pounds. *FEBS J.* 275: 5512–5526.

Bose, J. S., V. Gangan, R. Prakash, S. K. Jain, and S. K. Manna. 2009. A dihydrobenzofuran lignan induces cell death by modulating mitochondrial pathway and G2/M cell cycle arrest. *J. Med. Chem.* 52: 3184–3190.

Boudet, A.-M. 2007. Evolution and current status of research in phenolic compounds. *Phytochemistry* 68: 2722–2735.

Boumendjel, A., J. Boccard, P.-A. Carrupt, E. Nicolle, M. Blanc, A. Geze, L. Choisnard, D. Wouessidjewe, E.-L. Matera, and C. Dumontet. 2008. Antimitotic and antiprolifera-tive activities of chalcones: Forward structure–activity relationship. *J. Med. Chem.* 51: 2307–2310.

Boumendjel, A., A. McLeer-Florin, P. Champelovier, D. Allegro, D. Muhammad, F. Souard, M. Derouazi, V. Peyrot, B. Toussaint, and J. Boutonnat. 2009. A novel chalcone deriva-tive which acts as a microtubule depolymerising agent and an inhibitor of P-gp and BCRP in *in-vitro* and *in-vivo* glioblastoma models. *BMC Cancer* 9: 242.

Burjonroppa, S. and K. Fujise. 2006. Resveratrol as cardioprotective agent: Evidence from bench and bedside. In *Resveratrol in Health and Disease*, eds. B. B. Aggarwal and S. Shishodia, pp. 539–555. London, U.K.: Taylor & Francis.

Cao, T. M., D. Durrant, A. Tripathi, J. Liu, S. Tsai, G. E. Kellogg, D. Simoni, and R. M. Lee. 2008. Stilbene derivatives that are colchicine site microtubule inhibitors have antileuke-mic activity and minimal systemic toxicity. *Am. J. Hematol.* 83: 390–397.

Cardenas, M., M. Marder, V. C. Blank, L. P. Roguin, and P. Leonor. 2006. Antitumor activity of some natural flavonoids and synthetic derivatives on various human and murine can-cer cell lines. *Bioorg. Med. Chem.* 14: 2966–2971.

Cardile, V., R. Chillemi, L. Lombardo, S. Sciuto, C. Spatafora, and C. Tringali. 2007. Antiproliferative activity of methylated analogues of *E*- and *Z*-resveratrol. *Z. Naturforsch.* 62c: 189–195.

Cardile, V., L. Lombardo, C. Spatafora, and C. Tringali. 2005. Chemo-enzymatic synthesis and antiproliferative activity of resveratrol analogues. *Bioorg. Chem.* 33: 22–33.

Chang, J.-Y., M.-F. Yang, C.-Y. Chang, C.-M. Chen, C.-C. Kuo, and J.-P. Liou. 2006. 2-Amino and 2′-aminocombretastatin derivatives as potent antimitotic agents. *J. Med. Chem.* 49: 6412–6415.

Chaudhary, A., S. N. Pandeya, P. Kumar, P. P. Sharma, S. Gupta, N. Soni, K. K. Verma, and G. Bhardwaj. 2007. Combretastatin A-4 analogs as anticancer agents. *Mini Rev. Med. Chem.* 7: 1186–1205.

Chen, L. and H.-Y. Zhang. 2007. Cancer preventive mechanisms of the green tea polyphenol (−)-epigallocatechin-3-gallate. *Molecules* 12: 946–957.

Cheng, L., H. Tan, X. Wu, R. Hu, C. Aw, M. Zhao, H.-M. Shen, and Y. Lu. 2008. Novel syn-thetic luteolin analogue-caused sensitization of tumor necrosis factor-α-induced apopto-sis in human tumor cells. *Org. Biom. Chem.* 6: 4102–4104.

Chillemi, R., S. Sciuto, C. Spatafora, and C. Tringali. 2007. Anti-tumor properties of stilbene-based resveratrol analogues: Recent results. *Nat. Prod. Comm.* 2: 499–513.

Cushman, M., D. Nagarathnam, D. Gopal, A. K. Chakraborti, C. M. Lin, and E. Hamel. 1991. Synthesis and evaluation of stilbene and dihydrostilbene derivatives as potential antican-cer agents that inhibit tubulin polymerization. *J. Med. Chem.* 34: 2579–2588.

Daquino, C., A. Rescifina, C. Spatafora, and C. Tringali. 2009. Biomimetic synthesis of natu-ral and "unnatural" lignans by oxidative coupling of caffeic esters. *Eur. J. Org. Chem.* 6289–6300.

De Lemos, M. L. 2001. Effects of soy phytoestrogens genistein and daidzein on breast cancer growth. *Ann. Pharmacother.* 35: 1118–1121.

Delmas, D., A. Lançon, D. Colin, B. Jannin, and N. Latruffe. 2006. Resveratrol as a chemopre-ventive agent: A promising molecule for fighting cancer. *Curr. Drug Targets* 7: 423–442.

Dewick, P. M. 2009. *Medicinal Natural Products: A Biosynthetic Approach*, 3rd edn., pp. 1–539. Chichester, U.K.: J. Wiley.

Di Micco, S., C. Daquino, C. Spatafora, F. Mazué, D. Delmas, N. Latruffe, C. Tringali, R. Riccio, and G. Bifulco. 2011. Structural basis for the potential antitumor activity of DNA-interacting Benzo[*kl*]xanthene lignans. *Org. Biomol. Chem.* 9: 701–710.

Donnelly, A. C., J. R. Mays, J. A. Burlison, J. T. Nelson, G. Vielhauer, J. Holzbeierlein, and B. S. J. Blagg. 2008. The design, synthesis, and evaluation of coumarin ring derivatives of the novobiocin scaffold that exhibit antiproliferative activity. *J. Org. Chem.* 73: 8901–8920.

Ducki, S., G. MacKenzie, B. Greedy, S. Armitage, J. Fournier Dit Chabert, E. Bennett, J. Nettles, J. P. Snyder, and N. J. Lawrence. 2009a. Combretastatin-like chalcones as inhibitors of microtubule polymerisation. Part 2: Structure-based discovery of alpha-aryl chalcones. *Bioorg. Med. Chem.* 17: 7711–7722.

Ducki, S., G. Mackenzie, N. J. Lawrence, and J. P. Snyder. 2005. Quantitative structure–activity relationship (5D-QSAR) study of combretastatin-like analogues as inhibitors of tubulin assembly. *J. Med. Chem.* 48: 457–465.

Ducki, S., D. Rennison, M. Woo, A. Kendall, J. Fournier Dit Chabert, A. T. McGown, and N. J. Lawrence. 2009b. Combretastatin-like chalcones as inhibitors of microtubule polymerization. Part 1: Synthesis and biological evaluation of antivascular activity. *Bioorg. Med. Chem.* 17: 7698–7710.

East, A. J., W. D. Ollis, and R. E. Wheeler. 1969. Natural occurrence of 3-aryl-4-hydroxycoumarins. I. Phytochemical examination of *Derris robusta. J. Chem. Soc. C* 3: 365–374.

El-Refaei, M. F. and M. M. El-Naa. 2010. Inhibitory effect of caffeic acid phenethyl ester on mice bearing tumor involving angiostatic and apoptotic activities. *Chem. Bio. Interact.* 186: 152–156.

Etzenhouser, B., C. Hansch, S. Kapur, and C. D. Selassie. 2001. Mechanism of toxicity of esters of caffeic and dihydrocaffeic acids. *Bioorg. Med. Chem.* 9: 199–209.

Ferreira, M.-J. U., N. Duarte, N. Gyémánt, R. Radics, G. Cherepnev, A. Varga, and J. Molnar. 2006. Interaction between doxorubicin and the resistance modifier stilbene on multidrug resistant mouse lymphoma and human breast cancer cells. *Anticancer Res.* 26: 3541–3546.

Fiuza, S. M., C. Gomes, L. J. Teixeira, M. T. Girao da Cruz, M. N. D. S. Cordeiro, N. Milhazes, F. Borges, and M. P. M. Marques. 2004. Phenolic acid derivatives with potential anticancer properties: A structure–activity relationship study. Part 1: Methyl, propyl and octyl esters of caffeic and gallic acids. *Bioorg. Med. Chem.* 12: 3581–3589.

Flørenesd, V. A. and T. V. Hansen. 2008. 1,5-Disubstituted 1,2,3-triazoles as cis-restricted analogues of combretastatin A-4: Synthesis, molecular modelling and evaluation as cytotoxic agents and inhibitors of tubulin. *Bioorg. Med. Chem.* 16: 4829–4838.

Fujiki, H., M. Suganuma, K. Imai, and K. Nakachi. 2002. Green tea: Cancer preventive beverage and/or drug. *Cancer Lett.* 188: 9–13.

Go, M. L., X. Wu, and X. L. Liu. 2005. Chalcones: An update on cytotoxic and chemopreventive properties. *Curr. Med. Chem.* 12: 483–499.

Gordaliza, M., M. A. Castro, J. M. Miguel del Corral, and A. San Feliciano. 2000. Antitumor properties of podophyllotoxin and related compounds. *Curr. Pharm. Des.* 6: 1811–1839.

Gupta, D., B. Bleakley, and R. K. Gupta. 2008. Dragon's blood: Botany, chemistry and therapeutic uses. *J. Ethnopharmacol.* 115: 361–380.

Hackett, J. C., Y.-W. Kim, B. Su, and R. W. Brueggemeier. 2005. Synthesis and characterization of azole isoflavone inhibitors of aromatase. *Bioorg. Med. Chem.* 13: 4063–4070.

Hadden, M. K., L. Galam, R. L. Matts, and B. S. J. Blagg. 2007. Derrubone, an inhibitor of the Hsp90 protein folding machinery. *J. Nat. Prod.* 70: 2014–2018.

Hadfield, J. A., K. Gaukroger, N. Hirst, A. P. Weston, N. J. Lawrence, and A. T. McGown. 2005. Synthesis and evaluation of double bond substituted combretastatins. *Eur. J. Med. Chem.* 40: 529–541.

Haslam, E. 1998. *Practical Polyphenolics: From Structure to Molecular Recognition and Physiological Action*. Cambridge, U.K.: Cambridge University Press.

Hayes, C. J., B. P. Whittaker, S. A. Watson, and A. M. Grabowska. 2006. Synthesis and preliminary anticancer activity studies of C4 and C8-modified derivatives of catechin gallate (CG) and epicatechin gallate (ECG). *J. Org. Chem.* 71: 9701–9712.

Hill, S. A., G. M. Tozer, G. R. Pettit, and D. J. Chaplin. 2002. Preclinical evaluation of the antitumor activity of the novel vascular-targeting agent Oxi 4503. *Anticancer Res.* 22: 1453–1458. http://www.groupepolyphenols.com

Huang, X.-F., B.-F. Ruan, X.-T. Wang, C. Xu, H.-M. Ge, H.-L. Zhu, and R.-X. Tan. 2007. Synthesis and cytotoxic evaluation of a series of resveratrol derivatives modified in C2 position. *Eur. J. Med. Chem.* 42: 263–267.

Huo, C., S. B. Wan, W. H. Lam, L. Li, Z. Wang, K. R. Landis-Piwowar, D. Chen, Q. P. Dou, and T. H. Chan. 2008. The challenge of developing green tea polyphenols as therapeutic agents. *Inflammopharmacology* 16: 248–252.

Imai, K., K. Suga, and K. Nakachi. 1997. Cancer-preventive effects of drinking green tea among a Japanese population. *Prev. Med.* 26: 769–775.

Imbert, T. F. 1998. Discovery of podophyllotoxins. *Biochimie* 80: 207–222.

Itokawa, H., Q. Shi, T. Akiyama, S. L. Morris-Natschke, and K.-H. Lee. 2008. Recent advances in the investigation of curcuminoids. *Chin. Med.* 3: 11.

Jang, M., L. Cai, G. O. Udeani et al. 1997. Cancer chemopreventive activity of resveratrol, a natural product derived from grapes. *Science* 275: 218–220.

Jayaprakasam, B., M. Vanisree, Y. Zhang, D. L. Dewitt, and M. G. Nair. 2006. Impact of alkyl esters of caffeic and ferulic acids on tumor cell proliferation, cyclooxygenase enzyme, and lipid peroxidation. *J. Agric. Food Chem.* 54: 5375–5381.

Jazirehi, A. R. and B. Bonavida. 2006. Resveratrol as a sensitizer to apoptosis-inducing stimuli. In *Resveratrol in Health and Disease*, eds. B. B. Aggarwal and S. Shishodia, pp. 399–421. London, U.K.: Taylor & Francis.

Jeong, S. H., W. S. Jo, S. Song, H. Suh, S.-Y. Seol, S.-H. Leem, T. K. Kwon, and Y. H. Yoo. 2009. A novel resveratrol derivative, HS1793, overcomes the resistance conferred by Bcl-2 in human leukemic U937 cells. *Biochem. Pharmacol.* 77: 1337–1347.

Kanthou, C. and G. M. Tozer. 2007. Tumor targeting by microtubule-depolymerising vascular disrupting agents. *Expert Opin. Ther. Targets* 11: 1443–1457.

Katalinic, V., S. S. Mozina, D. Skroza, I. Generalic, H. Abramovic, M. Milos, I. Ljubenkov et al. 2009. Polyphenolic profile, antioxidant properties and antimicrobial activity of grape skin extracts of 14 *Vitis vinifera* varieties grown in Dalmatia (Croatia). *Food Chem.* 119: 715–723.

Katsori, A. M. and D. Hadjipavlou-Litina. 2009. Chalcones in cancer: Understanding their role in terms of QSAR. *Curr. Med. Chem.* 16: 1062–1081.

Kim, Y. W., J. C. Hackett, and R. W. Brueggemeier. 2004. Synthesis and aromatase inhibitory activity of novel pyridine-containing isoflavones. *J. Med. Chem.* 47: 4032–4040.

Kingston, D. G. I. 2009. Tubulin-interactive natural products as anticancer agents. *J. Nat. Prod.* 72: 507–515.

Kong, Y., J. Grembecka, M. C. Edler, E. Hamel, S. L. Mooberry, M. Sabat, J. Rieger, and M. L. Brown. 2005. Structure-based discovery of a boronic acid bioisostere of combretastatin A-4. *Chem. Biol.* 12: 1007–1014.

Kong, Y., K. Wang, M. C. Edler, E. Hamel, S. L. Mooberry, M. A. Paigeand, and M. L. Brown. 2010. A boronic acid chalcone analog of combretastatin A-4 as a potent anti-proliferation agent. *Bioorg. Med. Chem.* 18: 971–977.

Kumar, S. K., E. Hager, C. Pettit, H. Gurulingappa, N. E. Davidson, and S. R. Khan. 2003. Design, synthesis, and evaluation of novel boronic-chalcone derivatives as antitumor agents. *J. Med. Chem.* 46: 2813–2815.

Labbozzetta, M., M. Notarbartolo, P. Poma, A. Maurici, L. Inguglia, P. Marchetti, M. Rizzi, R. Baruchello, D. Simoni, and N. D'Alessandro. 2009. Curcumin as a possible lead compound against hormone-independent, multidrug-resistant breast cancer. Steroid enzymes and cancer. *Ann. N. Y. Acad. Sci.* 1155: 278–283.

Lambert, J. D., S. Sang, J. Hong, S.-J. Kwon, M.-J. Lee, C.-T. Ho, and C. S. Yang. 2006. Peracetylation as a means of enhancing in vitro bioactivity and bioavailability of epigallocatechin-3-gallate. *Drug Metab. Dispos.* 34: 2111–2116.

Lamy, S., D. Gingras, and R. Béliveau. 2002. Green tea catechins inhibit vascular endothelial growth factor receptor phosphorylation. *Cancer Res.* 62: 381–385.

Lançon, A., N. Hanet, B. Jannen, D. Delmas, J.-M. Heyuel, G. Lizard, M.-C. Chagnot, Y. Artur, and N. Latruffe. 2007. Resveratrol in human hepatoma HepG2 cells: Metabolism and inducibility of detoxifying enzymes. *Drug Metab. Dispos.* 35: 699–703.

Landis-Piwowar, K. R., C. Huo, D. Chen, V. Milacic, G. Shi, T. H. Chan, and Q. P. Dou. 2007. A novel prodrug of the green tea polyphenol (−)-epigallocatechin-3-gallate as a potential anticancer agent. *Cancer Res.* 67: 4303–4310.

Le Marchand, L., S. P. Murphy, J. H. Hankin, L. R. Wilkens, and L. N. Kolonel. 2000. Intake of flavonoids and lung cancer. *J. Natl. Cancer Inst.* 92: 154–160.

Lee, Y. J., P. H. Liao, W. K. Chen, and C. Y. Yang. 2000. Preferential cytotoxicity of caffeic acid phenethyl ester analogues on oral cancer cells. *Cancer Lett.* 153: 51–56.

Lee, K.-H. and Z. Xiao. 2003. Lignans in treatment of cancer and other diseases. *Phytochem. Rev.* 2: 341–362.

Lemiere, G., M. Gao, A. De Groot, R. Dommisse, J. Lepoivre, L. Pieters, and V. Buss. 1995. 3′,4-Di-*O*-methylcedrusin: Synthesis, resolution and absolute configuration. *J. Chem. Society, Perkin Trans.* 18: 11775–11779.

Li, H.-Q., H.-M. Ge, Y.-X. Chen, C. Xu, L. Shi, H. Ding, H.-L. Zhu, and R.-X. Tan. 2006. Synthesis and cytotoxic evaluation of a series of genistein derivatives. *Chem. Biodivers.* 3: 463–472.

Lin, H. S. and P. C. Ho. 2009. A rapid HPLC method for the quantification of 3,5,4′-trimethoxy-*trans*-stilbene (TMS) in rat plasma and its application in pharmacokinetic study. *J. Pharm. Biomed. Anal.* 49: 387–392.

Lin, T. S., A. S. Ruppert, A. J. Johnson, B. Fischer, N. A. Heerema, L. A. Andritsos, K. A. Blum et al. 2009. Phase II study of flavopiridol in relapsed chronic lymphocytic leukemia demonstrating high response rates in genetically high-risk disease. *J. Clin. Oncol.* 27: 6012–6018.

Lin, L., Q. Shi, A. K. Nyarko, F. Kenneth, C.-C. Wu, C.-Y. Su, C. C.-Y. Shih, and K.-H. Lee. 2006a. Antitumor agents. 250. Design and synthesis of new curcumin analogues as potential anti-prostate cancer agents. *J. Med. Chem.* 49: 3963–3972.

Lin, L., Q. Shi, C.-Y. Su, C. C.-Y. Shih, and K.-H. Lee. 2006b. Antitumor agents 247. New 4-ethoxycarbonylethyl curcumin analogs as potential antiandrogenic agents. *Bioorg. Med. Chem.* 14: 2527–2534.

Lin, H.-S., C. Tringali, C. Spatafora, Q.-Y. Choo, and P. C. Ho. 2010. LC determination of *trans*-3,5,3′,4′,5′-pentamethoxystilbene in rat plasma. *Chromatographia* 72: 827–832.

Liu, Y.-Q., L. Yang, and X. Tian. 2007. Podophyllotoxin: Current perspectives. *Curr. Bioact. Compd.* 3: 37–66.

Losiewicz, M. D., B. A. Carlson, G. Kaur, E. A. Sausville, and P. J. Worland. 1994. Potent inhibition of CDC2 kinase activity by the flavonoid L86–8275. *Biochem. Biophys. Res. Commun.* 201: 589–595.

Lv, P.-C., K.-R. Wang, Q.-S. Li, J. Chen, J. Sun, and H.-L. Zhu. 2010. Design, synthesis and biological evaluation of chrysin long-chain derivatives as potential anticancer agents. *Bioorg. Med. Chem.* 18: 1117–1123.

MacRae, W. D. and G. H. N. Towers. 1985. Non-alkaloidal constituents of *Virola elongate* bark. *Phytochemistry* 24: 561–566.

Magedov, I. V., M. Manpadi, E. Rozhkova, N. M. Przheval'skii, S. Rogelj, S. T. Shors, W. F. A. Steelant, S. Van Slambrouck, and A. Kornienko. 2007. Structural simplification of bioactive natural products with multicomponent synthesis: Dihydropyridopyrazole analogues of podophyllotoxin. *Bioorg. Med. Chem. Lett.* 17: 1381–1385.

Manthey, J. A. and N. Guthrie. 2002. Antiproliferative activities of *Citrus* flavonoids against six human cancer cell lines. *J. Agric. Food Chem.* 50: 5837–5843.

Matsubara, K., A. Saito, A. Tanaka, N. Nakajima, R. Akagi, M. Mori, and Y. Mizushina. 2007. Epicatechin conjugated with fatty acid is a potent inhibitor of DNA polymerase and angiogenesis. *Life Sci.* 80: 1578–1585.

Matsumura, K., K. Kaihatsu, S. Mori, H. H. Cho, N. Kato, and S. H. Hyon. 2008. Enhanced antitumor activities of (−)-epigallocatechin-3-O-gallate fatty acid monoester derivatives in vitro and in vivo. *Biochem. Biophys. Res. Comm.* 377: 1118–1122.

Mazué, F., D. Colin, J. Gobbo, M. Wegner, A. Rescifina, C. Spatafora, D. Fasseur et al. 2010. Structural determinants of resveratrol for cell proliferation inhibition potency: Experimental and docking studies of new analogs. *Eur. J. Med. Chem.* 45: 2972–2980.

Minutolo, F., G. Sala, A. Bagnacani, S. Bertini, I. Carboni, G. Placanica, G. Prota et al. 2005. Synthesis of a resveratrol analogue with high ceramide-mediated proapoptotic activity on human breast cancer cells. *J. Med. Chem.* 48: 6783–6786.

Modzelewska, A., C. Pettit, G. Achanta, N. E. Davidson, P. Huang, and S. R. Khan. 2006. Anticancer activities of novel chalcone and bis-chalcone derivatives. *Bioorg. Med. Chem.* 14: 3491–3495.

Mohanakumara, P., N. Sreejayan, V. Priti, B. T. Ramesha, G. Ravikanth, K. N. Ganeshaiah, R. Vasudeva et al. 2010. *Dysoxylum binectariferum* Hook.f (Meliaceae), a rich source of rohitukine. *Fitoterapia* 81: 145–148.

Moran, B. W., F. P. Anderson, A. Devery, S. Cloonan, W. E. Butler, S. Varughese, S. M. Draper, and P. T. M. Kenny. 2009. Synthesis, structural characterisation and biological evaluation of fluorinated analogues of resveratrol. *Bioorg. Med. Chem.* 17: 4510–4522.

Mosley, C. A., D. C. Liotta, and J. P. Snyder. 2007. Highly active anticancer curcumin analogues. *Adv. Exp. Med. Biol.* 595: 77–103.

Musa, M. A., J. S. Cooperwood, and M. O. F. Khan. 2008. A review of coumarin derivatives in pharmacotherapy of breast cancer. *Curr. Med. Chem.* 15: 2664–2679.

Nagle, D. G., D. Ferreira, and Y.-D. Zhou. 2006. Epigallocatechin-3-gallate (EGCG): Chemical and biomedical perspectives. *Phytochemistry* 67: 1849–1855.

Nichenametla, S. N., T. G. Taruscio, D. L. Barney, and J. H. Exon. 2006. A review of the effects and mechanisms of polyphenolics in cancer. *Crit. Rev. Food Sci. Nutr.* 46: 161–183.

Nicolosi, G., C. Spatafora, and C. Tringali. 2002. Chemo-enzymatic preparation of resveratrol derivatives. *J. Mol. Catal. B Enzym.* 16: 223–229.

Noble, M. E. M., J. A. Endicott, and L. N. Johnson. 2004. Protein kinase inhibitors: Insights into drug design from structure. *Science* 303: 1800–1805.

Nonomura, S., H. Kanagawa, and A. Makimoto. 1963. Chemical constituents of polygonaceous plants. I. Studies on the components of Ko-jo-kon (*Polygonum cuspidatum* Sieb. et Zucc.). *Yakugaku Zasshi* 83: 988–990.

Odlo, K., J. Fournier-Dit-Chabert, S. Ducki, O. A. B. S. M. Gani, I. Sylte, and T. V. Hansen. 2010. 1,2,3-Triazole analogs of combretastatin A-4 as potential microtubule-binding agents. *Bioorg. Med. Chem.* 18: 6874–6885.

Ohori, H., H. Yamakoshi, M. Tomizawa, M. Shibuya, Y. Kakudo, A. Takahashi, and S. Takahashi. 2006. Synthesis and biological analysis of new curcumin analogues bearing an enhanced potential for the medicinal treatment of cancer. *Mol. Cancer Ther.* 5: 2563–2571.

Pan, J.-Y., S.-L. Chen, M.-H. Yang, J. Wu, J. Sinkkonen, and K. Zous. 2009. An update on lignans: Natural products and synthesis. *Nat. Prod. Rep.* 26: 1251–1292.

Pan, M.-H., C.-L. Lin, J.-H. Tsai, C.-T. Ho, and W.-J. Chen. 2010. 3,5,3',4',5'-Pentamethoxystilbene (MR-5), a synthetically methoxylated analogue of resveratrol, inhibits growth and induces G1 cell cycle arrest of human breast carcinoma MCF-7 cells. *J. Agric. Food Chem.* 58: 226–334.

Pettit, G. R., N. Melody, A. Thornhill, J. C. Knight, T. L. Groy, and C. L. Herald. 2009a. Antineoplastic agents. 579. Synthesis and cancer cell growth evaluation of *E*-stilstatin 3: A resveratrol structural modification. *J. Nat. Prod.* 72: 1637–1642.

Pettit, R. K., G. R. Pettit, E. Hamel, F. Hogan, B. R. Moser, S. Wolf, S. Pon, J.-C. Chapuis, and J. M. Schmidt. 2009b. *E*-Combretastatin and *E*-resveratrol structural modifications: Antimicrobial and cancer cell growth inhibitory β-E-nitrostyrenes. *Bioorg. Med. Chem.* 17: 6606–6612.

Pettit, G. R., S. B. Singh, E. Hamel, C. M. Lin, D. S. Alberts, and K. D. Garcia. 1989. *Experientia* 45: 209–211.

Pettit, G. R., A. J. Thornhill, B. R. Moser, and F. Hogan. 2008. Antineoplastic agents. 552. Oxidation of combretastatin A-1: Trapping the *o*-quinone intermediate considered the metabolic product of the corresponding phosphate prodrug. *J. Nat. Prod.* 71: 1561–1563.

Pieters, L., S. Van Dyck, M. Gao, R. Bai, E. Hamel, A. Vlietinck, and G. Lemiere. 1999. Synthesis and biological evaluation of dihydrobenzofuran lignans and related compounds as potential antitumor agents that inhibit tubulin polymerization. *J. Med. Chem.* 42: 5475–5481.

Pinney, K. G., C. Jelinek, K. Edvardsen, D. J. Chaplin, and G. R. Pettit. 2005. The discovery and development of the combrestatins. In *Anticancer Agents from Natural Products*, eds. G. M. Cragg, D. G. I. Kingston, and D. J. Newman, pp. 23–46. Boca Raton, FL: CRC Press.

Pirali, T., S. Busacca, L. Beltrami, D. Imovilli, F. Pagliai, G. Miglio, A. Massarotti et al. 2006. Synthesis and cytotoxic evaluation of combretafurans, potential scaffolds for dual-action antitumoral agents. *J. Med. Chem.* 49: 5372–5376.

Quideau, S. 2006. Why bother with polyphenols? *Polyphénols Actualités* 24: 10–14.

Quintin, J., D. Buisson, S. Thoret, T. Cresteil, and G. Lewin. 2009. Semisynthesis and antiproliferative evaluation of a series of 3′- aminoflavones. *Bioorg. Med. Chem. Lett.* 19: 3502–3506.

Ramos, S. 2008. Cancer chemoprevention and chemotherapy: Dietary polyphenols and signalling pathways. *Mol. Nutr. Food Res.* 52: 507–526.

Renaud, S. and M. de Lorgeril. 1992. Wine, alcohol, platelets, and the French paradox for coronary heart disease. *Lancet* 339: 1523–1526.

Riveiro, M. E., N. De Kimpe, A. Moglioni, R. Vazquez, F. Monczor, C. Shayo, and C. Davio. 2010. Coumarins: Old compounds with novel promising therapeutic perspectives. *Curr. Med. Chem.* 17: 1325–1338.

Saiko, M., Z. Pemberger, I. Horvath, I. Savinc, M. Grusch, N. Handler, T. Erker, W. Jaeger, M. Fritzer-Szekeres, and T. Szekeres. 2008. Novel resveratrol analogs induce apoptosis and cause cell cycle arrest in HT29 human colon cancer cells: Inhibition of ribonucleotide reductase activity. *Oncol. Rep.* 19: 1621–1626.

Sánchez-del-Campo, L., F. Otón, A. Tárraga, J. Cabezas-Herrera, S. Chazarra, and J. Neptuno Rodríguez-López. 2008. Synthesis and biological activity of a 3,4,5-trimethoxybenzoyl ester analogue of epicatechin-3-gallate. *J. Med. Chem.* 51: 2018–2026.

Sánchez-del-Campo, L., A. Tarraga, M. F. Montenegro, J. Cabezas-Herrera, and J. N. Rodriguez-Lopez. 2009. Melanoma activation of 3-*O*-(3,4,5-trimethoxybenzoyl)-(−)-epicatechin to a potent irreversible inhibitor of dihydrofolate reductase. *Mol. Pharm.* 6: 883–894.

Schneider, Y., P. Chabert, J. Stutzmann, D. Coelho, A. Fougerousse, F. Gosse, J.-F. Launay, R. Brouillard, and F. Raul. 2003. Resveratrol analog (Z)-3,5,4′-trimethoxystilbene is a potent anti-mitotic drug inhibiting tubulin polymerization. *Int. J. Cancer* 107: 189–196.

Signorelli, P. and R. Ghidoni. 2005. Resveratrol as an anticancer nutrient: Molecular basis, open questions and promises. *J. Nutr. Biochem.* 16: 449–466.

Simoni, D., F. P. Invidiata, M. Eleopra, P. Marchetti, R. Rondanin, R. Baruchello, G. Grisolia et al. 2009. Design, synthesis and biological evaluation of novel stilbene-based antitumor agents. *Bioorg. Med. Chem.* 17: 512–522.

Simoni, D., M. Rizzi, R. Rondanin, R. Baruchello, P. Marchetti, F. P. Invidiata, M. Labbozzetta et al. 2008. Antitumor effects of curcumin and structurally β-diketone modified analogs on multidrug resistant cancer cells. *Bioorg. Med. Chem. Lett.* 18: 845–849.

Simoni, D., M. Roberti, F. P. Invidiata, E. Aiello, S. Aiello, P. Marchetti, R. Baruchello et al. 2006. Stilbene-based anticancer agents: Resveratrol analogues active toward HL60 leukemic cells with a non-specific phase mechanism. *Bioorg. Med. Chem. Lett.* 16: 3245–3248.

Singh, R. and H. Kaur. 2009. Advances in synthetic approaches for the preparation of combretastatin-based anti-cancer agents. *Synthesis* 15: 2471–2491.

Souza da Silva, S. A., A. L. Souto, M. de Fatima Agra et al. 2004. A new arylnaphthalene type lignan from *Cordia rufescens* A. DC (Boraginaceas). *Arkivok* 6: 54–58.

Spatafora, C., G. Basini, L. Baioni, F. Grasselli, A. Sofia, and C. Tringali. 2009. Antiangiogenic resveratrol analogues by mild *m*-CPBA aromatic hydroxylation of 3,5-dimethoxystilbenes. *Nat. Prod. Comm.* 4: 239–246.

Spatafora, C. and C. Tringali. 2007. Phenolic oxidative coupling in the biomimetic synthesis of heterocyclic lignans, neolignans and related compounds. In *Targets in Heterocyclic Systems. Chemistry and Properties*, Vol. 11, eds. O. A. Attanasi and D. Spinelli, pp. 284–312. Roma, Italy: Società Chimica Italiana.

Sporn, M. B., N. M. Dunlop, D. L. Newton, and J. M. Smith. 1976. Prevention of chemical carcinogenesis by vitamin A and its synthetic analogs (retinoids). *Fed. Proc.* 35: 1332–1338.

Srinivas, K. V., R. Y. Koteswara, I. Mahender, B. Das, K. V. S. Rama Krishna, K. Hara Kishore, and U. S. N. Murty. 2003. Flavonoids from *Caesalpinia pulcherrima*. *Phytochemistry* 63: 789–793.

Steward, W. P. and A. J. Gescher. 2008. Curcumin in cancer management: Recent results of analogue design and clinical studies and desirable future research. *Mol. Nutr. Food Res.* 52: 1005–1009.

Sun, A., Y. J. Lu, H. Hu, M. Shoji, D. C. Liotta, and J. P. Snyder. 2009. Curcumin analog cytotoxicity against breast cancer cells: Exploitation of a redox-dependent mechanism. *Bioorg. Med. Chem. Lett.* 19: 6627–6631.

Surh, Y. 1999. Molecular mechanisms of chemopreventive effects of selected dietary and medicinal phenolic substances. *Mutat. Res.* 428: 305–327.

Surh, Y. J. 2003. Cancer chemoprevention with dietary phytochemicals. *Nat. Rev. Cancer* 3: 768–780.

Takaoka, M. J. 1940. Of the phenolic substances of white hellebore (*Veratrum grandiflorum* Loes. fil.). *J. Fac. Agric. Hokkaido* 3: 1–16.

Thomasset, S. C., D. P. Berry, G. Garcea, T. Marczylo, W. P. Steward, and A. J. Gescher. 2006. Dietary polyphenolic phytochemicals—Promising cancer chemopreventive agents in humans? A review of their clinical properties. *Int. J. Cancer* 120: 451–458.

Tozer, G. M., C. Kanthou, and B. C. Baguley. 2005. Disrupting tumor blood vessels. *Nat. Rev. Cancer* 5: 423–435.

Tron, G. C., T. Pirali, G. Sorba, F. Pagliai, S. Busacca, and A. A. Genazzani. 2006. Medicinal chemistry of combretastatin A4: Present and future directions. *J. Med. Chem.* 49: 3033–3044.

Vogel, S., S. Ohmayer, G. Brunner, and J. Heilmann. 2008. Natural and non-natural prenylated chalcones: Synthesis, cytotoxicity and anti-oxidative activity. *Bioorg. Med. Chem.* 16: 4286–4293.

Walle, T. 2007. Methoxylated flavones, a superior cancer chemopreventive flavonoid subclass? *Semin. Cancer Biol.* 17: 354–362.

Weber, W. M., L. A. Hunsaker, C. N. B.-M. Roybal, V. Ekaterina, S. F. Abcouwer, R. E. Royer, L. M. Deck, and D. L. Vander Jagt. 2006. Activation of NF kappa B is inhibited by curcumin and related enones. *Bioorg. Med. Chem.* 14: 2450–2461.

Weng, C. J., Y. T. Yang, C. T. Ho, and G. C. Yen. 2009. Mechanisms of apoptotic effects induced by resveratrol, dibenzoylmethane, and their analogues on human lung carcinoma cells. *J. Agric. Food. Chem.* 57: 5235–5243.

William, N. W. Jr., J. V. Heymach, E. S. Kim, and S. M. Lippman. 2009. Molecular targets for cancer chemoprevention. *Nat. Rev. Drug Discov.* 8: 213–223.

Xia, C., L. H. Li, L. Feng, and W. -X. Hu. 2008. Synthesis of *trans*-caffeate analogues and their bioactivities against HIV-1 integrase and cancer cell lines. *Bioorg. Med. Chem. Lett.* 18: 6553–6557.

Yamakoshi, H., H. Ohori, C. Kudo et al. 2010. Structure–activity relationship of C5-curcuminoids and synthesis of their molecular probes thereof. *Bioorg. Med. Chem.* 18: 1083–1092.

Yang, C. S., S. Kim, G. Y. Yang, M.-J. Lee, J. Liao, J. Y. Chung, and C.-T. Ho. 1999. Inhibition of carcinogenesis by tea: Bioavailability of tea polyphenols and mechanisms of actions. *Proc. Soc. Exp. Biol. Med.* 220: 213–217.

Yerra, K. R., F. Shih-Hua, and T. Yew-Min. 2009. Synthesis and biological evaluation of 3′, 4′, 5′-trimethoxychalcone analogues as inhibitors of nitric oxide production and tumor cell proliferation. *Bioorg. Med. Chem.* 17: 7909–7914.

Yu, P.-F., H. Chen, J. Wang, C.-X. He, B. Cao, M. Li, N. Yang, Z.-Y. Lei, and M.-S. Cheng. 2008a. Design, synthesis and cytotoxicity of novel podophyllotoxin derivatives. *Chem. Pharm. Bull.* 56: 831–834.

Yu, Z., X. L. Qin, Y. Y. Gu, C. Di, Q. C. Cui, T. Jiang, S. B. Wan, and Q. P. Dou. 2008b. Prodrugs of fluoro-substituted benzoates of EGC as tumor cellular proteasome inhibitors and apoptosis inducers. *Int. J. Mol. Sci.* 9: 951–961.

Zaveri, N. T. 2006. Green tea and its polyphenolic catechins: Medicinal uses in cancer and noncancer applications. *Life Sci.* 78: 2073–2080.

10 Resveratrol against Major Pathologies

From Diet Prevention to Possible Alternative Chemotherapies with New Structural Analogues

Norbert Latruffe, Dominique Delmas, Gérard Lizard, Corrado Tringali, Carmela Spatafora, Dominique Vervandier-Fasseur, and Philippe Meunier

CONTENTS

10.1 INTRODUCTION

There are many evidences suggesting that dietary phenolic compounds lead to beneficial effects for health. In particular, resveratrol (E-3,5,4′-trihydroxystilbene = RSV) (**1**), a well-known wine grape component (Langcake and Pryce 1976), is one of the most promising natural polyphenols with anticancer properties (Aggarwal and Shishoida 2006; Delmas et al. 2006). Among such positive effects, resveratrol has been considered as an antioxidant with a direct impact on oxidative stress–related diseases (i.e., atherosclerotic cardiovascular diseases, cancer, neurodegenerative processes), as summarized in Figure 10.1.

Evidences for beneficial effects of phenol compounds come from in vitro, epidemiological, and some clinical studies (Levi et al. 2005). Mediterranean diet is now being considered as a protective factor for oxidative stress–associated diseases, as highlighted by the so-called French paradox (Renaud and de Longeril 1992). However, the knowledge of the role of single components and the interaction among them is required in order to perform future recommendations for primary and secondary prevention of disease. The efficacy, defined as the capacity to reduce secondary end points for disease, that is, oxidative stress, and individual component could be enhanced or reduced in the frame of a whole diet.

Concerning the effects of these polyphenolic compounds in humans, there are no controlled clinical trials showing an in vivo effect including dose-dependency analysis. Although phenol compounds are presumed to be beneficial for human health at any dose tested, in vitro studies suggest that this is not always the case and that paradoxical effects, that is, oxidative effects, may be observed at high

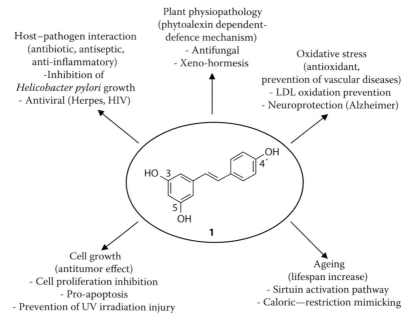

FIGURE 10.1 Biological effects of resveratrol (**1**) on living system.

doses. Moreover, the bioavailability, including bioabsorption and metabolism, and the biochemical action mechanisms of these molecules are not well understood. This last point is also essential to provide the scientific basis of therapeutic applications.

On the other hand, interestingly, animal and human cells react to plant polyphenols exposure as they behave toward chemical drugs, that is, by recognizing these molecules as xenobiotic compounds. As a consequence, the body cells will transform these compounds in order to eliminate them as quick and as extensively as possible. The pharmacokinetics studies allow evaluating the bioavailability of a polyphenol. This parameter is essential to select natural compounds showing biological activities, especially their possible anticancer properties. However, plant polyphenols, including RSV, often exhibit a poor bioavailability (Lançon et al. 2007; Niles 2006). To overcome this problem, possible improvements are the chemical modification of such natural polyphenols or the synthesis of structural analogues. Thus, this chapter will review selected results about the potency of RSV and some examples of the chemotherapeutic potential of its chemically modified analogues on cancer cell lines, especially those from colorectal origin. In fact colon cancer is a frequent cancer type and moreover often exhibits a resistance to classical therapies. The biochemical basis of colon cancer prevention and therapy is the main research field of some of us; thus, we focused part of this chapter on a summary of our recent results on colon cancer cell growth inhibition by RSV synthetic analogues. The plan of this review, ranging from diet prevention to alternative chemotherapy, will be the following: a presentation of control of cancer progression by RSV (Section 10.2); the RSV properties toward cell antioxidant status (Section 10.3); the RSV cell transport and metabolism (Section 10.4); the synthesis and biological potencies of some RSV structural analogues (Section 10.5); and finally the use of new RSV probes to target essential signaling cell pathways (Section 10.6). We will focus on the anti- and prooxidant properties of RSV, on the associated mechanisms, and on the potential clinical implications of this polyphenol in a major human pathology such as cancer. We will also discuss the RSV bioavailability which is a subject of only recent investigations.

10.2 CONTROL OF CANCER PROGRESSION

10.2.1 RESVERATROL AND MOLECULAR MECHANISMS IN CHEMOPREVENTION AND CHEMOSENSITIZATION

Dietary polyphenols are of great interest due to their antioxidative and anticarcinogenic activities. Indeed, polyphenols can have a chemoprotective effect which is the property of pharmacological or natural agents to promote the arrest or regression of a cancer process. Polyphenols such as RSV may inhibit carcinogenesis by affecting the molecular events at the initiation, promotion, and progression stages. An intricate network of signaling pathways is involved in these control mechanisms, especially the cell cycle and the induction of apoptosis. The cell cycle arrest is important for the cytostatic action and the induction of cell death in precancerous or malignant cells is considered to be a promising strategy for chemopreventive or chemotherapeutic purpose.

10.2.1.1 Resveratrol and Cell Cycle Perturbation

Natural compounds, like many cytotoxic agents, affect cell proliferation by disturbing the normal progression of the cell cycle. In fact, both stilbenes and flavonoids are able to block the cell cycle. This blockage depends on the cell type, the natural compound concentration, and the treatment duration. Various studies report that checkpoint at both G1/S and G2/M of the cell cycle is found to be perturbed by the phytochemicals (Hosokawa et al. 1990; Traganos et al. 1992; Zi et al. 1998). Checkpoints are controlled by a family of protein kinase complexes, and each complex is composed minimally of a catalytic subunit, cyclin-dependent kinases (cdks), and its essential activating partner, cyclin. Cyclins play a key regulatory role in this process by activating their partner cdks and targeting them to the respective protein substrates (Morgan 1995). Complexes formed in this way are activated at specific intervals during the cell cycle and their inhibition blocks the cell cycle at the corresponding control point. These key regulators can be affected by various stilbenes, especially RSV, leading to an arrest of the cell cycle. Interestingly, RSV was able to block different checkpoints of cell cycle in cancerous cell lines–dependent manner. However, in the colon cancer cells, treatment with RSV blocks only the cell cycle at the transition S/G2 (Table 10.1). Indeed, we and others have shown that RSV is able to act on the S phase with consequent effect on S/G2 transition in colon cancer cell lines or in animal models (Table 10.1). We have shown that RSV was able to induce proliferation arrest, which is associated with an accumulation of cancerous cells in the S phase. This arrest is reversible, but a continuous RSV treatment blocks the progression of colon cancer cells during the S/G2 transition (Delmas et al. 2002). An analysis of the deregulation of S phase in polyphenol-induced cell cycle arrest by

TABLE 10.1
Resveratrol Effects on Cell Cycle Progression in Colon Cancer Cells

Cell Systems	Cell Cycle Arrest	Resveratrol Effects	References
Human colonic adenocarcinoma cell line Caco-2	S/G2 transition	Cyclin D1 ↘; Cdk4 ↘ Cyclins A, E ↗ Hyperphosphorylated pRb ↘ Hyphosphorylated pRb ↗	Schneider et al. (2000)
Human colon carcinoma cell line HCT-116	S/G2 transition	Cyclin D1 ↘; Cdk4 ↘ Cyclins A, E ↗	Wolter et al. (2001)
Human colon adenocarcinoma cell line SW480	S/G2 transition	Cyclins A, B ↗ Hyperphosphorylated Cdks 1, 2 ↗ Nuclear localization of cyclin A Cyclin A/Cdk2 complex activity ↗	Colin et al. (2009); Delmas et al. (2002); Marel et al. (2008)
Human colon carcinoma cell line HT29	G2/M phase	Cdk1 phosphorylated ↗ Cdk1 kinase activity ↘ Cdk7 kinase activity ↘	Liang et al. (2003)

flux cytometer reveals three populations among the BrdU-incorporating cells (Colin et al. 2009). We demonstrated that RSV induced an accumulation of colon cancer cells in early S phase. In this early S phase, RSV treatment induces a time-dependent decrease in G_{1A} (which represents post-mitotic cells) and in G_{1B} (which represents cells ready to initiate DNA synthesis) phases.

Biochemical analysis of such phenomena shows a significant increase of cyclins A and B1 with the accumulation of cdks (cdk1 and cdk2), which are also increased in their inactive phosphorylated forms (Colin et al. 2009; Delmas et al. 2002). In fact, cdk1 is known to be a key regulator of the eukaryotic cell cycle and is believed to act in both G1 and G2 phases where the dephosphorylated form is required. In the same case, cdk2 plays an important part throughout the cell cycle where the cdk2 protein expression and its phosphorylation state are regulated with respect to cell cycle phase. Moreover, cdk2 kinase activity has been shown to be required for DNA synthesis (Pagano et al. 1993). Indeed, it has been shown that the accumulation of the inactive tyrosine 15-phosphorylated cdk1 form is associated to the cell division cycle arrest preventing the entry into G2/M phases (O'Connor et al. 1993, 1994). Cyclin A/cdk2 complex plays a key role during S phase progression, and cyclin B1/cdk1 complex controls the cell entry and progression of mitotic phase [M phase] (Hunter and Pines 1994; Nurse 1997). Since our results show that RSV provokes a hyperphosphorylation of cdk1 in SW480 human colorectal cells, one can suggest that RSV disrupts the dephosphorylation process of cdk1 leading to the arrest in the S phase (Figure 10.2). The same disruption through the cell cycle was observed in the epithelial cell during RSV treatment with a hyperphosphorylation of cdk1 (Ragione et al. 1998; Schneider et al. 2000; Wolter et al. 2001), and accumulation of p53 and p21 WAF1/CIP1 (Hsieh et al. 1999a).

Furthermore, consistently with the entry of cells into S phase, there is a dramatic increase in nuclear cdk2 activity associated with both cyclin A and cyclin E (Colin et al. 2009; Kuwajerwala et al. 2002). It seems that RSV treatment induces a specific response in a tissue-dependent manner and that this polyphenol may act as cell synchronizing agent. The S phase arrest was also shown in vivo where RSV exhibits antitumor activities on murine hepatoma H22 by a mechanism involving an arrest of the cell cycle by decreasing the expression of cyclin B1 and cdk1 protein (Yu et al. 2003). These results were also shown with the RSV analogues, piceatannol (E-3,5,3′,4′-tetrahydroxystilbene, **2**) and resveratrol triacetate (**3**), which induce an accumulation of colorectal cancer cells in the S phase (Colin et al. 2009; Wolter et al. 2002). This arrest is associated with an increase in cyclin A and cyclin E levels, whereas cyclin D1 and cdk4 are downregulated, and the abundance of p27 Kip1 is also reduced (Figure 10.3).

Most authors attribute the S phase arrest to an inhibition of ribonucleotide synthase and DNA synthesis. In fact, RSV is a scavenger of the essential tyrosyl radical of the small protein of ribonucleotide reductase and, consequently, inhibits deoxyribonucleotide synthesis during the S phase (Elleingand et al. 1998; Fontecave et al. 1998). Resveratrol is a much more effective inhibitor than hydroxyurea or hydroxyanisole, the only ribonucleotide reductase tyrosyl radical scavengers used in clinics, or indeed the potent p-propoxyphenol (Fontecave et al. 1998). It is also suggested that inhibition of DNA synthesis occurs at the level of DNA polymerase activity,

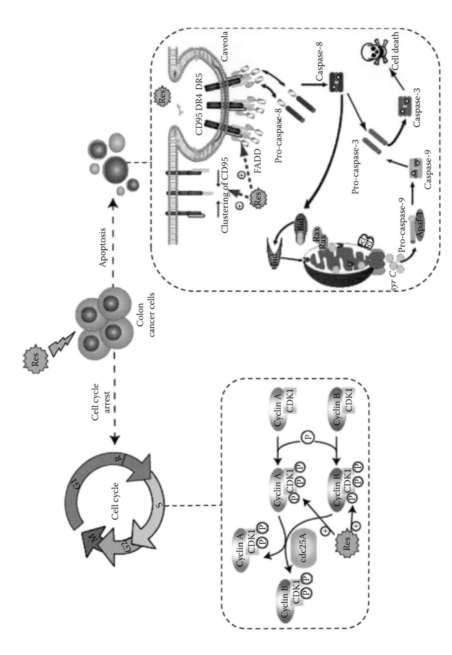

FIGURE 10.2 Molecular mechanisms of resveratrol in cell cycle arrest and apoptosis induction.

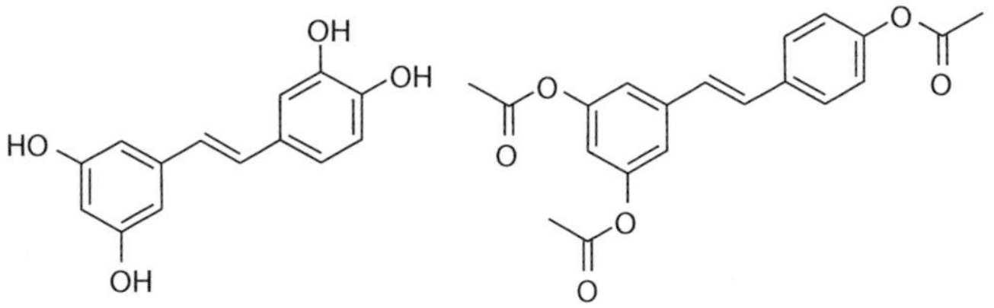

FIGURE 10.3   Structures of compounds **2** and **3**.

since the recruitment of PCNA and replication protein A (RPA) proteins to DNA replication sites is not affected by RSV (Stivala et al. 2001). More specifically, in vitro assays demonstrate that only *trans*-resveratrol significantly inhibits DNA polymerase α and δ (Stivala et al. 2001; Sun et al. 1998; Tsan et al. 2002). Stivala et al. have shown that the inhibition by RSV is found to be strictly specific for the B-type DNA polymerases α and δ (Stivala et al. 2001). Moreover, structure–activity relationships studies showed that the 4′-hydroxyl group in *trans*-conformation of resveratrol (hydroxystyryl moiety) is not the sole determinant for antioxidant properties, but acts synergistically with the 3- and 5-OH groups, and that the 4′-hydroxystyryl moiety of *trans*-resveratrol interacts with DNA polymerase δ (Stivala et al. 2001). Controversially, we and others have shown that RSV reduces viability but not DNA synthesis in cancer cells. Indeed, RSV increases the DNA synthesis associated with an accumulation of cells in S phase (Delmas et al. 2000, 2002; Marel et al. 2008). A possible mechanism is that RSV causes S phase arrest only when sister chromatid exchange is induced, as suggested by Matsuoka et al. in Chinese hamster lung cell lines (Matsuoka et al. 2001, 2002, 2004). The increase of cell cycle key protein regulators and the subsequent accumulation of colon tumor cells into S phase preceded apoptotic cell death (Colin et al. 2009).

10.2.1.2   Resveratrol and Apoptosis

Induction of apoptosis in precancerous or malignant cells is considered to be a promising strategy for chemopreventive or chemotherapeutic purposes. The induction of apoptosis triggered by polyphenolic compounds has been observed in various cell types with different pathways. Indeed it has been demonstrated that RSV is able to activate cell death by the mitochondrial pathway or by the death receptor pathway.

The mitochondrial pathway is activated in response to extracellular signals and internal disturbances such as DNA damages. We, like others, have shown that RSV (Delmas et al. 2003) induces apoptosis in various tumor cell lines by modulating pro-apoptotic Bcl-2 family proteins which are known as "BH3-only proteins" behaving as sensors of cellular damage and initiating the agents of death process. We have shown in colon cancer cells, and others in various cells types, that RSV downregulates Bcl-2 protein expression (Billard et al. 2002; Delmas et al. 2003; Hsieh et al. 1999b; Kim et al. 2004; Lee et al. 2008; Roman et al. 2002; Surh et al. 1999) and gene expression (Kaneuchi et al. 2003; Zhou et al. 2003, 2005), which normally stabilizes the mitochondrial potential of the membrane ($\Delta\varphi_m$), and inhibits ROS production. Contrary to Bcl-2, RSV has been shown to trigger an increase in Bax and Bak protein expression (Delmas et al. 2003; Jazirehi and Bonavida 2004; Kim et al. 2003a,b, 2004; Nakagawa et al. 2001) and gene expression (Kim et al. 2003b; Zhou et al. 2003, 2005). However, a bax-independent pathway to cell death has been identified in a HCT116 colon cancer cell clone in which both bax alleles had been inactivated (Mahyar-Roemer et al. 2002). The ability of RSV to trigger colon cancer cell apoptosis in the absence of Bax could be explained by the functional interchangeability of Bax and Bak. Cells from mice deficient in both Bax and Bak, but not cells deficient in only one of the two, are almost completely resistant to mitochondria-mediated apoptosis (Wei et al. 2001). We have shown that an exposure of adenocarcinoma colon cells to RSV induces conformational changes and mitochondrial redistribution of both Bax and Bak, suggesting that the two proteins are involved in RSV-induced cell death (Delmas et al. 2003). In addition, Bax has been shown to be involved in the chemopreventive effect of RSV in animal models of colon carcinogenesis, where Bax expression is enhanced in aberrant crypt foci (AFC) but not in the surrounding mucosa (Tessitore et al. 2000). So, we showed that RSV-induced apoptosis by this mechanism involves the release of molecules such as cytochrome c, Smac/Diablo present both in the intermembrane space of the mitochondria and in the cytosol under the control of Bcl-2 and Bcl-2-related proteins such as Bax. Cytochrome c, released in the cytosol, induces oligomerization of the adapter molecule Apaf-1 to generate a complex, the apoptosome, in which caspase-9 is activated. Active caspase-9 then triggers the catalytic maturation of caspase-3 and other resultant caspases, thus leading to cell death. Resveratrol induces other soluble molecules released from the mitochondria including Smac/Diablo (Delmas et al. 2003; Jazirehi and Bonavida 2004) that neutralizes caspase inhibitors of the IAP family such as XIAP (Figure 10.2) (Du et al. 2000; Srinivasula et al. 2000). Resveratrol itself is able to inhibit IAP family protein expression such as survivin expression (Fulda and Debatin 2004b; Hayashibara et al. 2002). A direct effect on apoptosis by downregulating bcl-2 expression and upregulating bax expression with p53 can occur and activate caspases (Miyashita et al. 1994). It appears that RSV can induce an increase of the tumor suppressor gene p53 in various cell types (Fulda and Debatin 2004a; Narayanan et al. 2003, 2004) and induce its phosphorylation (Haider et al. 2003; Lin et al. 2002; She et al. 2001; Shih et al. 2002, 2004; Zhang et al. 2004). This activation of the transcription factor p53 by RSV could contribute to death and cell cycle arrest (Huang et al. 1999; She et al. 2001, 2002), but the polyphenol can also induce apoptosis in p53-deficient cells (Mahyar-Roemer et al. 2001; Mahyar-Roemer and Roemer 2001), indicating that p53 is not an absolute requirement for the cytotoxic effect of the molecule.

An initial description of the death pathway triggered by RSV in tumor cells involved the upregulation of Fas-L mRNA and the Fas-L/Fas interaction in an autocrine or paracrine manner (Clement et al. 1998) but these results were subsequently challenged by several groups in various tumor models (Bernhard et al. 2000; Dorrie et al. 2001; Kuo et al. 2002; Tinhofer et al. 2001; Tsan et al. 2000; Wang et al. 2003), based on the observations that (i) Fas-L mRNA upregulation was not confirmed, (ii) Abs that prevent the FasL/Fas interaction did not prevent RSV-induced apoptosis, and (iii) cell lines resistant to Fas-mediated apoptosis, for example, leukemia cell lines, still underwent RSV-induced cell death. These arguments do not rule out the possibility of a Fas role in RSV-induced cell death. Receptor-mediated apoptosis was shown to depend upon prior activation of caspase-8, which was then capable of activating the other later caspases resulting in apoptosis. In addition, the Fas pathway may not be an absolute requirement for RSV-induced apoptosis, although it contributes to cell death when functional. The activation of caspase-8 identified in RSV-treated adenocarcinoma colon cells could occur earlier in the process at the level of death receptors or later in the process in the caspase cascade to amplify the apoptotic pathway. We have provided a potential explanation for these controversies by showing that RSV does not increase the expression of Fas and FasL at the surface of tumor cells but does induce a redistribution of Fas in the raft domains of the plasma membrane (Delmas et al. 2003). These lipid microdomains result from the preferential packing of complex sphingolipids and cholesterol in ordered plasma membrane structures and contain a variety of lipid-anchored and transmembrane proteins. Rafts play an important role in clustering or aggregating surface receptors, signaling enzymes and adaptor molecules into membrane complexes at specific sites and were shown to be essential for initiating signaling from a number of receptors. It appears that RSV treatment changes the homogeneous distribution of the protein existing in untreated colon cancer cells into a more clustered distribution. In addition, RSV induces a redistribution of Fas, together with FADD and procaspase-8, in the fractions enriched in cholesterol and sphingolipids (Delmas et al. 2003). The mechanisms trapping receptor molecules in membrane rafts have yet to be characterized. Whatever these mechanisms, RSV-induced redistribution of Fas in the rafts could contribute to the formation of the death-inducing signaling complex (DISC) observed in colon cancer cells treated with the polyphenol (Delmas et al. 2004).

10.2.1.3 Resveratrol and Chemosensitization

Recent evidence suggests that the use of RSV in combination with drugs, ionizing radiation or cytokines, can be effectively used for the sensitization to apoptosis. It appears that RSV can sensitize colon cancer cells to 5-fluorouracil, which is a classic drug used in colorectal and hepatoma chemotherapy. Indeed, it was reported that this natural stilbenoid can exert synergic effect with this drug to inhibit hepatocarcinoma and colon carcinoma cells proliferation by the induction of apoptosis (Colin et al. 2009; Fuggetta et al. 2004; Sun et al. 2002). An explanation of this result is that pretreatment induces the accumulation of colon cancer cells in early S phase, which may significantly account for the increased efficacy of 5-FU. This innovative combination may be more efficient than using a single drug at higher concentration.

Concerning cytokines, we and others have shown that RSV is able to sensitize to tumor necrosis factor-related apoptosis-inducing ligand (TRAIL)-induced apoptosis in cancer cells (Delmas et al. 2004; Fulda and Debatin 2004b). In human colon cancer cells that are resistant to the cytotoxic effect of RSV, we have shown that this polyphenol sensitizes these tumor cells to TNF, anti-CD95 antibodies, and TRAIL-mediated apoptosis and activates a caspase-dependent death pathway that escapes Bcl-2-mediated expression (Delmas et al. 2004). It appears that RSV pretreatment facilitates the formation of a functional DISC at plasma level. The cholesterol sequestering agent, nystatin prevents RSV-induced death receptor redistribution and cell sensitization to death receptor stimulation, suggesting that RSV-induced redistribution of death receptors in lipid rafts is an essential step in its sensitizing effect expression (Delmas et al 2004).

10.3 RESVERATROL AND CELL ANTIOXIDANT STATUS

It has also been shown that RSV can exhibit either antioxidant or prooxidant properties depending on its concentrations and the cell type (de la Lastra and Villegas 2007). Therefore, it has been proposed that such prooxidant activities could be a common mechanism for anticancer and chemopreventive properties of plant polyphenols. We will therefore focus on the anti- and prooxidant properties of RSV, on the associated mechanisms, and on the potential clinical implications of this polyphenol in major human pathologies such as cancer and degenerative diseases.

10.3.1 RESVERATROL AS AN ANTIOXIDANT AGENT: IN VIVO AND IN VITRO EVIDENCE OF GLUTATHIONE SYNTHESIS ACTIVATION AND OF ANTIOXIDANT ENZYMES STIMULATION

Oxidative stress is recognized as an important factor in the development of numerous pathologies, especially in liver pathologies, cardiovascular, and neurodegenerative diseases. Thus, in liver pathologies, the reactive oxygen species (ROS) endogenously generated or as a consequence of xenobiotic metabolism are eliminated by enzymatic and nonenzymatic cellular systems (Table 10.2). Besides endogenous defenses, the

TABLE 10.2
Potential Cellular Targets of Resveratrol Capable of Modulating the Redox Status in Normal and Tumoral Cells

Parameters Studied	Resveratrol and Normal Cells[a]	Resveratrol (without or with Cu^{2+}) and Tumor Cells
Oxidative markers	Lipid peroxidation ↘	DNA damages ↗
		Reactive oxygen species (ROS) ↗
Antioxidative markers	Glutathione (GSH) ↗	Glutathione (GSH) ↘
Prooxidative enzymes	NADPH-oxidase 1 (Nox1) ↘	NADPH-oxidase 1 (Nox1) ↗
	Xanthine oxidase ↘	NADPH-oxidase 4 (Nox4) ↗
Antioxidative enzymes	Glutathione peroxidase (GPx) ↗	
	Glutathione reductase ↗	
	Glutathione S-transferase ↗	
	Superoxide dismutase (SOD) ↗	Superoxide dismutase (SOD) ↘
	Catalase (CAT) ↗	

[a] Hepatocytes, platelets, smooth muscle cells, endothelial cells, neural cells.

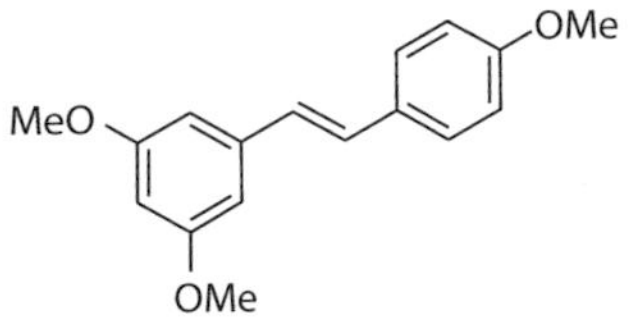

FIGURE 10.4   Structure of compound **4**.

antioxidant consumption in the diet has an important role in the protection against the development of diseases as a product of oxidative damage.

When liver damage was induced by acute oral administration of CCl_4 to Wistar rats, the GSH/GSSH ratio was decreased in the liver, and lipid peroxidation as well as gamma-glutamyl transpeptidase were significantly increased, RSV partially prevented the increase of these markers (Rivera et al. 2008). To this end, the OH groups of RSV are important for the antioxidant and hepatoprotective activities of the molecules of RSV. Indeed, authors report that the effects can be improved by replacing hydrogen by a methyl in these groups suggesting that trimethylresveratrol (i.e., E-3,5,4′-trimethoxystilbene **4**) could act like a prodrug with higher half-life than the original compound. Moreover, when hepatic cirrhosis in the rat was induced by repeated intraperitoneal administration of CCl_4, RSV has been shown to possess a strong anti-fibrinogenic activity (Chavez et al. 2008) (Figure 10.4).

In this condition, its action mechanism is probably associated with its ability to reduce NF-kappa B activation, which regulates the transcription of several genes including cytokines such as the pro-inflammatory and profibrogenic TGF-beta. When hepatotoxicity was induced by chronic intraperitoneal administration of ethanol fatty changes, necrosis, fibrosis, and inflammation were observed in liver sections (Kasdallah-Grissa et al. 2007). Ethanol also enhanced the formation of malondialdehyde (MDA) in the liver indicating an increase in lipid peroxidation, a major end point of oxidative damages, and caused drastic alterations of antioxidant defense systems: the activities of hepatic superoxide dismutase (SOD), glutathione peroxidase (GPx), and catalase (CAT) were reduced. Noteworthy, dietary supplementation with RSV inhibited ethanol-induced lipid peroxidation and ameliorated SOD, GPx, and CAT activities in the liver. Similarly, in acetaminophen-treated mice, the resulting hepatotoxicity was associated with a significant decrease of GSH which was counteracted by RSV (Sener et al. 2006). Conclusively, these different arguments obtained on different animal models support that RSV could have beneficial effects on hepatic damages associated with oxidative stress. In agreement with in vivo experiments, RSV has been shown to increase the activities of various antioxidant enzymes (CAT, SOD, GPx, glutathione-S-transferase) on primary cultures of rat hepatocytes and to activate biological pathways involving NF-E2-related factor 2 (Nrf2) signaling involved in the transcriptional regulation of glutamate cysteine ligase involved in GSH synthesis (Kluth et al. 2007; Kode et al. 2008; Rubiolo et al. 2008).

Concerning coronary artery diseases, some epidemiological and basic studies suggest that RSV increases the resistance to vascular oxidative stress which is one of the leading causes of these pathologies (Ungvari et al. 2007). As ROS and monocyte chemotactic protein-1 (MCP-1) contribute to the formation of foam cells (lipid-laden macrophages) playing key roles in the development of atherosclerosis, the effect of RSV on these parameters was investigated. Interestingly, RSV treatment of macrophages inhibited LPS-induced NADPH oxidase 1 (Nox1) expression as well as ROS generation, and also suppressed LPS-induced MCP-1 mRNA and protein expression. It was found that Akt–forkhead transcription factors of the O class (FoxO3a) is an important signaling pathway that regulates both Nox1 and MCP-1 genes. These inhibitory effects of RSV on Nox1 expression and MCP-1 production may target to the Akt and FoxO3a signaling pathways (Park et al. 2009). In addition, on the monocytic cells U937 (used as macrophage model) treated with various xenobiotics and frequently employed to investigate the biological effects of cholesterol oxide derivatives present at increased levels in atherosclerotic plaques, DNA damages associated with a decrease level of reduced glutathione (GSH) were counteracted by RSV (O'Brien et al. 2006). Moreover, on bovine aortic endothelial cells (BAEC) in primary culture, RSV protects from peroxinitrite-induced cell death by upregulating intracellular GSH levels (Brito et al. 2006), and on cultured aortic smooth muscle cells, this polyphenol induces cytoprotective factors against oxidative stress including SOD, CAT, glutathione reductase, glutathione peroxidase, GST, and NAD(P)H: quinine oxidoreductase-1 (NOQ-1) (Li et al. 2006). In blood platelets, playing also key roles in atherosclerosis, RSV counteracts lipid peroxidation and markedly reduces GSH decrease (Olas et al. 2008). In addition, in cardiomyocytes, RSV led to a great reduction of xanthine oxidase–induced intracellular accumulation of ROS (Cao and Li 2004). Taken together, these different results of RSV on cells of the vascular wall (endothelial cells, smooth muscle cells, and monocytes), and on cardiac cells highlight the antioxidant targets, potentially activated by RSV, which might contribute to its cardioprotective vascular effects.

Specific plasma membranes sites for polyphenol, including RSV, have been identified in the rat brain (Han et al. 2006); more specifically, grape polyphenols, including dietary RSV, are able to attenuate cognitive deterioration in a mouse model of Alzheimer's disease (Wang et al. 2008), to increase SOD expression and activity in mouse brain (Robb et al. 2008), to reduce ischemia induced toxicity, energy failure, and oxidative stress in rats (Ritz et al. 2008), to protect against kainic acid–induced seizures and oxidative stress in a rat model of epilepsy (Gupta et al. 2002), to reduce oxidative stress in tissue lesion area in rat after surgical brain injury (Ates et al. 2007) and experimental spinal cord injury (Ates et al. 2006), and to reverse colchicine-induced cognitive impairment (Kumar et al. 2007) and 3-nitropropionic acid–induced motor and cognitive impairment mimicking Huntington's disease in rats (Kumar et al. 2006). These different in vivo data support that polyphenols and RSV in particular have the ability to protect against some neurological diseases, and to contribute to functional recovery after brain injury. Noteworthy, these different observations performed on various animal models are supported by some in vitro investigations performed on cultured neuronal cells. Thus, in primary cortical astrocyte cultures, RSV has protective effects against H_2O_2-induced cell damage by improving glutamate uptake activity, increasing GSH content and stimulating S100B secretion, which all contribute to functional recovery after brain injury (Vieira de Almeida et al. 2007–2008). On sodium azide–treated dopaminergic neuron cultures, RSV prevents accumulation of ROS, and depletion of GSH (Okawara et al. 2007), and on beta amyloid–treated neuron cultures exerts an antioxidative action by enhancing the intracellular free radical scavenger GSH (Savaskan et al. 2003). These antioxidative effects of RSV in neuronal cells might involve the MAP kinase pathways (Tredici et al. 1999). Interestingly, some neuroprotective effects of RSV especially in ischemia could be triggered by sirtuin 1 (SIRT1), an NAD$^+$ (oxidized form of nicotinamide adenine dinucleotide)-dependent histone deacetylase related to increase life span in various species (Raval et al. 2008).

10.3.2 RESVERATROL AS A PROOXIDANT AGENT: APPLICATIONS IN CANCER CHEMOTHERAPY

Whereas numerous in vitro and in vivo studies investigating the biological activities of polyphenols, including RSV and its derivatives, revealed antioxidant and anti-inflammatory activities, it has also been demonstrated that these molecules are also able to block the multistep process of carcinogenesis at various stages: tumor initiation, promotion, and progression. In addition, depending on the concentration of the polyphenols and the cell type, it has been shown that RSV can exhibit prooxidant properties, leading to oxidative breakage of cellular DNA in the presence of transition metal ions such as copper (Hadi et al. 2000; de la Lastra and Villegas 2007). It has been suggested that anticancer mechanisms of plant polyphenols involve, at least in part, mobilization of endogenous copper, possibly chromatin-bound copper, which could contribute to prooxidant activities (Ahmad et al. 2006; Bhat et al. 2007). In agreement with this hypothesis, the prooxidant effect of RSV (**1**) and its synthetic analogues, that is, *E*-3,4,4′-trihydroxystilbene (3,4,4′-THS, **5**, *E*-3,4,5-trihydroxystilbene (3,4,5-THS, **6**), *E*-3,4-dihydroxystilbene (3,4-DHS, **7**), *E*-4,4′-dihydroxystilbene (4,4′-DHS, **8**), *E*-2,4-dihydroxystilbene (2,4-DHS, **9**), *E*-3,5-dihydroxystilbene (3,5-DHS, **10**), and *E*-3,5,4′-trimethoxystilbene (3,5,4′-TMS, **4**) on supercoiled pBR322 plasmid DNA strand breakage and calf thymus DNA damage in the presence of Cu (II) ions has been studied (Zheng et al. 2006). It was found that the compounds bearing *ortho*-dihydroxy groups (3,4-DHS, 3,4,4′-THS, and 3,4,5-THS) or bearing 4-hydroxyl groups (2,4-DHS, 4,4′-DHS, and RSV) exhibit remarkably higher activity in the DNA damage than the ones bearing no such functionalities. Moreover, kinetic analysis by UV-visible spectra demonstrates that the formation of ArOH-Cu (II) complexes, the stabilization of oxidative intermediate derived from ArOH and Cu (II)/Cu (I) redox cycles, might be responsible for the DNA damage. It has also been reported that the 4-hydroxystilbene structure is a major determinant for generation of ROS which are responsible for DNA strand scission (Fukuhara et al. 2006; Win et al. 2002) (Figure 10.5).

Using human peripheral lymphocytes and Comet assay, it has been confirmed that resveratrol-Cu (II) is indeed capable of causing DNA degradation in cells (Azmi et al. 2006). These different in vitro experiments strongly support that the cytotoxic activities of some polyphenols against cancer cells involve prooxidant activities leading to DNA damages. These molecules could constitute a new type of DNA cleaving agents (Fukuhara and Miyata 1998). Interestingly, when human prostate cancer

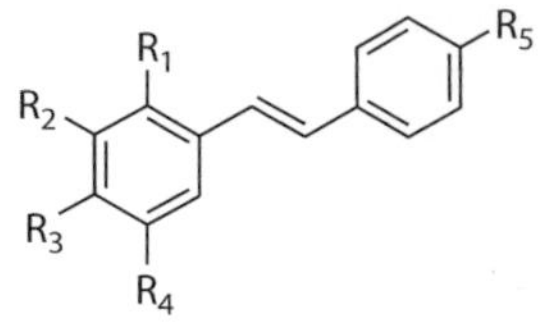

5 $R_1 = R_4 = H, R_2 = R_3 = R_5 = OH$
6 $R_1 = R_5 = H, R_2 = R_3 = R_4 = OH$
7 $R_1 = R_4 = R_5 = H, R_2 = R_3 = OH$
8 $R_1 = R_2 = R_4 = H, R_3 = R_5 = OH$
9 $R_1 = R_3 = OH, R_2 = R_4 = R_5 = H$
10 $R_1 = R_4 = R_5 = H, R_2 = R_4 = OH$

FIGURE 10.5   Structures of compounds **5–10**.

carcinoma cells or human carcinoma colorectal cells HT-29 are only treated with RSV, cytotoxic activities associated with overproduction of ROS and DNA damages are also observed (Cardile et al. 2003; Juan et al. 2008; Scifo et al. 2004). Moreover, it has been shown on endothelial cells that RSV induced an overproduction of ROS levels which was the leading cause of the accumulation of the cells in the S phase of the cell cycle. Indeed, using an siRNA approach, two NADPH oxidases, Nox1 and Nox4, were clearly identified as major targets of RSV and primary sources of ROS that act upstream of the observed S phase accumulation (Schilder et al. 2009). On human breast cancer cells ZR-75-1, MDA-MB-231 and T47D, it has also been reported that RSV-induced apoptosis was associated with a downregulation of mitochondrial SOD (Murias et al. 2008). Taken together, these different data support that polyphenol, including RSV, can induce ROS overproduction either by copper-dependent or copper-independent manners. In this latter condition, RSV acts on specific enzymes playing key roles in the equilibrium of the RedOx status.

10.4   RESVERATROL CELL TRANSPORT AND METABOLISM

Although the biological positive effects of RSV are largely admitted, in contrast, so far only little is known on the transport and the distribution of RSV through the body. At the organ and tissue level, scarce informations were brought about bioavailability mechanisms of RSV. Its intestinal absorption has been firstly studied by Andlauer et al. (2000) in the rat perfused small intestine where the digestive metabolic process may involve release of RSV aglycone after hydrolysis of its glycosylated forms by intestine microflora glucosidases and/or membrane surface–bound phorizine hydrolase of the enterocytes. To determine RSV level in plasma, HPLC methods have been validated (for instance, Adrian et al. 2000). Alternatively, absorption of labeled RSV in rats has been used by Soleas et al. (2001). According to its low water solubility (Belguendouz et al. 1997), RSV must be bound to proteins and/or conjugated to remain in a high serum concentration. In comparison, the efficiency of a therapeutic substance is related to its capacity (selectivity and affinity) to bind protein transporters (Jannin et al. 2004). Since albumin is well known to bind and carry out a large number of amphiphilic molecules, this protein is a good candidate as RSV plasmatic carrier. Jannin et al. (2004) reported the results of RSV binding assays with serum proteins, especially albumin, using different crossed methods including gel filtration chromatography coupled to UV absorption, fluorescence intensity measurements in presence of bovine serum albumin (BSA), quenching of tryptophanyl residues fluorescence by RSV, or Fourier transform infrared spectroscopy (FTIR) of BSA. This study afforded data on molecular interaction of RSV with albumin and highlighted a possible role of albumin as one of the plasmatic carriers of RSV through the blood circulation, allowing further delivery of this polyphenol to the cell surface before cell membrane uptake (Glatz and van der Vusse 1996) to produce its final intracellular biological effect. In the serum we showed that RSV binds to BSA and that fatty acids had a positive effect on this binding (Jannin et al. 2004). The role of fatty acids should be to ensure a lipophilic environment favorable to the binding of RSV. It is also known that the shape of the protein is modified by the presence of fatty acids (Curry et al. 1999). In the past, molecular interactions of serum albumin with small molecules have been extensively studied (Sengupta and Sengupta 2002). From our work, it appears that only a limited number of RSV molecules (4–8) bind the albumin tetramer. The interaction takes place at least at one of the two tryptophanyl residues of the albumin polypeptide chain. In addition, RSV binding induces small but significant conformational changes of the albumin tertiary structure, possibly alpha-helix, near peptide bonds (unpublished results). Proteins other than albumin may also be implicated in the fixation of RSV with higher affinity. For instance, Lamuela-Raventos et al. (1999) have reported binding of RSV to human LDL. Similarly, lactoferrin is able to bind flavonoids (Belguendouz et al. 1997). The binding of RSV by plasmatic albumin allows its transport and facilitates its accessibility to the cell surface. This binding should be reversible and not too strong in order to allow RSV delivery. Albumin-RSV interactions appear to answer to this requirement. In addition, the albumin receptor would allow interaction of RSV with plasma membrane and would make easier its dissociation with albumin (Jannin et al. 2004).

Taking into account that liver is a key step of the absorption and of the metabolism of polyphenols, the study of the RSV uptake, metabolism, and efflux mechanisms have been investigated in human hepatoblastoma cell line, HepG2. A transmembrane traffic of RSV (uptake and release as conjugated form) has been shown for the first time in these hepatic-derived cells (Lançon et al. 2004). These in vitro data are in agreement with in vivo results of Bertelli et al. (1998) that showed an accumulation of RSV in the liver of rats after oral administration. We reported (Lançon et al. 2004) that the fluorescent properties of RSV allowed an easy method to follow the transport of this polyphenol by fluorescence microscopy. Such a method was also convenient for the study of the transport of RSV derivatives such as resveratrol triacetate (3) and the resveratrol dimer ε-viniferin (11) (Colin et al. 2008). The microscopic observation of fluorescent cells showed that RSV is present essentially in cytoplasm and in nucleolus. A nucleolar localization was also reported for taxol (Guy et al. 1996), a natural product responsible for cell cycle arrest and used as chemotherapeutic agent. The authors showed a nucleolar localization of a fluorescent derivative of taxol during interphase of human fore-skin fibroblast cells. Therefore, this RSV localization may be related to the cell cycle perturbations described previously in HepG2-treated cells (Delmas et al. 2000) (Figure 10.6).

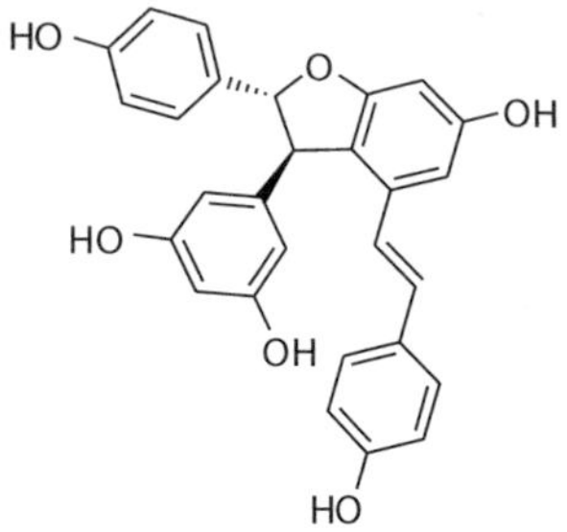

11

FIGURE 10.6   Structure of compound **11**.

The incubation with radiolabeled RSV showed a large uptake of the polyphenol by the hepatoblastoma cells, HepG2 (Lançon et al. 2004). The kinetics of this uptake was similar in human hepatocytes although with a higher capacity than hepatoma cells. No toxicity or hepatocyte lysis was observed with $30\,\mu M$ RSV. This observation is relevant because various antineoplastic agents cause a hepatotoxicity that limits their efficacy in anticancer therapy. Indeed, in accordance with previous studies (Gao et al. 2002; Lu et al. 2001), RSV appeared to have specific cytotoxic effects toward hepatic tumor cells, compared to normal human hepatocytes. This large uptake of RSV by human hepatocytes and its weak toxicity, at least in mouse tumors models (Jang et al. 1997), suggest that this polyphenol could have an important role in the prevention of liver damages. This hypothesis is supported by another studies on liver nonparenchymal cells (Chen et al. 1999) showing that RSV could protect from hepatocyte damage. Furthermore, an oral administration of this natural compound decreased hepatic metastatic invasion of B16 melanoma cells inoculated intrasplenically (Asensi et al. 2002).

Several authors (Goldberg et al. 2003) questioned about the significance of the in vitro studies performed with unconjugated RSV since it has been shown that this molecule is extensively conjugated in the intestine. However, experiments with a linked-rat model (Marier et al. 2002) showed that RSV aglycone is also bioavailable. Our results (Lançon et al. 2004) indicated that the uptake of RSV aglycone by liver cells involves both a passive diffusion process and a facilitated translocation. Indeed, we have shown that after a short time of incubation with RSV (1 h), the efflux of the polyphenol is fast. As previously found in influx experiments, the rates of release were slower at 4°C than at 37°C, showing that efflux was also temperature-dependent. This temperature dependence of uptake as well as of efflux suggests that at least one part of RSV enters the cells and exits via transporters. Concerning the identification of membrane carriers, it was shown that canalicular multispecific organic anion transporter (cMOAT), which is predominantly expressed on the canalicular membrane of hepatocytes under physiological conditions, exhibits transport activity (Maier-Salamon et al. 2008). This transport includes GSH-, glucuronide-, and sulfate-conjugated compounds, suggesting that it could be involved in the efflux of RSV metabolites. By comparative HPLC analysis of collected cell culture media, treated or not with beta-glucuronidase, we have found that at least one part of RSV released from the human hepatomas cells is in a conjugated form, glucuronide and possibly sulfate (Lançon et al. 2007). It is known

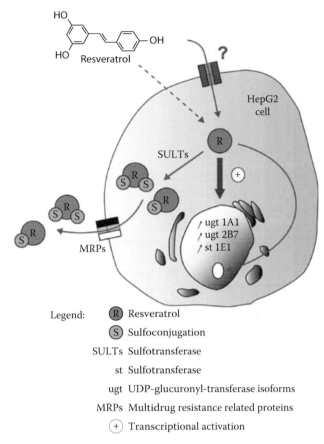

HO

HO

Resveratrol

OH

?

HepG2
cell

R

SULTs

S R S

S R S

S R

S R

MRPs

ugt 1A1
ugt 2B7
st 1E1

+

Legend: (R) Resveratrol

          (S) Sulfoconjugation

SULTs Sulfotransferase

   st Sulfotransferase

  ugt UDP-glucuronyl-transferase isoforms

MRPs Multidrug resistance related proteins

   (+) Transcriptional activation

FIGURE 10.7   Possible resveratrol intracellular transport.

that RSV is conjugated by human liver cells and that its glucuronidation occurs at positions 3 and 4′ and is mainly catalyzed by UGT1A1 isoform (Aumont et al. 2001). The uptake of conjugated RSV has been studied by Lançon et al. (2007) where the type of conjugation involved in HepG2 cells is mainly the sulfo-conjugation (Lançon et al. 2007). There are no or only poor effects of RSV metabolites on cell proliferation (Lançon et al. 2007). Vitrac et al. (2003) and Henry et al. (2005) showed that RSV uptake in Caco-2 cells involves MRP2/3. We have also seen the involvement of MRP3 in the efflux from HepG2 (Lançon et al. 2007). The possible overall RSV intracellular traffic is summarized in the Figure 10.7.

10.5   RESVERATROL STRUCTURAL ANALOGUES: SYNTHESIS AND BIOLOGICAL POTENCIES

We have seen above (Section 10.4) the rapid metabolism and efflux of resveratrol in the cells. One of the possible strategies to overcome the problem of resveratrol low bioavailability is to find new and potently active resveratrol analogues with higher

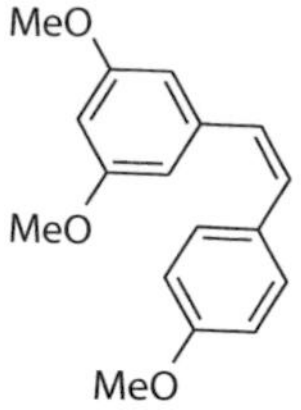

FIGURE 10.8   Structure of compound **12**.

half-life and possibly acting through a different biological mechanism. Many of these derivatives have been synthesized and are reported in a recent review (Chillemi et al. 2007). Among these, a series of acylated, methylated, and hydrogenated resveratrol analogues were prepared and subjected to MTT bioassay toward human tumor cell lines DU-145 (Cardile et al. 2005). Subsequently, four *E*- and *Z*-stereoisomeric couples of resveratrol and three further stilbenoids bearing two or three methoxy groups were evaluated toward a set of different human cancer cell lines, namely, DU-145 (androgen not responsive human prostate cancer), LNCaP (androgen responsive human prostate cancer tumor, M-14 (human melanoma), and KB (human mouth epidermoid carcinoma). In all cell lines, the permethylated resveratrol derivative *E*-3,5,4′-trimethoxystilbene (**4**) resulted more potent than resveratrol especially toward DU-145 cells. The corresponding *Z*-isomer (**12**) proved highly active toward KB cell line but with a poor effect in M-14 cells. (Cardile et al. 2007) (Figure 10.8).

Further literature data (Pettit et al. 2002; Roberti et al. 2003) and the above cited review (Chillemi et al. 2007) clearly indicate that polymethoxystilbenes and related compounds are a subgroup of resveratrol analogues showing promising antitumor properties such as potent antiproliferative and pro-apoptotic activity. In addition, some in vivo studies indicate that polymethoxystilbenes undergo different metabolic conversion and have a higher bioavailability with respect to resveratrol (Lin et al. 2010; Sale et al. 2004). In the majority of cases where pairs of *E*- and *Z*-stilbenoids were evaluated for antitumor activity, the *Z* isomers proved significantly more active than their *E* analogues; nevertheless, the antiproliferative/apoptotic activity ratio between the *E/Z* isomers reported showed wide variations and in some cases both either have comparable activities or the *E*-isomer may be even more active, as for *E*-resveratrol (**1**) and *Z*-resveratrol (**13**). Thus, the interpretation of the available data is not straightforward and the structure–activity relationships for *E* and *Z*-polymethoxystilbenes cannot be considered as completely established. However, a number of recent studies highlighted the structural analogy of polymethoxylated *Z*-stilbenoids with combretastatin A4 (**14**), a potent antimitotic agent which interacts with tubulin at colchicine binding site (de Lima et al. 2009), leading to a blockade of mitosis (M phase). This blockade provokes an increase of cell death. In addition, some observed biological properties, like the inhibition of tubulin polymerization by polymethoxystilbenes and in particular by *Z*-3,5,4′-trimethoxystilbene (**12**), considerably more active than resveratrol (Schneider et al. 2003, Seiler et al. 2004), supported the hypothesis that these resveratrol analogues could have a mechanism of action similar to that of **3**. On this basis, and in perspective for a possible application of resveratrol analogues in cancer therapy, we

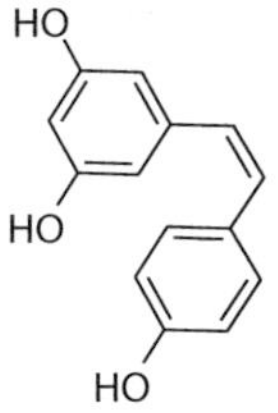

FIGURE 10.9 Structures of compounds **13** and **14**.

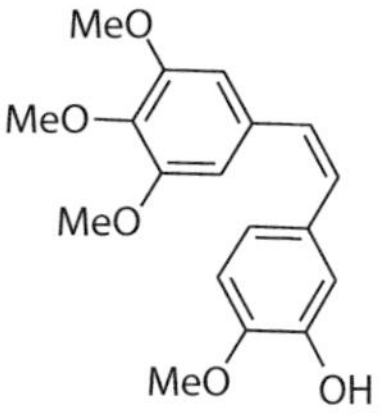

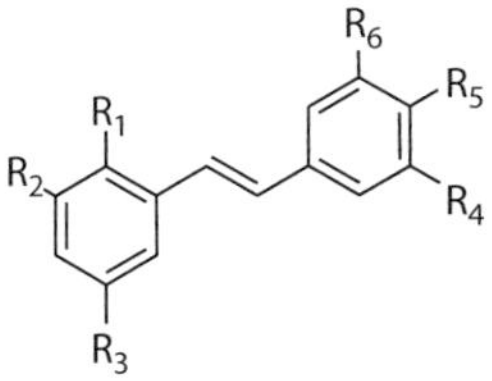

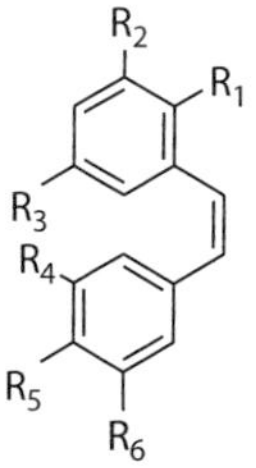

15 $R_1 = OH$, $R_2 = R_3 = R_5 = OMe$, $R_4 = R_6 = H$
17 $R_1 = R_5 = H$, $R_2 = R_3 = R_4 = R_6 = OMe$
19 $R_1 = OH$, $R_2 = R_3 = R_4 = R_6 = OMe$, $R_5 = H$
21 $R_1 = R_4 = H$, $R_2 = R_3 = R_5 = R_6 = OMe$
23 $R_1 = OH$, $R_2 = R_3 = R_5 = R_6 = OMe$, $R_4 = H$

16 $R_1 = OH$, $R_2 = R_3 = R_5 = OMe$, $R_4 = R_6 = H$
18 $R_1 = R_5 = H$, $R_2 = R_3 = R_4 = R_6 = OMe$
20 $R_1 = OH$, $R_2 = R_3 = R_4 = R_6 = OMe$, $R_5 = H$
22 $R_1 = R_4 = H$, $R_2 = R_3 = R_5 = R_6 = OMe$
24 $R_1 = OH$, $R_2 = R_3 = R_5 = R_6 = OMe$, $R_4 = H$

FIGURE 10.10 Structures of compounds **15–24**.

evaluated the potency of selected couples of *E*- and *Z*-polymethoxystilbenes toward human colorectal tumor cell line (SW 480) (Figure 10.9).

By a proper synthetic methodology, previously employed for some of the compounds (Spatafora et al. 2009), a library of resveratrol analogues (Figure 10.10) was obtained. These structural analogues of resveratrol were tested by in vitro approach as a prerequisite to perform studies at the in vivo level. In addition, we examined by a computational docking approach possible structure–activity relationships in connection with binding to the colchicine site of tubulin (Mazué et al. 2010).

In a first group, the 3,5,4′-hydroxyl groups of resveratrol were replaced by methoxy groups (compounds **4** and **12**) and a further hydroxyl group was inserted at position 2 (compounds **15** and **16**); a second group was represented by 3,5,3′,5′-tetramethoxystilbenes (compounds **17** and **18**) also bearing an hydroxyl group in C-2 (compounds **19** and **20**); in a third group of 3,5,3′,4′-tetramethoxystilbenes, the two rings were asymmetrically substituted (compounds **21** and **22**) and also in this case the 2-hydroxy analogues were prepared (compounds **23** and **24**). The antiproliferative potency of resveratrol synthetic analogues were compared to those of the two natural isomers of resveratrol, **1** and **13**; the *E*- and *Z*-isomers of the 3,5,4′-trimethoxystilbene (**2** and **14**) were also included in view of their previously reported high antiproliferative activity. The following parameters were evaluated: cell proliferation rate, cell cycle phase targets, type of death, state of cell polyploidy. IC_{50} values are summarized in Table 10.3.

TABLE 10.3

Half Inhibitory Concentrations (IC$_{50}$) of Resveratrol Derivatives toward SW480 Human Colorectal Cell Proliferation

Compound	IC$_{50}$ (µM)	Compound	IC$_{50}$ (µM)
1	20 ± 3	13	90 ± 12
4	54 ± 8	12	0.3 ± 0.04
15	43 ± 6	16	18 ± 2
17	48 ± 7	18	13 ± 2
19	17 ± 2	20	7 ± 2
21	100 ± 15	22	9.5 ± 2
23	42 ± 3	24	10 ± 2

The overall results, also in comparison with RSV properties, are shortly discussed below for each analogue.

E-3,5,4′-trihydroxystilbene (*E*-resveratrol, **1**) is the parent molecule used as reference. As expected, we observed antiproliferative activity (IC$_{50}$ = 20 µM) and a blockade of the cell cycle at the S/G2M phases transition. Only a little induction of cell polyploidy was observed.

E-3,5,4′-trimethoxystilbene (**4**) showed a weaker antiproliferative effect compared to RSV (**1**) giving ~40% of inhibition of cell growth after 48 h of treatment at 30 µM and an IC$_{50}$ = 54 µM. The proliferation blockade is related to a defect in the cell division in the M phase (50% of cells have several nuclei), leading to the generation of polyploid cells containing more than one nucleus suggesting a cytodieresis inhibition.

E-2-hydroxy-3,5,4′-trimethoxystilbene (**15**) showed effects similar to those exerted by **4**. No significant changes were caused by the presence of the hydroxyl group in position 2.

E-3,5,3′,5′-tetramethoxystilbene (**17**) showed a weak antiproliferative effect (IC$_{50}$ = 48 µM) and led to ~40% of inhibition of cell growth after 48 h of treatment at 30 µM. Neither a blockade of cell cycle at 48 h nor an increase of cell death was observed. The proliferation appears slow during the first 24 h of treatment.

Z-3,5,4′-trihydroxystilbene (*Z*-resveratrol, **13**) proved to be eightfold less potent than **1**, with an IC$_{50}$ of 90 µM.

Z-3,5,4′-trimethoxystilbene (**12**) largely showed the strongest inhibitory activity, in agreement with previously reported data (Cardile et al. 2007; Schneider et al., 2003). This *Z*-isomer is much more active than its *trans* counterpart, *E*-3,5,4′-trimethoxystilbene (**4**). At 0.2 µM of treatment, **14** inhibits cell growth at ~80%, thus resulting 66 times more efficient than **1** (IC$_{50}$ = 0.3 µM). The proliferation blockade is associated to a high accumulation of polyploid cells showing several nuclei (up to 16). After 48 h of treatment at 1 µM, 70% of cells have more than one nucleus.

Z-2-Hydroxy-3,5,4′-trimethoxystilbene (**16**) inhibits cell proliferation at 70% after 48 h of treatment at 30 μM. An accumulation of cells in S phase followed by an accumulation of polyploid cells (disappearance of the diploid G2 phase) was observed. Since the presence of more than one nucleus was difficult to establish, at this stage, it was not possible to conclude to a caryodieresis or to cytodieresis default.

Z-3,5,3′,5′-tetramethoxystilbene (**18**) proved significantly active, leading to a blockade of 90% of cell proliferation after 48 h of treatment at 30 μM with an IC_{50} of 13 μM. Its effect is to accumulate cells in S phase (disappearance of the diploid G1 phase), then to generate polyploid cells. However, since the presence of more than one nucleus was not evident it was not possible to conclude to a caryodieresis or cytodieresis default.

Z-2-Hydroxy-3,5,3′,5′-tetramethoxystilbene (**20**) proved to be a very active compound (IC_{50} = 7 μM). From 1 μM it inhibits cell proliferation at 60% after 48 h of treatment. Cells treated at 1 μM for 48 h are blocked in S phase of the diploid cycle. One-third of the cells are polyploid after this treatment, but they do not appear ongoing to the cycle.

E-tetramethoxystilbene analogues **21** and **23** proved less active than **1**. When **21** is compared to **17** (which is only different in the position of one methoxy group) it appears to be less active. The same tendency can be observed between E-tetramethoxy derivatives, that is, **23** compared to **18**.

Z-tetramethoxystilbene analogues **22** and **24** showed antiproliferative activity but they appear to be less efficient than the Z-trimethoxy analogue **12**.

This approach allowed studying the effect of critical methylation of RSV (**1**) on the proliferation inhibition of human colon cancer cells. From these results, it appears that methylation of RSV is crucial in the inhibitory potency and mechanism of action of analogues on human colorectal tumor cell proliferation and this effect is higher for the analogues with Z configuration. Some of the compounds proved more to much more active than the natural parent molecule. Compared to RSV, which leads to a cell growth arrest in S phase, the methylated derivatives stop the cell proliferation by inducing G2/M failures and also a polyploidization of the SW 480 cell line. E-RSV derivatives also induce apoptosis of cancer cells since we find sub-G1 peaks during the flow cytometry analysis. The cell polyploidization, which can occur naturally to repair damaged DNA, is induced by all of the methylated compounds. The inhibition leads to a default of mitosis and an impairment of cytodieresis. The destiny of tetraploidy-induced cells is unknown; these cells could still proliferate and become more resistant (Erenpreisa et al. 2005) or die by mitotic catastrophe (Mansilla et al. 2006).

In connection with the above cited results, it is worth noting here that recently the group of Szekeres (Saiko et al. 2008) has reported the influence of several E-resveratrol analogues on HT 29 human colon cancer cell proliferation inhibition and apoptosis and some results are similar to those above reported, for instance, the low activity of E-3,5,3′4′-tetramethoxystilbene (**21**) and blockade on G0-G1; some other are contrasting: the strong effect of E-3,5,4′-trimethylresveratrol (**4**) on HT 29 cells and the poor effect on SW480 cells (directly comparable to our data).

The presence of hydroxylated group does not significantly change the antiprolifera-tive effect, in agreement with results of Ovesna (2006) reporting that hydroxylation mainly protect against DNA damage. However here we are on established cell line which is already initiated. Thus, the only visible effect can be observed from the promotion step. More generally, methoxysubstituted resveratrol have stronger effect than the parent molecule. Indeed they inhibit the human tumor necrosis factor alpha-induced activation of transcription factor nuclear factor Kappa B (Heynekamp 2006).

In conclusion, while RSV is considered a promising molecule to fight cancer (Baur and Sinclair 2006), synthetic RSV analogues could offer a wide range of compounds potentially more active than E-RSV. In particular, some molecules tested by us seem to have a different way of delaying cancer cell growth. Resveratrol inhibits cells in S phase, while most of the other synthetic derivatives stop mitosis, or block cell growth in an unknown manner (**17** or **19**). We can consider that these methylated derivatives which are prevented of any hydroxyl groups-conjugation dependency would be less metabolized than RSV and potentially more bioavail-able. This has been recently confirmed by the above cited pharmacokinects study of Lin et al. (2010). In addition, the stronger effect of *cis* (Z) polymethoxy deriva-tives than their trans (E) counterparts is not linked to the lack of antioxidative effect (disappearance of hydroxyl groups) but due to a steric-dependent mecha-nism leading to interference to different pathways as compared to the *trans* deriva-tives. Our computational studies showed that almost all of the docked structures of (Z)-polymethoxy isomers are well superposed to the docked structure of com-bretastatin A-4 (**14**), in binding at colchicine-binding site of tubulin, while most of the (E)-polymethoxy isomers counterpart do not fit with **14**. These data sug-gest that tetramethoxystilbenes and hydroxylated polymethoxystilbenes are good candidates as tubulin inhibitors but probably do not reach the tubulin pocket with the same effectiveness of (Z)-3,5,4′-trimethoxystilbene (**12**), probably due to other factors such as hydrophobicity or metabolism.

Following hypothesis (Figure 10.11), the statement on critical resveratrol methyla-tion dependency of cell proliferation inhibition is presented below:

1. The relative weak effect of E-resveratrol is not due to its high metabolic rate and efflux since the masking of hydroxyl groups by methylation does not significantly improve resveratrol analogue efficacy.
2. The Z configuration is not required for resveratrol efficiency since the potency of Z-isomer is weaker than that of E-isomer.
3. In contrast to the preceding statement #2, the Z configuration, associated to the hydroxyl groups substitution by methoxy group is crucial and leads to a very potent inhibition efficiency.
4. The presence of more than three methoxy groups on Z-resveratrol core structure does not enhance efficiency but gives rather a weaker effect.
5. The presence of a new hydroxyl group on methylated Z-resveratrol ana-logues does not dramatically change inhibitory potency.
6. Concerning the inhibitory mechanism, the presence of methyl groups, whatever the resveratrol isomer (E-, Z-), induces a cell polyploidy resulting from the blockade of the cell divisions at the mitosis level.

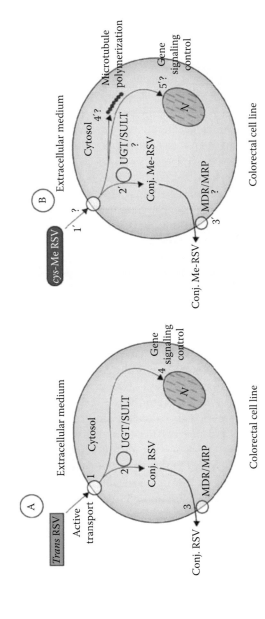

FIGURE 10.11   Working hypothesis.

10.6 RESVERATROL AND ANALOGUES AS PROBES TO TARGET ESSENTIAL SIGNALING CELL ELEMENTS

Resveratrol is well known for its multiple and varying therapeutic features which may be explained by the different ways in which RSV targets the essential signaling cell elements. Reported results of proteins, enzymes, or receptors targeted by RSV in different therapeutic fields are numerous. Furthermore, in the past 5 years, several research teams have turned to account the acidity of phenolic function in the goal of grafting RSV onto matrix beads or effecting chemical transformations on RSV which improve therapeutics features of the polyphenol.

One important target of RSV is the AMP-activated protein kinase. By activating the AMPK, RSV provides atheroprotective effects (Klinge et al. 2005), decreases lipid amounts and the acceleration of atherosclerosis (Zang et al. 2006), increases glucose uptake in myotubes (Park et al. 2007), and can induce apoptosis in drug-resistant cancer cells (Hwang et al. 2007). RSV is also an inactivator of COX-1 and COX-2, enzymes involved in the synthesis of vasoconstrictor and platelet aggregators (Szewczuk and Penning 2004; Szewczuk et al. 2004) and of prostaglandins (Murias et al. 2004). The human cytomegalovirus replication can be inhibited by RSV in which the mechanism is thought to be a blocking-up by the polyphenol of the epidermal growth factor receptor (EGF-R) and its downstream effects (Evers et al. 2004). RSV has been reported to slow the progression of bovine intervertebral disc (IVD) by modulating MMP-13 and PG production, proteins involved in disc degeneration (Li et al. 2008). Resveratrol is a well-known antitumoral agent and numerous reports have shown that this polyphenol is responsible for cell death in a wide range of tumors (see Section 10.2). Different signaling pathways in cancer cells are targeted by RSV, and we report some examples in the following. In prostate cancer, several genes are the targets of dietary phenolic compounds, suggesting that the growth inhibition of LNCaP cells is caused by the activation of multiple signaling pathways (Narayanan et al. 2002, 2003). Stat-3 protein is a signal transducer and activator of transcription initiated by a hyperactive tyrosine kinase Src; Stat-3 induces oncogenic processes which contribute to malignant transformation and progression (Yu and Jove 2004). Recently, it has been reported that both Src and Stat-3 are targets of RSV in human breast, pancreatic, prostate carcinoma cells (Kotha et al. 2006), and medulloblastoma cells (Yu et al. 2008). The binding of RSV on a receptor site of integrin $\alpha V\beta 3$ triggers the transduction of the stilbene signal into p53-dependent apoptosis of breast cancer cells (Lin et al. 2009). In different cancer cells, RSV may target mitochondria by inducing mitochondrial membrane potential that triggers the release of cytochrome c and the activation of caspases (Ma et al. 2007; van Ginkel et al. 2007); a sustained elevation of intracellular [Ca^{2+}] followed by the process of cell apoptosis (Ma et al. 2007) has been observed. In human colorectal carcinoma cells, RSV induces production of superoxide anions in the mitochondria of cells undergoing apoptosis in HT-29 (Juan et al. 2008). In breast cancer cells, the ability of RSV to bind the estrogen receptors is due to its structural resemblance to estrogen (Wang et al. 2006). Thus, estrogen receptors are very good targets of RSV whose agonistic and antagonistic properties lead to regulate mRNA expression of several genes involved in cell cycle control and estrogen receptors signaling pathway (Le Corre et al. 2006). A recent computational docking analysis of RSV and derivatives revealed for each polyphenol a distinct ERalpha binding orientation and estrogen target gene expression profile (Lappano et al. 2009). Aryl hydrocarbons such as dioxins generated by the chemical industry and present in cigarette smoke have been shown to cause diseases especially to favor tumor development. RSV has been reported to inhibit in vitro transcription of CYP1A1, responsible of the metabolism of aryl hydrocarbon in genotoxic metabolites (Ciolino et al. 1998) and to have an antagonist activity on the ArH receptor (Casper et al. 1999). The ability of the RSV to recognize ArH receptor has prompted Gao et al. to synthesize radiolabeled stilbene derivatives in order to use them as PET probes of ArH receptor in tumors (Gao et al. 2006). In addition to being a very good target of various kinds of tumors, RSV has fluorescent properties which allow observation of its cell uptake and its cellular localization in human hepatic cells (Lançon et al. 2004).

In a different therapeutic field, several reports indicate that natural polyphenols can reduce the incidence of age-related neurological disorders. Accumulation and aggregation of amyloid-β peptide have been identified as the core of Alzheimer's disease. The mechanism of action of RSV is yet ill defined (Singh et al. 2008); however, the degradation (Marambaud et al. 2005) and the inhibition of polymerization (Rivière et al. 2007) of amyloids by RSV have been reported. The powerful antioxidant properties of RSV seem to play an important role in neuroprotection (Jang and Surh 2001). Using [³H]-RSV, Bastianetto et al. have shown that the polyphenol bound on specific plasma membrane sites in the rat brain, which would indicate a mechanism type-receptor (Bastianetto et al. 2009; Han et al. 2006). However that may be, binding of RSV to amyloids has been examined by observing fluorescence emission spectra (Ahn et al. 2007); the amyloids enhanced the intrinsic fluorescence of RSV and the increase was proportional to their concentrations. RSV can be considered as a novel fluorescent probe which allows quantitative determination of amyloids. In addition to the ability of RSV to target a pletoric signaling cell elements, this polyphenol is a structurally simple small molecule. Thus, its synthesis can be easily carried out by the classical Wittig (Pettit et al. 2002), Wadsworth-Emmons (Wang et al. 1999), or Heck (Botella and Najera 2004) reactions; RSV may be transformed by alkylation of a phenolic function. Retaining the basic structure necessary for its biological activities, RSV may be used in different ways.

The reactivity of the phenolic function permitted the binding of the stilbene structure on agarose or sepharose matrix beads. In this manner, RSV is immobilized onto this solid matrix, forming a RSV-Affinity Column (RAC) and can qualitatively and quantitatively capture cellular targeting proteins from the eluted cell extracts. The development of this method was conducted by the Wang's group (Wang et al. 2004) which reported that protein bound with RSV immobilized was identified as quinone reductase 2 (QR2) after elution of extracts of prostate cells (Wang et al. 2004). This result is in agreement with the inhibition of QR2 by RSV in melanoma cells (Hsieh et al. 2005) and in vascular smooth muscle (Cai et al. 2008). A high-resolution structural analysis of a QR2–RSV complex has confirmed these results (Buryanovskyy et al. 2004): the analysis has revealed that RSV binds the active-site cleft and the three RSV hydroxyl groups form bonds with amino acids of QR2.

The results about the action of RSV on COX-2, enzyme highly expressed in various cancers are inconsistent. For example, Szewczuk et al. reported that RSV had a weak effect on COX-2 (Szewczuk et al. 2004) and in contrast, Murias et al. (2004) described RSV is a highly selective COX-2 inhibitor. This latest result has been corroborated by Zykova et al. (2008) using the method of RAC to show that RSV coupled to a sepharose matrix might directly bind with the COX-2 protein after elution of HT-29 colon cancer cells and thus decrease COX-2-mediated PGE_2 production.

By means of certain transformations on the RSV skeleton, new stilbene derivatives have been obtained and tested in different therapeutic yields. Hauss et al. (2007) have combined the neurodegenerative features of fatty alcohols structure and the neuroprotective effects of RSV; they have synthesized new hybrid compounds 25a–e called RSV fatty alcohols (RFAs). The stilbene skeletons were obtained by a Wadsworth–Emmons reaction but the preparation of the aromatic aldehydes bearing the long-chain fatty alcohol needed a Sonogashira reaction (Figure 10.12). The test of one of the new hybrid compounds ($n = 12$, RFA-12) on a mouse microglial cell line (MMGT-12) prove that the combination of the two biological properties in one molecule is a success. Indeed, RFA-12 can induce neuronal differentiation of neural stem cells while modulating microglial induction.

In cancer cells, mitochondria are targeted by RSV whose action is to modify the membrane potential and the intracellular $[Ca^{2+}]$ (Ma et al. 2007; van Ginkel et al. 2007). The coupling of RSV to a membrane-permeable lipophilic cation triphenylphosphonium provides a new mitochondriotropic derivative 26 (Biasutto et al. 2008). The polyphenol may be considered as a probe which leads the lipophilic cation selectively to the mitochondria of the cancer cells; in addition, the presence of the salt increases the solubility of the molecule in water (Figure 10.13).

The biological activities of the N-phosphoryl amino acid-modified RSV analogues 27a–d and 28a–d (Figure 10.14) were evaluated against human nasopharyngeal

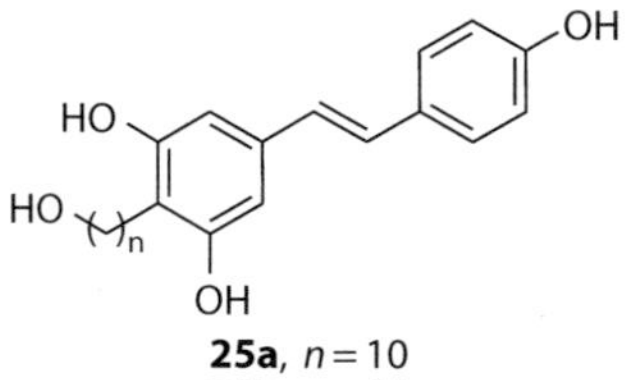

25a, $n = 10$
25b, $n = 12$
25c, $n = 14$
25d, $n = 16$
25e, $n = 18$

FIGURE 10.12   Structures of compounds 25a–e.

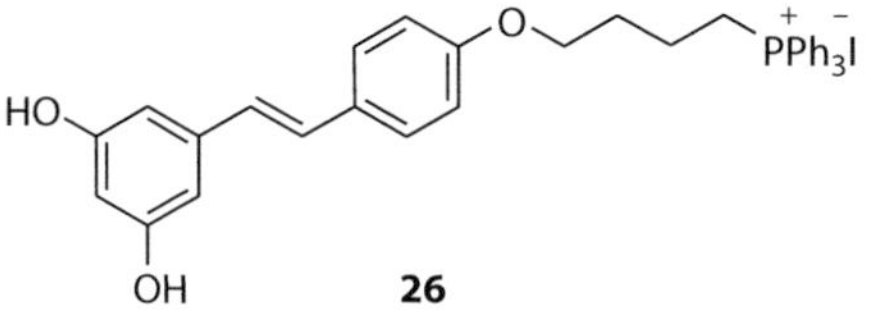

26

FIGURE 10.13   Structure of compound 26.

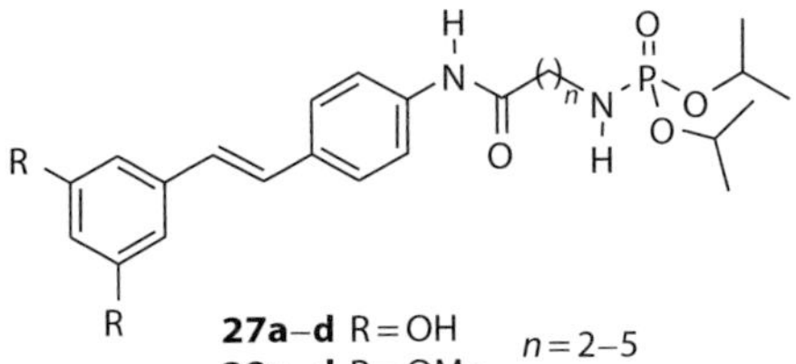

27a–d R=OH
28a–d R=OMe n=2–5

FIGURE 10.14 Structures of compounds **27a–d** and **28a–d**.

carcinoma cell lines CNE-1 and CNE-2 (Liu et al. 2008). The combination of RSV and *N*-phosphoryl amino acid offered a new series of potent growth inhibitors of CNE-1 and CNE-2 cell lines; when $n = 4$ and R_1 and R_2 are methyl groups, the inhibitory activity is 15-fold higher than RSV and is associated with caspase-9 activation, indicating that a mitochondrial pathway was involved in the apoptosis process. The characteristic in the structure of compounds **27a–d** and **28a–d** is the replacement of 4′-OH by an amido group; nevertheless, compound **27a–d** and **28a–d** still shows anticancer activity.

To carry RSV to tissues and human organs after consumption, the human serum albumin (HSA) must bind RSV. It has been shown by molecular modeling that a hydrophobic interaction exists between the plasma protein and the polyphenol, but only when its concentration is high (Lu et al. 2007).

Similar results from a study concerning the binding between RSV and BSA by fluorescence spectroscopy have been reported a few times later (Xiao et al. 2008). The HSA can also bind various other ligands including fatty acids. Thus, to improve the binding affinity between HSA and RSV and in the same time, to increase water solubility, the aliphatic acid **29** (Figure 10.15) was prepared by direct alkylation of RSV using brominated aliphatic ester, followed by saponification and acidification (Jiang 2008). In comparison to RSV, compound **29** is more soluble in water and shows a much more binding affinity to HSA. The results of fluorescence spectroscopy prove that **29** is tightly bound to HSA and the structure is very rigid. In another study (Chen et al. 2009) **29** has been reported to inhibit the TLR-2 mediated cell apoptosis. The cytotoxicity of the *O*-alkylated resveratrol analogue **29** (Figure 10.15) against human

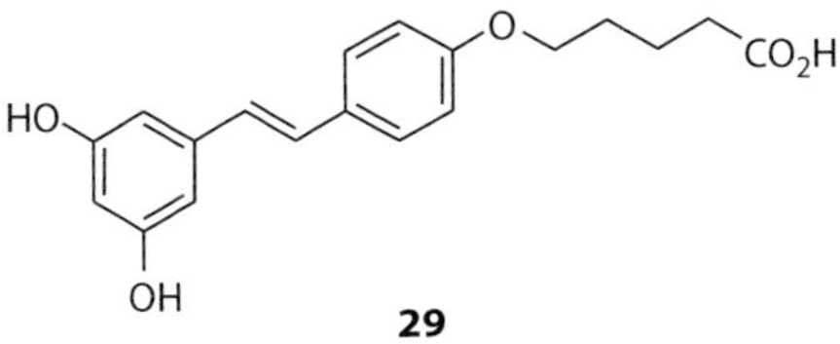

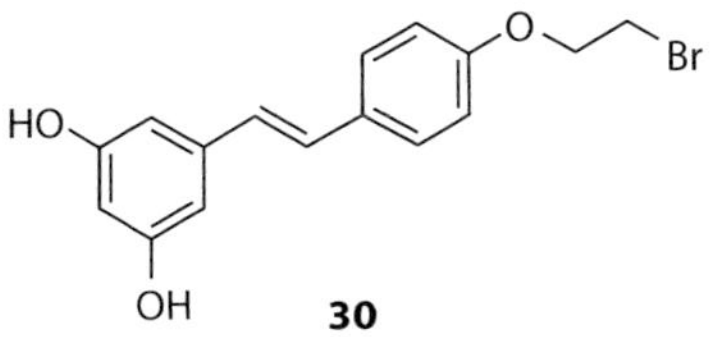

FIGURE 10.15 Structures of compounds **29** and **30**.

nasopharyngeal epidermoid tumor cell line KB was investigated (Ruan et al. 2006). This polyphenol showed a stronger anticancer activity than RSV. The compound **30** was easily synthesized by alkylation of RSV with 1,2-dibromoethane and K_2CO_3 as base in DMF.

10.7 CONCLUSIONS

Resveratrol (**1**) is currently considered a promising molecule to fight cancer (Baur and Sinclair 2006). Nevertheless, synthetic RSV derivatives could offer a wide range of compounds potentially more active than RSV: although the review of the resveratrol structural analogues cited in this chapter is not exhaustive and offers only some examples of compounds with potentiality in cancer chemotherapy, the case of Z-3,5,4'-trimethoxystilbene **12**, much more active than RSV, is worth of note. This and other derivatives lacking of readily conjugating hydroxyl groups would be less metabolized than RSV and consequently should result more bioavailable. In addition, the methylated analogues of RSV seem to have a different way of delaying cancer cell growth. Compared to RSV, which induces a cell growth arrest in S phase, these derivatives stop the cell proliferation by inducing G2/M failures and also a polyploidization of human colorectal cancer cell line (Mazue et al., 2010). RSV derivatives induce apoptosis of cancer cells as shown by the sub-G1 peaks observed by flow cytometry. Polyploidization could occur naturally to repair damaged DNA. Here, the polyploidization is induced by the compounds. *Cis*-trimethylresveratrol has been described as an inhibitor of tubulin polymerization (Schneider et al. 2003; Seiler et al. 2004). The destiny of polyploid induced cells is unknown; these cells could still proliferate and become more resistant (Erenpreisa et al. 2005) or die by mitotic catastrophe (Mansilla et al. 2006).

REFERENCES

Adrian, M., Jeandet, P., Douillet-Breuil, A.C., Tesson, L., and Bessis, R. 2000. Stilbene content of mature *Vitis vinifera* berries in response to UV-C elicitation. *J. Agric. Food Chem.* 48: 6103–6105.

Aggarwal, B.B. and Shishodia, S. 2006. Resveratrol in health and disease. In *Oxidative Stress and Disease*, Vol. 20, Taylor & Francis, London, U.K., pp. 1–712.

Ahmad, A., Syed, F.A., Singh, S., and Hadi, S.M. 2005. Prooxidant activity of resveratrol in the presence of copper ions: Mutagenicity in plasmid DNA. *Toxicol. Lett.* 159: 1–12.

Ahn, J.S., Lee, J.H., Kim, J.H., and Paik, S.R. 2007. Novel method for quantitative determination of amyloid fibrils of α-synuclein and amyloid β/A4 protein by using resveratrol. *Anal. Biochem.* 367: 259–265.

Andlauer, W., Kolb, J., Siebert, K., and Fürst, P. 2000. Assessment of resveratrol bioavailability in the perfused small intestine of the rat. *Drugs Exp. Clin. Res.* 26: 47–45.

Asensi, M., Medina, I., Ortega, A., Carretero, J., Bano, M.C., Obrador, E., and Estrela, J.M. 2002. Inhibition of cancer growth by resveratrol is related to its low bioavailability. *Free Radic. Biol. Med.* 33: 387–398.

Ates, O., Cayli, S., Altinoz, E., Gurses, I., Yucel, N., Kocak, A., Yologlu, S., and Turkoz, Y. 2006. Effects of resveratrol and methylprednisolone on biochemical, neurobehavioral and histopathological recovery after experimental spinal cord injury. *Acta Pharmacol. Sin.* 27: 1317–1325.

Ates, O., Cayli, S.R., Yucel, N., Altinoz, E., Kocak, A., Durak, M.A., Turkoz, Y., and Yologlu, S. 2007. Central nervous system protection by resveratrol in streptozotocin-induced diabetic rats. *J. Clin. Neurosci.* 14: 256–260.

Aumont, V., Krisa, S., Battaglia, E., Netter, P., Richard, T., Merillon, J.M., Magdalou, J., and Sabolovic, N. 2001. Regioselective and stereospecific glucuronidation of *trans*- and *cis*-resveratrol in human. *Arch. Biochem. Biophys.* 393: 281–289.

Azmi, A.S., Bhat, S.H., Hanif, S., and Hadi, S.M. 2006. Plant polyphenols mobilize endogenous copper in human peripheral lymphocytes leading to oxidative DNA breakage: A putative mechanism for anticancer properties. *FEBS Lett.* 580: 533–538.

Bastianetto, S., Dumont, Y., Han, Y., and Quirion, R. 2009. Comparative neuroprotective properties of stilbene and catechin analogs: Action via a plasma membrane receptor site? *CNS Neurosci. Ther.* 15: 76–83.

Baur, J.A. and Sinclair, D.A. 2006. Therapeutic potential of resveratrol: The *in vivo* evidence. *Nat. Rev. Drug Discov.* 5: 493–506.

Belguendouz, L., Fremont, L., and Linard, A. 1997. Resveratrol inhibits metal ion-dependent and independent peroxidation of porcine low density lipoproteins. *Biochem. Pharmacol.* 53: 1347–1355.

Bernhard, D., Tinhofer, I., Tonko, M., Hubl, H., Ausserlechner, M.J., Greil, R., Kofler, R., and Csordas, A. 2000. Resveratrol causes arrest in the S-phase prior to Fas-independent apoptosis in CEM-C7H2 acute leukemia cells. *Cell. Death Differ.* 7: 834–842.

Bertelli, A., Bertelli, A.A.E., Gozzini, A., and Giovannini, L. 1998. Plasma and tissue resveratrol concentrations and pharmacological activity. *Drug Exp. Clin. Res.* 24: 133–138.

Bhat, S.H., Azmi, A.S., and Hadi, S.M. 2007. Prooxidant DNA breakage induced by caffeic acid in human peripheral lymphocytes: Involvement of endogenous copper and a putative mechanism for anticancer properties. *Toxicol. Appl. Pharmacol.* 218: 249–255.

Biasutto, L., Mattarei, A., Marotta, E., Bradaschia, A., Sassi, N., Garbisa, S., Zoratti, M., and Paradisi, C. 2008. Development of mitochondria-targeted derivatives of resveratrol. *Bioorg. Med. Chem. Lett.* 18: 5594–5597.

Billard, C., Izard, J.C., Roman, V., Kern, C., Mathiot, C., Mentz, F., and Kolb, J.P. 2002. Comparative antiproliferative and apoptotic effects of resveratrol, epsilon-viniferin and vine-shots derived polyphenols (vineatrols) on chronic B lymphocytic leukemia cells and normal human lymphocytes. *Leuk. Lymphoma* 43: 1991–2002.

Botella, L. and Najera, C. 2004. Synthesis of methylated resveratrol and analogues by Heck reactions in organic and aqueous solvents. *Tetrahedron* 60: 5563–5570.

Brito, P.M., Mariano, A., Almeida, L.M., and Dinis, T.C. 2006. Resveratrol affords protection against peroxynitrite-mediated endothelial cell death: A role for intracellular glutathione. *Chem. Biol. Interact.* 164: 157–166.

Buryanovskyy, L., Fu, Y., Boyd, M., Ma, Y., Hsieh, T.C., Wu, J.M., and Zhang, Z. 2004. Crystal structure of quinone reductase 2 in complex with resveratrol. *Biochemistry* 43: 11417–11426.

Cai, J.B., Zhang, Z.H., Xu, D.J., Qian, Z.Y., Wang, Z.R., Huang, Y.Z., Zou, J.G., and Cao, K.J. 2008. Negative regulation of quinone reductase 2 by resveratrol in cultured vascular smooth muscle. *Clin. Exp. Pharm. Phys.* 35: 1419–1425.

Cao, Z. and Li, Y. 2004. Potent induction of cellular antioxidants and phase 2 enzymes by resveratrol in cardiomyocytes: Protection against oxidative and electrophilic injury. *Eur. J. Pharmacol.* 489: 39–48.

Cardile, V., Chillemi, R., Lombardo, L., Sciuto, S., Spatafora, C., and Tringali, C. 2007. Antiproliferative activity of methylated analogues of *E* and *Z*-resveratrol. *Z. Naturforsch. (C)* 62: 189–195.

Cardile, V., Lombardo, L., Spatafora, C., and Tringali, C. 2005. Chemoenzymatic synthesis and cell growth inhibition activity of resveratrol analogues. *Bioorg. Chem.* 33: 22–33.

Cardile, V., Scifo, C., Russo, A., Falsaperla, M., Morgia, G., Motta, M., Renis, M., Imbriani, E., and Silvestre, G. 2003. Involvement of HSP70 in resveratrol-induced apoptosis of human prostate cancer. *Anticancer Res.* 23: 4921–4926.

Casper, R.F., Quesne, M., Rogers, I.M., Shirota, T., Jolivet, A., Milgrom, E., and Savouret, J.F. 1999. Resveratrol has antagonist activity on the aryl hydrocarbon receptor: Implications for prevention of dioxin toxicity. *Mol. Pharmacol.* 56: 784–790.

Chávez, E., Reyes-Gordillo, K., Segovia, J., Shibayama, M., Tsutsumi, V., Vergara, P., Moreno, M.G., and Muriel, P. 2008. Resveratrol prevents fibrosis, NF-kappaB activation and TGF-beta increases induced by chronic CCl_4 treatment in rats. *J. Appl. Toxicol.* 28: 35–43.

Chen, T., Li, J., Cao, J., Xu, Q., Komatsu, K., and Namba, T. 1999. A new flavanone isolated from Rhizoma *smilacis glabrae* and the structural requirements of its derivatives for preventing immunological hepatocyte damage. *Planta Med.* 65: 56–59.

Chen, L., Zhang, Y., Sun, X., Li, H., Lesage, G., Javer, A., Zhang, X., Wei, X., Jiang, Y., and Yin, D. 2009. Synthetic resveratrol aliphatic acid inhibits TLR2-mediated apoptosis and an involvement of Akt/GSK3b pathway. *Bioorg. Med. Chem.* 17: 4378–4382.

Chillemi, R., Sciuto, S., Spatafora, C., and Tringali, C. 2007. Anti-tumor properties of stilbene-based resveratrol analogues: Recent results. *Nat. Prod. Commun.* 2: 1–15.

Ciolino, H.P., Daschner, P.J., and Chao Jeh, G. 1998. Resveratrol inhibits transcription of *CYP1A1 in vitro* by preventing activation of aryl hydrocarbon receptor. *Cancer Res.* 58: 5707–5712.

Clement, M.V., Hirpara, J.L., Chawdhury, S.H., and Pervaiz, S. 1998. Chemopreventive agent resveratrol, a natural product derived from grapes, triggers CD95 signaling-dependent apoptosis in human tumor cells. *Blood* 92: 996–1002.

Colin, D., Gimazane, A., Lizard, G., Izard, J.C., Solary, E., Latruffe, N., and Delmas, D. 2009. Effects of resveratrol analogs on cell cycle progression, cell cycle associated proteins and 5-fluorouracil sensitivity in human derived colon cancer cells. *Int. J. Cancer.* 124: 2780–2788.

Colin, D., Lançon, A., Delmas, D., Abrosssinow, J., Kahn, E., Lizard, G., Jannin, B., and Latruffe, N. 2008. Comparative study of the cell uptake and the antiproliferative effects of resveratrol, epsilon-viniferin and their acetates. *Biochimie* 90: 1674–1684.

Curry, S., Brick, P., and Franck, N.P. 1999. Fatty acid binding to human serum albumin: New insights from crystallographic studies. *Biochim. Biophys. Acta* 1441: 131–140.

de Almeida, L.M., Piñeiro, C.C., Leite, M.C., Brolese, G., Tramontina, F., Feoli, A.M., Gottfried, C., and Gonçalves, C.A. 2007. Resveratrol increases glutamate uptake, glutathione content, and S100B secretion in cortical astrocyte cultures. *Cell. Mol. Neurobiol.* 27: 661–668.

de la Lastra, C.A. and Villegas, I. 2005. Resveratrol as an anti-inflammatory and anti-aging agent: Mechanisms and clinical implications. *Mol. Nutr. Food. Res.* 49: 405–430.

de la Lastra, C.A. and Villegas, I. 2007. Resveratrol as an anti-oxidant and pro-oxidant agent: Mechanisms and clinical implications. *Biochem. Soc. Trans.* 35: 1156–1160.

de Lima, D.P., Rotta, R., Beatriz, A., Marques, M.R., Montenegro, R.C., Vasconcellos, M.C., Pessoa, C., de Moraes, M.O., Costa-Lotufo, L.V., Frankland Sawaya, A.C., and Eberlin, M.N. 2009. Synthesis and biological evaluation of cytotoxic properties of stilbenes based resveratrol analogs. *Eur. J. Med. Chem.* 44: 701–707.

Delmas, D., Jannin, B., Malki, M.C., and Latruffe, N. 2000. Inhibitory effect of resveratrol on the proliferation of human and rat hepatic derived cell lines. *Oncol. Rep.* 7: 847–852.

Delmas, D., Lançon, A., Colin, D., Jannin, B., and Latruffe, N. 2006. Resveratrol as a chemopreventive agent a promising molecule for fighting cancer. *Curr. Drug Targets* 7: 423–442.

Delmas, D., Passilly-Degrace, P., Jannin, B., Malki, M.C., and Latruffe, N. 2002. Resveratrol, a chemopreventive agent, disrupts the cell cycle control of human SW480 colorectal tumor cells. *Int. J. Mol. Med.* 10: 193–199.

Delmas, D., Rebe, C., Lacour, S., Filomenko, R., Athias, A., Gambert, P., Cherkaoui-Malki, M., Jannin, B., Dubrez-Daloz, L., Latruffe, N., and Solary, E. 2003. Resveratrol-induced apoptosis is associated with Fas redistribution in the rafts and the formation of a death-inducing signaling complex in colon cancer cells. *J. Biol. Chem.* 278: 41482–41490.

Delmas, D., Rebe, C., Micheau, O., Athias, A., Gambert, P., Grazide, S., Laurent, G., Latruffe, N., and Solary, E. 2004. Redistribution of CD95, DR4 and DR5 in rafts accounts for the synergistic toxicity of resveratrol and death receptor ligands in colon carcinoma cells. *Oncogene* 23: 8979–8986.

Dorrie, J., Gerauer, H., Wachter, Y., and Zunino, S.J. 2001. Resveratrol induces extensive apoptosis by depolarizing mitochondrial membranes and activating caspase-9 in acute lymphoblastic leukemia cells. *Cancer Res.* 61: 4731–4739.

Du, C., Fang, M., Li, Y., Li, L., and Wang, X. 2000. Smac, a mitochondrial protein that promotes cytochrome c-dependent caspase activation by eliminating IAP inhibition. *Cell* 102: 33–42.

Elleingand, E., Gerez, C., Un, S., Knupling, M., Lu, G., Salem, J., Rubin, H., Sauge-Merle, S., Laulhere, J.P., and Fontecave, M. 1998. Reactivity studies of the tyrosyl radical in ribonucleotide reductase from *Mycobacterium tuberculosis* and *Arabidopsis thaliana*— Comparison with *Escherichia coli* and mouse. *Eur. J. Biochem.* 258: 485–490.

Erenpreisa, J., Kalejs, M., and Cragg, M.S. 2005. Mitotic catastrophe and endomitosis in tumour cells: An evolutionary key to a molecular solution. *Cell Biol. Int.* 29: 1012–1018.

Evers, D.L., Wang, X., Huong, S.M., Huang, D.Y., and Huang, E.S. 2004. 3,4′,5-Trihydroxy-*trans*-stilbene (resveratrol) inhibits human cytomegalovirus replication and virus-induced cellular signaling. *Antiviral Res.* 63: 85–95.

Fontecave, M., Lepoivre, M., Elleingand, E., Gerez, C., and Guittet, O. 1998. Resveratrol, a remarkable inhibitor of ribonucleotide reductase. *FEBS Lett.* 421: 277–279.

Fuggetta, M.P., D'Atri, S., Lanzilli, G., Tricarico, M., Cannavo, E., Zambruno, G., Falchetti, R., and Ravagnan, G. 2004. In vitro antitumour activity of resveratrol in human melanoma cells sensitive or resistant to temozolomide. *Melanoma Res.* 14: 189–196.

Fukuhara, K. and Miyata, N. 1998. Resveratrol as a new type of DNA-cleaving agent. *Bioorg. Med. Chem. Lett.* 8: 3187–3192.

Fukuhara, K., Nagakawa, M., Nakanishi, I., Ohkubo, K., Imai, K., Urano, S., Fukuzumi, S., Ozawa, T., Ikota, N., Mochizuki, M., Miyata, N., and Okuda, H. 2006. Structural basis for DNA-cleaving activity of resveratrol in the presence of Cu(II). *Bioorg. Med. Chem.* 14: 1437–1443.

Fulda, S. and Debatin, K.M. 2004a. Sensitization for anticancer drug-induced apoptosis by the chemopreventive agent resveratrol. *Oncogene* 23: 6702–6711.

Fulda, S. and Debatin, K.M. 2004b. Sensitization for tumor necrosis factor-related apoptosis-inducing ligand-induced apoptosis by the chemopreventive agent resveratrol. *Cancer Res.* 6: 337–346.

Gao, M., Wang, M., Miller, K.D., Sledge, G.W., Hutchins, G.D., and Zheng, Q.H. 2006. Synthesis of radiolabelled stilbene derivatives as new potential PET probes for aryl hydrocarbon receptor in cancers. *Bioorg. Med. Chem. Lett.* 16: 5767–5772.

Gao, X., Xu, Y.X., Divine, G., Janakiraman, N., Chapman, R.A., and Gautam, S.C. 2002. Disparate in vitro and in vivo antileukemic effects of resveratrol, a natural polyphenolic compound found in grapes. *J. Nutr.* 132: 2076–2081.

Glatz, J.F.G. and van der Vusse, G.J. 1996. Cellular fatty acid binding protein: Their function and physiological significance. *Prog. Lip. Res.* 35: 243–282.

Goldberg, D.M., Yan, J., and Soleas, G.J. 2003. Absorption of three wine-related polyphenols in three different matrices by healthy subjects. *Clin. Biochem.* 36: 79–87.

Gupta, Y.K., Briyal, S., and Chaudhary, G. 2002. Protective effect of *trans*-resveratrol against kainic acid-induced seizures and oxidative stress in rats. *Pharmacol. Biochem. Behav.* 71: 245–249.

Guy, R., Scott, Z., Sloboda, R., and Nicolaou, K. 1996. Fluorescent taxoids. *Chem. Biol.* 3: 1021–1031.

Hadi, S.M., Asad, S.F., Singh, S., and Ahmad, A. 2000. Putative mechanism for anticancer and apoptosis-inducing properties of plant-derived polyphenolic compounds. *IUBMB Life* 50: 167–171.

Haider, U.G., Sorescu, D., Griendling, K.K., Vollmar, A.M., and Dirsch, V.M. 2003. Resveratrol increases serine15-phosphorylated but transcriptionally impaired p53 and induces a reversible DNA replication block in serum-activated vascular smooth muscle cells. *Mol. Pharmacol.* 63: 925–932.

Han, Y.S., Bastianetto, S., Dumont, Y., and Quirion, R. 2006. Specific plasma membrane binding sites for polyphenols, including resveratrol, in the rat brain. *J. Pharmacol. Exp. Therap.* 318: 238–245.

Hauss, F., Liu, J., Michelucci, A., Coowar, D., Morga, E., Heuschling, P., and Luu, B. 2007. Dual bioactivity of resveratrol fatty alcohols: Differentiation of neural stem cells and modulation of neuroinflammation. *Bioorg. Med. Chem. Lett.* 17: 4218–4222.

Hayashibara, T., Yamada, Y., Nakayama, S., Harasawa, H., Tsuruda, K., Sugahara, K., Miyanishi, T., Kamihira, S., Tomonaga, M., and Maita, T. 2002. Resveratrol induces downregulation in survivin expression and apoptosis in HTLV-1-infected cell lines: A prospective agent for adult T cell leukemia chemotherapy. *Nutr. Cancer.* 44: 193–201.

Henry, C., Vitrac, X., Decendit, A., Ennamany, R., Krisa, S., and Merillon, J.M. 2005. Cellular uptake and efflux of *trans*-piceid and its aglycone *trans*-resveratrol on the apical membrane of human intestinal Caco-2 cells. *J. Agric. Food Chem.* 53: 798–803.

Heynekamp, J.J. 2006. Substituted *trans*-stilbenes, including analogues of the natural product resveratrol, inhibit the human tumor necrosis factor alpha-induced activation of transcription factor nuclear factor Kappa B. *J. Med. Chem.* 49: 7182–7189.

Hosokawa, N., Hosokawa, Y., Sakai, T., Yoshida, M., Marui, N., Nishino, H., Kawai, K., and Aoike, A. 1990. Inhibitory effect of quercetin on the synthesis of a possibly cell-cycle-related 17-kDa protein, in human colon cancer cells. *Int. J. Cancer* 45: 1119–1124.

Hsieh, T.C., Burfeind, P., Laud, K., Backer, J.M., Traganos, F., Darzynkiewicz, Z., and Wu, J.M. 1999b. Cell cycle effects and control of gene expression by resveratrol in human breast carcinoma cell lines with different metastatic potentials. *Int. J. Oncol.* 15: 245–252.

Hsieh, T.C., Juan, G., Darzynkiewicz, Z., and Wu, J.M. 1999a. Resveratrol increases nitric oxide synthase, induces accumulation of p53 and p21(WAF1/CIP1), and suppresses cultured bovine pulmonary artery endothelial cell proliferation by perturbing progression through S and G2. *Cancer Res.* 59: 2596–2601.

Hsieh, T.C., Wang, Z., Hamby, C.V., and Wu, J.M. 2005. Inhibition of melanoma cell proliferation by resveratrol is correlated with upregulation of quinone reductase 2 and p53. *Biochem. Biophys. Res. Commun.* 334: 223–230.

Huang, C., Ma, W.Y., Goranson, A., and Dong, Z. 1999. Resveratrol suppresses cell transformation and induces apoptosis through a p53-dependent pathway. *Carcinogenesis* 20: 237–242.

Hunter, T. and Pines, J. 1994. Cyclins and cancer. II: Cyclin D and CDK inhibitors come of age. *Cell* 79: 573–582.

Hwang, J.T., Kwak, D.W., Lin, S.K., Kim, H.M., Kim, Y.M., and Park, O.J. 2007. Resveratrol induces apoptosis in chemoresistant cancer cells via modulation of AMPK signaling pathway. *Ann. N. Y. Acad. Sci.* 1095: 441–448.

Jang, M., Cai, L., Udeani, G.O., Slowing, K.V., Thomas, C.F., Beecher, C.W., Fong, H.H., Farnsworth, N.R., Kinghorn, A.D., Mehta, R.G., Moon, R.C., and Pezzuto, J.M. 1997. Cancer chemopreventive activity of resveratrol, a natural product derived from grapes. *Science* 275: 218–220.

Jang, J.H. and Surh, Y.J. 2001. Protective effects of resveratrol on hydrogen peroxide-induced apoptosis in rat pheochromocytoma (PC12) cells. *Mutat. Res.* 496: 181–190.

Jannin, B., Menzel, M., Berlot, J.-P., Delmas, D., Lançon, A., and Latruffe, N. 2004. Interactions of resveratrol, a chemopreventive agent, with serum albumin. *Biochem. Pharmacol.* 68: 1113–1118.

Jazirehi, A.R. and Bonavida, B. 2004. Resveratrol modifies the expression of apoptotic regulatory proteins and sensitizes non-Hodgkin's lymphoma and multiple myeloma cell lines to paclitaxel-induced apoptosis. *Mol. Cancer Ther.* 3: 71–84.

Jiang, Y.L. 2008. Design, synthesis and spectroscopic studies of resveratrol aliphatic acid ligands of human serum albumin. *Bioorg. Med. Chem.* 16: 6406–6414.

Juan, M.E., Wenzel, U., Daniel, H., and Planas, J.M. 2008. Resveratrol induces apoptosis through ROS-dependent mitochondria pathway in HT-29 human colorectal carcinoma cells. *J. Agric. Food Chem.* 56: 4813–4818.

Kaneuchi, M., Sasaki, M., Tanaka, Y., Yamamoto, R., Sakuragi, N., and Dahiya, R. 2003. Resveratrol suppresses growth of Ishikawa cells through down-regulation of EGF. *Int. J. Oncol.* 23: 1167–1172.

Kasdallah-Grissa, A., Mornagui, B., Aouani, E., Hammami, M., El May, M., Gharbi, N., Kamoun, A., and El-Fazaâ, S. 2007. Resveratrol, a red wine polyphenol, attenuates ethanol-induced oxidative stress in rat liver. *Life Sci.* 80: 1033–1039.

Kim, Y.A., Choi, B.T., Lee, Y.T., Park, D.I., Rhee, S.H., Park, K.Y., and Choi, Y.H. 2004. Resveratrol inhibits cell proliferation and induces apoptosis of human breast carcinoma MCF-7 cells. *Oncol. Rep.* 11: 441–446.

Kim, S., Min, S.Y., Lee, S.K., and Cho, W.J. 2003a. Comparative molecular field analysis study of stilbene derivatives active against A549 lung carcinoma. *Chem. Pharm. Bull.* 51: 516–521.

Kim, Y.A., Rhee, S.H., Park, K.Y., and Choi, Y.H. 2003b. Antiproliferative effect of resveratrol in human prostate carcinoma cells. *J. Med. Food* 6: 273–280.

Klinge, C.M., Blankenship, K.A., Risinger, K.E., Bhatnagar, S., Noisin, E.L., Sumanasekera, W.K., Zhao, L., Brey, D.M., and Keynton, R.S. 2005. Resveratrol and estradiol rapidly activate MAPK signaling through estrogen receptors alpha and beta in endothelial cells. *J. Biol. Chem.* 280: 7460–7468.

Kluth, D., Banning, A., Paur, I., Blomhoff, R., and Brigelius-Flohé, R. 2007. Modulation of pregnane X receptor- and electrophile responsive element-mediated gene expression by dietary polyphenolic compounds. *Free Radic. Biol. Med.* 42: 315–325.

Kode, A., Rajendrasozhan, S., Caito, S., Yang, S.R., Megson, I.L., and Rahman, I. 2008. Resveratrol induces glutathione synthesis by activation of Nrf2 and protects against cigarette smoke-mediated oxidative stress in human lung epithelial cells. *Am. J. Physiol. Lung Cell. Mol. Physiol.* 294: L478–L488.

Kotha, A., Sekharam, M., Cilenti, L., Siddiquee, K., Khaled, A., Zervos, A.S., Carter, B., Turkson, J., and Jove, R. 2006. Resveratrol inhibits Src and Stat3 signaling and induces the apoptosis of malignant cells containing activated Stat3 protein. *Mol. Cancer Ther.* 5: 621–629.

Kumar, A., Naidu, P.S., Seghal, N., and Padi, S.S. 2007. Neuroprotective effects of resveratrol against intracerebroventricular colchicine-induced cognitive impairment and oxidative stress in rats. *Pharmacology* 79: 17–26.

Kumar, P., Padi, S.S., Naidu, P.S., and Kumar, A. 2006. Effect of resveratrol on 3-nitropropionic acid-induced biochemical and behavioural changes: Possible neuroprotective mechanisms. *Behav. Pharmacol.* 17: 485–492.

Kuo, P.L., Chiang, L.C., and Lin, C.C. 2002. Resveratrol- induced apoptosis is mediated by p53-dependent pathway in Hep G2 cells. *Life Sci.* 72: 23–34.

Kuwajerwala, N., Cifuentes, E., Gautam, S., Menon, M., Barrack, E.R., and Reddy, G.P. 2002. Resveratrol induces prostate cancer cell entry into S phase and inhibits DNA synthesis. *Cancer Res.* 62: 2488–2492.

Lamuela-Raventos, R.M., Covas, M.I., Fito, M., Marrugat, J., and de la Torre-Boronat, M.C. 1999. Detection of dietary antioxidant phenolic compounds in human LDL. *Clin. Chem.* 45: 1870–1872.

Lançon, A., Delmas, D., Osman, H., Thénot, J.P., Jannin, B., and Latruffe, N. 2004. Human hepatic cell uptake of resveratrol: Involvement of both passive diffusion and carrier-mediated process. *Biochem. Biophys. Res. Commun.* 316: 1132–1137.

Lançon, A., Hanet, N., Jannin, B., Delmas, D., Heydel, J.M., Chagnon, M.C., Lizard, G., Artur, Y., and Latruffe, N. 2007. Resveratrol in human hepatoma HepG2 cells: Metabolism and inducibility of detoxifying enzymes. *Drug. Metab. Dispos.* 35: 699–703.

Langcake, P. and Pryce, R.J. 1976. The production of resveratrol by *Vitis vinifera* and other members of the vitaceae as a response to infection or injury. *Physiol. Plant Pathol.* 9: 77–86.

Lappano, R., Rosano, C., Madeo, A., Albanito, L., Plastina, P., Gabriele, B., Forti, L., Stivala, L.A., Iacopetta, D., Dolce, V., Andò, S., Pezzi, V., and Maggiolini, M. 2009. Structure-activity relationships of resveratrol and derivatives in breast cancer cells. *Mol. Nutr. Food Res.* 53: 845–858.

Le Corre, L., Chalabi, N., Delort, L., Bignon, Y.J., and Bernard-Gallon, D.J. 2006. Differential expression of genes induced by resveratrol in human breast cancer cell lines. *Nutr. Cancer* 56: 193–203.

Lee, D.H., Szczepanski, M., and Lee, Y.J. 2008. Role of Bax in quercetin-induced apoptosis in human prostate cancer cells. *Biochem. Pharmacol.* 75: 2345–2355.

Levi, F., Pasche, C., Lucchini, F., Ghidoni, R., Ferraroni, M., and La Vecchia, C. 2005. Epidemiological evidence that resveratrol from grape is inversely related to breast cancer risk. Study on 369 cases vs 602 controls of Swiss women from 1993 to 2003. *Eur. J. Cancer Prev.* 14: 139–142.

Li, X., Phillips, F.M., An, H.S., Ellman, M., Thonar, E.J., Wu, W., Park, D., and Im, H.J. 2008. The action of resveratrol, a phytoestrogen found in grapes, on the intervertebral disc. *Spine* 33: 2586–2595.

Li, Y., Cao, Z., and Zhu, H. 2006. Upregulation of endogenous antioxidants and phase 2 enzymes by the red wine polyphenol, resveratrol in cultured aortic smooth muscle cells leads to cytoprotection against oxidative and electrophilic stress. *Pharmacol. Res.* 53: 6–15.

Liang, Y.C., Tsai, S.H., Chen, L., Lin-Shiau, S.Y., and Lin, J.K. 2003. Resveratrol-induced G2 arrest through the inhibition of CDK7 and p34CDC2 kinases in colon carcinoma HT29 cells. *Biochem. Pharmacol.* 65: 1053–1060.

Lin, H.Y., Lansing, L., Merillon, J.M., Davis, F.B., Tang, H.Y., Shih, A., Vitrac, X., Krisa, S., Keating, T., Cao, H.J., Bergh, J., Quackenbush, S., and Davis, P.J. 2009. Integrin αVβ3 contains a receptor site for resveratrol. *FASEB J.* 20: 1133–1138.

Lin, H.Y., Shih, A., Davis, F.B., Tang, H.Y., Martino, L.J., Bennett, J.A., and Davis, P.J. 2002. Resveratrol induced serine phosphorylation of p53 causes apoptosis in a mutant p53 prostate cancer cell line. *J. Urol.* 168: 748–755.

Lin, H.-S., Tringali, C., Spatafora, C., Wu, C., and Ho, P.C. 2010. A simple and sensitive HPLC-UV method for the quantification of piceatannol analog 3,5,3′,4′-tetramethoxy-*trans*-stilbene in rat plasma and its application for a pre-clinical pharmacokinetic study. *J. Pharm. Biomed. Anal.* 51: 679–684.

Liu, H., Dong, A., Gao, C., Tan, C., Liu, H., Zu, X., and Jiang, Y. 2008. The design, synthesis, and anti-tumor mechanism study of *N*-phosphoryl amino acid modified resveratrol analogues. *Bioorg. Med. Chem.* 16: 10013–10021.

Lu, J., Ho, C.H., Ghai, G., and Chen, K.Y. 2001. Resveratrol analog, 3,4,5,4′-tetrahydroxystilbene, differentially induces pro-apoptotic p53/Bax gene expression and inhibits the growth of transformed cells but not their normal counterparts. *Carcinogenesis* 22: 321–328.

Lu, Z., Zhang, Y., Liu, H., Yuan, J., Zheng, Z., and Zou, G. 2007. Transport of a cancer chemopreventive polyphenol, resveratrol: Interaction with serum albumin and hemoglobin. *J. Fluoresc.* 17: 580–587.

Ma, X., Tian, X., Huang, X., Yan, F., and Qiao, D. 2007. Resveratrol-induced mitochondrial dysfunction and apoptosis are associated with Ca^{2+} and mCICR-mediated MPT activation in HepG2 cells. *Mol. Cell. Biochem.* 302: 99–109.

Mahyar-Roemer, M., Katsen, A., Mestres, P., and Roemer, K. 2001. Resveratrol induces colon tumor cell apoptosis independently of p53 and preceded by epithelial differentiation, mitochondrial proliferation and membrane potential collapse. *Int. J. Cancer* 94: 615–622.

Mahyar-Roemer, M., Kohler, H., and Roemer, K. 2002. Role of Bax in resveratrol-induced apoptosis of colorectal carcinoma cells. *BMC Cancer* 2: 27.

Mahyar-Roemer, M. and Roemer, K. 2001. p21 Waf1/Cip1 can protect human colon carcinoma cells against p53-dependent and p53-independent apoptosis induced by natural chemopreventive and therapeutic agents. *Oncogene* 20: 3387–3398.

Maier-Salamon, A., Hagenauer, B., Reznicel, G., Szekeres, T., Thalhammer, T., and Jäger, W. 2008. Metabolism and disposition of resveratrol in the isolated perfused rat liver: Role of mrp2 in the biliary excretion of glucuronides. *J. Pharma. Sci.* 97: 1615–1628.

Mansilla, S., Bataller, M., and Portugal, J. 2006. Mitotic catastrophe as a consequence of chemotherapy. *Anticancer Agents Med. Chem.* 6: 589–602.

Marambaud, P., Zhao, H., and Davies, P. 2005. Resveratrol promotes clearance of Alzheimer's disease amyloid-β peptides. *J. Biol. Chem.* 280: 37377–37382.

Marel, A.K., Lizard, G., Izard, J.C., Latruffe, N., and Delmas, D. 2008. Inhibitory effects of *trans*-resveratrol analogs molecules on the proliferation and the cell cycle progression of human colon tumoral cells. *Mol. Nutr. Food Res.* 52: 538–548.

Marier, J.-F., Vachon, P., Gritsas, A., Zhang, J., Moreau, J.-P., and Ducharme, M.P. 2002. Metabolism and disposition of resveratrol in rats: Extent of absorption, glucuronidation, and enterohepatic recirculation evidenced by a linked-rat model. *J. Pharmacol. Exp. Ther.* 302: 369–373.

Matsuoka, A., Furuta, A., Ozaki, M., Fukuhara, K., and Miyata, N. 2001. Resveratrol, a naturally occurring polyphenol, induces sister chromatid exchanges in a Chinese hamster lung (CHL) cell line. *Mutat. Res.* 494: 107–113.

Matsuoka, A., Lundin, C., Johansson, F., Sahlin, M., Fukuhara, K., Sjoberg, B.M., Jenssen, D., and Onfelt, A. 2004. Correlation of sister chromatid exchange formation through homologous recombination with ribonucleotide reductase inhibition. *Mutat. Res.* 547: 101–107.

Matsuoka, A., Takeshita, K., Furuta, A., Ozaki, M., Fukuhara, K., and Miyata, N. 2002. The 4'-hydroxy group is responsible for the in vitro cytogenetic activity of resveratrol. *Mutat. Res.* 521: 29–35.

Mazué, F., Colin, D., Gobbo, J., Wegner, M., Rescifina, A., Spatafora, C., Fasseur, D., Delmas, D., Meunier, P., Tringali, C., and Latruffe, N. 2010. Structural determination of resveratrol for cell proliferation inhibition potency: Experimental and docking studies of new analogs. *Eur. J. Med. Chem.* 45: 2972–2980.

Miyashita, T., Krajewski, S., Krajewska, M., Wang, H.G., Lin, H.K., Liebermann, D.A., Hoffman, B., and Reed, J.C. 1994. Tumor suppressor p53 is a regulator of bcl-2 and bax gene expression in vitro and in vivo. *Oncogene* 9: 1799–1805.

Morgan, D.O. 1995. Principles of CDK regulation. *Nature* 374: 131–134.

Murias, M., Handler, N., Erker, T., Pleban, K., Ecker, G., Saiko, P., Szekeres, T., and Jäger, W. 2004. Resveratrol analogues as selective cyclooxygenase-2 inhibitors: Synthesis and structure-activity relationship. *Bioorg. Med. Chem.* 12: 5571–5578.

Murias, M., Luczak, M.W., Niepsuj, A., Krajka-Kuzniak, V., Zielinska-Przyjemska, M., Jagodzinski, P.P., Jäger, W., Szekeres, T., and Jodynis-Liebert, J. 2008. Cytotoxic activity of 3,3',4,4',5,5'-hexahydroxystilbene against breast cancer cells is mediated by induction of p53 and downregulation of mitochondrial superoxide dismutase. *Toxicol. In Vitro* 22: 1361–1370.

Nakagawa, H., Kiyozuka, Y., Uemura, Y., Senzaki, H., Shikata, N., Hioki, K., and Tsubura, A. 2001. Resveratrol inhibits human breast cancer cell growth and may mitigate the effect of linoleic acid, a potent breast cancer cell stimulator. *J. Cancer Res. Clin. Oncol.* 127: 258–264.

Narayanan, B.A., Narayanan, N.K., Re, G.G., and Nixon, D.W. 2003. Differential expression of genes induced by resveratrol in LNCaP cells: P53-mediated molecular targets. *Int. J. Cancer* 104: 204–212.

Narayanan, B.A., Narayanan, N.K., Stoner, G.D., and Bullock, B.P. 2002. Interactive gene expression pattern in prostate cancer cells exposed to phenolic antioxidants. *Life Sci.* 70: 1821–1839.

Narayanan, N.K., Narayanan, B.A., and Nixon, D.W. 2004. Resveratrol-induced cell growth inhibition and apoptosis is associated with modulation of phosphoglycerate mutase B in human prostate cancer cells: Two-dimensional sodium dodecyl sulfate-polyacrylamide gel electrophoresis and mass spectrometry evaluation. *Cancer Detect. Prev.* 28: 443–452.

Niles, R.M. 2006. Resveratrol does not inhibit human melanoma xenograft tumor growth due to the rapid metabolism in athymic (nu/nu) mice. *J. Nutr.* 136: 2542–2546.

Nurse, P. 1997. Checkpoint pathways come of age. *Cell* 91: 865–867.

O'Brien, N.M., Carpenter, R., O'Callaghan, Y.C., O'Grady, M.N., and Kerry, J.P. 2006. Modulatory effects of resveratrol, citroflavan-3-ol, and plant-derived extracts on oxidative stress in U937 cells. *J. Med. Food* 9: 187–195.

O'Connor, P.M., Ferris, D.K., Hoffmann, I., Jackman, J., Draetta, G., and Kohn, K.W. 1994. Role of the cdc25C phosphatase in G2 arrest induced by nitrogen mustard. *Proc. Natl. Acad. Sci. USA* 91: 9480–9484.

O'Connor, P.M., Ferris, D.K., Pagano, M., Draetta, G., Pines, J., Hunter, T., Longo, D.L., and Kohn, K.W. 1993. G2 delay induced by nitrogen mustard in human cells affects cyclin A/cdk2 and cyclin B1/cdc2-kinase complexes differently. *J. Biol. Chem.* 268: 8298–8308.

Okawara, M., Katsuki, H., Kurimoto, E., Shibata, H., Kume, T., and Akaike, A. 2007. Resveratrol protects dopaminergic neurons in midbrain slice culture from multiple insults. *Biochem. Pharmacol.* 73: 550–560.

Olas, B., Wachowicz, B., Nowak, P., Stochmal, A., Oleszek, W., Glowacki, R., and Bald, E. 2008. Comparative studies of the antioxidant effects of a naturally occurring resveratrol analogue trans-3,3′,5,5′-tetrahydroxy-4′-methoxystilbene and resveratrol— Against oxidation and nitration of biomolecules in blood platelets. *Cell. Biol. Toxicol.* 24: 331–340.

Ovesna, Z. 2008. Antioxidant activity of resveratrol, piceatannol and 3,3′,4,4′5,5′-hexahydroxy-*trans*-stilbene in three leukemia cell lines. *Oncol. Rep.* 16: 617–624.

Pagano, M., Pepperkok, R., Lukas, J., Baldin, V., Ansorge, W., Bartek, J., and Draetta, G. 1993. Regulation of the cell cycle by the cdk2 protein kinase in cultured human fibroblasts. *J. Cell. Biol.* 121: 101–111.

Park, D.W., Baek, K., Kim, J.R., Lee, J.J., Ryu, S.H., Chin, B.R., and Baek, S.H. 2009. Resveratrol inhibits foam cell formation via NADPH oxidase 1-mediated reactive oxygen species and monocyte chemotactic protein-1. *Exp. Mol. Med.* 41: 171–179.

Park, C.E., Kim, M.J., Lee, J.H., Min, B.I., Bae, H., Choe, W., Kim, S.S., and Ha, J. 2007. Resveratrol stimulates glucose transport in C2C12 myotubes by activating AMP-activated protein kinase. *Exp. Mol. Med.* 39: 222–229.

Pettit, G.R., Grealish, M.P., Jung, M.K., Hamel, E., Pettit, R.K., Chapuis, J.C., and Schmidt, J.M. 2002. Antineoplastic agents. 465. Structural modification of resveratrol: Sodium resverastatin phosphate. *J. Med. Chem.* 45: 2534–2542.

Ragione, F.D., Cucciolla, V., Borriello, A., Pietra, V.D., Racioppi, L., Soldati, G., Manna, C., Galletti, P., and Zappia, V. 1998. Resveratrol arrests the cell division cycle at S/G2 phase transition. *Biochem. Biophys. Res. Commun.* 250: 53–58.

Raval, A.P., Lin, H.W., Dave, K.R., Defazio, R.A., Della Morte, D., Kim, E.J., and Perez-Pinzon, M.A. 2008. Resveratrol and ischemic preconditioning in the brain. *Curr. Med. Chem.* 15: 1545–1551.

Renaud, S. and de Longeril, M. 1992. Wine, alcohol, platelets, and the French paradox for coronary heart disease. *Lancet* 339: 1523–1526.

Ritz, M.F., Curin, Y., Mendelowitsch, A., and Andriantsitohaina, R. 2008. Acute treatment with red wine polyphenols protects from ischemia-induced excitotoxicity, energy failure and oxidative stress in rats. *Brain Res.* 1239: 226–234.

Rivera, H., Shibayama, M., Tsutsumi, V., Perez-Alvarez, V., and Muriel, P. 2008. Resveratrol and trimethylated resveratrol protect from acute liver damage induced by CCl_4 in the rat. *J. Appl. Toxicol.* 28: 147–155.

Rivière, C., Richard, T., Quentin, L., Krisa, S., Mérillon, J.M., and Monti, J.P. 2007. Inhibitory activity of stilbenes on Alzheimer's β-amyloid fibrils in vitro. *Bioorg. Med. Chem.* 15: 1160–1167.

Robb, E.L., Winkelmolen, L., Visanji, N., Brotchie, J., and Stuart, J.A. 2008. Dietary resveratrol administration increases MnSOD expression and activity in mouse brain. *Biochem. Biophys. Res. Commun.* 372: 254–259.

Roberti, M., Pizzirani, D., Simoni, D., Rondanin, R., Baruchello, R., Bonora, C., Buscami, F., Grimaudo, S., and Tolomeo, M. 2003. Synthesis and biological evaluation of resveratrol and analogues as apoptosis-inducing agents. *J. Med. Chem.*, 46: 3546–3554.

Roman, V., Billard, C., Kern, C., Ferry-Dumazet, H., Izard, J.C., Mohammad, R., Mossalayi, D.M., and Kolb, J.P. 2002. Analysis of resveratrol-induced apoptosis in human B-cell chronic leukaemia. *Br. J. Haematol.* 117: 842–851.

Ruan, B.F., Huang, X.F., Ding, H., Xu, C., Ge, H.M., Zhu, H.L., and Tan, R.X. 2006. Synthesis and cytotoxic evaluation of a series of resveratrol derivatives. *Chem. Biodivers.* 3: 975–981.

Rubiolo, J.A., Mithieux, G., and Vega, F.V. 2008. Resveratrol protects primary rat hepatocytes against oxidative stress damage: Activation of the Nrf2 transcription factor and augmented activities of antioxidant enzymes. *Eur. J. Pharmacol.* 591: 66–72.

Saiko, P., Szakmary, A., Jaeger, W., and Szekeres, T. 2008. Resveratrol and its analogs: Defense against cancer, coronary disease and neurodegenerative maladies or just a fad? *Mutat. Res.* 658: 68–94.

Sale, S., Verschoyle, R.D., Boocock, D., Jones, D.J.L., Wilsher, N., Ruparelia, K.C., Potter, G.A., Farmer, P.B., Steward, W.P., and Gescher, A.J. 2004. Pharmacokinetics in mice and growth-inhibitory properties of the putative cancer chemopreventive agent resveratrol and the synthetic analogue *trans* 3,4,5,4′-tetramethoxystilbene. *Brit. J. Cancer* 90: 736–744.

Savaskan, E., Olivieri, G., Meier, F., Seifritz, E., Wirz-Justice, A., and Müller-Spahn, F. 2003. Red wine ingredient resveratrol protects from beta-amyloid neurotoxicity. *Gerontology* 49: 380–383.

Schilder, Y.D., Heiss, E.H., Schachner, D., Ziegler, J., Reznicek, G., Sorescu, D., and Dirsch, V.M. 2009. NADPH oxidases 1 and 4 mediate cellular senescence induced by resveratrol in human endothelial cells. *Free Radic. Biol. Med.* 46: 1598–1606.

Schneider, Y., Chabert, P., Stutzmann, J., Coelho, D., Fougerousse, A., Gosse, F., Launay, J.-F., Brouillard, R., and Raul, F. 2003. Resveratrol analog (Z)-3,5,4′-trimethoxystilbene is a potent anti-mitotic drug inhibiting tubulin polymerization. *Int. J. Cancer* 107: 189–196.

Schneider, Y., Vincent, F., Duranton, B., Badolo, L., Gosse, F., Bergmann, C., Seiler, N., and Raul, F. 2000. Anti-proliferative effect of resveratrol, a natural component of grapes and wine, on human colonic cancer cells. *Cancer Lett.* 158: 85–91.

Scifo, C., Cardile, V., Russo, A., Consoli, R., Vancheri, C., Capasso, F., Vanella, A., and Renis, M. 2004. Resveratrol and propolis as necrosis or apoptosis inducers in human prostate carcinoma cells. *Oncol. Res.* 14: 415–426.

Seiler, N., Schneider, Y., Gosse, F., Schleiffer, R., and Raul, F. 2004. Polyploidisation of metastatic colon carcinoma cells by microtubule and tubulin interacting drugs: Effect on proteolytic activity and invasiveness. *Int. J. Oncol.* 25: 1039–1048.

Sener, G., Toklu, H.Z., Sehirli, A.O., Velioğlu-Oğünç, A., Cetinel, S., and Gedik, N. 2006. Protective effects of resveratrol against acetaminophen-induced toxicity in mice. *Hepatol. Res.* 35: 62–68.

Sengupta, B. and Sengupta, P.K. 2002. The interaction of quercetin with human serum albumin: A fluorescence spectroscopic study. *Biochem. Biophys. Res. Commun.* 299: 400–403.

She, Q.B., Bode, A.M., Ma, W.Y., Chen, N.Y., and Dong, Z. 2001. Resveratrol-induced activation of p53 and apoptosis is mediated by extracellular-signal-regulated protein kinases and p38 kinase. *Cancer Res.* 61: 1604–1610.

She, Q.B., Huang, C., Zhang, Y., and Dong, Z. 2002. Involvement of c-jun NH(2)-terminal kinases in resveratrol-induced activation of p53 and apoptosis. *Mol. Carcinog.* 33: 244–250.

Shih, A., Davis, F.B., Lin, H.Y., and Davis, P.J. 2002. Resveratrol induces apoptosis in thyroid cancer cell lines via a MAPK- and p53-dependent mechanism. *J. Clin. Endocrinol. Metab.* 87: 1223–1232.

Shih, A., Zhang, S., Cao, H.J., Boswell, S., Wu, Y.H., Tang, H.Y., Lennartz, M.R., Davis, F.B., Davis, P.J., and Lin, H.Y. 2004. Inhibitory effect of epidermal growth factor on resveratrol-induced apoptosis in prostate cancer cells is mediated by protein kinase C-alpha. *Mol. Cancer Ther.* 3: 1355–1364.

Singh, M., Arseneault, M., Sanderson, T., Murthy, V., and Ramassamy, C. 2008. Challenges for the research on polyphenols from foods in Alzheimer's disease: Bioavailability, metabolism and cellular and molecular mechanisms. *J. Agric. Food Chem.* 56: 4855–4873.

Spatafora, C., Basini, G., Baioni, L., Grasselli, F., Sofia, A., and Tringali, C. 2009. Antiangiogenic resveratrol analogues by mild m-CPBA aromatic hydroxylation of 3,5-dimethoxystilbenes. *Nat. Prod. Comm.* 4: 239–246.

Soleas, G.J., Angelini, M., Grass, L., Diamandis, E.P., and Goldberg, D.M. 2001. Absorption of *trans*-resveratrol in rats. *Methods Enzymol.* 335: 145–154.

Srinivasula, S.M., Datta, P., Fan, X.J., Fernandes-Alnemri, T., Huang, Z., and Alnemri, E.S. 2000. Molecular determinants of the caspase-promoting activity of Smac/DIABLO and its role in the death receptor pathway. *J. Biol. Chem.* 275: 36152–36157.

Stivala, L.A., Savio, M., Carafoli, F., Perucca, P., Bianchi, L., Maga, G., Forti, L., Pagnoni, U.M., Albini, A., Prosperi, E., and Vannini, V. 2001. Specific structural determinants are responsible for the antioxidant activity and the cell cycle effects of resveratrol. *J. Biol. Chem.* 276: 22586–22594.

Sun, Z.J., Pan, C.E., Liu, H.S., and Wang, G.J. 2002. Anti-hepatoma activity of resveratrol in vitro. *World J. Gastroenterol.* 8: 79–81.

Sun, N.J., Woo, S.H., Cassady, J.M., and Snapka, R.M. 1998. DNA polymerase and topoisomerase II inhibitors from *Psoralea corylifolia*. *J. Nat. Prod.* 61: 362–366.

Surh, Y.J., Hurh, Y.J., Kang, J.Y., Lee, E., Kong, G., and Lee, S.J. 1999. Resveratrol, an antioxidant present in red wine, induces apoptosis in human promyelocytic leukemia (HL-60) cells. *Cancer Lett.* 140: 1–10.

Szewczuk, L.M., Forti, L., Stivala, L.A., and Penning, T.M. 2004. Resveratrol is a peroxidase-mediated inactivator of COX-1 but not COX-2: A mechanistic approach to the design of COX-1 selective agents. *J. Biol. Chem.* 279: 22727–22737.

Szewczuk, L.M. and Penning, T.M. 2004. Mechanism-based inactivation of COX-1 by red wine *m*-hydroquinones: A structure-activity relationship study. *J. Nat. Prod.* 67: 1777–1782.

Tessitore, L., Davit, A., Sarotto, I., and Caderni, G. 2000. Resveratrol depresses the growth of colorectal aberrant crypt foci by affecting bax and p21(CIP) expression. *Carcinogenesis* 21: 1619–1622.

Tinhofer, I., Bernhard, D., Senfter, M., Anether, G., Loeffler, M., Kroemer, G., Kofler, R., Csordas, A., and Greil, R. 2001. Resveratrol, a tumor-suppressive compound from grapes, induces apoptosis via a novel mitochondrial pathway controlled by Bcl-2. *FASEB J.* 15: 1613–1615.

Traganos, F., Ardelt, B., Halko, N., Bruno, S., and Darzynkiewicz, Z. 1992. Effects of genistein on the growth and cell cycle progression of normal human lymphocytes and human leukemic MOLT-4 and HL-60 cells. *Cancer Res.* 52: 6200–6208.

Tredici, G., Miloso, M., Nicolini, G., Galbiati, S., Cavaletti, G., and Bertelli, A. 1999. Resveratrol, map kinases and neuronal cells: Might wine be a neuroprotectant? *Drugs Exp. Clin. Res.* 25: 99–103.

Tsan, M.F., White, J.E., Maheshwari, J.G., Bremner, T.A., and Sacco, J. 2000. Resveratrol induces Fas signaling-independent apoptosis in THP-1 human monocytic leukaemia cells. *Br. J. Haematol.* 109: 405–412.

Tsan, M.F., White, J.E., Maheshwari, J.G., and Chikkappa, G. 2002. Anti-leukemia effect of resveratrol. *Leuk. Lymphoma* 43: 983–987.

Ungvari, Z., Orosz, Z., Rivera, A., Labinskyy, N., Xiangmin, Z., Olson, S., Podlutsky, A., and Csiszar, A. 2007. Resveratrol increases vascular oxidative stress resistance. *Am. J. Physiol. Heart. Circ. Physiol.* 292: H2417-H2424.

van Ginkel, P.R., Sareen, D., Subramanian, L., Walker, Q., Darjatmoko, S.R., Lindstrom, M.J., Kulkarni, A., Albert, D.M., and Polans, A.S. 2007. Resveratrol inhibits tumor growth of human neuroblastoma and mediates apoptosis by directly targeting mitochondria. *Clin. Cancer Res.* 13: 5162–5169.

Vieira de Almeida, L.M., Piñeiro, C.C., Leite, M.C., Brolese, G., Leal, R.B., Gottfried, C., and Gonçalves, C.A. 2008. Protective effects of resveratrol on hydrogen peroxide induced toxicity in primary cortical astrocyte cultures. *Neurochem. Res.* 33: 8–15.

Vitrac, X., Desmoulière, A., Brouillaud, B., Krisa, S., Deffieux, G., Barthe, N., Rosenbaum, J., and Mérillon, J.-M. 2003. Distribution of [^{14}C]-*trans*-resveratrol, a cancer chemopreventive polyphenol, in mouse tissues after oral administration. *Life Sci.* 72: 2219–2233.

Wang, J., Ho, L., Zhao, W., Ono, K., Rosensweig, C., Chen, L., Humala, N., Teplow, D.B., and Pasinetti, G.M. 2008. Grape-derived polyphenolics prevent Abeta oligomerization and attenuate cognitive deterioration in a mouse model of Alzheimer's disease. *J. Neurosci.* 28: 6388–6392.

Wang, Z., Hsieh, T.C., Zhang, Z., Ma, Y., and Wu, J.M. 2004. Identification and purification of resveratrol targeting proteins using immobilized resveratrol affinity chromatography. *Biochem. Biophys. Res. Commun.* 323: 743–749.

Wang, M., Jin, Y., and Ho, C.T. 1999. Evaluation of resveratrol derivatives as potential antioxidants and identification of a reaction product of resveratrol and 2,2-diphenyl-1-picryhydrazyl radical. *J. Agric. Food Chem.* 47: 3974–3977.

Wang, Y., Lee, K.W., Chan, F.L., Chen, S., and Leung, L.K. 2006. The red wine polyphenol resveratrol displays bilevel inhibition on aromatase in breast cancer cells. *Toxicol. Sci.* 92: 71–77.

Wang, Q., Li, H., Wang, X.W., Wu, D.C., Chen, X.Y., and Liu, J. 2003. Resveratrol promotes differentiation and induces Fas-independent apoptosis of human medulloblastoma cells. *Neurosci. Lett.* 351: 83–86.

Wei, M.C., Zong, W.X., Cheng, E.H., Lindsten, T., Panoutsakopoulou, V., Ross, A.J., Roth, K.A., MacGregor, G.R., Thompson, C.B., and Korsmeyer, S.J. 2001. Proapoptotic BAX and BAK: A requisite gateway to mitochondrial dysfunction and death. *Science* 292: 727–730.

Win, W., Cao, Z., Peng, X., Trush, M.A., and Li, Y. 2002. Different effects of genistein and resveratrol on oxidative DNA damage in vitro. *Mutat. Res.* 513: 113–120.

Wolter, F., Akoglu, B., Clausnitzer, A., and Stein, J. 2001. Downregulation of the cyclin D1/Cdk4 complex occurs during resveratrol-induced cell cycle arrest in colon cancer cell lines. *J. Nutr.* 131: 2197–2203.

Wolter, F., Clausnitzer, A., Akoglu, B., and Stein, J. 2002. Piceatannol, a natural analog of resveratrol, inhibits progression through the S phase of the cell cycle in colorectal cancer cell lines. *J. Nutr.* 132: 298–302.

Xiao, J.B., Chen, X.Q., Jiang, X.Y., Hilczer, M., and Tachiya, M. 2008. Probing the interaction of trans-resveratrol with bovine serum albumin: A fluorescence quenching study with Tachiya model. *J. Fluoresc.* 18: 671–678.

Yu, H. and Jove, R. 2004. The STATS of cancer-new molecular targets come of age. *Nat. Rev. Cancer* 4: 97–105.

Yu, L., Sun, Z.J., Wu, S.L., and Pan, C.E. 2003. Effect of resveratrol on cell cycle proteins in murine transplantable liver cancer. *World J. Gastroenterol.* 9: 2341–2343.

Yu, L.J., Wu, M.L., Li, H., Chen, X.Y., Wang, Q., Sun, Y., Kong, Q.Y., and Liu, J. 2008. Inhibition of STAT3 expression and signaling in resveratrol-differentiated medulloblastoma cells. *Neoplasia* 10: 736–744.

Zang, M., Xu, S., Maitland-Toolan, K.A., Zuccollo, A., Hou, X., Jiang, B., Wierzbicki, M., Verbeuren, T.J., and Cohen, R.A. 2006. Polyphenols stimulate AMP-activated protein kinase, lower lipids, and inhibit accelerated atherosclerosis in diabetic LDL receptor-deficient mice. *Diabetes* 55: 2180–2191.

Zhang, S., Cao, H.J., Davis, F.B., Tang, H.Y., Davis, P.J., and Lin, H.Y. 2004. Oestrogen inhibits resveratrol-induced post-translational modification of p53 and apoptosis in breast cancer cells. *Br. J. Cancer* 91:178–185.

Zheng, L.F., Wei, Q.Y., Cai, Y.J., Fang, J.G., Zhou, B., Yang, L., and Liu, Z.L. 2006. DNA damage induced by resveratrol and its synthetic analogues in the presence of Cu (II) ions: Mechanism and structure-activity relationship. *Free Radic. Biol. Med.* 41: 1807–1816.

Zhou, H.B., Chen, J.J., Wang, W.X., Cai, J.T., and Du, Q. 2005. Anticancer activity of resveratrol on implanted human primary gastric carcinoma cells in nude mice. *World J. Gastroenterol.* 11: 280–284.

Zhou, H.B., Yan, Y., Sun, Y.N., and Zhu, J.R. 2003. Resveratrol induces apoptosis in human esophageal carcinoma cells. *World J. Gastroenterol.* 9: 408–411.

Zi, X., Grasso, A.W., Kung, H.J., and Agarwal, R. 1998. A flavonoid antioxidant, silymarin, inhibits activation of erbB1 signaling and induces cyclin-dependent kinase inhibitors, G1 arrest, and anticarcinogenic effects in human prostate carcinoma DU145 cells. *Cancer Res.* 58: 1920–1929.

Zykova, T.A., Zhu, F., Zhai, X., Ma, W.Y., Ermakova, S.P., Lee, K.W., Bode, A.M., and Dong, Z. 2008. Resveratrol directly targets COX-2 to inhibit carcinogenesis. *Mol. Carcinog.* 47: 797–805.

11 Drug Discovery from Natural Substances

A Case Study—Camptothecins

Gabriele Fontana and Lucio Merlini

CONTENTS

11.1 INTRODUCTION

Many books and papers have been written to tell the tale of the discovery of camptothecin (CPT) (1) (Potmesil and Pinedo 1995, Pantaziz and Giovanella 1996, Wall et al. 1966, Liehr et al. 2000, Adams and Burke 2005). This chapter is meant to be only a modest contribution to this tale and to remind us how we have to be determinate day by day if we want to discover something useful for saving human lives.

The story of the discovery of CPT begins almost 60 years ago when, at the beginning of the 1950s, something very important happened in the fight against cancer in the United States. The Memorial Sloan-Kettering Cancer Center (New York), a prestigious cancer research institute founded in 1884, promoted the concept that cancer could be cured with drugs as any other disease. The replacement of invasive treatments like surgery and radiotherapy with chemotherapy was something very desirable and the idea found immediately a wide number of supporters. The Memorial Sloan-Kettering Center, which was based on the industrial organizational model, started then an ambitious program of screening with the aim to find substances active against cancer. The program was designed to test over 3000 compounds per year in in vitro cytotoxicity assay (essentially leukemia cell lines). The compounds that proved active in the primary screening passed to in vivo murine tumor models. In a few years, the campaign was embraced also by the

National Cancer Institute (NCI) at the National Institute of Health (NIH), which in 1955 launched an even more massive campaign. The number of compounds to be screened per year was increased 10-fold (from 3,000 to 30,000). Any kind of substance, either synthetic or isolated from natural sources, could be admitted into the screening. At the US NIH, the scientists were already familiar with screening programs and knew the tremendous potential of such a resource. For example, thanks to a similar approach, they were successful in making cortisone, a drug, available to a large population. At that time, cortisone was known to be a powerful remedy against rheumatoid arthritis and inflammation. The real problem was that the compound was of animal origin and therefore of scarce abundance. NIH initiated then a program of screening aimed to find in nature other sources of cortisone or suitable precursors for semisynthesis. The program ended with the identification of diosgenin, isolated from the Mexican yam, as a cheap precursor in the semisynthesis of cortisone.

In the screening program for anticancer agents, anybody from academic, industrial, private institution, or common people were entitled to submit compounds for testing, which was actually great news. In other words, the screening was open to public contributions. Everybody got involved and everybody could get compounds tested free of charge in a screening completely funded by the government.

The primary screening was conceptually simple, since the compounds were to be assayed in in vitro cytotoxicity tumor models. The tumor cell lines chosen were representative of simple tumor and common tumors (again leukemia). In the primary screening, NCI was not looking for outstanding activity, but for any signal of activity. Therefore, what was needed were tumor cell lines stable over time and sensitive to even weak activities. All the compounds that resulted positive to the primary screening would have been passed to a secondary screening, based essentially on in vivo testing aimed to measure the increased span life of mice with transplanted tumors.

It is important to note that the screening program launched by NCI has evolved over time, with the continuous review of the approach of primary and secondary screening, and of the filters for the promotion of a drug to the following research phase. The program is still active today. For example, the primary screening has been renewed from the original assay on three cell lines of tumors to the "sixty-line panel screening" that is not only able to detect activity but also to tell if the mechanism of action is already known (Paull et al. 1989, Shoemaker 2006).

The NCI program sounds today like a legend since the most important anticancer drugs, like paclitaxel and CPT, were discovered within that initiative. And what is amazing to note is that the people behind the discovery of diosgenin, CPT, paclitaxel, and many other active substances were always the same even if, at least at the beginning, they worked independently. Names like Monroe Wall, Mansukh Wani, and Jonathan Hartwell are written in bold in the history of medicinal chemistry. At the beginning, they presumably did not know each other, but for sure they shared a common passion for natural products. Jonathan Hartwell was the head of the National Products Section at the NCI. He joined NCI in 1938, after being employed by both DuPont and Interchemical Corporation. He was the person that started the screening program on extracts and natural products. Monroe Wall was a natural product chemist at the U.S. Department of Agriculture (USDA), Eastern Utilization and Development Division, Philadelphia. Within the USDA, he was responsible for the qualitative and quantitative screening of sterols, alkaloids, tannins, and flavonoids present in the botanical species living on the U.S. ground that brought to the discovery of diosgenin. From the USDA, Monroe Wall collaborated with Jonathan Hartwell sending samples of botanical extracts for anticancer screening. In fact, in 1957, Hartwell convinced Wall to send to the NCI for antitumor activity testing about 1000 ethanolic plant extracts that Wall had obtained during his work at USDA. In 1958, Hartwell informed Wall that, of all the 1000 extracts, only *Camptotheca acuminata* had demonstrated activity in the CA-755 and L1210 mouse leukemia assay. Wall became increasingly interested in drug discovery. Given its focus on agriculture, the USDA did not share Wall's interest in studying cancer-fighting plants. So, in 1962, Wall moved to the Research Triangle Institute (RTI), Durham, North Carolina, where with Mansukh Wani he established the Natural Products Laboratory to isolate antitumor compounds under a contract with NCI. Wall and Wani were engaged in a work of bioguided fractionation of botanical extracts on behalf of the NCI, including that from *C. acuminata*.

C. *acuminata* is not native to the United States. It is a rapidly growing, deciduous tree (up to 25 m) that occurs at elevations from 150 to 2400 m in southeastern China where it is known as *xi shu* ("tree of joy"). It may also grow in Burma and northern Thailand. The tree forms part of the Chinese-mixed mesophytic forest in warm, moist, temperate regions. The northern limits to this habitat lie along the Tsinling mountains, which divide the watersheds of the Yellow and Yangtze Rivers. In 1911, Arnold Arboretum, on behalf of the USDA's Plant Introduction Division, for the first time classified the seeds procured by a private collector. But the plant was grown for the first time in 1934 at the Chico Plant Introduction Station (California) from seeds collected by A.N. Steward at the College of Agriculture and Forestry, Nanking University, and germinated at the Gleen Dale Plant Introduction Station. It is exactly from the Chico station that in 1959 Wall received a sample of leaves and bark. The plant was known to Wall as used in traditional Chinese medicine.

It is often said that the discovery of drugs by chance are "serendipitous." The discovery of paclitaxel, penicillin, vinblastine, and CPT are classified as "serendipitous." Anyway, behind the selection of *C. acuminata,* as well as behind the isolation of *Penicillium* or *Vinca rosea*, more than luckily events, we can see the studies, the observations, the perseverance, and the intuition of clever scientists.

The activity of *C. acuminata* extract on L1210 mouse leukemia assay was unusual since most plants did not exhibit such activity. That fact obviously attracted the interest of NCI. Then in 1963, 5 years after the discovery of the activity, Wall, who was already established at RTI, received 19 kg of *C. acuminata* stem bark from Chico Station. In 1966, upon an impressive fractionation guided by in vitro 9 KB cytotoxicity test and in vivo L1210 mouse leukemia life prolongation assay, the active compound was isolated and named NSC-94600 (Figure 11.1). On the basis of the concentration and of the activity, it was estimated that NSC-94600 had a concentration of 0.01% in the plant. Wani obtained crystals of the substance that were sent to Andrew McPhail and George Sim at the University of Illinois for x-ray analysis (Wall et al. 1966).

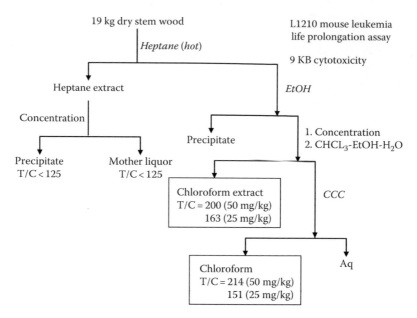

FIGURE 11.1 Bioactivity-guided fractionation of *C. acuminata* extract. (Reconstructed from Wall, M.E. et al., *J. Am. Chem. Soc.*, 88, 3888, 1996.)

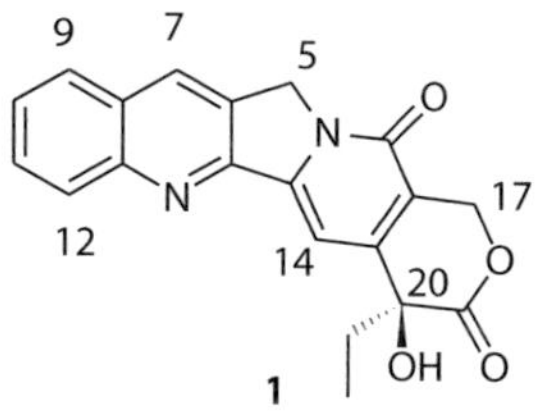

FIGURE 11.2 Structure of camptothecin.

Camptothecin (**1**, Figure 11.2), is a pentacyclic compound in which a quinoline unit (ring A and B) is joined to a pyridone ring (ring D) through a five-membered ring (ring C). The fifth ring (ring E) is a lactone, containing the only stereogenic center C-20, which just slightly disrupts the planarity of the molecule. Although the structure was unusual and not basic, CPT was immediately related to monoterpene indole alkaloids and, as a consequence, numbered like them (Shamma 1968).

CPT proved effective against a number of tumors other than L1210, like P388 and Walker WM, with a dose in the range of 0.5–4.0 mg/kg. The promising results obtained convinced the NCI scientists to accelerate preclinical and formulative studies in order to quickly promote the drug in human trials. As it happens with compounds like CPT, it was calculated that the minimum amount of drug needed to complete preclinical evaluation, formulative studies, and Phase I trial in humans was 100 g. To any organic chemist, 100 g does not sound anything special, but to a natural product chemist, that sounds like a source of trouble. In fact, with a content of 0.01% in the bark and an average water loss of 45%, 100 g of drug correspond to 1 metric ton of dry stem bark, that is, about 250 trees. A major concern arose when it was calculated that to proceed from Phase I to the approval of the drug, the amount needed was 2 kg (in average 7500 trees). It was obvious that such number of trees was not available in the United States, and that the access to biomass interior in China or India was difficult for collectors at that time. Immediately, 1300 seedlings were planted at the Chico Station in California, and the following year further 5000 seedlings to guarantee further harvests. The problem was solved just for the moment, since it was clear that the poor availability and renewability of the biomass would have limited the future manufacture of the drug.

Another issue became evident during the in vivo experiments in animal models. CPT, as the majority of its derivatives or analogs, has sparing solubility in physiological media like water for injection. That was actually a huge limitation in the administration of the drug and in dose-ranging studies. However, from chemical observations, it was noted that CPT is very sensitive to the pH. In particular, at physiological pH (7.4), the ring-E lactone gets hydrolyzed to generate the open hydroxycarboxylate form (Figure 11.3). The idea was then not to administer CPT itself, but the sodium salt of the hydroxycarboxylate derivative (CPT Salt, NSC-100880, CS). CS was much more soluble than CPT, although 10-fold less active. However, the loss of activity was not an issue since the reaction of hydrolysis is reversible; therefore, CPT can be restored in vivo through a rapid equilibrium. Everything was set and under the enthusiasm of the preclinical results the drug was quickly (too quickly) promoted to clinical investigations.

In 1970, CS entered Phase I at the cancer institutes of Baltimore and Washington on patients suffering gastrointestinal or bone-marrow cancer. Although the study was aimed to establish the tolerance of a weekly and daily ×5 schedule, the efficacy of the drug was kept under control. In gastrointestinal cancer, 5 out of 18 patients gave partial response (response rate = 28%), while in bone-marrow cancer only 2 patients out of 10 gave partial response, practically at the limit of the NCI threshold to continue clinical investigations (RR > 20%). The dose limiting toxicities were diarrhea, vomiting, and hematological depression. But the public expectations on the drug were so high that it was pushed to Phase II. It was a disaster. Besides only 2 partial responses in 68 patients, the toxicities were so severe that the trials were terminated in 1972. In particular, anemia and bleeding cystitis were the most severe toxicities upon repeated administration (Pizzolato and Saltz 2003).

Only later it was understood that what was thought to be a solution to solubility issues—the administration as sodium salt—was actually the detail that killed the drug. In fact, patients administered with CS were exposed to high concentration of

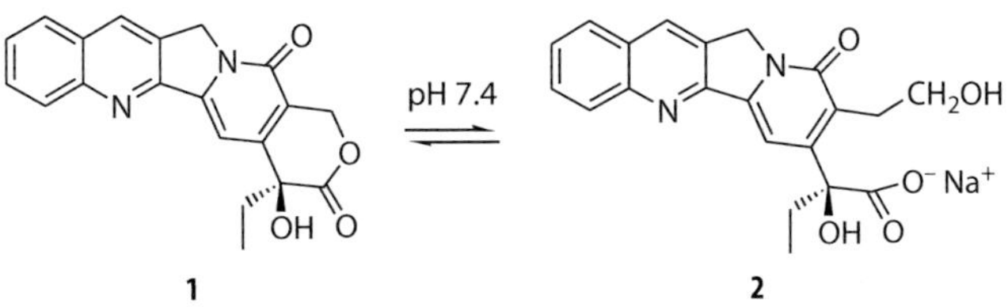

FIGURE 11.3   Lactone–hydroxyacid equilibrium in camptothecin.

carboxylate, which is rapidly excreted in urine, whose pH is acidic and pushes the equilibrium toward the lactone form. Due to its low solubility, CPT precipitates and accumulates in kidneys and bladder, resulting in hemorrhagic cystitis. The phenomenon was not predicted by the preclinical toxicological studies, conducted essentially in rodents. Later an explanation was found. The hydroxyl carboxylate form has great affinity for serum albumin. Serum albumin becomes a major way of elimination or compartimentalization of CS. But CS has greater affinity for human serum albumin (HSA) than for mouse serum albumin (MSA); therefore, the toxicities in rodents were underestimated and did not predict correctly the behavior of the drug in humans.

11.2   DISCOVERY OF THE MECHANISM OF ACTION

CPT was officially abandoned, albeit some academic scientists continued their studies on this interesting compound. Amongst them was Susan B. Horwitz whose efforts helped to elucidate the peculiar mechanism of action of CPT, still unknown up to that time (Horwitz and Horwitz 1971; Horwitz et al. 1971, 1972, 1973, Liebskind 1974). Horwitz observed that the drug was antimitotic, with the ability to interfere with DNA in S-phase of cell division cycle, inducing DNA breaks and interfering with DNA and RNA synthesis (Figure 11.4).

The real pulse to the mechanism elucidation came from John Randall at SmithKline Beecham. In 1980, SmithKline Beecham, John Hopkins University, University of Florida, and University of Virginia joined their efforts in the National Cooperative Development Drug Group. Their aim was to find novel topoisomerase II inhibitors as anticancer agents. Topoisomerases II are involved in the DNA

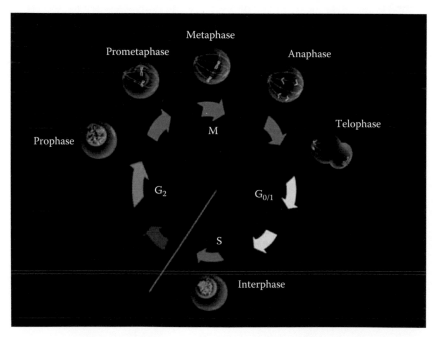

FIGURE 11.4   Arrest at the S/G_2 phase of the cell cycle by CPT.

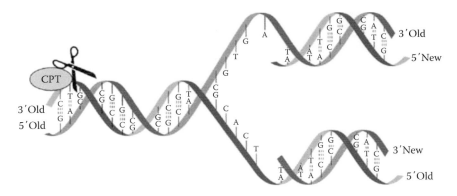

FIGURE 11.5 CPT inhibition of topoisomerase I.

replication as responsible for relaxation, religation, and damage reparation. Their inhibition blocks DNA (and RNA) replication, inducing apoptosis. John Randall worked at NCI before moving to SmithKline Beecham. He knew and had a long-standing interest in CPT because the compound shared many common properties with Topo II inhibitors. In 1985, he asked for some CPT for mechanism-based screening. The drug was used in an assay for topoisomerase II activity that happened to use topoisomerase I as a cofactor. The data obtained clearly showed that CPT acts as a potent poison of topoisomerase Ib (Hsiang et al. 1985).

The details of the mechanism were discovered in the following years. CPT acts forming a reversible ternary complex (known also as "cleavable complex") with DNA and topoisomerase I (Figure 11.5), preventing the religation of DNA strand, cut by topoisomerase I to allow relaxation, during the DNA replication in cell division, thus inducing apoptosis (Figure 11.6) (Thomas et al. 2004).

After so many years, in any case, how CPT forms the ternary complex is still not well understood. It is established that CPT does not significantly bind to either DNA or Topo I alone, but it binds to the covalent binary complex between DNA and Topo I. In any case, there is no evidence of the formation of a covalent bond between CPT and the binary complex. The x-ray structure of crystals of such a complex of a 22-base DNA fragment with topoisomerase I and topotecan has been reported (Staker et al. 2002) and molecular models of the interaction have been proposed (Fan et al. 1998, Redinbo et al. 1998, Kerrigan and Pilch 2001).

CPTs, like other TopoI inhibitors, bind at the interface of the Top1-DNA complex by intercalating at the cleavage site (π–π interactions) and by forming a network of H-bonds with critical Top1 residues involved in CPT resistance (Pommier 2009).

11.3 SAR STUDIES

The discovery that the primary cellular target of CPT is topoisomerase I was the breakthrough that revived interest in the drug. Immediately, the research activities restarted with the purpose to find new derivatives or analogs of CPT endowed with less toxicity and more specificity to the target cells. At that time, the research based its efforts on some assumptions derived from the history above described.

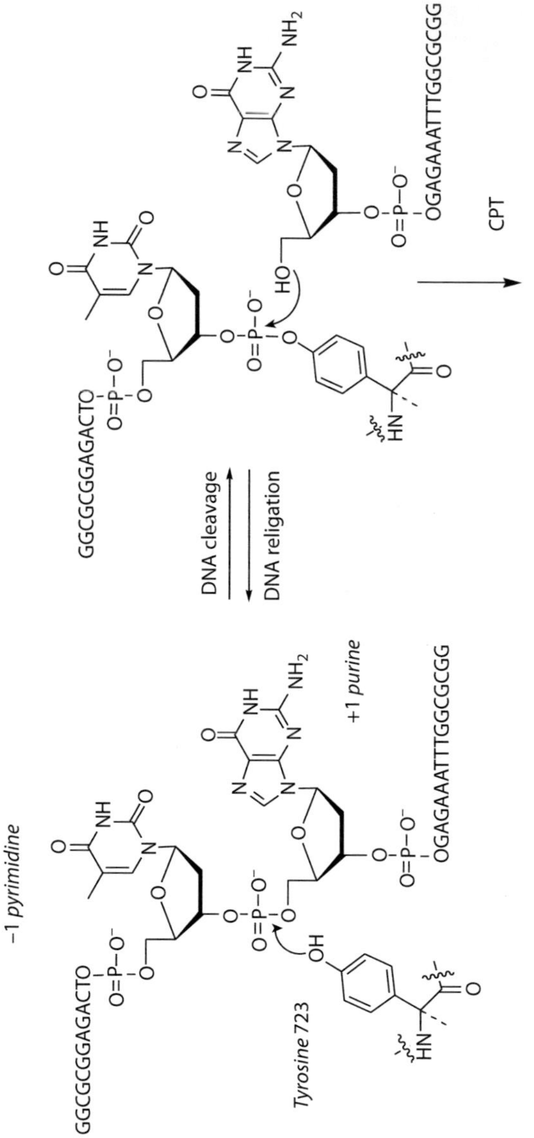

FIGURE 11.6 Mechanism of action of topoisomerase I.

For example, a desirable CPT analog or derivative should be endowed with good solubility, in order to be administered by i.v. infusion. The best CPT should have an intact and stable lactone ring-E, since the hydroxycarboxylate form was associated to HSA affinity and therefore to adverse effects. The best CPT analog/derivative should be readily available and therefore preferably totally synthetic. The best CPT should be specific to the target cell.

The efforts were then directed to increase solubility, to stabilize lactone E-ring, and to draw a structure–activity relationship (SAR) in order to design simple and specific synthetic analogs.

It is usually difficult to build an SAR when the biological target is a cellular target rather than a protein or an enzyme. That is what usually happens in cancer research when the primary bioassay is a cytotoxicity test and the parameter measured is essentially the average dose able to kill 50% of the cultured tumor cells. In functional tests, the efficacy of a drug is determined by a larger number of parameters, like the cell permeability and the stability of the compound over the exposure time, than in inhibition assays. Therefore, a large number of analogs is needed to consolidate the data and understand the impact of a chemical change on the efficacy of a compound.

With an impressive chemical work, Wall and Wani introduced and studied the effect of structural changes on the biological activity of CPTs, resulting in the seminal pages on the SAR of CPTs at the end of the 1960s. Of course, they took advantage of their studies on the extracts of *C. acuminata*, submitting to the bioassays all the analogs isolated from the plant like the ubiquitous 9 and/or 10 hydroxy/methoxy derivatives (Figure 11.7).

With this work, they established that substitution in position 10 and 9 is tolerated, with the preference for position 10, while substitution in 11 and 12 abolishes the activity of the compound with the notable exception of the 10,11-dimethoxy derivatives. The results were confirmed also by derivatization of CPT. By chemical derivatization, Wall and Wani also demonstrated that the hydroxyl group in position 20 is essential to the activity and it can only be replaced, with decrease of activity, by electron-withdrawing groups or hydrogen bond donors/acceptors. Furthermore, during their studies on the total synthesis of CPT, Wall and Wani also established that 20*S* configuration is essential to the activity of the drug (Wani et al. 1987).

The other fundamentals in the SAR of CPT were achieved by different groups in the following years; amongst them we cannot forget to mention Sawada's group at Yakult

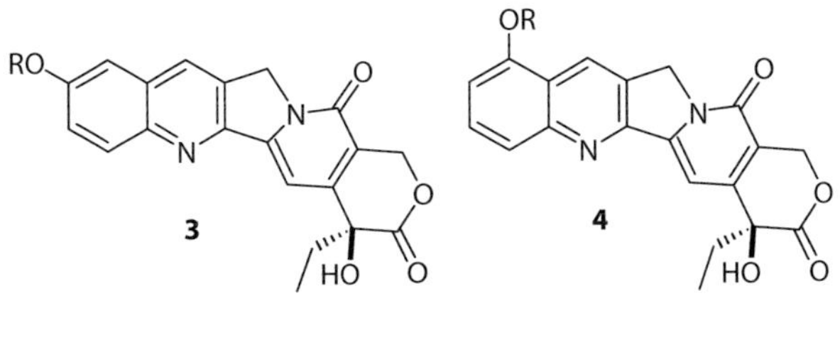

R = H, CH$_3$

FIGURE 11.7   Camptothecin-related alkaloids in the plant.

(Sawada et al. 1991). In particular, it was shown that substitution in position 7 is not only tolerated but can also potentiate the potency of the compound. Providing that position 7 lies in a large pocket on the ternary complex (presumably coincident with the minor groove of DNA), the type and the number of tolerated substituents are quite large.

The substitution in position 5 is tolerated as long as it does not disrupt the planarity of the molecule, although it is not generally convenient because it is usually difficult to control and stabilize the configuration of the forming stereocenter. The transformation into a 5-methylidene derivative has not given better results.

The simplification of the molecule with the replacement of pyridone ring D and lactone ring E with carbocyclic residues resulted in the loss of activity. A recent review by Hansch (Verma and Hansch 2009) summarizes the SAR studies (Figure 11.8) and reports some Quantitative Structure-Activity Relationship (QSAR) models.

The assumption that lactone is good and hydroxyl carboxylate is undesirable has been the rationale for important structural modification that found their best realizations in the homocamptothecins series (in which the ring E has been enlarged from 6- to 7-membered lactone) or in the Pierre Fabre ring E cyclopentanone analogs, for example, S39625 (Hautefaye et al. 2003). Homocamptothecins lactones are reported to be more resistant to hydrolysis than the 6-membered lactones, due to the shift of the activating OH from the α to the β position with respect to the CO group (Lavergne et al. 1998). Anyway, this effect did not produce any particular advantage in terms of efficacy so that none of the compounds of the series (e.g., homosilatecan, diflomotecan) has been promoted to clinical trials. All other attempts to remove the lactone functionality abolished the activity of the compounds.

The scientific community has been arguing for years on the importance of the stability of the lactone ring, elaborating a second dogma in this class: that the more lipophilic is the CPT the more stable is the lactone and therefore the more active is the drug. In the elaboration of this syllogism, maybe it has not been considered that lipophilic CPTs are very well absorbed in lipophilic tissues where the hydrolysis to hydroxycarboxylate is very slow or nil. In that meaning, the lipophilicity stabilizes the lactone and makes the compound more suitable for passing cell membranes. But that is also a double-edged sword since the compound is accumulated and slowly released (with side effects). Moreover, it leads us to a kind of paradox: we need lipophilicity to get activity, but lipophilicity is opposite to druggability.

That is the reason why so many derivatives/analogs, including prodrugs, have been designed to balance the lipophilicity with the solubility of the drug (Figures 11.9 and 11.10). Amongst them, we can mention the three launched CPT derivatives: irinotecan, topotecan, and belotecan. Topotecan and belotecan are two semisynthetic derivatives of CPT that have been solubilized by the introduction of a tertiary amino residue in position 9 and 7, respectively. Actually, irinotecan is a prodrug of the active 7-ethyl-10-hydroxycamptothecin (SN38) by N-piperidino-4-piperidinyl carbamoyl derivative in position 10. Irinotecan is today the top-selling CPT, with a total turnover beyond 1000 million US$/year in the regulated countries. Irinotecan is prescribed as first-line treatment for colon cancer and gastric tumors, especially in combination with fluorouracil and vascular disrupting agents. Despite its great success, irinotecan produces a significant number of side effects, particularly gastrointestinal disorders. The carbamoyl unit is easily hydrolyzed by enzymes of the

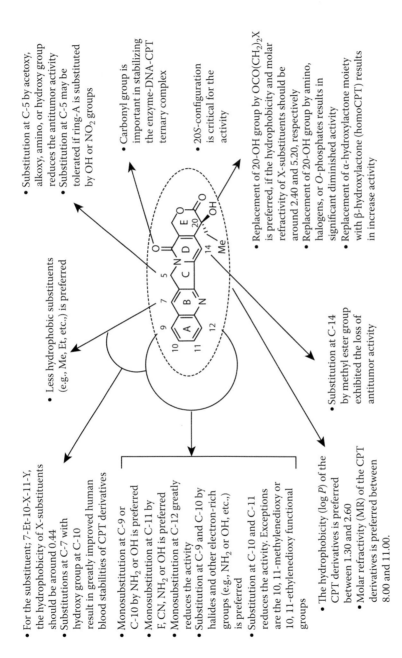

- Substitution at C-5 by acetoxy, alkoxy, amino, or hydroxy group reduces the antitumor activity
- Substitution at C-5 may be tolerated if ring-A is substituted by OH or NO_2 groups

- Carbonyl group is important in stabilizing the enzyme-DNA-CPT ternary complex

- 20S-configuration is critical for the activity

- Less hydrophobic substituents (e.g., Me, Et, etc.,) is preferred

- Replacement of 20-OH group by $OCO(CH_2)_2X$ is preferred, if the hydrophobicity and molar refractivity of X-substituents should be around 2.40 and 5.20, respectively
- Replacement of 20-OH group by amino, halogens, or O-phosphates results in significant diminished activity
- Replacement of α-hydroxylactone moiety with β-hydroxylactone (homoCPT) results in increase activity

- For the substituent; 7-Et-10-X-11-Y, the hydrophobicity of X-substituents should be around 0.44
- Substitutions at C-7 with hydroxy group at C-10 result in greatly improved human blood stabilities of CPT derivatives

- Monosubstitution at C-9 or C-10 by NH_2 or OH is preferred
- Monosubstitution at C-11 by F, CN, NH_2 or OH is preferred
- Monosubstitution at C-12 greatly reduces the activity
- Substitution at C-9 and C-10 by halides and other electron-rich groups (e.g., NH_2 or OH, etc.,) is preferred
- Substitution at C-10 and C-11 reduces the activity. Exceptions are the 10, 11-methylenedioxy or 10, 11-ethylenedioxy functional groups

- The hydrophobicity (log P) of the CPT derivatives is preferred between 1.30 and 2.60
- Molar refractivity (MR) of the CPT derivatives is preferred between 8.00 and 11.00.

- Substitution at C-14 by methyl ester group exhibited the loss of antitumor activity

FIGURE 11.8 QSAR of camptothecins from Verma and Hansch (2009). (Reproduced from Verma, R.P. and Hansch, C. Camptothecins: A SAR/QSAR study, *Chem. Rev.*, 109, 213–235, 2009. Copyright 2009 American Chemical Society.)

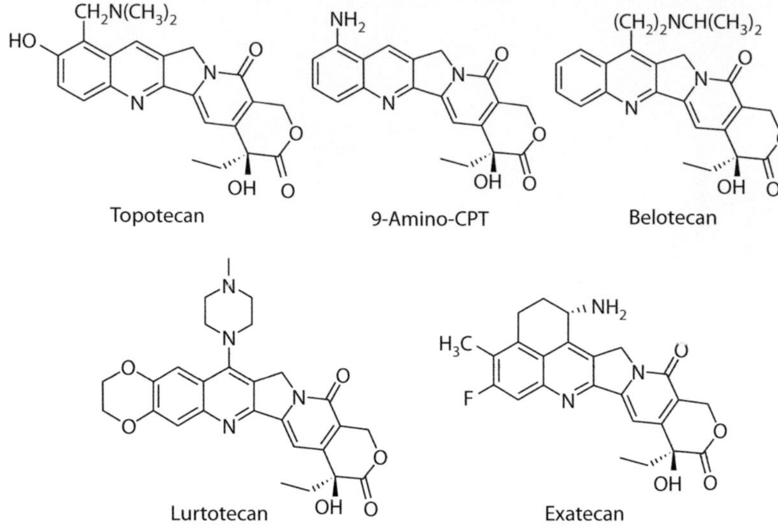

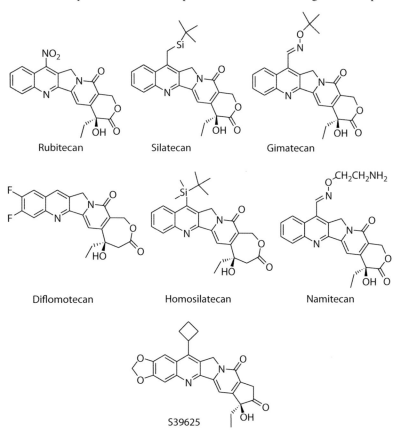

FIGURE 11.9 Camptothecins in clinical practice or in advanced stage of development (A).

FIGURE 11.10 Camptothecins in clinical practice or in advanced stage of development (B).

carboxylesterase class, but these enzymes are expressed in an unhomogeneous way in humans, so that the response to the drug is not always predictable.

The card of making prodrugs has been played from the very beginning of the story with the purpose of making CPTs soluble or more absorbable. Wall and Wani put particular efforts in the preparation of esters in position 20. The propionate was promoted to clinical trials at the beginning of the 1990s. Particularly interesting was 9-aminocamptothecin, obtained by nitration of CPT, followed by reduction of the nitro group. This compound could be administered as a salt for injection. But the precursor 9-nitrocamptothecin, known as rubitecan, became much more interesting, being transformed in vivo to 9-aminocamptothecin itself. Rubitecan is still today under investigation in Phase III clinical studies.

The derivatives with cleavable functionalities were used not only to enhance the kinetic properties, but also to improve the selectivity of the drug against tumors. Notable examples of targeting compounds are afeletecan and conjugates mentioned below (Figure 11.11).

The assumption on the stability of the lactone ring remained a dogma until the x-ray structure of crystals of the complex of a 22-base DNA fragment with topoisomerase I and topotecan was published (Staker et al. 2002). In fact, in that image, topotecan seems to be an averaged population between the lactone and the hydroxycarboxylate derivative. It can be argued that for the activity of the compound, hydroxycarboxylate cannot be eliminated. Maybe the lipophilic lactone is important for passing through the cell membranes and entering into the tumor, but the hydroxyl carboxylate is

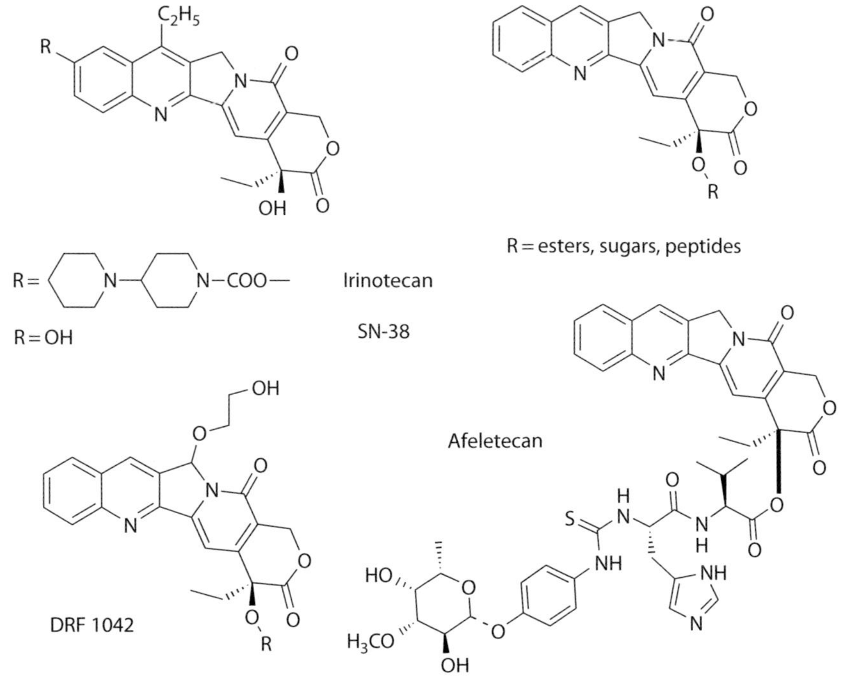

FIGURE 11.11 Prodrugs of camptothecins.

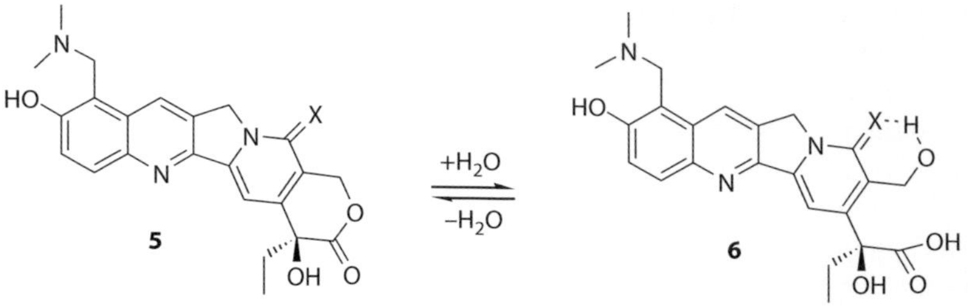

	ΔG° (KCal/mol)	
	O	S
B3YLP/6-31G(d)	−2.38	−0.83
B3YLP water	−1.31	+0.15

FIGURE 11.12 Calculated free energy for the lactone–hydroxyacid equilibrium in thiotopotecan.

fundamental for the distribution and can enhance the stability of the ternary complex. This rationale was the basis for the design and development of the class of thiocamptothecins, in which the equilibrium between the lactone ring and the hydroxycarboxylate is maintained very dynamically by the weakening of the 17-OH-16a-carbonyl hydrogen bond, which in the CPT series shifts the population of compounds toward the hydroxycarboxylate form. The model is well exemplified by the results of the *in silico ab initio* comparison between topotecan and thiotopotecan, for which the free energy of the equilibrium approaches zero for the thio couple indicating that in this case the population of the lactone is much more represented (Figure 11.12) (Samorì et al. 2010).

11.4   SYNTHETIC EFFORTS AROUND CAMPTOTHECIN

The SAR studies have been obviously supported by synthetic efforts oriented toward the total synthesis of CPT and to the design of simplified analogs.

Several reviews have been published on the synthesis of this class of compounds, with notable summaries by Du (Du 2003), Hecht, (Thomas et al. 2004) and, more recently, by Dallavalle (Dallavalle and Merlini 2008).

In a plain model of discussion, the CPT synthesis can be summarized in four basic disconnections: the route based on the formation of the B-ring from suitable synthons, the route based on the formation of C-ring, the one bringing to the formation of rings B and C, and the formation of C-D-E rings (Figure 11.13).

The synthetic approach based on B-ring retro-synthesis was the first published by Monroe Wall (Wall et al. 1986). The core of the approach (Figure 11.14) is the Friedländer condensation of anthranylaldehyde with a tricyclic pyrrolidinone derivative. This last synthon can be obtained in 12 steps from cyanoacetamide and ethyl acetylpyruvate. After the initial condensation to get the carbethoxycyanopyrimidine and its N-alkylation with methyl ω-halopropionate, the pyrrolidinone is formed by intramolecular Claisen condensation. The formation of the future E-ring is quite straightforward, but the real key-step of the whole approach is the introduction of the 20*S* stereocenter. In Wall's approach, the 20-OH is introduced by oxidation α

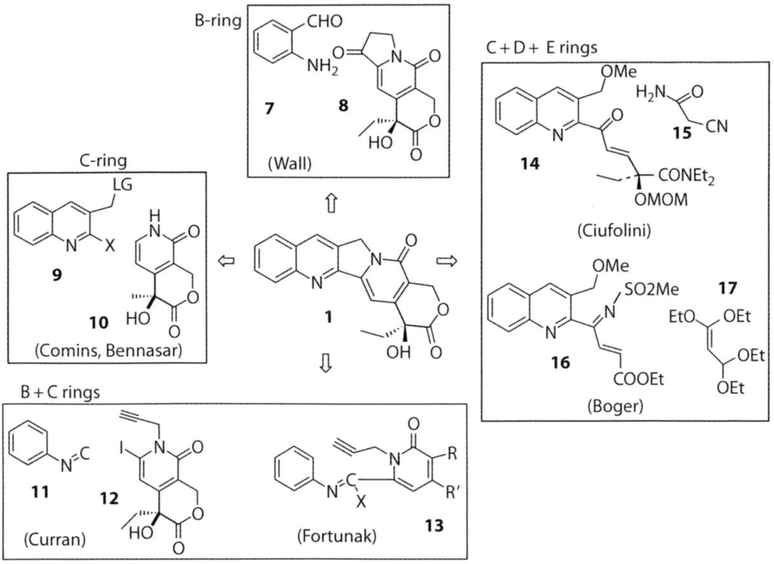

FIGURE 11.13 Main routes of synthesis of camptothecin.

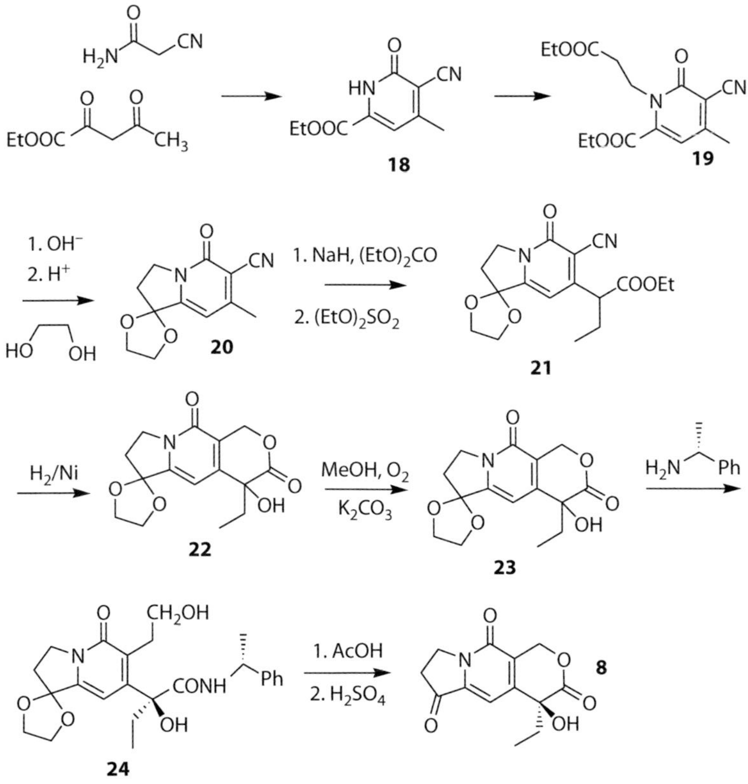

FIGURE 11.14 Wall synthesis of rings CDE.

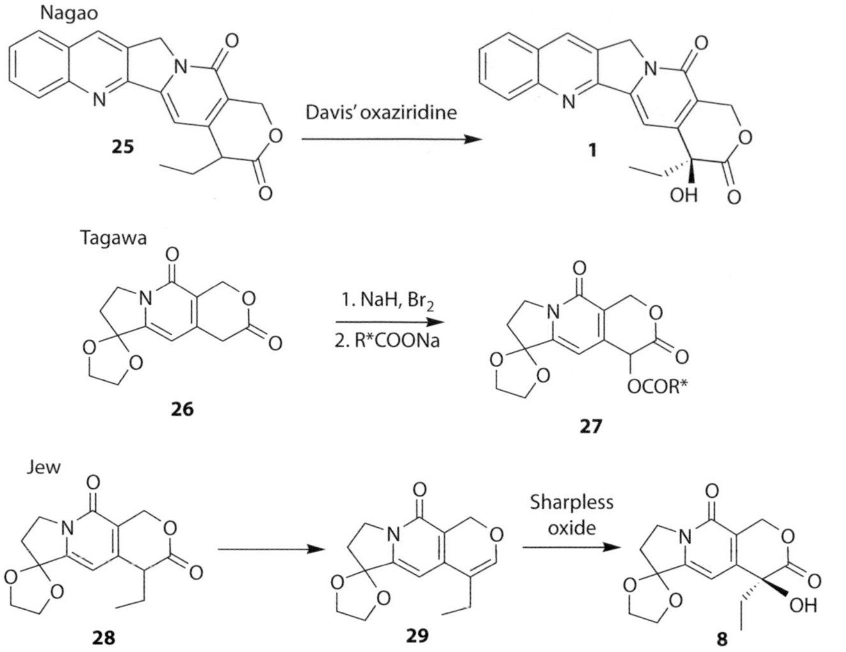

FIGURE 11.15 Different approaches to 20S-camptothecin.

to the carboxylate in alkali medium. The racemic mixture is then resolved with phenylethylamine. In the following years, different authors approached the problem suggesting interesting alternatives.

The quickest evolution of Wall method was the oxidation by means of Davis' oxaziridine as suggested by Nagao (Tagami et al. 2000). Tagawa proposed the α-bromination of the lactone carboxyl group and following nucleophilic substitution with a chiral carboxylate (Ejima et al. 1990). The hydrolysis of the so obtained alpha ester gave the 20-OH in the right configuration. More elegant was Jew's approach who introduced the 20-S-hydroxyl by Sharpless epoxidation of a chromene-like system to get a 20-S lactol that is then oxidized to lactone (Jew et al. 1995) (Figure 11.15).

The introduction of the 20S-hydroxyl group is a common problem to any of the synthetic approaches to CPT. An original methodology was described by Ciufolini in his scheme (Figure 11.16) based on the C + D + E rings disconnection (Ciufolini and Roschangar 1997). According to this author, the 20-S-hydroxyl is introduced as a preformed stereocenter from optically pure diethyl ethyl tartronate by kinetic resolution with pig liver esterase.

In the Comins' approach (Figure 11.17), the first representative of the C-ring pathway, the 20-S stereocenter, is generated by stereo-directed alkylation of the pyridone precursor with a chiral ester of propanoic acid (Comins and Nolan 2001).

In the same disconnection pathway, the more concise Bennasar's approach (Figure 11.18) introduces the 20S moiety as a preformed stereocenter by alkylation with t-butyl ethyl dioxolanone, obtained according to a well-established method for the desymmetrization of glycolic acid (Bennasar et al. 2002).

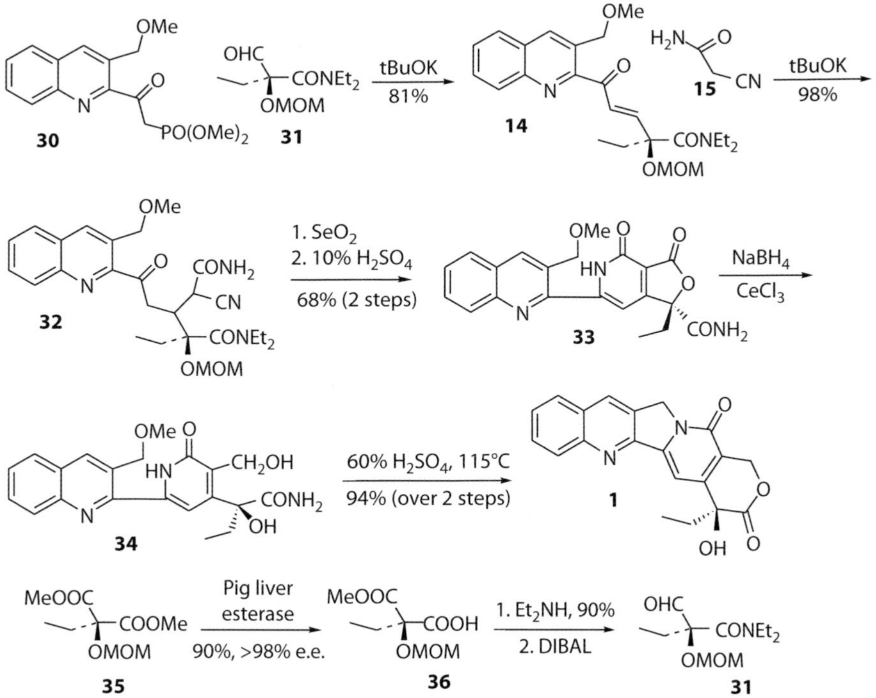

FIGURE 11.16 Ciufolini's synthesis of camptothecin.

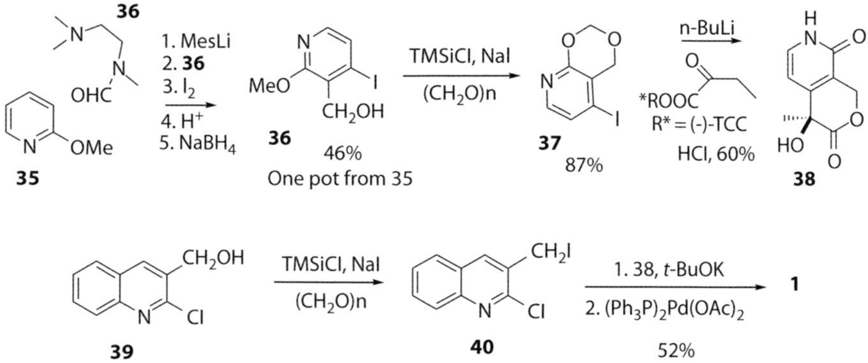

FIGURE 11.17 Comins' approach to camptothecin.

A new concept for the synthesis of this class of compounds has been introduced by Fortunak with the pericyclic synthesis of B + C rings and then of E-ring (Fortunak et al. 1996). Within this disconnective approach, the Curran's radical cascade cyclization is by far the more concise method for the preparation of CPT (Josien et al. 1998). Here, the 20S stereocenter is introduced by Sharpless epoxidation of the chromene as suggested by Jew (Figure 11.13).

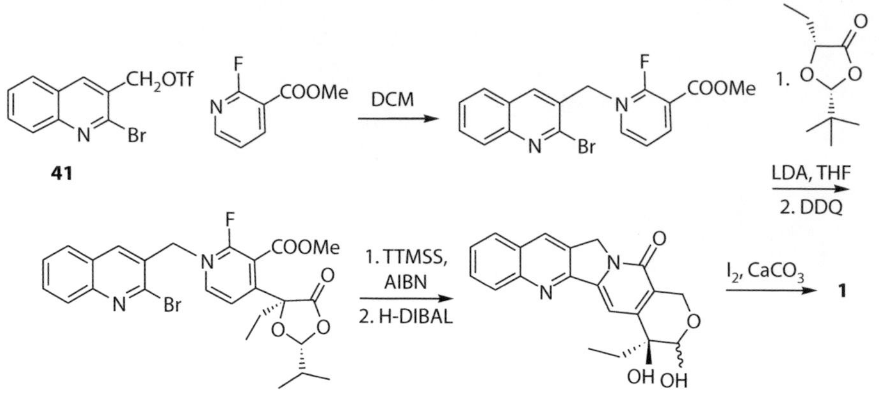

FIGURE 11.18   Bennasar's route to camptothecin.

Of course we have mentioned only few of the dozens of methodologies, and we apologize to all the scientists who have devoted their efforts to the synthetic studies on CPTs, including the never-enough-mentioned Danishefsky's group.

11.5   SEMI-SYNTHESIS AS COMPROMISE

In the SAR studies of a class of compounds, the total synthesis offers wide possibilities to explore the chemical space of the molecular target. However, especially when an asymmetric synthesis is required—as usually happens in the field of natural products—the number of steps involved is often incompatible with the cost/benefit ratio of the drug and sometimes the technology required is a limitation for the industrialization of the process. In the case of CPTs, the difficulties associated with the reproducibility and renewability of the botanical source were of less concern than the efficiency of the total synthesis. Therefore it is not suprising that the compounds registered and launched for clinical use (irinotecan, topotecan, and belotecan) are prepared by semisynthesis starting either from CPT or natural precursors (Figure 11.19).

11.6   SOURCING OF CAMPTOTHECIN: SUPPLY OF BIOMASS

The reproducibility and the renewability of the biomass are two problems that are relevant for the production of any drug of natural origin. In the case of CPTs, due to the high potency of the drugs, pharmacological doses are quite low, so that the amount of CPT required is in the range of hundreds of kilograms per year (almost nothing if we compare it to the tons of salicylic acid produced per year). Furthermore, *Camptotheca spp.* are plants growing at a reasonable rate, so that the plantation can be planned with just a few years advance notice. An interesting paper by Lorence and Nessler (Lorence and Nessler 2004) summarizes also how CPT can be isolated from plants of unrelated orders or families of angiosperms: Order Celastrales: *Nothapodytes foetida*, *Pyrenacantha klaineana*, and *Merrilliodendron megacarpum* (Icacinaceae); Order Cornales: *Camptotheeca acuminata*, *C. lowreyana*, and *C. yunnanensis* (Nyssaceae); Order Gentianales: *Ophiorrhiza mungos*, *O. pumila*, *O. filistipula* (Rubiaceae), *Ervatamia heyneana* (Apocynaceae), and *Mostuea brunonis* (Gelsemiaceae) (Table 11.1).

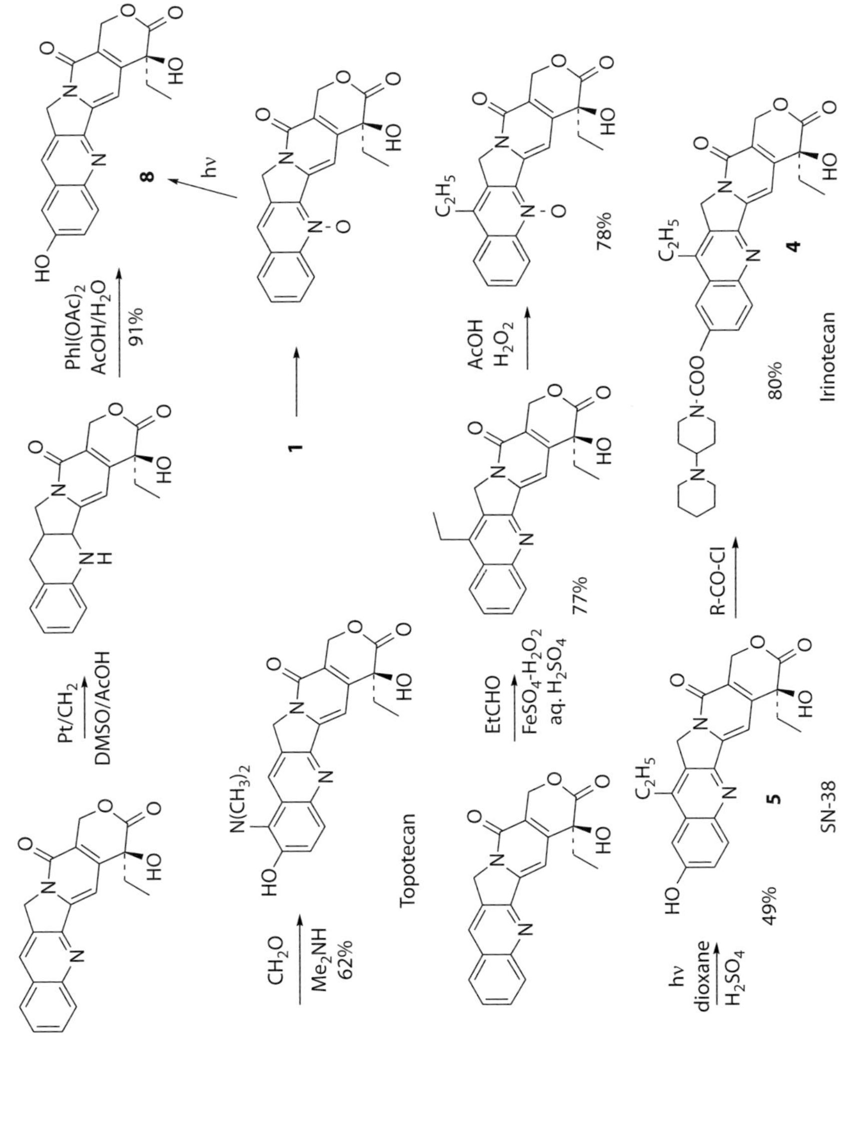

FIGURE 11.19 Semisynthesis of topotecan and irinotecan.

TABLE 11.1

Sites of Accumulation of the Antitumor Alkaloid CPT and Its Natural Derivatives in Several Natural Sources

Species	Tissue Analyzed	Sample Origin	Camptothecinoid Content (µg/g Dry Weight)
Camptotheca	Young leaves	Texas, United States	CPT 4000 HCPT 20–30
acuminata	Seeds		CPT 3000 HCPT 25
Decaisne	Bark		CPT 1800–2000 HCPT 2–90
	Roots		CPT 400 HCPT 13–20
	Young leaves	Texas, United States	CPT 2421–3022
	Old leaves		CPT 482
	Young fruit		CPT 842
	Old fruit		CPT 2362
	Hairy roots	Texas, United States	CPT 1000 HCPT 150
	Callus	Shanghai, China	CPT 2040–2360 HCPT 80–100
	Cell cultures		CPT 2. 5-4
Camptotheca	Young leaves	Texas, United States	CPT 3913–5537
lowreyana Li	Old leaves		CPT 909–1184
Camptotheca	Young leaves	Texas, United States	CPT 2592–4494
yunnanensis Dode	Old leaves		CPT 590
Ervatamia heyneana	Wood and stem	India	CPT 1300 MCPT 400
(Wall) T. Cooke	bark		
Nothapodytes foetida	Stem wood	Okinawa, Japan	CPT 1400–2400 dCPT 19
(Wight)			
Sleumer	Stem	Taiwan	ACPT 0.24
	Shoot	Mahabaleshwar, India	CPT 750 MCPT 130
	Plantlet culture		MCPT 7
	Callus		MCPT 1
	Stem	Godavari, India	MACPT 2.5
	Callus	Ooty, India	CPT 9.5 MCPT traces
	Cell culture	Satara, India	CPT 1.1 MCPT 0.81
Merriliodendron	Leaves and stem	Guam	CPT 530 MCPT 170
megacarpum			
(Hemsl.) Sleumer			
Mostuea brunonis	Entire plant	Lope, Gabon	CPT-20-O-$\tilde{\beta}$ glucoside 100
Didr.			Deoxypumiloside 100
			Strictosamide 600
Ophiorrhiza mungos	Entire plant	Colombo, Sri Lanka	CPT 12 MCPT 10.41
Linn.			

TABLE 11.1 (continued)

Sites of Accumulation of the Antitumor Alkaloid CPT and Its Natural Derivatives in Several Natural Sources

Species	Tissue Analyzed	Sample Origin	Camptothecinoid Content (µg/g Dry Weight)
Ophiorrhiza pumila	Leaves	Japan	CPT 300–400
Champ.	Young roots		CPT 1000
	Hairy roots		CPT 1000
	Entire plant	Kagoshima, Japan	CPT 300–510 MCPT 70–140
			Chaboside 300–690
	Hairy roots		CPT 240
	Cell cultures	Japan	None
Pyrenacantha	Stems	Ankasa Game	CPT 4.8 MCPT 1.6
klaineana		Reserve, Ghana	
Pierre ex Exell &			
Mendoca			

Source: Reproduced from Lorence, A. and Nessler, C.L. *Phytochemistry,* 65, 2735, 2004. With permission.

CPT, camptothecin; ACPT, O-acetyl-CPT; dCPT, (20S)-18,19-dehydro CPT; HCPT, 10-hydroxy CPT; MACPT, 9-methoxy-20-O-acetyl-CPT; MCPT, 9-methoxy CPT.

These observations opened the way to find in Nature other sources of CPT. A further important finding was that the highest level of CPT is found in young leaves (4–5 mg g^{-1} dry weight), approximately 50% higher than in seeds and 250% higher than in the bark, previously used for commercial CPT production (Lopez-Meyer et al. 1994). This observation has been confirmed by other groups, while studying other members of the genus *Camptotheca* (Li et al. 2002) and *Ophiorrhiza pumila* (Yamazaki et al. 2003). That sounded as good news to supply chain managers whose job is to identify renewable, cheap, and fast-growing biomasses containing CPT. From a scientific standpoint, the observation that the highest content of CPT is found in young leaves and seedlings (particularly of *C. acuminata*) suggested that the compound may represent a chemical defense to deter attacks by herbivores, pathogens, or both. This can be true because the genes that encode the enzyme for the production of CPT were not lost during phylogeny, and *C.* trees do not suffer pathogenic damages by insects.

For defense, plants produce many toxic compounds that harm other organisms. However, if the target of these compounds is a fundamental biological process then the producing plant may also be harmed. In such cases, self-resistance strategies must coevolve with the biosynthetic pathway of toxic metabolites. In the case of CPT-producing plants, the target protein of CPT, topoisomerase 1, has been mutated in order to overcome the toxicity of the compound (Sirikantaramas et al. 2009).

Interestingly, a patent application for the use of CPT and its analogs in termite control was filed in the United States (Li 2002). In addition, CPT has been reported to inhibit the sprouting of potato and root growth of radishes (Wang et al. 1980).

The problem of CPT supply prompted finding new sources. Such an effort had to pass through a clear understanding of the biochemical origin of the compound.

The first study on CPT biosynthesis in *C. acuminata* was reported in the early 1970s (Hutchinson et al. 1974). However, after 35 years, the biosynthetic pathway has not been completely elucidated yet. From its discovery, CPT was reconducted to the synthesis of monoterpenoid indole alkaloids starting from tryptamine and secologanin (Figure 11.20). The key intermediate strictosamide, which forms by intramolecular cyclization of strictosidine, was identified by Hutchinson in 1979 by incorporation of radiolabeled precursors (Hutchinson et al. 1979). Pumiloside, which is supposed to be the intermediate precursor between strictosamide and CPT, was isolated from *Ophiorriza spp* and from *C. acuminata* (Lorence and Nessler 2004), but the pathway between strictosamide and CPT remains still today largely unexplored. The mechanism of the rearrangement of the indole moiety into quinoline, which most probably involves oxidation and recyclization of B and C rings, as well as the conversion of ring E into lactone, which most probably involves removal of C-21 glucose unit and oxidation, are only hypotheses to be confirmed.

The biosynthetic pathway prior to the formation of strictosidine has been extensively studied including cDNA cloning of several biosynthetic enzymes. Strictosidine is produced from the condensation of secologanin and tryptamine by strictosidine synthase. Such an enzyme has been detected in roots and stems, suggesting that those are the parts of the plant where the biosynthesis takes place. From there CPT is distributed to the other parts through an unknown mechanism. One possible mechanism for such an insoluble compound is its conversion into a more water soluble form such as a glucoside (Aimi et al. 1990) and transport to other parts of the plant (Yamazaki et al. 2003).

More discussion has been dedicated to the origin of secologanin. For many years, the mevalonate pathway (MVA) was thought to be the only source of building blocks for all plant isoprenoids. Recently it has been found that isopentenyl diphosphate can be obtained through a second route, called $2C$-methyl-D-erythritol-4-phosphate pathway (MEP), operating in plastids of higher plants (Cane 1999). The two pathways operate simultaneously and their separation is not absolute. Recently, feeding studies with *O. pumila* hairy roots and [1-^{13}C]-glucose have shown that the secologanin moiety of CPT was synthesized via the MEP pathway (Yamazaki et al. 2004). In *C. acuminata*, the contribution of MVA and MEP routes to CPT production needs to be studied in further detail. The tryptophan moiety is clearly derived from the shikimate pathway.

The information above was something fundamental to the scientists oriented in finding new renewable sources of CPT, especially with biotechnological methods. We know in fact that the great amount of CPT required is still supplied exclusively from intact plants, mainly *C. acuminata and N. foetida*. The chemical synthesis has been extensively studied (Du 2003, Thomas et al. 2004, Dallavalle and Merlini 2008), but no large-scale synthesis has been reported yet. Shortage of the plants and environmental issues force finding alternative sources.

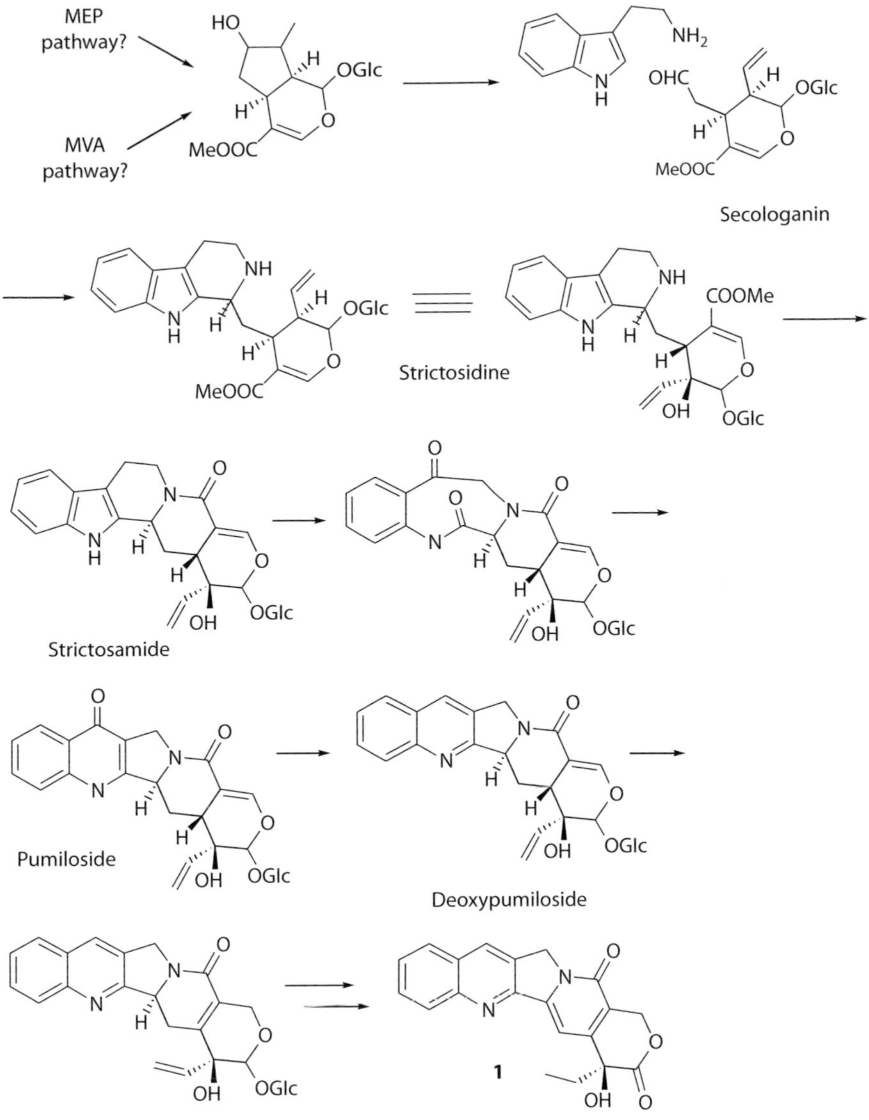

FIGURE 11.20 Synthesis of monoterpenoid indole alkaloids starting from tryptamine and secologanin.

The production of secondary metabolites of botanical origin from cell suspension culture has always represented a dream to all the scientists working in natural product chemistry. Also, in this case, we assisted in a large number of studies oriented to produce CPT *from C. acuminata, N. foetida,* and *O. pumila* cell cultures. The great limitation of this kind of technique is mainly the instability of the cell lines and the vulnerability to attack by microbial and viral agents. Nevertheless, some excellent cases have found an industrial application like the production of paclitaxel by cultivation of *Taxus brevifolia* cells.

In the case of CPT, the production by botanical cells cultivation is not available on industrial scale yet. The best results were obtained on bench scale in callus cultures of *C. acuminata* (0.2% of dry weight) (Wiedenfeld et al. 1997). In addition to CPT, the culture was also able to yield 10-hydroxy CPT as 0.08% on dry weight. However, such a way of production is still not convenient considering that young leaves contains CPT in an average 0.4% of dry weight. Callus and suspension cultures of *N. foetida and O. pumila* are by far less convenient since they do not accumulate CPT and CPT-related alkaloids (Kitajima et al. 1998). Interestingly, *Agrobacterium rhizogenes*-transformed hairy root cultures of *O. pumila* gave results applicable for industrial development. In fact, such a cultivation produces CPT up to 0.1% of dry weight, and the metabolite does not accumulate in the tissues but is excreted into the culture medium where it can be captured on an adsorbent resin (like Diaion HP-20) (Sudo et al. 2002). *A. rhizogenes* is an attractive experimental system as it consists of long-term aseptic root clones, genetically stable, with growth rates comparable to those of the fastest growing cell suspension cultures. Remarkably, these cultures were able to synthesize the alkaloids at levels equal to, or sometimes greater than, the roots in planta (1.0 mg/g dry weight). Recently, the company ROOTec reported production of CPT and 10-OHCPT from hairy roots of *C. acuminata* (Guillon et al. 2006).

Interestingly, CPT has been found to be produced also by endophytic fungi isolated from *N. foetida*, *C. acuminata*, and *Apodytes dimidiata* (Shweta et al. 2010), which may become a new renewable source of CPT. In an experiment aimed at ascertaining the potential of production, the yield of CPT by fermentation of *Entrophosphora infrequens* was 4.96% of dry cell mass in 48 h in a bioreactor (Amna et al. 2006).

Biotechnology, with the intended cultivation of callus, modified hairy roots and fungi, is a promising source of supply from nonplant origin. The encouraging results mentioned above indicate good opportunities for the future, but the technical know-how and the costs of the product make this way of production still noncompetitive with the extraction from intact plants. Further studies are needed to clarify completely the biosynthesis of CPT, and therefore to modify or force the biosynthetic pathway.

The extraction from intact plant remains the way to be exploited. Lorence and Nessler (Lorence and Nessler 2004) have described two methods to increase the amount of CPT harvested available for derivatization. One way is through the clonal propagation of elite cultivars via shoot bud culture, which ensures that highly productive trees can be rapidly propagated. The other way is the repeated harvest of *C. acuminata* young leaves, without the destruction of the tree. Under greenhouse conditions, a 6 week interval in the harvest of young leaves produced a high amount of CPT per gram of fresh weight.

11.7   CLINICAL APPLICATIONS OF CAMPTOTHECINS

The studies in vivo (Venditto and Simanek 2010) and the clinical application of the CPTs (Garcia-Carbonero and Supko 2002, Zunino and Pratesi 2004) have been discussed in several excellent reviews.

11.8 FUTURE OF CAMPTOTHECIN DERIVATIVES

It is beyond any doubt that CPTs have fundamental importance in clinical practice for the treatment of tumors. After so many years of their discovery, some CPTs are the only first-line treatment for some type of tumors like colon and lung cancer.

Of course, the research of new derivatives or methods to increase the therapeutic index of the drug is ongoing. For antimitotic drugs, like CPTs, selective tumor targeting is a long-standing goal to improve the therapeutic index.

Various methods have been devised to target the drug to the tumor, via covalent or noncovalent attachment to the carrier. Among the noncovalent drug delivery systems developed to improve solubility and lactone stability of CPTs, there are micelles, liposomes, dendrimers, nanoparticle drug formulations, and hydrogels. The covalent conjugation of CPTs to macromolecular architectures has shown great potential for improving pharmacokinetics and increasing tumor efficacy. Most commonly, CPT is attached to the polymer through an ester bond with the 20-hydroxy moiety. This linkage not only conveys solubility through conjugation with a water-soluble polymer, but also improves lactone stability. Some linkages are chosen as specific substrates for enzymatic cleavage, while others are used due to their pH sensitivity, but may also undergo hydrolysis.

The tremendous amount of work dedicated to the targeting of CPTs cannot be discussed here in detail, so that the reader is directed to some recent reviews, for example, the special issue of Advanced Drug Delivery for the use of polymers (Vicent et al. 2009), a review on dendrimers (Tekade et al. 2009), and the recent review by Simanek (Venditto and Simanek 2010) for the application of many of these devices for the in vivo studies of CPTs.

Another strategy for intracellular delivery of CPTs is that of active receptor-mediated tumor targeting systems. In that case, the selective delivery of the drug is driven by a moiety of the antitumor drug able to bind specifically to some receptor expressed by the cancer cell, like, for example, Arg-Gly-Asp (RGD) peptides (Dal Pozzo et al. 2010), which target integrins. The method is theoretically powerful but demands some requirements. First, the moiety that binds to the receptor does not have to interfere with the mechanism of action of the anticancer compound or, in a prodrug-like approach, it has to be cleaved off immediately after cell targeting. Then, the tumor has to over-express the targeting receptor. This could be a great limitation since some tumors can be phenotypically different from patient to patient.

More recently, a new method for selective delivery of anticancer drugs to tumors has been discovered and applied in some cases with success. This methodology is based on the enhanced permeability and retention (EPR) effect of lipids and macromolecules in tumors. According to most scientists, EPR provides a versatile and nonsaturable opportunity for tumor-selective delivery (Matsumura 2009).

The EPR effect is the property by which certain sizes of molecules, typically liposomes, lipids, or macromolecular drugs, tend to accumulate in tumor tissue much more than they do in normal tissues. The general explanation given for this phenomenon is that, in order for tumor cells to grow quickly, they must stimulate the production of blood vessels. Tumor cell aggregates of size as small as 150–200 μm become dependent on blood supply carried out by neovasculature for their nutritional and oxygen supply. These newly formed tumor vessels are usually abnormal in form and architecture. They are poorly aligned defective endothelial cells with wide fenestrations, lacking a smooth muscle layer, or innervation with a wider lumen, and impaired secretion of vascular permeability factors stimulating extravasation within cancer tissue. Furthermore, tumor tissues usually lack effective lymphatic drainage with impaired clearance of the accumulated macromolecules. All these factors will lead to abnormal molecular and fluid transport dynamics, especially for macromolecular drugs.

Several techniques like modification of drug structures and development of drug carriers have been adopted to maximally exploit the EPR effect. Usually, polymeric micelles, especially nanoparticles, able to incorporate anticancer drugs are expected to generate accumulation in solid tumors. For an optimal accumulation in tumor cells versus normal cells, the size of the micelles have to be in the range of 20–100 nm so as to ensure that they do not penetrate normal vessel walls. The distribution of the drug through EPR effect is expected to decrease the incidence of drug-induced side effects owing to reduced drug distribution in normal tissues.

Recently, two drugs based on the natural antimitotics doxorubicin and paclitaxel have been registered and launched under the name of Doxil® and Abraxane®, respectively. The SN-38 PEG-conjugate NKTR-102 (Von Hoff et al. 2008) and the SN-38-incorporating polymeric micelles NK-012 (Saito et al. 2010) look very promising and are in Phase II of clinical development.

A strategically interesting way to achieve site-selective delivery is by chemical attachment of therapeutic agents to ultrasmall superparamagnetic iron oxide nanoparticles (USPIO) via a cell-specific labile linkage, and steering the drug–USPIO assembly to specific diseased areas in the body under the influence of an external magnetic field. This implies that the drug–nanoparticle assembly must be internalized by cells, that this uptake be enhanced by an external magnetic field, and subsequently, that the drug be released intracellularly to exert its expected therapeutic effects. The design, preparation, and evaluation of magnetic nanoparticles displaying drugs at their surface are only in their early phase of development, but an application to CPTs has already been reported (Cengelli et al. 2009).

REFERENCES

Adams, V.R. and T.G. Burke, eds. 2005. *Camptothecins in Cancer Therapy*. Totowa: Humana Press.

Aimi, N., Hoshino, H., Nishimura, M., Sakai, S.I., and J. Haginiwa. 1990. Chaboside, first natural glycocamptothecin found from *Ophiorriza pumila*. *Tetrahedron Lett.* 31:5169–5172.

Amna, T., Puri, S.C., Verma, V., Sharma, J.P., Khajuria, R.K., Musarrat, J., Spiteller, M., and G.N. Qazi. 2006. Bioreactor studies on the endophytic fungus *Entrophospora infrequens* for the production of an anticancer alkaloid camptothecin. *Can. J. Microbiol.* 52:189–196.

Bennasar, M.-L., Zulaica, E., Juan, C., Alonso, Y., and J. Bosch. 2002. Addition of ester enolates to *N*-alkyl-2-fluoropyridinium salts: Total synthesis of (±)-20-deoxycamptothecin and (+)-camptothecin. *J. Org. Chem.* 67:7465–7474.

Cane, D.E., ed. 1999. *Comprehensive Natural Product Chemistry*, vol. 2. Amsterdam, the Netherlands: Elsevier, pp. 1–13 and 45–67.

Cengelli, F., Grzyb, J.A., Montoro, A., Hofmann, H., Hanessian, S., and L. Juillerat-Jeanneret. 2009. Surface-functionalized ultrasmall superparamagnetic nanoparticles as magnetic delivery vectors for camptothecin. *ChemMedChem* 4:988–997.

Ciufolini, M.A. and F. Roschangar. 1997. Practical total synthesis of (+)-camptothecin: The full story. *Tetrahedron* 53:1049–1060.

Comins, D.L. and J.M. Nolan. 2001. A practical six-step synthesis of (*S*)-camptothecin. *Org. Lett.* 3:4255–4257 and refs. quoted therein.

Dal Pozzo, A., Ni, M.-H., Esposito, E., Dallavalle, S., Musso, L., Bargiotti, A., Pisano, C., Vesci, L., Bucci, F., Castorina, M., Foderà, R., Giannini, G., Aulicino, C., and S. Penco. 2010. Novel tumor-targeted RGD peptide–camptothecin conjugates: Synthesis and biological evaluation. *Bioorg. Med. Chem.* 18:64–72.

Dallavalle, S. and L. Merlini. 2008. Camptothecin and analogs: Structure and synthetic efforts. In *Modern Alkaloids*, Fattorusso, E. and O. Taglialatela-Scafati, eds. Weinheim, Germany: Wiley-VCH, pp. 503–520.

Du, W. 2003. Towards new anticancer drugs: A decade of advances in synthesis of camptothecins and related alkaloids. *Tetrahedron* 59:8649–8687.

Ejima, A., Terasawa, H., Sugimori, H., and H. Tagawa. 1990. Antitumour agents. Part 2. Asymmetric synthesis of (*S*)-camptothecin. *J. Chem. Soc., Perkin Trans.* I:27–31.

Fan, Y., Shi, L.M., Kohn, K.W., Pommier, Y., and J.N. Weinstein. 1998. Quantitative structure-antitumor activity relationships of camptothecin analogues: Cluster analysis and genetic algorithm-based studies. *J. Med. Chem.* 41:2216–2226.

Fortunak, J.M.D., Kitteringham, J., Mastrocola, A.R., Mellinger, M., Sisti, N.J., Wood, J.L., and Z.-P. Zhuang. 1996. Novel syntheses of camptothecin alkaloids, part 2. Concise synthesis of (*S*)-camptothecins. *Tetrahedron Lett.* 37:5683–5686.

Garcia-Carbonero, R. and J.G. Supko. 2002. Current perspectives on the clinical experience, pharmacology, and continued development of the camptothecins. *Clin. Cancer Res.* 8:641–661.

Guillon, S., Trémouillaux-Guiller, J., Pati, P.K., Rideau, M., and P. Gantet. 2006. Hairy root research: Recent scenario and exciting prospects. *Curr. Opin. Plant Biol.* 9:341–346.

Hautefaye, P., Cimetière, B., Pierré, A., Léonce, S., Hickman, J., Laine, W., Bailly, C., and G. Lavielle. 2003. Synthesis and pharmacological evaluation of novel non-lactone analogues of camptothecin. *Bioorg. Med. Chem. Lett.* 13:2731–2735.

Horwitz, S.B., Chang, C.-K., and A.P. Grollman. 1971. Studies on camptothecin: I. Effects on nucleic acid and protein synthesis. *Mol. Pharmacol.* 7:632–644.

Horwitz, S.B., Chang, C.K., and A.P. Grollman. 1972. Antiviral Action of Camptothecin. *Antimicrob. Agents Chemother.* 2:395–401.

Horwitz, M.S. and S.B. Horwitz. 1971. Intracellular degradation of HeLa and adenovirus type 2 DNA induced by camptothecin. *Biochem. Biophys. Res. Comm.* 45:723–727.

Horwitz, S.B. and M. Horwitz. 1973. Effects of camptothecin on the breakage and repair of DNA during the cell cycle. *Cancer Res.* 33:2834–2836.

Hsiang, Y., Hertzberg, R., Hecht, S., and L.F. Liu. 1985. Camptothecin induces protein-linked DNA breaks via mammalian DNA topoisomerase I. *J. Biol. Chem.* 260:14873–14888.

Hutchinson, C.R., Heckendorf, A.H., Daddona, P.E., Hagaman, E., and E. Wenkert. 1974. Biosynthesis of camptothecin. I. Definition of the overall pathway assisted by carbon-13 nuclear magnetic resonance analysis. *J. Am. Chem. Soc.* 96:5609–5611.

Hutchinson, C.R., Heckerdorf, A.H., Straughn, J.L., Daddona, P.E., and D.E. Cane. 1979. Biosynthesis of camptothecin. 3. Definition of strictosamide as the penultimate biosynthetic precursor assisted by carbon-13 and deuterium NMR spectroscopy. *J. Am. Chem. Soc.* 101:3358–3369.

Jew, S.S., Ok, K.D., Kim, H.J., Kim, M.G., Kim, J.M., Hah, J.M., and Y.S. Cho. 1995. Enantioselective synthesis of 20(S)-camptothecin using sharpless catalytic asymmetric dihydroxylation. *Tetrahedron Asymm.* 6:1245–1248.

Josien, H., Ko, S.B., Bom, D., and D.P. Curran. 1998. A general synthetic approach to the (20S)-camptothecin family of antitumor agents by a regiocontrolled cascade radical cyclization of aryl isonitriles. *Chem. Eur. J.* 4:67–83 and refs. quoted therein.

Kerrigan, J.E. and D.S. Pilch. 2001. A structural model for the ternary cleavable complex formed between human topoisomerase I, DNA, and camptothecin. *Biochemistry* 40:9792–9798.

Kitajima, M., Fischer, U., Nakamura, M., Ohsawa, M., Ueno, M., Takayama, H., Unger, M., Stöckigt, J., and N. Aimi. 1998. Anthraquinones from *Ophiorrhiza pumila* tissue and cell cultures. *Phytochemistry* 48:107–111.

Lavergne, O., Lesueur-Ginot, L., Rodas, F.P., Kasprzyk, P.G., Pommier, J., Demarquay, D., Prevost, G., Ulibarri, G., Rolland, A., Schiano-Liberatore, A., Harnett, J., Pons, D., Camara, J., and D.C.H. Bigg. 1998. Homocamptothecins: Synthesis and antitumor activity of novel E-ring-modified camptothecin analogues. *J. Med. Chem.* 41:5410–5419.

Li, S. 2002. *Camptotheca* lowreyana tree named 'Katie' U.S. Pat. 2002 0018762.

Li, S., Yi, Y., Wang, Y., Zhang, Z., and R.S. Beasley. 2002. Camptothecin accumulation and variations in *Camptotheca*. *Planta Med.* 68:1010–1016.

Liebskind, D., Horwitz, S.B., Horwitz, M.S., and K.C. Hsu. 1974. Immunoreactivity to antinucleoside antibodies in camptothecin treated HeLa cells. *Expert Cell Res.* 86:174–178.

Liehr, J.G., Giovanella, B.C., and C.F. Verschraegen, eds. 2000. The camptothecins: Unfolding their anticancer potential. *Ann. NY Acad. Sci.* 922:1–360.

Lopez-Meyer, M., Nessler, C.L., and T.D. Mcknight. 1994. Sites of accumulation of the antitumor alkaloid camptothecin in *Camptotheca acuminata*. *Planta Med.* 60:558–560.

Lorence, A. and C.L. Nessler. 2004. Camptothecin, over four decades of surprising findings. *Phytochemistry* 65:2735–2749.

Matsumura, Y. 2009. NK-012. *Drugs Future* 34:276–281.

Pantaziz, P. and B.C. Giovanella, eds. 1996. The camptothecins: From discovery to patient. *Ann. NY Acad. Sci.* 803:1–335.

Paull, K.D., Shoemaker, R.H., Hodes, L., Monks, A., Scudiero, D.A., Rubinstein, L., Plowman, J., and M. Boyd. 1989. Display and analysis of patterns of differential activity of drugs against human tumor cell lines: Development of mean graph and COMPARE algorithm. *J. Natl. Cancer Inst.* 81:1088–1092.

Pizzolato, J.F. and L.B. Saltz. 2003. The camptothecins. *Lancet* 361:2235–2242 and references cited therein.

Pommier, Y. 2009. DNA topoisomerase I inhibitors: Chemistry, biology, and interfacial inhibition. *Chem. Rev.* 109:2894–2902.

Potmesil, M. and H. Pinedo, eds. 1995. *Camptothecins: New Anticancer Agents*. Boca Raton, FL: CRC Press.

Redinbo, M.R., Stewart, L., Kuhn, P., Champoux, J.J., and W.G. Hol. 1998. Crystal structures of human topoisomerase I in covalent and noncovalent complexes with DNA. *Science* 279:1504–1513.

Saito, Y., Yasunaga, M., Kuroda, J.-I., Koga, Y., and Y. Matsumura. 2010. Antitumour activity of NK012, SN-38-incorporating polymeric micelles, in hypovascular orthotopic pancreatic tumour. *Eur. J. Cancer* 46:650–658.

Samorì, C., Beretta, G.L., Varchi, G., Guerrini, A., Di Micco, S., Basili, S., Bifulco, G., Riccio, R., Moro, S., Bombardelli, E., Zunino, F., and Fontana, G. 2010. Structure-activity relationship study of 16a-thiocamptothecins: An integrated in vitro and in silico approach. *ChemMedChem* 5:2006–2015.

Sawada, S., Okajima, S., Aiyama, R., Nokata, K., Furuta, T., Yokokura, T., Sugino, E., Yamaguchi, K., and T. Miyasaka. 1991. Synthesis and antitumor activity of 20(S)-camptothecin derivatives: Carbamate-linked, water-soluble derivatives of 7-ethyl-10-hydroxycamptothecin. *Chem. Pharm. Bull.* 39:1446–1450.

Shamma, M. 1968. A numbering system for camptothecin based on its biogenesis. *Experientia* 24:107.

Shoemaker, R.H. 2006. The NCI60 human tumour cell line anticancer drug screen. *Nat. Rev. Cancer* 6:813–823.

Shweta, S., Zuehlke, S., Ramesha, B.T., Priti, V., Kumar, P.M., Ravikanth, G., Spiteller, M., Vasudeva, R., and R.U. Shaanker. 2010. Endophytic fungal strains of *Fusarium solani,* from *Apodytes dimidiata* E. Mey. ex Arn (Icacinaceae) produce camptothecin, 10-hydroxycamptothecin and 9-methoxycamptothecin. *Phytochemistry* 71:117–122, and refs. quoted therein.

Sirikantaramas, S., Yamazaki, M., and K. Saito. 2009. A survival strategy: The coevolution of the camptothecin biosynthetic pathway and self-resistance mechanism. *Phytochemistry* 70:1894–1898.

Staker, B.L., Hjerrild, K., Feese, M.D., Behnke, C.A., Burgin Jr., A.B., and L. Stewart. 2002. The mechanism of topoisomerase I poisoning by a camptothecin analog. *Proc. Natl. Acad. Sci.* 99:15387–15392.

Sudo, H., Yamakawa, T., Yamazaki, M., Aimi, N., and K. Saito. 2002. Bioreactor production of camptothecin by hairy root cultures of *Ophiorrhiza pumila. Biotechnol. Lett.* 24:359–363.

Tagami, K., Nakazawa, N., Sano, S., and Y. Nagao. 2000. Asymmetric syntheses of (+)-camptothecin and (+)-7-ethyl-10-methoxycamptothecin. *Heterocycles.* 53:771–775.

Tekade, R.K., Kumar, P.V., and N.K. Jain. 2009. Dendrimers in oncology: An expanding horizon. *Chem. Rev.* 109:49–87.

Thomas, C.J., Rahier, N.J., and S.M. Hecht. 2004. Camptothecin: Current perspectives. *Bioorg. Med. Chem.* 12:1585–1604.

Venditto, V.J. and E.E. Simanek. 2010. Cancer therapies utilizing the camptothecins: A review of the *in vivo* literature. *Mol. Pharm.* 7:307–349.

Verma, R.P. and C. Hansch. 2009. Camptothecins: A SAR/QSAR study. *Chem. Rev.* 109:213–235.

Vicent, M.J., Ringsdorf, H., and R. Duncan. 2009. *Adv. Drug. Deliv. Rev.* 61:1117–1120 and following papers.

Von Hoff, D.D., Jameson, G.S., Borad, M.J., Rosen, L.S., Utz, J., Dhar, S., Acosta, L., Barker, T., Walling, J., and J.T. Hamm. 2008. First phase I trial of NKTR-102 (PEG-irinotecan) reveals early evidence of broad anti-tumor activity in 595 three schedules. *20th EORTC-NCI-AACR Symp.*, Geneva, Switzerland, poster # 595.

Wall, M.E., Wani, M.C., Cooke, C.E., Palmer, K.T., McPhail, A.T., and G.A. Sim. 1966. Plant antitumor agents. I. The isolation and structure of camptothecin, a novel alkaloidal leukemia and tumor inhibitor from *Camptotheca acuminata. J. Am. Chem. Soc.* 88:3888–3890.

Wall, M.E., Wani, M.C., Natschke, S.M., and A.W. Nicholas. 1986. Plant antitumor agents. 22. Isolation of 11-hydroxycamptothecin from *Camptotheca acuminata* decne: Total synthesis and biological activity. *J. Med. Chem.* 29:1553–1555.

Wang, C.Y., Buta, J.G., Moline, H.E., and H.W. Hruschka. 1980. Potato sprout inhibition by camptothecin, a naturally occurring plant growth regulator. *J. Am. Soc. Hort. Sci.* 105:120–124.

Wani, M.C., Nicholas, A.W., Manikumar, G., and M.E. Wall. 1987. Plant antitumor agents. 25. Total synthesis and antileukemic activity of ring A substituted camptothecin analogs. Structure–activity correlations. *J. Med. Chem.* 30:1774–1779.

Wiedenfeld, H., Furmanowa, M., Roeder, E., Guzewska, J., and W. Gustowski. 1997. Camptothecin and 10-hydroxycamptothecin in callus and plantlets of *Camptotheca acuminata. Plant Cell Tissue Org. Cult.* 49:213–218.

Yamazaki, Y., Kitajima, M., Arita, M., Takayama, H., Sudo, H., Yamazaki, M., Aimi, N., and K. Saito. 2004. Biosynthesis of camptothecin. In silico and in vivo tracer study from [1-^{13}C]glucose. *Plant Physiol.* 134:161–170.

Yamazaki, Y., Urano, A., Sudo, H., Kitajima, M., Takayama, H., Yamazaki, M., Aimi, N., and K. Saito. 2003. Metabolite profiling of alkaloids and strictosidine synthase activity in camptothecin producing plants. *Phytochemistry* 62:461–470.

Zunino, F. and G. Pratesi. 2004. Camptothecins in clinical development. *Exp. Opin. Investig. Drugs* 13:269–284.

12 Plant Compounds and Derivatives as Inhibitors of Cancer Cell Multidrug Resistance

Virginia Lanzotti, Orazio Taglialatela-Scafati, Ernesto Fattorusso, and Attilio Di Pietro

CONTENTS

12.1 INTRODUCTION

Cancer cells often become resistant to structurally unrelated chemotherapeutics by overexpressing within their plasma membranes adenosine-5'-triphosphate (ATP)-binding cassette (ABC) transporters behaving as drug-efflux pumps at the expense of ATP hydrolysis. Three main transporters have been successively identified to confer such a multidrug-resistance (MDR) phenotype: first, P-glycoprotein, also called MDR1 (Endicott and Ling 1989), then MRP1 ("multidrug resistance-[associated] protein 1") (Cole et al. 1992), and more recently BCRP ("breast cancer resistance protein") (Doyle et al. 1998, Miyake et al. 1999).

These different multidrug transporters belong to the same ABC family since they contain the seven consensus motifs involved in ATP binding: Walker-A and Walker-B, ABC signature (or motif C), A-loop, Q-loop, D-loop, and H-loop. They also display a catalytic glutamate immediately after the Walker-B. These motifs are located within a cytosolic nucleotide-binding domain. The minimal functional unit is composed of two nucleotide-binding domains (NBDs) associated to two membrane domains, each containing six transmembrane alpha-helical spans. Differences in topology and sequence of the different domains explain why the three transporters belong to different classes of the HUGO (HUman classification Genome Organization): B for P-glycoprotein/MDR1 (ABCB1), C for MRP1 (ABCC1), and G for BCRP (ABCG2).

Since the ABCB1 substrates are largely hydrophobic, access to their drug-binding site(s) seemingly takes place from the lipid bilayer rather than the aqueous phase. Accordingly, both transport and binding studies suggest that ABCB1 intercepts ligands from the inner leaflet of the membrane, extruding them directly into the extracellular medium through a "hydrophobic vacuum cleaner" process (Bolhuis et al. 1997), or alternatively with a "flippase" mechanism (Higgins 1994).

These multidrug transporters are polyspecific, and ABCB1 and ABCG2 display highly overlapping patterns of drug substrates including epipodophyllotoxins, camptothecins, anthracyclines, and other anthraquinones such as mitoxantrone (Szakács et al. 2006). Interestingly, the transporters appear much more specific for inhibitors. Since ABCB1 was identified first, considerable efforts have been made to find out efficient inhibitors: series of first-generation inhibitors previously known for their calcium antagonist (verapamil, nifedipine, dexniguldipine), immunosuppressive (cyclosporin A), antiarythmic (amiodarone), or antiparasitic (quinine, quinidine) activity and second-generation derivatives, allowed to model pharmacophores finally leading to the synthesis of several third-generation leads, such as LY335979 (Zosuquidar), GF120918 (Elacridar), R-101933 (Laniquidar), VX-710 (Biricodar), and XR-9576 (Tariquidar), which are under advanced phase 2/3 clinical trials (Bates et al. 2002, Bardelmeijer et al. 2004) (Figure 12.1). While new ABCB1 inhibitors will essentially bring additional information about the molecular and cellular mechanisms of interaction and inhibition, potent ABCG2-specific inhibitors still need to be found out in order to completely abolish the MDR phenotype related to drug efflux by the different transporters.

ABCC1 effluxes different substrates, which are anionic and conjugated, mainly with glutathione such as leukotriene C4 and aflatoxin, or alternatively with

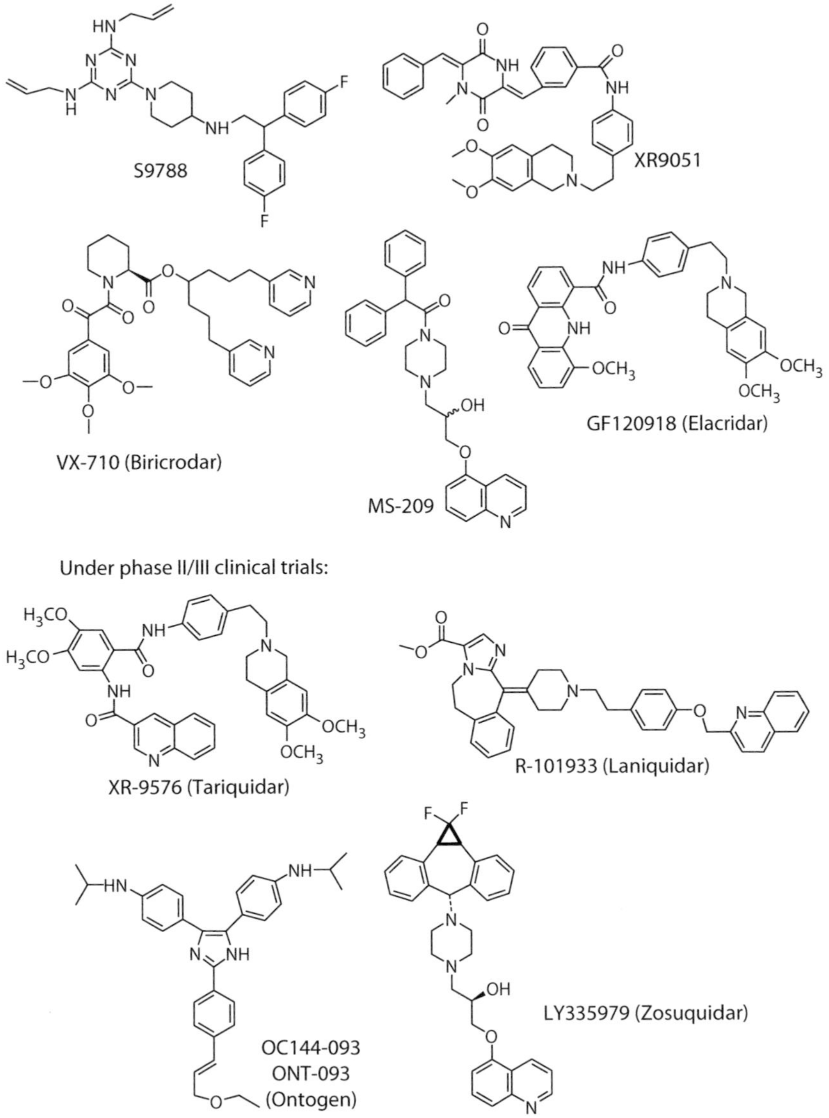

FIGURE 12.1 Second- and third-generation inhibitors of ABCB1, designed from 3D-QSARs.

glucuronide or sulfate. Such a dependence on charged substrates makes difficult the identification of competitive inhibitors, displaying low membrane diffusion and therefore limited accessibility to the transporter. Interestingly, hydrophobic compounds like vincristine may also be transported, by cotransport with glutathione (Cole and Deeley 2006). Hydrophobic, penetrating compounds, such as verapamil, are able, maybe by mimicking vincristine but without being transported, to stimulate glutathione efflux producing a fast and massive intracellular glutathione depletion, and inducing a selective apoptosis of ABCB1-positive cells (Trompier et al. 2004,

Boumendjel et al. 2005, Laberge et al. 2007). This phenomenon is highly stereospecific since only S-verapamil is efficient (Perrotton et al. 2007); it might constitute
a new therapeutic strategy to selectively eliminate ABCB1-expressing multidrug
resistant cancer cells without requiring any treatment by conventional chemotherapy.

The physiological role of these ABC transporters, especially ABCB1 and
ABCG2, is critical for protection and detoxification, especially in sensitive tissues
such as brain (within the blood-brain barrier), testis (blood-testis barrier), and fetus
(placenta), as well as stem cells (plasma membrane). They also play an important
role of excretion at the apical membrane of polarized cells (hepatocytes, kidney,
enterocytes) (Van Herwaarden and Schinkel 2006). Despite initial controversies, the
clinical relevance of ABCB1 overexpression has been clearly demonstrated in adult
acute myeloid leukemia (AML) (Benderra et al. 2004). More recently, ABCG2 was
also shown to be clinically involved in both childhood and adult AMLs (Steinbach
and Legrand 2007, Ho et al. 2008). The involvement of ABCC1 is clearly evident
in lung cancer, related to both small cells (SCLC) and non-small cells (NSCLC)
(Berger et al. 2005, Cole and Deeley 2006).

It is therefore clear that inhibitory compounds able to efficiently overcome multidrug resistance of tumor cells will be considered as valuable drug candidates of
clinical interest for cancer-suffering patients.

12.2   NATURAL COMPOUNDS AS ABCB1 INHIBITORS

12.2.1   FLAVONOIDS

Flavonoids have been extensively studied as modulators of ABCB1 and other multidrug ABC transporters, as summarized in various reviews (Boumendjel et al. 2002,
Di Pietro et al. 2002, Deferme and Augustijns 2003, Morris and Zang 2006, Aszalos
2008, Alvarez et al. 2009, Bansal et al. 2009). Some examples of commonly used
classes and substituents are illustrated in Figure 12.2.

Early contradictory results were reported for flavonoid effects on multidrug
resistance and its reversal, depending on the type of cancer cells and the chemotherapeutic drug used. Flavonols such as quercetin, kaempferol, and galangin were
reported to increase adriamycin efflux from HCT-15 colon cancer cells (Critchfield
et al. 1994), whereas a hydrophobic quercetin derivative inhibited rhodamine 123
efflux from MCF-7 breast cancer cells and abolished their multidrug resistance
phenotype (Scambia et al. 1994), and various flavonols inhibited drug efflux from
ABCB1-overexpressing hepatocytes (Chieli et al. 1995). The isoflavone genistein, at
higher concentration, was also reported to inhibit drug efflux (Castro and Altenberg
1997). Such a discrepancy might be, at least partly, related to multiple cellular targets. An alternative possibility could be the existence of different flavonoid-binding
sites within the same multidrug transporter. Indeed, quercetin was found not only
to inhibit ABCB1 ATPase activity (Shapiro and Ling 1997a), but also to either prevent the binding of transported drugs such as colchicin or Hoechst 33342 or, on
the contrary, to activate the binding of rhodamine 123 (Shapiro and Ling 1997b),
further demonstrating the involvement of several interacting binding-sites for drugs
and modulators.

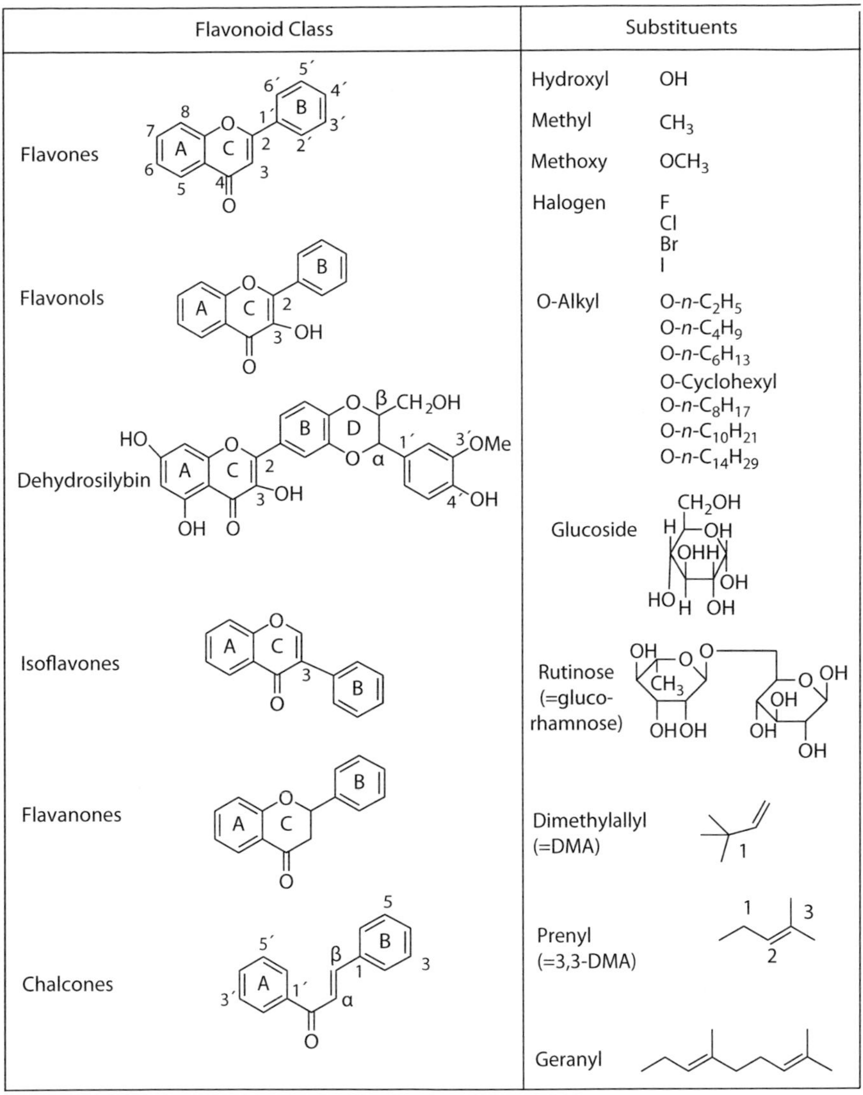

Flavonoid Class	Substituents		

FIGURE 12.2 Different classes and substituents of flavonoids studied as ABCB1 modulators.

12.2.1.1 Molecular Interactions with ABCB1 and Other Multidrug Transporters

We have investigated a large number of flavonoids for their direct binding either to the NBD2 cytosolic domain of mouse P-glycoprotein or Leishmania Ltrmdr1 or to the whole Pdr5p yeast transporter by quenching of the protein intrinsic fluorescence essentially due to tryptophan residues. An extensive study was performed with mouse ABCB1 NBD2. First, different classes of flavonoids were investigated for their binding ability, as estimated by the determined K_D while the maximal quenching of fluorescence was generally high (80%–100%). The following sequence in affinity was obtained, dehydrosilybin > chalcone > flavonol > flavone > isoflavone > flavanone, when comparing, for example, dehydrosilybin to 2′,4′,6′-triOH-chalcone, galangine or kaempferol, chrysin or apigenin, genistein, and naringenin or silybin, respectively (Bois et al. 1998, Conseil et al. 1998, Maitrejean et al. 2000). Second, inside the same class, the following efficiency of substituents was observed: alkoxyl, geranyl > dimethylallyl > halogen > monolignol > methoxy > hydroxyl > glycosyl. Hydrophobicity of the substituents was an important parameter since (1) alkoxylation up to 8–10 carbon atoms gradually increased the chalcone binding affinity (Bois et al. 1999), (2) geranylation was more efficient than prenylation in both chrysin and dehydrosilybin (Maitrejean et al. 2000, Comte et al. 2001), and (3) halogens were better than H or OH in chalcones, and the sequence I > Br > Cl > F correlated the degree of lipophilicity (Bois et al. 1999). The positive effects of n-octyl and iodine substituents were also observed in galangin derivatives (Boumendjel et al. 2001). The positive effects of prenylation were observed in all classes, including flavones (Comte et al. 2001) and chalcones (Daskiewicz et al. 1999), as well as in other polyphenol compounds such as xanthones (Noungoué Tchamo et al. 2000). Other hydrophobic substituents were studied, such as isopropyl and benzyl, but gave more variable effects (Comte et al. 2001). A positive effect was produced by the high-size monolignol unit, when comparing silybin to taxifolin and dehydrosilybin to galangin (Maitrejean et al. 2000). Slightly positive effects were produced by methyl/methoxy groups, for example, at either position 6, 7, or 8 of chrysin, position 4′ of galangin, or position 4 of chalcone, but not at position 3 of galangin. The hydroxyl groups appeared to be important at position 3, when comparing flavonols to flavones, and at position 5 (Conseil et al. 1998). In contrast, all forms of glycosylation at different positions dramatically altered the binding affinity, as observed for rutin and for apigenin-7-O-glucoside and vitexin as well. Molecular models of flavonoidic inhibitors were recently constructed by both ligand-based drug design from 3D quantitative structure–activity relationship (QSARs) of a number of flavonoids and derivatives (Boccard et al. 2009), and structure-based drug design by inhibitors docking into a model NBD2 structure (Badhan and Penny 2006).

Similar structure–activity relationships (SARs) were obtained with parasite Ltrmdr1 NBD2, despite a lower solubility of the recombinant protein (Pérez-Victoria et al. 1999). In addition, the 1,1-dimethylallyl (DMA) isomer appeared to be better than the 3,3- one (prenyl) in both chrysin and galangin, prenylation at position 8 was found slightly more efficient than at position 6, and prenylation appeared better than geranylation (Pérez-Victoria et al. 2001). The prenylation effects produced in full-length Pdr5p from yeast were qualitatively comparable, but quantitatively much lower (Conseil et al. 2000). This might be attributable, at least partly, to the presence of residual detergent, 0.02% n-dodecyl β-D-maltoside, required to keep the transporter soluble, which is expected to lower protein interactions with hydrophobic ligands.

Both N-terminal NBD1 (Dayan et al. 1997) and C-terminal NBD2 (Conseil et al. 1998) from ABCB1, as well as NBD2 from Ltrmdr1 (Pérez-Victoria et al. 1999), contain a region interacting with hydrophobic steroid derivatives such as RU486. This region is probably located in close proximity to the ATP-binding site since RU486 completely prevented or displaced the hydrophobic nucleotide derivative 2′(3′)-methylanthraniloyl-ATP (MANT-ATP) (Dayan et al. 1997). The binding of kaempferide to ABCB1 NBD2 was partly prevented by preincubation with either ATP or RU486, or additively by both ATP and RU486, which suggests that kaempferide displays bifunctional interactions with the ATP-binding site and the hydrophobic steroid-interacting region (Conseil et al. 1998). The binding of flavonoids to the ATP-binding site was also monitored by studying their ability to prevent photoaffinity labeling by [γ-^{32}P]TNP-8-azido-ATP. Kaempferide indeed bound to the ATP-binding site since it prevented NBD2 photolabeling with an IC$_{50\%}$ value similar to the K$_D$ of direct binding (De Wet et al. 2001). However, among a set of 29 flavonoids tested, only three, all being flavonols, were found to bind to the ATP site. This indicates that the hydroxyl group at position 3 is critical; the higher IC$_{50\%}$ value for dehydrosilybin as compared to the K$_D$ of direct binding suggests that extension of B-ring is possible but that the large size of the monolignol unit might produce some steric hindrance. In addition, the 2–3 double bond was critical since silybin was inefficient. These requirements are similar to those observed for quercetin binding to the Hck tyrosine kinase as demonstrated by cocrystallization (Sicheri et al. 1997) and to other ATPases by inhibition kinetics (Murakami et al. 1999). In contrast, they differ from those concerning the cyclin-dependent kinase CDK2, the crystal structure of which was determined with bound chrysin derivatives (De Azevedo et al. 1996). Interestingly, hydrophobic substitution by prenylation at either position 6 or 8, which considerably increased the binding affinity for ABCB1 NBD2, shifted flavonol binding outside the ATP-binding site, since it no longer competed with nucleotides. It is likely that the binding of prenyl-flavonols might better overlap the hydrophobic steroid-interacting region than the binding of unsubstituted ones.

The effects of flavonoids on nucleoside triphosphate hydrolysis was studied on the yeast Pdr5p transporter (Conseil et al. 2000) within enriched plasma membranes. Chrysin and quercetin behaved as poor inhibitors, but flavonoid prenylation markedly increased the efficiency of inhibition, up to an IC$_{50\%}$ value of 4.9 µM for 6-prenyl-galangin toward uridine 5′-triphosphatase (UTPase) activity; the inhibition appeared to be noncompetitive. In the case of cyclic adenosine 5′-triphosphate (AMP)-dependent kinase, a prenylated derivative, waranglone, also behaved as a much more potent inhibitor than the corresponding unsubstituted flavonoid, and produced a noncompetitive inhibition with respect to ATP (Wang et al. 1997).

The same flavonoids and prenylated derivatives were very efficient for inhibiting the energy-dependent interaction of rhodamine 6G with the Pdr5p-enriched plasma membranes (Conseil et al. 2000). Here also, the prenylated derivatives were much potent inhibitors than the unsubstituted flavonoids for chrysin, kaempferide, 3-methylgalangin, and galangin. The best affinity was also obtained for 6-prenyl-galangin, with an IC$_{50\%}$ value of 0.24 µM, indicating a 200-fold higher inhibition efficiency than for ATPase activity. In addition, the inhibition was always found to be competitive with respect to rhodamine 6G. Therefore, prenyl-flavonoids exhibit a marked preference for interacting at the drug-binding site(s) of Pdr5p over the ATP-binding site(s).

12.2.1.2 Cellular Effects of Flavonoids

Prenyl-flavonoids produced an efficient inhibition of ABCB1-mediated drug efflux within leukemic K562/R7 cells (Comte et al. 2001), as monitored by flow cytometry. The highest effect was produced by 8-prenyl-chrysin, but significant effects were also produced by 6-prenyl, 8-geranyl, 6-geranyl, and 6,8-diprenyl derivatives. Clearly, although hydrophobicity was a critical parameter for both binding affinity toward NBD2 and inhibition of cellular ABCB1 activity, other determinants were also important for the inhibition. Indeed, the most efficient inhibitor, 8-prenyl-chrysin, is a moderately hydrophobic compound, displaying a high-affinity binding to, and a high maximal fluorescence quenching (>80%) of, ABCB1 NBD2. In contrast, more hydrophobic compounds producing a lower inhibition, such as isopropyl-, diprenyl-, and geranyl- derivatives, probably bound differently to NBD2 since the maximal quenching was significantly lower (Comte et al. 2001). Hydrophobic substitution at position 7 with an *N*-benzylpiperazine chain was also found to increase the potency of ABCB1 inhibition, and the overall lipophilicity was also concluded to be an important, although not a lone, determinant (Ferté et al. 1999).

Prenylation was also critical in vivo to improve flavonoid inhibition in the case of the *Leishmania tropica* multidrug transporter Ltrmdr1 since 8-DMA-kaempferide was a better modulator than either cyclosporin A or verapamil (Pérez-Victoria et al. 1999), whereas apigenin produced a significant but much lower effect, and no inhibition at all was observed with rutin. Therefore, prenylation appears to be important for both increasing the binding affinity toward the cytosolic domain of the parasite transporter and inhibiting drug efflux to the extracellular medium. Finally, prenylation was quite determinant for chemosensitizing the parasite growth (Pérez-Victoria et al. 2001) to the presence of cytotoxic drugs such as daunomycin at high concentration. This was true for any flavonoid tested such as chrysin, galangin, or dehydrosilybin, as well for DMA, prenyl, and geranyl, at either position 6 or 8. The highest effect was produced by 8-prenyl-dehydrosilybin at 10 µM, since almost no growth inhibition was observed for the wild-type strain, whereas a marked inhibition was observed with the drug-resistant strain. However, the DMA isomer substituent was more efficient in chrysin and galangin, suggesting that DMA-dehydrosilybin would be expected to behave as an even stronger chemosensitizer. Unsubstituted dehydrosilybin appeared to be highly cytotoxic, as compared to silybin; this might be attributable to its ability to interact with ATP-binding sites that are present on a number of other cellular targets. Therefore, we observed a strong correlation between the affinity of in vitro binding to the *Leishmania* Ltrmdr1 NBD2, and the efficiency of both in vivo modulation of drug accumulation and reversion of the resistant phenotype in the MDR *L. tropica* line (Pérez-Victoria et al. 1999, 2001, 2006).

A tentative mechanism for the interaction of flavonoids with ABCB1 and related multidrug transporters was proposed (Di Pietro et al. 2002). Unsubstituted flavonols, such as galangin, kaempferol, kaempferide, or dehydrosilybin, appear to bifunctionally interact with cytosolic NBDs: the hydroxyl groups at positions 3 and 5, in addition to the ketone at position 4, would bind to the ATP-binding site while other parts of the molecule would bind to a vicinal region also able to interact with hydrophobic steroid derivatives. Prenylation at either position 6 or 8 of the A-ring would increase hydrophobic interactions with both the cytosolic steroid-interacting region and the membrane drug-binding site of the full-length transporter. This would produce a significant shift in flavonoid positioning, in such a way that overlapping of the ATP-binding site would no longer occur. Such a prenyl-flavonoid positioning appears to be quite efficient to directly inhibit drug binding and transport, and possibly to indirectly interfere with ATP hydrolysis and/or energy transduction. In this way, prenyl-flavonoids appear to be promising modulators of multidrug resistance, mediated by either ABCB1 in cancer cells or related ABC transporters in various species.

Since ABCB1 is located in the apical membrane of a number of epithelial cells, including jejunum and colon, its activity is recognized to limit the oral absorption of drugs by mediating their secretion from blood to intestinal lumen (Hunter and Hirst 1997, Van Asperen et al. 1998). The herbal modulation of ABCB1 by curcumin, ginsenosides, piperine, catechins, silymarin and natural flavonols such as kaempferol, quercetin, and galangin (Zhou et al. 2004), as well as by biochanin A and naringenin (Chung et al. 2005) has therefore been questioned (Brand et al. 2006) and recently reviewed (Aszalos 2008). Flavonoids activity has been observed in vivo in animal studies: naringenin increased the oral bioavailability of quinine and baicalein of cyclosporine (Lai et al. 2004) in rats; quercetin subcutaneously increased moxidectin bioavailability in lambs (Dupuy et al. 2003), and the oral bioavailability of paclitaxel in rats (Choi et al. 2004) and of digoxin in pigs (Wang et al. 2004b), which however resulted in serious toxicity due to pharmacokinetic interactions.

Drug bioavailability also depends on the activity of Cytochrome P450 (CYP), phase I enzymes, which are expressed in the intestine (Lin et al. 1999). Interestingly, ABCB1 and CYPs share many substrates (Wacher et al. 1995) and inhibitors (Lin et al. 1999), and appear to be both regulated by similar compounds (Schuetz et al. 1996). This raised the question whether some diet components could act as ABCB1 and CYP inhibitors, and enhance drug bioavailability (Evans 2000). An increased concentration of many drugs has been demonstrated when coadministrated with grapefruit juice. However, this might be due to inhibition of either CYP activity, presumably by naringenin (Fuhr and Kummert 1995), or ABCB1 by another compound (Takanaga et al. 1998). On the contrary, it has been shown that grapefruit juice significantly activates the efflux of some drugs that are ABCB1 substrates (Soldner et al. 1999). Thus, the activating effect of grapefruit juice on ABCB1 would partially counteract its CYP3A-inhibitory effect, which contrasts with the earlier assumption that intestinal CYP3A share common inhibitors with ABCB1. In conclusion, the involvement of food components on drug bioavailability certainly deserves more attention. For example, flavonoids like naringenin may in fact more strongly inhibit CYPs than ABCB1. In addition, the link between CYP and ABCB1 activities is still unclear. There are reports on independent regulations of the two enzymes (Wang et al. 2004), and increasing evidences for differential inhibitors (Dantzig et al. 1999, Edwards et al. 1999) such as polymethoxylated flavones from orange juice, which inhibit ABCB1 but not CYP3A4 (Takanaga et al. 2000).

12.2.2 Macrocyclic Diterpenes

Jatrophane diterpenes from *Euphorbiae* have been recently discovered as a new class of plant powerful inhibitors of ABCB1, thus opening new frontiers for research studies on this genus (Appendino et al. 1998, Hohmann et al. 2002, Corea et al. 2003a,b, 2004a,b, 2009, Barile et al. 2007). In addition, the isolated compounds are based on the same core, only differing on the substitution pattern. Thus, the isolation of structurally related analogs allowed us to establish SARs giving information on the

key pharmacophoric elements of these new class of promising drugs, without need-
ing any synthesis or semisynthetic modification. In this way, we got a better idea of
ideal ABCB1-inhibitors.

Diterpenes are constituents of *Euphorbia* plants that are present in both latex and
the whole plant. Their isolation was performed by exhaustive extraction with ethylacetate (EtOAc), and filtration on silica gel in order to eliminate gummy compounds.
Then, soluble materials, after taking to dryness, were separated by medium pressure
liquid chromatography (MPLC) on silica gel column using a linear gradient system
solvent from 100% hexane to 100% EtOAc. Major metabolites were constituted by
polyol diterpenes, mainly jatrophanes, polyesterified with short-chain organic acids.
The diterpenes-containing fractions were repeatedly chromatographed by high performance liquid chromatography (HPLC) on silica-gel semipreparative column, thus
affording the isolation of pure compounds.

Structure elucidation of the isolated diterpenes was performed by spectroscopic
methods, mainly mass spectrometry (MS) and nuclear magnetic resonance (NMR)
techniques. Thus, MS spectra gave key information on the molecular weight (MW)
of the analyzed compounds while NMR experiments (^{1}H, ^{13}C and ^{2}D) allowed building up the chemical structure and relative configuration of chiral centers. In particular, ^{1}H and ^{13}C NMR spectra allowed detecting all substituents, while COrrelation
SpectroscopY (COSY) and homonuclear Hartmann-Hahn (HOHAHA) spectra
identified the spin systems of the skeletal core. Then, heteronuclear single quantum coherence (HSQC) allowed the correlation of directly linked proton and carbon
atoms, thus constructing chemical substructures. At this point, extensive study of
the $^{2,3}J_{C-H}$ correlations, inferred from the heteronuclear multiple-bond correlation
(HMBC) spectrum, allowed the connection of all substructures through quaternary carbons and located the susbstituents on methine carbons. Finally, a rotational
nuclear overhauser effect spectroscopy (ROESY) spectrum allowed connection of
the substituents on quaternary carbons and also defined the relative stereochemistry
of chiral centers, thus completing the stereostructure of the studied metabolite. This
procedure had to be applied to each isolated metabolites to obtain the exact location
of substituents on the skeletal core that could not be obtained by simple comparison
of ^{1}H and ^{13}C NMR data.

By this procedure, we have isolated and characterized a number of jatrophane
diterpene analogs (**1–47**, Figures 12.3 through 12.5) from spurges that are characteristic of the Mediterranean landscape: *Euphorbia dendroides* (Corea et al. 2003a,b),
Euphorbia characias (Corea et al. 2004a), *Euphorbia peplus* (Corea et al. 2004b)
and *Euphorbia paralias* (Barile et al. 2007). Diterpenes isolated from *E dendroides,* named by Dioscorides (40–90 AD) as female spurge because of a characteristic
gentle shape, are constituted by polyol diterpenes, mainly jatrophanes, polyesterified
with short-chain organic acids. The jatrophane diterpenes, isolated for the first time
by our group, have been named euphodendroidins A (**1**), B (**2**), C (**3**), D (**4**), E (**5**),
F (**6**) G (**8**), H (**9**), and I (**10**) (Figure 12.3) (Corea et al. 2003a), in addition to the known
compound **7** previously described from *Euphorbia terracina* (Marco et al. 1998).

Inspection of ^{1}H NMR and ^{13}C NMR spectra of **1** initially suggested the presence of a diterpene skeleton pentaesterified with three acetyls, an isobutyryl, and a
benzoyl group and with four additional methyls, one methylene, and two methines.

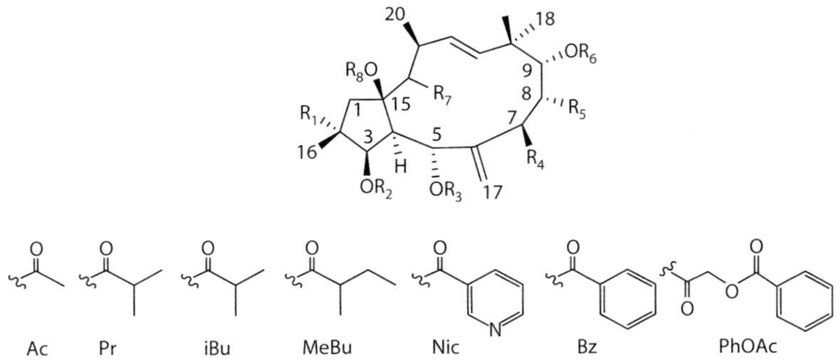

Name	R_1	R_2	R_3	R_4	R_5	R_6	20-Me	R_7	R_8	Inhibition
Euphodendroidin A (**1**)	OAc	H	iBu	OBz	OAc	Ac	b	O	H	97 ± 13.5%
Euphodendroidin B (**2**)	OAc	H	MeBu	OBz	OAc	Ac	b	O	H	105 ± 12.7%
Euphodendroidin C (**3**)	OAc	H	Nic	OBz	OAc	Ac	b	O	H	56 ± 4.8%
Euphodendroidin D (**4**)	H	H	iBu	OBz	OAc	Ac	b	O	H	183 ± 17%
Euphodendroidin E (**5**)	H	Ac	iBu	OBz	OAc	Ac	b	O	H	88 ± 6.7%
Euphodendroidin F (**6**)	OH	Ac	iBu	OBz	OAc	Ac	b	O	H	38 ± 4.1%
Compound **7**	OAc	PhOAc	Ac	OiBz	OAc	Ac	b	O	H	75 ± 12.2%
Euphodendroidin G (**8**)	OAc	Nic	Ac	OiBz	OAc	Ac	b	O	H	99 ± 6.5%
Euphodendroidin H (**9**)	H	Bz	Ac	OiBz	OAc	Ac	b	O	H	74 ± 9.5%
Euphodendroidin I (**10**)	ONic	Ac	iBu	OAc	OAc	Nic	b	O	Ac	42 ± 7.4%
Euphocharacin A (**11**)	OH	Bz	Ac	H	OAc	Nic	b	O	Ac	59 ± 1%
Euphocharacin B (**12**)	OH	Bz	Ac	H	OAc	Nic	b	O	H	72 ± 1%
Euphocharacin C (**13**)	OH	Bz	Ac	H	OAc	Bz	b	O	H	123 ± 2%
Euphocharacin D (**14**)	OH	MeBu	Ac	H	OAc	Nic	b	O	Ac	52 ± 3%
Euphocharacin E (**15**)	H	Bz	Ac	H	OAc	Nic	b	O	H	105 ± 3%
Euphocharacin F (**16**)	H	Bz	Ac	H	OAc	Nic	b	O	Ac	86 ± 2%
Euphocharacin G (**17**)	H	iBu	Ac	H	OAc	Nic	b	O	H	61 ± 2%
Euphocharacin H (**18**)	H	iBu	Ac	H	OAc	Nic	b	O	Ac	62 ± 4%
Euphocharacin I (**19**)	H	Pr	Ac	H	OAc	Nic	b	O	Ac	123 ± 3%
Euphocharacin J (**20**)	H	Ac	Ac	H	OAc	Nic	b	O	Ac	62 ± 2%
Euphocharacin K (**21**)	H	iBu	H	H	OAc	Nic	b	O	Ac	47 ± 5%
Euphocharacin L (**22**)	OH	Bz	H	H	OAc	Nic	b	O	Ac	79 ± 4%
Pepluanin A (**23**)	OAc	Bz	Ac	OAc	OAc	Nic	b	b OAc	H	207 ± 17%
Pepluanin B (**24**)	OAc	Bz	Ac	OMeBu	OH	Nic	b	b OAc	H	60 ± 4%
Pepluanin C (**25**)	OAc	Ac	iBu	OBz	OAc	Ac	b	b OAc	H	54 ± 5%
Pepluanin D (**26**)	H	Ac	Ac	OAc	H	Ac	b	O	Ac	76 ± 7%
Pepluanin E (**27**)	OAc	Bz	Ac	OiBu	OAc	Nic	b	O	H	94 ± 5%
Compound **28**	OAc	Bz	Ac	OAc	H	Ac	b	b OAc	H	80 ± 6%
Compound **29**	OAc	Ac	Ac	OAc	H	Nic	a	O	Ac	47 ± 7%

FIGURE 12.3 Chemical structures of jatrophane diterpenes **1–29** isolated from *E. dendroides*, *E. characias,* and *E. peplus,* and MDR reversing effects evaluated by their inhibition, comparatively to cyclosporin A (CsA) taken as a reference (100%), of cellular ABCB1-mediated daunomycin efflux, monitored by flow cytometry.

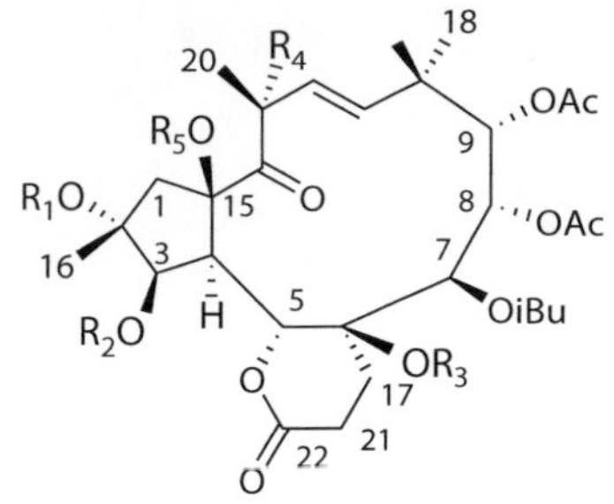

Name	R_1	R_2	R_3	R_4	R_5	Daunomycin-Efflux Inhibition
Terracinolide J (**30**)	Ac	Ac	iBu	H	H	101% ± 19%
Terracinolide K (**31**)	H	Ac	Ac	H	Ac	58% ± 6%
Terracinolide L (**32**)	H	Ac	iBu	H	Ac	0%
Terracinolide B (**33**)	Ac	Ac	Ac	H	Ac	22% ± 8%
Terracinolide C (**34**)	Ac	H	Ac	H	Ac	52% ± 11%
Terracinolide F (**35**)	Ac	Ac	iBu	H	Ac	0%
Terracinolide H (**36**)	Ac	H	iBu	H	Ac	138% ± 27%
13a-OH Terracinolide F (**37**)	Ac	Ac	iBu	OH	Ac	Not assayed
13a-OH Terracinolide B (**38**)	Ac	Ac	Ac	OH	Ac	27% ± 13%
13a-OH Terracinolide G (**39**)	Ac	Ac	Ac	OH	H	102 ± 12%

FIGURE 12.4   Chemical structures of terracinolides **30–39** isolated from *E. dendroides,* and MDR reversing effects evaluated by their inhibition, relatively to CsA, of cellular ABCB1-mediated daunomycin efflux.

Then, the 2D HSQC spectra showed that, in addition to the signals attributable to the above ester groups, euphodendroidin A (**1**) contained a ketone carbonyl group, two double bonds, one exocyclic and one trans disubstituted, and seven (five methines and two unprotonated) oxygenated sp³ carbon atoms. Application of ¹H-¹H COSY allowed us to sequence the multiplets of the core diterpene structure into three spin systems: C-3 to C-5, C-7 to C-9, and C-11 to C-20. It should be noted that although the methine protons of the fragment C-7 to C-9 appeared as broad singlets, pointing to a preferred conformation with dihedral angles near 90°, correlation peaks of H-8 with both H-7 and H-9 were evident in the COSY spectrum of **1**. Extensive study of the $^{2,3}J_{C-H}$ correlations, inferred from the HMBC spectrum, allowed us to determine the connection of all the above deduced moieties and the location of the oxygenated carbons. With these data in our hands, the final task to define the planar structure of euphodendroidin A (**1**) was the correct location of the five acyl groups. The relative configuration of the stereogenic centers of euphodendroidin A (**1**) was deduced from coupling constant values, spatial proximities evidenced through a ROESY experiment and comparison with literature data (Jakupovic et al. 1998a,b). In particular, the ROESY cross-peak of OH-3 with OH-15 and those of H-4 with both H-3 and Ac-2 were instrumental in defining both the relative stereochemistry of

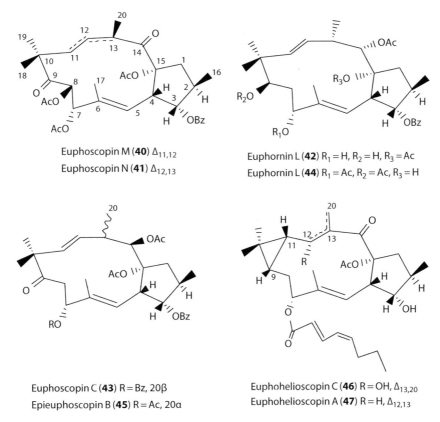

FIGURE 12.5 Chemical structures of jatrophane (**40–45**) and lathyrane (**46–47**) diterpenes isolated from *E. helioscopia*.

the five-membered ring and the *trans* geometry of the ring junction. On the other hand, the very small value of the coupling constants $J_{H-7/H-8}$ and $J_{H-8/H-9}$ was consistent with the *cis*-orientation of C-8 and C-9 acyl groups and the trans-orientation of C-7 and C-8 acyl groups. The key ROESY cross-peak of H-4 with H-7 connected the relative stereochemistry of this fragment with that described above for the five-membered ring. Finally, the spatial proximities H-5/H-8 and H-4/H-13, revealed through the ROESY experiment, allowed the assignment of the relative orientation to H-5 and to H-13, respectively, thus completely defining the relative stereochemistry pattern of the jatrophane core of **1**. Assuming the *S* configuration at C-4, typical of all the members of the jatrophane family isolated to date (deduced by x-ray analysis of some derivatives) (Hohmann et al. 1997), the absolute stereochemistry reported in structure **1** could be confidently assigned to euphodendroidin A. Moreover, it should be noted that the small value of $J_{H-4/H-5}$ (1.2 Hz) has been correlated with a conformation of the 12-member ring with the *exo*-methylene group pointing outward and H-5 pointing inward, named *exo*-type conformation (Corea et al. 2005). Accordingly, H-5 showed ROESY cross-peak with both OH-15 and with H-12.

Euphodendroidins B-I (**2–6, 8–10**) were closely related to compound **1**. In particular, with the only exception of **10**, they differed from euphodendroidin A (**1**) only for the acylation pattern at C-2, C-3, C-5, or C-7. Consequently, stereostructure elucidation of these molecules was greatly aided by comparison of their spectroscopic data with those obtained for **1**. Anyway, the same set of spectra have been acquired for all compounds in order to unambiguously elucidate their structure. Euphodendroidins represent the first series of compounds isolated by our research group, and have been used to give a first insight in SAR studies for this class of inhibitors. Among them, euphodendroidin D (**4**, Figure 12.3) was found to be the highly potent inhibitor since it was almost twofold more efficient at 5 µM (183% ± 17%) than cyclosporin A (CsA), used as a reference in the experiments, by monitoring daunomycin accumulation inside the cells. Comparison of the modulatory activity of euphodendroidins in terms of chemical structure evidenced that substitution pattern at C-2, C-3, and C-5 strongly affected the activity.

Thus, the relevance of a free hydroxyl at C-3 was highlighted by comparison of **4** (183% ± 17%) versus **5** (88% ± 6.7%), which showed a marked decrease in activity upon acylation of the 3-hydroxyl. A second important feature was the lack of oxygenation at C-2, since hydroxylation or acyloxylation at this position was also detrimental for activity. Comparison of **4** with **1–3** clearly demonstrates this point. Consequently, the presence of both components, acetylation of the 3-hydroxyl and oxygenation of C-2, caused an additional decrease in efficiency as observed by comparing **4** (183% ± 17%) to both **6** (38% ± 4.1%) and **10** (42% ± 7.4%). Besides, the acyloxyl at C-5 could also modulate the inhibitory efficiency, since the nicotinyl derivative euphodendroidin C (**3**) was twofold less active (56% ± 4.8%) by comparison to its 2-methylbutyryl (euphodendroidin B, **2**) and isobutyryl (euphodendroidin A, **1**) analogs.

Chemical investigation of *E. characias*, collected in Campania and described by Dioscorides as male spurge for its vigorous aspect, afforded 12 new jatrophane diterpenes, named euphocharacins A (**11**), B (**12**), C (**13**), D (**14**), E (**15**), F (**16**) G (**17**), H (**18**), I (**19**), J (**20**), K (**21**) and L (**22**) (Figure 12.3) (Corea et al. 2004a).

A common characteristic of the euphocharacin skeleton is the absence of any substituent at C-7. In particular, ROESY correlations of H_2-17/H-5 and of H-17b/H-8 indicated the presence of a favorite *endo*-type conformation (Corea et al. 2005), with the exomethylene perpendicular to the plane of the molecule. This was also confirmed by the high value observed for $J_{H-4/H-5}$ of 10.0 Hz. Euphocharacins A-L (**11–22**) are close analogs, differing only in the acylation pattern; therefore, their inhibition of multidrug resistance allowed getting interesting SAR data.

Among the series of euphocharacins, both C and I were strong inhibitors (123%) (Figure 12.3). Among compounds without hydroxyl group at C-2, the results showed the importance of a propionyl group at C-3 (euphocharacin I, 123%), which was much more efficient than benzoyl (euphocharacin F, 86% ± 2%) and isobutyryl (euphocharacin H, 62% ± 4%) or acetyl (euphocharacin J, 62% ± 2%). Thus, the following sequence in efficiency at C-3 was proposed: propionyl > benzoyl > acetoyl or isobutyryl. By contrast, among this series of compounds bearing a hydroxyl at position 2, the substitution of benzoyl (euphocharacin A, 59% ± 1%) with a 2-methylbutyryl (euphocharacin D, 52% ± 3%) did not affect the activity. Another positive role was played by benzoyl at C-9 (euphocharacin C, 123% ± 2%), which was better than nicotinyl in euphocharacin B (72% ± 1%).

In this set of compounds, without substitution at C-7, two aromatic rings did not damage the inhibitory activity, as for euphodendroidin I (**10**, 42% ± 7.4%), possibly due to a major flexibility of the macrocyclic ring. Furthermore, substitution at C-5 seemed not to be critical since limited and variable effects of euphocharacin L (79% ± 4%) versus A (59% ± 1%) or of K (47% ± 5%) versus H (62% ± 4%) were produced by either hydroxyl or acetyl. In contrast, a negative effect of hydroxyl, by comparison to the absence of any substituent at C-2, was observed for both euphocharacin B (72% ± 1%) versus E (105% ± 3%) and euphocharacin A (59% ± 1%) versus F (86% ± 2%); this confirms the observation in euphodendroidins where both acetyl and hydroxyl groups were detrimental to inhibition.

Analysis of *E. peplus*, commonly named petty spurge for its size of 5–30 cm height, afforded compounds **23–29** (Figure 12.3). The isolated compounds belong to two series: the first one having an acetoxyl at C-9 (**23**, **24**, **25**, **28**), and the second one characterized by a carbonyl at the same position (**26**, **27**, **29**), as found for the other metabolites here described. The new compounds have been named pepluanins A-E (**23–27**), while compounds **28** and **29** were already described in the literature (Jakupovic et al. 1998).

Pepluanin A (**23**), according to the NMR profiles, was indicative of a jatrophane polyol bearing several ester groups. The nature and location of the ester groups has been identified by NMR spectra that showed the presence of five acetates (at positions 2, 5, 7, 8 and 14), one benzoate (at position 3), and one nicotinate (at position 9). The remaining oxygenated function has been ascribed to a hydroxyl group located at C-15. The relative configuration of **23**, as depicted in the formula, was determined by analyzing the *J* pattern and the results of nuclear overhauser effect (NOE) correlations. In addition, the small coupling constant values (0–2 Hz) found for the macrocyclic ring protons suggested an orthogonal relationship between them and therefore an *exo*-type conformation as found for *E. dendroides* diterpenes (Corea et al. 2003a).

Pepluanins B-E (**24–27**) are closely related to pepluanin A. In particular, apart from C-14, they differ from one another by their acylation pattern. Consequently, also for these compounds, stereostructure elucidation was greatly aided by comparison of their spectroscopic data with those obtained for both pepluanin A and other known jatrophanes. The same relative stereochemistry previously assigned to pepluanin A was also confirmed for pepluanins B-E by 2D NMR ROESY spectroscopy.

The ABCB1-binding ability of this set of compounds was used to extend the SAR to the other oxygenated carbon atoms of the medium-sized ring (C-7/C-15), highlighting the importance of an acetoxyl at C-8, by comparison to a free hydroxyl, and of a free hydroxyl at C-15. The obtained percentages of inhibition for this compound series at 5 μM are reported in Figure 12.3. Our previous studies on *E. dendroides* diterpenes (Appendino et al. 1998) evidenced for the first time the involvement of ring A in binding to ABCB1. The present set of compounds has allowed us to go beyond these data, underlining the importance of the substitution on other carbons of the medium-sized ring for modulating the activity. Pepluanin A was found here to be an even more powerful inhibitor, with an efficiency at least twofold higher (207% ± 17%, Figure 12.3) than the conventional modulator CsA. This appears quite interesting also in the light of some structure–activity considerations that can be drawn. First, comparison of the activity of pepluanin A and B, the only structural differences of which being confined to C-8 substitution, evidenced a collapse of the inhibitory potency (207% ± 17% in pepluanin A versus 60% ± 4% in pepluanin B). This clearly indicates that also the substitution pattern at the carbons of the medium-sized ring was of great importance for the activity.

Another point to clarify was that pepluanin A, unlike the majority of the other tested jatrophanes and compounds 26, 27, 29, has an acetoxyl function at C-14 instead of a carbonyl. It was rather difficult at this point to draw a definitive conclusion on this structural detail. Anyway, by comparing the activity of 27 (94% ± 5%) with that of 24 (60% ± 4%), essentially differing in this structural detail, we could hypothesize that the observed inhibitory potency of pepluanin A might be further increased by a carbonyl at position 14. Similarly, a positive effect might be expected by substituting nicotinyl by acetoxyl at C-9, since 28 (80% ± 6%) was more efficient than 24 (60% ± 4%). As found in the most powerful compounds tested (see euphocharacin jatrophanes), an additional key point was the presence of a free hydroxyl group at C-15. Indeed, compounds 26 (76% ± 7%) and 29 (47% ± 7%), presenting an ester function at this position, exhibited the lowest efficiency.

We completed the characterization of the diterpenoid fraction of *E. dendroides* by describing the structural elucidation and biological evaluation of 10 additional compounds of the modified jatrophane type, terracinolides J, K, L (30–32) and 13a-OH terracinolide F (37), described for the first time (Figure 12.4). Compounds 33–36 and 38–39 were identified as known terracinolides B, C, F, H, and 13a-OH terracinolide B and G, respectively, by comparison of their MS and NMR data with those reported for the original compounds (Marco et al. 1996, 1997, 1999, Jakupovic et al. 1998).

Terracinolides are based on 17-ethyl bis-homojatrophane framework, a skeleton previously found in *E. terracina* diterpenes (Marco et al. 1996), which according to (Marco et al. 1997) could arise through opening of a 5,17-epoxide by nucleophilic attack of a C_2 unit (either acetate or malonate) followed by lactone-ring closure (Scheme 12.1). Within this set of compounds, terracinolide H showed the highest activity, being even more efficient (138% ± 27%) than CsA. SAR studies showed the effects of substitutions at positions 3, 6, and 15 (Figure 12.4). First, an hydroxyl group at either position 3 or 15 appeared of major importance since its substitution by an acetyl completely abolished the inhibitory effect in compound 36 versus 35,

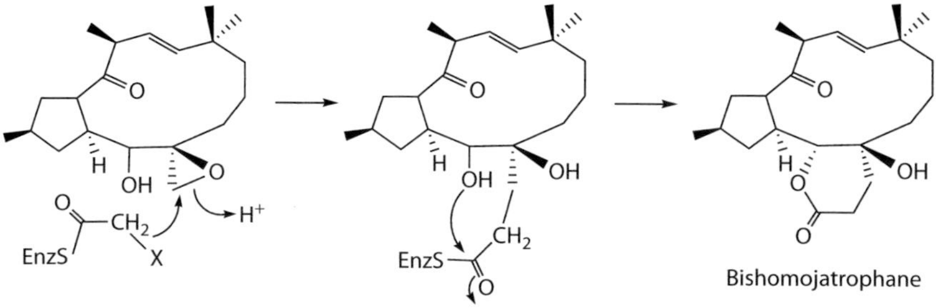

SCHEME 12.1 Possible biogenesis of bishomojatrophanes (named terracinolides) by incorporation of a C_2 unit (from acetate or malonate) into jatrophane precursor.

and dramatically reduced the inhibitory effect in compound **34** (52% ± 11%) versus **33** (22% ± 8%) and in **39** (102% ± 12%) versus **38** (27% ± 13%). Like euphodendroidins, an unfavorable feature for the activity was the presence of a free hydroxyl at C-2, in addition to the lack of any substituent at C-15 (see compound **30** versus **32**). On the other hand, a hydroxyl group at C-13 did not affect the activity as we could see in two pairs of compounds: **30** versus **39,** and **33** versus **38**.

We finally analyzed *Euphorbia helioscopia*, whose name derives from the Greek word "helioscopion" used first by Dioscorides to describe this species and meaning "sun gazer" because the plant moves its heads round to follow the sun. Our analysis afforded compounds **40–47** (Figure 12.5). Four of them were based on new chemical structures, and were named euphoscopin M (**40**), euphoscopin N (**41**), euphornin L (**42**), and the lathyrane euphohelioscopin C (**46**) (Figure 12.5). The remaining were determined as the known compounds euphoscopin C (**43**), euphornin (**44**), epieuphoscopin B (**45**), and euphohelioscopin A (**47**), first isolated from samples of *E. helioscopia* (Yamamura et al. 1989) (compound **43**, **45** and **47**) and *Euphorbia maddeni* (compound **44**) (Ahmed-Belkacem et al. 2005).

The biological activity of *E. helioscopia* jatrophane (**40**, **41**, **43**, **44**, and **45**) and lathyrane (**47**) diterpenes was monitored through their concentration-dependent ability to inhibit ABCB1-mediated mitoxantrone efflux leading to drug accumulation, as measured by flow cytometry (Figure 12.6, top). It appeared that epieuphoscopin B was twofold more potent than the reference inhibitor cyclosporin A, with an IC_{50} of 1.71 ± 0.83 μM, by comparison to 3.37 ± 1.39 μM. In contrast, euphornin and euphohelioscopin A were much less efficient, with respective values of 8.46 ± 3.51 and 14.0 ± 2.4 μM. In the latter case, concerning the lathyrane euphohelioscopin A, a high concentration of 100 μM was required to get a maximal inhibition. Finally, the remaining compounds euphoscopin M, N, and C appeared similar to cyclosporin A, with respective IC_{50} values of 3.78 ± 2.18, 3.47 ± 1.88, and 3.58 ± 1.78 μM.

Three main SARs of ABCB1 inhibition could be drawn for substituents from the differential inhibitory effects observed: (1) a marked, fivefold, positive effect played by a carbonyl versus an O-acetyl group at position 9 when comparing compounds **45** and **44**, while the same group was detrimental for the activity at position 14 (compare compounds **41** and **45**), (2) a twofold positive effect of an O-acetyl versus O-benzyl substituent at position 7 when comparing compounds **45** and **43**, and (3) a neutral effect of the double bond at either position 11–12 or 12–13 in compounds **41** and **40**. These new important data about jatrophanes isolated from *E. helioscopia* are complementary to those previously drawn for jatrophanes isolated from other *Euphorbiae* species.

Interestingly, the same compounds were not able to alter the mitoxantrone efflux mediated under similar conditions by the other multidrug transporter ABCG2 (Figure 12.6, bottom). Indeed, no significant inhibition was observed for any compound up to a 20 μM concentration, whereas the control inhibitor GF120918 gave a submicromolar IC_{50} value, in agreement with previous observations (Sahai et al. 1981, Szakács et al. 2006).

Taken together, all these observations suggest that jatrophanes and modified compounds share a common gross pharmacophore, which is dramatically affected by changes of the oxygenation pattern, but is surprisingly tolerant in terms of

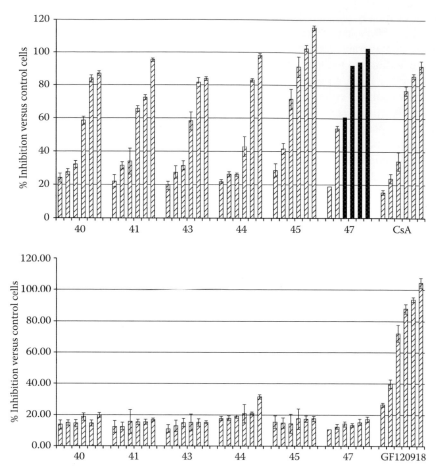

FIGURE 12.6 Concentration-dependent inhibitions of the purified compounds on mitoxantrone efflux by either MDR1-transfected NIH3T3 cells (top) or ABCG2-transfected HEK-293 cells (bottom). The inhibitors were used at increasing concentrations of 0.1, 0.5, 2.5, 10, and 20 μM, except for compound **47** tested at 0, 20, 40, 60, 80, 100 μM (last four concentrations in black because higher than for the other compounds) in top experiment. Mitoxantrone accumulation was assayed by flow cytometry, and the inhibition efficiency determined by comparison to control cells. CsA (top) and GF120918 (bottom) were used as reference inhibitors of ABCB1 and ABCG2, respectively.

modifications of connectivity (Figure 12.7). Finally, the last set of jatrophanes and derivatives tested by our group, individually investigated for their ABCB1- and ABCG2-inhibiting properties, appeared to be specific inhibitors of ABCB1 since they showed no significant activity against ABCG2. A similar specificity was observed with Celastraceae sesquiterpenes, which affected various functions of ABCB1 (reversion of resistance to daunomycin and vinblastine, inhibition of drug efflux, noncompetitive inhibition of ATPase activity) but not of either ABCC1, ABCC2, or ABCG2 (Munoz-Martinez et al. 2004). The compounds were found to be not transported by ABCB1 (Munoz-Martinez et al. 2006).

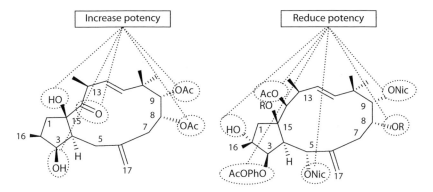

FIGURE 12.7 Key pharmacophoric elements, either positive or negative, for the anti-MDR activity of ABCB1.

Such a specificity toward ABCB1 was observed with other types of inhibitors, such as amiodarone, nifedipine, dexniguldipine, LY335979 (Zosuquidar), and R101933 (Laniquidar) as reviewed (Szakács et al. 2006). Conversely, other types of inhibitors, such as fumitremorgin C, Ko143, tectochrysin, and 6-prenylchrysin, were found to selectively inhibit ABCG2 and not ABCB1 (*cf.* here below). All these third-generation specific inhibitors constitute convenient and efficient tools to evaluate the contribution of each transporter to cell drug-efflux and related MDR phenotype, in various cell lines or tissues, under either normal or pathological conditions.

12.2.3 STEROIDAL SAPONINS AND OTHER NATURAL INHIBITORS

Bioguided fractionation of the roots of *Paris polyphylla* (Trilliaceae), based on inhibition of ABCB1-mediated daunorubicin efflux, led to the isolation and identification of the three saponins 3-*O*-Rha(1 → 2)[Ara(1 → 4)]Glc-pennogenine (**1**), gracillin (**2**), and polyphyllin D (**3**), and the two ecdysteroids 20-hydroxyecdysone (**4**) and pinnatasterone (**5**) (Figure 12.8). By contrast to a very weak efficiency on ABCG2, the three saponins displayed significant effects as inhibitors of ABCB1-mediated drug efflux (Nguyen et al. 2009). Interestingly, slight modifications in the sugar moiety of compounds **1–3** were able to significantly impact the inhibitory potency. Such a specificity of steroidal saponines for ABCB1 contrasts with the wider effects of triterpenoid saponines from *Panax ginseng* rhizomes, ginsenosides, reported to inhibit both ABCB1 (Choi et al. 2003, Kim et al. 2003) and ABCG2 (Jin et al. 2006).

Finally, various other classes of plant compounds have been reported to bind and inhibit ABCB1, including hypericin and hyperforin found in St John's Wort (Wang et al. 2004), dietary phytochemicals such as sesamin, ginkgolic acid, matairesinol, glycyrrhetinic acid, glabridin and phyllodulcin (Nabekura et al. 2008), cnidiadin (Barthomeuf et al. 2005), conferone (Barthomeuf et al. 2006) and other coumarins (Raad et al. 2006), quinine and other antiparasitic drugs (Hayeshi et al. 2006), eudesmin, a bicyclic lignan from *Haplophyllum perforatum* (Lim et al. 2005), triterpenoids such as amooranin from *Amoora rohituka* (Ramachandran et al. 2003), a number of diterpenes and carotenoids (Molnar et al. 2006), tetandrine (Zhu et al. 2005), and bisbenzylisoquinoline (Fu et al. 2001) and bisbibenzyl (Li et al. 2009) derivatives,

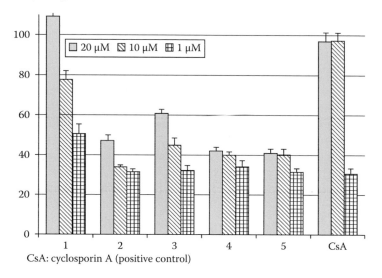

1. $R_1 = OH$, $R_2 = 3$-O-Rha(1 → 2)[Ara(1 → 4)]Glc   **4.** $R_1 = H$, $R_2 = OH$, 20-hydroxyecdysone

2. $R_1 = H$, $R_2 = 3$-O-Rha(1 → 2)[Glc(1 → 3)]Glc   **5.** $R_1 = OH$, $R_2 = H$, Pinnatasterone

3. $R_1 = H$, $R_2 = 3$-O-Rha(1 → 2)[Ara(1 → 4)]Glc

CsA: cyclosporin A (positive control)

FIGURE 12.8   Steroidal saponines: Structure of the different saponines studied, and ability to prevent ABCB1-mediated daunomycin efflux, by comparison to CsA.

and indole alkaloids of the ibogan-type (Kam et al. 2004). Recently, dimers of naturals compounds such as quinine homodimers (Pires et al. 2009) and flavonoid dimers (Chan et al. 2006) have been shown to constitute potent inhibitors.

12.3   ABCG2 INHIBITORS

12.3.1   Different Types of Known ABCG2 Inhibitors

The ABCG2 inhibitors may be classified by different ways: according to either (a) their specificity, (b) their chemical class, or (c) their origin (natural compounds, clinical use as drugs, design,…). The first type of classification is used here, with special emphasis on natural compounds.

12.3.1.1   Nonselective Inhibitors

Known ABCG2 inhibitors are structurally unrelated and therefore belong to different chemical classes. ABCB1 inhibitors were among the first compounds to be examined for their ability to also inhibit ABCG2 (Ahmed-Belkacem et al. 2006, Florin et al. 2008).

The pattern of substrate drugs transported by ABCG2 highly overlaps that of ABCB1, which is clearly not the case for inhibitors. In addition, the effects of the hot-spot mutation R482G/T, which are well known to change the pattern of transported substrates, have not been systematically characterized for the inhibitors. One of the first compounds to be studied was GF120918 (elacridar), which was designed as an ABCB1 inhibitor (De Bruin et al. 1999), as described above, and later demonstrated to be a lower-affinity, but efficient, inhibitor of ABCG2. Following these major results, the structure of GF120918 was used as a template to conceive new ABCG2 inhibitors. In this regard, structural analogs of GF120918 (Figure 12.9), has been synthesized by shortening the aryl side-chain, and found to be slightly more active than GF120918, as shown on mitoxantrone efflux from human wild-type (R482) transfected cells (Boumendjel et al. 2007). Tariquidar (XR9576), a structural analog of GF120918, has also been shown to inhibit ABCG2-mediated drug efflux (Robey et al. 2004). A number of derivatives have allowed the elaboration of a molecular model from three-dimensional 3D-QSARs, with minor differences between ABCB1 and ABCG2 (Pick et al. 2008), whereas other derivatives appeared to be much more efficient and specific for ABCG2 (Kühnle et al. 2009). The group of S.E. Bates (Lee et al. 1997) demonstrated that the immunosuppressive drug cyclosporin A was an inhibitor of rhodamine transport in MCF-7/AdVp cells, which was subsequently confirmed in various ABCG2-expressing cell lines (Qadir et al. 2005, Gupta et al. 2006).

Various compounds of natural origin were recently reported to inhibit ABCG2, but their specificity and inhibition potency were not always characterized in details. Various stilbenoids from *Bletilla striata* (Orchidaceae) potentiated SN-38 cytotoxicity in ABCG2-transfected K562 cells (Morita et al. 2005). Metabolites of ginsenosides from *Panax ginseng* significantly increased mitoxantrone cytotoxicity in human

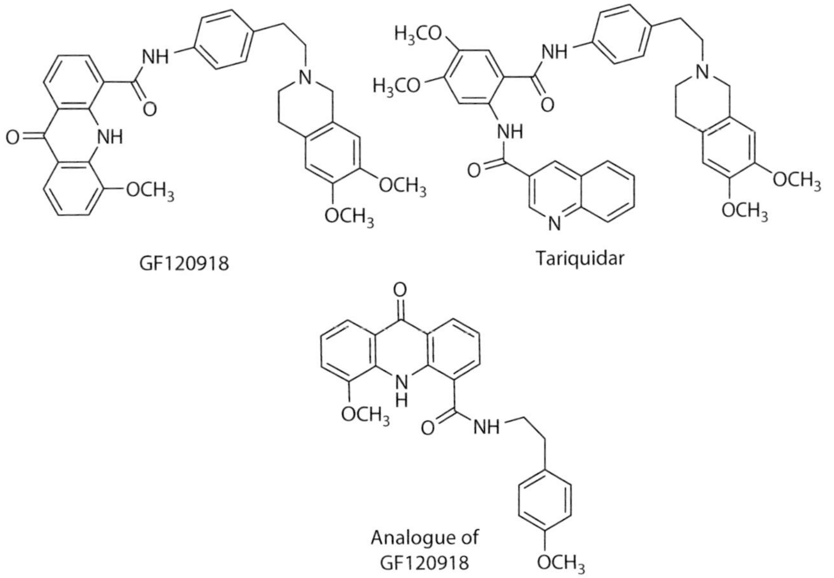

GF120918

Tariquidar

Analogue of
GF120918

FIGURE 12.9   Derivatives of GF12018 (elacridar) as inhibitors of ABCG2.

MCF-7/MX cells but the efficacy appeared rather moderate (Jin et al. 2006). Some terpenoids present in herbal medicines, such as glycyrrhetic acid and abietic acid, were shown to inhibit ABCG2 in addition to ABCC2, while ABCB1 appeared less sensitive (Yoshida et al. 2008). Curcuminoids, extracted from turmeric, which were previously known to modulate the function of ABCC1 (Wortelboer et al. 2003, Cherwae et al. 2006) and ABCB1 (Limtrakul et al. 2007), have been recently shown to inhibit the in vivo transport activity of ABCG2 in knock-out mice, by increasing the bioavailability of sulfasalazine (Shukla et al. 2009). Hyperforin, a phloroglucinol present in St John's Wort and known to stimulate apoptosis in B cell chronic lymphocytic leukemia cells and to display antiangiogenic properties, was shown to inhibit not only ABCB1 (Wang et al. 2004) but also ABCG2 activity (Quiney et al. 2007) of drug efflux. Very recently, some botrillamides, from the marine ascidian *Botryllus tyreus*, have been found to both inhibit ABCG2-mediated efflux of boron-dipyrromethene (BODIPY)-prazosin, prevent iodoarylazidoprazosin labeling, stimulate ATPase activity, and reverse cell growth resistance (Henrich et al. 2009).

Taxoids have been recently reported as promising ABCG2 inhibitors. SARs studies have shown that the presence of the C-13 side chain is critical for activity. Recently, a series of noncytotoxic taxoids where the C-13 side chain is replaced by an acetyloxy group has been reported (Brooks et al. 2003). Structural diversity was obtained by introducing substituents on the hydroxyl groups at positions 1, 2, 7, 10, and 14, leading to tRA98006 (Figure 12.10), which was active against both wild-type R482 and mutant R482G/T ABCG2. The basic amino group is essential for the inhibitory activity and strengthens the importance of basic nitrogen atoms among the structure of ABCG2 inhibitors.

Another class of drugs extensively studied as reversers of ABCG2-dependent multidrug resistance was constituted by anti-HIV drugs. The expression of ABCG2 in a CD4$^+$ T-cell line was found to confer cellular resistance to nucleoside reverse-transcriptase inhibitors, and ABCG2 was then considered to constitute a potential target for improving HIV treatments (Wang and Baba 2005). The effect of a set of anti-HIV drugs on ABCG2 activity was assessed by increased pheophorbide A accumulation in MDCKII-ABCG2 cells, as compared to the corresponding parental MDCKII cell line. The two most active drugs were found to be lopinavir and nelfinavir (Figure 12.11), while a weak inhibitory activity was observed with the well known anti-HIV drug AZT (Weiss et al. 2007).

Tyrosine kinase inhibitors (TKIs) have emerged as a new class of ABCG2 inhibitors. Hydrophobic TKIs have a great potential to interact with ATP-dependent active

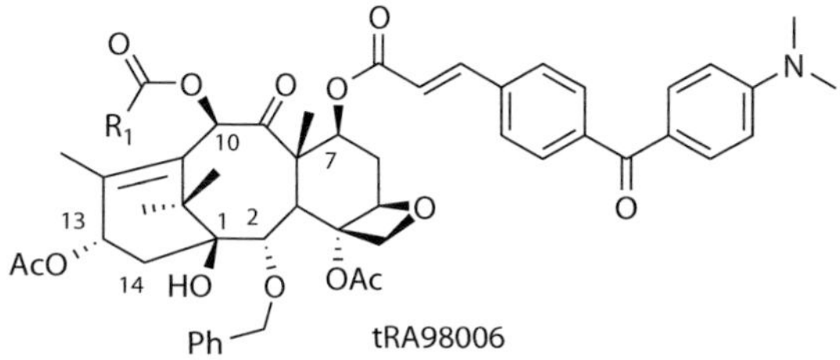

FIGURE 12.10 Evaluation of taxoids.

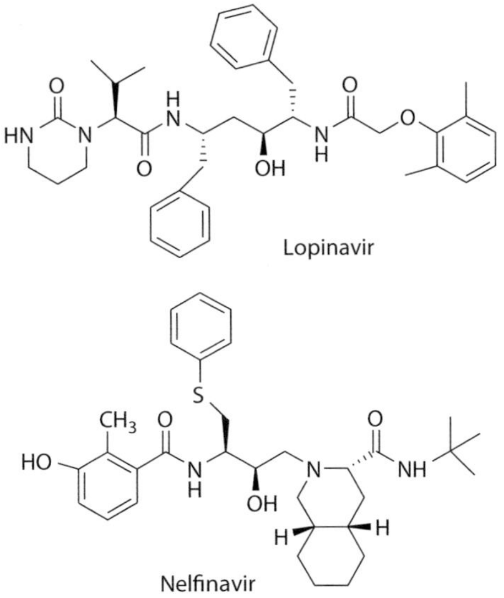

FIGURE 12.11 Antiviral drugs as inhibitors of ABCG2.

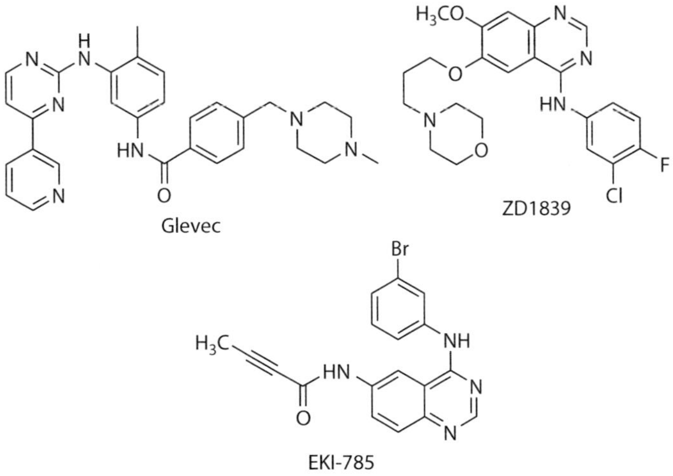

FIGURE 12.12 Tyrosine kinase inhibitors as potent ABCG2 inhibitors.

transporters, including ABCG2. Several compounds have been studied in more detail (Figure 12.12). Gleevec (Imatinib mesylate, STI-571) is a well-known tyrosine kinase inhibitor widely used for the treatment of various types of cancer. Overexpression of ABCG2 has been related to resistance to Gleevec. In human osteosarcoma Saos2 ABCG2#4 cells, Gleevec was able to reverse resistance to camptothecins (IC$_{50}$ = 0.17 µM) and increased topotecan accumulation (Houghton et al. 2004). By contrast to the initial conclusion that STI-571 was not transported, some resistance was indeed observed and related to STI-571 efflux (Burger et al. 2004). ZD1839 (Iressa, Gefitinib) and EKI-785 are two TKIs reported to affect the ABCG2 function (Ozvegy-Laczka et al. 2004). At low concentration (0.1 μM), ZD1839 and EKI-785 stimulated ATPase activity, whereas they inhibited at higher concentrations, indicating that ZD1839 and EKI-785 are probably substrates for ABCG2. CI1033, another TKI was shown to reverse ABCG2-mediated resistance to SN-38 and topotecan in transfected cells. However, it also acts as a substrate, as demonstrated by its low accumulation in ABCG2-expressing cells.

12.3.1.2 ABCG2-Specific Inhibitors

12.3.1.2.1 Fumitremorgin C and Related Analogs

Fumitremorgin C (FTC) was isolated from the fermentation broth of *Aspergillus fumigatus* (Rabindran et al. 1998). FTC was tested on a mitoxantrone-resistant colon carcinoma cell line and found to be specific and a potent inhibitor of ABCG2; however, it induced neurotoxic effects, excluding any possible clinical use. In order to provide less toxic ABCG2 inhibitors, FTC analogs became attractive targets. In this regard, a large series of synthetic FTC analogs, namely indolyldiketopiperazines, were screened on ABCG2-transfected cells and led to the identification of three compounds, namely Ko132, Ko134, and Ko143, as promising candidates (Figure 12.13) (Van Loevezijn et al. 2001, Allen et al. 2002). In addition, Ko143 significantly increased the topotecan oral availability in mice. Tryprostatin A (TPS-A) is another natural analog of FTC, as a fungal secondary metabolite. Using an assay test based on mitoxantrone uptake and cytotoxicity, Woehlecke et al. (2003) demonstrated that TPS-A inhibited wild-type ABCG2. The structural difference between FTC and TPS-A essentially concerns the noncohering diketopiperazine C ring in TPS-A, allowing an alternative conformation. It could also be hypothesized that rings A, B, and D, as well as the C-3 side chain, are necessary for ABCG2 inhibition.

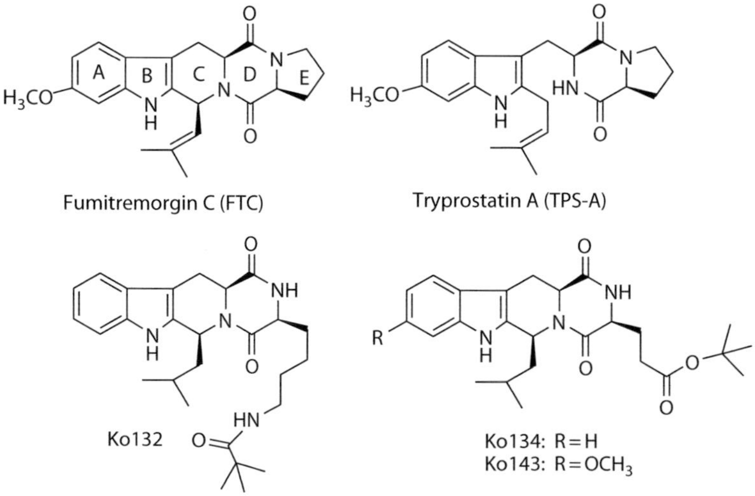

FIGURE 12.13   FTC and related derivatives as specific inhibitors.

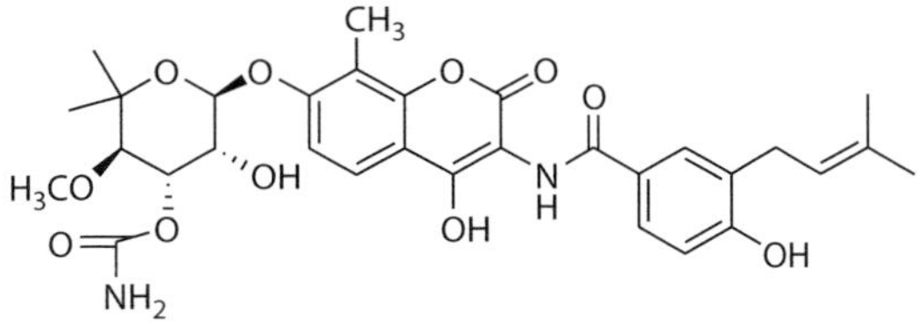

FIGURE 12.14   Structure of novobiocin.

12.3.1.2.2   Novobiocin and Other Compounds

Novobiocin is an antibiotic that is produced by *Streptomyces niveus* (Figure 12.14). This prenylated aminocoumarin is an effective antibacterial agent used in the treatment against methicillin-resistant *Staphylococcus aureus* (Walsh et al. 1993), and is also active against *Staphylococcus epidermidis*. Novobiocin was found to increase cellular accumulation of topotecan and to inhibit its efflux from MCF7/TPT300 cells, as well as to enhance topotecan and mitoxantrone toxicity in the same cells (Yang et al. 2003). It also increased the intracellular topotecan accumulation in PC-6/SN2-5H2 small cell lung cancer cells and inhibited the topotecan transport into membrane vesicles prepared from these cells (Shiozawa et al. 2004). The inhibition was clearly dependent on the cell line used, since in a recent report Su and Sinko (2006) showed that mitoxantrone resistance of CPT-K5 cells (human acute lymphoblastic leukemia cells) was not reversed by novobiocin. In addition, even when active, novobiocin requires rather high concentrations, likely due to limited membrane diffusion related to the rather hydrophilic character of the molecule.

12.3.2   FLAVONOIDS

Flavonoids constitute a growing class of ABCG2 inhibitors. Silymarin, hesperetin, quercetin, and daidzein were shown to increase intracellular accumulation of mitoxantrone in ABCG2-expressing cells (Cooray et al. 2004). Among the naturally occurring flavonoids (Figure 12.15), biochanin A, chrysin, and tectochrysin were the most active on mitoxantrone-resistant MCF7 and NCI-H460 cell lines (Ahmed-Belkacem et al. 2005, Zang et al. 2004a,b). The flavonol quercetin was found to inhibit SN-38 efflux (Yoshikawa et al. 2004), whereas the isoflavone genistein was transported. A first comparison of different flavonoids subclasses showed the following efficiency order against mitoxantrone efflux: flavones > flavonols > isoflavones > flavanones (Ahmed-Belkacem et al. 2005). Structure–activity studies led to the identification of novel ABCG2 inhibitors such as 6-prenylchrysin (Ahmed-Belkacem et al. 2005) exhibiting an IC_{50} of 0.3 μM similar to that of GF120918. Interestingly, the R482T hot-spot mutation altered the positive impact of prenylation on the inhibitory potency, the natural compound tectochrysin being then the best compound with an IC_{50} of 1.9 μM. The relatively low toxicity of both tectochrysin and 6-prenylchrysin and their efficient sensitization of cell growth to mitoxantrone made these compounds promising for future potential use in clinical trials. In a recent study, the inhibitory effect of naturally occurring flavonoids on ABCG2 was correlated with their positive effects on the pharmacokinetics of anticancer drugs. A panel of 32 flavonoids was screened by using topotecan accumulation and cytotoxicity assays, and led to the identification 3′,4′,7-trimethoxyflavone and apigenin as the most

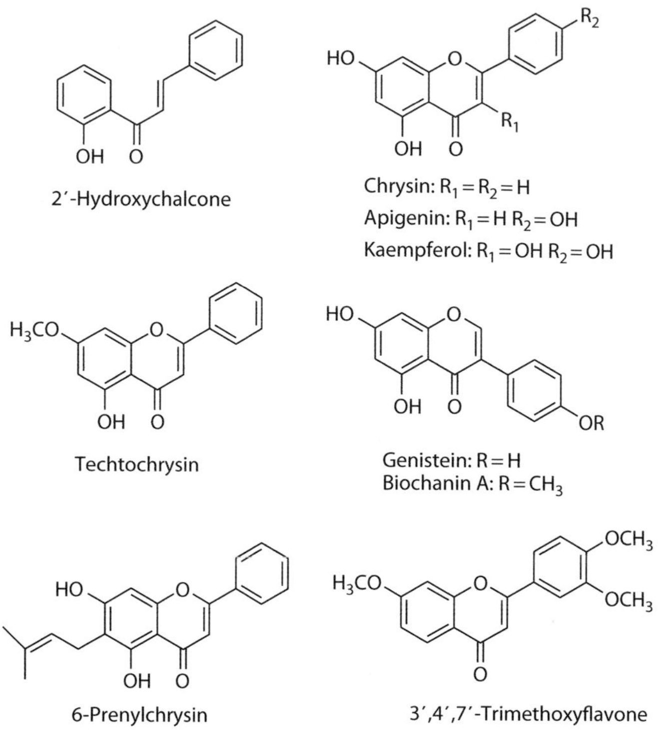

FIGURE 12.15 Representative flavonoids investigated for their ABCG2 inhibition.

potent inhibitors of ABCG2. Very recently, chalcones were found to inhibit differentially ABCB1 and ABCG2, basic chalcones being more efficient on the first transporter (Liu et al. 2008), and nonbasic chalcones on the second one (Han et al. 2008).

The acridone derivatives, initially designed for mimicking GF120918 (Boumendjel et al. 2007), were in fact quite selective for ABCG2. This indicated that they bound to distinct sites on ABCG2, although producing a similar inhibition. A similar behavior was recently observed between XR9576 (tariquidar), a third-generation ABCB1 inhibitor also inhibiting ABCG2, of which some derivatives were found to be quite specific for ABCG2 (Kühnle et al. 2009), While the flavonoid efficiency in vitro is now well recognized, their in vivo activity needs to be further optimized.

12.3.3 ROTENOIDS

The phytochemical analysis of *Boerhaavia diffusa* (Nyctaginaceae), a popular herbal remedy in the Indian Ayurvedic medicine (known with the name punarnava), provided an attractive opportunity to extend the SARs available from flavonoids. Indeed, chromatographic purification of the organic extract obtained from the roots of *B. diffusa* afforded ten rotenoid derivatives and a coumaronochromone as pure compounds. The names and structures of the isolated compounds are reported in Figure 12.16. All compounds except boeravinones C, E, and J are racemic mixtures at the single stereogenic center C-6.

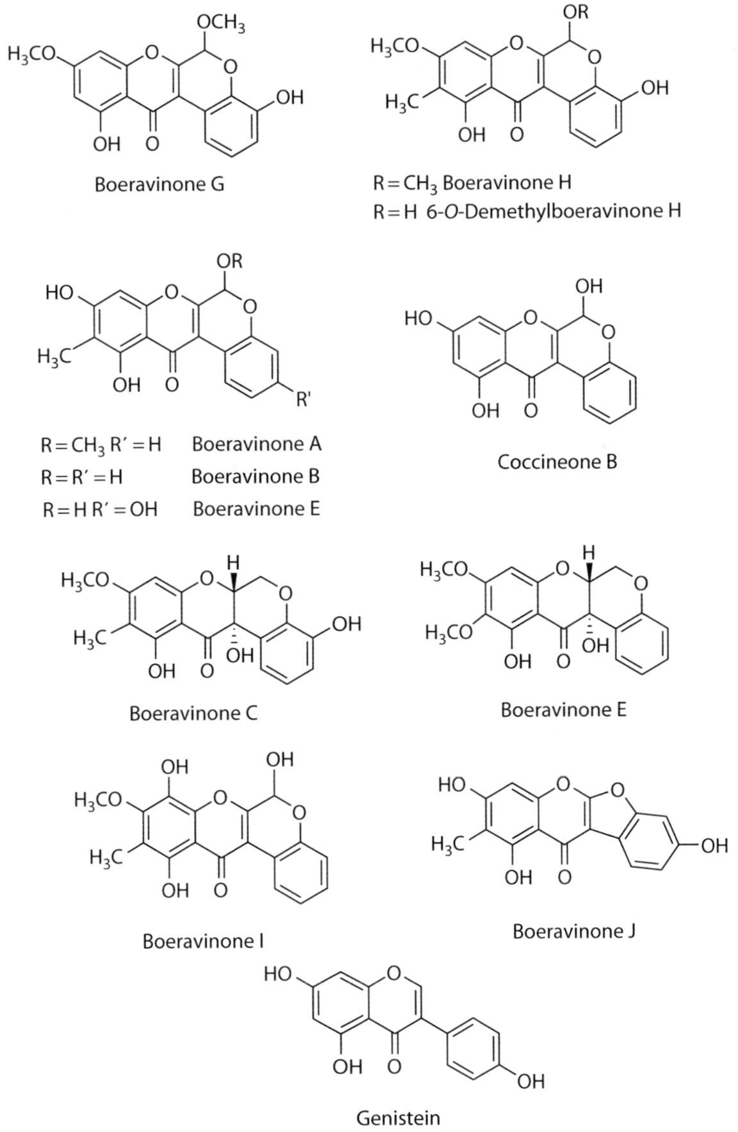

FIGURE 12.16 Chemical structures of rotenoids from *Boerhaavia diffusa*, and of the isoflavone genistein.

Rotenoids are isoflavonoid derivatives (3-phenylchromen-4-one in place of 2-phenylchromen-4-one backbone) typically found as secondary metabolites of Leguminosae and Fabaceae plants. Biogenetically, isoflavonoids derive from flavonoids through a 1,2-aryl migration, likely involving radical intermediates. The oxidation of a 5-OMe isoflavonoid, followed by an intramolecular pseudo-Diels-Alder reaction (Figure 12.17), leads to the formation of a fourth ring characteristic of the rotenoid skeleton. Thus, compared to tricyclic isoflavonoids (as genistein and

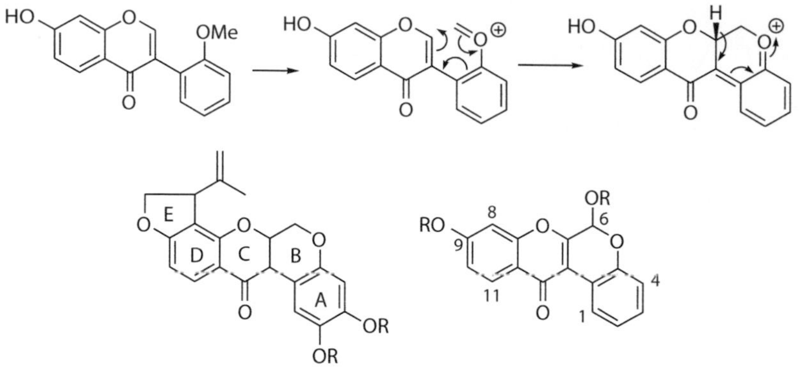

FIGURE 12.17 Above: Biosynthetical steps connecting isoflavonoids with rotenoids. Below, Left: General structure of rotenoids from Leguminosae and Fabaceae. Right: General structure of rotenoids from Nyctaginaceae.

daidzein), the peculiar structural features of rotenoids are: (1) an additional ring (ring B) including an oxygen atom and, almost invariably, (2) a 2,3-disubstitution on ring A, and (3) a prenyl group attached at position 8, commonly cyclized to form a five- or a six-membered ring (ring E) (Figure 12.17). The skeleton of coumarono-chromones differs from that of rotenoids in possessing a five-membered ring B, with loss of C-6. Most likely, coumaronochromones derive biogenetically from a direct cyclization of isoflavonoids, without incorporation of an additional carbon atom.

Rotenoids and analogs are frequently toxic through inhibition of mitochondrial electron-transport chain at complex I, and some of them have found application as potent insecticides (Gutman et al. 1970). However, an interesting study aimed at establishing the "toxophore" of the rotenoid molecules revealed that both the prenyl-derived ring and the dimethoxy substitution on ring A were essential requirements for toxicity (Crombie et al. 1992). Accordingly, the group of rotenoids isolated from *B. diffusa* and other plants belonging to Nyctaginaceae, which lack the isoprenoid residue on ring D and show a monosubstituted or unsubstituted ring A (Figures 12.16 and 12.17), proved to be noncytotoxic (IC$_{50}$ > 20 µM) on several cell lines.

The rotenoid derivatives obtained from *B. diffusa* were assayed for ABCG2 inhibition through evaluation of alteration of mitoxantrone efflux in HEK-293 human cells transfected by wild-type (R482) ABCG2 (Ahmed-Belkacem et al. 2007). The inhibitory activity of boeravinones was then evaluated by flow cytometry. The obtained results are summarized in Table 12.1.

The most potent compounds were boeravinones G and H, for which a subsequent concentration-dependence study allowed the determination of IC$_{50}$ values of 0.7 and 2.5 µM, respectively. The other boeravinones were less active and required higher concentrations to produce appreciable inhibition, with coccineone E and boeravinones I and J being the less active compounds (Table 12.1).

The close similarities among the structures of tested compounds allowed the drawing of some SARs, identifying a couple of important substituent positive effects: (1) absence of methyl substituent at position 10 (compare boeravinones G and H); (2) presence of a methoxy group over a hydroxy group at position 6 (compare

TABLE 12.1

Inhibition of ABCG2-Mediated Mitoxantrone Efflux by *Boerhaavia* Rotenoids

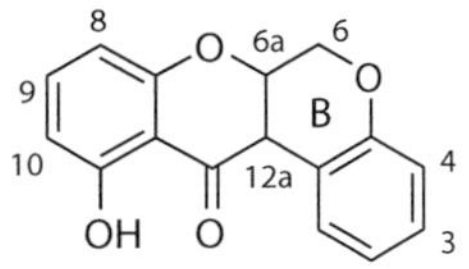

Compound	\multicolumn{6}{c}{Substituents}	Concentration	% Maximal Mitoxantrone Accumulation
	3	4	6	9	10	6a/12a	(µM)	
Boeravinone G	H	OH	OMe	OMe	H	DB	5	92 ± 6.5
Boeravinone H	H	OH	OMe	OMe	Me	DB	5	68 ± 6.1
Boeravinone E	OH	H	OH	OH	Me	DB	10	56 ± 5.0
Boeravinone B	H	H	OH	OH	Me	DB	10	55 ± 5.8
Boeravinone C	H	OH	H	OMe	Me	12a-OH	10	31 ± 4.2
Coccineone B	H	H	OH	OH	H	DB	10	29 ± 5.3
Boeravinone A	H	H	OMe	OH	Me	DB	10	27 ± 5.1
Coccineone E	H	H	H	OMe	OMe	12a-OH	10	15 ± 5.2
DemethylBoer. H	H	OH	OH	OMe	Me	DB	20	15 ± 3.1
Boeravinone I	H	H	OH	OMe	Me	DB	20	12 ± 5.4
Boeravinone J	OH	H	Contracted ring B	OH	Me	DB	20	15 ± 3.1

Substituent at position 8: Always H, except for boeravinone I (OH).
DB, double bond.

boeravinone H and its 6-O-demethyl derivative). However, when position 4 was unsubstitued and a hydroxyl replaced the methoxy group at both positions 9 and 6, the two above effects appeared to be reversed. In addition, an intact ring B was evidently preferable, since its contraction to a five-membered ring, giving a coumaronochromone derivative, resulted to be clearly detrimental (compare boeravinones E and J). Similarly, the tetrasubstitution of ring D, present in boeravinone I, appeared to be responsible for a marked decrease in activity. In contrast, the presence of a hydroxyl group at position 3 seemed not to have significant effects on activity.

Interestingly, although boeravinones C and E had no direct counterpart, it seemed that loss of the double bond $\Delta^{6a(12a)}$ between rings B and C, and consequently of the molecular planarity, was deleterious for activity. As reported above, a similar effect, at the same positions, was evidenced for flavanones, which resulted to be much less active than the corresponding flavones. Moreover, the comparison with data reported in a different work for the isoflavonoid genistein (Figure 12.16) (Ahmed-Belkacem et al. 2005), which exhibited $IC_{50} = 24\,\mu M$, highlighted the importance of ring B (Figure 12.17), which is characteristic of the rotenoid skeleton, by comparison to isoflavonoids.

The above delineated SARs extended and gave support to those previously deduced for flavonoids. Figure 12.18 evidences that the complete "western" moiety (including

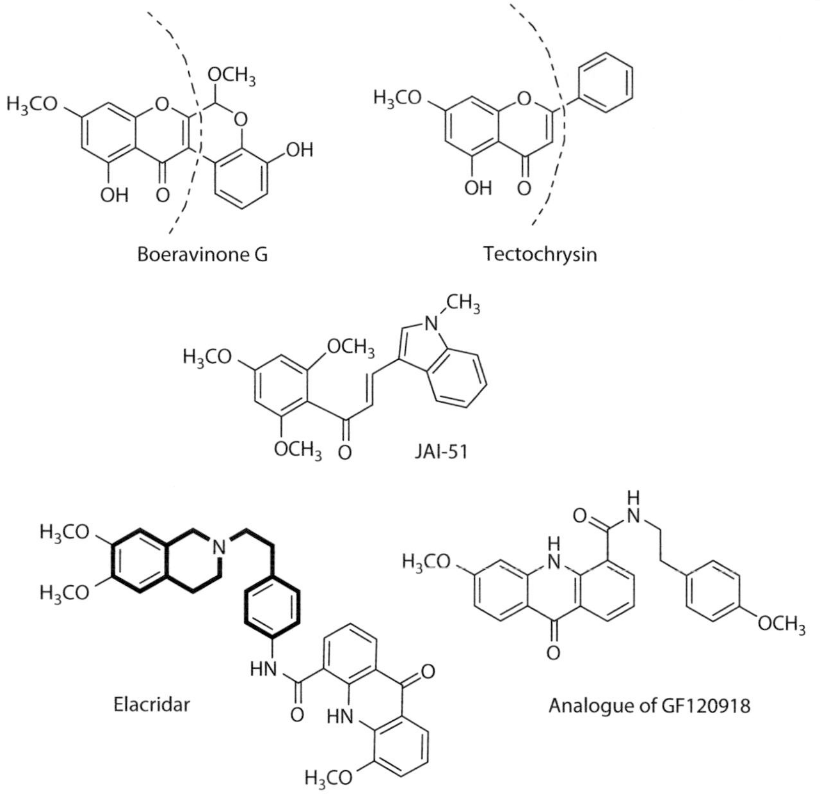

FIGURE 12.18 Above: Structure similarity between the boeravinone G and tectochrysin. Below: Chemical structures of JAI-51, elacridar (similarities with the rotenoid scaffold highlighted in bold), and a synthetic acridone.

rings C and D) of boeravinone G, the best rotenoid identified to date, was identical to that of the previously reported tectochrysin, the best unprenylated flavonoid (Ahmed-Belkacem et al. 2005). However, boeravinone G was more potent than tectochrysin, with an IC_{50} of 0.7 μM as compared to 3.0 μM; these data, once again, highlighted the role of the "eastern" moiety of the molecule, which in rotenoid is stabilized by the additional heterocyclic and substituted ring B. It is interesting to notice that the introduction of a noncyclized prenyl group at C-6 in chrysin derivatives (corresponding to C-10 in boeravinone numbering) was found to increase the inhibitory efficiency (IC_{50} from 3.0 to 0.3 μM for 6-prenylchrysin [Ahmed-Belkacem et al. 2005]). Thus, it would be quite interesting to test the ABCG2-inhibitory activity of natural or synthetic boeravinone G derivatives bearing a noncyclized prenyl group at the same position.

Finally, it is also worthwhile underlining that a number of synthetic compounds reported to be active in ABCG2 inhibition also exhibited some of the above delineated structural features. For example, JAI-51 and other synthetic chalcone-based derivatives (Figure 12.18) (Liu et al. 2008, Boumendjel et al. 2009), recently reported to inhibit ABCG2, showed a methoxylated ring D and an organization of the other substitutents, which is clearly reminiscent of those of flavones and rotenoids.

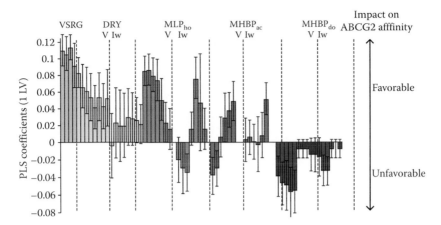

FIGURE 12.19 Partial least-squares (PLS) coefficients of the molecular model of ABCG2-specific inhibitors (V: Volume, Iw: Integy). VSRG, shape parameters: Volume/surface/rugosity/globularity; DRY, polarizability; MLPho, molecular lipophilic potential; MHBP, molecular hydrogen bound potential as acceptor (ac) or donor (do).

Analogously, Figure 12.18 shows that a portion of the structure of elacridar (highlighted in bold), a reference compound for ABCG2 inhibition (although probably mediated by a different mechanism by comparison to flavonoids), includes a number of structural motifs, which are similar to the rotenoid structural organization. Moreover, a synthetic acridone derivative (**17**, *cf.* Figure 12.9), related to elacridar and showing an IC$_{50}$ value quite similar to that of boeravinone G, shares with this rotenoid also a number of structural features, including methoxylation at the peripheral aromatic ring (Nicolle et al. 2009a).

At this stage, the parallel data obtained from flavones, chalcones, and rotenoids, which have been systematically analyzed in different 3D-QSAR studies (see next paragraphs) (Nicolle et al. 2009a,b), appeared to be sufficient to guide a number of chemical modifications on these simple planar structures in order to pave the way for the development of even more potent MDR modulators. The lack of cytotoxicity of these compounds is also encouraging and makes them candidates for in vivo tests as adjuvant in the antitumor chemotherapy.

12.3.4 MOLECULAR MODEL OF ABCG2-SPECIFIC INHIBITORS: LIGAND-BASED DRUG DESIGN

A series of compounds derived from naturally occurring flavonoids and synthetic analogs, including both flavones and benzopyrane/benzofurane derivatives, boeravinones and acridone derivatives have been evaluated under the same conditions for their inhibition of mitoxantrone efflux from human ABCG2-transfected HEK-293 cells. Five new natural compounds obtained from *Morus mesozygia* Stapf and one synthetic chromone, comprising a flavonoidic scaffold, were also evaluated. Based on the results obtained with a total of 34 compounds, a 3D linear solvation energy QSAR was investigated by VolSurf descriptors of molecular interaction fields (MIFs) related to hydrophobic interaction forces, polarizability, and hydrogen-bonding capacity. Accuracy of the constructed 3D-QSAR model was attested by a correlation coefficient of 0.77. Figure 12.19 shows that shape parameters and hydrophobicity were revealed to be major physicochemical parameters responsible for the inhibition activity of flavonoid derivatives and synthetic analogs toward ABCG2, whereas hydrogen-bond acceptor capacity had limited effect and hydrogen-bond donor capacity appeared highly unfavorable (Nicolle et al. 2009a). The reliability of the model generated by the training test was validated by a test set of seven other related compounds. Such a molecular model will allow the design and synthesis of even more potent and specific new inhibitors of ABCG2.

12.4   FUTURE DEVELOPMENTS

12.4.1   MOLECULAR MODELS FOR COMPETITIVE AND NONCOMPETITIVE INHIBITORS

Due to their polyspecificity, multidrug ABC transporters are known to have large, poorly defined, regions involved in binding transported substrate drugs and inhibitors. Up to three distinct substrate-binding sites have been identified by the group of Victor Ling in ABCB1: two catalytic, cooperative, transport sites revealed by rhodamine 123 (R site) and Hoechst 33342 (H site) (Shapiro and Ling 1997b), and one regulatory, allosteric, site binding progesterone and prazosin (Shapiro et al. 1999). Interestingly, only one catalytic site, the R site, appears to function for transport in wild-type ABCG2, whereas the H site is unmasked by the hot-spot mutation R482G/T occurring within the third transmembrane span (Sarkadi et al. 2004). In addition, the BODIPY derivative of prazosin is indeed transported by ABCG2. All these common and divergent parameters should be taken into account for the competitive inhibitors, and allow the generation of distinct, but possibly partially overlapping, molecular models established by ligand-based drug design.

Similarly, noncompetitive inhibitors might interact with different, possibly large, areas outside the catalytic sites, including transmembrane spans and/or intracellular loops. The latter ones, known to connect and transduce conformational changes between the membrane and nucleotide-binding domains, might as well contribute to inhibitor (and substrate?) recognition and binding. Again, distinct molecular models would be expected for different classes of noncompetitive inhibitors. The specificity of a given transporter for different noncompetitive inhibitors should be correlated to their interaction to various sequences within the binding area expected to be significantly divergent between ABCB1 and ABCG2.

12.4.2   FEASIBILITY OF DOCKING: STRUCTURE-BASED DRUG DESIGN

A high-resolution crystal structure has been recently published for mouse mdr1a P-glycoprotein, either free or with bound enantiomers of a cyclic hexapeptide inhibitor, QZ59 (Aller et al. 2009). This should constitute a good template to construct a molecular model of human ABCB1, in order to investigate docking of the different types of substrates and inhibitors. This might allow a structural localization of the different binding sites, and give a better idea of their relative overlapping if any.

Partial molecular models could also be generated for ABCG2 and ABCC1, and similarly help to visualize the different binding sites for substrates and modulators, either as inhibitors for ABCG2 or apoptogenic compounds for MRP1, since verapamil is a common ligand to the different multidrug ABC transporters.

ACKNOWLEDGMENTS

This work was supported by the Centre National de la Recherche Scientifique (CNRS) and Université Lyon 1 (UMR 5086) and grants from the Agence Nationale de la Recherche (ANR-06-BLAN-0420), the Association pour la Recherche sur le Cancer (ARC 3942), and the Ligue Nationale contre le Cancer (Comités du Rhône, de la Drôme et de la Savoie, et Labellisation Ligue 2009).

REFERENCES

Ahmed-Belkacem A., Macalou S., Borrelli F., Capasso R., Fattorusso E., Taglialatela-Scafati O., and Di Pietro A. 2007. Nonprenylated rotenoids, a new class of potent breast cancer resistance protein inhibitors. *J. Med. Chem.* 50:1933–1938.

Ahmed-Belkacem A., Pozza A., Macalou S., Perez-Victoria J.M., Boumendjel A., and Di Pietro A. 2006. Inhibitors of cancer cell multidrug resistance mediated by breast cancer resistance protein (BCRP/ABCG2). *Anticancer Drug.* 17:239–244.

Ahmed-Belkacem A., Pozza A., Munoz-Martinez F., Bates S.E., Castanys S., Gamarro F., Di Pietro A., and Perez-Victoria J.M. 2005. Flavonoid structure–activity studies identify 6-prenylchrysin and tectochrysin as potent and specific inhibitors of breast cancer resistance protein ABCG2. *Cancer Res.* 65:852–4860.

Allen J.D., van Loevezijn A., Lakhai J.M., van der Valk M., van Tellingen O., Reid G., Schellens J.H.M., Koomen G.J., and Schinkel A.H. 2002. Potent and specific inhibition of the breast cancer resistance protein multidrug transporter in vitro and in mouse intestine by a novel analogue of fumitremorgin C. *Mol. Cancer Ther.* 1:417–425.

Aller S.G., Yu J., Ward A., Weng Y., Chittaboina S., Zhuo R., Harrell P.M., Tringh Y.T., Zhang Q., Urbatsch I.L., and Chang G. 2009. Structure of P-glycoprotein reveals a molecular basis for a poly-specific drug binding. *Science* 323:1718–1722.

Alvarez A.I., Real R., Pérez M., Mendoza G., Prieto J.G., and Merino G. 2009. Modulation of the activity of ABC transporters (P-glycoprotein, MRP2, BCRP) by flavonoids and drug response. *J. Pharm. Sci.* 99:598–617.

Appendino G., Jakupovic S., Tron G.C., Jakupovic J., Milon V., and Ballero M. 1998. Macrocyclic diterpenoids from *Euphorbia semiperfoliata. J. Nat. Prod.* 61:749–756.

Arnaud O., Boumendjel A., Gèze A., Honorat M., Matera E.L., Guitton J., Stein W.D., Bates S.E., Falson P., Dumontet C., Di Pietro A., and Payen L. 2011. The acridone derivative MBLI-87 sensitizes breast cancer resistance protein-expressing xenografts to irinotecan. *Eur. J. Cancer* 47:640–648.

Aszalos A. 2008. Role of ATP-binding cassette (ABC) transporters in interactions between natural products and drugs. *Curr. Drug Metab.* 9:1010–1018.

Badhan R. and Penny J. 2006. In silico modeling of the interaction of flavonoids with human P-glycoprotein nucleotide-binding domain. *Eur. J. Med. Chem.* 41:285–295.

Bansal T., Jaggi M., Khar R.K., and Talegaonkar S. 2009. Emerging significance of flavonoids as P-glycoprotein inhibitors in cancer chemotherapy. *J. Pharm. Pharm. Sci.* 12:46–78.

Bardelmeijer H.A., Ouwehand M., Beijnen J.H., Schellens J.H., and van Tellingen O. 2004. Efficacy of novel P-glycoprotein inhibitors to increase the oral uptake of paclitaxel in mice. *Invest. New Drug.* 22:219–229.

Barile E., Fattorusso E., Ialenti A., Ianaro A., and Lanzotti V. 2007. Paraliane and pepluane diterpenes as anti-inflammatory agents: First insights in structure–activity relationships. *Bioorg. Med. Chem. Lett.* 17:4196–4200.

Barthomeuf C., Demeule M., Grassi J., Saikhodjaev A., and Béliveau R. 2006. Conferone from Ferula schtschurowskiana enhances vinblastine cytotoxicity in MDCK-MDR1 cells by competitively inhibiting P-glycoprotein transport. *Planta Med.* 72:634–639.

Barthomeuf C., Grassi J., Demeule M., Fournier C., Boivin D., and Béliveau R. 2005. Inhibition of P-glycoprotein transport function and reversion of MDR1 multidrug resistance by cnidiadin. *Cancer Chemother. Pharmacol.* 56:173–181.

Bates S.E., Chen C., Robey R., Kang M., Figg W.D., and Fojo T. 2002. Reversal of multidrug resistance: Lessons from clinical oncology. *Novartis Found Symp.* 243:83–96; discussion 96–102, 180–185.

Benderra Z., Faussat A.-M., Sayada L., Perrot J.-Y., Chaoui D., Marie J.-P., and Legrand O. 2004. Breast cancer resistance protein and P-glycoprotein in 149 adult acute myeloid leukemias. *Clin. Cancer Res.* 10:7896–7902.

Berger W., Setinek U., Hollaus P., Zidek T., Steiner E., Elbling L., Cantonati H., Attems J., Gsur A., and Micksche M. 2005. Multidrug resistance markers P-glycoprotein, multidrug resistance protein 1, and lung resistance protein in non-small cell lung cancer: Prognostic implications. *J. Cancer Res. Clin. Oncol.* 131:355–363.

Boccard J., Bajot F., Di Pietro A., Rudaz S., Boumendjel A., Nicolle E., and Carrupt P.A. 2009. A 3D linear solvation energy model to quantify the affinity of flavonoid derivatives towards P-glycoprotein. *Eur. J. Pharm. Sci.* 36:254–264.

Bois F., Beney C., Boumendjel A., Mariotte A.-M., Conseil G., and Di Pietro A. 1998. Halogenated chalcones with high-affinity binding to P-glycoprotein: Potential modulators of multidrug resistance. *J. Med. Chem.* 41:4161–4164.

Bois F., Boumendjel A., Mariotte A.-M., Conseil G., and Di Pietro A. 1999. Synthesis and biological activity of 4-alkoxy chalcones: Potential hydrophobic modulators of P-glycoprotein-mediated multidrug resistance. *Bioorg. Med. Chem.* 7:2691–2695.

Bolhuis H., van Veen H.W., Poolman B., Driessen A.J.M., and Konings W.N. 1997. Mechanisms of multidrug transporters. *FEMS Microbiol. Rev.* 21:55–84.

Boumendjel A., Baubichon-Cortay H., Trompier D., Perrotton T., and Di Pietro A. 2005. Anticancer multidrug resistance mediated by MRP1: Recent advances in the discovery of reversal agents. *Med. Res. Rev.* 25:453–472.

Boumendjel A., Bois F., Beney C., Mariotte A.-M., Conseil G., and Di Pietro A. 2001. B-ring substituted 5,7-dihydroxyflavonols with high-affinity binding to P-glycoprotein responsible for cell multidrug resistance. *Bioorg. Med. Chem. Lett.* 11:75–77.

Boumendjel A., Di Pietro A., Dumontet C., and Barron D. 2002. Recent advances in the discovery of flavonoids and analogs with high-affinity binding to P-glycoprotein responsible for cancer cell multidrug resistance. *Med. Res. Rev.* 22:512–529.

Boumendjel A., Macalou S., Ahmed-Belkacem A., Blanc M., and Di Pietro A. 2007. Acridone derivatives: Design, synthesis, and inhibition of breast cancer resistance protein ABCG2. *Bioorg. Med. Chem.* 15:2892–2897.

Boumendjel A., McLeer-Florin A., Champelovier P., Allegro D., Muhammad D., Souard F., Derouazi M., Peyrot V., Toussaint B., and Boutonnat J. 2009. A novel chalcone derivatives which acts as a microtubule depolymerising agent and an inhibitor of Pg-P and BCRP in in-vitro and in-vivo glioblastoma models. *BMC Cancer* 9:242–252.

Brand W., Schutte M.E., Williamson G., van Zanden J.J., Cnubben N.H., Groten J.P., van Bladeren P.J., and Rietjens I.M. 2006. Flavonoid-mediated inhibition of intestinal ABC transporters may affect the oral bioavailability of drugs, food-borne toxic compounds and bioactive ingredients. *Biomed. Pharmacother.* 60:508–519.

Brooks T.A., Minderman H., O'Loughlin K.L., Pera P., Ojima I., Baer M.R., and Bernacki R.J. 2003. Taxane-based reversal agents modulate drug resistance mediated by P-glycoprotein, multidrug resistance protein, and breast cancer resistance protein. *Mol. Cancer Ther.* 2:1195–1205.

Burger H., van Tol H., Boersma A.W., Brok M., Wiemer E.A., Stoter G., and Nooter K. 2004. Imatinib mesylate (STI571) is a substrate for the breast cancer resistance protein (BCRP)/ABCG2 drug pump. *Blood* 104:2940–2942.

Castro A.F. and Altenberg G.A. 1997. Inhibition of drug transport by genistein in multidrug-resistant cells expressing P-glycoprotein. *Biochem. Pharmacol.* 53:89–93.

Chan K.F., Zhao Y., Burkett B.A., Wong I.L., Chow L.M., and Chan T.H. 2006. Flavonoid dimers as bivalent modulators for P-glycoprotein-based multidrug resistance: Synthetic apigenin homodimers linked with defined-length poly(ethylene glycol) spacers increase drug retention and enhance chemosensitivity in resistant cancer cells. *J. Med. Chem.* 49:6742–6759.

Cherwae W., Wu C.P., Chu H.Y., Lee T.R., Ambudkar S.V., and Limtrakul P. 2006. Curcuminoids purified from turmeric powder modulate the function of human multidrug resistance protein 1 (ABCC1). *Cancer Chemother. Pharmacol.* 57:376–388.

Chieli E., Romiti N., Cervelli F., and Tongiani R. 1995. Effects of flavonols on P-glycoprotein activity in cultured rat hepatocytes. *Life Sci.* 57:1741–1751.

Choi J.S., Jo B.W., and Kim Y.C. 2004. Enhanced paclitaxel bioavailability after oral administration of paclitaxel or prodrug to rats pretreated with quercetin. *Eur. J. Pharm. Biopharm.* 57:313–318.

Choi C.H., Kang G., and Min Y.D. 2003. Reversal of P-glycoprotein-mediated multidrug resistance by protopanaxatriol ginsenosides from Korean red ginseng. *Planta Med.* 69:235–240.

Chung S.Y., Sung M.K., Kim N.H., Jang J.O., Go E.J., and Lee H.J. 2005. Inhibition of P-glycoprotein by natural products in human breast cancer cells. *Arch. Pharm. Res.* 28:823–828.

Clark R., Kerr I.D., and Callaghan R. 2006. Multiple drugbinding sites on the R482G isoform of the ABCG2 transporter. *Brit. J. Pharmacol.* 149:506–515.

Cole S.P., Bhardwaj G., Gerlach J.H., Mackie J.E., Grant C.E., Almquist K.C., Stewart A.J., Kurz E.U., Duncan A.M., and Deeley R.G. 1992. Overexpression of a transporter gene in a multidrug-resistant human lung cancer cell line. *Science* 258:1650–1654.

Cole S.P. and Deeley R.G. 2006. Transmembrane transport of endo- and xenobiotics by mammalian ATP-binding cassette multidrug resistance proteins. *Physiol. Rev.* 86:849–899.

Comte G., Daskiewicz J.-B., Bayet C., Conseil G., Viornery-Vanier A., Dumontet C., and Di Pietro A. 2001. *C*-isoprenylation of flavonoids enhances binding affinity towards P-glycoprotein and modulation of cancer cell chemoresistance. *J. Med. Chem.* 44:763–768.

Conseil G., Baubichon-Cortay H., Dayan G., Jault J.-M., Barron D., and Di Pietro A. 1998. Flavonoids: A class of modulators with bifunctional interactions at vicinal ATP- and steroid-binding sites on mouse P-glycoprotein. *Proc. Natl. Acad. Sci. USA* 95:9831–9836.

Conseil G., Decottignies A., Jault J.-M., Comte G., Barron D., Goffeau A., and Di Pietro A. 2000. Prenyl flavonoids as potent inhibitors of the Pdr5p multidrug ABC transporter from *Saccharomyces cerevisiae*. *Biochemistry* 39:6910–6917.

Cooray H.C., Janvilisri T., van Veen H.W., Hladky S.B., and Barrand M.A. 2004. Interaction of the breast cancer resistance protein with plant polyphenols. *Biochem. Biophys. Res. Commun.* 317:269–275.

Corea G., Di Pietro A., Dumontet C., Fattorusso E., and Lanzotti V. 2009. Jatrophane diterpenes from *Euphorbia* spp. as modulators of multidrug resistance in cancer therapy. *Phytochem. Rev.* 8:431–448.

Corea G., Fattorusso C., Fattorusso E., and Lanzotti V. 2005. Amygdaloidins A-L, twelve new 13 α-OH jatrophane diterpenes from *Euphorbia amygdaloides* L. *Tetrahedron* 61:4485–4494.

Corea G., Fattorusso E., Lanzotti V., Motti R., Simon P.N., Dumontet C., and Di Pietro A. 2004a. Structure–activity relationships for euphocharacins A-L, a new series of jatrophane diterpenes, as inhibitors of cancer cell P-glycoprotein. *Planta Med.* 70:657–665.

Corea G., Fattorusso E., Lanzotti V., Motti R., Simon P.N., Dumontet C., and Di Pietro A. 2004b. Jatrophane diterpenes as modulators of multidrug resistance. Advances of structure–activity relationships and discovery of the potent lead pepluanin A. *J. Med. Chem.* 47:988–992.

Corea G., Fattorusso E., Lanzotti V., Taglialatela-Scafati O., Appendino G., Ballero M., Simon P.N., Dumontet C., and Di Pietro A. 2003a. Jatrophane diterpenes as P-glycoprotein inhibitors. First insights of structure–activity relationships and discovery of a new, powerful lead. *J. Med. Chem.* 46:3395–3402.

Corea G., Fattorusso E., Lanzotti V., Taglialatela-Scafati O., Appendino G., Ballero M., Simon P.N., Dumontet C., and Di Pietro A. 2003b. Modified jatrophane diterpenes as modulators of multidrug resistance from *Euphorbia dendroides* L. *Bioorg. Med. Chem.* 11:5221–5227.

Critchfield J.W., Welsh C.J., Phang J.M., and Yeh G.C. 1994. Modulation of adriamycin in accumulation and efflux by flavonoids in HCT-15 colon cells. Activation of P-glycoprotein as a putative mechanism. *Biochem. Pharmacol.* 48:1437–1445.

Crombie L., Josephs J.L., Cayley J., Larkin J., and Weston J.B. 1992. The rotenoid core structure: Modifications to define the requirements of the toxophore. *Bioorg. Med. Chem. Lett.* 2:13–16.

Dantzig A.H., Shepard R.L., Law K.L., Tabas L., Pratt S., Gillespie J.S., Binkley S.N., Kuhfeld M.T., Starling J.J., and Wrighton S.A. 1999. Selectivity of the multidrug resistance modulator, LY335979, for P-glycoprotein and effect on cytochrome P-450 activities. *J. Pharmacol. Exp. Ther.* 290:854–862.

Daskiewicz J.-B., Comte G., Barron D., Di Pietro A., and Thomasson F. 1999. Organolithium-mediated synthesis of prenylchalcones as potential inhibitors of chemoresistance. *Tetrahedron Lett.* 40:7095–7098.

Dayan G., Jault J.-M., Baubichon-Cortay H., Baggetto L.G., Renoir J.M., Baulieu E.E., Gros P., and Di Pietro A. 1997. Binding of steroid modulators to recombinant cytosolic domain from mouse P-glycoprotein in close proximity to the ATP site. *Biochemistry* 36:15208–15215.

De Azevedo W. F. Jr., Mueller-Dieckmann H.-J., Schulze-Gahmen U., Worland P. J., Sausville E., and Kim S.-H. 1996. Structural basis for specificity and potency of a flavonoid inhibitor of human CDK2, a cell cycle kinase. *Proc. Natl. Acad. Sci. USA* 93:2735–2740.

De Bruin M., Miyake K., Litman T., Robey R.W., and Bates S.E. 1999. Reversal of resistance by GF120918 in cell lines expressing the ABC half-transporter, MXR. *Cancer Lett.* 146:117–126.

De Wet H., McIntosh D.B., Conseil G., Baubichon-Cortay H., Krell T., Jault J.-M., Daskiewicz J.-B., Barron D., and Di Pietro A. 2001. Sequence requirements of the ATP-binding site within the C-terminal nucleotide-binding domain of mouse P-glycoprotein: Structure–activity relationships for flavonoid binding. *Biochemistry* 40:10382–10391.

Deferme S. and Augustijns P. 2003. The effect of food components on the absorption of P-gp substrates: A review. *J. Pharm. Pharmacol.* 55:153–162.

Di Pietro A., Conseil G., Pérez-Victoria J.M., Dayan G., Baubichon-Cortay H., Trompier D., Steinfels E., Jault J.M., de Wet H., Maitrejean M., Comte G., Boumendjel A., Mariotte A.M., Dumontet C., McIntosh D.B., Goffeau A., Castanys S., Gamarro F., and Barron D. 2002. Modulation by flavonoids of cell multidrug resistance mediated by P-glycoprotein and related ABC transporters. *Cell. Mol. Life Sci.* 59:307–322.

Doyle L.A., Yang W., Abruzzo L.V., Krogmann T., Gao Y., Rishi A.K., and Ross D.D. 1998. A multidrug resistance transporter from human MCF-7 breast cancer cells. *Proc. Natl. Acad. Sci. USA* 95:15665–15670.

Dupuy J., Larrieu G., Sutra J.F., Lespine A., and Alvinerie M. 2003. Enhancement of moxidectin bioavailability in lambs by a natural flavonoid: Quercetin. *Vet. Parasitol.* 112:337–347.

Edwards D.J., Fitzsimmons M.E., Schuetz E.G., Yasuda K., Ducharme M.P., Warbasse L.H., Woster P.M., Schuetz J.D., and Watkins P. 1999. 6′,7′-Dihydroxybergamottin in grapefruit juice and Seville orange juice: Effects on cyclosporin disposition, enterocyte CYP3A4, and P-glycoprotein. *Clin. Pharmacol. Ther.* 65:237–244.

Endicott J.A. and Ling V. 1989. The biochemistry of P-glycoprotein-mediated multidrug resistance. *Annu. Rev. Biochem.* 58:137–171.

Evans A.M. 2000. Influence of dietary components on the gastrointestinal metabolism and transport of drugs. *Ther. Drug. Monit.* 22:131–136.

Ferté J., Kuhnel J.M., Chapuis G., Rolland Y., Lewin G., and Schwaller M.A. 1999. Flavonoid-related modulators of multidrug resistance: Synthesis, pharmacological activity, and structure–activity relationships. *J. Med. Chem.* 42:478–489.

Florin A., Boutonnat J., and Boumendjel A. 2008. Overcoming BCRP-mediated multidrug resistance. *Drugs Future* 33:533–542.

Fu L.W., Deng Z.A., Pan Q.C., and Fan W. 2001. Screening and discovery of novel MDR modifiers from naturally occurring bisbenzylisoquinoline alkaloids. *Anticancer Res.* 21:2273–2280.

Fuhr U. and Kummert A.L. 1995. The fate of naringin in humans: A key to grapefruit juice-drug interactions? *Clin. Pharmacol. Ther.* 58:365–373.

Gupta A., Dai Y., Vethanayagam R.R., Hebert M.F., Thummel K.E., Unadkat J.D., Ross D.D., and Mao Q. 2006. Cyclosporin A, tacrolimus and sirolimus are potent inhibitors of the human breast cancer resistance protein (ABCG2) and reverse resistance to mitoxantrone and topotecan. *Cancer Chemother. Pharmacol.* 58:374–383.

Gutman M., Singer T.P., and Casida J.E. 1970. Studies on the respiratory chain-linked reduced nicotinamide adenine dinucleotide dehydrogenase. XVII. Reaction sites of piericidin A and rotenone. *J. Biol. Chem.* 245:1992–1997.

Han Y., Riwanto M., Go M.L., and Ee P.L. 2008. Modulation of breast cancer resistance protein (BCRP/ABCG2) by non-basic chalcone analogues. *Eur. J. Pharm. Sci.* 35:30–41.

Hayeshi R., Masimirembwa C., Mukangayama S., and Ungell A.L. 2006. The potential inhibitory effect of antiparasitic drugs and natural products on P-glycoprotein mediated efflux. *Eur. J. Pharm. Sci.* 29:70–81.

Henrich C.J., Robey R.W., Takada K., Bokesch H.R., Bates S.E., Shukla S., Ambudkar S.V., McMahon J.B., and Gustafson K.R. 2009. Botryllamides: Natural product inhibitors of ABCG2. *ACS Chem. Biol.* 4:637–647.

Higgins C.F. 1994. P-glycoprotein. To flip or not to flip? *Curr. Biol.* 4:259–260.

Ho M.M., Hogge D.E., and Ling V. 2008. MDR1 and BCRP1 expression in leukemic progenitors correlates with chemotherapy response in acute myeloid leukemia. *Exp. Hematol.* 36:433–442.

Hohmann J., Molnar J., Redei D., Evanics F., Forgo P., Kalman A., Argay G., and Szabo P. 2002. Discovery and biological evaluation of a new family of potent modulators of multidrug resistance: Reversal of multidrug resistance of mouse lymphoma cells by new natural jatrophane diterpenoids isolated from Euphorbia species. *J. Med. Chem.* 45:2425–2431.

Hohmann J., Vasas A., Gunther G., Mathe I., Evanics F., Dombi G., and Jerkovich G. 1997. Macrocyclic diterpene polyester of the jatrophane type from *Euphorbia esula*. *J. Nat. Prod.* 60:331–335.

Houghton P.J., Germain G.S., Harwood F.C., Schuetz J.D., Stewart C.F., Buchdunger E., and Traxler P. 2004. Imatinib mesylate is a potent inhibitor of the ABCG2 (BCRP) transporter and reverses resistance to topotecan and SN-38 in vitro. *Cancer Res.* 64:2333–2337.

Hunter J. and Hirst B.H. 1997. Intestinal secretion of drugs. The role of P-glycoprotein and related drug efflux systems in limiting oral drug absorption. *Adv. Drug. Deliv. Rev.* 25:129–157.

Imai Y., Tsukahara S., Asada S., and Sugimoto Y. 2004. Phytoestrogens/flavonoids reverse breast cancer resistance protein/ABCG2-mediated multidrug resistance. *Cancer Res.* 64:4346–4352.

Jakupovic J., Jeske F., Morgenstern T., Tsichritzis F., Marco J.A., and Berendshon W. 1998a. Diterpenes from *Euphorbia segetalis*. *Phytochemistry* 47:1583–1600.

Jakupovic J., Morgenstern T., Bittner M., and Silva M. 1998b. Diterpenes from *Euphorbia peplus*. *Phytochemistry* 47:1601–1609.

Jin J., Shahi S., Kang H.K., van Veen H.W., and Fan T.P. 2006. Metabolites of gingenosides as novel BCRP inhibitors. *Biochem. Biophys. Res. Commun.* 345:1308–1314.

Kam T.S., Sim K.M., Pang H.S., Koyano T., Hayashi M., and Komiyama K. 2004. Cytotoxic effects and reversal of multidrug resistance by ibogan and related indole alkaloids. *Bioorg. Med. Chem. Lett.* 14:4487–4489.

Katayama K., Masuyama K., Yoshioka S., Hasegawa H., Mitsuhashi J., and Sugimoto Y. 2007. Flavonoids inhibit breast cancer resistance protein-mediated drug resistance: Transporter specificity and structure–activity relationship. *Cancer Chemother. Pharmacol.* 60:789–797.

Kim S.W., Kwon H.Y., Chi D.W., Shim J.H., Park J.D., and Lee Y.H. 2003. Reversal of P-glycoprotein-mediated multidrug resistance by ginsenoside Rg(3). *Biochem. Pharmacol.* 65:75–82.

Kühnle M., Egger M., Müller C., Mahringer A., Bernhardt G., Fricker G., König B., and Buschauer A. 2009. *J. Med. Chem.* 52:1190–1197.

Laberge R.M., Karwatsky J., Lincoln M.C., Leimanis M.L., and Georges E. 2007. Modulation of GSH levels in ABCC1 expressing tumor cells triggers apoptosis through oxidative stress. *Biochem Pharmacol.* 73:1727–1737.

Lai M.Y., Hsiu S.L., Hou Y.C., Tsai S.Y., and Chao P.D. 2004. Significant decrease of cyclosporine bioavailability in rats caused by a decoction of the roots of *Scutellaria baicalensis*. *Planta Med.* 70:132–137.

Lee J.S., Scala S., Matsumoto Y., Dickstein B., Robey R.W., Zhan Z., Altenberg G., and Bates S.E. 1997. Reduced drug accumulation and multidrug resistance in human breast cancer cells without associated P-glycoprotein or MRP overexpression. *J. Cell. Biochem.* 65:513–526.

Li X., Sun B., Zhu C.J., Yuan H.Q., Shi Y.Q., Goa J., Li S.J., and Lou H.X. 2009. Reversal of P-glycoprotein-mediated multidrug resistance by macrocyclic bisbibenzyl derivatives in adriamycin-resistant human myelogenous leukemia (K562/A02) cells. *Toxicol. In Vitro* 23:29–36.

Lim S., Grassi J., Akhmedjanova V., Debiton E., Balansard G., Béliveau R., and Barthomeuf C. 2005. Reversal of P-glycoprotein-mediated drug efflux by eudesmin from Haplophyllum perforatum and cytotoxicity pattern versus diphyllin, podophyllotoxin and etoposide. *Planta Med.* 73:1563–1567.

Limtrakul P., Cherwae W., Shukla S., Phisalphong C., and Ambudkar S.V. 2007. Modulation of function of three ABC transporters, P-glycoprotein (ABCB1), mitoxantrone resistance protein (ABCG2) and multidrug resistance protein 1 (ABCC1) by tetrahydrocurcumin, a major metabolite of curcumin. *Mol. Cell. Biochem.* 296:85–95.

Lin J.H., Chiba M., and Baillie T.A. 1999. Is the role of the small intestine in first-pass metabolism overemphasized? *Pharmacol. Rev.* 51:135–157.

Liu X.L., Tee H.W., and Go M.L. 2008. Functionalized chalcones as selective inhibitors of P-glycoprotein and breast cancer resistance protein. *Bioorg. Med. Chem.* 16:171–180.

Maitrejean M., Comte G., Barron D., El Khirat K., Conseil G., and Di Pietro A. 2000. The flavanolignan silybin and its hemisynthetic derivatives, a novel series of potential modulators of P-glycoprotein. *Bioorg. Med. Chem. Lett.* 10:157–160.

Marco J.A., Sanz-Cervera J.F., Yuste A., and Jakupovic J. 1997. Terracinolides from *Euphorbia terracina*. *Phytochemistry* 45:137–140.

Marco J.A., Sanz-Cervera J.F., Yuste A., and Jakupovic J. 1999. Isoterracinolides A and B, novel bishomoditerpene lactones from euphorbia terracina. *J. Nat. Prod.* 62:110–113.

Marco J.A., Sanz-Cevera J.F., Yuste A., Jakupovic J., and Jeske F. 1998. Jatrophane derivatives and rearranged jatrophane from *Euphorbia terracina. Phytochemistry* 47:1621–1630.

Marco J.A., Sanz-Cervera J.F., Yuste A., Jakupovic J., and Lex J. 1996. Terracinolides A and B, two bishomoditerpene lactones with a novel carbon framework from *Euphorbia terracina. J. Org. Chem.* 61:1707–1709.

Miyake K., Mickley L., Litman T., Zhan Z., Robey R., Cristensen B., Brangi M., Greenberger L., Dean M., Fojo T., and Bates S.E. 1999. Molecular cloning of cDNAs which are highly overexpressed in mitoxantrone-resistant cells: Demonstration of homology to ABC transport genes. *Cancer Res.* 59:8–13.

Molnar J., Gyemant N., Tanaka M., Hohmann J., Bergmann-Leitner E., Molnar P., Deli J., Didiziapetris R., and Ferreira M.J. 2006. Inhibition of multidrug resistance of cancer cells by natural diterpenes, triterpenes and carotenoids. *Curr. Pharm. Des.* 12:287–311.

Morita H., Koyama K., Sugimoto Y., and Kobayashi J. 2005. Antimitotic activity and reversal of breast cancer resistance protein-mediated drug resistance by stilbenoids from *Bletilla striata. Bioorg. Med. Chem. Lett.* 15:1051–1054.

Morris M.E. and Zhang S. 2006. Flavonoid-drug interactions: Effects of flavonoids on ABC transporters. *Life Sci.* 78:2116–2130. *Biomed. Pharmacother.* 60:508–519.

Munoz-Martinez F., Lu P., Cortes-Selva F., Perez-Victoria J.M., Jimenez I.A., Ravelo A.G., Sharom F.J., Gamarro F., and Castanys S. 2004. Celastraceae sesquiterpenes as a new class of modulators that bind specifically to human P-glycoprotein and reverse cellular multidrug resistance. *Cancer Res.* 64:7130–7138.

Munoz-Martinez F., Reyes C.P., Perez-Lomas A.L., Jimenez I.A., Gamarro F., and Castanys S. 2006. Insights into the molecular mechanism of action of Celastraceae sesquiterpenes as specific, non-transported, inhibitors of human P-glycoprotein. *Biochim. Biophys. Acta* 1758:98–110.

Murakami S., Muramatsu M., and Tomisawa K. 1999. Inhibition of gastric H^+,K^+-ATPase by flavonoids: A structure–activity study. *J. Enzyme Inhib.* 14:151–166.

Nabekura T., Yamaki T., Ueno K., and Kitagawa S. 2008. Inhibition of P-glycoprotein and multidrug resistance protein 1 by dietary phytochemicals. *Cancer Chemother. Pharmacol.* 62:867–873.

Nguyen V.T.B., Darbour N., Bayet C., Doreau A., Raad I., Phung B.H., Dumontet C., Di Pietro A., Dijoux-Franca M.-G., and Guilet D. 2009. Selective modulation of P-glycoprotein activity by steroidal saponines from *Paris polyphylla. Fitoterapia* 80:39–42.

Nicolle E., Boccard J., Guilet D., Dijoux-Franca M.-G., Zelefac F., Macalou S., Grosselin J., Schmidt J., Carrupt P.A., Di Pietro A., and Boumendjel A. 2009a. Breast cancer resistance protein (BCRP/ABCG2): New inhibitors and QSAR studies by a 3D linear solvation energy approach. *Eur. J. Pharm. Sci.* 38:39–46.

Nicolle E., Boumendjel A., Macalou S., Genoux E., Ahmed-Belkacem A., Carrupt P.-A., and Di Pietro A. 2009b. QSAR analysis and molecular modeling of ABCG2-specific inhibitors. *Adv. Drug Deliver. Rev.* 61:34–46.

Noungoué Tchamo D., Dijoux-Franca M.-G., Mariotte A.-M., Tsamo E., Daskiewicz J.-B., Bayet C., Barron D., Conseil G., and Di Pietro A. 2000. Prenylated xanthones as potential P-glycoprotein modulators. *Bioorg. Med. Chem. Lett.* 10:1343–1345.

Ozvegy-Laczka C., Hegedus T., Varady G., Ujhelly O., Schuetz J.D., Varadi A., Keri G., Orfi L., Nemet K., and Sarkadi B. 2004. High-affinity interaction of tyrosine kinase inhibitors with the ABCG2 multidrug transporter. *Mol. Pharmacol.* 65:1485–1495.

Pérez-Victoria J.M., Chiquero M.J., Conseil G., Dayan G., Di Pietro A., Barron D., Castanys S., and Gamarro F. 1999. Correlation between the affinity of flavonoids binding to cytosolic site of *Leishmania tropica* multidrug transporter and their efficiency to revert parasite resistance to daunomycin. *Biochemistry* 38:1736–1743.

Pérez-Victoria J.M., Cortés-Selva F., Parodi-Talice A., Bavchvarov B.I., Pérez-Victoria F.J., Muñoz-Martínez F., Maitrejean M., Costi M.P., Barron D., Di Pietro A., Castanys S., and Gamarro F. 2006. Combination of suboptimal doses of inhibitors targeting different domains of LtrMDR1 efficiently overcomes resistance of *Leishmania* spp. to Miltefosine by inhibiting drug efflux. *Antimicrob. Agents Chemother.* 50:3102–3110.

Pérez-Victoria J.M., Pérez-Victoria F.J., Conseil G., Maitrejean M., Comte G., Barron D., Di Pietro A., Castanys S., and Gamarro F. 2001. High-affinity binding of silybin derivatives to the nucleotide-binding domain of a *Leishmania tropica* P-glycoprotein-like transporter and chemosensitization of a multidrug resistant parasite to daunomycin. *Antimicrob. Agents Chemother.* 45:439–446.

Perrotton T., Trompier D., Chang X.-B., Di Pietro A., and Baubichon-Cortay H. 2007. S- and R-verapamil differentially modulate the multidrug resistance protein MRP1. *J. Biol. Chem.* 282:31542–31548.

Pick A., Müller H., and Wiese M. 2008. Structure–activity relationships of new inhibitors of breast cancer resistance protein (ABCG2). 2008. *Bioorg. Med. Chem.* 16:8224–8236.

Pires M.M., Emmert D., Hrycyna C.A., and Chmielewski J. 2009. Inhibition of P-glycoprotein-mediated paclitaxel resistance by reversibly linked quinine homodimers. *Mol. Pharmacol.* 75:92–100.

Qadir M., O'Loughlin K.L., Fricke S.M., Williamson N.A., Greco W.R., Minderman H., and Baer M.R. 2005. Cyclosporin A is a broad-spectrum multidrug resistance modulator. *Clin. Cancer Res.* 11:2320–2326.

Quiney C., Billard C., Faussat A.M., Salanoubat C., and Kolb J.P. 2007. Hyperforin inhibits P-gp and BCRP activities in chronic lymphocytic leukemia cells and myeloid cells. *Leuk. Lymphoma* 48:1587–1599.

Raad I., Terreux R., Richomme P., Matera E.L., Dumontet C., Raynaud J., and Guilet D. 2006. Structure–activity relationships of natural and synthetic coumarins inhibiting the multidrug transporter P-glycoprotein. *Bioorg. Med. Chem.* 14:6979–6987.

Rabindran S.K., He H., Singh M., Brown E., Collins K.I., Annable T., and Greenberger L.M. 1998. Reversal of a novel multidrug resistance mechanism in human colon carcinoma cells by fumitremorgin C. *Cancer Res.* 58:5850–5858.

Ramachandran C., Rabi T., Fonseca H.B., Melnick S.J., and Escalon E.A. 2003. Novel plant triterpenoid drug amooranin overcomes multidrug resistance in human leukemia and colon carcinoma cell lines. *Int. J. Cancer* 105:784–789.

Robey R.W., Steadman K., Polgar O., Morisaki K., Blaynay M., Mistry P., and Bates S.E. 2004. Pheophorbide a is a specific probe for ABCG2 function and inhibition. *Cancer Res.* 64:1242–1246.

Sahai R., Rastogi R.P., Jakupovic J., and Bohlmann F. 1981. A diterpene from *Euphorbia maddeni. Phytochemistry* 20:1665–1667.

Sarkadi B., Ozvegy-Laczka C., Német K., and Váradi A. 2004. ABCG2—A transporter for all seasons. *FEBS Lett.* 567:116–120.

Scambia G., Ranelletti F.O., Panici P.B., De Vincenzo R., Bonanno G., Ferrandina G., Piantelli M., Bussa S., Rumi C., Cianfriglia M., and Mancuso S. 1994. Quercetin potentiates the effect of adriamycin in a multidrug-resistant MCF-7 human breast-cancer cell line: P-glycoprotein as a possible target. *Cancer Chemother. Pharmacol.* 34:459–464.

Schuetz E.G., Beck W.T., and Schuetz J.D. 1996. Modulators and substrates of P-glycoprotein and cytochrome P4503A coordinately up-regulate these proteins in human colon carcinoma cells. *Mol. Pharmacol.* 49:311–318.

Shapiro A.B., Fox K., Lam P., and Ling V. 1999. Stimulation of P-glycoprotein-mediated drug transport by prazosin and progesterone. Evidence for a third drug-binding site. *Eur. J. Biochem.* 259:841–850.

Shapiro A.B. and Ling V. 1997a. Effect of quercetin on Hoechst 33342 transport by purified and reconstituted P-glycoprotein. *Biochem. Pharmacol.* 53:587–596.

Shapiro A.B. and Ling V. 1997b. Positively cooperative sites for drug transport by P-glycoprotein with distinct drug specificities. *Eur. J. Biochem.* 250:130–137.

Shiozawa K., Oka M., Soda H., Yoshikawa M., Ikegami Y., Tsurutani J., Nakatomi K., Nakamura Y., Doi S., Kitazaki T., Mizuta Y., Murase K., Yoshida H., Ross D.D., and Kohno S. 2004. Reversal of breast cancer resistance protein (BCRP/ABCG2)-mediated drug resistance by novobiocin, a coumermycin antibiotic. *Int. J. Cancer* 108:146–151.

Shukla S., Zaher H., Hartz A., Bauer B., Ware J.A., and Ambudkar S.V. 2009. Curcumin inhibits the activity of ABCG2/BCRP1, a multidrug resistance-linked ABC drug transporter in mice. *Pharm. Res.* 26:480–487.

Sicheri F., Moarefi I., and Kuriyan J. 1997. Crystal structure of the Src family tyrosine kinase Hck. *Nature* 385:602–609.

Soldner A., Christians U., Susanto M., Wacher V.J., Silverman J.A., and Benet L.Z. 1999. Grapefruit juice activates P-glycoprotein-mediated drug transport. *Pharm. Res.* 16:478–485.

Steinbach D. and Legrand O. 2007. ABC transporters and drug resistance in leukemia: Was P-gp nothing but the first head of the Hydra? *Leukemia* 21:1172–1176.

Su Y. and Sinko P.J. 2006. Inhibition of efflux transporter ABCG2/BCRP does not restore mitoxantrone sensitivity in irinotecan-selected human leukemia CPT-K5 cells: Evidence for multifactorial multidrug resistance. *Eur. J. Pharm. Sci.* 29:102–110.

Szakács G., Paterson J.K., Ludwig J.A., Booth-Genthe C., and Gottesman M.M. 2006. Targeting multidrug resistance in cancer. *Nat. Rev. Drug Discov.* 5:219–234.

Takanaga H., Ohnishi A., Matsuo H., and Sawada Y. 1998. Inhibition of vinblastine efflux mediated by P-glycoprotein by grapefruit components in Caco-2 cells. *Biol. Pharm. Bull.* 21:1062–1066.

Takanaga H., Ohnishi A., Yamada S., Matsuo H., Morimoto S., Shoyama Y., Ohtani H., and Sawada Y. 2000. Polymethoxylated flavones in orange juice are inhibitors of P-glycoprotein but not cytochrome P450 3A4. *J. Pharmacol. Exp. Ther.* 293:230–236.

Trompier D., Chang X.-B., Barattin R., du Moulinet d'Hardemare A., Di Pietro A., and Baubichon-Cortay H. 2004. Verapamil and its derivative trigger apoptosis through glutathione extrusion by multidrug resistance protein MRP1. *Cancer Res.* 64:4950–4956.

Van Asperen J., Van Tellingen O., and Beijnen J. 1998. The pharmacological role of P-glycoprotein in the intestinal epithelium. *Pharmacol. Res.* 37:429–435.

Van Herwaarden A.E. and Schinkel A.H. 2006. The function of breast cancer resistance protein in epithelial barriers, stem cells and milk secretion of drugs and xenotoxins. *Trends Pharmacol. Sci.* 27:10–16.

Van Loevezijn A., Allen J.D., Schinkel A.H., and Koomen G.J. 2001. Inhibition of BCRP-mediated drug efflux by fumitremorgin-type indolyl diketopiperazines. *Bioorg. Med. Chem. Lett.* 11:29–32.

Wacher V.J., Wu C.Y., and Benet L.Z. 1995. Overlapping substrate specificities and tissue distribution of cytochrome P450 3A and P-glycoprotein: Implications for drug delivery and activity in cancer chemotherapy. *Mol. Carcinog.* 13:129–134.

Walsh T.J., Standiford H.C., Reboli A.C., John J.F., Mulligan M.E., Ribner B.S., Montgomerie J.Z., Goetz M.B., Mayhall C.G., Rimland D., Stevens D.A., Hansen S.L., Gerard G.C., and Ragual R.J. 1993. Randomized double-blinded trial of rifampin with either novobiocin or trimethoprim-sulfamethoxazole against methicillin-resistant *Staphylococcus aureus* colonization: Prevention of antimicrobial resistance and effect of host factors on outcome. *Antimicrob. Agents Chemother.* 37:1334–1342.

Wang X. and Baba M. 2005. The role of breast cancer resistance protein (BCRP/ABCG2) in cellular resistance to HIV-1 nucleoside reverse transcriptase inhibitors. *Antivir. Chem. Chemother.* 16:213–216.

Wang E.J., Barecki-Roach M., and Johnson W.W. 2004a. Quantitative characterization of direct P-glycoprotein inhibition by St John's Wort constituents hypericin and hyperforin. *J. Pharm. Pharmacol.* 56:123–128.

Wang Y.H., Chao P.D., Hsiu S.L., Wen K.C., and Hou Y.C. 2004b. Lethal quercetin-digoxin interaction in pigs. *Life Sci.* 74:1191–1197.

Wang B.H., Ternai B., and Polya G. 1997. Specific inhibition of cyclic AMP-dependent protein kinase by waranglone and robustic acid. *Phytochemistry* 44:787–796.

Weiss J., Rose J., Storch C.H., Ketabi-Kiyanvash N., Sauer A., Haefeli W.E., and Efferth T. 2007. Modulation of human BCRP (ABCG2) activity by anti-HIV drugs. *J. Antimicrob. Chemother.* 59:238–245.

Woehlecke H., Osada H., Herrmann A., and Lage H. 2003. Reversal of breast cancer resistance protein-mediated drug resistance by tryprostatin A. *Int. J. Cancer* 107:721–728.

Wortelboer H.M., Usta M., van der Velde A.E., Boersma M.G., Spenkelink B., van Zanden J.J., Rietjens I.M., van Bladeren P.J., and Cnubben N.H. 2003. Interplay between MRP inhibition and metabolism of MRP inhibitors: The case of curcumin. *Chem. Res. Toxicol.* 16:1642–1651.

Yamamura S., Shizuri Y., Kosemura S., Ohtsuka J., Tayama T., Ohiba S., Ito M., Saito Y., and Terada Y. 1989. Diterpenes from *Euphorbia helioscopia. Phytochemistry* 28:3421–3436.

Yang C.H., Chen Y.C., and Kuo M.L. 2003. Novobiocin sensitizes BCRP/MXR/ABCP over-expressing topotecan-resistant human breast carcinoma cells to topotecan and mitoxantrone. *Anticancer Res.* 23:2519–2523.

Yoshida N., Takada T., Yamamura Y., Adachi I., Suzuki H., and Kawakami J. 2008. Inhibitory effects of terpenoids on multidrug resistance-associated protein 2- and breast cancer resistance protein-mediated transport. *Drug Metab. Dispos.* 36:1206–1211.

Yoshikawa M., Ikegami Y., Sano K., Yoshida H., Mitomo H., Sawada S., and Ishikawa T. 2004. Transport of SN-38 by the wild-type ABC transporter ABCG2 and its inhibition by quercetin, a natural flavonoid. *J. Exp. Ther. Oncol.* 4:25–35.

Zhang S., Wang X., Sagawa K., and Morris M.E. 2005. Flavonoids chrysin and benzoflavone, potent breast cancer resistance protein inhibitors, have no significant effect on topotecan pharmacokinetics in rats or mdr1a/b (-/-) mice. *Drug Metab. Dispos.* 33:341–348.

Zhang H., Wong C.W., Coville P.F., and Wanwimolruk S. 2000. Effect of the grapefruit flavonoid naringenin on pharmacokinetics of quinine in rats. *Drug Metab. Drug Inter.* 17:351–363.

Zhang S., Yang X., and Morris M.E. 2004a. Flavonoids are inhibitors of breast cancer resistance protein (ABCG2)-mediated transport. *Mol. Pharmacol.* 65:1208–1216.

Zhang S., Yang X., and Morris M.E. 2004b. Combined effects of multiple flavonoids on breast cancer resistance protein (ABCG2)-mediated transport. *Pharm. Res.* 21:1263–1273.

Zhou S., Lim L.Y., and Chowbay B. 2004. Herbal modulation of P-glycoprotein. *Drug Metab. Rev.* 36:57–104.

Zhu X., Sui M., and Fan W. 2005. In vitro and in vivo characterizations of tetandrine on the reversal of P-glycoprotein-mediated drug resistance to paclitaxel. *Anticancer Res.* 25:1953–1962.

13 Recent Advances in the Search of Novel Calmodulin Inhibitors from Selected Mexican Plants and Fungi

Rachel Mata, Mario Figueroa, Isabel Rivero-Cruz, and Martín González-Andrade

CONTENTS

13.1 INTRODUCTION

A mega-diverse country like Mexico, rich in organisms not yet explored, offers excellent prospects for the discovery of new bioactive compounds. Potentially, these organisms are sources of novel chemical entities representing appropriate templates for the development of new drugs and pesticide agents. Therefore, a few years ago, we initiated a program to discover new calmodulin (CaM) inhibitors from natural sources considering that agents that inhibit its regulatory properties might play an important role in the development of new chemotherapeutic agents for treating several diseases. For example, recent findings have shown that most cancers are associated with elevated levels of calcium-bound CaM (Ca^{2+}-CaM) and that CaM antagonists inhibit tumor cell invasion in vitro and metastasis in vivo (Shim et al. 2007). Then, CaM is an important target for the development of new anticancer agents.

On the other hand, many CaM inhibitors are valuable research tools for the study of the complex CaM messenger system in plants and animals (Chin and Means 2000; Bouche et al. 2005). Finally, CaM could be also a specific target for natural phytotoxic compounds with potential as herbicides agents due to the following reasons: First, in higher plants, CaM is responsible for modulating the activity of several enzymes involved in plant development. A few of these proteins are found only in plants; others, although not exclusive from plants, are regulated by CaM only in plants. Finally, higher plants express multiple divergent CaM isoforms, some of which differentially control CaM-modulated enzymes (Bouche et al. 2005; Sheng 2008; Kim et al. 2009), representing specific targets for plant-growth inhibitors.

It is worth mentioning that two decades ago a few natural products with anti-CaM properties were already discovered. Thereafter, the interest in the subject declined. More recently, nevertheless, research in several laboratories worldwide revealed a renewed interest for discovering new CaM inhibitors. This concern is probably due to the significant progress in the knowledge of this protein, the discovery of new CaM target enzymes and isoforms, as well as its role in the regulation of several physiological processes (Chin and Means 2000; Bouche et al. 2005) such as cell growth and proliferation (Shim et al. 2007), muscle contraction and relaxation (Somlyo and Somlyo 2003), learning and memory, immune responses (Horikawa et al. 2005), osteoclastogenesis (Seales et al. 2006), and anxiety physiology (Du et al. 2004), to mention the most important.

In the next sections, a concise overview on CaM and natural anti-CaM compounds will be considered. Recent examples of CaM inhibitors stemming from our own work will be also described in detail including some aspects of their natural sources, isolation, and structure elucidation.

13.2   CALMODULIN: AN OVERVIEW

To show the state of the art regarding CaM, it would be necessary to cite an impressive number of works; however, the interested reader is referred elsewhere for this information (Chin and Means 2000; Vetter and Leclerc 2003; Kortvely and Gulya 2004; Gardiner et al. 2005; Choi and Husain 2006; Luan 2008; Means 2008; Sheng 2008; Kim et al. 2009; *inter alia*).

CaM is the major ubiquitous Ca^{2+}-binding protein of all eukaryotes fundamental to the regulation of several cellular events. It becomes activated upon the transitory elevation of Ca^{2+} inside the cell and regulates several Ca^{2+}-dependent physiological processes by interacting with a heterogeneous group of target enzymes, both in plant and animal cells (Chin and Means 2000).

From the structural point of view, vertebrate CaM is an acidic small protein (17 kD) composed by 148 amino acids organized in two distinct N- and C-terminal globular domains. The 3D structures of apo-CaM and Ca^{2+}-CaM have been determined by NMR and x-ray crystallography methods. The crystal structure of Ca^{2+}-CaM (PDB code 1CLL) disclosed a protein with dumbbell structure with a flexible α-helical central linker region between residues Lys 77 and Asp 80

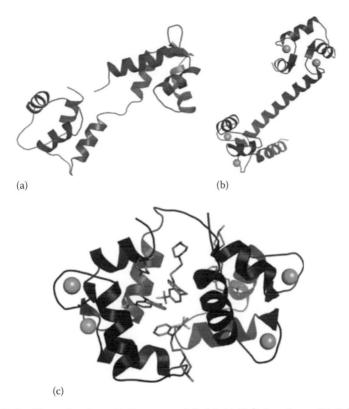

(a)   (b)

(c)

FIGURE 13.1   (See color insert.) Structure of CaM: (a) Ca^{2+}-free form; (b) Ca^{2+}-binding; and (c) Ca^{2+}-TFP complex. (PDB entry codes: 1CFD, 1CLL and 1LIN, respectively.)

(Figure 13.1). The flexibility of this central linker region allows CaM to adopt a variety of conformations upon binding to different targets. Each lobe of CaM contains a pair of EF-hand motifs (sites I–IV) allowing it to bind four Ca^{2+} ions. Once Ca^{2+} binds to the EF-hand motifs in both domains, the conformation of CaM changes into an open form permitting the protein to interact with and activate a surprisingly diverse set of target enzymes. In the 3D structures of Ca^{2+}-CaM complexes of these enzymes, the peptides are wrapped by the two domains of CaM, forming several contacts with the residues of the hydrophobic pockets located at the inner parts of each domain and with the surrounding acidic side chains throughout polar interactions (Figure 13.1). The conformation of the two domains of CaM is unaffected by target binding, but their relative orientation changes drastically, bringing the two previously independent domains into contact. In contrast, apo-CaM forms a closed conformation by its two homologous domains (Bouche et al. 2005; Johnson 2006).

Distinctive structural features of CaM are its high methionine content and the large number of negatively charged acidic amino acid residues. The preponderance of methionine in its target peptide binding sites creates two highly bendable surfaces, which can accommodate the binding of peptides with different amino acid sequences.

Accordingly, small molecule antagonists of CaM generally possess not only a hydrophobic part but also a charged side chain. The hydrophobic part of the antagonists is required for binding to N- and/or C-terminal hydrophobic pockets of the protein, while the charged side chain is necessary for antagonistic functions, which prevent CaM target peptides from accessing the hydrophobic pockets of CaM.

As such, CaM has no enzymatic activity, but acting in response to Ca^{2+}-mediated signaling events, it plays a key role in modulating gene regulation, protein synthesis, fast axonal transport, smooth muscle contraction, secretion, growth cone elongation, organelle tubulation, ion channel function, cell motility, and chemotaxis. CaM controls these processes through the activation of a series of CaM-dependent enzymes such as nitric oxide synthases, adenylate cyclases, phosphodiesterases, several kinases, calcium-ATPase pumps, ion channels, phosphatases as well as cytoskeletal structural proteins. Consequently, this protein is implicated in a variety of cellular functions, including cell growth and proliferation, inflammation, short-term memory, viral penetration, cell growth and differentiation, the immune response, and the cell cycle implicating it in AIDS, Alzheimer's, some cancers, and other diseases. Because of the relevant role of Ca^{2+} and CaM in cellular processes and degenerative diseases, there is an urgent need for drugs that affect the binding of certain target proteins to CaM (Chin and Means 2000; Du et al. 2004; Bouche et al. 2005; Seales et al. 2006).

13.3 METHODS FOR DETECTING CALMODULIN INHIBITORS

CaM inhibitors have been detected by means of several methods. The most widely used includes affinity chromatography (Ovadi 1989; Molnar et al. 1995), native gels electrophoresis (Leung et al. 1985; Rivero-Cruz et al. 2003), UV and CD spectroscopies (Molnar et al. 1995; Vertessy et al. 1998; Harmat et al. 2000), fluorescence techniques (González-Andrade et al. 2009), fluorescence resonance energy transfer (FRET) (Molnar et al. 1995; Au et al. 2000; Sharma et al. 2005), nuclear magnetic resonance (NMR) (Craven et al. 1996; Lalor et al. 2003), X-ray diffraction (Vertessy et al. 1998; Harmat et al. 2000; Lalor et al. 2003), functional enzymatic assays (Sharma and Wang 1979; Martínez-Luis et al. 2007 and references cited therein), site-directed mutagenesis (Odom et al. 1997; Au et al. 2000), isothermal titration calorimetry (ITC) (Gilli et al. 1998; Brokx et al. 2001), and computational methods (Gabdoulline and Wade 2002; Ladbury and Williams 2004).

For most assays, the antipsychotics phenothiazines, trifluoperazine (TFP) and chlorpromazine (CPZ), the naphthalenesulfonamide W-7 [N-(6-aminohexyl)-5-chloro-1-napththalenesulfonamide], the antibreast cancer agents tamoxifen and its analogs, as well as calmidazolium, and felodipine have been extensively used as positive controls. These antagonists bind to CaM, with few exceptions (e.g., calmidazolium with binding constant, K_d, value of 3 nM), much weaker (K_d values in the µM range) than its target proteins (K_d values in the nM range) (Harmat et al. 2000). Detailed structural information of their binding nature to CaM is available for TFP and W7. These two compounds are regarded as classics antagonists of CaM and exert their anti-CaM effect by a competitive mechanism of action preventing the formation of the complexes CaM-target enzymes. There are also compounds that bind to CaM (fendiline and the arylalkylamine derivative, KHL-8430) but do not avoid the complexes formation, interfering only with the ternary enzyme-CaM-ligand complexes functions; these compounds are considered as functional inhibitors (Harmat et al. 2000).

13.4   NATURAL CALMODULIN INHIBITORS

As a follow-up of our previous reviews (Martínez-Luis et al. 2007; Mata et al. 2008), we have updated in Table 13.1 the information regarding natural CaM inhibitors detected using different assays. Table 13.1 includes the names of the inhibitors, their natural sources, methods employed for detecting their anti-CaM activity, and suitable references. Figures 13.2 through 13.9 show the structures of some of these inhibitors (**1–42**) belonging to different types of secondary metabolites.

13.5   SELECTED CaM INHIBITORS FROM MEXICAN BIODIVERSITY

For pursuing our research to discover natural compounds with anti-CaM activity, we have followed the strategy described in the following lines. Thus, in the case of fungi, once the microorganisms are selected according to the conventional criteria, appropriated fermentation conditions are established. Then, organic soluble extracts are prepared from the fermentation media and mycelia. For plants, usually medicinal species are selected and organic extracts obtained. The resulting extracts (fungal or vegetal) are submitted for testing using a nondenaturing homogeneous electrophoresis (PAGE) in polyacrylamide gels. In this experiment, the change in the electrophoretic mobility of CaM treated with the extracts is analyzed. Those active extracts are fractionated until active pure compounds are isolated. The active isolates are subjected to structure elucidation using standard methodology. Finally, their anti-CaM activity is assessed using one or more assays and CPZ as positive control (Martínez-Luis et al. 2005, 2006, 2007; Figueroa et al. 2009). The application of this methodology allowed us to isolate several CaM inhibitors from several plants species, including *Hofmeisteria schaffneri* (Asteraceae), *Prionosciadium watsoni* (Umbelliferae), and the orchids *Maxillaria densa* and *Epidendrum rigidum*. In particular, the orchid bibenzyls and the northymol hofmeisterin turned out to be good lead compounds for the development of herbicides agents. The fungi *Guanomyces polythrix* and *Phoma herbarum* contain also potent phytotoxic agents with CaM-inhibitor properties. In this chapter, we have reviewed the anti-CaM activity of other natural products isolated from two ascomycetes (*Emericella* 25379 and *Malbranchea aurantiaca*), and one medicinal plant from the Asteraceae family (*Brickellia veronicifolia*).

13.5.1   Prenylated Xanthones from Emericella 25379

The genus *Emericella*, one of the anamorphous of *Aspergillus*, biosynthesizes a wide variety of bioactive compounds representing valuable leads for the development of new pharmaceutical agents. Therefore, the new species *Emericella* 25379 (Figure 13.10), isolated from the coral species *Pacifigorgia rutilia*, was selected as a potential source of novel CaM inhibitors (Figueroa et al. 2009).

TABLE 13.1

Natural Anti-CaM Metabolites Reported to Date

Compound	Natural Source	Method of Detection	References
Alkaloids			
K-252a	*Nocardiopsis* sp. (Nocardiopsacea)	Functional enzymatic assay (PDE1)	Kase et al. (1986)
K-252b			Nakanishi et al. (1986)
K-252c			
K-252d			
Eudistomidin A	*Eudistoma glaucus* (Polycitoridae)	Functional enzymatic assay (PDE1)	Kobayashi et al. (1986)
Eudistomidin C			Kobayashi et al. (1990a)
Pyridindolol	*Streptoverticillium album* K-251 (Streptomycetaceae)	Functional enzymatic assay (PDE1)	Matsuda et al. (1988)
Rigidin	*Eudistoma* cf. *rigida* (Polycitoridae)	Functional enzymatic assay (PDE1)	Kobayashi et al. (1990b)
Stellettamide A	*Stelletta* sp. (Stellettidae)	Functional enzymatic assay (PDE1)	Abe et al. (1997)
Stellettazole A			Tsukamoto et al. (1999a,b)
Vinblastine (**1**)	*Catharanthus roseus* (Apocynaceae)	Fluorescence resonance energy transfer CD measurements	Molnar et al. (1995)
Vincristine (**2**)			Makarov et al. (2007)
Catharantine			
Vindoline			
Vinflunine			
Leptosin M	*Leptosphaeria* sp. OUPS-4 (Phaeosphaeriaceae) or *Sargassum tortil* (Sargassaceae)	Functional enzymatic assay (CaMKII, PTK)	Yamada et al. (2002)
Verruculogen	*Penicillium* (Trichocomaceae)	Functional enzymatic assay (PDE1) Fluorescence resonance energy transfer	Pala et al. (1999)
Melatonin	Wide (from algae to human)	Computational methods	Del Rio et al. (2004)

Compound	Source	Assay	References
Malbrancheamide (3) Malbrancheamide B (4) Iso-malbrancheamide B (5) Pre-malbrancheamide (6)	*Malbranchea aurantiaca* (Myxotrichaceae)	Functional enzymatic assay (PDE1)	Martínez-Luis et al. (2006) Figueroa et al. (2008) Ding et al. (2008)
Berbamine Dauricine Daurisoline	*Menispermun dauricum* (Menispermaceae) or *Berberis poiretii* (Berberidaceae)	Functional enzymatic assay (PDE1)	Hu et al. (1988) Chen et al. (1990)
Papaverine	*Papaver somniferum* (Papaveraceae)	Functional enzymatic assay (PDE1, PDE10)	Ronca-Testoni et al. (1985) Boswell-Smith et al. (2006) Siuciak et al. (2006)
Peptides			
Domoic acid	*Mytilus edulis* overfed with the algae *Nitzschia pungens* (Bacillariophyceae)	Functional enzymatic assay (adenylate cyclase)	Nijjar and Nijjar (2000)
Sanjoinine A (7) Sanjoinine Ahl (8) Sanjoinine B (9) Sanjoinine F (10) Daechuine S4 (11) Sanjoinine (12) Dihydrosanjoinine A (13) Sanjoinine D (14) Sanjoinine G1 (15) Sanjoinine G2 Sanjoinine A dialdehyde Daechuine S10 Daechuine S27	*Zizyphus vulgaris* var. spinosus and *Z. jujube* var inermis (Rhamnaceae)	Functional enzymatic assay (PDE1, PKII)	Han et al. (1993, 2005) Hwang et al. (2001)
Konbamide	*Theonella* sp. (Spongiidae)	Functional enzymatic assay (PDE1)	Kobayashi et al. (1991)

(continued)

TABLE 13.1 (continued)
Natural Anti-CaM Metabolites Reported to Date

Compound	Natural Source	Method of Detection	References
Polymyxin B (16)	*Bacillus polymyxa* (Bacillaceae)	Functional enzymatic assay (PDE1); Affinity chromatography Native gel electrophoresis	Hegemann et al. (1991)
Bistellettadines A Bistellettadines B	*Stelletta* sp. (Stellettidae)	Functional enzymatic assay (PDE1)	Tsukamoto et al. (1999a)
Melittin (17) Apamin (18)	Bee venom and Insect venom	Functional enzymatic assay (PDE1, CaMKII) computational methods	Barnette et al. (1983) Kataoka et al. (1989) Wang et al. (2009)
Lesueurin Aurein 1.1 Citropin 1.1 Aurein 2.2 Aurein 2.3 Aurein 2.4 Not named Frenatin 3 Splendipherin Caerin 1.1 Caerin 1.10 Caerin 1.6 Caerin 1.8 Caerin 1.9	*Litoria* genus (Hylidae)	Functional enzymatic assay (*n*NOS, calcineurin)	Doyle et al. (2002) Pukala et al. (2008)

Compound	Source	Assay	References
Polyamines			
Spermine	Mammalian tissues	Functional enzymatic assay (MLCK)	Qi et al. (1983)
1,12-Diaminododecane			
Spermidine			
Cadaverine			
Putrescine			
1,10-Diaminodecane			
Terpenoids			
Ophiobolin A (**19**)	*Bipolaris oryzae* (Pleosporaceae)	Functional enzymatic assay (PDE1) Intrinsic fluorescence UV spectroscopy site-directed mutagenesis	Leung et al. (1984) Au and Leung (1998) Au et al. (2000) Evidente et al. (2006)
Fasciculic acid A	*Naematoloma fasciculare* (Strophariaceae)	Functional enzymatic assay (PDE1)	Takahashi et al. (1989) Kubo et al. (1985)
Fasciculic acid B			
Fasciculic acid C			
Fasciculic acid D			
Fasciculic acid E			
Fasciculic acid F			
KS-505a	*Streptomyces argenteolus* A-2 (Streptomycetaceae)	Functional enzymatic assay (PDE1) Native gel electrophoresis	Nakanishi et al. (1994) Ichimura et al. (1996)
Dehydroflourensic acid	*Flourensia cernua* (Asteraceae)	Functional enzymatic assay (PDE1), SDS gel electrophoresis	Mata et al. (2003b)
Flourensadiol			
1-Hydroxy-2-oxoeremophil-1(10),7(11), 8(9)-trien-12(8)-olide (**20**)	*Malbranchea aurantiaca* (Myxotrichaceae)	Functional enzymatic assay (PDE1)	Martínez-Luis et al. (2005)

(*continued*)

TABLE 13.1 (continued)
Natural Anti-CaM Metabolites Reported to Date

Compound	Natural Source	Method of Detection	References
Ergosta-4,6,8(14),22-tetraen-3-one	*Guanomyces polytrix* (Chaetomiaceae)	Functional enzymatic assay (PDE1, NADK)	Mata et al. (2003a)
Hofmeisterin 3′,4′,4a′,9a′-Tetrahydro-6,7′-dimethylspiro[benzofuran-3(2*H*), 2′-pyrano[2,3-*b*]benzofuran]-2,4a′-diol	*Hofmeisteria schaffneri* (Asteraceae)	Functional enzymatic assay (PDE1)	Pérez-Vásquez et al. (2005)
Forskolin (**21**)	*Coleus forskohlii* (Lamiaceae)	Functional enzymatic assay (CaMKII)	Grey and Burrell (2008)
Gossypol	Cotton plants	Functional enzymatic assay (ATPase, Ser/Thr phosphatase)	Baumgrass et al. (2001)
Quinovic acid Quinovic glycoside C	*Mitragyna stipulosa* (Rubiaceae)	Functional enzymatic assay (PDE1)	Fatima et al. (2002)
Jujuboside A	Chinese herbal medicine sanzaoren	NMR and EPR spectroscopies	Zhou et al. (1994)
Helenalin (**22**)	*Arnica Montana* (Asteraceae)	Functional enzymatic assay (CaMKII)	Olofsson et al. (2008)
Coumarins, γ-Pyrones and Xanthones			
Reticulol	*Streptoverticillium album* K-251 (Streptomycetaceae)	Functional enzymatic assay (PDE1)	Matsuda et al. (1988)
Secalonic acid	*Penicillium oxalicum* (Trichocomaceae)	Functional enzymatic assay (PDE1)	Pala et al. (1999)

Compound	Source	Method	Reference
15-Chlorotajixanthone hydrate (**23**) 14-Methoxytajixanthone (**24**) Shamixanthone (**25**) Tajixanthone hydrate (**26**) Emericellin (**27**)	*Emericella* 25379	Functional enzymatic assay (PDE) Native gel electrophoresis Computational methods	Figueroa et al. (2009)
(9R,10R)-9-Acetoxy-8,8-dimethyl-9,10-dihydro-2H,8H-benzo[1,2-b:3,4-b']dipyran-2-one-10-yl ester Isobutyric acid (9R,10R)-9-hydroxy-8,8-dimethyl-9,10-dihydro-2H,8H-benzo[1,2-b:3,4-b']-dipyran-2-one-10-yl ester 2-Methylbut-(2Z)-enoic acid (9R,10R)-10-hydroxy-8,8-dimethyl-9,10-dihydro-2H,8H-benzo[1,2-b:3,4-b']dipyran-2-one-9-yl ester Seravshanin Quianhucoumarin D (+)-cis-Khellactone Jatamansin Isobutyric acid (9R)-8,8-dimethyl-9,10-dihydro-2H,8H-benzo[1,2-b:3,4-b']dipyran-2-one-9-yl ester (+)-Lomatin 2-Methylbut-(2Z)-enoic acid (3R)-5-methoxy-3,4-dihydro-2,2,8-trimethyl-6-oxo-2H,6H-benzo[1,2-b:5,4-b']dipyran-3-yl ester	*Prionosciadium watsoni* (Umbelliferae)	SDS gel electrophoresis	Valencia-Islas et al. (2002)

(continued)

TABLE 13.1 (continued)
Natural Anti-CaM Metabolites Reported to Date

Compound	Natural Source	Method of Detection	References
Isobutyric acid (3R)-5-methoxy-3,4-dihydro-2,2,8-trimethyl-6-oxo-2H,6H-benzo[1,2-b:5,4-b']dipyran-3-yl ester	Guanomyces polythrix (Chaetomiaceae)	Functional enzymatic assay (PDE1, NADK) SDS gel electrophoresis Affinity chromatography	Macías et al. (2000, 2001)
(+)-5-Methoxyhamaudol			
(2S,3S)-5-Hydroxy-6,8-dimethoxy-2,3-dimethyl-4H-2,3-dihydronaphtho[2,3-b]-pyran-4-one			
(2S,3S)-5-Hydroxy-6,8,10-trimethoxy-2,3-dimethyl-4H-2,3-dihydronaphtho[2,3-b]-pyran-4-one			
(2S)-5-Hydroxy-6,8-dimethoxy-2-methyl-4H-2,3-dihydronaphtho[2,3-b]-pyran-4-one			
(2S)-5-Hydroxy-6,8,10-trimethoxy-2-methyl-4H-2,3-dihydronaphtho[2,3-b]-pyran-4-one			
(2S,3R)-5-Hydroxy-6,8-dimethoxy-2,3-dimethyl-2,3-dihydro-4H-naphtho[2,3-b]-pyran-4-one			
(2S,3R)-5-Hydroxy-6,8,10-trimethoxy-2,3-dimethyl-2,3-dihydro-4H-naphtho[2,3-b]-pyran-4-one			
5-Hydroxy-6,8-dimethoxy-2,3-dimethyl-4H-naphtho[2,3-b]-pyran-4-one			
Rubrofusarin B			
8-Hydroxy-6-methyl-9-oxo-9H-xanthene-1-carboxylic acid methyl ester			

Lignans

Compound	Source	Assay	Reference
2'-Methoxykobusin	Leucophyllum ambiguum (Scrophulariaceae)	Functional enzymatic assay (PDE1) SDS gel electrophoresis	Rojas et al. (2003)
2'-Methoxy-4''-hydroxydemethoxykobusin Kobusin			
2',2''-Dimethoxysesamin			
Honokiol	Magnolia officinalis	Functional enzymatic assay (CaMKII)	Lo et al. (1994) Zhai et al. (2005)

Anthracyclins and Anthraquinones

Compound	Source	Assay	Reference
K-259-2	Micromonospora olivasterospora K-259 (Micromonosporaceae)	Functional enzymatic assay (PDE1)	Matsuda et al. (1987, 1990)
KS-619-1	Streptomyces californicus (Streptomycetaceae)	Functional enzymatic assay (PDE1)	Matsuda and Kase (1987) Yasuzawa et al. (1987)
Adriamycin	Streptomyces sp (Streptomycetaceae)	Functional enzymatic assay (PDE1)	Matsuda and Kase (1987)
Emodin	Guanomyces polythrix (Chaetomiaceae)	Functional enzymatic assay (PDE1)	Macías et al. (2001)
Daunorubicin	Streptomyces peucetius or Streptomyces galilaeus (Streptomycetaceae)	Functional enzymatic assay (PDE1)	Nwankwoala and West (1988)
Aclacinomycin A			

Stilbenoids

Compound	Source	Assay	Reference
3,5-Dihydroxy-2-(3-methyl-2-butenyl)-bibenzyl	Radula kojana (Radulaceae)	Functional enzymatic assay (PDE1)	Asakawa et al. (1991)
2-Geranyl-3,5-dihydroxy-bibenzyl			
Nidemone (28)	Nidema boothii (Orchidaceae)	Functional enzymatic assay (PDE1) SDS gel electrophoresis	Hernández-Romero et al. (2004)
Ephemeranthoquinone			
Gigantol (29)			
Batatasin III (30)			
Marchantin A	Marchantia polymorpha (Marchantiaceae)	NMR experiments	Kamory et al. (1995) Keseru and Nogradi (1996)

(continued)

TABLE 13.1 (continued)
Natural Anti-CaM Metabolites Reported to Date

Compound	Natural Source	Method of Detection	References
3′,4-Dihydroxy-3,5′-dimethoxybibenzyl	*Epidendrum rigidum* (Orchidaceae)	Functional enzymatic assay (PDE1) SDS gel electrophoresis	Hernández-Romero et al. (2005)
3′,4-Dihydroxy-5′-methoxybibenzyl			
3′,5′-Dihydroxy-bibenzyl			
3-Hydroxy-4-methoxybibenzyl			
3,4,3′,5′-Tetrahydroxybibenzyl			
3′-(β-D-Glucopyranosyloxy)-benzyl-2,6-dimethoxybenzoate (**31**)	*Brickellia veronicifolia* (Asteraceae)	Functional enzymatic assay (PDE1)	Rivero-Cruz et al. (2005)
3″-Hydroxybenzyl-2,6-dimethoxybenzoate (**32**)			
2′-Methoxybenzyl-2-hydroxybenzoate (**33**)			
Benzyl 2,6-dimethoxybenzoate (**34**)			
3′-Methoxybenzyl-2-hydroxy-6-methoxybenzoate (**35**)			
Benzyl 2-hydroxy-6-methoxybenzoate (**36**)			
Benzyl 2,5,6-trimethoxybenzoate (**37**)			
Benzyl 2-hydroxy-5,6-dimethoxybenzoate (**38**)			
3′-Methoxybenzyl 2,6-dimethoxybenzoate (**39**)			
Miscellaneous polyketides			
KS-501	*Sporothrix* sp. KAC-1985 (Ophiostomataceae)	Functional enzymatic assay (PDE1)	Nakanishi et al. (1990a); Yasuzawa et al. (1990)
KS-502			
Methyl orsellinate	*Flourensia cernua* (Asteraceae) And several lichens	Functional enzymatic assay (PDE1)	Mata et al. (2003b)

Compound	Source	Assay	Reference
MS-282a MS-282b	*Streptomyces tauricus* ATCC 27470 (Streptomycetaceae)	Functional enzymatic assay (MLKC)	Nakanishi et al. (1994)
Obscurolide A_1 Obscurolide A_2 Obscurolide A_3 Obscurolide A_4 Obscurolide $B_{2\alpha}$ Obscurolide $B_{2\beta}$ Obscurolide B_3 Obscurolide B_4 Obscurolide D_2 Obscurolide $C_{2\alpha}$ Obscurolide $C_{2\beta}$ Obscurolide C_2 methyl esther	*Streptomyces viridochromogenes* (Streptomycetaceae)	Functional enzymatic assay (PDE1)	Hoff et al. (1992) Ritzau et al. (1993)
KS-504a KS-504b KS-504c KS-504d	*Mollisia ventosa* kac-1148 (Cortinariaceae)	Functional enzymatic assay (PDE1, MLCK)	Nakanishi et al. (1990b)
Herbarumin I Herbarumin II Herbarumin III	*Phoma herbarum* (Sphaeropsidaceae)	Functional enzymatic assay (PDE1) SDS gel electrophoresis	Rivero-Cruz et al. (2003)
Citreoviridin	*Penicillium* sp. (Trichocomaceae)	Functional enzymatic assay (PDE1) Intrinsic fluorescence	Pala et al. (1999)

(*continued*)

TABLE 13.1 (continued)
Natural Anti-CaM Metabolites Reported to Date

Compound	Natural Source	Method of Detection	References
Flavonoids and Other Phenolics Compounds			
Quercetin	Several plants	Intrinsic fluorescence functional enzymatic assay (PDE1)	Paliyath and Poovaiah (1985)
Genistein			Goto et al. (1987)
Biochanin A			
Naringenin			
Glycoside rutine			
Catechin			
epi-Catechin			
Hesperidin			
Chrysin (**40**)	*Oroxylum indicum* (Bignoniaceae)	Computational studies	Li et al. (2007)
Tectorigenin	*Streptoverticillium album* K-251 (Streptomycetaceae)	Functional enzymatic assay (PDE1)	Matsuda et al. (1988)
Daidzein			
4-Hydroxyderricin	*Angelica keiskei* (Umbelliferae)	Intrinsic fluorescence	Okuyama et al. (1991)
Xanthoangelol			
Cafeic acid	Several plants	Functional enzymatic assay (PDE1)	Paliyath and Poovaiah (1985)
Vitexin (**41**)	*Aloysia citriodora* Palau (Verbenaceae)	Functional enzymatic assay (PDE1)	Ragone et al. (2007)
Isovitexin			
Dioclein (**42**)	*Dioclea grandiflora* (Fabaceae)	Functional enzymatic assay (PDE1)	Bhattacharyya et al. (1995) Goncalves et al. (2009)

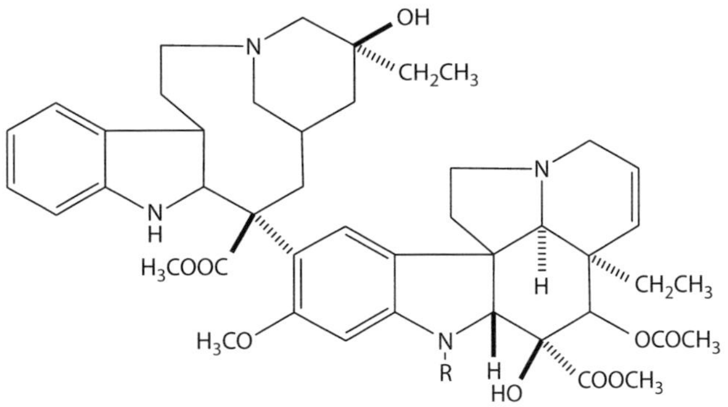

R

(1) CH₃

(2) CHO

FIGURE 13.2 Alkaloids with CaM inhibitor properties.

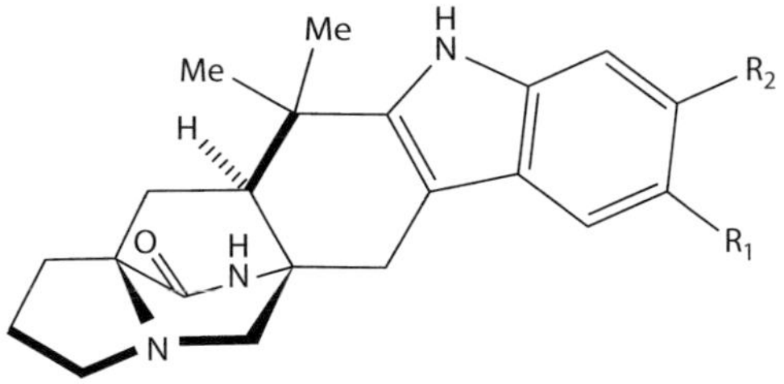

	R₁	R₂
(3)	Cl	Cl
(4)	H	Cl
(5)	Cl	H
(6)	H	H

FIGURE 13.3 Malbrancheamides with CaM inhibitor properties.

From the combined mycelia and culture extracts of *Emericella* 25379, we isolated two novel CaM inhibitors, namely, 15-chlorotajixanthone hydrate (**23**) and 14-methoxytajixanthone (**24**) (Figure 13.6), along with three known compounds characterized as shamixanthone (**25**), tajixanthone hydrate (**26**), and emericellin (**27**) (Figueroa et al. 2009).

In all cases, the molecular formulae were established by high resolution mass spectrometry (HRMS) and in the case of **23**, the mass spectrum was consistent with the presence of one chlorine atom in the molecule since the isotopic pattern around the molecular ion showed the typical M/M + 2 ratio of 100:35. The five natural products exhibited similar UV, IR, and NMR spectra (Table 13.2) suggesting the presence of a hydroxyxanthone chromophore. As an example, the NMR spectra of compound **23** are illustrated in Figure 13.11. The substitution pattern along the xanthone and

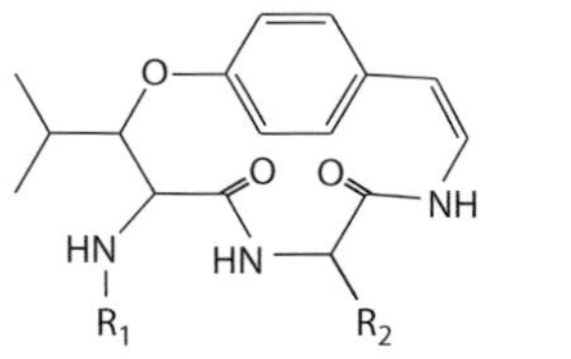

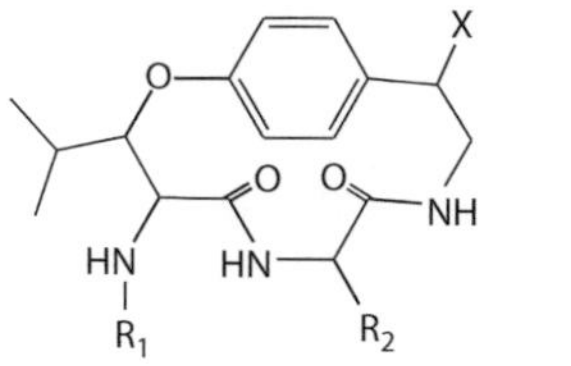

	R$_1$	R$_2$		R$_1$	R$_2$	X
(7)	N(CH$_3$)$_2$Phe(S)-	(CH$_3$)$_2$CHCH$_2$-	(13)	N(CH$_3$)$_2$Phe(S)-	(CH$_3$)$_2$CHCH$_2$-	H
(8)	N(CH$_3$)$_2$Phe(R)-	(CH$_3$)$_2$CHCH$_2$-	(14)	N(CH$_3$)$_2$Phe(R)-	(CH$_3$)$_2$CHCH$_2$-	OCH$_3$
(9)	N(CH$_3$)Phe-	(CH$_3$)$_2$CHCH$_2$-	(15)	N(CH$_3$)Phe-	(CH$_3$)$_2$CHCH$_2$-	OH
(10)	N(CH$_3$)$_2$Phe-	(CH$_3$)$_2$CHCH(OH)-				
(11)	N(CH$_3$)$_2$Leu-	(CH$_3$)$_2$CHCH$_2$-				
(12)	t-cinnamoyl-	(CH$_3$)$_2$CHCH$_2$-				

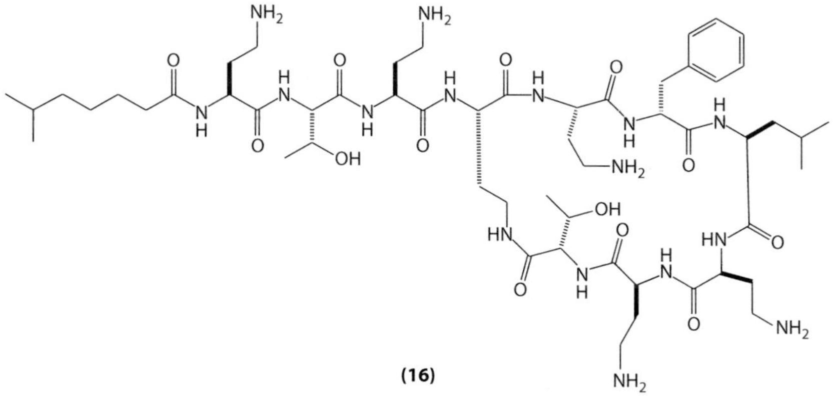

(16)

GIGAVLKVLTTGLPALISWILRLRQQ-NH$_2$ CNCKAPETALCARRCQQ-NH$_2$

(17)   (18)

FIGURE 13.4 Peptides with CaM inhibitor properties.

dihydropyran residues was corroborated by detailed analysis of the correlation spectroscopy (COSY), heteronuclear multiple bond coherence (HMBC), and nuclear Overhauser effect spectroscopy (NOESY) experiments. In the case of compounds **23**, and **25–27**, the configuration at the chiral centers was assigned based on biogenetic grounds, NOESY correlations, and by the correspondence between the Cotton effects observed in their CD spectra and those of other related xanthones. The CD spectra of **24**, however, revealed some significant differences with respect to those of **23**, and **25–27**, which were attributed to the presence of the new chiral center at C-14.

The absolute configuration at the stereogenic centers in **24** was established by a combination of molecular modeling calculations and advanced Mosher's methodology. Analysis of the $\Delta\delta H_{S-R}$ of the per-(*S*)- and per-(*R*)-MTPA Mosher esters derivatives **24s** and **24r** prepared from **24** showed that $\Delta\delta H_{S-R}$ for H-19a, H-19b, and H-20 were +0.0037, +0.0025, and +0.0040, respectively. Therefore, the absolute configuration at C-20 and C-25 was established as *S* and *R*, respectively; this finding was in agreement with the stereochemistry at these centers for compounds **23**, and **25–27**, as well as for other

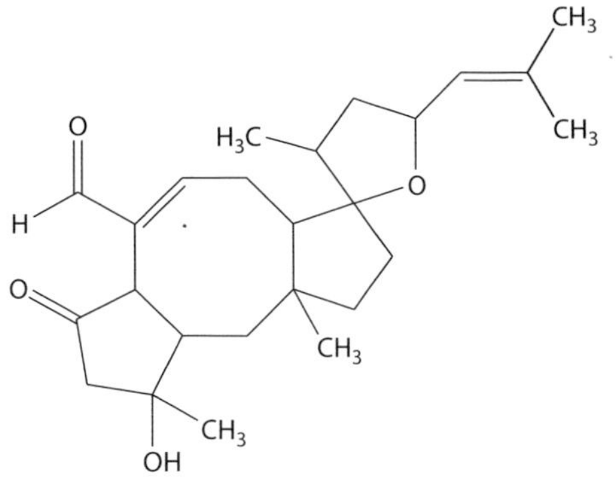

FIGURE 13.5 Terpenoids with CaM inhibitor properties.

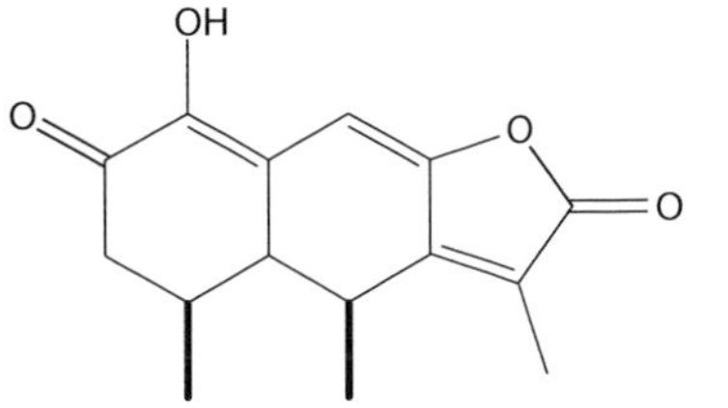

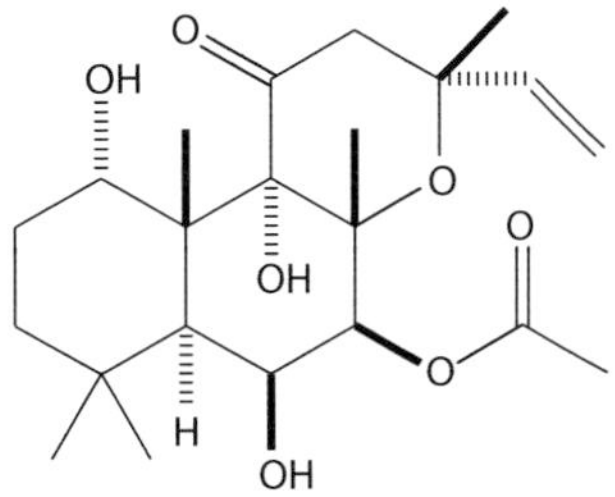

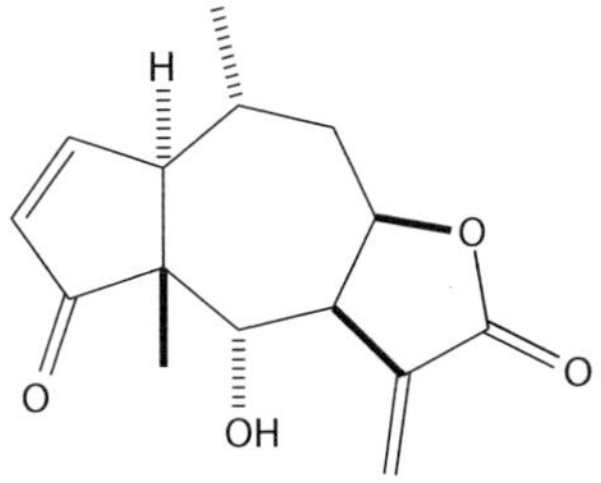

related xanthones. Finally, the configuration at C-14 and C-15 of compound **24** was determined by comparing the optical rotation and ^{1}H-^{1}H coupling constant experimental values with those obtained through molecular modeling calculations at DFT B3LYP/ DGDZVP level of theory for diastereoisomers **24a–24d**. Table 13.3 summarizes the remarkable differences between the experimental optical rotation of **24** ($[\alpha]_D = -38$) and the calculated values for **24a**, **24c**, and **24d**, revealing that the (14S,15S,20S,25R)-stereoisomer (**24b**) represents the correct structure for 14-methoxytajixanthone (**24**). Moreover, the agreement between the calculated and observed coupling constants $J_{19a,20}$, $J_{19b,20}$, and $J_{20,25}$ confirmed the *trans*-relationship between the substituents at C-20 and C-25 in the dihydropyran ring (Figueroa et al. 2009).

The effect of **23–27** on CaM was first assessed with the PDE1 assay, which is commonly employed to detect CaM antagonists; a human recombinant-CaM was employed as the activator. In vitro PDE1 activity was established through a functional coupled enzymatic reaction by measuring spectrophotocolorimetrically at 655 nm the amount of inorganic phosphorous released by the hydrolysis of 5′-AMP. The latter compound is generated by the hydrolysis of *c*AMP by the action of PDE1. The resulting 5′-AMP is then converted to adenosine and inorganic phosphate in the presence of a 5′-nucleotidase (Sharma and Wang 1979). The results are expressed as 50% inhibition concentration (IC$_{50}$) values. The overall enzymatic process is performed following the procedure of Sharma and Wang with modifications of our own (Rivero-Cruz et al. 2007; Figueroa et al. 2009).

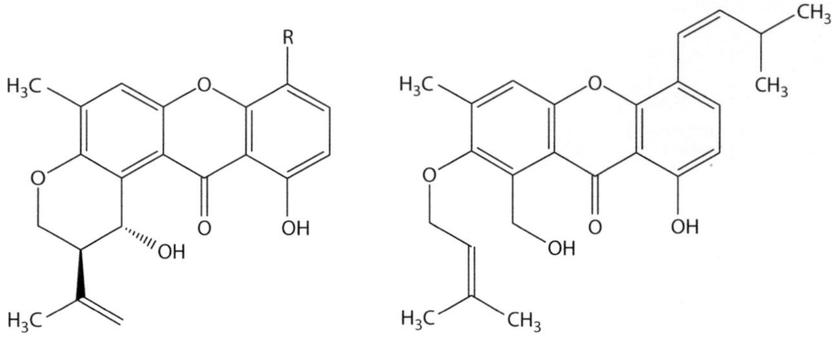

FIGURE 13.6 Xanthones with CaM inhibitor properties.

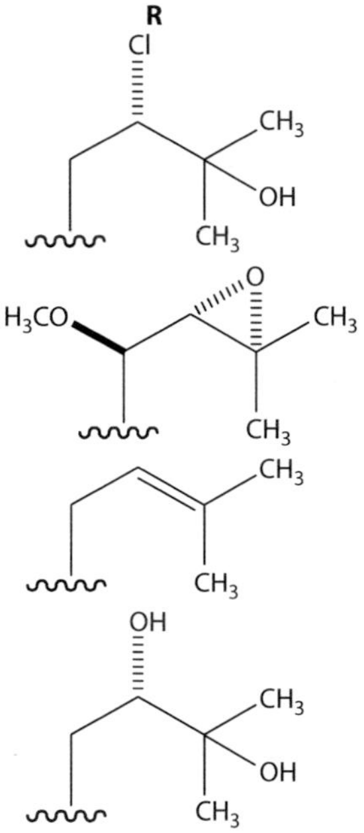

The results showed that the activation of PDE1 was better inhibited in the presence of **24** [IC_{50} = 5.54 µM] and **26** [IC_{50} = 5.62 µM]. The activity was comparable to that of CPZ [IC_{50} = 7.26 µM]. Compounds **23**, **25**, and **27** showed IC_{50} values of 9.59, 29.16, and 68.17 µM, respectively. In all cases, the effect was concentration dependent. In addition, a kinetic analysis using different amounts of CaM in the presence of different concentrations of **24** and **26** indicated that both compounds act as competitive antagonists of CaM, thus interfering with the formation of the CaM-PDE1 active complex. The estimated K_i (inhibition constant) values were 25.38 and 13.92 µM, respectively (Figueroa et al. 2009).

The change of the electrophoretic behavior of CaM treated with the xanthones **23–27**, as detected in a PAGE electrophoresis, provided an additional evidence of their interaction with the protein. The results indicated that CaM treated with

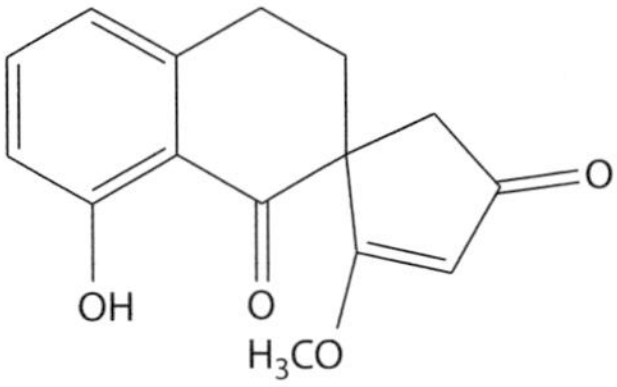

(28)

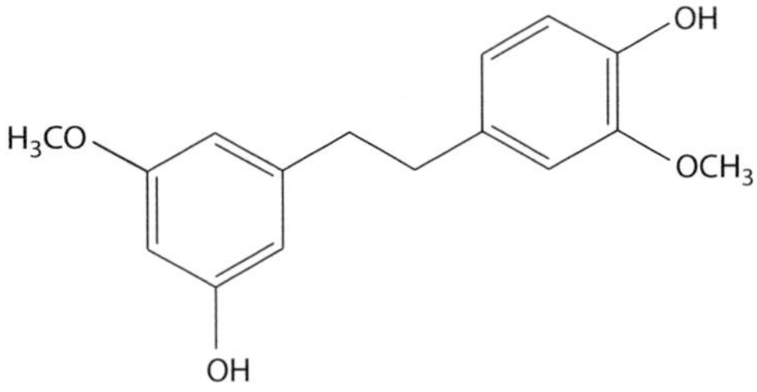

(29)

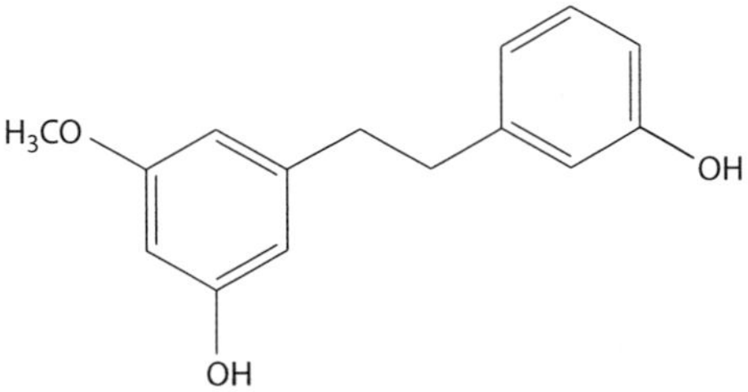

(30)

FIGURE 13.7 Stilbenoids with CaM inhibitor properties.

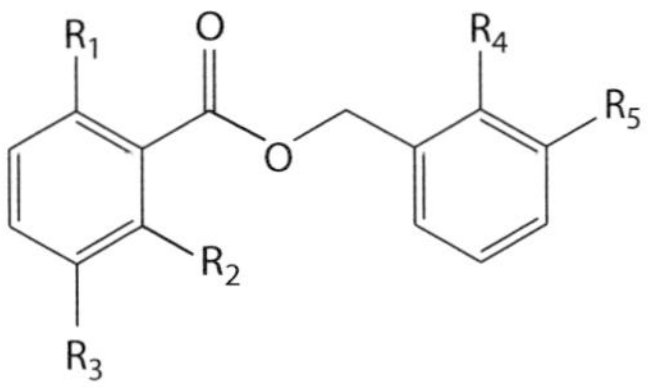

	R_1	R_2	R_3	R_4	R_5
(31)	OCH₃	OCH₃	H	H	β-D-glucopiranosyloxy
(32)	OCH₃	OCH₃	H	H	OH
(33)	OH	H	H	OCH₃	H
(34)	OCH₃	OCH₃	H	H	H
(35)	OH	OCH₃	H	H	OCH₃
(36)	OH	OCH₃	H	H	H
(37)	OCH₃	OCH₃	OCH₃	H	H
(38)	OH	OCH₃	OCH₃	H	H
(39)	OCH₃	OCH₃	H	H	OCH₃

FIGURE 13.8 Benzyl benzoates with CaM inhibitor properties.

xanthones **23**–**27** has a lower electrophoretic mobility than the untreated protein. As in the PDE1 assay, the best effect was observed with compounds **24** and **26**.

Xanthones **23**–**27** quenched also the extrinsic fluorescence of a fluorescent-engineered human CaM (*h*CaMM124C–mBBr) recently developed by González-Andrade and coworkers (2009). The protein was engineered by a rational design, replacing Met124 by cysteine using site-directed mutagenesis; the resulting protein was purified and the fluorophore monobromobimane (mBBr) attached covalently to Cys124

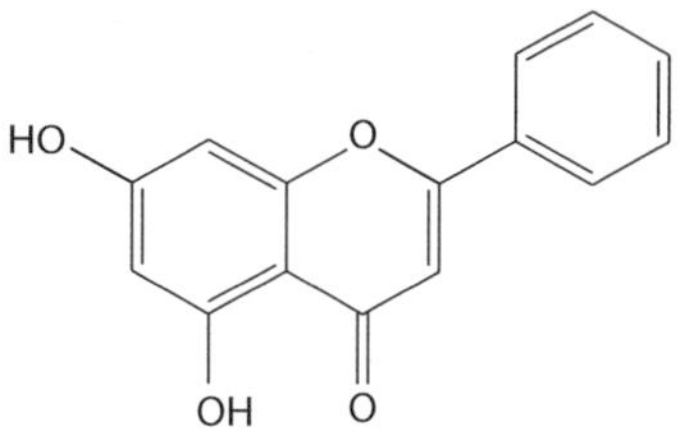

(40)

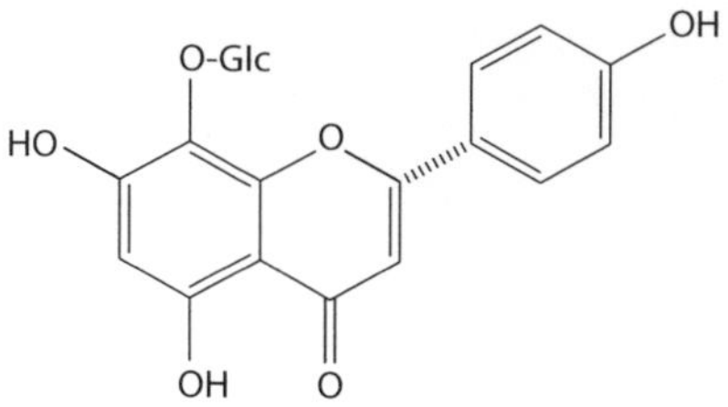

(41)

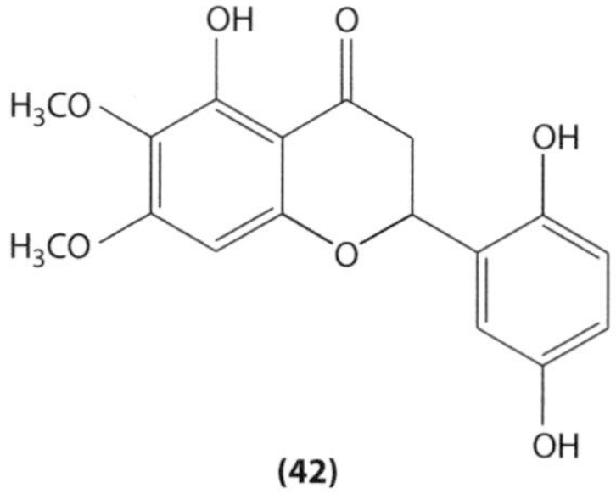

(42)

FIGURE 13.9   Flavonoids with CaM inhibitor properties.

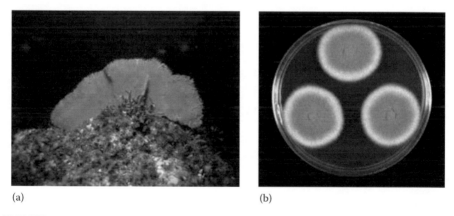

(a)   (b)

FIGURE 13.10   *Emericella* 25379: (a) *Emericella* host, soft coral *Pacifigorgia rutilia*; and (b) fungi growing in potato dextrose agar.

as fluorescent probe. The microenvironment surrounding position 124 is very susceptible to classical CaM inhibitors, for that reason this position was selected for attaching the fluorophore. The data obtained by titrating labeled Ca^{2+}-CaM (5 µM) with different amounts of compounds **23–27** revealed that the fluorescence intensity changed with increasing concentrations of compounds **23–27**. Figure 13.12 shows quenching of the fluorescence of CPZ (A) and compound **24** (B) as well as nonlinear fit plot for calculating the K_d values. The K_d of metabolites **23–27** were 33, 30, 65, 57, and 92 nM, respectively while that of CPZ was 1 µM. This information clearly revealed that the binding of xanthones **23–27** to CaM is stronger than that of CPZ.

TABLE 13.2

NMR Data of Isolates 23–27 from *Emericella* 25379

Position	δ_C/δ_H (ppm), Mult. (J in Hz)				
	23	24	25	26	27
1	160.5	162.2	159.7	160.3	159.7
2	137.1/7.53 d (8.0)	135.6/7.72 d (8.5)	109.8/6.74 d (8.4)	109.9/6.72 d (8.4)	106.2/6.85 d (8.4)
3	119.3/6.80 d (8.0)	119.0/6.88 d (8.5)	136.6/7.44 d (8.4)	138.3/7.49 d (8.4)	136.6/7.48 dd (0.45, 8.4)
4	115.2	109.0	109.2	109.2	109.8
5	109.3/7.33 d (1.0)	110.9/7.23 d (1.0)	119.4/7.30 s	119.1/7.19 s	119.4/7.28 d (0.6)
6	138.6	138.9	138.4	138.5	138.4
7	149.6	149.8	149.4	149.5	149.4
8	119.3	119.1	120.9	120.8	123.9
9	109.3	116.8	118.9	116.3	118.8
10	152.1	152.7	152.8	153.1	152.8
11	152.0	151.9	152.3	151.9	152.3
12	121.1	115.9	116.7	116.8	118.7
13	184.4	184.4	184.5	184.3	184.5
14	28.6/a 3.05 d (16.6); b 3.13 dd (8.0, 16.3)	66.7/3.18 d (8.0)	27.5/a 3.49 d (5.5); b 3.50 d (6.0)	32.0/a 3.16 dd (1.2, 14.0); b 2.63 dd (10.8, 14.0)	27.1/3.39 d (7.2)
15	63.4/3.03 dd (2.0, 10.5)	76.1/4.67 d (8.0)	121.7/5.31 dd (7.2, 7.6)	77.7/5.31 dd (7.2, 7.6)	122.1/5.34 dd (1.5, 7.5)
16	116.9	56.9	133.3	72.9	133.3
17	24.8/1.34 s	24.7/1.25 s	25.8/1.75 s	26.5/1.34 s	25.3/1.74 s
18	19.0/1.46 s	19.8/1.32 s	17.9/1.79 s	23.6/1.39 s	18.4/1.74 s
19	64.6/a 4.44 ddd (1.0, 3.5, 11.0); b 4.37 dd (3.0, 11.0)	64.7/a 4.43 dd (3.6, 10.8); b 4.35 dd (3.0, 10.8)	64.6/a 4.43 dd (2.8, 11.2); b 4.34 dd (2.8, 10.8)	64.5/a 4.41 dd (2.0, 10.8); b 4.31 dd (2.8, 10.8)	71.6/4.45 dd (7.5)
20	44.9/2.74 ddd (3.0, 3.0, 3.5)	45.0/2.74 ddd (2.1, 3.0, 3.6)	44.9/2.73 s	44.8/2.69 s	119.5/5.61 dd (1.5, 7.2)
21	142.6	142.5	142.6	142.5	142.6
22	112.3/a 4.82 ddd (0.5, 1.5, 2.5); b 4.60 dd (1.5, 2.5)	112.3/a 4.81 d (2.5); b 4.60 d (1.0)	112.3/a 4.80 s; b 4.58 s	112.3/a 4.77 s; b 4.53 s	25.8/1.81 s
23	22.5/1.87 t (0.5)	22.7/1.85 s	22.5/1.85 s	22.6/1.82 s	18.2/1.71 s
24	17.4/2.37 d (1.0)	17.5/2.36 s	17.5/2.36 s	17.4/2.28 s	17.9/2.41 s

(continued)

TABLE 13.2 (continued)
NMR Data of Isolates 23–27 from *Emericella* 25379

	δ_C/δ_H (ppm), Mult. (*J* in Hz)				
Position	23	24	25	26	27
25	63.2/5.43 ddd (1.0, 3.0, 3.5)	63.3/5.43 d (2.1)	63.2/5.41 s	63.2/5.34 s	56.2/5.41 s
OH-1	12.66 d (0.5)	12.86 s	12.60 s	12.54 s	12.92 d (0.3)
CH₃O-14		57.8/3.40 s			
OH-15				2.44 s	
OH-16	2.47 s			2.37 s	
OH-25	5.03 d (4.0)	4.94 sa	5.09 s	4.98 d (4.0)	4.50 d (8.1)

Note: Spectra recorded in CDCl₃ at 500 MHz (¹H) or 125 MHz (¹³C).

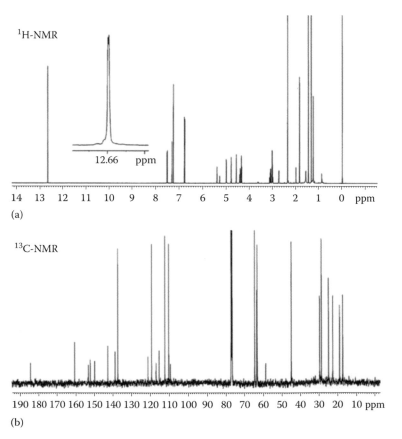

FIGURE 13.11 Selected NMR data of 15-chlorotajixanthone hydrate (**23**) in CDCl₃: (a) ¹H-NMR (500 MHz); (b) ¹³C-NMR (125 MHz);

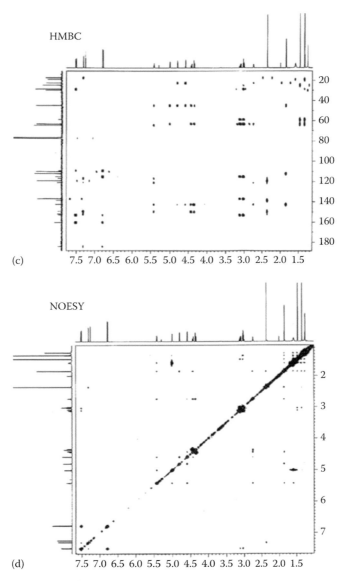

(c)   7.5 7.0 6.5 6.0 5.5 5.0 4.5 4.0 3.5 3.0 2.5 2.0 1.5

(d)   7.5 7.0 6.5 6.0 5.5 5.0 4.5 4.0 3.5 3.0 2.5 2.0 1.5

FIGURE 13.11 (continued)   (c) HMBC; and (d) NOESY.

Finally, to assess the putative binding mode of **23–27** with CaM, a docking study analysis was performed using the advanced docking AutoDock4.0 software (http:// autodock.scripps.edu). Briefly, the docking protocol was validated for the CaM crystal structure (PDB code 1a29) predicting the binding mode of TFP, which was removed from the active site and docked back into the pocket in the same confor- mation found in its crystal structure. AutoDock successfully predicted the bind- ing mode not only of TFP but also of CPZ and with a root mean square deviation (RMSD) of 1.931 and 1.837 Å, respectively (Figure 13.13). The results of molecular

TABLE 13.3

Absolute Differences between the Experimental Optical Rotation of 14-methoxytajixanthone (24) and the DFT B3LYP/DGDZVP Calculated Values for Stereoisomers 24a–24d

	24a (14R,15R)	24b (14S,15S)	[α]D_calc 24c (14R,15S)	24d (14S,15R)
24 (20S,25R) [α]D_exp = −38.0	−102.30	−16.11	−233.48	95.95
Δ|([α]D_exp − [α]D_calc)|	64.30	**21.89**	195.48	57.95

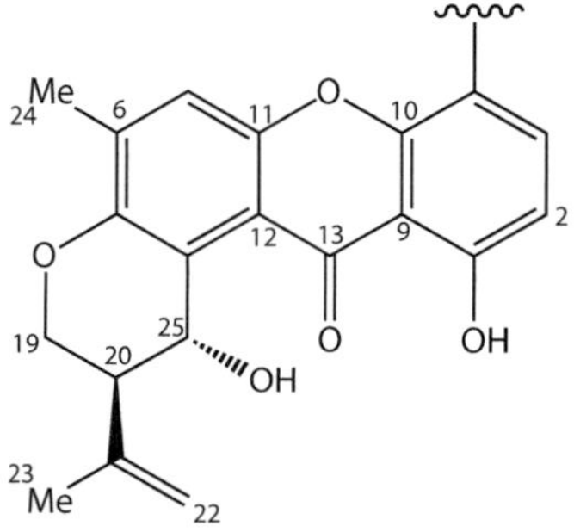

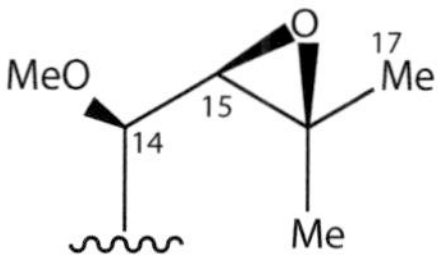

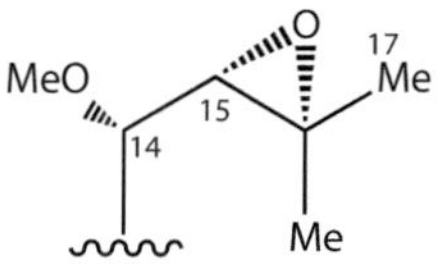

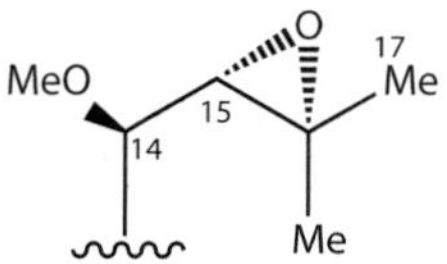

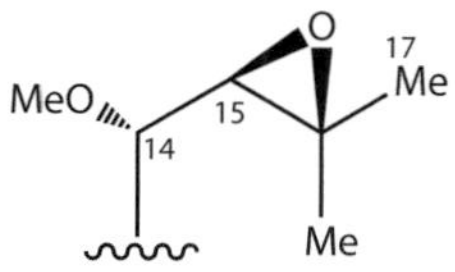

Note: The small difference for (14S,15S)-**24b** allows the stereochemical assignment of **24**.

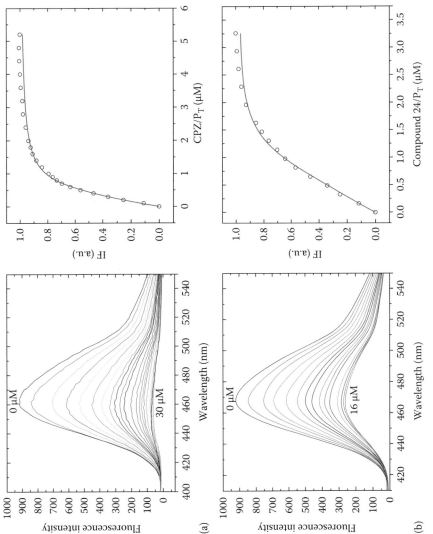

FIGURE 13.12 Titration by fluorescence of hCaM engineered hCaM with CPZ and compound **24**.

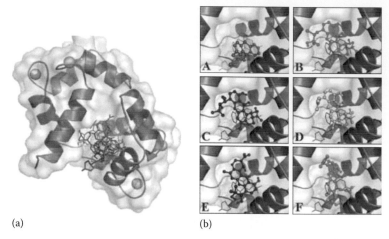

(a)   (b)

FIGURE 13.13   (See color insert.) (a) Docking results obtained using AutoDock 4.0 inside the active site of CaM (orange): (b) (A) Docked CPZ ligand (green, sticks) into CaM appears superimposed on the cocrystallized TPF (red, lines). Top ranked binding mode of the most populated cluster of **23** (B, pale-green, sticks), **24** (C, blue, sticks), **25** (D, pale-yellow, sticks), **26** (E, gray, sticks), and **27** (F, cyano, sticks) into the binding site. Compounds **24** and **26** attach to CaM at the same position as TFP. The metal atoms (Ca^{2+}) are shown as pale-yellow color balls. Hydrogens are omitted for clarity.

docking study revealed that **23–27** interacted with CaM in the same pocket of TFP and CPZ (Figueroa et al. 2009).

13.5.2 ALKALOIDS FROM *MALBRANCHEA AURANTIACA*

Malbranchea aurantiaca Singler and Carmich (Myxotrichaceae) (Figure 13.14) is an ascomycete obtained from bat guano at the Juxtlahuaca caves, State of Guerrero, Mexico. The fungus was initially selected as a source of phytotoxic agents considering the noted phytogrowth inhibitory activity against *Amaranthus hypochondriacus* seedlings of the extracts prepared from broths and mycelia of the fungus fermented in different conditions (Martínez-Luis et al. 2005).

The first fermentation was accomplished in dark conditions at 25°C. The resulting active organic extract (mycelial and broth) (IC$_{50}$ = 195.0 µg/mL against *A. hypochondriacus*) yielded 1-hydroxy-2-oxoeremophil-1(10),7(11),8(9)-trien-12(8)-olide (**20**), a novel natural product, and penicillic acid, a well-known phytotoxic compound (Martínez-Luis et al. 2005).

Compound **20** [IC$_{50}$ = 6.57 µM] inhibited radicle growth of seedlings of *A. hypochondriacus* using a Petri dish bioassay with a similar potency to 2,2-dichlorophenoxyacetic acid [2,4-D; IC$_{50}$ = 18 µM], used as a positive control.

The effect of compound **20** on CaM was investigated. The results showed that activation of PDE1 was inhibited in the presence of the compound (IC$_{50}$ = 10.2 µM) and CaM. Its effect was higher than that of CPZ (IC$_{50}$ = 18.4 µM) (Martinez-Luis et al. 2005).

Subsequent fermentations of this fungal species were carried out in potato dextrose broth for 30 days at 25°C and under natural daylight cycles. The resulting extracts from the mycelium and fermentation media were extensively chromatographed by

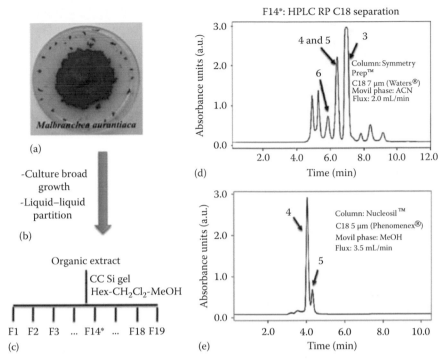

FIGURE 13.14 Isolation of novel malbrancheamide analogs (**3–6**) from *Malbranchea aurantiaca*: (a) Specimen growth in PDA plate; (b) fermentation and extraction process; (c) open column fractionation; (d) HPLC preparative separation of compounds **3**, **4**, and the mixture of **5** and **6**; and (e) HPLC semi-preparative separation of compounds **4** and **5**.

open column chromatography and high performance liquid chromatography (HPLC) (Figure 13.14) to yield four novel alkaloids (**3–6**, Figure 13.2). The alkaloids were given the trivial names of malbrancheamide (**3**), malbrancheamide B (**4**), isomalbrancheamide B (**5**), and premalbrancheamide (**6**) (Martínez-Luis et al. 2006; Figueroa et al. 2008)

The malbrancheamide (**3–6**) family of alkaloids belongs to a unique and rare group of prenylated indole alkaloids containing a bicycle [2.2.2] diazooctane core. These compounds are related to the brevianamides, aspergamides, macfortines, paraherquamides, sclerotamides, and stephacidins (Williams and Cox 2003). These natural products have been isolated periodically from different strains of fungi of the genera *Aspergillus* and *Penicillium* since their discovery in 1969. It has been proposed that they are biosynthesized by an intramolecular hetero-Diels-Alder reaction of a suitable 5-hydroxypyrazinone intermediate.

The four alkaloids (**3–6**) showed a positive color reaction with Dragendroff's and Erlich's reagents and were identified by spectroscopic and spectrometric methods. Their molecular formulae were determined as $C_{21}H_{23}ON_3Cl_2$, $C_{21}H_{24}ON_3Cl$, $C_{21}H_{24}ON_3Cl$, and $C_{21}H_{25}ON_3$, respectively by HRMS. This information as well as the UV absorption maxima at ~233 and ~293 nm suggested that the alkaloids were indole derivatives. The presence of one or two chlorine atoms in the molecules were consistent with the relative abundance of the [M + 2] peak, in the case of malbrancheamide B (**4**) and isomalbrancheamide B (**5**), and of the [M + 2] and [M + 4] peaks, in the case of malbrancheamide (**3**), with respect to the molecular ions [M$^+$] in their mass spectra. In all cases, the IR spectra showed typical absorption bands for lactams at ~3298 and ~1658 cm^{-1}.

In the proton NMR spectra, the profile and coupling pattern of the aromatic protons (Figure 13.15, Table 13.4) were consistent with the position of the substituents

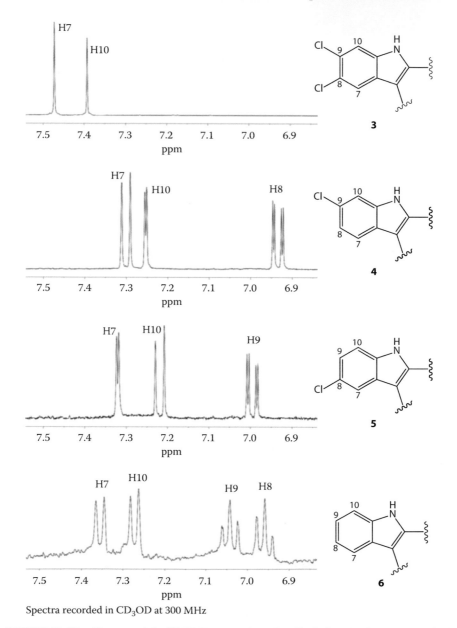

Spectra recorded in CD$_3$OD at 300 MHz

FIGURE 13.15 Close up of the ^{1}H-NMR aromatic region (δ_H 6–8 ppm) of compounds **3–6**.

TABLE 13.4
NMR Data of Isolates 3–6 from *M. aurantiaca*

| Position | δ_C/δ_H (ppm), Mult. (*J* in Hz) | | | |
	3	4	5	6
1	28.1/a 2.452 m; b 1.46 m	28.5/a 2.42 m; b 1.34 m	28.2/a 2.51 m; b 1.46 m	28.2/a 2.43 m; b 1.49 m
2	23.6/1.87 m	22.4/1.72 m	23.6/1.86 m	23.5/1.90 m
3	55.3/a 3.05 m; b 2.15 q (2.0, 5.0)	55.2/a 2.93 ddd (2.1, 6.0, 9.6); b 2.41 ddd (2.1, 6.0, 9.6)	52.9/a 3.03 m; b 2.15 q (2.0, 5.0)	55.4/a 3.07 m; b 2.13 q (2.0, 5.0)
5	59.4/a 2.25 dd (2.0, 10.0); b 3.42 d (10.0)	59.5/a 2.13 d (9.9); b 3.25 d (9.9)	59.5/a 2.26 dd (2.0, 10.0); b 3.42 d (10.0)	59.5/a 2.28 dd (2.0, 10.0); b 3.36 d (10.0)
5a	57.5	58.5	55.3	57.7
6	30.0/a 2.86 m; b 2.85 m	29.9/a 2.79 d (15.9); b 2.72 d (15.9)	30.3/a 2.86 m; b 2.85 m	30.4/a 2.96 m; b 2.80 m
6a	104.8	103.7	104.9	104.5
6b	123.3	122.7	123.0	122.0
7	119.6/7.47 s	120.7/7.31 d (8.7)	120.0/7.32 d (2.0)	122.0/7.35 ddd (1.0, 1.8, 7.6)
8	125.4	120.6/6.94 dd (1.8, 8.4)	123.3	119.4/6.97 dd (1.0, 7.5)
9	128.2	126.6	125.3/7.00 dd (2.6, 8.6)	119.5/6.97 dd (1.0, 7.5)
10	113.1/7.39 s	112.1/7.26 d (1.7)	111.6/7.22 d (8.6)	128.2/7.26 ddd (0.8, 1.6, 8.0)
10a	137.3	135.4	135.6	138.4
11	11.04 s			
11a	145.2	144.4	146.7	146.7
12	35.5	34.1	32.6	35.4
12a	48.5/2.14 m	46.9/2.07 m	49.8/2.14 m	49.6/2.09 m
13	32.5/a 1.99 m; b 1.94 m	33.9/a 1.98 m; b 1.91 m	32.6/a 1.99 m; b 1.94 m	32.5/a 1.99 m; b 1.98 m
13a	66.1	64.1	66.1	66.2
14	176.7	173.1	176.7	176.7
15	8.37 s			
16	30.6/1.32 s	31.1/1.26 s	30.8/1.32 s	30.9/1.32 s
17	24.2/1.42 s	26.6/1.31 s	24.2/1.42 s	24.4/1.46 s

Note: Spectra recorded in CD_3OD at 300 MHz (^1H) or 100 MHz (^{13}C).

or the absence of any, as in the case of premalbrancheamide (**6**). Also apparent in the proton NMR spectra (Table 13.4) were the resonances for the methyl, aliphatic methine and methylene groups. The ^{13}C-NMR spectra revealed 21 carbon resonances and the presence of the lactam moiety and indole nuclei in the alkaloids.

Detailed 2D NMR spectra analyses led to the establishment of the connectivity of functional groups and, in turn, of the molecular structures. The position of the

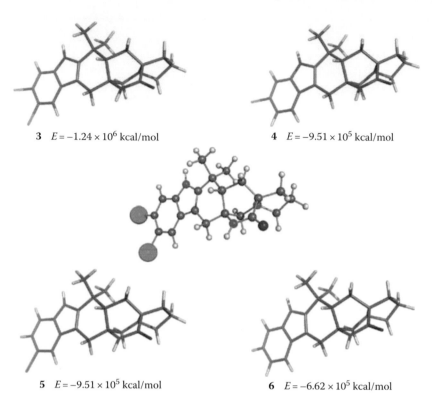

3 $E = -1.24 \times 10^6$ kcal/mol   4 $E = -9.51 \times 10^5$ kcal/mol

5 $E = -9.51 \times 10^5$ kcal/mol   6 $E = -6.62 \times 10^5$ kcal/mol

FIGURE 13.16 **(See color insert.)** X-ray structure of **3** (center) and optimized DFT structures of **3–6** at the B3LYP/631G+ level of theory.

functional groups along the hexacyclic moieties was corroborated by the HMBC experiments. The structure of malbrancheamide (**3**) could be corroborated by an x-ray analysis (Figure 13.16) and the absolute stereochemistry was secured by Flack's method. Thus, the first alkaloid of the series was fully characterized as (5a*S*,12a*S*,13a*S*)-8,9-dichloro-12,12-dimethyl-2,3,11,12,12a,13-hexahydro-1*H*,6*H*-5a,13a (epiminomethano) indolizino[7,6b]carbazol-14-one (**3**). Since the CD spectra of all four analogs were similar, we assumed that the absolute configuration at the stereogenic centers was identical in all compounds.

The conformational behavior of the four alkaloids was analyzed using DFT (B3LYP/631G+ level of theory). In the four optimized structures, ring E displays a normal envelop conformation and the atoms N4-C13a-C1-C2 are located in the plane while C-3 is out of the ring-plane (Figure 13.16). However, in the solid state, malbrancheamide (**3**) showed ring C in the bicyclo [2.2.2] diazaoctane system with a boat conformation whereas ring F (lactame) with an envelope conformation. Ring D was also observed as an envelope while ring E adopts a normal envelope conformation. The atoms C13a-C1-C2-C3 are located in the ring plane while N4 is out of the plane.

Recently, Williams and coworkers (Miller et al. 2008a), from the University of Colorado, reported the total synthesis of racemic malbrancheamide (*d,l* **3**) and malbrancheamide B (*d,l* **4**), which were achieved by a biomimetic hetero-Diels-Alder cycloaddition strategy. Of course the NMR, IR, and UV spectra of the synthetic and natural products were identical. Also the group of Professor Simpkins at the University of Birmingham just synthesized *ent*-malbrancheamide B (*d,l* **4**) (Frebault et al. 2009).

Malbrancheamides (**3–6**) inhibited the activation of PDE1 in a concentration-dependent manner. The IC_{50} values were 19.3, 60.3, 41.6, and 183.3 µM, respectively. Only **3** displayed similar activity to CPZ ($IC_{50} = 17$ µM). Furthermore, a kinetic analysis of the inhibition of the activity of PDE1 using different amounts of CaM in the presence of different concentrations of malbrancheamide (**3**) suggested that the alkaloid acts also as a competitive antagonist of CaM. The estimated K_i value was 47 µM.

In collaboration with Prof. Williams, a number of malbrancheamides analogs **43–55** (Figure 13.17) were tested for their anti-CaM properties (Miller et al. 2008b). These compounds were prepared while attempting the synthesis of malbrancheamide (**3**) and malbrancheamide B (**4**). For each analog, the activity of the complex CaM-PDE1 was measured and a number of active compounds were identified, including **46**, **44**, **54**, and racemic malbrancheamide (*d,l* **3**). Their relative potencies in relation to CPZ (RPCPZ) were 0.9, 0.9, 0.7, and 1.1, respectively.

The effect of indole chlorination, C-12a relative stereochemistry, and the oxidation state of the bicyclodiazaoctane core on the enzymatic inhibition was analyzed. The overall results revealed that malbrancheamide (**3**) and *d,l* **3**, both the natural (+)-enantiomer and a racemate were the most active inhibitors. Apparently, the unnatural (−)-enantiomers of malbrancheamide (**3**) and malbrancheamide B (**4**) are slightly more active than the naturally occurring ones since the RPCPZ of racemic **3** and **4** is higher than that of the natural enantiomerically pure substances. As with malbrancheamides **3–6**, the indole chlorine substitution was important on PDE1 activity among the *syn*-dioxopiperazines (**49–52**); the dichloro substituted species **49** (RPCPZ = 0.9) displayed the highest potency, but compound **52**, which lacks chlorine substitution, was virtually inactive (RPCPZ = 0.1).

In the antidioxopiperazines (**43**, **44**, and **49–55**) series, compound **55**, which completely lacks chlorine substitution, was the most active substance (RPCPZ = 0.7) of this group.

Comparing the monooxopiperazines **45–48**, and *d,l* **3**-*d,l* **6**, synthetic *d,l* **3** (RPCPZ=1.1), and **46** (RPCPZ=0.9) were the most potent whereas *epi*-malbrancheamides **45** and **47** were the least potent (RPCPZ = 0.1 in both cases). It is particularly striking that the monooxopiperazine compound **45** (C-12a-*epi*-malbrancheamide) was essentially inactive, whereas the corresponding C-5-oxo-isomer **46** was very potent (RPCPZ = 0.9). In any case, the relative lack of activity of compounds **45** and **47** reveals that the relative stereochemistry at C-12a is quite important, and also reveals the significance of the C-5 and C-14 amide carbonyl residues.

13.5.3 BENZYL BENZOATES FROM *BRICKELLIA VERONICIFOLIA*

The aerial parts of *Brickellia veronicifolia* (Kunth) Gray (Asteraceae) are broadly commercialized in Mexico to cure gastrointestinal disorders including stomach aches, biliary colic, gastritis, and dyspepsia. In addition, the species is highly valued for the treatment of arthritis, diabetes, infectious diseases, as well as painful inflammatory complaints (Rivero-Cruz et al. 2005; Palacios-Espinosa et al. 2008).

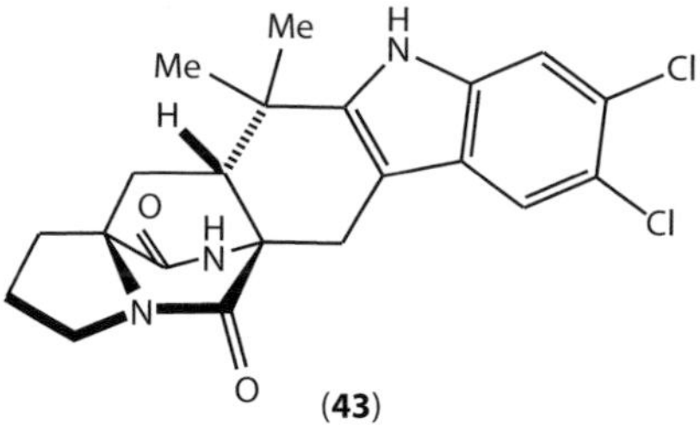
(**43**)

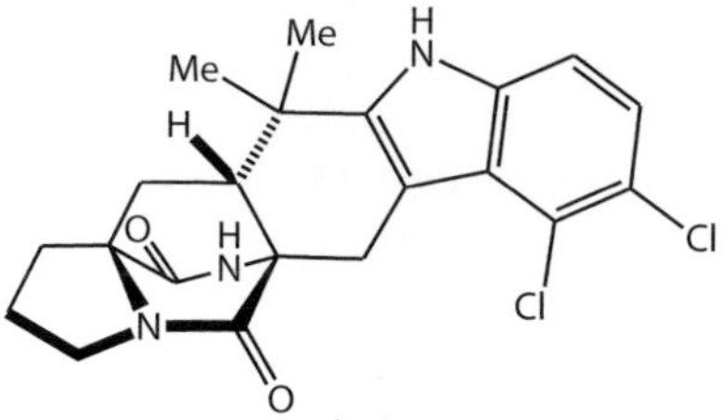
(**44**)

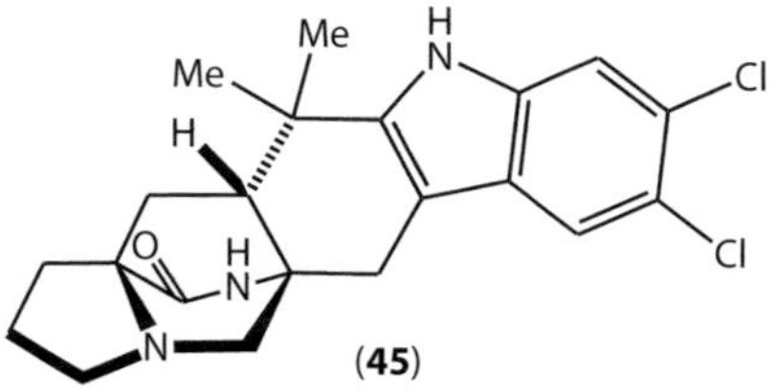
(**45**)

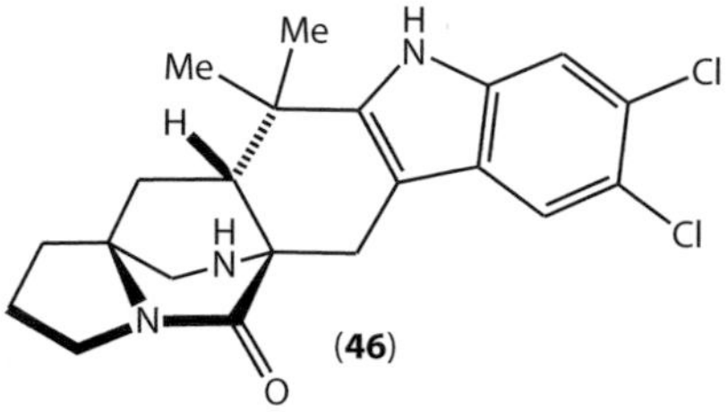
(**46**)

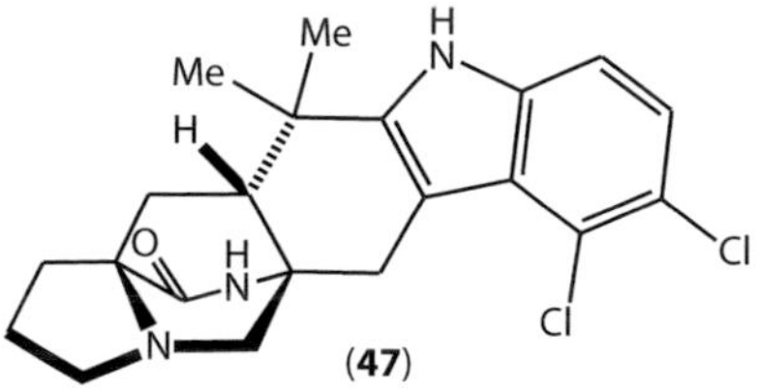
(**47**)

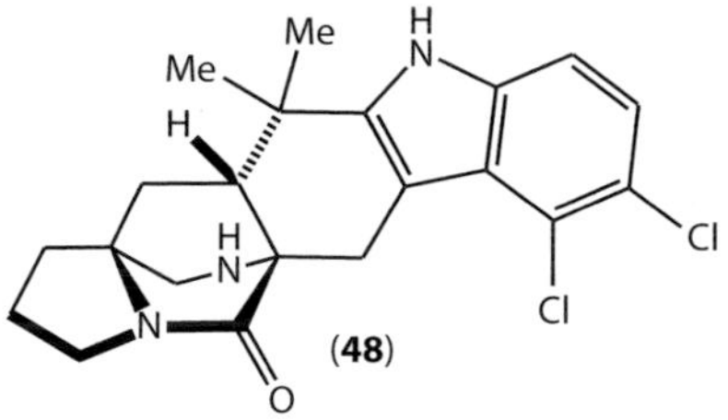
(**48**)

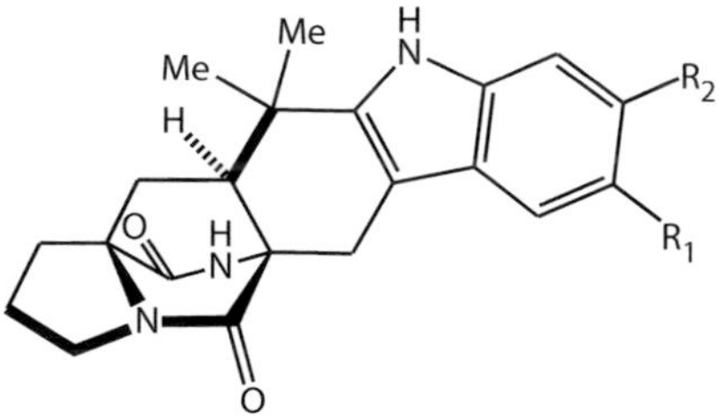

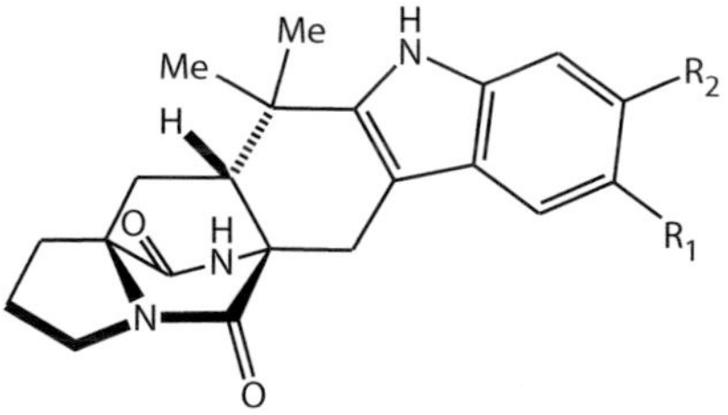

	R_1	R_2
(**49**)	Cl	Cl
(**50**)	Cl	H
(**51**)	H	Cl
(**52**)	H	H

	R_1	R_2
(**53**)	Cl	H
(**54**)	H	Cl
(**55**)	H	H

FIGURE 13.17 Synthetic malbrancheamides derivatives with CaM inhibitors properties.

Classical chemical studies conducted also with the aerial parts of the plant led to the isolation of several 6-methoxy-flavonols and labdane-type diterpenes (Roberts et al. 1980; Calderón et al. 1983; Inhuma et al. 1985).

Pharmacological investigation of different extracts and essential oil prepared from the aerial parts of the plant have demonstrated its antidiarrhoeal (Pérez-Gutiérrez 1996), antioxidant, hypoglycemic (Pérez-Gutiérrez et al. 1998; Pérez et al. 2000;

Pérez et al. 2004; Palacios-Espinosa et al. 2008), spasmolytic (Rivero-Cruz et al. 2005), and antinociceptive (Palacios-Espinosa et al. 2008) activities. In addition, the species was not toxic to mice, nor mutagenic according to the Lorke and Ames tests, respectively (Déciga-Campos et al. 2007).

Bioassay-guided fractionation of a CH_2Cl_2-MeOH (1:1) spasmolytic extract from the aerial parts of the plant led to the isolation of several benzyl benzoates (**31–39**, Figure 13.8), some of them (**31–35**, **38**, and **39**) with noted in vitro smooth muscle relaxant properties (Rivero-Cruz et al. 2005).

The smooth muscle relaxant action of the benzyl benzoates **31–39** was demonstrated measuring their effect on the tone and amplitude of the spontaneous contraction of the guinea-pig ileum. Table 13.5 lists the corresponding IC_{50} values of these compounds. Compounds **31–35**, **38**, and **39** induced a significant concentration-dependent inhibition of the spontaneous contractions of the guinea-pig ileum. These compounds were more potent than papaverine, used as a positive control. The data presented in the Table 13.5 show that the presence of a glucosyl or methoxyl residue at C-3′ in the benzyl unit as well as the 2,6-dimethoxybenzoyloxy moiety, as in the case of compounds **31** and **34**, are important structural requirements for the biological activity. Compound **32** has good activity, although lower than that of **31**, indicating how the presence of a free hydroxyl group at C-3′ provoked a significant decrease in the spasmolytic action (Rivero-Cruz et al. 2005). In addition, benzyl benzoate **34**

TABLE 13.5

Smooth Muscle Relaxant and CaM-PDE1 Activities In Vitro of Natural Benzyl Benzoates 31–39

Compound	Inhibition on the Spontaneous Contraction of Isolated Guinea Pig Ileum		Inhibition of PDE1-CaM Complex	
	IC_{50} (µM)	Potency	IC_{50} (µM)	Potency
31	1.49 ± 0.17	2.8	>100	
32	2.61 ± 0.24	1.6	67.51 ± 12.36	0.2
33	2.07 ± 0.16	2.0	10.61 ± 2.84	1.4
34	1.84 ± 0.11	2.3	12.34 ± 4.67	1.2
35	2.29 ± 0.28	1.8	14.59 ± 5.27	1.0
36	ND	ND	>100	
37	ND	ND	23.73 ± 9.02	0.6
38	4.96 ± 0.08	0.8	21.01 ± 3.19	0.6
39	1.52 ± 0.11	2.8	46.94 ± 11.18	0.3
Papaverine[a]	4.23 ± 0.68	1.0		
CPZ[a]			14.61 ± 3.55	1.0

Note: Values to mean ± SEM; n = 6; $p < 0.05$; ND not determined; potency was obtained by the formula: IC_{50} (µM) positive control[a]/IC_{50} (µM) compound, assuming a value of 1.0 for positive controls.

showed a good antinociceptive response when tested by the writhing test in mice at the dose of 100 mg/kg (Palacios-Espinosa et al. 2008).

Several events that take place during smooth muscle contraction–relaxation events are regulated by CaM-dependent mechanisms (Webb 2003). On the other hand, CaM modulates the activity of other important enzymes participating in the signal transduction events during the smooth muscle contraction–relaxation events; some of these proteins include CaM kinase II, CaM sensitive-phosphodiesterase (PDE1), the nitric oxide synthases, adenylate cyclases 1 and 8, and several ion channels, notably voltage-gated Ca^{2+} channels (Chin and Means 2000). Therefore, the effect of benzyl benzoates **31–39** (Figure 13.8) on CaM using the PDE1 enzymatic assay was evaluated to establish, at least in vitro, if their spasmolytic action could be mediated by CaM. According to the data in Table 13.5, which summarized the IC_{50} values of the natural benzyl benzoates, including their RPCPZ, the most active natural inhibitors of the system CaM-PDE1 were benzyl benzoates **33–35**. Compound **33** was almost twice as potent (RPCPZ = 1.8) as the positive control, but **34** and **35** were as active as CPZ with an RPCPZ of 1. The three compounds inhibited the activity of PDE1 in a concentration-dependent manner. However, there was not a direct correlation between the potency of the compounds as smooth muscle relaxant agents and CaM inhibitors.

Since compound **34** showed similar efficacy in vitro to CPZ, a series of analogs (**56–78**, Figure 13.18) were prepared in order to increment its activity against the complex CaM-PDE1 using the same functional assay. The compounds were designed considering the commercial availability of suitable acids, acid chlorides, or benzyl alcohols. The modifications were planned to generate valuable information regarding the weight of the position and number of oxygenated substituents of ring A and the presence of different types of substituents in C-3′ of ring B on the enzymatic activity. Thus, the first series (**56–70**, Figure 13.18) possesses no substituents in ring B but different number and position of hydroxyl or methoxy groups in ring A. The second group (**71–78**, Figure 13.18), on the other hand, possesses an A ring identical to that on compound **34**, but different substituents in ring B. Compounds **56–63** were prepared by condensation of benzyl alcohol with the appropriate acid chlorides in the presence of triethylamine. Benzyl benzoates **61–67** were obtained by the reaction of a suitable benzoic acid derivative with benzyl alcohol in the presence of 1,1′-carbonyldiimidazole (DCC) while **71–74** were prepared following the same strategy but using 2,6-dimethoxy benzoyl chloride and a suitable *m*-substituted benzyl alcohol derivative. Compound **75** was obtained by acetylation with pyridine and acetic anhydride (Ac_2O) of natural product **32**. Finally, products **76–78** possessing an ester group at C-3′ in the benzyl alcohol moiety were prepared by esterification of the 3′-hydroxybenzyl-2,6-dimethoxy benzoate, and the appropriate benzoyl, 2,6-dimethoxy benzoyl or *o*-anisoyl chlorides, respectively, in the presence of triethylamine.

The most active compounds of the first series were **60**, **61**, **65**, **67**, and **70** with RPCPZ of 3.33, 1.79, 1.18, 1.13, and 1.42, respectively. As the lead molecule **34** (RPCPZ = 1), these compounds have two oxygenated substituents in ring A, except in the case of **70**, which has three oxygenated substituents; it is worth mentioning that when the substituents were methoxy groups and were located in C-2/C-4 (i.e., **60**) or in C-3/C-4 (i.e., **61**), the higher inhibitory activity was found.

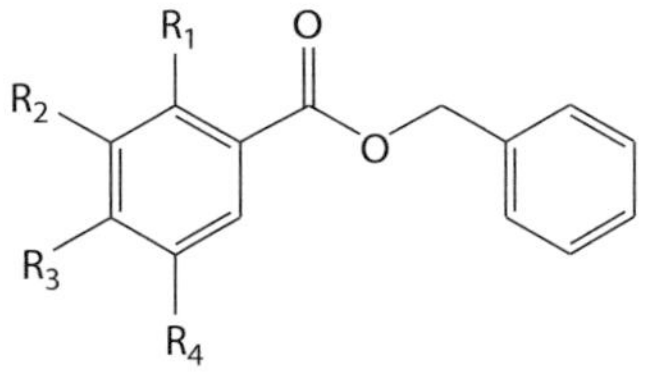

	R_1	R_2	R_3	R_4
(56)	H	H	H	H
(57)	OCH$_3$	H	H	H
(58)	H	OCH$_3$	H	H
(59)	H	H	OCH$_3$	H
(60)	OCH$_3$	H	OCH$_3$	H
(61)	H	OCH$_3$	OCH$_3$	H
(62)	H	OCH$_3$	H	OCH$_3$
(63)	H	OCH$_3$	OCH$_3$	OCH$_3$
(64)	OH	H	H	H
(65)	H	OH	H	H
(66)	H	H	OH	H
(67)	OH	OH	H	H
(68)	H	OH	OH	H
(69)	H	OH	H	OH
(70)	H	OH	OH	OH

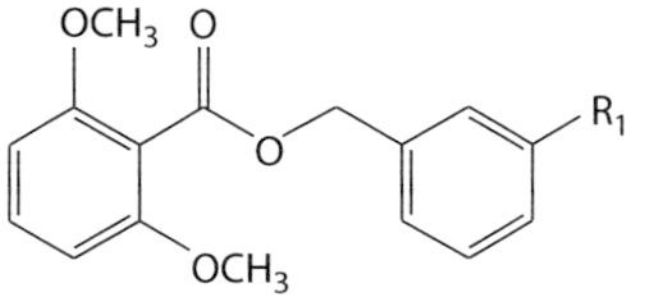

	R_1
(71)	Cl
(72)	NH$_2$
(73)	NO$_2$
(74)	OCH$_3$
(75)	OCOCH$_3$

(76)

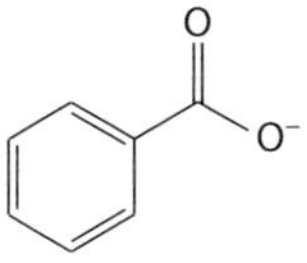

(77)

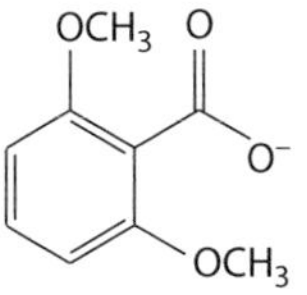

(78)

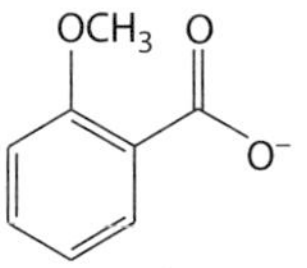

FIGURE 13.18 Synthetic benzyl benzoates with CaM inhibitors properties.

In the second series, the most active benzyl benzoates were **76** (RPCPZ = 6.5) and **77** (RPCPZ = 0.9), which possess benzoyloxy moiety at C-3′.

A kinetic analysis of the most active compounds (**33, 34, 60, 61,** and **76**) was also carried out. The K_i values of **33, 34, 60, 61,** and **76** were calculated using a nonlinear fit of the data. In this case, the Dixon plots approach was not used because after subtracting the activity found at saturating inhibitor concentration for each Ca^{2+}-CaM concentration employed, the curvature in the Dixon plot was not completely eliminated. The adjusted K_i values for **33**, **34**, **60**, **61**, and **76** were 13.4, 18.4, 6.5, 15.3, and 1.6 µM, respectively. The calculated values are consistent with benzyl benzoates, **33**, **34**, **60**, **61**, and **76** being antagonists of the ternary complex Ca^{2+}-CaM-PDE1.

To obtain more information regarding the interaction of the benzyl benzoates with CaM, quenching of the extrinsic fluorescence of the fluorescent-engineered human CaM hCaM124C–mBBr was determined. The results after titrating the labeled Ca^{2+}-CaM (5 µM) with different amounts of benzyl benzoates revealed, however, that in any case the fluorescence intensity changed with increasing concentrations of the benzyl benzoates (up to 160 µM) tested. Figure 13.19 illustrates the effect of compound **73**. These results are in agreement with benzyl benzoates being not classical

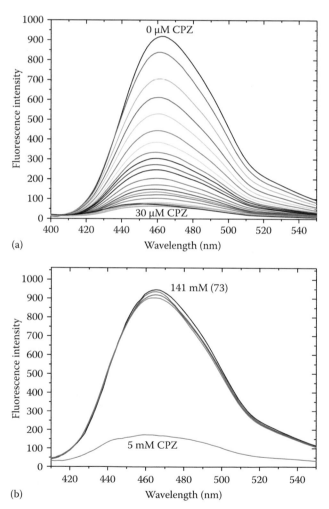

FIGURE 13.19 **(See color insert.)** Titration by fluorescence of hCaM M124C-mBBr engineered hCaM with (a) CPZ and (b) compound **73**.

CaM inhibitors but functional and should not be regarded as competitive antagonist as previously described considering only the kinetic experiments.

13.6 CONCLUSIONS

As shown in this review, CaM inhibitor properties have been described for numerous natural products from plants, animals, and microorganisms. However, few of them have been further investigated in preclinical studies for the development of new drugs useful for treating neurodegenerative diseases.

The novel xanthones isolated from *Emericella* 25379 are potent CaM inhibitors, and interact with the protein at the same binding site of TFP and CPZ. The level of activity exhibited by these compounds suggests that these are promising leads for the development of therapeutics agents.

From the coprophylus fungus *Malbranchea aurantiaca*, four novel prenylated indol alkaloids (**3–6**) possessing the bicyclo [2.2.2] diazaoctane ring system were isolated and characterized. Compounds **3–5** are the first chlorinated alkaloids in this family. Related compounds such as brevianamides, paraherquamides, aspergamides, and macfortines possess a spiro-ψ-indoxyl system, while malbrancheamides do not. On the other hand, the relative configuration at C-12a in the bicyclo [2.2.2] diazaoctane ring system is different to that of previous reported analogs such as (−) VM55599 and stephacidin A (Williams and Cox 2003). Altogether, these features make malbrancheamides unique among these complex indole alkaloid derivatives. These compounds represent a new type of CaM inhibitors being malbrancheamide (**3**) the most relevant. The best activity of this compound in comparison with its natural analogs is the presence of two chlorine atoms that confer a more hydrophobic character to the molecule, which in turn could facilitate its interaction with CaM.

The smooth muscle relaxant benzyl benzoates from *Brickellia veronicifolia* are functional CaM inhibitors. Some more active synthetic analogs were obtained; the most active analog was **76** (six times more potent than CPZ), which possesses a benzoyloxy moiety at C-3′ in ring B but a ring A identical to the natural products.

ACKNOWLEDGMENTS

This work has been conducted under the auspices of several project grants funded by Consejo Nacional de Ciencia y Tecnología (CONACyT) and Dirección General de Asuntos del Personal Académico, UNAM (DGAPA). R. Mata is very grateful to all colleagues, in particular Rogelio Rodríguez Sotres, and PhD students whose collaboration has contributed to the results described in this chapter, in particular to Blanca Rivero-Cruz, Sergio Martínez-Luis, and Araceli Pérez-Vásquez.

REFERENCES

Abe, Y., S. Saito, M. Hori, H. Ozaki, N. Fusetani, and H. Karaki. 1997. Stellettamide-A, a novel inhibitor of calmodulin, isolated from a marine sponge. *British Journal of Pharmacology* 121:1309–1314.

Asakawa, K., K. Kondo, M. Tori, T. Hashimoto, and S. Ogawa. 1991. Prenyl bibenzyls from the liverwort *Radula kojana. Phytochemistry* 30:219–234.

Au, T. K., W. S. Chick, and P. C. Leung. 2000. The biology of ophiobolins. *Life Sciences* 67:733–742.

Au, T. K. and P. C. Leung. 1998. Identification of the binding and inhibition sites in the calmodulin molecule for ophiobolin A by site-directed mutagenesis. *Plant Physiology* 118:965–973.

Barnette, M. S., R. Daly, and B. Weiss. 1983. Inhibition of calmodulin activity by insect venom peptides. *Biochemical Pharmacology* 32:2929–2933.

Baumgrass, R., M. Weiwad, F. Erdmann, J. O. Liu, D. Wunderlich, S. Grabley, and G. Fischer. 2001. Reversible inhibition of calcineurin by the polyphenolic aldehyde gossypol. *Journal of Biological Chemistry* 276:47914–47921.

Bhattacharyya, J., J. S. Batista, and R. N. Almeida. 1995. Dioclein, a flavanone from the roots of *Dioclea grandiflora. Phytochemistry* 38:277–278.

Boswell-Smith, V., D. Spina, and C. P. Page. 2006. Phosphodiesterase inhibitors. *British Journal of Pharmacology* 147:S252–S257.

Bouche, N., A. Yellin, W. A. Snedden, and H. Fromm. 2005. Plant-specific calmodulin-binding proteins. *Annual Plant Reviews* 56:435–466.

Brokx, R. D., M. M. Lopez, H. J. Vogel, and G. I. Makhatadze. 2001. Energetics of target peptide binding by calmodulin reveals different modes of binding. *Journal of Biological Chemistry* 276:14083–14091.

Calderón, J., L. Quijano, M. Cristia, F. Gómez, and T. Ríos. 1983. Labdane diterpenes from *Brickellia veronicaefolia. Phytochemistry* 22:1783–1785.

Chen, S., Z. Hu, Z. Hao, H. Cai, and W. Huang. 1990. Calmodulin antagonist-study on structure–activity relationship of bis-benzylisoquinoline compounds. *Shengwu Huaxue Zazhi* 6:413–416.

Chin, D. and A. R. Means. 2000. Calmodulin: A prototypical calcium sensor. *Trends in Cell Biology* 10:322–328.

Choi, J. and M. Husain. 2006. Calmodulin-mediated cell cycle regulation: New mechanisms for old observations. *Cell Cycle* 5:2183–2186.

Craven, C. J., B. Whitehead, S. K. A. Jones, E. Thulin, G. M. Blackburn, and J. P. Waltho. 1996. Complexes formed between calmodulin and the antagonists J-8 and TFP in solution. *Biochemistry* 35:10287–10299.

Déciga-Campos, M., I. Rivero-Cruz, M. Arriaga-Alba, G. Castañeda-Corral, G. E. Angeles-Lopez, A. Navarrete, and R. Mata. 2007. Acute toxicity and mutagenic activity of Mexican plants used in traditional medicine. *Journal of Ethnopharmacology* 110:334–342.

Del Rio, B., J. M. Garcia Pedrero, C. Martinez-Campa, P. Zuazua, P. S. Lazo, and S. Ramos. 2004. Melatonin, an endogenous-specific inhibitor of estrogen receptor alpha via calmodulin. *Journal of Biological Chemistry* 279:38294–38302.

Ding, Y., T. J. Greshock, K. A. Miller, D. H. Sherman, and R. M. Williams. 2008. Premalbrancheamide: Synthesis, isotopic labeling, biosynthetic incorporation, and detection in cultures of *Malbranchea aurantiaca. Organic Letters* 10:4863–4866.

Doyle, J., L. E. Llewellyn, C. S. Brinkworth, J. H. Bowie, K. L. Wegener, T. Rozek, P. A. Wabnitz, J. C. Wallace, and M. J. Tyler. 2002. Amphibian peptides that inhibit neuronal nitric oxide synthase. Isolation of lesueurin from the skin secretion of the Australian Stony Creek frog *Litoria lesueuri. European Journal of Biochemistry* 269:100–109.

Du, J., S. T. Szabo, N. A. Gray, and H. K. Manji. 2004. Focus on CaMKII: A molecular switch in the pathophysiology and treatment of mood and anxiety disorders. *International Journal of Neuropsychopharmacology* 7:243–248.

Evidente, A., A. Andolfi, A. Cimmino, M. Vurro, M. Fracchiolla, and R. Charudattan. 2006. Herbicidal potential of ophiobolins produced by *Drechslera gigantea. Journal of Agricultural and Food Chemistry* 54:1779–1783.

Fatima, N., L. A. Tapondjou, D. Lontsi, B. L. Sondengam, R. Atta Ur, and M. I. Choudhary. 2002. Quinovic acid glycosides from *Mitragyna stipulosa*-first examples of natural inhibitors of snake venom phosphodiesterase I. *Natural Product Letters* 16:389–393.

Figueroa, M., M. C. González, and R. Mata. 2008. Malbrancheamide B, a novel compound from the fungus *Malbranchea aurantiaca*. *Natural Products Research* 22:709–714.

Figueroa, M., M. C. González, R. Rodríguez-Sotres, A. Sosa-Peínado, M. González-Andrade, C. M. Cerda-García-Rojas, and R. Mata. 2009. Calmodulin inhibitors from the fungus *Emericella sp. Bioorganic & Medicinal Chemistry* 17:2167–2174.

Frebault, F., N. S. Simpkins, and A. Fenwick. 2009. Concise enantioselective synthesis of *ent*-malbrancheamide B. *Journal of the American Chemical Society* 131:4214–4215.

Gabdoulline, R. R. and R. C. Wade. 2002. Biomolecular diffusional association. *Current Opinion in Structural Biology* 12:204–213.

Gardiner, E. E., J. F. Arthur, M. C. Berndt, and R. K. Andrews. 2005. Role of calmodulin in platelet receptor function. *Current Medicinal Chemistry Cardiovascular & Hematological Agents* 3:283–287.

González-Andrade, M., M. Figueroa, R. Rodriguez-Sotres, R. Mata, and A. Sosa-Peinado. 2009. An alternative assay to discover potential calmodulin inhibitors using a human fluorophore-labeled CaM protein. *Analytical Biochemistry* 387:64–70.

Gilli, R., D. Lafitte, C. Lopez, M. Kilhoffer, A. Makarov, C. Briand, and J. Haiech. 1998. Thermodynamic analysis of calcium and magnesium binding to calmodulin. *Biochemistry* 37:5450–5456.

Goncalves, R. L., C. Lugnier, T. Keravis, M. J. Lopes, F. A. Fantini, M. Schmitt, S. F. Cortes, and V. S. Lemos. 2009. The flavonoid dioclein is a selective inhibitor of cyclic nucleotide phosphodiesterase type 1 (PDE1) and a *c*GMP-dependent protein kinase (PKG) vasorelaxant in human vascular tissue. *European Journal of Pharmacology* 620:78–83.

Goto, J., Y. Matsuda, K. Asano, I. Kawamoto, T. Yasuzawa, K. Shirahata, H. Sano, and H. Kase. 1987. K-254-I (genistein), a new inhibitor of Ca^{2+} and calmodulin-dependent cyclic nucleotide phosphodiesterase from *Streptosporangium vulgare. Agricultural and Biological Chemistry* 51:3003–3009.

Grey, K. B. and B. D. Burrell. 2008. Forskolin induces NMDA receptor-dependent potentiation at a central synapse in the leech. *Journal of Neurophysiology* 99:2719–2724.

Han, Y. N., K. H. Hwang, and B. H. Han. 2005. Inhibition of calmodulin-dependent protein kinase II by cyclic and linear peptide alkaloids from *Zizyphus* species. *Archives of Pharmacal Research* 28:159–163.

Han, Y. N., G.-Y. Kim, K. H. Hwang, and B. H. Han. 1993. Binding of sanjoinine-A (frangufoline) to calmodulin. *Archives of Pharmacal Research* 16:289–294.

Harmat, V., Z. Bocskei, G. Naray-Szabo, I. Bata, A. S. Csutor, I. Hermecz, P. Aranyi, B. Szabo, K. Liliom, B. G. Vertessy, and J. Ovadi. 2000. A new potent calmodulin antagonist with arylalkylamine structure: Crystallographic, spectroscopic and functional studies. *Journal of Molecular Biology* 297:747–755.

Hegemann, L., L. A. van Rooijen, J. Traber, and B. H. Schmidt. 1991. Polymyxin B is a selective and potent antagonist of calmodulin. *European Journal of Pharmacology* 207:17–22.

Hernández-Romero, Y., L. Acevedo, L. Sanchez Mde, W. T. Shier, H. K. Abbas, and R. Mata. 2005. Phytotoxic activity of bibenzyl derivatives from the orchid *Epidendrum rigidum. Journal of Agricultural and Food Chemistry* 53:6276–6280.

Hernández-Romero, Y., J. I. Rojas, R. Castillo, A. Rojas, and R. Mata. 2004. Spasmolytic effects, mode of action, and structure-activity relationships of stilbenoids from *Nidema boothii. Journal of Natural Products* 67:160–167.

Hoff, H., H. Drautz, H. P. Fiedler, H. Zahner, J. E. Schultz, W. Keller-Schierlein, S. Philipps, M. Ritzau, and A. Zeeck. 1992. Metabolic products of microorganisms. 261. Obscurolides, a novel class of phosphodiesterase inhibitors from streptomyces. I. Production, isolation, structural elucidation and biological activity of obscurolides A1 to A4. *Journal of Antibiotics* 45:1096–1107.

Horikawa, N., T. Suzuki, T. Uchiumi, T. Minamimura, K. Tsukada, N. Takeguchi, and H. Sakai. 2005. Cyclic AMP-dependent Cl- secretion induced by thromboxane A2 in isolated human colon. *Journal of Physiology* 562:885–897.

Hu, Z., S. Chen, Z. Hao, W. Huang, and S. Peng. 1988. Benzylisoquinoline compounds inhibit the ability of calmodulin to activate cyclic nucleotide phosphodiesterase. *Cellular Signalling* 1:181–185.

Hwang, K. H., Y. N. Han, and B. H. Han. 2001. Inhibition of calmodulin-dependent calcium-ATPase and phosphodiesterase by various cyclopeptides and peptide alkaloids from the *Zizyphus* species. *Archives of Pharmacal Research* 24:202–206.

Ichimura, M., R. Eiki, K. Osawa, S. Nakanishi, and H. Kase. 1996. KS505a, an isoform-selective inhibitor of calmodulin-dependent cyclic nucleotide phosphodiesterase. *Biochemical Journal* 316:311–316.

Inhuma, M., M. Roberts, S. Matlin, V. Stacey, B. Timmerman, T. Mabry, and R. Brown. 1985. Synthesis and revised structure of the flavone brickellin. *Phytochemistry* 24:1367–1368.

Johnson, C. K. 2006. Calmodulin, conformational states, and calcium signaling. A single-molecule perspective. *Biochemistry* 45:14233–14246.

Kamory, E., G. M. Keseru, and B. Papp. 1995. Isolation and antibacterial activity of marchantin A, a cyclic bis(bibenzyl) constituent of Hungarian *Marchantia polymorpha*. *Planta Medica* 61:387–388.

Kase, H., K. Iwahashi, and Y. Matsuda. 1986. K-252a, a potent inhibitor of protein kinase C from microbial origin. *Journal of Antibiotics* 39:1059–1065.

Kataoka, M., J. F. Head, B. A. Seaton, and D. M. Engelman. 1989. Melittin binding causes a large calcium-dependent conformational change in calmodulin. *Proceedings of the National Academy of Sciences* 86:6944–6948.

Keseru, G. M. and M. Nogradi. 1996. Molecular similarity analysis on biologically active macrocyclic bis(biphenyls). *Journal of Molecular Recognition* 9:133–138.

Kim, M. C., W. S. Chung, D. J. Yun, and M. J. Cho. 2009. Calcium and calmodulin-mediated regulation of gene expression in plants. *Molecular Plant* 2:13–21.

Kobayashi, J., J. Cheng, Y. Kikuchi, M. Ishibashi, S. Yamamura, Y. Ohizumi, T. Ohta, and S. Nozoe. 1990b. Rigidin, a novel alkaloid with calmodulin antagonistic activity from the Okinawan marine tunicate *Eudistoma* cf. *rigida*. *Tetrahedron Letters* 31:4617–4620.

Kobayashi, J., J. Cheng, T. Ohta, S. Nozoe, Y. Ohizumi, and T. Sasaki. 1990a. Eudistomidins B, C, and D: Novel antileukemic alkaloids from the Okinawan marine tunicate *Eudistoma glaucus*. *Journal of Organic Chemistry* 55:3666–3670.

Kobayashi, J., H. Nakamura, Y. Ohizumi, and Y. Hirata. 1986. Eudistomidin-A, a novel calmodulin antagonist from the Okinawan tunicate *Eudistoma glaucus*. *Tetrahedron Letters* 27:1191–1194.

Kobayashi, J., M. Sato, T. Murayama, M. Ishibashi, M. R. Walchi, M. Kanai, J. Shoji, and Y. Ohizumi. 1991. Konbamide, a novel peptide with calmodulin antagonistic activity from the Okinawan marine sponge *Theonella* sp. *Journal of the Chemical Society, Chemical Communications* 15:1050–1052.

Kortvely, E. and K. Gulya. 2004. Calmodulin, and various ways to regulate its activity. *Life Sciences* 74:1065–1070.

Kubo, I., A. Matsumoto, M. Kozuka, and W. F. Wood. 1985. Calmodulin inhibitors from the bitter mushroom *Naematoloma fasciculare* (Fr.) Karst. (Strophariaceae) and absolute configuration of fasciculols. *Chemical and Pharmaceutical Bulletin* 33:3821–3825.

Ladbury, J. E. and M. A. Williams. 2004. The extended interface: Measuring non-local effects in biomolecular interactions. *Current Opinion in Structural Biology* 14:562–569.

Lalor, D. J., T. Schnyder, V. Saridakis, D. E. Pilloff, A. Dong, H. Tang, T. S. Leyh, and E. F. Pai. 2003. Structural and functional analysis of a truncated form of *Saccharomyces cerevisiae* ATP sulfurylase: C-terminal domain essential for oligomer formation but not for activity. *Protein Engineering* 16:1071–1079.

Leung, P. C., W. A. Taylor, J. H. Wang, and C. L. Tipton. 1984. Ophiobolin A. A natural product inhibitor of calmodulin. *Journal of Biological Chemistry* 259:2742–2747.

Leung, P. C., W. A. Taylor, J. H. Wang, and C. L. Tipton. 1985. Role of calmodulin inhibition in the mode of action of ophiobolin A. *Plant Physiology* 77:303–308.

Li, L., D. Q. Wei, J. F. Wang, and K. C. Chou. 2007. Computational studies of the binding mechanism of calmodulin with chrysin. *Biochemical and Biophysical Research Communications* 358:1102–1107.

Lo, Y. C., C. M. Teng, C. F. Chen, C. C. Chen, and C. Y. Hong. 1994. Magnolol and honokiol isolated from *Magnolia officinalis* protect rat heart mitochondria against lipid peroxidation. *Biochemical Pharmacology* 47:549–553.

Luan, S. 2008. Paradigms and networks for intracellular calcium signaling in plant cells. *Annual Plant Reviews* 33:163–188.

Macías, M., A. Gamboa, M. Ulloa, R. A. Toscano, and R. Mata. 2001. Phytotoxic naphthopyranone derivatives from the coprophilous fungus *Guanomyces polythrix*. *Phytochemistry* 58:751–758.

Macías, M., M. Ulloa, A. Gamboa, and R. Mata. 2000. Phytotoxic compounds from the new coprophilous fungus *Guanomyces polythrix*. *Journal of Natural Products* 63:757–761.

Makarov, A. A., P. O. Tsvetkov, C. Villard, D. Esquieu, B. Pourroy, J. Fahy, D. Braguer, V. Peyrot, and D. Lafitte. 2007. Vinflunine, a novel microtubule inhibitor, suppresses calmodulin interaction with the microtubule-associated protein STOP. *Biochemistry* 46:14899–14906.

Martínez-Luis, S., M. C. González, M. Ulloa, and R. Mata. 2005. Phytotoxins from the fungus *Malbranchea aurantiaca*. *Phytochemistry* 66:1012–1016.

Martínez-Luis, S., A. Perez-Vasquez, and R. Mata. 2007. Natural products with calmodulin inhibitor properties. *Phytochemistry* 68:1882–1903.

Martínez-Luis, S., R. Rodríguez, L. Acevedo, M. C. González, A. Lira-Rocha, and R. Mata. 2006. Malbrancheamide, a new calmodulin inhibitor from the fungus *Malbranchea aurantiaca*. *Tetrahedron* 62:1817–1822.

Mata, R., R. Bye, E. Linares, M. Macias, I. Rivero-Cruz, O. Perez, and B. N. Timmermann. 2003b. Phytotoxic compounds from *Flourensia cernua*. *Phytochemistry* 64:285–291.

Mata, R., A. Gamboa, M. Macias, S. Santillan, M. Ulloa, and M. C. Gonzalez. 2003a. Effect of selected phytotoxins from *Guanomyces polythrix* on the calmodulin-dependent activity of the enzymes cAMP phosphodiesterase and NAD-kinase. *Journal of Agricultural and Food Chemistry* 51:4559–4562.

Mata, R., S. Martínez-Luis, and A. Pérez-Vásquez. 2008. Phytotoxic compounds with calmodulin inhibitor properties from selected Mexican fungi and plants. *In Selected Topics in the Chemistry of Natural Products*, ed. Raphael Ikan. World Scientific, 427–428. Singapore.

Matsuda, Y., K. Asano, I. Kawamoto, and H. Kase. 1987. K-259-2, a new inhibitor of Ca^{2+} and calmodulin-dependent cyclic nucleotide phosphodiesterase from *Micromonospora olivasterospora*. *Journal of Antibiotics* 40:1092–1100.

Matsuda, Y., K. Asano, I. Kawamoto, T. Yasuzawa, K. Shirahata, H. Sano, and H. Kase. 1988. K-251 compounds, inhibitors of Ca^{2+} and calmodulin-dependent cyclic nucleotide phosphodiesterase from *Streptoverticillium album*. *Agricultural and Biological Chemistry* 52:3211–3213.

Matsuda, Y. and H. Kase. 1987. KS-619-1, a new inhibitor of Ca^{2+} and calmodulin-dependent cyclic nucleotide phosphodiesterase from *Streptomyces californicus*. *Journal of Antibiotics* 40:1104–1110.

Matsuda, Y., S. Nakanishi, K. Nagasawa, and H. Kase. 1990. Inhibition by new anthraquinone compounds, K-259-2 and KS-619-1, of calmodulin-dependent cyclic nucleotide phosphodiesterase. *Biochemical Pharmacology* 39:841–849.

Means, A. R. 2008. The year in basic science: Calmodulin kinase cascades. *Molecular Endocrinology* 22:2759–2765.

Miller, K. A., M. Figueroa, M. W. N. Valente, T. J. Greshock, R. Mata, and R. M. Williams. 2008b. Calmodulin inhibitory activity of the malbrancheamides and various analogs. *Bioorganic and Medicinal Chemistry Letters* 18:6479–6481.

Miller, K. A., T. R. Welch, T. J. Greshock, Y. Ding, D. H. Sherman, and R. M. Williams. 2008a. Biomimetic total synthesis of malbrancheamide and malbrancheamide B. *Journal of Organic Chemistry* 73:3116–3119.

Molnar, A., K. Liliom, F. Orosz, B. G. Vertessy, and J. Ovadi. 1995. Anti-calmodulin potency of indole alkaloids in in-vitro systems. *European Journal of Pharmacology* 291:73–82.

Nakanishi, S., A. Katsuhiko, I. Kawamoto, and H. Kase. 1990a. KS-501 and KS-502, a new inhibitors of Ca2+ and calmodulin-dependent cyclicnucleotide phosphodiesterase from *Sporothrix* sp. *Journal of Antibiotics* 42:1049–1055.

Nakanishi, S., K. Kita, Y. Uosaki, M. Yoshida, Y. Saitoh, A. Mihara, I. Kawamoto, and Y. Matsuda. 1994. MS-282a and MS-282b, new inhibitors of calmodulin-activated myosin light chain kinase from *Streptomyces tauricus* ATCC 27470. *Journal of Antibiotics* 47:855–861.

Nakanishi, S., K. Kuroda, K. Osawa, and H. Kase. 1990b. Calmodulin antagonistic action of KS-504a, a novel metabolite of the fungus *Mollisia ventosa*. *Agricultural and Biological Chemistry* 54:2697–2702.

Nakanishi, S., Y. Matsuda, K. Iwahashi, and H. Kase. 1986. K-252b, c and d, potent inhibitors of protein kinase C from microbial origin. *Journal of Antibiotics* 39:1066–1071.

Nijjar, M. S. and S. S. Nijjar. 2000. Domoic acid-induced neurodegeneration resulting in memory loss is mediated by Ca^{2+} overload and inhibition of Ca^{2+} calmodulin-stimulated adenylate cyclase in rat brain (review). *International Journal of Molecular Medicine* 6:377–389.

Nwankwoala, R. N. and W. L. West. 1988. Inhibition of alpha-tocopherol and calcium calmodulin-stimulated phosphodiesterase activity in vitro by anthracyclines. *Clinical and Experimental Pharmacology and Physiology* 15:805–814.

Odom, A., M. Del Poeta, J. Perfect, and J. Heitman. 1997. The immunosuppressant FK506 and its nonimmunosuppressive analog L-685, 818 are toxic to *Cryptococcus eoformans* by inhibition of a common target protein. *Antimicrobial Agents in Chemotherapy* 41:156–161.

Okuyama, T., M. Takata, J. Takayasu, T. Hasegawa, H. Tokuda, A. Nishino, H. Nishino, and A. Iwashima. 1991. Anti-tumor-promotion by principles obtained from *Angelica keiskei*. *Planta Medica* 57:242–246.

Olofsson, M. H., A. M. Havelka, S. Brnjic, M. C. Shoshan, and S. Linder. 2008. Charting calcium-regulated apoptosis pathways using chemical biology: Role of calmodulin kinase II. *BMC Chemical Biology* 8:2.

Ovadi, J. 1989. Effects of drugs on calmodulin-mediated enzymatic actions. *Progress in Drug Research* 33:353–395.

Pala, I., A. Srinivasan, P. Vig, and D. Desaiah. 1999. Modulation of calmodulin and protein kinase C activities by *Penicillium mycotoxins*. *International Journal of Toxicology* 18:91–96.

Palacios-Espinosa, F., M. Deciga-Campos, and R. Mata. 2008. Antinociceptive, hypoglycemic and spasmolytic effects of *Brickellia veronicifolia*. *Journal of Ethnopharmacology* 118:448–454.

Paliyath, G. and B. W. Poovaiah. 1985. Identification of naturally occurring calmodulin inhibitors in plants and their effects on calcium- and calmodulin-promoted protein phosphorylation. *Plant and Cell Physiology* 26:201–209.

Pérez, R. M., H. Cervantes, M. A. Zavala, J. Sanchez, S. Perez, and C. Perez. 2000. Isolation and hypoglycemic activity of 5,7,3′-trihydroxy-3,6,4′-trimethoxyflavone from *Brickellia veronicaefolia*. *Phytomedicine* 7:25–29.

Pérez, G. R., S. R. Vargas, M. F. Martínez, and R. I. Cordova. 2004. Antioxidant and free radical scavenging activities of 5,7,3′-trihydroxy-3,6,4′-trimethoxyflavone from *Brickellia veronicaefolia*. *Phytotherapy Research* 18:428–430.

Pérez-Gutiérrez, R. 1996. Effect of aqueous extract of *Brickellia veronicaefolia* on the gastro-intestinal tract of guinea-pig, rats and mice. *Phytotherapy Research* 10:677–679.

Pérez-Gutiérrez, R., M. C. Pérez-González, M. A. Zavala-Sánchez, and S. Pérez-Gutiérrez. 1998. Hypoglycemic activity of *Bouvardia terniflora*, *Brickellia veronicaefolia*, and *Parmentiera edulis*. *Salud Pública Mexicana* 40:354–358.

Pérez-Vásquez, A., A. Reyes, E. Linares, R. Bye, and R. Mata. 2005. Phytotoxins from hof-meisteria schaffneri: Isolation and synthesis of 2′-(2′-hydroxy-4′-methylphenyl)-2′-oxoethyl acetate1. *Journal of Natural Products* 68:959–962.

Pukala, T. L., T. Urathamakul, S. J. Watt, J. L. Beck, R. J. Jackway, and J. H. Bowie. 2008. Binding studies of nNOS-active amphibian peptides and Ca^{2+} calmodulin, using nega-tive ion electrospray ionisation mass spectrometry. *Rapid Communications in Mass Spectrometry* 22:3501–3509.

Qi, D. F., R. C. Schatzman, G. J. Mazzei, R. S. Turner, R. L. Raynor, S. Liao, and J. F. Kuo. 1983. Polyamines inhibit phospholipid-sensitive and calmodulin-sensitive Ca^{2+}-dependent protein kinases. *Biochemical Journal* 213:281–288.

Ragone, M. I., M. Sella, P. Conforti, M. G. Volonte, and A. E. Consolini. 2007. The spasmolytic effect of *Aloysia citriodora*, Palau (South American cedron) is partially due to its vitexin but not isovitexin on rat duodenums. *Journal of Ethnopharmacology* 113:258–266.

Ritzau, M., S. Philipps, A. Zeeck, H. Hoff, and H. Zahner. 1993. Metabolic products of micro-organisms. 268. Obscurolides, a novel class of phosphodiesterase inhibitors from strep-tomyces. II. Minor components belonging to the obscurolide B to D series. *Journal of Antibiotics* 46:1625–1628.

Rivero-Cruz, J. F., M. Macias, C. M. Cerda-Garcia-Rojas, and R. Mata. 2003. A new phy-totoxic nonenolide from *Phoma herbarum*. *Journal of Natural Products* 66:511–514.

Rivero-Cruz, B., I. Rivero-Cruz, R. Rodriguez-Sotres, and R. Mata. 2007. Effect of natural and synthetic benzyl benzoates on calmodulin. *Phytochemistry* 68:1147–1155.

Rivero-Cruz, B., M. A. Rojas, R. Rodríguez-Sotres, C. M. Cerda-Garcia-Rojas, and R. Mata. 2005. Smooth muscle relaxant action of benzyl benzoates and salicylic acid derivatives from *Brickellia veronicaefolia* on isolated guinea-pig ileum. *Planta Medica* 71:320–325.

Roberts, M., B. Timmermann, and T. Mabry. 1980. 6-Methoxyflavonols from *Brickellia veron-icaefolia*. *Phytochemistry* 19:127–129.

Rojas, S., L. Acevedo, M. Macias, R. A. Toscano, R. Bye, B. Timmermann, and R. Mata. 2003. Calmodulin inhibitors from *Leucophyllum ambiguum*. *Journal of Natural Products* 66:221–224.

Ronca-Testoni, S., S. Hrelia, G. Hakim, and C. A. Rossi. 1985. Interaction of smooth muscle relax-ant drugs with calmodulin and cyclic nucleotide phosphodiesterase. *Experientia* 41:75–76.

Seales, E. C., K. J. Micoli, and J. M. McDonald. 2006. Calmodulin is a critical regulator of osteo-clastic differentiation, function, and survival. *Journal of Cellular Biochemistry* 97:45–55.

Sharma, B., S. K. Deo, L. G. Bachas, and S. Daunert. 2005. Competitive binding assay using fluorescence resonance energy transfer for the identification of calmodulin antagonists. *Bioconjugate Chemistry* 16:1257–1263.

Sharma, R. K. and J. H. Wang. 1979. Preparation and assay of the Ca^{2+}-dependent modulator protein. *Advances in Cyclic Nucleotide Research* 10:187–198.

Sheng, L. 2008. Paradigms and networks for intracellular calcium signaling in plant cells. *Annual Plant Reviews* 33:163–188.

Shim, J. S., J. Lee, K. N. Kim, and H. J. Kwon. 2007. Development of a new Ca^{2+}/calmodulin antagonist and its anti-proliferative activity against colorectal cancer cells. *Biochemical and Biophysical Research Communications* 359:747–751.

Siuciak, J. A., D. S. Chapin, J. F. Harms, L. A. Lebel, S. A. McCarthy, L. Chambers, A. Shrikhande, S. Wong, F. S. Menniti, and C. J. Schmidt. 2006. Inhibition of the stria-tum-enriched phosphodiesterase PDE10A: A novel approach to the treatment of psycho-sis. *Neuropharmacology* 51:386–396.

Somlyo, A. P. and A. V. Somlyo. 2003. Ca^{2+} sensitivity of smooth muscle and nonmuscle myosin II: Modulated by G proteins, kinases, and myosin phosphatase. *Physiological Reviews* 83:1325–1358.

Takahashi, A., G. Kusano, T. Ohta, Y. Ohizumi, and S. Nozoe. 1989. Fasciculic acids A, B and C as calmodulin antagonists from the mushroom *Naematoloma fasciculare*. *Chemical and Pharmaceutical Bulletin* 37:3247–3250.

Tsukamoto, S., T. Yamashita, S. Matsunaga, and N. Fusetani. 1999a. Bistellettadines A and B: Two bioactive dimeric stellettadines from a marine sponge *Stelletta* sp. *Journal of Organic Chemistry* 64:3794–3795.

Tsukamoto, S., T. Yamashita, S. Matsunaga, and N. Fusetani. 1999b. Stellettazole A: An antibacterial guanidinoimidazole alkaloid from a marine sponge *Stelletta* sp. *Tetrahedron Letters* 40:737–738.

Valencia-Islas, N., H. Abbas, R. Bye, R. Toscano, and R. Mata. 2002. Phytotoxic compounds from *Prionosciadium watsoni*. *Journal of Natural Products* 65:828–834.

Vertessy, B. G., V. Ramat, Z. Böcskei, G. Náray-Szabó, F. Orosz, and J. Ovádi. 1998. Simultaneous binding of drugs with different chemical structures to Ca^{2+}-calmodulin: Crystallographic and spectroscopic studies. *Biochemistry* 37:15300–15310.

Vetter, S. W. and E. Leclerc. 2003. Novel aspects of calmodulin target recognition and activation. *European Journal of Biochemistry* 270:404–414.

Wang, C., T. Chen, N. Zhang, M. Yang, B. Li, X. Lu, X. Cao, and C. Ling. 2009. Melittin, a major component of bee venom, sensitizes human hepatocellular carcinoma cells to tumor necrosis factor-related apoptosis-inducing ligand (TRAIL)-induced apoptosis by activating CaMKII-TAK1-JNK/p38 and inhibiting IkappaBalpha kinase-NFkappaB. *Journal of Biological Chemistry* 284:3804–3813.

Webb, R. C. 2003. Smooth muscle contraction and relaxation. *Advances in Physiology Education* 27:201–206.

Williams, R. M. and R. J. Cox. 2003. Paraherquamides, brevianamides, and asperparalines: Laboratory synthesis and biosynthesis. An interim report. *Accounts of Chemical Research* 36:127–139.

Yamada, T., C. Iwamoto, N. Yamagaki, T. Yamanouchi, K. Minoura, T. Yamori, Y. Uehara, T. Andoh, K. Umemura, A. Numata, and T. Nasahara. 2002. Leptosins M-N1, cytotoxic metabolites from a *Leptosphaeria* species separated from a marine alga. Structure determination and biological activities. *Tetrahedron* 58:479–487.

Yasuzawa, T., Y. Saitoh, and H. Sano. 1990. Structures of KS-501 and KS-502, the new inhibitors of Ca^{2+} and calmodulin-dependent cyclic nucleotide phosphodiesterase. *Journal of Antibiotics* 43:336–343.

Yasuzawa, T., M. Yoshida, K. Shirahata, and H. Sano. 1987. Structure of a novel Ca^{2+} and calmodulin-dependent cyclic nucleotide phosphodiesterase inhibitor KS-619-1. *Journal of Antibiotics* 40:1111–1114.

Zhai, H., K. Nakade, N. Oda, Y. Mitsumoto, M. Akagi, J. Sakurai, and Y. Fukuyama. 2005. Honokiol-induced neurite outgrowth promotion depends on activation of extracellular signal-regulated kinases (ERK1/2). *European Journal of Pharmacology* 516:112–117.

Zhou, Y., Y. Li, Z. Wang, Y. Ou, and X. Zhou. 1994. ^{1}H NMR and spin-labeled EPR studies on the interaction of calmodulin with jujuboside A. *Biochemical and Biophysical Research Communications* 202:148–154.

14 Neuroprotective Effects of Natural Products from Traditional Chinese Herbs

Hai Yan Zhang and Xi Can Tang

CONTENTS

14.1 INTRODUCTION

Population ageing is one of humanity's greatest challenges. World Health Organization (WHO) projections indicate that by 2050, there will be a total of about 2 billion people over the age of 60. Given the expected dramatic increase in the incidence and prevalence of many age-related neurodegenerative diseases (NDs), the identification of successful prevention and treatment strategies is critical. However, the available therapeutic strategies can only modestly improve symptoms and cannot cure or prevent the disease progression. Currently, the major obstacle to the development of new therapies is our inadequate knowledge of the etiology of the NDs. Nevertheless, increasing evidence suggest that NDs, such as Alzheimer's disease (AD) and Parkinson's disease (PD), likely represent a syndrome instead of a single disorder with the same primary cause in all cases. Several common abnormalities have been observed in the pathogenesis of NDs, which include oxidative stress, mitochondrial dysfunction, protein aggregation and misfolding, inflammation, excitotoxicity, as well as apoptosis (Halliwell 1992; Halliwell 2001; Liu and Hong 2003; Lee et al. 2009; Yacoubian and Standaert 2009). These pathogenic alterations likely act synergistically through complex interactions to promote neurodegeneration. Thus, it becomes evident that syndromes, such as AD and PD, will require multi-target drug therapy to address the varied pathological aspects of the disease. Moreover, studies have indicated that bi- or multifunctional compounds might provide greater symptomatic efficacy as potential anti-ND drugs than single-target compound (Weinreb et al. 2009).

China is a country with abundant storage of various species of Chinese herbs, decoctions of which have been proved to be effective for many specific symptoms in Chinese literatures like *A Chinese Bestiary, the Medical Classic of the Yellow Emperor*. Therefore, the traditional Chinese medicines, natural products, as well as plant-derived compounds (phytochemicals) are the potential sources for the development of effective drug for the treatment of NDs. In recent years, scientists in China isolated a lot of novel compounds from Chinese herbs. Many of them were demonstrated to have poly-pharmacological neuroprotective activities with few side effects and, thus, recognized as promising anti-ND drug candidates. In this review, we will introduce some of the pure active components from natural sources as well as the possible mechanisms involved.

14.2 *HUPERZIA SERRATA* AND HUPERZINE A

Huperzia serrata (Qian Ceng Ta) (Figure 14.1a) is mainly distributed in the northeastern, southern, southwestern parts of China and Changjiang River zone. It is traditionally used for traumatic injury, detoxification, pain relief, as well as schizophrenia. In the 1970s, schizophrenia patients who used this herbal extraction to treat relative ailments showed major or minor signs of cholinergic stimulation, which prompted scientists to explore the active principles involved in it. Among all the isolated principles, (-)-huperzine A (Hup A) was proved to be the most potent acetylcholinesterase (AChE) inhibitor.

(a)   (b)

(c)   (d)

FIGURE 14.1 **(See color insert.)** Representative plants from traditional Chinese medicine producing neuroprotective metabolites. (a) *Huperzia serrata*, (b) Caulis Sinomenii, (c) *Panax ginseng*, (d) *Apium graveolens*.

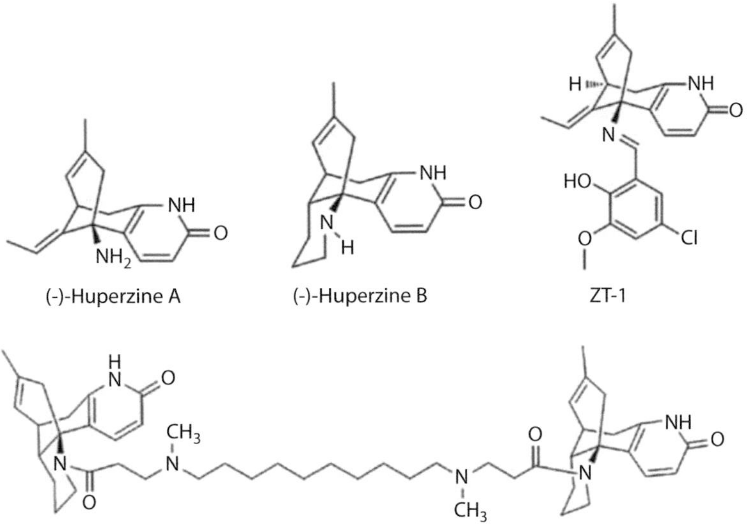

FIGURE 14.2 Structures of huperzine A, huperzine B, ZT-1, and FS-0311.

Hup A is a mixed-competitive and reversible AChE inhibitor (Figure 14.2). It shows higher potency and selectivity of AChE inhibition in vivo as compared with galanthamine, donepezil, and rivastigmine, which are approved by food and drug administration (FDA) for AD treatment. Moreover, recent study showed that Hup A has an eight- and twofold higher potency for increasing cortical ACh level than donepezil and rivastigmine, respectively, with a longer duration of action and fewer side effects. These effects could be well explained by a study about 3D x-ray crystal structure of Hup A—*Tc*AChE complex, which revealed that the "ingenious design" of Hup A has a tighter binding and greater specificity to the AChE enzyme than other known AChE inhibitors, such as tacrine (reviewed in Wang et al. 2006).

Hup A is widely proved to reverse or attenuate cognitive deficits in several behavior models and different animal species as well as nonhuman primates (Table 14.1). Moreover, clinical trials in China also indicated that Hup A treatment could significantly improve the memory, cognitive skills, and daily life abilities of AD patients with mild to no side effects (reviewed in Wang et al. 2006). Hup A is now widely used clinically for AD treatment in China and in a phase II clinical trial in the United States (http://clinicaltrials.gov/ct2/show/NCT00083590). Interestingly, a number of studies suggested that, aside from the potent AChE inhibitory effects, Hup A has a range of neuroprotective activities such as regulating β-amyloid precursor protein (APP) metabolism, counteracting neurotoxicities induced by many neurotoxins, including β-amyloid (Aβ) peptide, and improving mitochondrial function (reviewed in Wang et al. 2006; Zhang and Tang 2006). This may provide adequate evidence for the clinical and experimental cognitive improvement gained from Hup A.

Hup A has been proved to enhance nonamyloidogenic pathway by increasing the levels of sAPPα and PKCα (Zhang et al. 2004), which is mediated via two divergent transduction pathways that converge at a step of tyrosine phosphorylation: a

TABLE 14.1

Effects of Active Compounds on Cognitive Improvement in Multiple Animal Models

Animal Models of Cognitive Deficits	Species	Hup A	Hup B	ZT-1	Rg1
			mg/mL		
Scopolamine	Rat	0.1		0.1	20
	Mice	0.1	1.25	0.05	50
	Monkey	0.01		0.015	
Aged	Rat	0.1			
	Mice	0.125	1.25		
	Monkey	0.001		0.001	
CO_2	Rat	0.1			
	Mice	0.075	0.6		
$A\beta$	Rat	0.1			
	Mice				5
Alcohol	Mice				200
AF64A	Rat	0.4		0.5	
Cycloheximide	Mice	0.1	1.0		
Electroconvulsive shock	Mice	0.125	1.5		
Hemicholinium	Rat	0.1			
Ibotenic acid	Rat	0.3			
Ischemia	Rat	0.1			
	Gerbil				5
Kainic acid	Rat	0.2			
$NaNO_2$	Mice	0.1	1.25		200

PKC-dependent pathway and a Ras/MAPK dependent pathway (Yan et al. 2007). Consequently, Hup A might be able to indirectly decrease amyloidogenic pathway, since increased sAPPα secretion is associated with the reduction of Aβ generation. Moreover, pretreatment of cells with Hup A prior to Aβ or hydrogen peroxide (H_2O_2) exposure was found to enhance the cell survival and the activities of anti-oxidant enzymes, including glutathione peroxidase (GSH-Px), superoxide dismutase (SOD), and catalase (CAT), and decrease the level of lipid peroxidation product—malondialdehyde (MDA) (Xiao et al. 1999, 2000a,b), these anti-oxidative effects were also confirmed in chronic cerebral hypoperfusion rats and aged rats (Shang et al. 1999; Wang et al. 2000). Furthermore, the anti-apoptotic effects of Hup A have also been well proved in many cellular and animal models, which might be mediated by reversing the down-regulation of the expression of Bcl-2 and the up-regulation of the expressions of Bax and P53, as well as attenuating the increase of Caspase-3 activity (Wang et al. 2001, Wang et al. 2001b; Xiao et al. 2002; Zhou et al. 2001; Zhou and Tang 2002; Zhang and Tang 2003).

The precise mechanisms of the multiple neuroprotective effects of Hup A still remain unclear. The noncholinergic effect of Hup A may at least partly be involved, since synthetic (+)-Hup A and natural (−)-Hup A have similar potencies against

Aβ-induced toxicities (Zhang et al. 2002), while their potencies on AChE inhibition are very different based on in vitro and in vivo data (Tang et al. 1994). Moreover, recent studies demonstrated the protective effects of Hup A on mitochondrial dysfunction in Aβ-exposed PC12 cells or isolated mitochondria, which occur through reducing the levels of reactive oxygen species (ROS) as well as increasing the activities of key components of the respiratory chain and key enzymes in TCA cycle (Gao and Tang 2006; Gao et al. 2009). The study on isolated mitochondria directly demonstrates the protective effects of Hup A on mitochondrial function, which occurs through increasing ATP production, reducing mitochondrial swelling, and maintaining membrane integrity. In addition, the direct effects of Hup A on isolated mitochondria extend its noncholinergic functions as well, and provide a potential template for the design of anti-AD drugs.

Besides the clinical efficacy in treating AD, considerable evidence suggested that Hup A can also markedly improve the memory impairment in patients with vascular dementia (VaD), mild cognitive deficits, as well as schizophrenia. These beneficial effects in clinical trials for VaD are in agreement with previous studies in animal and cell models related to VaD (reviewed in Wang et al. 2006). Studies showed that Hup A can ameliorate acute inflammation in transient focal cerebral ischemic rats or in oxygen-glucose-deprived C6 cells (Wang and Tang 2007; Wang et al. 2008b) and chronic inflammation in cerebral hypoperfusion rats induced by occlusion of bilateral common carotid arteries or in hypoxia-induced BV-2 cells (Wang et al. 2010). The neuroprotective effects of Hup A might be due in part to a cholinergic anti-inflammatory pathway, involving nicotinic acetylcholine receptor (nAChR). Previous study have proved that Hup A has no direct effect on the amplitude or kinetics of nAChRs (Fayuk and Yakel 2004), the above-mentioned anti-inflammatory effects of Hup A may be partly mediated by the indirect activation of nAChR due to the increased synapse ACh level through AChE inhibition. In addition, Hup A can slow the rate of recovery from desensitization of the α7 containing nAChRs in rat hippocampal interneurons and increase both the amplitude and decay time of non-α7 nAChRs-mediated responses through an indirect mechanism involving the inhibition of the breakdown of ACh, resulting in an enhanced amount of duration of ACh exposed to these channels (Fayuk and Yakel 2004). All these advantages might contribute to the clinical effect of Hup A in the treatment of VaD and other neurologic disorders, although further studies will be needed for making this conclusion.

ZT-1, a semisynthetic derivative of Hup A, was originally synthesized and selected from other over 100 Hup A derivatives in Shanghai Institute of Materia Medica, Chinese Academy of Sciences (Ma et al. 2007). The AChE inhibitory effect of ZT-1 is almost as potent as that of Hup A, while its potency in inhibiting human serum butyrylcholinesterase (BuChE) is less than that of Hup A (Table 14.2). Maximal AChE inhibition in rat whole brain was reached at 60 min and maintained for 360 min following oral administration of ZT-1. It shows higher potency and selectivity of AChE inhibition in vivo as compared with donepezil (Figure 14.3), which is consistent with the cortical acetylcholine (ACh) release as measured by microdialysis and high-performance liquid chromatography (HPLC) (Figure 14.4). It can significantly alleviate scopolamine or AF64A-induced spatial learning deficits in mice, rats, or monkeys (Figures 14.5 through 14.7) (Table 14.1). All of the above information

TABLE 14.2

In Vitro Inhibitory Effects of ZT-1, Hup A, and Donepezil on AChE and BuChE

	IC_{50} (nM)[a]		Ratio of IC_{50}
AChEIs	AChE	BuChE	(BuChE/AChE)
ZT-1	82.9 ± 1.5	131000.0 ± 11000.0	1580
Hup A	61.6 ± 3.0	52780.0 ± 1530.0	857
Donepezil	13.8 ± 0.7	7663.3 ± 346.4	555

[a] The IC_{50} was defined as the concentration of inhibitor necessary to yield 50% inhibition of enzyme activity. Rats' cortex was homogenized and used as AChE resources. Rats' serum was used as BuChE resources.

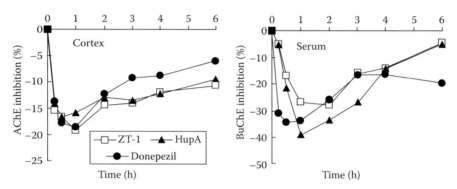

FIGURE 14.3 Time course of cholinesterase inhibition following oral administration of ZT-1 (1 μmol/kg), Hup A (1 μmol/kg), and donepezil (24 μmol/kg). Results are expressed as percentage increases above baseline (averaged from three consecutive values obtained before stimulation). Mean values ± SEM are shown.

allows ZT-1 to be a promising new drug candidate in treating AD. A subcutaneous sustained-release implant formulation of ZT-1 was developed in Europe and was tested in a phase II clinical trial recently. In mild to moderate AD patients, both ZT-1 and donepezil proved to be similarly effective in cognitive improvement

Another active principle from *Huperzia serrata* is huperzine B (Hup B). It is also a potent, reversible AChE inhibitor exhibiting more selective inhibition on AChE activity than galanthamine in vitro (Xu and Tang 1987; Tang et al. 1988). X-ray crystal structure of HupB—TcAChE complex revealed that Hup B binds to the enzyme with dissociation constants of 0.33 μM, compared to 0.18 μM for Hup A (Dvir et al. 2002). Hup B is widely proved to improve cognition in both adult and aged mice as well as reverse learning and memory deficits induced by scopolamine, NaNO_2, electroconvulsive shock, and cycloheximide (Zhu and Tang 1988) (Table 14.1). Hup B could protect cells against H_2O_2 and oxygen glucose deprivation, which may be partly due to elevating antioxidant enzyme activities and decreasing

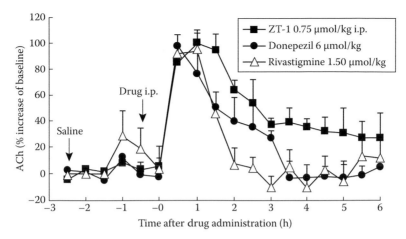

FIGURE 14.4 Effects of ZT-1, donepezil, and rivastigmine (i.p. administration) on the ACh levels of cortex in conscious rats. Brain cortical CSF was collected by microdialysis procedure, and ACh levels in dialysis samples were determined with an HPLC-ECD system (ESA Bedford, MA). Results were expressed as the increase percentages of the baseline and shown as mean ± SEM.

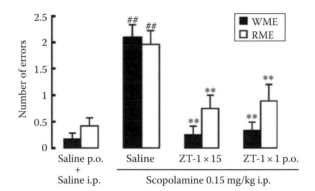

FIGURE 14.5 Acute and chronic effects of ZT-1 on scopolamine-induced working memory (WME) and reference memory (RME) disruption of the partially baited radial maze performance in rats. The behavior protocol was described in our previous paper (Wang and Tang 1998). ZT-1 was administered 30 min before the behavioral testing. Data represent mean ± SEM. $^{##}P < 0.01$ vs. saline group, $^{**}P < 0.01$ vs. scopolamine group.

MDA level (Zhang and Tang 2000; Wang et al. 2002). In addition, Hup B has been demonstrated to antagonize cerebral NMDA receptor (Wang et al. 1999), besides its inhibitory effect on AChE. However, further investigations in clinic will be needed to determine the effect of Hup B on dementia. Moreover, in light of previous successful cholinesterase inhibitors, a series of bis-Hup B derivatives have been designed and synthesized (Feng et al. 2005; He et al. 2007). Among these derivatives, FS-0311 showed the highest potency and selectivity as an AChE inhibitor. FS-0311 had a high oral bioavailability and a long duration of AChE inhibitory action in vivo. FS-0311 was found to antagonize cognitive deficits induced by scopolamine or transient brain

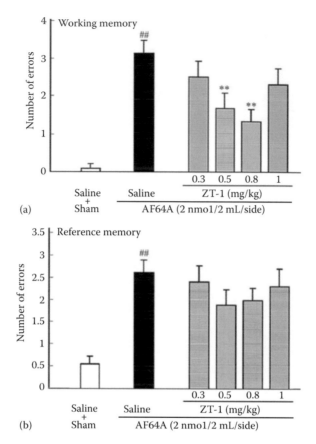

FIGURE 14.6   Effects of ZT-1 on AF64A-induced working and reference memory disruption of the partially baited radial maze performance in rats. The behavior protocol was described in our previous paper (Cheng and Tang 1998). ZT-1 was administered orally 30 min before the behavioral testing. Data represent mean ± SEM. ##$P < 0.01$ vs. sham group, **$P < 0.01$ vs. AF64A group.

ischemia and reperfusion in a water maze task. FS-0311 also possessed the ability to protect PC12 cells against Aβ peptide induced toxicity, OGD insult, and staurosporine-induced apoptosis. The neuroprotective effects of FS-0311 appeared to reflect an attenuation of oxidative stress (Wang et al. 2008a).

14.3   CAULIS SINOMENII AND SINOMENINE

Caulis Sinomenii (Figure 14.1b) was clinically used for the treatment of rheumatoid arthritis due to its anti-inflammation and analgesic activity. One of the well-reported active principles from Caulis Sinomenii is sinomenine (SN) (Liu et al. 2006) (Figure 14.8). SN has been used clinically as an anti-inflammatory agent in several inflammation-related diseases (Feng et al. 1965). It can markedly reduce the production of pro-inflammatory mediators, such as TNF-α, IL-1, PGE2, and NO from macrophages (Wang et al. 2003; Wang et al. 2005), which might be due to the inhibition

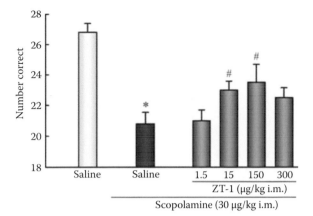

FIGURE 14.7 Effects of ZT-1 on the deficit of delayed-response performance induced by scopolamine in young adult monkeys. Saline or ZT-1 was administered i.m. 20 min before testing. Scopolamine was administered 30 min before testing. ZT-1 produced a dose-related improvement in the delayed-response performance of young monkeys. Values represent mean ± SEM, number of trials correct out of a possible 30 trials. [#]$P < 0.05$ vs. saline control, [*]$P < 0.05$ vs. scopolamine control.

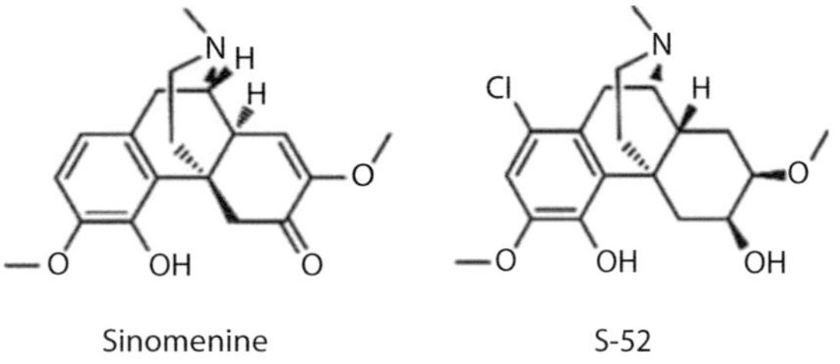

Sinomenine   S-52

FIGURE 14.8   Structure of sinomenine from Caulis Sinomenii.

of superoxide production through the inhibition of microglial PHOX activity. It is also reported to reduce the inflammation in many ischemic injury models (Wu et al. 2009; Zhou et al. 2009).

S-52, one of the derivatives of SN synthesized in this institute by Prof. Guowei Qin, was found to exert many neuroprotective effects. This semisynthetic compound could also markedly elevate cortical ACh level (Figure 14.9) with no AChE inhibitory effect (Table 14.3). Moreover, S-52 could significantly increase the cell viability in Aβ or glutamate-induced cell injury models (Figure 14.10), which might be due to regulating the activities of anti-oxidative enzymes, including SOD, GSH-PX, CAT, reducing the lipid peroxidation products—MDA (Figure 14.11), alleviating ROS production, increasing ATP level, as well as regulating the mitochondrial function through increasing the activities of respiratory chain complexes and TCA enzymes (unpublished data). S-52 could significantly alleviate scopolamine or ischemia-induced spatial learning deficits in mice (Figure 14.12). As mentioned earlier, S-52 is proved to be an active SN derivative with multiple molecular targets. However, the precise mechanisms and efficacy in clinical trial still need further investigation.

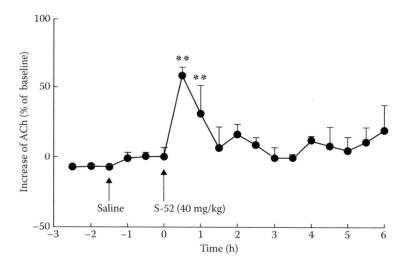

FIGURE 14.9 Effects of S-52 on the ACh levels of cortex in conscious rats. The brain corti-cal CSF was collected by microdialysis procedure, and ACh levels in dialysis samples were determined with an HPLC-ECD system (ESA Bedford, MA). Results were expressed as the increase percentages of the baseline and shown as mean ± SEM. S-52 (40 mg/kg) was given by i.p. administration. **$P < 0.01$ vs. baseline.

TABLE 14.3

In Vitro Inhibitory Effect of S-52 on AChE and BuChE

IC_{50}[a]		Ratio of IC_{50}
AChE	BuChE	(BuChE/AChE)
3.20 mM	0.63 mM	0.20

[a] The IC_{50} was defined as the concentration of inhibitor necessary to yield 50% inhibition of enzyme activity. Rats' cortex was homog-enized and used as AChE resources. Rats' serum was used as BuChE resources.

14.4 *AMARYLLIDACEAE* FAMILY AND GALANTHAMINE

Galanthamine was first isolated from the bulbs of *Galanthus woronowii*, plant of *Amaryllidaceae* family, in 1952. Shortly after that, Chinese phytochemists in Shanghai Institute of Materia Medica, Chinese Academy of Sciences, began to isolate this alka-loid from *Lycoris squamigera maxim*. Galanthamine is being studied as a possible ther-apeutic agent in treating ND, especially AD, because of its central cholinergic effects.

Galanthamine was reported to be a selective, reversible, and competitive inhibitor of AChE. It inhibits AChE activity at submicromolar concentrations, which is less than Hup A and donepezil but more potent than rivastigmine (Wang et al. 2006).

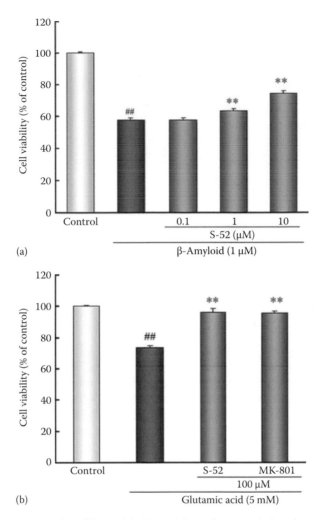

(a)

(b)

FIGURE 14.10 Protective effects of S-52 on Aβ or glutamate induced neurotoxicities in PC12 cells. Data are percentage of control values and expressed as mean ± SEM. Three independent experiments were carried out in triplicate. $^{\#\#}P < 0.01$ vs. control group, $^{**}P < 0.01$ vs. Aβ or glutamate group.

Although its action is mostly directed at the regulation of cholinergic transmission, galanthamine can also afford neuroprotection against H_2O_2-induced toxicity, which may involve alleviating oxidative stress (Ezoulin et al. 2008), potentiating NMDA receptor (Wang et al. 1999; Cao et al. 2006), upregulating anti-apoptotic proteins expression (Arias et al. 2004), as well as blocking delayed rectifier, but not transient outward potassium current (Pan et al. 2003). As already mentioned, galanthamine is proved to be a natural product with multiple molecular targets. Besides, the beneficial effects of galanthamine may involve an additional mechanism, most likely allosteric modulation of nicotinic acetylcholine receptors (nAChRs) (Woodruff-Pak et al. 2002). In fact, protective effects of galanthamine have been found in many brain ischemic

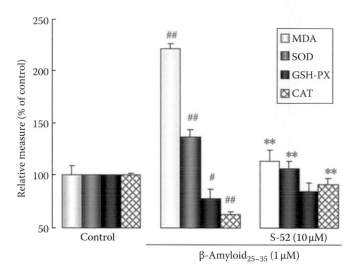

FIGURE 14.11   Effects of S-52 on $A\beta_{25-35}$-induced antioxidant enzyme activities and level of lipid peroxidation in PC12 cells. S-52 was added 2 h prior to $A\beta$ addition. Data are expressed as MDA: malondialdehyde; SOD: superoxide dismutase; GSH-Px: glutathione peroxidase; CAT: catalase. $^{\#}P < 0.05$, $^{\#\#}P < 0.01$ vs. control, $^{*}P < 0.01$ vs. $A\beta_{25-35}$ group.

models, which may involve attenuation of the release of inflammatory cytokines, and possibly mediated by nAChRs (Giunta et al. 2004; Ji et al. 2007; Lorrio et al. 2007).

Large amount of studies have demonstrated that galanthamine could significantly improve learning and memory deficits induced by scopolamine, streptozotocin, and ischemia (Ji et al. 2007), which is consistent with the other studies reported in foreign countries (Harvey 1995; Iliev et al. 1999). It is proved to be effective in treating moderate AD, VaD, and age-associated memory impaired patients in China (Xie et al. 2003; Sun et al. 2005; Hong et al. 2006) and in foreign countries (Erkinjuntti et al. 2002, 2003). The use of galanthamine for the treatment of mild to moderate AD has been approved by the SFDA in China and the US FDA. In addition, galanthamine could also significantly ameliorate peripheral nerve injury in patient by renovation of nerve (Gui and You 2008). All the above-mentioned clinical data together with the multiple neuroprotective effects of galanthamine support the potency of the compound in alleviating the progress of ND.

14.5   *PANAX GINSENG* AND GINSENOSIDES Rg1 AND Rb1

Panax ginseng (Figure 14.1c) has been used as medicine among Native Americans and the peoples of China and Korea for thousands of years. The herb is traditionally used as preventive medicine, but has exceptional therapeutic benefits for several medical conditions. If taken regularly, it enhances vitality and is reputed to lengthen life span. Recently, clinical trials in China suggested that ginseng decoction could significantly improve the cognitive function in AD patients (Cao and Hu 2008), which may predict *P. ginseng* to be a promising source of principles for the treatment of dementia.

The main active components of *Panax ginseng* are ginsenosides, which are triterpene saponins rich in stalks and leaves of the herb. Recently, studies have shown

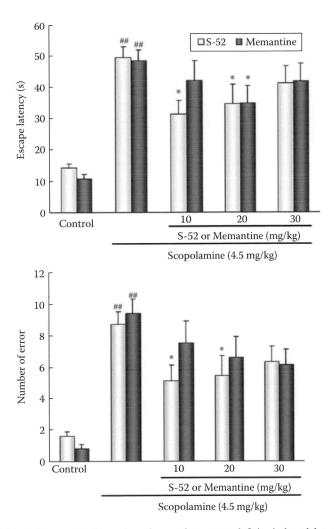

FIGURE 14.12 Effects of S-52 on learning and memory deficits induced by scopolamine. Mice were orally administrated with S-52 1 h before testing. Scopolamine was i.p. administrated 20 min before testing. Data represent mean ± SEM. ##$P < 0.01$ vs. control group, *$P < 0.05$ vs. scopolamine group.

that one of the most active ginsenosides—Rg1 (Figure 14.13) is able to significantly enhance cholinergic function through increasing ACh level and density of muscarinic ACh receptor (Cheng et al. 2005). It also has other potential neurotrophic and neuroprotective effects (Yang et al. 2007), which may involve the blockade of inflammation cascade (Joo et al. 2005), calcium channels and cell apoptosis (Wu et al. 2007; Wang et al. 2008), the regulation of NGF (Rausch et al. 2006), reduction of oxidative stress (Wang et al. 2005a; Zhao et al. 2009), as well as the amelioration of mitochondrial dysfunction. Rg1 could improve learning and memory in normal, aged animals, as well as in various animal models with impaired memory induced by alcohol, Aβ, scopolamine, and ischemia (Wu et al. 1999; Wang et al. 2001a; Shen and Zhang 2004;

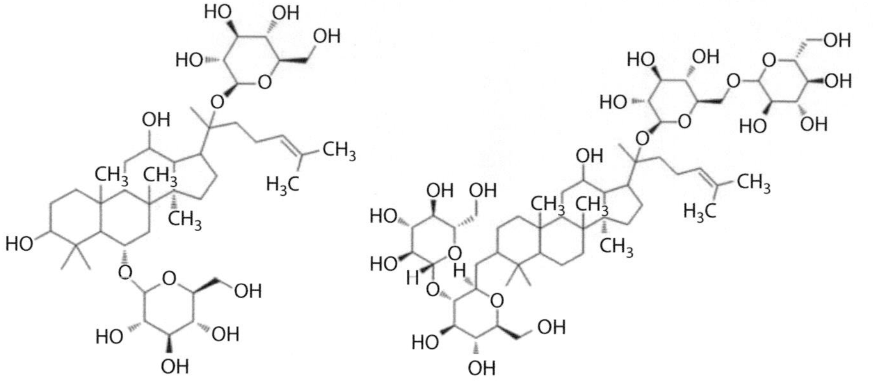

FIGURE 14.13 Extracts Rg1 and Rb1 from the Chinese herb *Panax ginseng.*

Chen and Zhu 2005; Cheng et al. 2005; Zhang et al. 2008) (Table 14.1). Besides the protective effects mentioned earlier, recent studies also indicated that Rg1 could attenuate dopamine-induced apoptosis in PC12 cells (Chen et al. 2001) and modulate COX-2 expression through the P38 signaling pathway in the substantia nigra of PD mice induced by MPTP (Wang et al. 2008b). Most recently, pretreatment with Rg1 was found to significantly inhibit MPP(+)-induced up-regulation of DMT1-IRE in MES23.5 cells, which was associated with ROS production and translocation of nuclear factor-κB (NFκB) to nuclei (Xu et al. 2009). Moreover, Rgl has been also proved to have estrogen-like activities and exert neuroprotective effects on the dopaminergic neurons in the 6-OHDA-induced ovariectomized (ovx) rat model of PD, which may at least, in part, be attributed to its effects on attenuating iron overload and apoptosis (Xu et al. 2008). Furthermore, it is also reported that the neuroprotective effects of Rg1 on dopaminergic neurons in the 6-OHDA model of nigrostriatal injury could involve the activation of the IGF-IR signaling pathway (Wang et al. 2009a). These earlier mentioned effects of Rg1 may play a beneficial role in treating ND, especially AD and PD.

One of the other well-studied active ginsenosides is Rb1 (Figure 14.13). It could significantly improve learning and memory deficits (Ying et al. 1994; Mook-Jung et al. 2001; Cheng et al. 2005), which may involve enhancement of cholinergic metabolism (Zhang et al. 1990), increase of hippocampal synaptic density (Ying et al. 1994) and nerve growth factor expression (Liu et al. 2009), maintenance of intracellular calcium homeostasis (Zhang et al. 2005), inhibition of tau phosphorylation (Li et al. 2009), as well as anti-apoptosis (Yang et al. 2008). However, more investigations will be needed to determine the efficacies of Rg1 and Rb1 in the therapy for ND.

14.6　*APIUM GRAVEOLENS* AND L-NBP

Apium graveolens (Figure 14.1d), herb of Europe and Asia, is a plant species in the family *Apiaceae*, commonly known as celery (*var. dulce*) or celeriac (*var. rapaceum*). It was traditionally believed in Chinese folk to alleviate hypertension, vertigo,

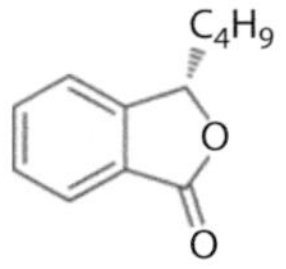

L-NBP

FIGURE 14.14 Structure of L-NBP from *Apium graveolens.*

headache, redness in the face and eyes, swelling and pains. L-3-n-butylphthalide (L-NBP) (Figure 14.14) is the active component extracted from the seeds of *Apium graveolens.* The multi-target action might be involved in the neuroprotective effects of L-NBP. In cultured neuronal cells, L-NBP was found to markedly reverse Aβ-induced nuclear fragmentation and early apoptosis, probably through inhibiting tau protein hyperphosphorylation (Peng et al. 2008). In agreement with the cellular data, daily treatments of L-NBP significantly improved Aβ-induced spatial learning deficits and attenuated working memory deficits in Morris water maze task, which may involve disturbing oxidative stress insults via reversing the reduction of GSH-Px activities and over production of MDA levels in the cortex and hippocampus, and inhibiting neuronal apoptosis possibly by blocking caspase-3 activation (Peng et al. 2009). Moreover, it was reported that L-NBP could ameliorate learning and memory deficits in aged rats or after bilateral permanent occlusion of the common carotid arteries in rats (Peng et al. 2007). Furthermore, long-term treatment of L-NBP significantly improves learning and memory deficits in aged rats, which may be mediated through enhancing ChAT activity, reducing expression of BACE and tau hyperphosphorylation (Ma et al. 2009). Taken together, besides inhibiting BACE expression and tau hyperphosphorylation, the underlying mechanism might also involve microvessel improvement possibly via increasing the ratio of PGI_2/TXA_2 (Chong and Feng 1997; Liu et al. 2007), enhancing the activity of antioxidative enzymes, improving mitochondrial function, inhibiting lipidperoxidation (Dong and Feng 2002), blocking platelet aggregation (Peng et al. 2004), increasing regional cerebral blood flow (Yan et al. 1998), as well as reducing ischemia ipsilateral cortical calcineurin and calpain activities (Dong and Feng 2000). Therefore, L-NBP might be a new drug candidate for long-term treatment of age-related ND, especially dementia. It has been approved for phase I clinic trial by SFDA in China.

DL-3-n-butylphthalide (DL-NBP), a racemic mixture containing both L- and D-isomers of butylphthalide, has been developed from L-NBP. The beneficial effects of NBP have been well established in a variety of in vivo and in vitro models of stroke (Li et al. 2009a). NBP has been shown to exert its protective actions against ischemic brain tissue through multiple mechanisms, such as protection of mitochondria and regulation of energy metabolism, decrease of oxidative damage (Li et al. 2009a), reduction of neuronal apoptosis (Dong and Feng 2000) and inhibition of the inflammatory response (Xu and Feng 2000), inhibition of platelet aggregation (Peng et al. 2004), as well as improvement of rat brain microcirculation (Chong and Feng 1997; Yan et al. 1998; Liu et al. 2007). Therefore, DL-NBP is developed as an anticerebral ischemic agent, which was approved by SFDA in 2002 for clinical use in stroke patients in China.

14.7   OTHER ACTIVE PRINCIPLES FROM CHINESE HERBS

There are many other well-studied active components that possess multiple neuro-protective effects. The beneficial profits of active components from *Gingko biloba, Camellia sinensis, Tripterygium wilfordii Hook F,* and *Carthamus tinctorius* are briefly described in the following paragraphs, and the others are summarized in Table 14.4.

Gingko biloba has been used in China as a traditional medicine in ameliorating blood circulation for thousand years. The widely known extracts from *Ginkgo biloba* leaf (EGB761) contains 24% flavonoids and 6% terpene lactones, which is a mixture of many active components including ginkgolides, bilobalide, ginkgolic acids, as well as KYNA and 6-HKA (Chen 1996; Chen and Yin 1998). These extracts have been widely used in China for treating cardiovascular diseases, strokes, and complications caused by brain trauma. The neuroprotective effects of EGb761 may involve amelioration of axon damage (Li et al. 2009b) and mitochondrial dysfunction (Zhang et al. 2007; Li et al. 2008), inhibition of monoamine oxidase (Wu and Zhu 1999), reduction of oxidative stress and platelets aggregation, and anti-apoptosis (Zhang et al. 2007; Li et al. 2008).

Green tea is made from the dried leaves of *Camellia sinensis.* It is well known that green tea allows people to keep health, with reduced onset of cardiovascular disease, obesity, and cancer, as well as a slowing down of the aging process. The most active component of *Camellia sinensis* is (-)-epigallocatechin-3-gallate (EGCG). EGCG was proved to be effective in many cellular and animal models of neurological disorders (Mandel et al. 2008; Li and Li 2009). The neuroprotective effect of EGCG could be achieved through complementary mechanisms involving down-regulation of pro-apoptotic genes (Chen et al. 2000; Yang et al. 2007a; Hou et al. 2008), scavenge of ROS (Li et al. 2006; Yang et al. 2007a) and iron chelation (Higuchi et al. 2003), anti-inflammation (Li et al. 2004), as well as reducing Aβ levels (Xu and Zhang 2008).

The most abundant and active component isolated from *Tripterygium wilfordii Hook F* is triptolide, which was proved to be promising for the treatment of ND, especially PD. Its beneficial effects might be mediated through reduction of inflammation (Zhou et al. 2005; Matta et al. 2009), inhibition of apoptosis (Wei et al. 2004), reduction of oxidative stress (Wu et al. 2006), stimulation of the expression and release of NGF (Xue et al. 2007), as well as inhibition of NF-κB activation (Wu et al. 2006; Pan and Chen 2008).

Carthamus tinctorius (safflower) has long been used in the prevention and treatment of cardiovascular disease and thrombosis in China. The major active component of safflower is nicotiflorin. It can significantly improve learning and memory deficits in many ischemic animal models (Li et al. 2006a), which could be attributed to its potency in protecting energy metabolism and alleviating oxidative stress (Huang et al. 2007). Based on the promising effect of safflower on cardiovascular system, it is worthy to put more efforts in discovering novel active principles for the treatment of dementia, especially VaD.

TABLE 14.4
Other Promising Herbal Components with Multiple Neuroprotective Effects

Active Components (Herb)	Improve Cholinergic Function	Mechanisms Against			
		Inflammation	Oxidation	Apoptosis	Others
Puerarin (Radix puerariae)	Yan et al. (2006)	Wang et al. (2006a); Zhang et al. (2008a)	Pan et al. (2008); Xu and Zhao (2002)	Wu et al. (2009a); Yan et al. (2006)	Regulate BDNF expression (Zhang and Wang 2007)
Ligustrazine (Rhizoma chuanxiong)	Zhang et al. (2008b)	Chang et al. (2007); Hu et al. (2005)		Qin et al. (2007); Xiao and Liu (2007)	Cell proliferation (Qiu et al. 2006)
Resveratrol (Polygonum cuspidatum sieb. et zucc.)		Liu et al. (2007b); Tsai et al. (2007)	Liu et al. (2007a); Lu et al. (2006)	Alvira et al. (2007); Morin et al. (2003)	Energy metabolism (Liu et al. 2007a; Yousuf et al. 2009)
Rhodosin (Rhodiola rosae)	Jiang et al. (2001)		Jiang et al. (2001); Xie et al. (2003a)	Xie et al. (2003a)	
Paeoniflorin (Radix paeoniae Alba)			Chen et al. (2008); Sun et al. (2006)	Sun et al. (2005a)	Energy metabolism (Sun et al. 2006)
Asarone (Acori graminei rhizome)	Mukherjee et al. (2007)		Jiang et al. (2007)	Chen et al. (2007)	Block NMDA receptor function (Li et al. 2007b);
Gastrodine (Gastrodia elata Bl.)	Zhang et al. (2008c)		Li et al. (2003; Liao et al. (2006)	Wu and Ren (2008)	
Ganoderma lucidum polysaccharide (Lucid ganoderma)		Wang et al. (2007)	Guo et al. (2006)	Zhang et al. (2008d)	
Tenuigenin/Presemgemin (Polygala tenuifolia)	Chen et al. (2002)		Sun et al. (2007)	Chen et al. (2006)	Inhibit Aβ secretion (Jia et al. 2004)
Icariin (Epimedium brevicornum Maxim)		Xia et al. (2008)	Li et al. (2005); Luo et al. (2007); Shen et al. (2002); Shen et al. (1999)		
Osthol (Fructus cnidii)	Shen et al. (1999)				Decrease Aβ level (Luo et al. 2007)

14.8   CONCLUSIONS

It has been suggested that multiple abnormalities, including mitochondrial dys-function, oxidative stress, inflammation, and apoptosis, play a central role in the pathogenesis of many NDs. Due to the complex symptomatology of NDs, the use of multifunctional drugs will probably be more effective. Over the years, it has become evident that some combinations do induce a favorable clinical response, not achieved by each of the drugs given alone (Keith et al. 2005). However, drugs with multiple functions might be more proficient as compared with multidrug combination, since they will simplify administering procedures and obviate multiple single drugs with potentially different degrees of bioavailability, pharmacokinetics, etc. Interestingly, while precise mechanisms for the beneficial effects of the natural active principles discussed in the previous sections have yet to be discerned, it is clear that they involve multiple neuroprotective effects. However, further studies will be needed to determine whether these effects can be successfully translated into prospect human studies to affect the progression of ND.

Although the significant number of plant extracts has yielded positive results in the therapy of ND, great efforts are still needed to isolate, purify, and characterize the active ingredients. It is our hope, that the data collated and reviewed here will provide useful information and serve to stimulate further research on this topic.

ACKNOWLEDGMENTS

The work was supported in part by the Ministry of Science and Technology of China (G199805110, G1998051115, No. 2004CB518907), National Natural Science Foundation of China (39170860, 39770846, 3001161954, 30123005, 30271494, 30572169 and 30801402), National Science & Technology Major Project "Key New Drug Creation and Manufacturing Program" of China (Number 2009 ZX09301-001 and 063) and the Knowledge Innovation Program of the Chinese Academy of Sciences (No. SIMM0709QN-12). The authors are grateful to Dr. Xiaoqiu Xiao, Dr. Jin Zhou, Dr. Han Yan, Dr. Zhifei Wang, Dr. Xin Gao, Ms. Ying Wang et al. for their contributions to Hup A, ZT-1 or S-52 studies, and to Prof. Luqi Wang (Institute of Chinese Materia Medica, Chinese Academy of Chinese Medical Sciences) for providing the color pictures of *Huperzia serrata, Caulis Sinomenii, Panax ginseng, and Apium graveolens*.

REFERENCES

Alvira, D., M. Yeste-Velasco, J. Folch, E. Verdaguer, A.M. Canudas, M. Pallas, and A. Camins. 2007. Comparative analysis of the effects of resveratrol in two apoptotic models: Inhibition of complex I and potassium deprivation in cerebellar neurons. *Neuroscience* 147(3): 746–756.

Arias, E., E. Ales, N.H. Gabilan, M.F. Cano-Abad, M. Villarroya, A.G. Garcia, and M.G. Lopez. 2004. Galantamine prevents apoptosis induced by beta-amyloid and thapsigargin: Involvement of nicotinic acetylcholine receptors. *Neuropharmacology* 46(1): 103–14.

Cao, L.M. and Z.C. Hu. 2008. Clinical analysis on senile dementia treated by ginseng tonic decoction. *J. Pract. Trad. Chin. Med.* 24: 207.

Cao, X.Z., L.M. Zhang, Y.E. Shen, S.S. Zhou, and G.H. Jiang. 2006. NMDA receptors mediate the effect of galanthamine on the cognitive function of Alzheimer's disease in rats. *J. Apopl. Nerv. Dis.* 23: 678–680.

Chang, Y., G. Hsiao, S.H. Chen, Y.C. Chen, J.H. Lin, K.H. Lin, D.S. Chou, and J.R. Sheu. 2007. Tetramethylpyrazine suppresses HIF-1alpha, TNF-alpha, and activated caspase-3 expression in middle cerebral artery occlusion-induced brain ischemia in rats. *Acta Pharmacol. Sin.* 28(3): 327–333.

Chen, Z. 1996. The chemical constituents and quality of ginkgo preparations. *J. Chin. Pharm.* 31: 326–329.

Chen, Q., Y.G. Cao, and C.H. Zhang. 2002. Effect of tenuigenin on cholinergic decline induced by beta-amyloid peptide and ibotenic acid in rats. *Acta Pharm. Sin.* 37: 913–917.

Chen, G.B., H.P. Chen, T. Wu, J.T. Lin, L. Cui, and F.C. Shen. 2008. Effect of paeoniflorin on malondlaldehyde and superoxide dismutase levels in neonatal rats with hypoxic-ischemic brain damage. *J. Appl. Clin. Pediatr.* 23(16): 1272–1273.

Chen, Y.Z., Y.Q. Fang, Y. Lian, and Y.W. Wang. 2007. Effect of β-asarone on mitochondrial membrane potentials and ultrastructural changes induced apoptosis by glutamate in cultured rat cortical neurons. *Stroke Nerv. Dis.* 14(5): 263–266.

Chen, Q., C.X. Gao, and L.H. Ge. 2006. The effect of tenuigenin on neuromorphopathological changes of AD rats with beta-amyloid1-40 injection into right nucleus basalis. *Acta Laser Biol. Sin.* 15(3): 294–298.

Chen, Z. and M. Yin. 1998. Comment on various manufacturing processes of EGb. *Chin. Trad. Herbal Drugs* 29(suppl.): 1–4.

Chen, C., R. Yu, E.D. Owuor, and A.N. Kong. 2000. Activation of antioxidant-response element (ARE), mitogen-activated protein kinases (MAPKs) and caspases by major green tea polyphenol components during cell survival and death. *Arch. Pharmacal. Res.* 23(6): 605–612.

Chen, X.M. and G.B. Zhu. 2005. Effects and mechanisms of ginsenoside Rg1 on learning and memory impairment induced by scopolamine hydrobromide. *Chin. J. Clin. Pharmacol. Ther.* 10: 898–902.

Chen, X.C., Y.G. Zhu, X.Z. Wang, L.A. Zhu, and C. Huang. 2001. Protective effect of ginsenoside Rg1 on dopamine-induced apoptosis in PC12 cells. *Acta Pharmacol. Sin.* 22(8): 673–678.

Cheng, Y., L.H. Shen, and J.T. Zhang. 2005. Anti-amnestic and anti-aging effects of ginsenoside Rg1 and Rb1 and its mechanism of action. *Acta Pharmacol. Sin.* 26(2): 143–149.

Chong, Z.Z. and Y.P. Feng. 1997. Effects of dl-3-n-butylphthalide on production of TXB2 and 6-keto-PGF1 alpha in rat brain during focal cerebral ischemia and reperfusion. *Zhongguo Yao Li Xue Bao* 18(6): 505–508.

Dong, G.X. and Y.P. Feng. 2000. Effects of 3-N-butylphthalide on cortical calcineurin and calpain activities in focal cerebral ischemia rats. *Yao Xue Xue Bao* 35(10): 790–792.

Dong, G.X. and Y.P. Feng. 2002. Effects of NBP on ATPase and anti-oxidant enzymes activities and lipid peroxidation in transient focal cerebral ischemic rats. *Zhongguo Yi Xue Ke Xue Yuan Xue Bao* 24(1): 93–97.

Dvir, H., H.L. Jiang, D.M. Wong, M. Harel, M. Chetrit, X.C. He, G.Y. Jin, G.L. Yu, X.C. Tang, I. Silman, D.L. Bai, and J.L. Sussman. 2002. X-ray structures of *Torpedo californica* acetylcholinesterase complexed with (+)-huperzine A and (−)-huperzine B: Structural evidence for an active site rearrangement. *Biochemistry* 41(35): 10810–10818.

Erkinjuntti, T., A. Kurz, S. Gauthier, R. Bullock, S. Lilienfeld, and C.V. Damaraju. 2002. Efficacy of galantamine in probable vascular dementia and Alzheimer's disease combined with cerebrovascular disease: A randomised trial. *Lancet* 359(9314): 1283–1290.

Erkinjuntti, T., A. Kurz, G.W. Small, R. Bullock, S. Lilienfeld, and C.V. Damaraju. 2003. An open-label extension trial of galantamine in patients with probable vascular dementia and mixed dementia. *Clin. Ther.* 25(6): 1765–1782.

Ezoulin, M.J., J.E. Ombetta, H. Dutertre-Catella, J.M. Warnet, and F. Massicot. 2008. Antioxidative properties of galantamine on neuronal damage induced by hydrogen peroxide in SK-N-SH cells. *Neurotoxicology* 29(2): 270–277.

Fayuk, D. and J.L. Yakel. 2004. Regulation of nicotinic acetylcholine receptor channel function by acetylcholinesterase inhibitors in rat hippocampal CA1 interneurons. *Mol. Pharmacol.* 66(3): 658–666.

Feng, C.I., Y. Chin, N.C. Wang, and S.S. Chang. 1965. The pharmacology of sinomenine. VII. Effect of sinomenine on the gastro-intestinal movement and its mechanism. *Acta Pharm. Sin.* 12(8): 492–495.

Feng, S., Z. Wang, X. He, S. Zheng, Y. Xia, H. Jiang, X. Tang, and D. Bai. 2005. Bis-huperzine B: Highly potent and selective acetylcholinesterase inhibitors. *J. Med. Chem.* 48(3): 655–657.

Gao, X. and X.C. Tang. 2006. Huperzine A attenuates mitochondrial dysfunction in beta-amyloid-treated PC12 cells by reducing oxygen free radicals accumulation and improving mitochondrial energy metabolism. *J. Neurosci. Res.* 83(6): 1048–1057.

Gao, X., C.Y. Zheng, L. Yang, X.C. Tang, and H.Y. Zhang. 2009. Huperzine A protects isolated rat brain mitochondria against beta-amyloid peptide. *Free Radic. Biol. Med.* 46(11): 1454–1462.

Giunta, B., J. Ehrhart, K. Townsend, N. Sun, M. Vendrame, D. Shytle, J. Tan, and F. Fernandez. 2004. Galantamine and nicotine have a synergistic effect on inhibition of microglial activation induced by HIV-1 gp120. *Brain Res. Bull.* 64(2): 165–170.

Gui, T. and Z.M. You. 2008. Estimate the curative effect of the use of galanthamin on peripheral nerve injury. *Chin. New Med. Forum.* 8: 4–6.

Guo, Y.J., H. Yuan, L.N. Zhang, and S.W. Gan. 2006. Effects of *Ganoderma lucidum* polysaccharides on antioxidant ability and modality in the hippocampus of AD rats. *Acta Anat. Sin.* 37(5): 509–513.

Halliwell, B. 1992. Reactive oxygen species and the central nervous system. *J. Neurochem.* 59(5): 1609–1623.

Halliwell, B. 2001. Role of free radicals in the neurodegenerative diseases: Therapeutic implications for antioxidant treatment. *Drugs Aging* 18(9): 685–716.

Harvey, A.L. 1995. The pharmacology of galanthamine and its analogues. *Pharmacol. Ther.* 68(1): 113–128.

He, X.C., S. Feng, Z.F. Wang, Y. Shi, S. Zheng, Y. Xia, H. Jiang, X.C. Tang, and D. Bai. 2007. Study on dual-site inhibitors of acetylcholinesterase: Highly potent derivatives of bis- and bifunctional huperzine B. *Bioorg. Med. Chem.* 15(3): 1394–1408.

Higuchi, A., K. Yonemitsu, A. Koreeda, and S. Tsunenari. 2003. Inhibitory activity of epigallocatechin gallate (EGCg) in paraquat-induced microsomal lipid peroxidation—A mechanism of protective effects of EGCg against paraquat toxicity. *Toxicology* 183(1–3): 143–149.

Hong, X., Z.X. Zhang, L.N. Wang, F.Y. Shao, S.F. Xiao, Y.H. Wang, C.Y. Qian, L. Shu, S.D. Chen, and X.H. Xu. 2006. A randomized study comparing the effect and safety of galantamine and donepezil in patients with mild to moderate Alzheimer's disease. *Chin. J. Neurol.* 39: 379–382.

Hou, R.R., J.Z. Chen, H. Chen, X.G. Kang, M.G. Li, and B.R. Wang. 2008. Neuroprotective effects of (-)-epigallocatechin-3-gallate (EGCG) on paraquat-induced apoptosis in PC12 cells. *Cell Biol. Int.* 32(1): 22–30.

Hu, Y.J., G.H. Hu, B. Xiao, and W.P. Gu. 2005. Dynamic changes of the expression of intercellular adhesion molecule-1 in brain of rat with focal cerebral ischemia reperfusion and the intervention of ligustrazine. *Chin. J. Clin. Rehabil.* 9(25): 149–151.

Huang, J.L., S.T. Fu, Y.Y. Jiang, Y.B. Cao, M.L. Guo, Y. Wang, and Z. Xu. 2007. Protective effects of nicotiflorin on reducing memory dysfunction energy metabolism failure and oxidative stress in multi-infarct dementia model rats. *Pharmacol. Biochem. Behav.* 86(4): 741–748.

Iliev, A., V. Traykov, D. Prodanov, G. Mantchev, K. Yakimova, I. Krushkov, and N. Boyadjieva. 1999. Effect of the acetylcholinesterase inhibitor galanthamine on learning and memory in prolonged alcohol intake rat model of acetylcholine deficit. *Methods Find. Exp. Clin. Pharmacol.* 21(4): 297–301.

Ji, X., C. Li, Y. Lu, Y. Chen, and L. Guo. 2007. Post-ischemic continuous administration of galantamine attenuates cognitive deficits and hippocampal neurons loss after transient global ischemia in gerbils. *Neurosci. Lett.* 416(1): 92–95.

Jia, H., Y. Jiang, Y. Ruan, Y. Zhang, X. Ma, J. Zhang, K. Beyreuther, P. Tu, and D. Zhang. 2004. Tenuigenin treatment decreases secretion of the Alzheimer's disease amyloid beta-protein in cultured cells. *Neurosci. Lett.* 367(1): 123–128.

Jiang, Y., Y.Q. Fang, and Y.Y. Zou. 2007. Effects of the β-asarone on learning and memory ability and SOD, GSH-Px and MDA levels in AD mice. *Chin. J. Gerontol.* 27(12): 1126–1127.

Jiang, W.H., X.R. Meng, L.M. Hao, L. Cui, Z.Y. Dong, and S.L. Wang. 2001. Study of anti-aging and anti-dementia effects of rhodosin on aging rats and experimental dementia rats. *J. Norman Bethune Univ. Med. Sci.* 27(2): 127–129.

Joo, S.S., T.J. Won, and D.I. Lee. 2005. Reciprocal activity of ginsenosides in the production of proinflammatory repertoire, and their potential roles in neuroprotection in vivo. *Planta Med.* 71(5): 476–481.

Keith, C.T., A.A. Borisy, and B.R. Stockwell. 2005. Multicomponent therapeutics for networked systems. *Nat. Rev. Drug Discov.* 4(1): 71–78.

Lee, J., J.H. Boo, and H. Ryu. 2009. The failure of mitochondria leads to neurodegeneration: Do mitochondria need a jump start? *Adv. Drug Deliv. Rev.* 61(14): 1316–1323.

Li, Y.M., F.P. Chen, and G.Q. Liu. 2003. Studies on inhibitive effect of gastrodin on PC12 cell damage induced by glutamate and H_2O_2. *J. Chin. Pharm. Univ.* 34: 456–460.

Li, R., M. Guo, G. Zhang, X. Xu, and Q. Li. 2006a. Neuroprotection of nicotiflorin in permanent focal cerebral ischemia and in neuronal cultures. *Biol. Pharm. Bull.* 29(9): 1868–1872.

Li, R., Y.G. Huang, D. Fang, and W.D. Le. 2004. (-)-Epigallocatechin gallate inhibits lipopoly-saccharide-induced microglial activation and protects against inflammation-mediated dopaminergic neuronal injury. *J. Neurosci. Res.* 78(5): 723–731.

Li, X., X.L. Jing, J.Y. Lin, X.M. Cui, H.Y. Wei, C.L. Hu, and X.X. Liao. 2009b. Protective effect of ginkgo biloba extract on neuronal axon injury after cerebral infarction. *Chin. J. Integ. Trad. Western Med. Int. Crit. Care* 16(3): 149–151.

Li, Q. and Y. Li. 2009. Green tea polyphenols prevents the age-related decline of learning and memory ability of C57BL/6J mice. *Sci. Tech. Rev.* 27(22): 26–31.

Li, Y.K., C.H. Ouyang, and S.J. Shu. 2006. Protective effect of EGCG on focal cerebral ischemia-reperfusion injury in rats. *J. Xianning Coll. Med. Sci.* 20: 373–376.

Li, G.D., B.M. Yuan, W.H. Yan, and Y. Xing. 2009. Interventional effect of ginsenosides Rb1 on Tau protein phosphorylation of neuronal cells induced by endogenous Aβ peptide. *Shandong Med. J.* 49: 26–28.

Li, L., B. Zhang, Y. Tao, Y. Wang, H. Wei, J. Zhao, R. Huang, and Z. Pei. 2009a. DL-3-n-butylphthalide protects endothelial cells against oxidative/nitrosative stress, mitochondrial damage and subsequent cell death after oxygen glucose deprivation in vitro. *Brain Res.* 1290: 91–101.

Li, J.J., X.L. Zhang, X.S. Wang, Q. Sun, and L.R. Zhang. 2008. Influence of *Ginkgo biloba* extract on neural cell apoptosis and ultrastructure in ischemic penumbra after cerebral ischemia/reperfusion in rats. *J. Bengbu Med. Coll.* 33(4): 400.

Li, L., Q.X. Zhou, and J.S. Shi. 2005. Protective effects of icariin on neurons injured by cerebral ischemia/reperfusion. *Chin. Med. J.* 118(19): 1637–1643.

Liao, Q.B., X.Q. Liu, J.P. Liu, Y.X. Zhang, F.Y. Nie, and K. Zou. 2006. Relationship between antioxidant activity and gastrodin content of *Gastrodia elata Blume*. *J. Chin. Three Gorges Univ. Nat. Sci.* 28(1): 80–82.

Liu, B. and J.S. Hong. 2003. Role of microglia in inflammation-mediated neurodegenerative diseases: Mechanisms and strategies for therapeutic intervention. *J. Pharmacol. Exp. Ther.* 304(1): 1–7.

Liu, Y.G., F.J. Li, and S.L. Xie. 2007a. Anti-oxidative and mitochondria protective effects of resveratrol on focal cerebral ischemia-reperfusion injury in rats. *Chin. Trad. Pat. Med.* 29: 1274–1277.

Liu, C.L., S.J. Liao, J.S. Zeng, J.W. Lin, C.X. Li, L.C. Xie, X.G. Shi, and R.X. Huang. 2007. Dl-3n-butylphthalide prevents stroke via improvement of cerebral microvessels in RHRSP. *J. Neurol. Sci.* 260(1–2): 106–113.

Liu, Y.G., X.D. Wang, and X.B. Zhang. 2007b. Effects of resveratrol on inflammatory process induced by focal cerebral ischemia-reperfusion in rats. *Zhongguo Zhong Yao Za Zhi* 32: 1792–1795.

Liu, L.J., L. Yang, J.D. Xiao, and D.P. Wang. 2009. Effects of ginsenoside Rb1 and Rgl on nerve growth factor expression in Schwann cells. *J. Clin. Rehabil. Tissue Eng.* 13: 6393–6396.

Liu, L., J. Zhang, Q. Fu, L.C. He, and Y. Li. 2006. Concentration and extraction of sinomenine from herb and plasma using a molecularly imprinted polymer as the stationary phase. *Anal. Chim. Acta* 561(1–2): 178–182.

Lorrio, S., M. Sobrado, E. Arias, J.M. Roda, A.G. Garcia, and M.G. Lopez. 2007. Galantamine postischemia provides neuroprotection and memory recovery against transient global cerebral ischemia in gerbils. *J. Pharmacol. Exp. Ther.* 322(2): 591–599.

Lu, K.T., R.Y. Chiou, L.G. Chen, M.H. Chen, W.T. Tseng, H.T. Hsieh, and Y.L. Yang. 2006. Neuroprotective effects of resveratrol on cerebral ischemia-induced neuron loss mediated by free radical scavenging and cerebral blood flow elevation. *J. Agric. Food Chem.* 54(8): 3126–3131.

Luo, Y., J. Nie, Q.H. Gong, Y.F. Lu, Q. Wu, and J.S. Shi. 2007. Protective effects of icariin against learning and memory deficits induced by aluminium in rats. *Clin. Exp. Pharmacol. Physiol.* 34(8): 792–795.

Ma, X., C. Tan, D. Zhu, D.R. Gang, and P. Xiao. 2007. Huperzine A from Huperzia species—An ethnopharmacolgical review. *J. Ethnopharmacol.* 113(1): 15–34.

Ma, S., S. Xu, B. Liu, J. Li, N. Feng, L. Wang, and X. Wang. 2009. Long-term treatment of l-3-n-butylphthalide attenuated neurodegenerative changes in aged rats. *Naunyn Schmiedebergs Arch. Pharmacol.* 379(6): 565–574.

Mandel, S.A., T. Amit, O. Weinreb, L. Reznichenko, and M.B. Youdim. 2008. Simultaneous manipulation of multiple brain targets by green tea catechins: A potential neuroprotective strategy for Alzheimer and Parkinson diseases. *CNS Neurosci. Ther.* 14(4): 352–365.

Matta, R., X. Wang, H. Ge, W. Ray, L.D. Nelin, and Y. Liu. 2009. Triptolide induces anti-inflammatory cellular responses. *Am. J. Transl. Res.* 1(3): 267–282.

Mook-Jung, I., H.S. Hong, J.H. Boo, K.H. Lee, S.H. Yun, M.Y. Cheong, I. Joo, K. Huh, and M.W. Jung. 2001. Ginsenoside Rb1 and Rg1 improve spatial learning and increase hippocampal synaptophysin level in mice. *J. Neurosci. Res.* 63(6): 509–515.

Morin, C., R. Zini, E. Albengres, A.A. Bertelli, A. Bertelli, and J.P. Tillement. 2003. Evidence for resveratrol-induced preservation of brain mitochondria functions after hypoxia-reoxygenation. *Drugs Exp. Clin. Res.* 29(5–6): 227–233.

Mukherjee, P.K., V. Kumar, M. Mal, and P.J. Houghton. 2007. In vitro acetylcholinesterase inhibitory activity of the essential oil from *Acorus calamus* and its main constituents. *Planta Med.* 73(3): 283–285.

Pan, X.D. and X.C. Chen. 2008. Advances in the study of immunopharmacological effects and mechanisms of extracts of *Tripterygium wilfordii* Hook. f. in neuroimmunologic disorders. *Acta Pharm. Sin.* 43(12): 1179–1185.

Pan, Z.Y., X.M. Wang, Z.Y. Wang, F.Y. Wang, and Y.L. Shen. 2008. The effects of puerarin on learning memory ability and cerebral cortex cholinergic system in D-galactose model mice. *Chin. J. Gerontol.* 28(23): 2308–2309.

Pan, Y.P., X.H. Xu, and X.L. Wang. 2003. Galantamine blocks delayed rectifier, but not transient outward potassium current in rat dissociated hippocampal pyramidal neurons. *Neurosci. Lett.* 336(1): 37–40.

Peng, Y., C. Xing, C.A. Lemere, G. Chen, L. Wang, Y. Feng, and X. Wang. 2008. 1-3-n-Butylphthalide ameliorates beta-amyloid-induced neuronal toxicity in cultured neuronal cells. *Neurosci. Lett.* 434(2): 224–229.

Peng, Y., C. Xing, S. Xu, C.A. Lemere, G. Chen, B. Liu, L. Wang, Y. Feng, and X. Wang. 2009. 1-3-n-Butylphthalide improves cognitive impairment induced by intracerebroventricular infusion of amyloid-beta peptide in rats. *Eur. J. Pharmacol.* 621(1–3): 38–45.

Peng, Y., S. Xu, G. Chen, L. Wang, Y. Feng, and X. Wang. 2007. 1-3-n-Butylphthalide improves cognitive impairment induced by chronic cerebral hypoperfusion in rats. *J. Pharmacol. Exp. Ther.* 321(3): 902–910.

Peng, Y., X. Zeng, Y. Feng, and X. Wang. 2004. Antiplatelet and antithrombotic activity of L-3-n-butylphthalide in rats. *J. Cardiovasc. Pharmacol.* 43(6): 876–881.

Qin, A.J., Z.L. Di, Y. Tian, A.T. Li, G.J. Zhang, and Y. Wang. 2007. The effect and significance of ligustrazine on the expression of NF-κB in primary cultured neural cells after ischemia-reperfusion injury. *J. Diff. Compl. Cases.* 6(11): 644–645.

Qiu, F., Y. Liu, P.B. Zhang, Y.F. Tian, J.J. Zhao, Q.Y. Kang, C.F. Qi, and X.L. Chen. 2006. The effect of ligustrazine on cells proliferation in cortex and striatum after focal cerebral ischemia in adult rats. *Zhong Yao Cai* 29: 1196–1200.

Rausch, W.D., S. Liu, G. Gille, and K. Radad. 2006. Neuroprotective effects of ginsenosides. *Acta Neurobiol. Exp (Wars).* 66(4): 369–375.

Shang, Y.Z., J.W. Ye, and X.C. Tang. 1999. Improving effects of huperzine A on abnormal lipid peroxidation and superoxide dismutase in aged rats. *Acta Pharmacol. Sin.* 20(9): 824–828.

Shen, L.X., L.Q. Jin, D.S. Zhang, and X.G.P. 2002. Effect of osthol on memory impairment of mice in AlCl3-induced acute senile model. *Yao Xue Xue Bao* 37(3): 178–180.

Shen, L.H. and J.T. Zhang. 2004. Ginsenoside Rg1 increases the survival rate of hippocampal neural stem cells and improves learning and memory in gerbils suffered from transient global ischemia. *Central South Pharm.* 1: 6–9.

Shen, L.X., D.S. Zhang, L. Zhang, L.M. Ren, and C. J.Q. 1999. Action of Osthol on learning and memory and its mechanism analysis. *Acta Pharm. Sin.* 34: 405–409.

Sun, G.B., X.C. Deng, and C.H. Li. 2007. The protective effects of tenuigenin on the PC12 cells injury induced by H_2O_2. *Zhong Yao Cai* 30(8): 991–993.

Sun, R., L.L. Lv, and G.Q. Liu. 2006. Effects of paeoniflorin on cerebral energy metabolism, nitric oxide and nitric oxide synthase after cerebral ischemia in mongoliagerbils. *China J. Chin. Mater. Med.* 31(10): 832–835.

Sun, L.Y., Z.E. Wang, Y. Wang, H. Ai, and H.Z. Yu. 2005a. Effects of galanthamine in 20 vascular dementia patients. *Chin. J. Clin. Rehabil.* 9: 190.

Sun, R., D.D. Wu, and G.Q. Liu. 2005. Nitric oxide-induced PC12 cells apoptosis and protective effect of paeoniflorin. *Chin. J. Clin. Pharmacol. Therap.* 10(11): 1266–1269.

Tang, X.C., G.H. Kindel, A.P. Kozikowski, and I. Hanin. 1994. Comparison of the effects of natural and synthetic huperzine-A on rat brain cholinergic function in vitro and in vivo. *J. Ethnopharmacol.* 44(3): 147–155.

Tang, X.C., X.D. Zhu, and W.H. Lu. 1988. Studies on the nootropic effects of huperzine A and B: Two selective AchE inhibitors. *Curr. Res. Alz. Ther.* 289–293.

Tsai, S.K., L.M. Hung, Y.T. Fu, H. Cheng, M.W. Nien, H.Y. Liu, F.B. Zhang, and S.S. Huang. 2007. Resveratrol neuroprotective effects during focal cerebral ischemia injury via nitric oxide mechanism in rats. *J. Vasc. Surg.* 46(2): 346–353.

Wang, M.B., W.F. Chen, and J.X. Xie. 2008. The protective effect of ginsenoside Rg1 against 6-OHDA-induced neurotoxicity in MES23.5 cells. *Acta Acad. Med. Qingdao Univ.* 44: 377–378.

Wang, X.D., X.Q. Chen, H.H. Yang, and G.Y. Hu. 1999. Comparison of the effects of cholinesterase inhibitors on [3H]MK-801 binding in rat cerebral cortex. *Neurosci. Lett.* 272(1): 21–24.

Wang, X.Y., J. Chen, and J.T. Zhang. 2001a. Effect of ginsenoside Rg1 on learning and memory impairment induced by beta-amyloid peptide(25–35) and its mechanism of action. *Acta Pharm. Sin.* 36: 1–4.

Wang, Y., Y. Fang, W. Huang, X. Zhou, M. Wang, B. Zhong, and D. Peng. 2005. Effect of sinomenine on cytokine expression of macrophages and synoviocytes in adjuvant arthritis rats. *J. Ethnopharmacol.* 98(1–2): 37–43.

Wang, L., M.X. Gong, X. Chen, and X.C.Y. Xiao. 2005a. The protective effects of ginsenoside Rg1 on hydrogen peroxide-induced oxidative damage in PC12 Cells. *Shanghai J. Trad. Chin. Med.* 39: 56–58.

Wang, L.M., Y.F. Han, and X.C. Tang. 2000. Huperzine A improves cognitive deficits caused by chronic cerebral hypoperfusion in rats. *Eur. J. Pharmacol.* 398(1): 65–72.

Wang, Z.F. and X.C. Tang. 2007. Huperzine A protects C6 rat glioma cells against oxygen-glucose deprivation-induced injury. *FEBS Lett.* 581(4): 596–602.

Wang, W.J., P.X. Wang, and X.J. Li. 2003. The effect of sinomenine on cyclooxygenase activity and the expression of COX-1 and COX-2 mRNA in human peripheral monocytes. *Zhongguo Zhong Yao Za Zhi* 28(4): 352–355.

Wang, Z.F., J. Wang, H.Y. Zhang, and X.C. Tang. 2008b. Huperzine A exhibits anti-inflammatory and neuroprotective effects in a rat model of transient focal cerebral ischemia. *J. Neurochem.* 106(4): 1594–1603.

Wang, Z.Y., X.B. Wei, B. Zhang, R. Sun, X. Sun, Y. Zhong, C.X. Zuo, and X.M. Zhang. 2006a. Reducing the levels of ET-1 and IL-6 in isehemia-reperfusion injury rats with hydroxyethylpuerarin. *Chin. J. Biochem. Pharm.* 27(5): 280–282.

Wang, R., X.Q. Xiao, and X.C. Tang. 2001b. Huperzine A attenuates hydrogen peroxide-induced apoptosis by regulating expression of apoptosis-related genes in rat PC12 cells. *Neuroreport.* 12(12): 2629–2634.

Wang, Z.F., J. Yan, Y. Fu, X.C. Tang, S. Feng, X.C. He, and D.L. Bai. 2008a. Pharmacodynamic study of FS-0311: A novel highly potent, selective acetylcholinesterase inhibitor. *Cell Mol. Neurobiol.* 28(2): 245–261.

Wang, R., H. Yan, and X.C. Tang. 2006. Progress in studies of huperzine A, a natural cholinesterase inhibitor from Chinese herbal medicine. *Acta Pharmacol. Sin.* 27(1): 1–26.

Wang, B.H., H. Yuan, Z.H. Lv, and G. Li. 2007. Effects of ganoderma lucidum polysaccharide on the expression of interleukin-6 in the hippocampus of the rats with the learning and memory dysfunction. *Stroke Nerv. Dis.* 14(4): 206–209.

Wang, R., H.Y. Zhang, and X.C. Tang. 2001. Huperzine A attenuates cognitive dysfunction and neuronal degeneration caused by beta-amyloid protein-(1–40) in rat. *Eur. J. Pharmacol.* 421(3): 149–156.

Wang, J., H.Y. Zhang, and X.C. Tang. 2010. Huperzine A improves chronic inflammation and cognitive decline in rats with cerebral hypoperfusion. *J. Neurosci. Res.* 88(4): 807–815.

Wang, Q., H. Zheng, Z.F. Zhang, and Y.X. Zhang. 2006b. Ginsenoside Rg1 modulates COX-2 expression in the substantia nigra of mice with MPTP-induced Parkinson disease through the P38 signalling pathway. *J. South Med. Univ.* 28(9): 1594–1598.

Wang, Z.F., J. Zhou, and X.C. Tang. 2002. Huperzine B protects rat pheochromocytoma cells against oxygen-glucose deprivation-induced injury. *Acta Pharmacol. Sin.* 23(12): 1193–1198.

Wei, D.M., G.Z. Huang, Y.G. Zhang, and G.X. Rao. 2004. Influence of triptolide on neuronal apoptosis in rat with cerebral injury after focal ischemia reperfusion. *Chin. Trad. Herbal Drugs* 29(11): 1089–1091, 1116.

Weinreb, O., S. Mandel, O. Bar-Am, M. Yogev-Falach, Y. Avramovich-Tirosh, T. Amit, and M.B. Youdim. 2009. Multifunctional neuroprotective derivatives of rasagiline as anti-Alzheimer's disease drugs. *Neurotherapeutics* 6(1): 163–174.

Woodruff-Pak, D.S., C. Lander, and H. Geerts. 2002. Nicotinic cholinergic modulation: Galantamine as a prototype. *CNS Drug Rev.* 8(4): 405–426.

Wu, L., Y.B. Chen, Q. Wang, S.Y. Cheng, W.X. Liang, and Z.H. Wen. 2007. In-vitro inhibitory effect of ginsenoside Rg1 on Aβ25–35-induced NG108–15 apoptosis in Alzheimer's disease cellular model. *J. Guangzhou Univ. Trad. Chin. Med.* 24: 126–131.

Wu, Y., J. Cui, X. Bao, S. Chan, D.O. Young, D. Liu, and P. Shen. 2006. Triptolide attenuates oxidative stress, NF-kappaB activation and multiple cytokine gene expression in murine peritoneal macrophage. *Int. J. Mol. Med.* 17(1): 141–150.

Wu, H.Q., H.N. Guo, H.Q. Wang, M.Z. Chang, G.L. Zhang, and Y.X. Zhao. 2009a. Protective effects and mechanism of puerarin on learning-memory disorder after global cerebral ischemia-reperfusion injury in rats. *Chin. J. Integr. Med.* 15(1): 54–59.

Wu, L., K.X. Liu, J.L. Feng, A.Y. Zeng, and H. Li. 2009. Effects of sinomenine on expressions of cyclooxygenase-2 and prostaglandin E2 content following cerebral ischemic reperfusion injury in rats. *China J. Rehabil. Med.* 4: 293–296.

Wu, L.O., Y. Lu, R. Yang, K. Tian, P. Wu, and J. Chen. 1999. Improvement of impaired memory in mice by notoginsenoside-Rg1. *Acad. J. Kunming Med. Coll.* 20: 29–31.

Wu, Z.L. and N. Ren. 2008. Effect of gastrodin injection on the neuronal apoptosis of the brain after focal cerebral ischemia reperfusion. *Pharm. J. Chin. People's Liber. Army* 24(3): 204–207.

Wu, W.R. and X.Z. Zhu. 1999. Involvement of monoamine oxidase inhibition in neuroprotective and neurorestorative effects of Ginkgo biloba extract against MPTP-induced nigrostriatal dopaminergic toxicity in C57 mice. *Life Sci.* 65(2): 157–164.

Xia, S.J., Z.Y. Shen, J.C. Dong, X.Y. Liu, J.H. Huang, H.D. Wang, B. Wu, and W.H. Chen. 2008. Inflamm-aging related genes expressions in aged rat hippocampus and Icariin intervention outcome. *Geriatr. Health Care* 14(6): 340–344.

Xiao, C.Y. and J.T. Liu. 2007. The effect of tetramethyipyrazine on Bcl- 2,c- fos,Caspase- 3 in rat cerebral ischemia reperfusion. *J. Henan Univ. Chin. Med.* 22(4): 28–29.

Xiao, X.Q., R. Wang, Y.F. Han, and X.C. Tang. 2000a. Protective effects of huperzine A on beta-amyloid(25–35) induced oxidative injury in rat pheochromocytoma cells. *Neurosci. Lett.* 286(3): 155–158.

Xiao, X.Q., R. Wang, and X.C. Tang. 2000b. Huperzine A and tacrine attenuate beta-amyloid peptide-induced oxidative injury. *J. Neurosci. Res.* 61(5): 564–556.

Xiao, X.Q., J.W. Yang, and X.C. Tang. 1999. Huperzine A protects rat pheochromocytoma cells against hydrogen peroxide-induced injury. *Neurosci. Lett.* 275(2): 73–76.

Xiao, X.Q., H.Y. Zhang, and X.C. Tang. 2002. Huperzine A attenuates amyloid beta-peptide fragment 25–35-induced apoptosis in rat cortical neurons via inhibiting reactive oxygen species formation and caspase-3 activation. *J. Neurosci. Res.* 67(1): 30–36.

Xie, S.Z., Y.X. Ma, X.Y. Zhu, S.F. Wu, F.D. Liu, J.Y. Yang, and Y.D. Gu. 2003. Galantamine in treating age-associated memory impairment and Alzheimer's disease by double blinded method. *Geriatr. Health Care* 9: 223–226.

Xie, G.Q., X.L. Sun, S.P. Tian, and Q.S. Chen. 2003a. Studies on the preventive and therapeutic effects of rhodosin on rats with AIzheimer's disease. *Chin. J. Behav. Med. Sci.* 12(1): 18–20.

Xu, L., W.F. Chen, and M.S. Wong. 2009a. Ginsenoside Rg1 protects dopaminergic neurons in a rat model of Parkinson's disease through the IGF-I receptor signalling pathway. *Br. J. Pharmacol.* 158(3): 738–748.

Xu, H.L. and Y.P. Feng. 2000. Inhibitory effects of chiral 3-n-butylphthalide on inflammation following focal ischemic brain injury in rats. *Acta Pharmacol. Sin.* 21(5): 433–438.

Xu, H., H. Jiang, J. Wang, and J. Xie. 2009. Rg1 protects the MPP(+)-treated MES23.5 cells via attenuating DMT1 up-regulation and cellular iron uptake. *Neuropharmacology* 58(2): 488–494.

Xu, L., L.X. Liu, W.F. Chen, J.X. Xie, and W.X. Huang. 2008. The protective effect of ginsenoside Rg1 on dopaminergic neurons of substantia in the ovariectomized rat model of Parkinson's disease. *Chin. J. Appl. Physiol.* 24: 1–5.

Xu, H. and X.C. Tang. 1987. Cholinesterase inhibition by huperzine B. *Acta Pharm. Sin.* 8(1): 18–22.

Xu, Y. and J.J. Zhang. 2008. Protective effects of green tea and green tea polyphenols on Alzheimer's disease. *Chin. J. Clin. Neurosci.* 16: 328–331.

Xu, X.H. and T.Q. Zhao. 2002. Effects of puerarin on D-galactose-induced memory deficits in mice. *Acta Pharmacol. Sin.* 23(7): 587–590.

Xue, B., J. Jiao, L. Zhang, K.R. Li, Y.T. Gong, J.X. Xie, and X.M. Wang. 2007. Triptolide upregulates NGF synthesis in rat astrocyte cultures. *Neurochem. Res.* 32(7): 1113–1119.

Yacoubian, T.A. and D.G. Standaert. 2009. Targets for neuroprotection in Parkinson's disease. *Biochim. Biophys. Acta* 1792(7): 676–687.

Yan, C.H., Y.P. Feng, and J.T. Zhang. 1998. Effects of dl-3-n-butylphthalide on regional cerebral blood flow in right middle cerebral artery occlusion rats. *Zhongguo Yao Li Xue Bao* 19(2): 117–120.

Yan, F.L., Y.Q. Wang, G. Lu, and Z. Hong. 2006. Effects of puerarin on protein expression of hyperphosphorylated tau and ChAT in hippocampus of Alzheimer's disease rats. *J. Clin. Neurol.* 19(3): 191–193.

Yan, H., H.Y. Zhang, and X.C. Tang. 2007. Involvement of M1-muscarinic acetylcholine receptors, protein kinase C and mitogen-activated protein kinase in the effect of huperzine A on secretory amyloid precursor protein-alpha. *Neuroreport.* 18(7): 689–692.

Yang, K.H., S.X. Ge, B.Y. Xu, J.L. Yan, and L.O. Wu. 2007. Variation of BDNF mRNA on focal cerebral ischemia reperfusion injury in rats with notogisenoside-Rg1. *Chin. Trad. Herbal Drugs* 30(3): 313–316.

Yang, C.X., L.J. Li, X.Q. Cao, B. Chen, L. Deng, Z.L. Sun, and Q.L. Yuang. 2008. Ginsenoside Rb1 prohibits cell apoptosis and modulates the expression of apoptotic-related genes in rats subjected to focal cerebral ischemia. *J. Apoplexy Nerv. Dis.* 25: 12–16.

Yang, L.B., S.L. Li, S.Q. Wang, and S.Q. Zhang. 2007b. Effect of *Acorus gramimeus* and its active component α-asarone on N-methyl-D-asperate receptor 1 of hippocamp neurons in epileptic young rats. *Chin. Trad. Herbal Drugs* 38(11): 1670–1673.

Yang, S., G.H. Zhu, and D.H. Xie. 2007a. Epigallocatechin gallate ameliorates accumulated β-amyloid protein caused by oxidative free radicals in cortex neuron in vitro. *J. Clin. Res.* 24: 1065–1067.

Ying, Y., J.T. Zhang, C.Z. Shi, Z.W. Qu, and Y. Liu. 1994. Study on the nootropic mechanism of ginsenoside Rb1 and Rg1–influence on mouse brain development. *Acta Pharm. Sin.* 29(4): 241–245.

Yousuf, S., F. Atif, M. Ahmad, N. Hoda, T. Ishrat, B. Khan, and F. Islam. 2009. Resveratrol exerts its neuroprotective effect by modulating mitochondrial dysfunctions and associated cell death during cerebral ischemia. *Brain Res.* 1250: 242–253.

Zhang, L.D., X.J. Gong, M.M. Hu, Y.M. Li, and G.Q. Liu. 2008c. Anti-vascular dementia effect of gastrodin and its mechanisms of action. *Chin. J. Nat. Med.* 6(2): 130–134.

Zhang, X.C., B. He, P. Chen, J.J. Bai, L. Yang, and Z.Q. Shen. 2008. Effects of notogisenoside Rg1 on learning and memory function in multiple models. *Pharmacol. Clin. Chin. Mater. Med.* 24: 13–16.

Zhang, Y.F., Z.L. Jiang, M.H. Cao, and K.F. Ke. 2005. Effects of GSRb1 on free intracellular calcium concentrations in ischemic neurons of rats. *J. Clin. Neurol.* 18: 440–442.

Zhang, S., X.J. Li, and T.T. Wei. 2007. Protective effects of *Ginkgo biloba* extract on hydrogen peroxide-induced oxidative damage in astrocytes. *Chin. J. Histochem. Cytochem.* 16(1): 92–98.

Zhang, P., T. Li, B. Zhu, and B. Yuan. 2008a. The experimental study of puerarin on TNF-α and TGF-β1 expression afert focal cerebral ischenmia in rats. *Shaanxi Med. J.* 37(4): 398–401.

Zhang, H.Y., Y.Q. Liang, X.C. Tang, X.C. He, and D.L. Bai. 2002. Stereoselectivities of enantiomers of huperzine A in protection against beta-amyloid(25–35)-induced injury in PC12 and NG108–15 cells and cholinesterase inhibition in mice. *Neurosci. Lett.* 317(3): 143–146.

Zhang, J.T., Z.W. Qu, Y. Liu, and H.L. Deng. 1990. Preliminary study on antiamnestic mechanism of ginsenoside Rg1 and Rb1. *Chin. Med. J. Engl.* 103(11): 932–938.

Zhang, H.Y. and X.C. Tang. 2000. Huperzine B, a novel acetylcholinesterase inhibitor, attenuates hydrogen peroxide induced injury in PC12 cells. *Neurosci. Lett.* 292(1): 41–44.

Zhang, H.Y. and X.C. Tang. 2003. Huperzine A attenuates the neurotoxic effect of staurosporine in primary rat cortical neurons. *Neurosci. Lett.* 340(2): 91–94.

Zhang, H.Y. and X.C. Tang. 2006. Neuroprotective effects of huperzine A: New therapeutic targets for neurodegenerative disease. *Trends Pharmacol. Sci.* 27(12): 619–625.

Zhang, B.Q. and Y.L. Wang. 2007. Effects of puerarin on hippocampal pyramidal cell and the expression of BDNF in vascular dementia rats. *Chin. J. Neuroanat.* 23(6): 615–620.

Zhang, C., S.Z. Wang, and T. Wang. 2008b. Effects of tetramethylpyrazine on the hippocampal cholinergic system in D-galactose induced mice model with Alzheimer's disease. *J. Capital Univ. Med. Sci.* 29(1): 15–18.

Zhang, H.Y., H. Yan, and X.C. Tang. 2004. Huperzine A enhances the level of secretory amyloid precursor protein and protein kinase C-alpha in intracerebroventricular beta-amyloid-(1–40) infused rats and human embryonic kidney 293 Swedish mutant cells. *Neurosci. Lett.* 360(1–2): 21–24.

Zhang, Y.P., H. Yuan, L. Li, and X.Y. Wang. 2008d. Effects of *Ganoderma lucidum* polysaccharides on Caspase-3 and FasL expressions in the hippocampus of Alzheimer disease model rats. *Chin. J. Histochem. Cytochem.* 17(5): 484–489.

Zhao, Z.M., H.S. Pan, and Y.C. Feng. 2009. Experimental studies on the antioxidant capacity of ginsenoside Rg1. *J. Jiangxi Coll. Trad. Chin. Med.* 21: 36–38.

Zhou, J., Y. Fu, and X.C. Tang. 2001. Huperzine A protects rat pheochromocytoma cells against oxygen-glucose deprivation. *Neuroreport.* 12(10): 2073–2077.

Zhou, H.F., X.Y. Liu, D.B. Niu, F.Q. Li, Q.H. He, and X.M. Wang. 2005. Triptolide protects dopaminergic neurons from inflammation-mediated damage induced by lipopolysaccharide intranigral injection. *Neurobiol. Dis.* 18(3): 441–449.

Zhou, S.X., K. Su, J. Yu, Y. Peng, Y. Zhou, F. Lin, and H. Li. 2009. Effect of sinomenine on expression of P- selectin and ICAM-1 following cerebral ischemic reperfusion injury in diabetic rats. *Lish. Med. Mater. Med. Res.* 20: 1593–1595.

Zhou, J. and X.C. Tang. 2002. Huperzine A attenuates apoptosis and mitochondria-dependent caspase-3 in rat cortical neurons. *FEBS Lett.* 526(1–3): 21–25.

Zhu, X.D. and X.C. Tang. 1988. Improvement of impaired memory in mice by huperzine A and huperzine B. *Acta Pharmacol. Sin.* 9(6): 492–497.

15 Drugs for the Neglected Disease Malaria Based on Natural Products

S. Brøgger Christensen and Ib Christian Bygbjerg

CONTENTS

15.1 NEGLECTED DISEASES

Orphan diseases and neglected diseases may both be defined as diseases for which discovering and development of new treatments are assumed not to provide substantial financial return to companies willing to take on the costs. In general, the term "orphan diseases" is used for rare diseases in the industrialized world, meaning that the limited number of patients prevents sufficient return (Editorial 2007; Heemstra et al. 2008; Joppi et al. 2009), whereas the term "neglected diseases" is used for diseases in the developing world, where the poor income of the patients prevents sufficient return (Oprea et al. 2009; Trouiller et al. 2001). In both cases, pharmaceutical companies will refrain from investing in drug development. Initiatives in the European Union and the United States have resulted in registration of drugs for orphan diseases (Heemstra et al. 2008; Joppi et al. 2009). Incentives to develop drugs for neglected diseases are made through public–private partnerships (PPP), which are public-health-driven not-for-profit organizations that drive neglected-disease drug development in conjugation with industry groups. An example is Medicine for Malaria Venture (Moran 2005). In 2005, half the activity in this field was performed in PPP with small companies, whereas the other half was performed in PPP with large companies such as Glaxo SmithKline, Novartis, Astra Zeneca, and Sanofi-Aventis (Moran 2005). PPP has resulted in drugs having a major impact on global health such as ivermectin (Mectizan), which halved the global burden of onchocerciasis between 1990 and 2000, praziquantel (Biltricide), which has helped to control schistosomiasis in Brazil and other endemic countries, and artemether-lumefantrine (Coartem), which has delivered Africa its first safe effective new antimalarial drug for many years (Moran 2005). An unusual program of combination therapy is presently running, in which the pharmaceutical company Merck & Co. donates Mectizan and GlaxoSmithKline donates Albendazole (Zentel) to combat river blindness in exposed areas in Africa (The Mectizan donation program). A successful outcome of this program might eliminate the disease from the infected areas. As an appreciation of the pharmaceutical company Novartis' involvement in the development of Coartem, the American Food and Drug Agency (FDA) has awarded the company quick access to register new drugs (Anderson 2009).

In spite of these efforts, neglected diseases still is a major burden for humanity affecting more than one-sixth of the world population (more than 1.3 billion people, with malaria alone contributing to more than 0.3 billion [WHO 2008]) and cause more than 1.5 millions deaths each year, malaria alone causing more than 1 million deaths (Greenwood et al. 2005; Hotez et al. 2006).

15.2 MALARIA

15.2.1 MALARIA DISEASE

Malaria is a disease caused by protozoan parasites belonging to the genus *Plasmodium*. The parasites are transmitted by night-biting mosquitoes of the genus *Anopheles*. Until recently, it was believed that only four could infect humans: *P. falciparum, P. vivax, P. malariae,* or *P. ovale.* The use of genetic techniques, however, has revealed that a cluster of cases in Kapic (Malaysia), was caused by infection with the monkey parasite *P. knowlesi* and not, as expected, by *P. malariae* (Singh et al. 2004). *P. falciparum* is by far the most important species, causing most of the deaths and disease in the tropics, in particular in nonimmune young children, pregnant women, and migrants and travelers to endemic areas.

15.2.1.1 Life Cycle of the Malaria Parasite

The life cycle of the malaria parasite (Figure 15.1) involves a sexual stage in mosquitoes and an asexual stage in humans. When an infected mosquito bites a

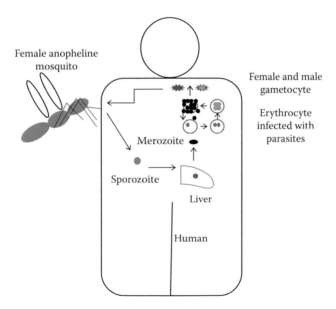

FIGURE 15.1 Life cycle of malaria parasites. When a female infected anopheline mosquito bites a human, it injects sporozoites into the bloodstream. The sporozoites invade liver cells and proliferate in liver cells. After 5–15 days, merozoites are released to the blood and infect red blood cells (erythrocytes). In the blood cells, they are transformed into first trophozoites, which develop into multinucleated schizonts (brown). Finally, after 2–3 days the parasites rupture the cells as either merozoites, which invade new red blood cells, where they divide into new schizonts etc. After few days, some merozoites become male or female gametocytes, which do not multiply nor give symptoms. When taken up by a mosquito they mate and multiply (sexual life cycle) in the mosquito and eventually develop into sporozoites, which can infect a new human when bitten by the mosquito.

human, it will inoculate sporozoites (Figure 15.1). After inoculation, the sporozoites invade the liver, where they proliferate in the hepatocytes. The liver stage, in which the patient feels no symptoms, lasts between 5 (*P. falciparum*) and 15 days (*P. malariae*). After this period, merozoites released into the bloodstream invade red blood cells (erythrocytes). In the blood cell, the parasite is transformed into trophozoites, which develop into multinucleated schizonts. After 48 h (*P. falciparum, P. vivax*, and *P. ovale*) or 72 h (*P. malariae*), the cell ruptures and releases between 6 and 32 merozoites. A subpopulation of the parasites develops into male and female gametocytes, which may persist in the blood as long as the life of the red cell, for example, 120 days. Gametocytes give no symptoms, but if taken up by mosquitoes, they mate, multiply, and become infective sporozoites. The characteristic intermittent fever is a consequence of the rupture of the red blood cells (White 2003). Deaths are mainly caused by cerebral malaria and anemia, caused by *P. falciparum*.

15.3 SPECIES AND STAGES OF MALARIA PARASITES AFFECTED BY ANTIMALARIALS

Schizontocides are antimalarial drugs that act on the asexual, dividing parasites in the erythrocytes. These include quinine, chloroquine and derivatives, and folic acid antagonists. For the benign species, *Plasmodium vivax, P. ovale,* and *P. malariae,* the sexual stages (gametocytes) are also affected, but in the malignant, potentially deadly *P. falciparum* only primaquine and artemisinins hit this stage, as well as the asexual blood stages. Primaquine also works on the liver-stages of *P. vivax* and *P. ovale,* which may otherwise remain dormant up to three years after infection, and give rise to delayed malaria attacks. Primaquine is therefore given after the *P. vivax* or *P. ovale* attack. It is most unfortunate that most antimalarials are not eradicating the sexual stage of *P. falciparum*, and drug pressure may even induce gametocytogenesis, that is, cure the patient from fever, but make him more infectious. Another problem of high drug pressure, in falciparum malaria in particular, is development of resistance. Therefore, the World Health Organization since 2006 recommends combination of antimalarials, preferably as artemisinin combination therapy (ACT) for treatment of *P. falciparum* infections (WHO 2006).

15.4 PHARMACOLOGICAL TARGETS FOR TREATMENT OF MALARIA

15.4.1 Digestive Vacuole

During the growth in the red blood cells (erythrocytes), the malaria parasite digests in the acidic food digestive vacuoles (DV) the protein strings of hemoglobin (Figure 15.2).

The major part of the released amino acid is used by the parasites in the protein synthesis (Egan 2004). Attempts are made to develop inhibitors of the proteolytic enzymes involved in the degradation of hemoglobin and of other proteases

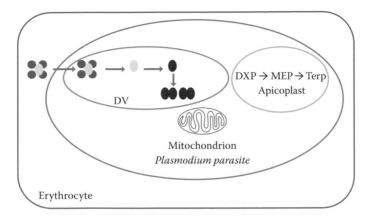

FIGURE 15.2 Pharmacological targets for drugs against malaria.

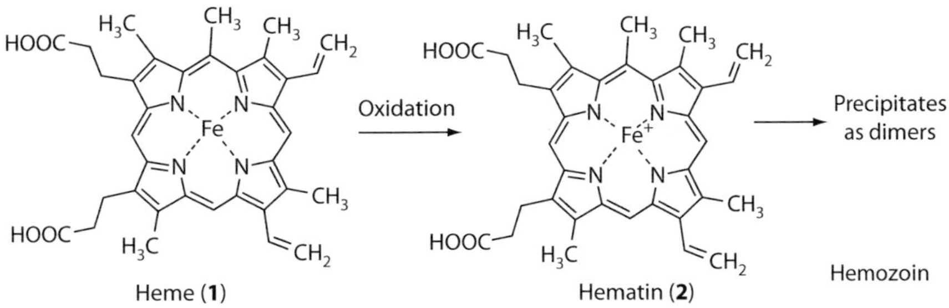

SCHEME 15.1 Conversion of heme (**1**) into hemozoin (**2**) in the digestive vacuole of the parasites.

characteristic for the *Plasmodium* parasites (Choi et al. 2008; Rosenthal 2001; Yeh and Altman 2006). After degradation of the protein part of haemoglobin heme (**1**), which is an iron(II) complex with a porphyrin skeleton (Scheme 15.1), remains. Heme is spontaneously oxidized to hematin (**2**), an iron(III) complex with a porphyrin skeleton. The parasite overcomes the cytotoxicity of hematin by converting it into dimers, which precipitate in the DV as microcrystalline hemozoin.

The exact mechanism whereby the parasite converts heme into a dimer is not known in detail, but it is assumed that some histidine-rich proteins characteristic for the parasite are involved in the detoxification of the hematin (Kannan et al. 2005).

15.4.2 APICOPLAST AND TERPENOID BIOSYNTHESIS

In plants, photosynthesis occurs in chloroplasts, which are believed to originate in cyanobacteria domesticated by eukaryote cells. Molecular and biochemical data have revealed that a reminiscence of the chloroplast, the apicoplast, remains in the *Plasmodium* parasite even though the ability to perform photosynthesis has been lost (Köhler et al. 1997; Lichtenthaler et al. 2000). The metabolic functions of the

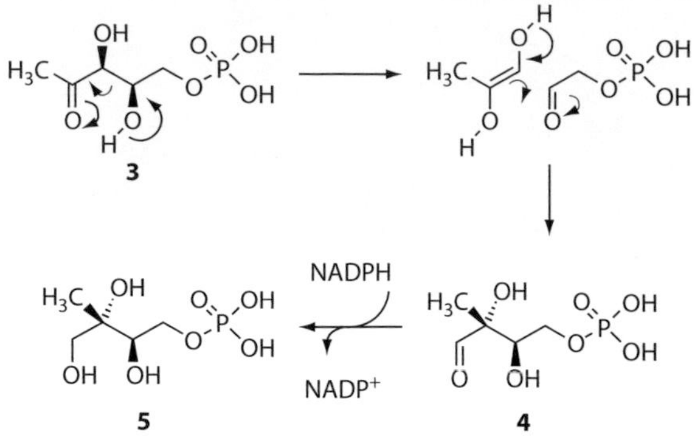

SCHEME 15.2 Suggested mechanism for rearrangement of 1-deoxy-D-xylulose 5-phosphate (**3**) to 2-C-methyl-D-erythritol 4-phosphate (**4**). The reaction is catalyzed by 1-deoxy-D-xylulose reductoisomerase. (From Dewick, P.M., *Medicinal Natural Products*, John Wiley & Sons Ltd., Chichester, U.K.; Dewick, P.M., The mevalonate and methylerythritol phosphate pathways: Terpenoids and steroids, in *Medicinal Natural Products*, John Wiley & Sons Ltd., Chichester, U.K., 2009b, pp. 187–310.)

apicoplast are essential for the survival of the parasite. The apicoplast performs synthesis of fatty acids, terpenoids, and heme, and malfunction of this organelle or inhibition of the reductoisomerase might be lethal for the parasite (Choi et al. 2008). The apicoplast still contains its own genome consisting of a circular 35-kb genome encoding prokaryote like ribosomal RNAs, tRNAs, and for some proteins.

Plants and microorganisms possess two pathways for terpenoid synthesis: The mevalonate pathway and the methylerythritol phosphate pathway, also known as the deoxyxylulose pathway. In plants, the deoxyxylulose pathway is the major source of mono-, di-, and tetraterpenoids but there are examples where both pathways supply different parts of the molecule (Dewick 2009). Humans are only able to synthesize terpenoids via the mevalonate pathway. In contrast, the terpenoids formed in the apicoplasts of *Plasmodium* parasites are only formed via the deoxyxylose pathway, making the methylerithritol pathway an interesting pharmacological target for treatment of malaria. The methylerithritol pathway involves a rearrangement of 1-deoxy-D-xylulose 5-phosphate (**3**) to 2-C-methyl-D-erythritol 4-phosphate (**4**) followed by a reduction of the aldehyde to the alcohol (Hunter 2007). The reaction, which is catalyzed by 1-deoxy-D-xylulose reductoisomerase, might proceed as shown in Scheme 15.2 (Dewick 2009).

15.4.3 ELECTRON TRANSPORT

In most organisms, the reduction of oxygen to generate energy for the cell occurs in the mitochondria. In mammalian cells, the energy is formed as ATP, but apparently enzymes for this process are missing in the mitochondria of *Plasmodium* parasites (Mather et al. 2007). In spite of this, the mitochondrial electron flow is still essential for (1) generating an electron sink for cellular metabolism such as ubiquinone-dependent dehydrogenases for dihydroorotate dehydrogenase and succinate,

(2) for maintenance of the transmembrane proton gradient required for transport of metabolites and proteins over the mitochondrial membrane, and (3) to reduce the formation of reactive oxygen species (Mather et al. 2007).

Like all other cytochrome bc_1 complexes, the *Plasmodium* complex contains di-heme cytochrome b, containing heme b-566 and b-560, cytochrome c_1, and an iron-sulfur protein known as the Rieske protein (Trumpower and Gennis 1994). The path of electrons flows from ubiquinol to cytochrome c through the bc_1 complex. The overall reaction is that one molecule of ubiquinol is oxidized to one molecule of ubiquinone and two molecules of oxidized cytochrome c are reduced. In addition, the two protons from ubiquinol and two protons from the negative side of the membrane are transferred to the positive side (Equation 15.1) (Trumpower and Gennis 1994):

$$QH_2 + 2\ cyt\ c_{ox} + 2H^+ \rightarrow Q + 2\ cyt\ c_{red} + 4H^+ \qquad (15.1)$$

The differences between the cytochrome bc_1 complexes electron transferring enzymes, in particular a different amino acid in position 275 of cytochrome b in human and *Plasmodium* mitochondria, enable blocking of the electron flow in mitochondria of the *Plasmodium* parasite without affecting the electron flow in the mitochondria of the mammalian host.

15.4.4 OTHER PHARMACOLOGICAL TARGETS

In contrast to the human host, which takes up folic acid as vitamin B_9, parasites are dependent on *de novo* folate synthesis. This leads to the important difference between the host and the parasite, for example, dihydropteroate synthase is absent in humans. Structural differences between the dihydrofolate reductases (DHFR) in the host and the parasite also have allowed development of drugs selectively targeting the essential tetrahydrofolic acid synthesis in parasites. Sulfonamides like sulfadoxine and dapsone are typical examples of drugs for treatment of malaria, which are antimetabolites to p-aminobenzoic acid, a critical building block for dihydrofolic acid. Other important drugs for malaria, proguanil, cycloguanil, and pyrimethamine are inhibitors of DHFR (Schlitzer 2007). These drugs are not natural products or derived from natural products and therefore beyond the scope of this chapter.

15.5 DRUGS FOR TREATMENT OF MALARIA

When the British major Ronald Ross discovered the vector function of mosquitoes belonging to the genus *Anopheles* in 1897, he enthusiastically wrote a poem:

I know this little thing.
A myriad men will save.
O death, where is thy sting.
Thy victory. O grave.

Unfortunately, the optimism of Ross has not been met. Comparison of maps on the distribution of malaria in 1945, 1977, and 2007 reveals that although malaria is under control in many parts of Asia and the Americas, and extinct from Europe, only limited progress has been made in Africa (WHO 2008). Also it is worth noticing that among the many natural products that were mentioned in the previous review on antiplasmodial natural products in this series (Christensen and Kharazmi 2001), none has made its way into the clinic.

Only a limited number of drugs are used for treatment of malaria. From the 1940s to the 1970s, chloroquine was the mainstay for prevention and treatment of malaria. Development of resistant parasites, however, has severely limited the use of the drug (Dorsey et al. 2001; Egan 2006). Artemisinin-based combination therapies (ACT) are safe and efficient and are also highly effective in reducing parasite transmission owing to the antigametocytic-activity of the artemisinin component. At the present, no clinical important resistance toward ACT is reported but artemisinin resistant parasites have been seen in the laboratory (White 2008), and delayed clearance of parasites have been observed in Cambodia (Dondorp et al. 2009). If resistance makes artemisinins obsolete, few if any replacements are available in large parts of the world.

Besides artemisinins, mefloquine, quinine, primaquine, and tetracyclines are used for treatment of malaria and a combination of atovaquone and proguanil (Malarone®), as well as combinations of sulfadoxine and pyrimethamine (Fansidar®) is used for treatment as well as prophylaxis. Among these agents, quinine, tetracycline, and artemisinins are natural products and atovaquone and mefloquine are developed from natural products. The present review, consequently, will focus on these compounds. Fosmidomycin will be included since this molecule reveals a new potential pharmacological target. In this text, drugs for treatment of malaria will be grouped according to their mechanism of action: Drugs that affect the polymerization of heme (quinine, chloroquine, artemisinin), drugs that affect the electron transport in mitochondria (atovaquone), and drugs that affects the protein synthesis of the apicoplasts, tetracyclines. The mechanism of action of most of these drugs, however, is still debated.

15.5.1 Drugs Affecting Targets in the Digestive Vacuole

15.5.1.1 Cinchona Bark

The first efficient drug the Europeans learned to use for treatment of malaria was the bark of trees belonging to the genus *Cinchona* (Rutaceae) (Rocco 2003). It might be considered a manifestation of the irony of history that this drug was discovered on a continent, South America, in which malaria only became a major problem after the arrival of the Europeans. The Jesuits during their missionary journeys in the conquered Inca Empire noticed that the Incas chewed the bark to prevent shivering. In the hope that the bark also might prevent the shivering caused by malaria, they brought the drug to Rome, which in the seventeenth century was malarious. Indeed, alcoholic extracts of the bark not only prevented shivering it also cured some cases of malaria. Gradually, the drug became recognized as a cure for malaria (Meshnick and Dobson 2001; Rocco 2003). The crude bark, however, possesses a number of disadvantages. A major problem being that the content of quinine present in the different species of Cinchona may vary between 4% and 17% (Dewick 2009). The small therapeutic window of quinine thus makes the irreproducible doses a major problem.

The isolation of quinine in a reasonably pure state in 1820 (Cooper 1970; Greenwood 1992) thus meant a breakthrough in the treatment, since the drug now could be offered in reproducible doses. The importance of the access to pure quinine is illustrated by four expeditions on the river Niger. In 1805, 39 of 44 Europeans died from malaria when sailing on the river and in 1833, 32 of 40 men died. In 1834, MacWilliam used quinine for treatment of crew-members infected with malaria. The crew consisted of 62 men, of which 55 were infected with malaria but in 39 cases the treatment was successful meaning that only 16 died. The final proof that Central and Western Africa could be colonized by Europeans was obtained by Baikie in 1854. Of the 12 Europeans and 54 Africans participating in the expedition on the Niger lead by him, no one died of malaria; among the two who died, one was murdered and the other drowned. The success was obtained by prophylactic treatment of the members of the crew with three or four grains (194 or 260 mg) of quinine dissolved in sherry each morning and evening (Rocco 2003).

In 1885, the European colonial imperial powers regulated African colonization and trade at a conference in Berlin.

15.5.1.2   Constituents of Cinchona Bark

Quinine (Figure 15.3, **6**) is the dominating alkaloid of the four major alkaloids present in *Cinchona* bark (Dewick 2009). The most important sources for quinine are *Cinchona calisaya* containing 4%–7% of alkaloids, *C. succiruba*, 5%–7%, and *C. ledgeriana*, 5%–14% (Dewick 2009).

From a chemical and medicinal point of view an interesting feature of quinine and quinidine (**8**) [and cinchonidine (**7**) and cinchonine (**9**)] is the opposite configuration at C-8 and C-9. The preserved configuration at C-3 makes the compounds diastereoismers. The opposite configurations at C-8 and C-9 enable the use of derivatives of quinine and quinidine for introduction of opposite chirality in alkenes by Sharpless asymmetric dihydroxylation (Kolb et al. 1994). Besides being found in species belonging to *Cinchona,* quinine is also found in species belonging to *Remijia,* (Rutaceae), which like *Cinchona* species have been used for treatment of malaria in traditional South American medicine (Kvist et al. 2006).

15.5.1.3   Accessibility of Quinine

Today the demand of quinine is met by isolation from the bark of *Cinchona* trees cultivated in many parts of the world, including Bolivia, Guatemala, India, Indonesia,

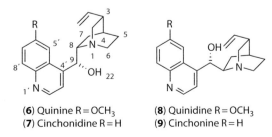

(**6**) Quinine R = OCH$_3$ (**8**) Quinidine R = OCH$_3$
(**7**) Cinchonidine R = H   (**9**) Cinchonine R = H

FIGURE 15.3   Structures of the four major *Cinchona* alkaloids.

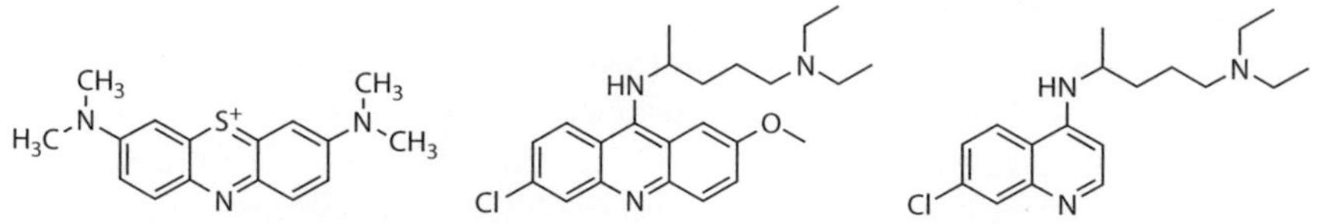

Methylene blue (**10**)   Mepacrine (racemic, **11**)   Chloroquine (racemic, **12**)

FIGURE 15.4 Structures of methylene blue (**10**), mepacrine (**11**), and chloroquine (**12**).

Zaire, Tanzania, and Kenya. The major three species cultivated are *C. succirubra, C. ledgeriana,* and *C. calisaya* (Dewick 2009). The *Cinchona* species are 15–20 m tall trees native to Colombia, Ecuador, and Peru (Samuelsson 2004). A number of protocols for total synthesis of quinine have been developed (Nicolaou and Snyder 2003), but none of these are of commercial interest.

Several attempts, however, were and are made to develop antimalarial drugs, which may be synthesized by a commercial feasible procedure.

During the Second World War, the Japanese occupation of Java cut off the major supply of quinine for treatment of malaria in infected soldiers in the allied forces. In order to overcome this problem considerably, efforts were undertaken in order to produce synthetic drugs, in particular to reconstruct the synthetic pathway to mepacrine. Mepacrine (**11**, Figure 15.4) became the main drug for prophylaxis and treatment of malaria during the Pacific war (Schlitzer 2007).

Paul Ehrlich noticed that methylene blue (**10**) was particularly efficient in staining malaria parasites and concluded that because the parasite took up the dye so efficiently it might be poisoned. In 1891, Ehrlich cured two malaria patients using methylene blue as a drug (Meshnick and Dobson 2001). Studies to optimize the drug led to development of chloroquine (**12**), which for many years was suspected of being too toxic for clinical use. After the Second World War, chloroquine became the foundation of malaria therapy for at least two decades.

15.5.1.4 Medicinal Chemistry of the Cinchona Alkaloids

Quinine was the first Cinchona alkaloid to be isolated in a pure state and consequently the first pure chemical entity to be used as an antimalarial drug. Even though quinidine is twice as active in vitro (Karle and Bhattacharjee 1999), quinine is still the preferred dug for treatment. The importance of the stereochemistry of the two chirality centers C-8 and C-9 is emphasized by only a residual activity of the two *threo*-isomers epiquinine and epiquinidine (Karle and Bhattacharjee 1999). The antiplasmodial activity of the quinine isomers parallels the affinity of the compounds for hematin, which is taken as an evidence for the binding of quinine to hematin as the mechanism of action for quinine (de Villiers et al. 2008). Whereas it is generally accepted that the effect of the quinoline drug chloroquine is mediated via its binding to hematin (Egan 2004; Schlitzer 2007), other pharmacological targets have been suggested for the quinoline drugs like quinine and mefloquine (Figure 15.5, **13**). Mefloquine was developed because of emerging chloroquine resistance (Trenholme et al. 1975), and because quinine must be taken thrice a day for 7 days, which is often not completed because of side effects (cinchonism). These drugs might

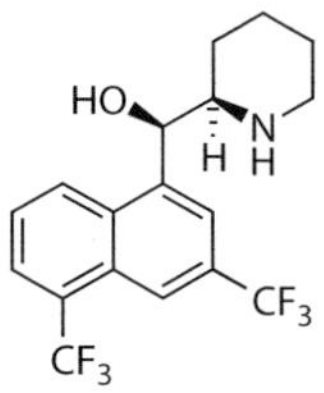

Mefloquine (**13**)

FIGURE 15.5   The structure of mefloquine (**13**). The clinically used mefloquine is the *erythro* racemate.

mediate their effects by inhibition of vacuole-vesicle fusion (Hoppe et al. 2004) or binding to essential proteins in the parasite (Foley and Tilley 1997).

15.5.1.5   *Artemisia annua*

The plant *Artemisia annua,* is a 1–1.5 m tall herb belonging to Asteraceae. The plant is cultivated in many different clime zones to get varieties or chemical races with an optimum content of the active constituent artemisinin. Up to 42% of the total artemisinin content is found in the glandular trichomes of the upper leaves (Bhakuni et al. 1988, 1990). Ge Hong, a Chinese scholar (284–363), was the first to describe that an emulsion obtained by wringing humide herbs of qinghao could be used for treatment of intermittent fever (Hsu 2006a,b). In the early 1970s, during the Vietnam War, a Chinese research team investigated the old remedy. The results were not published until 1979, but in a remarkable publication including the discovery of the antimalarial principle and clinical trials (Qinhaosu Antimalaria Coordinating Research Group 1979). Even though a number of species belonging to the genus *Artemisia* have been studied, artemisinin has only been detected in minor amounts in *A. apiacea* and *A. lancea* besides *A. annua* (Hsu 2006a,b). These observations also explain why *A. vulgaris* irresponsibly promoted as an antimalarial for travelers appeared to have no effect in vivo or in vitro (Kurzhals 2005).

15.5.1.6   Constituents of *Artemisia annua*

Artemisinin (Scheme 15.3, **14**) is a naturally occurring sesquiterpene lactone isolated from the plant. The stability of the peroxide bridge is surprising. Actually the lactone group artemisinin might be reduced using sodium borohydride to the corresponding semiacetal without degradation of the peroxide (Casteel 1997; Klayman 1985).

In vitro dihydroartemisinin (**15**) is almost one order of magnitude more potent in killing parasites (Casteel 1997). The real advantage, however, of this derivative is that it enables preparation of ethers, for example, arteether (**17**) and artemether (**16**), which are soluble in vegetable oils and therefore can be given intramuscularly, or esters like artesunate (**18**), which is soluble in water and can be formulated intravenously or by rectal administration (Barnes et al. 2004). The later method of administration is beneficial in patients in whom vomiting prevents per oral administration. Besides artemisinin, a series of other sesquiterpenes have been isolated from *A. annua,* including artemisinic acid (Figure 15.6, **19**), which is about 10 times more abundant than is artemisinin. None of these compounds shows antimalarial activity (Bhakuni et al. 2002).

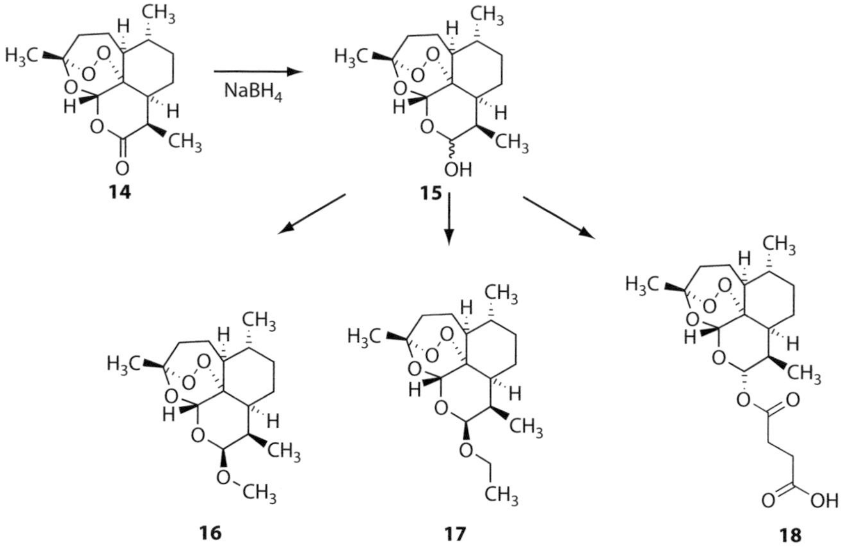

SCHEME 15.3 Conversion of artemisinin (**14**) into artemether (**16**), arteether (**17**), and arte-sunate (**18**) via dihydroatemisinin (**15**).

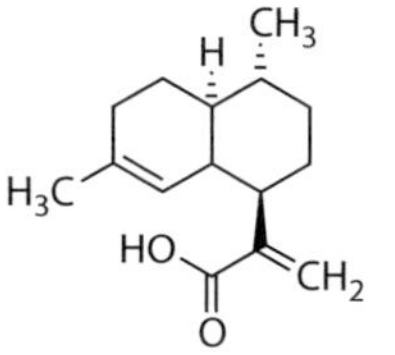

Artemisinic acid (**19**)

FIGURE 15.6 Structure of artemisinic acid (**19**).

15.5.1.7 Mechanism of Action of Artemisinin and Derivatives

Artemisinin is assumed to kill the parasites by reacting with the porphyrin skeleton of heme and thereby prevent the dimerization (Fidock et al. 2008; Kannan et al. 2002, 2005). Model studies have revealed that the complex of mangan(II) *meso*-tetraphenylporphyrin (**21**) reacts with artemisinin and other antimalarial trioxanes and forms a product which by a sodium borohydride reduction is converted into a C-alkylated porphyrin (**21**) (Robert and Meunier 1998) (Scheme 15.4). The initial C-radical formation might proceed as indicated in Scheme 15.5 (Robert and Meunier 1998).

The reaction is initiated by a transfer of an electron from Mn(II) to the perxoy-group to form a C-radical, which reacts with the porphyrin skeleton (Scheme 15.5). An intramolecular electron transfer converts the Mn(III) radical to a Mn(II) carbocation, which is reduced with borohydride. The acetalic acetoxy group is also lost during the reduction.

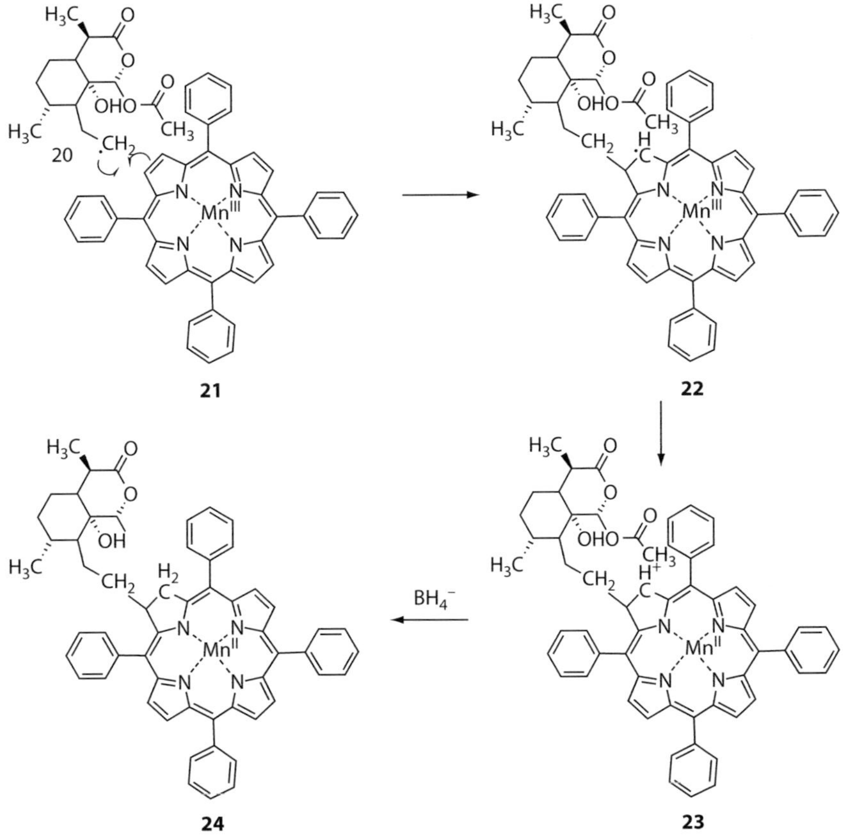

SCHEME 15.4 Formation of a C-alkylated porphyrin (**24**) after reaction between mangan(II) tetraphenylporphyrin (**21**) and a C-radical of artemisinin (**23**).

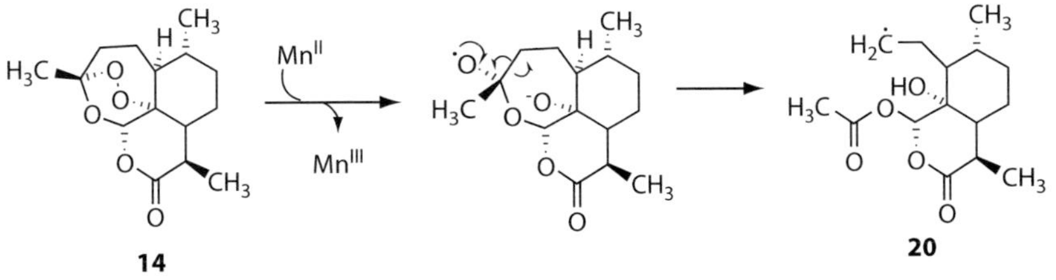

SCHEME 15.5 A suggested mechanism for formation of the C-radical (**20**) from artemisinin (**14**).

Model studies have revealed that artemisinin is consumed in the presence of heme (**1**) (Hong et al. 1994) or hemoglobin (Kannan et al. 2005). Studies using ^{14}C-labelled artemisnin revealed that the radioactivity follows the porphyrin skeleton supporting the suggestion that a covalent bond is formed between the porphyrin skeleton and artemisinin (Hong et al. 1994). The alkyl substituents at the vinylic carbon atoms

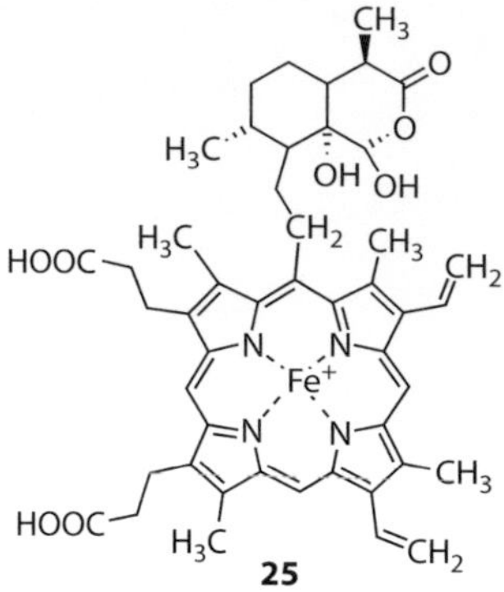

FIGURE 15.7   Possible product (**25**) formed from hematin and artemisinin.

in the dihydropyrrole rings of heme makes an alkylation as outlined in Scheme 15.4 less likely. Substitution at one of the methine groups between the pyrrole rings followed by hydrolysis of the acetate ester could give a product like compound **25** (Figure 15.7). An interesting feature by this structure is that the m/z value is 856. In the electrospray mass spectrum of the reaction product between hematin (**2**) and artemisinin, a peak at this m/z value was observed but no further analysis of the structure was performed (Hong et al. 1994). From the reaction mixture of hematin and artemisinin, peaks at m/z 838 and 1060 were observed (Kannan et al. 2005). Loss of water from the structure **25** would give a structure with m/z 838. Reaction between hematin and two C-radicals of artemisinin followed by hydrolysis of the acetate esters and loss of two molecules of water from the reaction product would give a structure with m/z 1060. Obviously these speculations do not prove how the artemisinin residue is attached to the porphyrin skeleton, but they might indicate the building blocks present in the reaction products.

Besides affecting malaria parasites, artemisinin and derivatives induce apoptosis in cultured human cancer cell lines and have been used successfully for the treatment of uveal melanoma after standard chemotherapy had failed (Efferth 2007). Artemisinins may also be used against another parasite disease, namely, schistosomiasis (Utzinger et al. 2007).

15.5.1.8   Artemisinin Supply

The increasing use of artemisinin or derivatives thereof causes a growing demand for efficient production of affordable drugs, which for being applicable in severely infected countries can be marketed at a price below $1 for an adult treatment course (White 2008). The artemisinin yield of wild type *A. annua* is 6 kg per hectare. If three harvests each year are possible, a yield of 18 kg per hectare may be obtained (Efferth 2007). If metabolically engineered *Escherichia coli* or *Saccharomyces cerevisiae* are used, up to 25 kg of the artemisinin precursor dihydroartemisinic acid can be produced in 8 h (Arsenault et al. 2008; Efferth 2007; Ro et al. 2006). Amyris Biotechnologies Inc. and the Institute for One World Health have announced an arrangement with Sanofi-Aventis to produce semisynthetic artemisinin from precursors obtained by a microbial source developed at the University of California at Berkeley (Covello 2008). It is expected that amounts of artemisinin corresponding to 200 million treatments can be produced by a combination of microbiological sources and semisynthesis (White 2008). The present metabolically engineered organisms all produce artemisinic acid and not the target compound artemisinin (Arsenault et al. 2008; Ro et al. 2006). A possible explanation for this finding is that the conversion of artemisinic acid to artemisinin is a spontaneous reaction occurring in *A. annua* without involvement of enzymes (Brown and Sy 2007; Covello 2008). Photooxidation of dihydroartemisinic acid (**26**) into artemisinin can be performed in preparative scale (Roth and Acton 1989) even though mechanism of the reaction is not known in detail (Scheme 15.6). The reaction might proceed via the aldehyde (Covello 2008), but other mechanisms have been suggested (Bhakuni et al. 2002).

In principle, total synthesis is a possibility for sustainable supply of artemisinin. Several procedures have been reported (Bhakuni et al. 2002; Covello 2008), but none of these are feasible for large-scale production, especially not in an economically feasible way. An alternative for sustainable supply of drugs possessing the pharmacological target of artemisinin is design of an agent, which in an economically feasible way may be produced by total synthesis, using the pharmacophore of artemisinin as a scaffold. A number of antimalarial synthetic trioxolanes inspired by the endoperoxide moiety in artemisinin and based on the performed structure–activity relationships have been made (Bhakuni et al. 2002; Casteel 1997; Schlitzer 2007) and one of these (RBX11160 or OZ277, Figure 15.8, **27**) entered clinical trials. Unfortunately, the development was discontinued because of its instability in blood (White 2008).

Considering the prices for antimalarial drugs based on artemisinin isolated from the herb, the possibility of using locally grown *A. annua* as a self-reliant treatment for malaria in developing countries has been discussed (de Ridder et al. 2008).

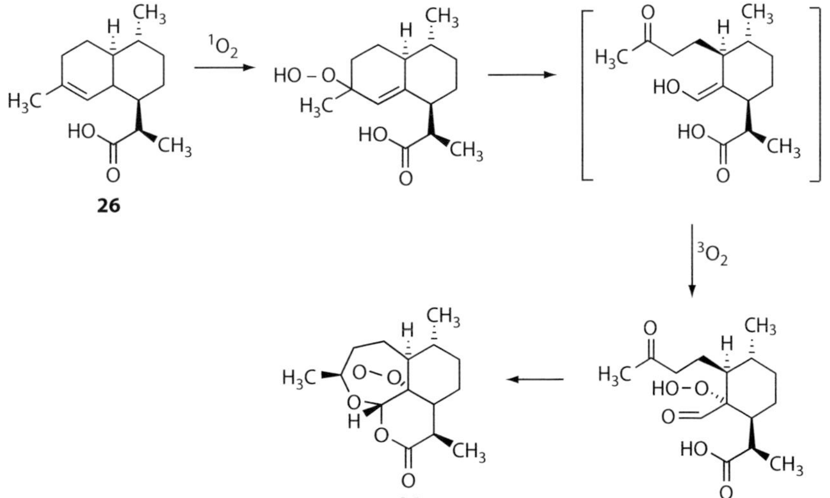

SCHEME 15.6 Suggested mechanism for conversion of dihydroartemisinic acid (**26**) into artemisinin (**14**) by photooxidation (Covello 2008).

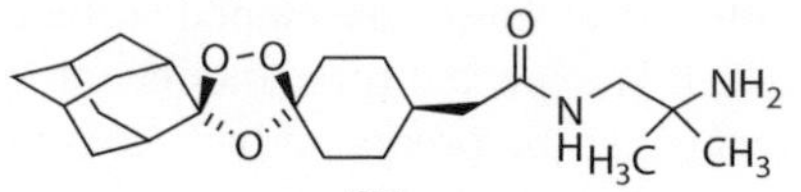

27

FIGURE 15.8   Structure of RBX11160 (**27**).

15.5.1.9 Artemisinin Combination Therapy

If artemisinin is given alone, a seven day regimen is required for obtaining maximal cure rates (White 2008), as is the case for quinine. Such an extended treatment is unrealistic in many endemic areas and consequently artemisinin in generally given in combination with slowly eliminated drugs allowing a complete treatment within 3 days. The first ACT was a combination of artemisinin with mefloquine. WHO currently recommends four combination therapies: Artesunate-sulfadoxine-pyremethamine, arteusunate-amodiaquine, artesunate-mefloquine, and artemether-lumefantrine and to ensure adherence encourage the use of fixed dose combinations. A potential risk by combination therapy in particular when using drugs with very different half-lives in the body and in areas with high transmission is that resistance might develop against the long living drug. Selection of resistant parasites in Uganda (Dokomajilar et al. 2006) and Zanzibar (Sisowath et al. 2005) has been reported. The drawback of using a preparation containing two agents has inspired the development of drugs containing two pharmacophores, for example, the pharmacophore from chloroquine and that from artemisinin (Benoit-Vical et al. 2007). Some verification for the mechanism of action for these so-called trioxaquines has been obtained by establishing that the compounds do alkylate heme in malaria-infected mice (Garah et al. 2008).

15.5.2 Drugs Affecting the Terpenoid Biosynthesis

15.5.2.1 Phosphonic Acids Including Fosmidomycin

A search for antibiotics in species belonging to the genus *Streptomyces rubellomurinus* leads to the isolation of an antibiotic phosphonic acid FR900098 (Figure 15.9, **28**) (Okuhara et al. 1980a). An additional search in *S. lavendulae* and other strains of *S. rubellomurinus* leads to the isolation of three additional phosphonic acids (Kuroda et al. 1980), namely, compounds **29–31**. (Figure 15.9).

Compound **29** has been given the trivial name fosmidomycin. Compounds **29** and **31** show significant activities against *Pseudomonas aeruginosa*, *Proteus vulgaris*, *Salmonella typhi*, and *Bacillus subtilis* but much less activity against *Klebsiella pneumoniae* and no activity against *Staphyllococus aureus*. In contrast, **30** showed almost no activity against these bacteria and **28** only a poor activity (Okuhara et al. 1980b).

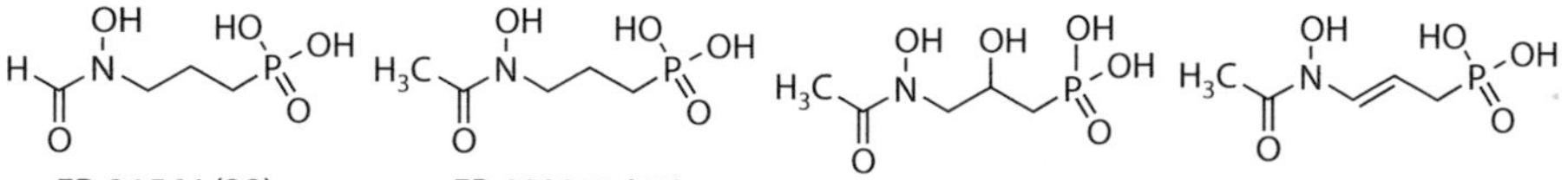

FR-31564 (**29**)   FR-900098 (**28**)   FR-33289 (**30**)   FR-32863 (**31**)

FIGURE 15.9   Structures of phosphonic acid antibiotics (**28–31**) isolated from *Streptomyces* species. (From Kuroda, Y. et al., *J. Antibiot.*, 33, 29, 1980.)

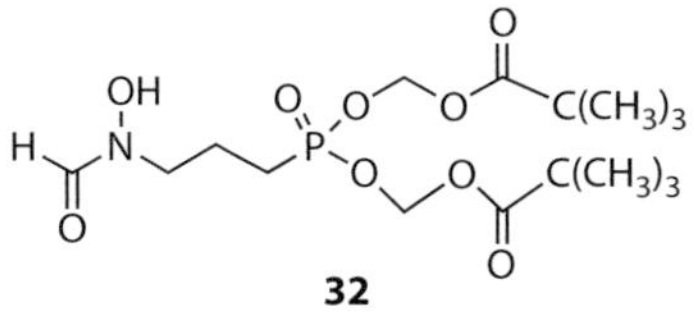

FIGURE 15.10 Structure of a fosmidomycin prodrug (**32**).

Fosmidomycin (**29**) is an analogue of 2-C-methyl-D-erythritol 4-phosphate (**4**, see Scheme 15.2) and the antibiotic effect is based on an inhibition of the enzyme 1-deoxy-D-xylulose reductoisomerase (Lichtenthaler et al. 2000).

A strong conservation of the amino acid sequences of 1-deoxy-D-xylulose reductoisomerase in plants, algae, bacteria, and parasites explains the ability of fosmidomycin to control the growth of all these organisms (Lichtenthaler et al. 2000). A further advantage is that humans do not posses this pathway (Dewick 2009) making drugs targeting the reductoisomerase selective toward the parasite. In vitro and in vivo assays revealed that the homologue **28** was twice as active as fosmidomycin. Prodrugs such as bispivalyloxymethyl esters (Figure 15.10, **32**) have been shown to possess improved in vitro and in vivo activity (Ortmann et al. 2003; Schluter et al. 2006; Wiesner et al. 2007).

Some clinical trials for *P. falciparum* malaria have been performed with fosmidomycin either as monotherapy or in combination with artesunate or clindamycin (Borrmann et al. 2004, 2005, 2006; Missinou et al. 2002). In spite of encouraging clinical results, the attempts to develop a drug based on fosmidomycin have been stopped for the time being.

15.5.3 DRUGS AFFECTING THE ELECTRON TRANSPORT

15.5.3.1 Naphtoquinones

The naphtoquinone lapachol (Figure 15.11, **33**) is widely distributed in plants belonging to the order Lamiales, in particular in plants belonging to the family Bignoniaceae, but has also been found in plants belonging to the families Proteaceae, Fabiaceae, Saptoceae, and Malvaceae (Hussain et al. 2007). Screening studies revealed that naphtoquinones possessed a potential for treatment of *P. lophurae* in ducks (Fieser et al. 1948). The studies resulted in the development of lapinone, which was used for curing vivax-malaria in two patients by intravenous injection of high doses of lapinone (**34**, 2 g per day for 4 days).

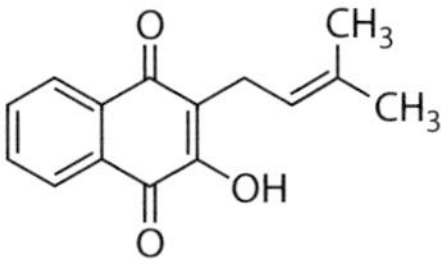

Lapachol (**33**) 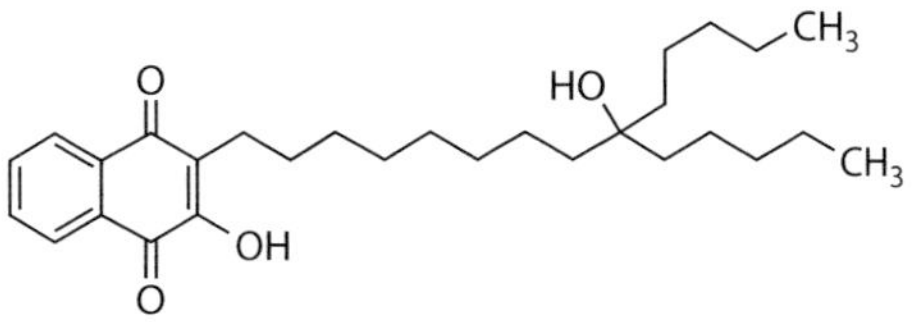 Lapinone (**34**)

FIGURE 15.11 Structures of lapachol (**33**) and lapinone (**34**).

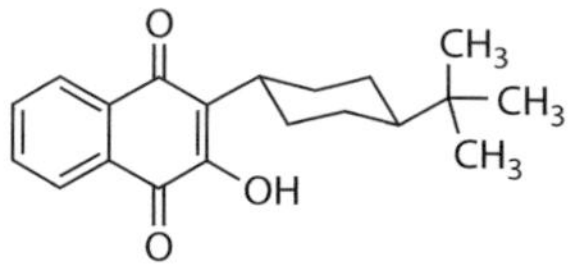

BW58C80 (**35**) 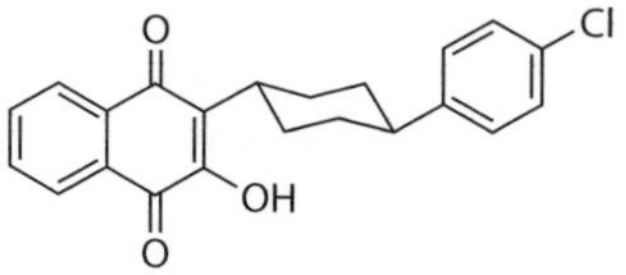   Atovaquone (**36**)

FIGURE 15.12   Structures of BW58C80 (**35**) and atovaquone (**36**).

The appearance of the cheap and effective chloroquine, however, resulted in a poor interest for the naphtoquinones.

15.5.3.2   Atovaquone

The appearance of resistance against chloroquine and other drugs for treatment of malaria renewed the interest for naphtoquinones. A major drawback of the available naphtoquinones was a fast catabolism of the side chain, inactivating the compounds. Fortunately, the naphtoquinones showed activity against *Theilera parva* infections in cattle leading to extensive structural variations (Boehm et al. 1981). A major outcome of this study was the importance of a cycloalkyl group in the side chain leading to wide exploration of a cyclohexyl moiety (Hudson et al. 1985). The fast metabolism of the tertiary butyl group in BW58C80 (Figure 15.12, **35**), which has broad antiparasitic activity, prevented its use in humans. Variations at the 4-position of the cyclohexyl group led to introduction of a 4-chlorophenyl group. The developed agent has been introduced as a drug under the name atovaquone (**36**) (Hudson et al. 1991; Hughes et al. 1991). The advent of the AIDS epidemic and the need to prevent and treat *Pneumocystis carinii* pneumonia, the most important opportunistic infection in AIDS patients from the Western world, prompted clinical trials of atovaquone, where it is now having a place as an alternative to sulfamethoxazole-trimethoprim, in patients not tolerating that combination. *Pneumocystis carinii*, now *Pneumocystis jirovecii*, was originally considered a protozoa, but is now classified as a fungus (Russian and Kovacs 1998).

15.5.3.3   Mechanism of Action of the Naphtoquinones

Atovaquone and other naphtoquinones block the electron flow in the cytochrome bc_1 complex by interchelating the Glu272 of the C helix of cytochrome c through a water bridge and the His181 of the Rieske protein (Figure 15.13). The model is based on studies of the cytochrome bc_1 complex from yeast (*Saccharomyces cerevisiae*), which has a high sequence identity with those of *Plasmodium* (Trumpower and Gennis 1994). The interchelation prevents electron transfer to the Rieske iron–sulfur complex (Mather et al. 2007).

The selectivity illustrated by a yeast bc_1 complex is inhibition with an IC_{50} value of 50 nM, whereas the corresponding value for bovine enzyme is 400 nM. A characteristic difference between the two enzymes is the presence of a leucine at position 275 of yeast cytochrome b, whereas this position in the bovine enzyme is occupied by phenylalanine. *In silico* modeling and studies using a L175F mutated yeast enzyme has revealed that this difference can explain the different affinities of atovaquone for

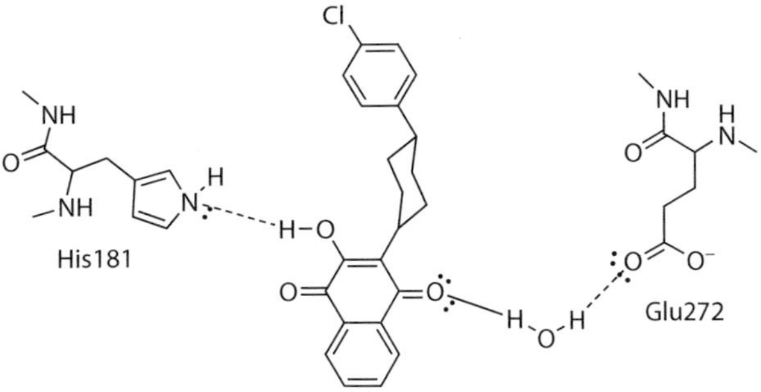

FIGURE 15.13 Interchelation of the Riekse protein through a hydrogen bond from histidine 181 to the hydroxygroup of atovaquone and from the carbonyl group of atovaquone to glutaric acid 272 of the C helix of cytochrome c.

the cytochromes (Kessl et al. 2003). The bovine and the human proteins are 80% identical. Both yeast and *Plasmodium* parasites have leucine in position 275, whereas both human and bovine proteins have phenylalanine in this position. In malaria, atovaquone is only used in combination with proguanil, as Malarone®, to avoid resistance, both for chemoprophylaxis and therapy. However, resistance has already been reported in travelers (David et al. 2003).

15.6　ANTIBIOTICS

15.6.1　Tetracyclines

The tetracyclines (compounds **37–39**, Figure 15.14) originally were introduced as broad spectrum antibiotics for treatment of infections with Gram-positive and Gram-negative bacteria (Corey et al. 2007). The antibiotics are produced by soil dwelling bacteria belonging to the *Streptomyces* species. The mechanism of action relates to the ability to bind to the 30S subunit of bacterial ribosomes and thereby preventing binding of amino-acyl tRNA to the acceptor site of the mRNA-ribosome complex. It is suggested that the antimalaria effect is based on the ability of the tetracyclines to disrupt the expression of apicoplast genes. Even though the apicoplast is not prevented from functioning, the effect of the tetracycline will contain insufficient levels of apicoplast encoded proteins for importation and processing of the nuclear genes needed for normal function. Ultimately this results in

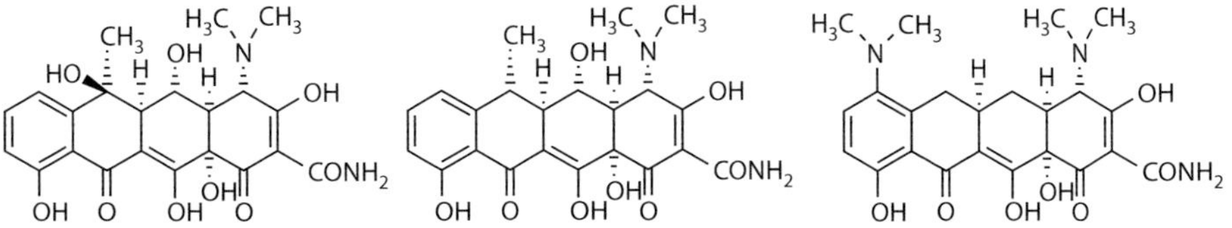

FIGURE 15.14 Structures of tetracycline (**37**), doxycycline (**38**), and minocycline (**39**).

parasite death (Dahl et al. 2006). However, teracyclines work slowly, and should only be used for prophylaxis, most commonly as doxycycline, or in combination with, for example, quinine, for therapy. Here, the combination may work even for treatment of *P. falciparum* infections with lowered sensitivity to quinine (Chin and Intraprasert 1973).

15.7 RESISTANCE

The molecular basis for the development of resistance against drugs is now increasingly understood. In the case of chloroquine, resistance may be caused by mutations in the genes coding for *P. falciparum* chloroquine resistance transporter (PfCRT) and for Pgh1, an ATP binding transporter involved in food vacuole import that causes resistance to the drug (Turschner and Efferth 2009; Valderramos and Fidock 2006). Controversy exists over the role of the Plasmodial pfmdr1, which is a member of the superfamily of ATP-binding cassette transporters (van Es et al. 1994). The pfmdr1-encoded protein, Pgh1, in the wild type, can mediate an increased intracellular accumulation of chloroquine, and that function is impaired in chloro-quine-resistant mutant forms of the protein. Interestingly, pfmdr1 mutations may contribute to quinine resistance but enhance mefloquine and artemisinin sensitiv-ity in *P. falciparum* (Sidhu et al. 2005). In the case of drugs acting on the folate pathway, resistance follows mutations of the enzymes involved in the folate synthe-sis such as dihydrofolate reductase and dihydropteroate synthetase (Gregson and Plowe 2005; Hyde 2002).

Alarming field isolates have revealed that mutations the intracellular calcium pump PfATP6 result in reduced susceptibility toward artemisinin in *P. falciparum* (Jambou et al. 2005).

The increased understanding of the mechanisms behind resistance might enable development of therapies for overcoming the problem. However, at the present no drug for circumventing resistance is available.

15.8 CONCLUSION

In spite of many initiatives over the years, malaria remains a major public health problem in the poorest parts of the world, sub-Saharan Africa in particular. In those areas, the disease may be responsible for more than one-third of deaths among chil-dren under the age of five and up to one-fifth of the death of pregnant women. In addition, malaria may cause decreased learning capacity in children, students, and trainees and in loss of workforce (Mboera et al. 2007). These facts impose the ques-tion whether the population is poor because it suffers from malaria, or whether it suf-fers from malaria because it is poor? At the present and probably in the near future, no efficient vaccine exists for treatment of malaria or will be available, meaning that treatment of malaria still depends on drugs based on small molecules, supplemented by prevention by long-lasting insecticide impregnated bed-nets (The global fund to fight aids, tuberculosis and malaria 2010). Interestingly, the original substances used for impregnation are also natural products, namely, pyrethrins from *Chrysanthemum cinerariaefolium* (Asteraceae) (Dewick 2009) (Figure 15.15).

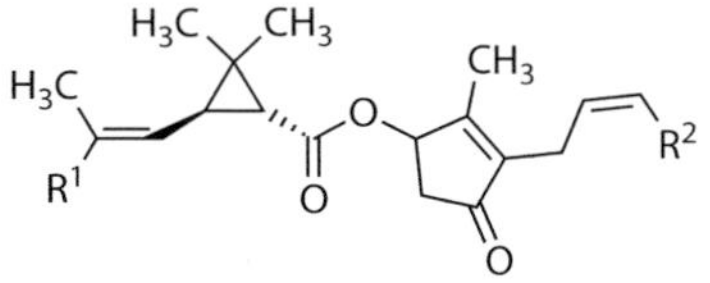

Pyrethrin I (**40**) R^1 = –CH$_3$ R^2 = –CH=CH$_2$

Cinerin I (**41**) R^1 = –CH$_3$ R^2 = –CH$_3$

Jasmolin II (**42**) R^1 = –CH$_3$ R^2 = –CH$_2$CH$_3$

Pyrethrin II (**43**) R^1 = –COOCH$_3$ R^2 = –CH=CH$_2$

Cinerin II (**44**) R^1 = –COOCH$_3$ R^2 = –CH$_3$

Jasmolin II (**45**) R^1 = –COOCH$_3$ R^2 = –CH$_2$CH$_3$

FIGURE 15.15   Constitutions of the major insecticides present in an extract of *Chrysanthemum* flowers.

Natural products have been and still are an important source for new drugs and in particular drugs against parasitic diseases like malaria, drugs against other infectious diseases, and drugs against cancer (Newman and Cragg 2007). At the present, WHO recommends combinations therapies using artemisinin and derivatives thereof as one of the components as drug of first choice. Hopefully, the problem of a sustainable supply of artemisinin will be solved giving access to sufficient amounts of the drug. However, in the light of the first alarming indications that resistance is developing, even though artemisinin is given in combination therapy, efforts for the development of drugs, which can be used in parts of the world where malaria is endemic, and which are cheap and simple to administer—preferentially as a single dose—should be encouraged.

REFERENCES

Anderson T. 2009. Novartis under fire for accepting new reward for old drug. *Lancet* 373: 1414.

Arsenault PR, Wobbe KK, and Weathers PJ. 2008. Recent advances in artemisinin production through heterologous expression. *Curr. Med. Chem.* 15: 2886–2896.

Barnes KI, Mwenechanya J, Tembo M, Mcilleron H, Folb PI, Ribeiro I, Little F, Gomes M, and Molyneux ME. 2004. Efficacy of rectal artesunate compared with parenteral quinine in initial treatment of moderately severe malaria in African children and adults: A randomised study. *Lancet* 363: 1598–1605.

Benoit-Vical F, Lelievre J, Berry A, Deymier C, Dechy-Cabaret O, Cazelles J, Loup C, Robert A, Magnaval JF, and Meunier B. 2007. Trioxaquines are new antimalarial agents active on all erythrocytic forms, including gametocytes. *Antimicrob. Agents Chemother.* 51: 1463–1472.

Bhakuni DS, Goel AK, Goel AK, Jain S, Mehrotra BN, and Srimal RC. 1990. Screening of Indian plants for biological activity: Part XIV. *Indian J. Exp. Biol.* 28: 619–637.

Bhakuni DS, Goel AK, Jain S, Mehrotra BN, Patnaik GK, and Prakash V. 1988. Screening of Indian plants for biological activity: Part XIII. *Indian J. Exp. Biol.* 26: 883–904.

Bhakuni RS, Jain BDC, and Sharma RP. 2002. Phytochemistry of *Artemisia annua* and the development of artemisinin-derived antimalarial agents. In *Artemisia*, ed. CW Wright, pp. 211–248. London, U.K.: Taylor & Francis.

Boehm P, Cooper K, Hudson AT, Elphick JP, and Mchardy N. 1981. In vitro activity of 2-alkyl-3-hydroxy-1,4-naphthoquinones against *Theileria parva*. *J. Med. Chem.* 24: 295–299.

Borrmann S, Adegnika AA, Matsiegui PB, Issifou S, Schindler A, Mawili-Mboumba DP, Baranek T, Wiesner J, Jomaa H, and Kremsner PG. 2004. Fosmidomycin-clindamycin for *Plasmodium falciparum* infections in African children. *J. Infect. Dis.* 189: 901–908.

Borrmann S, Adegnika AA, Moussavou F, Oyakhirome S, Esser G, Matsiegui P, Ramharter M, Lundgren I, Kombila M, Issifou S, Hutchinson D, Wiesner J, Jomaa H, and Kremsner PG. 2005. Short-course regimens of artesunate-fosmidomycin in treatment of uncomplicated *Plasmodium falciparum* malaria. *Antimicrob. Agents Chemother.* 49: 3749–3754.

Borrmann S, Lundgren I, Oyakhirome S, Impouma B, Matsiegui P, Adegnika AA, Issifou S, Kun Jurgen FJ, Hutchinson D, Wiesner J, Jomaa H, and Kremsner PG. 2006. Fosmidomycin plus clindamycin for treatment of pediatric patients aged 1 to 14 years with *Plasmodium falciparum* malaria. *Antimicrob. Agents Chemother.* 50: 2713–2718.

Brown GD and Sy LK. 2007. In vivo transformations of artemisinic acid in *Artemisia annua* plants. *Tetrahedron* 63: 9548–9566.

Casteel DA. 1997. Antimalarial agents. In *Burger's Medicinal Chemistry and Drug Discovery*, ed. ME Wolff, pp. 3–91. New York: John Wiley & Sons, Inc.

Chin W and Intraprasert R. 1973. The evaluation of quinine alone or in combination with tetracycline and pyrimethamine against falciparum malaria in Thailand. *South East Asian J. Trop. Med. Public Health* 4: 245–249.

Choi SR, Mukherjee P, and Avery MA. 2008. The fight against drug-resistant malaria: Novel plasmodial targets and antimalarial drugs. *Curr. Med. Chem.* 15: 161–171.

Christensen SB and Kharazmi A. 2001. Antimalarial natural products. In *Bioactive Compounds from Natural Sources*, ed. C Tringali, pp. 381–431. London, U.K.: Taylor & Francis.

Cooper P. 1970. Pelletier and caventou: Discoveries of quinine. *Pharmaceutical J.* 205: 536–537.

Corey EJ, Czakó B, and Kürti L. 2007. *Molecules and Medicine*. Hoboken, NJ: John Wiley & Sons.

Covello PS. 2008. Making artemisinin. *Phytochemistry* 69: 2881–2885.

Dahl EL, Shock JL, Shenai BR, Gut J, DeRisi JL, and Rosenthal PJ. 2006. Tetracyclines specifically target the apicoplast of the malaria parasite plasmodium falciparum. *Antimicrob. Agents Chemother.* 50: 3124–3131.

David KP, Alifrangis M, Salanti A, Vestergaard LS, Ronn A, and Bygbjerg IB. 2003. Atovaquone/proguanil resistance in Africa: A case report. *Scand. J. Infect. Dis.* 35: 897–898.

de Ridder S, van der Kooy F, and Verpoorte R. 2008. *Artemisia annua* as a self-reliant treatment for malaria in developing countries. *J. Ethnopharmacol.* 120: 302–314.

de Villiers KA, Marques HM, and Egan TJ. 2008. The crystal structure of halofantrine-ferriprotoporphyrin IX and the mechanism of action of arylmethanol antimalarials. *J. Inorg. Biochem.* 102: 1660–1667.

Dewick PM. 2009a. *Medicinal Natural Products*. Chichester, U.K.: John Wiley & Sons Ltd.

Dewick PM. 2009b. The mevalonate and methylerythritol phosphate pathways: Terpenoids and steroids. In *Medicinal Natural Products*, pp. 187–310. Chichester, U.K.: John Wiley & Sons Ltd.

Dokomajilar C, Nsobya SL, Greenhouse B, Rosenthal PJ, and Dorsey G. 2006. Selection of *Plasmodium falciparum* pfmdr1 alleles following therapy with artemether-lumefantrine in an area of Uganda where malaria is highly endemic. *Antimicrob. Agents Chemother.* 50: 1893–1895.

Dondorp AM, Nosten F, Yi P, Das D, Phyo AP, Tarning J, Lwin KM, Ariey F, Hanpithakpong W, Lee SJ, Ringwald P, Silamut K, Imwong M, Chotivanich K, Lim P, Herdman T, An SS, Yeung S, Singhasivanon P, Day NP, Lindegardh N, Socheat D, and White NJ. 2009. Artemisinin resistance in *Plasmodium falciparum* malaria. *N. Engl. J. Med.* 361: 455–467.

Dorsey G, Fidock DA, Wellems TE, and Rosenthal AS. 2001. Mechanisms of quinoline resistance. In *Antimalarial Chemotherapy*, ed. AS Rosenthal, pp. 153–172. Totowa, NJ: Humana Press.

Editorial. 2007. Drugs for rare diseases: Mixed assessment in Europe. *Prescrire Int.* 16: 36–42.

Efferth T. 2007. Willmar Schwabe award 2006: Antiplasmodial and antitumor activity of artemisinin—From bench to bedside. *Planta Med.* 73: 299–309.

Egan TJ. 2004. Haemozoin formation as a target for the rational design of new antimalarials. *Drug Des. Rev. Online* 1: 93–110.

Egan TJ. 2006. Chloroquine and primaquine: Combining old drugs as a new weapon against falciparum malaria? *Trends Parasitol.* 22: 235–237.

Fidock DA, Eastman RT, Ward SA, and Meshnick SR. 2008. Recent highlights in antimalarial drug resistance and chemotherapy research. *Trends Parasitol.* 24: 537–544.

Fieser LF, Berlinger E, Bondhus FJ, Chang FC, Dauben WG, Ettlinger MG, Fawaz G, Fields M, Fieser M, Heidelberger C, Heymann H, Seligman AM, Vaughan WR, Wilson AG, Wilson E, Wu M, Leffler MT, Hamlin KE, Hathaway RJ, Matson EJ, Moore EE, Moore MB, Rapala RT, and Zaugg HE. 1948. Naphtoquinone antimalarials. I. General survey. *J. Am. Chem. Soc.* 70: 3151–3155.

Foley M and Tilley L. 1997. Quinoline antimalarials: Mechanisms of action and resistance. *Int. J. Parasitol.* 27: 231–240.

Garah FBE, Claparols C, Benoit-Vical F, Meunier B, and Robert A. 2008. The antimalarial trioxaquine DU1301 alkylates heme in malaria-infected mice. *Antimicrob. Agents Chemother.* 52: 2966–2969.

Greenwood D. 1992. The quinine connection. *J. Antimicrob. Chemother.* 30: 417–427.

Greenwood BM, Bojang K, Whitty CJM, and Targett GAT. 2005. Malaria. *Lancet* 365: 1487–1498.

Gregson A and Plowe CV. 2005. Mechanisms of resistance of malaria parasites to antifolates. *Pharmacol. Rev.* 57: 117–145.

Heemstra HE, de Vrueh RLA, van Weely S, Bueller HA, and Leufkens HGM. 2008. Orphan drug development across Europe: Bottlenecks and opportunities. *Drug Discov. Today* 13: 670–676.

Hong YL, Yang YZ, and Meshnick SR. 1994. The interaction of artemisinin with malarial hemozoin. *Mol. Biochem. Parasit.* 63: 121–128.

Hoppe HC, van Schalkwyk DA, Wiehart UI, Meredith SA, Egan J, and Weber BW. 2004. Antimalarial quinolines and artemisinin inhibit endocytosis in *Plasmodium falciparum*. *Antimicrob. Agents Chemother.* 48: 2370–2378.

Hotez PJ, Molyneux DH, Fenwick A, Ottesen E, Sachs SE, and Sachs JD. 2006. Incorporating a rapid-impact package for neglected tropical diseases with programs for HIV/AIDS, tuberculosis, and malaria—A comprehensive pro-poor health policy and strategy for the developing world. *Plos Medicine* 3: 576–584.

Hsu E. 2006a. Reflections on the 'discovery' of the antimalarial qinghao. *Br. J. Clin. Pharmacol.* 61: 666–670.

Hsu E. 2006b. The history of qing hao in the Chinese materia medica. *Trans. R. Soc. Trop. Med. Hyg.* 100: 505–508.

Hudson AT, Dickins M, Ginger CD, Gutteridge WE, Holdich T, Hutchinson DB, Pudney M, Randall AW, and Latter VS. 1991. 566C80: A potent broad spectrum anti-infective agent with activity against malaria and opportunistic infections in AIDS patients. *Drugs Exp. Clin. Res.* 17: 427–435.

Hudson AT, Randall AW, Fry M, Ginger CD, Hill B, Latter VS, Mchardy N, and Williams RB. 1985. Novel anti-malarial hydroxynaphthoquinones with potent broad spectrum anti-protozoal activity. *Parasitology* 90 (Pt 1): 45–55.

Hughes WT, Kennedy W, Shenep JL, Flynn PM, Hetherington SV, Fullen G, Lancaster DJ, Stein DS, Palte S, Rosenbaum D et al. 1991. Safety and pharmacokinetics of 566C80, a hydroxynaphthoquinone with anti-*Pneumocystis carinii* activity: A phase I study in human immunodeficiency virus (HIV)-infected men. *J. Infect. Dis.* 163: 843–848.

Hunter WN. 2007. The non-mevalonate pathway of isoprenoid precursor biosynthesis. *J. Biol. Chem.* 282: 21573–21577.

Hussain H, Krohn K, Ahmad VU, Miana GA, and Green IR. 2007. Lapachol: An overview. *Arkivoc* ii: 145–171.

Hyde JE. 2002. Mechanisms of resistance of *Plasmodium falciparum* to antimalarial drugs. *Microbes Infect.* 4: 165–174.

Jambou R, Legrand E, Niang M, Khim N, Lim P, Volney B, Ekala MT, Bouchier C, Esterre P, Fandeur T, and Mercereau-Puijalan O. 2005. Resistance of *Plasmodium falciparum* field isolates to in vitro artemether and point mutations of the SERCA-type PfATPase6. *Lancet* 366: 1960–1963.

Joppi R, Bertele V, and Garattini S. 2009. Orphan drug development is not taking off. *Br. J. Clin. Pharmacol.* 67: 494–502.

Kannan R, Kumar K, Sahal D, Kukreti S, and Chauhan AK. 2005. Reaction of artemisinin with haemoglobin: Implications for antimalarial activity. *Biochem. J.* 385: 409–418.

Kannan R, Sahal D, and Chauhan VS. 2002. Heme-artemisinin adducts are crucial mediators of the ability of artemisinin to inhibit heme polymerization. *Chem. Biol.* 9: 321–332.

Karle JM and Bhattacharjee AK. 1999. Stereoelectronic features of the cinchona alkaloids determine their differential antimalarial activity. *Bioorg. Med. Chem.* 7: 1769–1774.

Kessl JJ, Lange BB, Merbitz-Zahradnik T, Zwicker K, Hill P, Meunier B, Palsdottir H, Hunte C, Meshnick S, and Trumpower BL. 2003. Molecular basis for atovaquone binding to the cytochrome bc(1) complex. *J. Biol. Chem.* 278: 31312–31318.

Klayman DL. 1985. Qinhaosu (Artemisinin): An antimalarial drug from China. *Science* 228: 1045–1055.

Köhler S, Delwiche CF, Denny PW, Tilney LG, Webster P, Wilson RJ, Palmer JD, and Roos DS. 1997. A plastid of probable green algal origin in apicomplexan parasites. *Science* 275: 1485–1489.

Kolb HC, van Nieuwenhze MS, and Sharpless KB. 1994. Catalytic asymmetric dihydroxylation. *Chem. Rev.* 94: 2483–2547.

Kuroda Y, Okuhara M, Goto T, Okamoto M, Terano H, Kohsaka M, Aoki H, and Imanaka H. 1980. Studies on new phosphonic acid antibiotics.4. Structure determination of Fr-33289, Fr-31564 and Fr-32863. *J. Antibiot.* 33: 29–35.

Kurzhals JA. 2005. Ineffective change of antimalaria prophylaxis to *Artemisia vulgaris* in a group travelling to west Africa. *Ugeskr. Læger* 167: 4082–4083.

Kvist LP, Christensen SB, Rasmussen HB, Mejia K, and Gonzalez A. 2006. Identification and evaluation of Peruvian plants used to treat malaria and leishmaniasis. *J. Ethnopharmacol.* 106: 390–402.

Lichtenthaler HK, Zeidler J, Schwender J, and Muller C. 2000. The non-mevalonate isoprenoid biosynthesis of plants as a test system for new herbicides and drugs against pathogenic bacteria and the malaria parasite. *Z. Naturforsch. C* 55: 305–313.

Mather MW, Henry KW, and Vaidya AB. 2007. Mitochondrial drug targets in apicomplexan parasites. *Curr. Drug Targets* 8: 49–60.

Mboera LE, Makundi EA, and Kitua AY. 2007. Uncertainty in malaria control in Tanzania: Crossroads and challenges for future interventions. *Am. J. Trop. Med. Hyg.* 77: 112–118.

Meshnick SR and Dobson MJ. 2001. The history of antimalarial drugs. In *Antimalarial Chemitherapy*, ed. PJ Rosenthal, pp. 15–25. Totowa, NJ: Humana Press.

Missinou MA, Borrmann S, Schindler A, Issifou S, Adegnika AA, Matsiegui P, Binder R, Lell B, Wiesner J, Baranek T, Jomaa H, and Kremsner PG. 2002. Fosmidomycin for malaria. *Lancet* 360: 1941–1942.

Moran M. 2005. A breakthrough in R&D for neglected diseases: New ways to get the drugs we need. *PLoS Med.* 2: e302.

Newman DJ and Cragg GM. 2007. Natural products as sources of new drugs over the last 25 years. *J. Nat. Prod.* 70: 461–477.

Nicolaou KC and Snyder SA. 2003. Quinine. In *Classics in Total Synthesis II*, pp. 443–462. Weinheim, Germany: Wiley-VCH.

Okuhara M, Kuroda Y, Goto T, Okamoto M, Terano H, Kohsaka M, Aoki H, and Imanaka H. 1980a. Studies on new phosphonic acid antibiotics. 1. Fr-900098, isolation and characterization. *J. Antibiot.* 33: 13–17.

Okuhara M, Kuroda Y, Goto T, Okamoto M, Terano H, Kohsaka M, Aoki H, and Imanaka H. 1980b. Studies on new phosphonic acid antibiotics. III. Isolation and characterization of FR-31564, FR-32863 and FR-33289. *J. Antibiot.* 33: 24–28.

Oprea L, Braunack-Mayer A, and Gericke CA. 2009. Ethical issues in funding research and development of drugs for neglected tropical diseases. *J. Med. Ethics* 35: 310–314.

Ortmann R, Wiesner J, Reichenberg A, Henschker D, Beck E, Jomaa H, and Schlitzer M. 2003. Acyloxyalkyl ester prodrugs of FR900098 with improved in vivo anti-malarial activity. *Bioorg. Med. Chem. Lett.* 13: 2163–2166.

Qinhaosu Antimalaria Coordinating Research Group. 1979. Antimalaria studies on qinhaosu. *Chin. Med. J.* 92: 811–816.

Ro DK, Paradise EM, Ouellet M, Fisher KJ, Newman KL, Ndungu JM, Ho KA, Eachus RA, Ham TS, Kirby J, Chang MC, Withers ST, Shiba Y, Sarpong R, and Keasling JD. 2006. Production of the antimalarial drug precursor artemisinic acid in engineered yeast. *Nature* 440: 940–943.

Robert A and Meunier B. 1998. Alkylating properties of antimalarial artemisinin derivatives and synthetic trioxanes when activated by a reduced heme model. *Chem. Eur. J.* 4: 1287–1296.

Rocco F. 2003. *The Miraculous Fever-Tree. Malaria, Medicine and the Cure That Changed the World*. London, U.K.: HarperCollins Publisher.

Rosenthal PJ. 2001. Protease inhibitors. In *Antimalarial Chemotherapy*, ed. PJ Rosenthal, pp. 325–345. Totowa, NJ: Humana Press.

Roth RJ and Acton N. 1989. A simple conversion of artemisinic acid into artemisinin. *J. Nat. Prod.* 52: 1183–1185.

Russian DA and Kovacs JA. 1998. *Pneumocystis carinii*: A fungus resistant to antifungal therapies—Mechanisms of action of antipneumocystis drugs. *Drug Resist. Updat.* 1: 16–20.

Samuelsson G. 2004. *Drugs of Natural Origin*. Kristianstad, Sweden: Apotekarsocieteten.

Schlitzer M. 2007. Malaria chemotherapeutics part I: History of antimalarial drug development, currently used therapeutics, and drugs in clinical development. *Chem. Med. Chem.* 2: 944–986.

Schluter K, Walter RD, Bergmann B, and Kurz T. 2006. Arylmethyl substituted derivatives of fosmidomycin: Synthesis and antimalarial activity. *Eur. J. Med. Chem.* 41: 1385–1397.

Sidhu AB, Valderramos SG, and Fidock DA. 2005. Pfmdr1 mutations contribute to quinine resistance and enhance mefloquine and artemisinin sensitivity in *Plasmodium falciparum*. *Mol. Microbiol.* 57: 913–926.

Singh B, Sung LK, Matusop A, Radhakrishnan A, Shamsul SSG, Cox-Singh J, Thomas A, and Conway DJ. 2004. A large focus of naturally acquired *Plasmodium knowlesi* infections in human beings. *Lancet* 363: 1017–1024.

Sisowath C, Stromberg J, Martensson A, Msellem M, Obondo C, Bjorkman A, and Gil JP. 2005. In vivo selection of *Plasmodium falciparum* pfmdr1 86N coding alleles by artemether-lumefantrine (Coartem). *J. Infect. Dis.* 191: 1014–1017.

The global fund to fight aids, tuberculosis and malaria. 2010. Early evidence of substantial impact on malaria. http://www.theglobalfight.org/view/resources/uploaded/global_fund_impact_on_malaria.pdf, accessed on November.

Trenholme CM, Williams RL, Desjardins RE, Frischer H, Carson PE, Rieckmann KH, and Canfield CJ. 1975. Mefloquine (WR 142,490) in the treatment of human malaria. *Science* 190: 792–794.

Trouiller P, Torreele E, Olliaro P, White N, Foster S, Wirth D, and Pecoul B. 2001. Drugs for neglected diseases: A failure of the market and a public health failure? *Trop. Med. Int. Health* 6: 945–951.

Trumpower BL and Gennis RB. 1994. Energy transduction by cytochrome complexes in mitochondrial and bacterial respiration: The enzymology of coupling electron transfer reactions to transmembrane proton translocation. *Ann. Rev. Biochem.* 63: 675–716.

Turschner S and Efferth T. 2009. Drug resistance in plasmodium: Natural products in the fight against malaria. *Mini-Rev. Med. Chem.* 9: 206–214.

Utzinger J, Xiao SH, Tanner M, and Keiser J. 2007. Artemisinins for schistosomiasis and beyond. *Curr. Opin. Investig. Drugs* 8: 105–116.

Valderramos SG and Fidock DA. 2006. Transporters involved in resistance to antimalarial drugs. *Trends Pharmacol. Sci.* 27: 594–601.

van Es HH, Karcz S, Chu F, Cowman AF, Vidal S, Gros P, and Schurr E. 1994. Expression of the plasmodial pfmdr1 gene in mammalian cells is associated with increased susceptibility to chloroquine. *Mol. Cell Biol.* 14: 2419–2428.

White NJ. 2003. Malaria. In *Manson's Tropical Diseases*, eds. GC Cook and AI Zumla, pp. 1205–1293. China: Elsevier Science Limited.

White NJ. 2008. Qinghaosu (Artemisinin): The price of success. *Science* 320: 330–334.

WHO. 2006. *Guidelines for the Treatment of Malaria*. Geneva, Switzerland: World Health Organization.

WHO. 2008. *Global Malaria Control and Elimination: Report of a Technical Review*. Geneva, Switzerland: World Health Organization.

Wiesner J, Ortmann R, Jomaa H, and Schlitzer M. 2007. Double ester prodrugs of FR900098 display enhanced in vitro antimalarial activity. *Arch. Pharm.* 340: 667–669.

Yeh I and Altman RB. 2006. Drug targets for *Plasmodium falciparum*: A post-genomic review/survey. *Mini. Rev. Med. Chem.* 6: 177–202.

16 Endophytic Microorganisms as a Source of Bioactive Compounds

Yoshihito Shiono and Ken-ichi Kimura

CONTENTS

16.1   INTRODUCTION

Natural products have been discovered from a wide variety of living organisms such as plants, microorganisms, marine lives, and insects; and many of them showed fascinating structures or remarkable biological activities. In fact, 60% of anticancer drugs and 75% of anti-infective drugs developed between 1981 and 2002 are natural products themselves or derivatives developed from natural products (Newman et al., 2003). Substances such as cyclosporine, FK506, mevastatin, micafungin, and avermectin are derived from natural products produced by microbes; and they are major examples that gave a significant impact to basic and clinical researches (Ganesan, 2008). Furthermore, natural products not only contribute greatly to the development of medical drug and agrichemical discovery researches etc., but also play an important role as bioprobes for elucidating life phenomena, studies of which have rapidly advanced recently (Osada, 2000). Thus, compounds derived from natural products are, quite simply, repository of drug types with limitless possibilities in the field of drug discovery research.

In the previous decade, the search for drug discovery from natural sources has been extended to a wide diversity of living species. It has been reported that, out of all the higher plants resources existing on the earth, the species whose biological activity has been investigated occupy a little more than 5%–20% (McChesney et al., 2002), which is a small percentage of all higher plants. As for microbial resources, less than 1% of the entire fungi and less than 5% of true fungi are known, and the rest remain uninvestigated (Abel et al., 2002). Therefore, we still do not fully utilize biological resources existing on the earth. The trend of exploration sources and obtainment methods of biologically active substances from microbes during the last several years include a wide variety of approaches such as obtainment of the substances from bacteria of marine origin (Bugni and Ireland, 2004), symbiotic fungi, unculturable microbes, and unused bacteria whose substance productivity has not been studied (Huang et al., 1995), as well as production of nonnatural type natural products using biosynthetic gene engineering (Yanai et al., 2004, Watanabe et al., 2009) and the use of a new assay system that targets subjects on the molecular level (Rydzewki, 2008). These approaches and methods have all been developed by researchers based on their expectations for the superb substance producing capabilities of microbes in an effort to maximize their potentials. On this basis, we focused this chapter on "endophytic fungi," which are microorganisms whose productivity has been investigated only partly. Thus, we report here selected examples of bioactive products recently isolated by us from cultured mycelia, through the combination of antimicrobial activity, behavior on the thin layer chromatography (TLC) method, and characteristic structural information obtained mainly from NMR spectral measurement. While this method cannot treat many extracts at one time, since it scrutinizes cultured extracts one by one, it can capture substances that are not captured by activity-based screening, thereby possibly leading to the discovery of new substances. Moreover, we have also studied the characteristics of isolated substances by conducting Ca^{2+}-signal transduction inhibition or apoptosis inducing assays (Shitamukai et al., 2000, Nishikawa et al., 2008). These recent results of research and some other examples of research in this field will be reported below.

16.2   ENDOPHYTES

The term "endophyte" was first used by the German scientist De Bary in 1884 to describe endosymbionts, that is, microorganisms living inside plants affording some beneficial biological characteristic to the host. Nowadays, the definition of endophytes has been broadened by many researchers, and the term endophyte applies to a fungal or bacterial microorganism that can reside in the intercellular space of living plant tissues without causing discernible symptoms of plant disease. Endophytes live in various plants, everywhere ranging from the tropical rainforest jungle to cacti in the desert, in a close symbiotic relationship with the plants. In most cases, the endophytes play important roles in plant development and health because they are capable of synthesizing bioactive compounds that can be involved in plants–endophytes relationship and useful for the plant for defense against pathogenic fungi and bacteria, and herbivorous insects (Schulz and Boyle, 2005, Wicklow et al., 2005, Mejia et al., 2008, Vega et al., 2008). However, it has not been fully clarified how endophytes penetrate inside the host, how they live in coexistence with it, and what they do inside it. In addition, once the balance of coexistence is lost, the fungi seem to turn into phytopathogenic species or saprophytes (Basidiomycetes). Endophytic microorganisms have recently been recognized as a rich and untapped source of structurally novel and biologically active secondary metabolites. Indeed, a lot of bioactive and chemically novel compounds were discovered from the culture of these microorganisms in recent years (Strobel et al., 2004, Zhang et al., 2006). Endophytes discussed herein are endophytic fungi living in healthy trees without causing any disease to them.

16.3   SCREENING FOR ENDOPHYTIC METABOLITES

16.3.1   Isolation of Endophytic Fungi

As first step of endophytes collection, a bark sample from Yamagata prefecture was disinfected using 70% ethanol and sodium hypochlorite, and subsequently placed on agar media (Schulz et al., 1993). After a few days, filamentous fungi that developed around the plants were collected on slants. At this point, in addition to the conventional method, by making the nutrient concentration of the isolation media extremely low and focusing attention to slowly growing colonies, isolation of fungi that had not been studied to date was attempted.

16.3.2   Fermentation Media

Numerous kinds of different media in microorganism cultures have been used for the production of biologically active secondary metabolites and enzymes, and it is well known that different culture conditions may allow the production of unprecedented products. In industry, the submerged culture conditions have been used successfully for the production of enzyme and secondary metabolites (Robinson et al., 2001). It is also worth mentioning here that for thousands of years many traditional foods, such as *shoyu*, *miso*, and *tempeh* in oriental countries, are produced under the solid state fermentation. Although the use of solid state fermentation technology has not been commonly employed for industrial applications in the past, more recently solid state culture conditions have proved to be efficient in the production of a number of microbial enzymes and products at the laboratory scale (Singhania et al., 2009). In addition, the solid state fermentation are normally carried out on various agricultural by-products, including millet, barely, wheat straw, rice, maize, rice grains, and corncobs. These agro-industrial wastes are cheap and easily available, so the production of secondary metabolites using solid state fermentation offers an economic advantage over liquid fermentation. Therefore, in our collection, we carried out the investigation of bioactive compounds by endophytic fungi using moist unpolished rice as a solid substrate. An estimated total of over 200 fungal strains isolated from plant materials were fermented with unpolished rice media for 4 weeks at 25°C.

16.3.3 SCREENING

In addition to a microorganism-based biological screening, we employed a chemical screening approach in order to select the fungus producing interesting secondary metabolites. Chemical screening was based on the analysis of the characteristic behavior on TLC after spraying with chromogenic reagents or examination under UV light (Grabley and Thiericke, 1999). The biological screening criterion is focused to define possible pharmaceutical application. Although chemical screening does not have a direct relationship with specific biological activities, it frequently allows discovery of novel, and frequently interesting, secondary metabolites: actually a lot of compounds have been isolated and reported using chemical screening methods, and some of them showed promising biological properties. In order to apply the chemical screening, the crude mycelium extracts were analyzed by TLC, normally eluted with 10% MeOH in CHCl$_3$. In the subsequent step, the secondary metabolites pattern of each strain was analyzed by the Rf value on TLC, the color after spraying with chromogenic reagents and heating at 120°C.

About 200 strains screened in this way were inoculated on brown-rice solid culture media, and then extracts of mold were screened using antimicrobial activity test and/or chemical screening using TLC. This means that we attempted to produce a new biologically active substance by combining new microorganism sources and the characteristic individual culture method. As a result, four filamentous fungi strains, YUA-026, YST-55, KS 37-2, and KS-246 were obtained. Using DNA base sequence analysis, these fungi were identified as xylariaceous endophytic fungus, *Anthracobia* sp., *Phomopsis* sp., and *Neonectria ramulariae* Wollenw, respectively.

16.4   EREMOXYLARINS FROM XYLARIACEOUS ENDOPHYTIC FUNGUS YUA-026

16.4.1   ISOLATION AND STRUCTURE DETERMINATION

The fungus YUA-026 was stationarily cultured at 25°C for 4 weeks in unpolished rice. Following fermentation, the MeOH extract of the moldy unpolished rice was concentrated and partitioned with EtOAc. The purification of metabolites was guided by their antimicrobial activity against *Pseudomonas aeruginosa* and intense blue characteristic coloration with vanillin-sulfuric acid solution on TLC plates. The EtOAc extract was chromatographed on a column of silica gel. Two

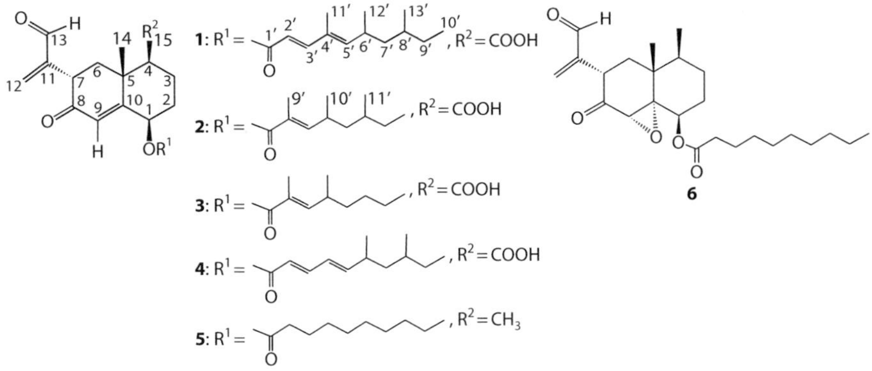

FIGURE 16.1 Structures of **1–6**.

fractions obtained were chosen for further purification using octadecylsilane (ODS) column chromatography to afford new compounds, eremoxylarins A (**1**) and B (**2**) (Shiono and Murayama, 2005) (Figure 16.1).

The molecular formula of **1**, $C_{28}H_{38}O_6$, was determined by high resolution fast atom bombardment mass spectrometry (HRFABMS) measurement. The IR spectrum exhibited bands at 1733, 1718, 1685, and 1653 cm^{-1}, characteristic of multiple carbonyl groups. ^{13}C-NMR (δc 176.6) of **1** and formation of a monomethyl ester by treatment with trimethylsilyldiazomethane supported the presence of a carboxyl group. Its UV spectrum showed an absorption maximum at 269 nm, which suggested the presence of a conjugated dienoyl moiety. The gross structure of **1** was deduced from detailed analyses of ^{1}H (400 MHz, CD$_3$OD) and ^{13}C-NMR (100 MHz, CD$_3$OD) data aided by 2D NMR experiments. The ^{13}C-NMR spectrum showed a total of 28 carbons atoms, and their multiplicity assignments using distortionless enhancement by polarization transfer (DEPT) established the presence of five methyls, five sp^3 methylene, five sp^3 methines, one sp^3 quaternary carbon, three sp^2 quaternary carbons, four sp^2 methines, one sp^2 methylene, one ester, one conjugated ketone, one carboxylic acid, and one aldehyde. The 10 unsaturation equivalents required by the molecular formula indicated this compound has two rings. The ^{1}H-NMR and heteronuclear multiple quantum coherence (HMQC) spectra of **1** revealed the presence of one primary methyl [δ_H 0.84 (3H, t, $J = 7.3$ Hz, H$_3$-10′)], two secondary methyls [δ_H 0.83 (3H, d, $J = 6.8$ Hz, Me-13′) and 0.97 (3H, d, $J = 6.8$ Hz, Me-12′)], one tertiary methyl [δ_H 1.48 (3H, s, Me-14)], one methyl attached to an olefinic carbon [δ_H 1.81 (3H, s, Me-11′)], one disubstituted [δ_H 5.80 (1H, d, $J = 15.6$ Hz, H-2′) and 7.31 (1H, d, $J = 15.6$ Hz, H-3′)] and two trisubstituted double bonds [δ_H 6.02 (1H, s, H-9) and 5.70 (1H, d, $J = 9.8$ Hz, H-5′)], vinylidene protons [δ_H 6.29 (1H, s, H-12) and 6.42 (1H, s, H-12)], five methines [δ_H 1.23–1.36 (overlapped, H-8′), 2.44 (1H, dd, $J = 13.2, 2.9$ Hz, H-4), 2.69 (1H, m, H-6′), 3.73 (1H, dd, $J = 14.5, 5.5$ Hz, H-7), 5.49 (1H, br. s, H-1)]. Detailed analyses of the ^{1}H-^{1}H correlation spectroscopy (COSY) spectrum established the presence of partial structures (Figure 16.2).

A combination of HMQC and heteronuclear multiple bond connectivity (HMBC) experiments let us conclude that **1** may be an eremophilane sesquiterpene skeleton substituted by a decadienoic acid. The HMBC correlations between Me-11′ and

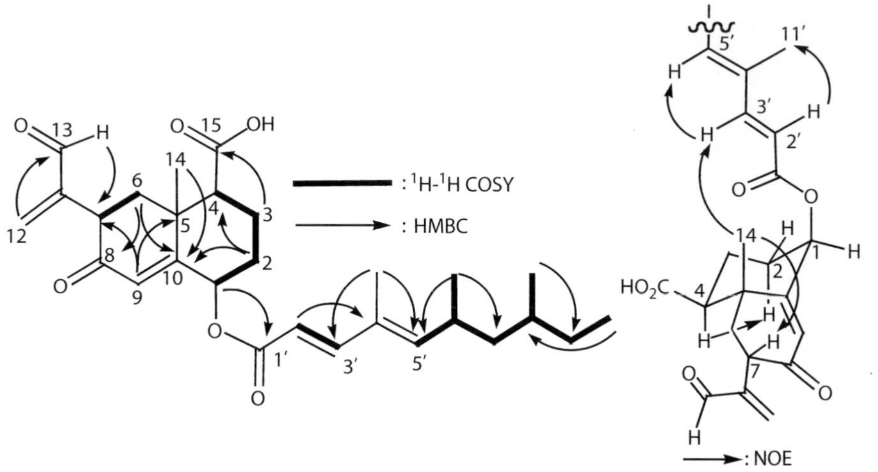

FIGURE 16.2 Important ¹H-¹H COSY, HMBC correlations, and selected NOEs observed for **1**.

C-3′ and between Me-11′ and C-5′ suggested that there is a 4,6,8-trimethyldeca-2,4-dienoyl moiety in this molecule. In fact, acidic hydrolysis of **1** with 6 N HCl, followed by methylation with trimethylsilyldiazomethane afforded methyl 4,6,8-tri-methyldeca-2,4-dienoate (Figure 16.3).

Furthermore, the HMBC correlation between H-1 and C-1′ and substantial downfield shift for H-1 revealed that the location of the decadienoyl was at C-1. Thus, the planar structure of eremoxylarin A was assigned as **1**. The relative configurations of C-1, C-4, C-5, and C-7 in **1** were deduced from nuclear Overhauser effect (NOE) experiments in CD₃OD (Figure 16.2). NOE correlations from Me-14 to H-3′, Me-14 to H-7 indicated that Me-14, H-7 and ester side chain moiety at C-1 were all β-oriented. Furthermore, NOE correlations were observed from H-4 to H-2α, suggesting that the carboxylic acid moiety was β-oriented. The configurations of the C-7 and C-4 were also supported by the coupling constants between H₂-6 and H-7, and H₂-3 and H-4, respectively. However, the configurations of methyl group at C-6′ and C-8′ could not be determined by NOE experiments.

Eremoxylarin B (**2**) had a molecular formula of $C_{26}H_{36}O_6$, based on HRFABMS. The IR absorption bands were very similar to those of **1**. The presence of a carboxyl group was also supported by the methylation of **2** with trimethylsilyldiazomethane giving a mono methyl ester. The ¹H- and ¹³C-NMR data for **2** were very similar to those of **1** except for the signals of substitution group at C-1. Identification of the

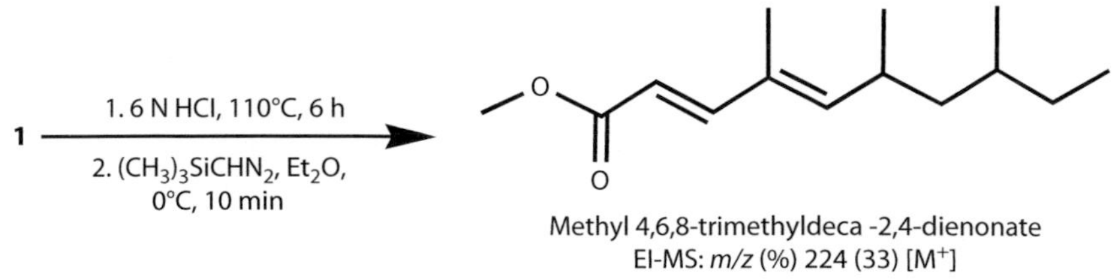

Methyl 4,6,8-trimethyldeca -2,4-dienonate
EI-MS: *m/z* (%) 224 (33) [M⁺]

FIGURE 16.3 Hydrolysis of **1**.

substitution group in **2** as 2,4,6-trimethyloct-2-enoyl was deduced from data of ^{1}H-^{1}H COSY, HMBC, and NOE experiments. In addition, hydrolysis of **2** with 6 N HCl, followed by methylation with trimethylsilyldiazomethane afforded methyl-2,4,6-trimethyloct-2-enoate. The esterification site was confirmed by an HMBC correlation from H-1 to the ester carbonyl at δ_C 169.0. Based on the NOE difference experiments, the relative stereochemistry of eremophilane sesquiterpene moiety in **2** were determined. The configuration of the trisubstituted $\Delta^{2'}$ double bond was assigned as *E* on the basis of the NOE correlation between Me-9′ and H-4′. The relative configurations at C-4′ and C-6′ remains uncertain.

16.4.2 Biological Activity

The minimum inhibitory concentration (MIC) of **1** and **2** was 12.5 and 25 μg/mL against *Staphylococcus aureus*, and 6.25 and 12.5 μg/mL against *Pseudomonas aeruginosa*, respectively. However, **1** and **2** had little or no activity against *Candida albicans* and *Aspergillus clavatus*. Although generally many eremophilane-type sesquiterpens are reported from plant sources, several compounds structurally related with eremoxylarins have been isolated from xylariaceous fungus. For example, integric acid (**3**), an HIV-1 integrase inhibitor, has been reported from *Xylaria* sp. (Singh et al., 1999), the cytotoxic agent 07H239-A (**4**), selective for the NPY Y5 receptors, was obtained from the marine-derived xylariaceous fungus LL-07H239 (McDonald et al., 2004), and xylarenals A (**5**) and B (**6**) were isolated from *Xylaria persicaria* (Smith et al., 2002) (Figure 16.1).

As described in Introduction, one promising research area at the interface between chemistry and biology is the use of biologically active natural products (bioprobes) for the exploration of cell biology. It is important for finding a unique natural product used in chemical biology to select appropriate natural sources and a biochemical or cell-based screening system. We used endophytic microorganisms as natural sources and the mutant yeast as screening tools for isolation of unique compounds.

The Ca^{2+}-signal transduction pathways have important roles in the regulation of diverse cellular processes such as T-cell activation, muscle contraction, neurotransmitter release, and secretion (Clapham, 1995). The budding yeast *Saccharomyces cerevisiae* has been especially useful for isolation of mutants. Genetic- and molecular genetic studies by using the budding yeast greatly contributed to understand the cell-cycle regulation in eukaryotes. In addition to its characteristics as a model eukaryotic cell for molecular- and cell-biological research, yeast has technical advantages such as simple growth conditions, rapid cell division, and the development of a wealth of genetic tools for analysis of biological functions. These characteristics have expanded the application of yeast as screening tool to the field of drug discovery (Simon and Bedalov, 2004). Thus, the bioactive compounds isolated by the yeast screening system may be applicable to mammalian cell systems, especially to humans. The only cause for a growth defect in the G2 phase of the *zds1Δ* cells of *Saccharomyces cerevisiae* in medium with CaCl$_2$ depends on the hyperactivation of cellular Ca^{2+}-signal (Mizunuma et al., 1998). The inhibitors of Ca^{2+}-signal transduction are detected by their ability to stimulate the growth of the cells as a growth zone around a paper disc containing the active compound (Figure 16.4) (Shitamukai et al., 2000, Miyakawa and Mizunuma, 2007).

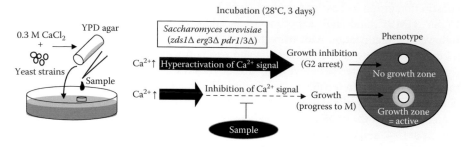

FIGURE 16.4 Phenotypic screening method using the mutant yeast strain.

However, the drug permeability through the cell membrane of *S. cerevisiae* is not especially good. Thus, we improved a drug-sensitive strain with disrupted genes of *erg3*, *pdr1*, and *pdr3* except *zds1* (called *zds1Δ erg3Δ pdr1Δ pdr3Δ* strain as "the mutant yeast" in this section) and used for the evaluation of Ca²⁺-signal transduction inhibitor (Figure 16.5) (Chanklan et al., 2008, Ogasawara et al., 2008).

The Ca²⁺-signaling pathways for growth regulation (cell cycle) are composed of several signaling molecules such as the Ca²⁺ channel (target of antihypertension drugs), calcineurin (CN) (target of immunosuppressant drugs), Pkc1 protein kinase C (target of anticancer drugs), Mpk1 MAPK (mitogen-activated protein kinase) (target of anticancer drugs), and Mck1 GSK-3 (target of antidiabetes and Alzheimer's disease drugs). In fact, calcineurin inhibitors FK506 and cyclosporine A, which are important clinical medicines as immunosuppressants, showed a growth zone of the cells in this screening (Ogasawara et al., 2008). The inhibition mechanism of FK506 and cyclosporine A to calcineurin is remarkable, because both compounds are bound by their respective binding proteins (immunophilins), named FKBP12 and cyclophilin A, and their complexes inhibit calcineurin, leading to suppressed T cell activation

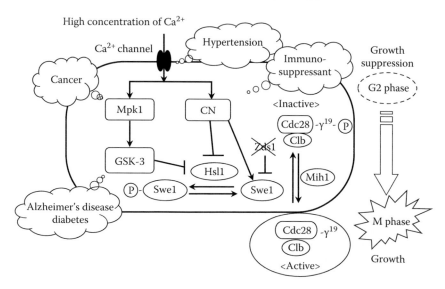

FIGURE 16.5 Ca²⁺-signal transduction and the molecular target in the mutant yeast.

(Liu et al., 1991, Mann, 2001). In addition, we have already found that the Ca^{2+} channel blocker diltiazem showed a growth zone at high concentration (4 mg/disc) and the HSP90 inhibitor radicicol showed a growth zone at low concentration (1.6 μg/disc). The synthetic GSK-3βinhibitor, GSK-3β inhibitor I, showed a growth zone around a clear inhibitory zone. Conversely, calcineurin inhibitors of clinical immunosuppressants FK506 and cyclosporine A showed only a growth zone without an inhibitory zone against the mutant yeast at low concentration (Ogasawara et al., 2008).

In many molecular targets, calcineurin is a Ca^{2+}/CaM-dependent serine/threonine protein phosphatase and is a fascinating drug target for immunosuppressants and antiinflammation drugs. To find a new type of calcineurin inhibitor different from FK506 and cyclosporine A, we used a unique screening system and found **1** and **2** have calcineurin inhibition activity (Ogasawara et al., 2008).

The phenotype of a growth zone has a different character depending on each molecular target and/or inhibitor. Under the same conditions, **1** and **2** showed restored growth activity against the mutant strain (*zds1Δ erg3Δ pdr1Δ pdr3Δ*) in a dose-dependent manner with a small inhibition zone (Figure 16.6). Eremoxylarin A (**1**) has less potent activity than that of **2**.

The cell-cycle regulation by Ca^{2+} is executed through the activation of two parallel pathways, calcineurin and the Mpk1 MAPK cascade, and the deletion of both genes showed a lethal phenotype (Nakamura et al., 1996). Isogenic strains differing only in the presence or absence of functional calcineurin or MAPK were tested for sensitivity to eremoxylarins in growth inhibition assays. Eremoxylarins A (**1**) and B (**2**) showed the growth inhibition zone against the *mpk1Δ* strain specifically having the same character as FK506. It was shown that both compounds apparently acted on the pathway of calcineurin (Figure 16.7).

Mutations in the catalytic or regulatory subunits of calcineurin caused a defect of tolerance to salts such as NaCl and LiCl (Nakamura et al., 1993). Using this character, calcineurin could be confirmed as the sample target. As we expected, eremoxylarins showed an inhibition zone to the wild type of *S. cerevisiae* with 0.16 M LiCl and it is similar in character to FK506. The wild-type strain exposed to eremoxylarins

FIGURE 16.6 Growth restored activity of **1** against the mutant yeast. No. 1–6: 40, 20, 10, 5.0, 2.5, 1.3 μg/disc, respectively; 7: FK506 0.02 μg/disc.

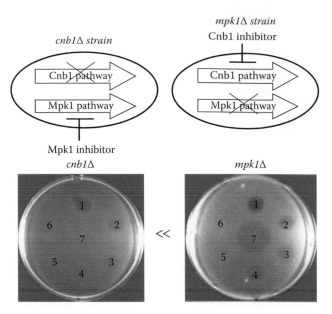

FIGURE 16.7 Presumption of the molecular target of **1** using the character of synthetic lethal effect. No. 1–6: 10, 5.0, 2.5, 1.3, 0.63, 0.31 μg/spot, respectively; 7: FK 506 0.1 μg/spot.

and 0.16 M LiCl lost the salt tolerance activity; therefore, it was shown that both compounds apparently acted on calcineurin. The immunosuppressant drugs FK506 and cyclosporine A bind to specific immunophilins and their complexes selectively inhibit calcineurin, leading to the suppression of T-cell proliferation (Liu et al., 1991). Thus, using the deletion mutant strain of each immunophilin, *fkb1Δ* and *cph1Δ* with the combined character of *zds1Δ erg3Δ*, it can be examined whether eremoxylarins need any immunophilin or not for the calcineurin inhibition. The immunosuppressant FK506 (0.02 μg/disc) showed a clearly restored growth zone only to the strain of *zds1Δ erg3Δ cph1Δ* and the immunosuppressant cyclosporine A (40 μg/disc) showed a clearly restored growth zone only to the strain of *zds1Δ erg3Δ fkb1Δ*. Eremoxylarins A (**1**) (40 μg/disc) and B (**2**) (80 μg/disc) showed faint growth zones compared to those of FK506 and cyclosporine A on both plates. These results also support the idea that eremoxylarins inhibit calcineurin directly without known immunophilin FKB1 and CPH1 in *S. cerevisiae*.

To determine the molecular target, the inhibition activity of **1** and **2** on calcineurin in vitro was directly examined. Eremoxylarins A (**1**) and B (**2**) inhibited calcineurin in a dose-dependent manner and showed $IC_{50} = 2.7$ and $1.4 \mu M$, whereas the control compound trifluoperazine (calmodulin antagonist) showed $IC_{50} = 20.8 \mu M$. The inhibition mechanisms of both **1** and **2** was competitive inhibition against calcineurin with a synthetic substrate in the Dixon plot ($Ki = 1.1$ and $0.7 \mu M$, respectively) as shown in Figure 16.8.

Although an increase in calmodulin could affect the inhibition activity of the calmodulin antagonist trifluoperazine, it did not affect the inhibition activities of **1** and **2**. FK506 and cyclosporine are excellent therapeutic immunosuppressants, but

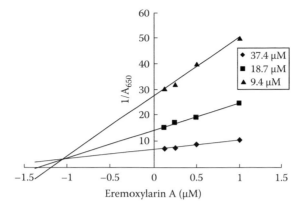

FIGURE 16.8 Dixon plot of **1** against PP2B (calcineurin).

those compounds alone cannot inhibit calcineurin and each needs an immunophilin that is a binding protein for this inhibition. Because immunophilins have been shown to be involved in various other biological processes, some of the side effects of FK506 and cyclosporine might be caused by the inhibition of these processes. However, a direct and selective inhibitor has not been reported so far (Baba et al., 2003). Eremophilane sesquiterpenoid compounds belong to a different type with respect to the known compound FK506 (macrolide), cyclosporine A (cyclic peptide), and cantharidin derivatives. Eremoxylarins A (**1**) and B (**2**) were proved to inhibit calcineurin without any immunophilin and showed potent inhibition activity toward human calcineurin. Although they have other biological activities such as antimicrotubule, HIV-1 integrase inhibition, and antitumor activities, which have already reported in related compounds (Singh et al., 1999, Smith et al., 2002, McDonald et al., 2004), the eremophilane sesquiterpenoid structure of eremoxylarins could represent a good lead compound for the development of immunosuppressants and antiallergy drugs.

16.5 ANTHRACOBIC ACIDS FROM *ANTHRACOBIA* SP. YST-55

16.5.1 ISOLATION AND STRUCTURE DETERMINATION

The fermented unpolished rice inoculated with *Anthracobia* sp. YST-55 was soaked in MeOH. The purification of these metabolites was guided by their antimicrobial activity and intense blue characteristic coloration with vanillin-sulfuric acid solution on TLC plates. The MeOH extract was purified by silica gel and ODS column chromatography to afford anthracobic acids A (**7**) and B (**8**) (Figure 16.9) (Shiono, 2006). Both compounds were unstable under light and aerobic conditions at room temperature and their color gradually turned brown.

Anthracobic acid A (**7**) was obtained as yellow powder with molecular formula of $C_{25}H_{32}O_4$ as determined by HRFABMS. The UV spectrum of **7** showed the strong absorption at 337 nm in MeOH due to a conjugated olefine system. IR absorption bands at 3402 and 1716 cm^{-1} implied the presence of a carboxyl group. Furthermore, a ^{13}C-NMR peak at δ_C 169.4 (100 MHz, C_5D_5N) as well as the formation of a mono methyl ester by treatment with trimethylsilyldiazomethane also

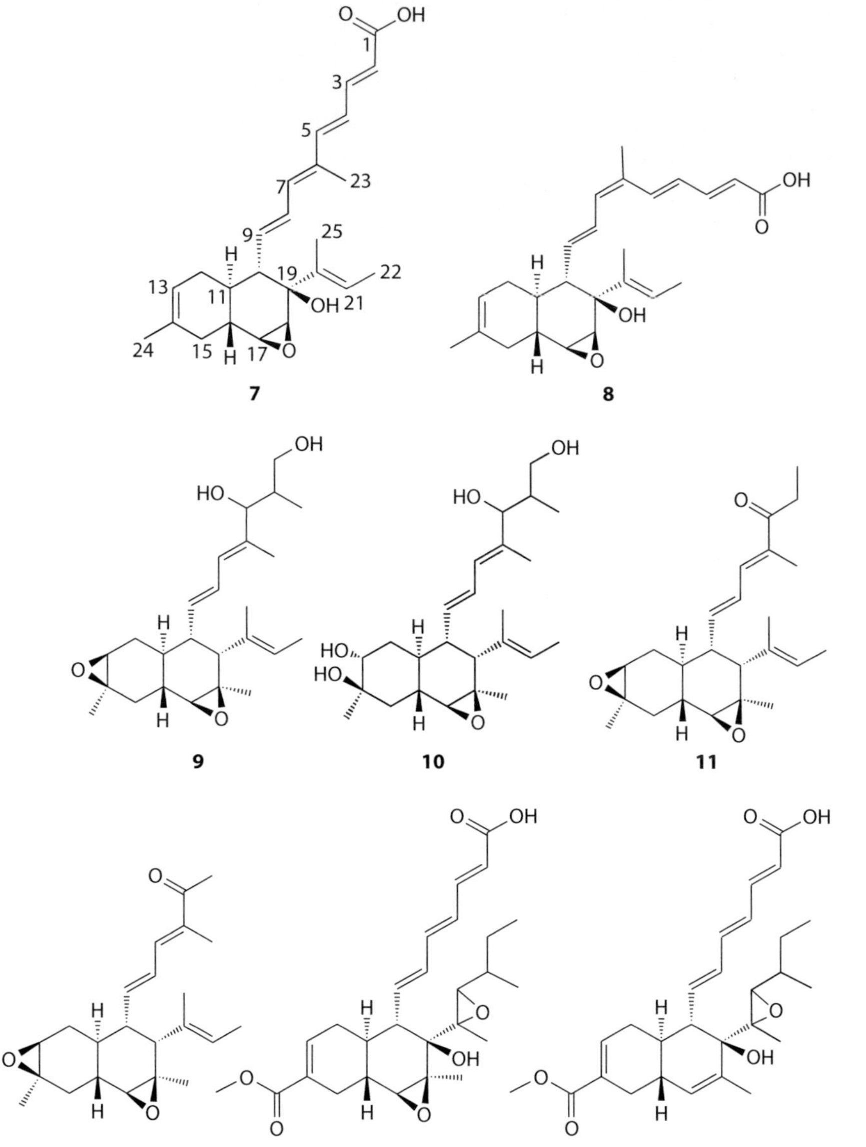

FIGURE 16.9   Structures of **7–14**.

supported the presence of a carboxyl group. The ^{13}C-NMR spectrum of **7** showed altogether the presence of 25 carbons and confirmed the molecular formula. A combined analysis of both the DEPT and the HSQC spectra together indicated the presence of four methyls, two methylenes, nine olefinic methines, three methines, two oxygenated methines, an oxygenated quaternary carbon, three olefinic quaternary carbons, a carboxyl carbon. The ^{1}H-NMR spectrum (400 MHz, C$_5$D$_5$N) showed signals for four methyl groups at δ_H 1.62 (3H, s, Me-24), 1.79 (3H, s, Me-23),

1.96 (3H, s, Me-25) and a doublet at δ_H 1.71 (3H, d, $J = 6.8$ Hz, Me-22), two oxygenated methines at δ_H 3.23 (1H, d, $J = 3.4$ Hz, H-17) and 3.47 (1H, d, $J = 3.4$ Hz, H-18), three methines at δ_H 1.63–1.67 (1H, m, H-11), 1.82–1.89 (1H, m, H-16) and 2.66 (1H, t, $J = 10.7$ Hz, H-10), nine olefinic methines at δ_H 5.43 (1H, br. s, H-13), 5.65 (1H, dd, $J = 15.2, 10.7$ Hz, H-9), 5.68–5.73 (1H, m, H-21), 6.32 (1H, d, $J = 15.1$ Hz, H-2), 6.43 (1H, d, $J = 11.2$ Hz, H-7), 6.53 (1H, dd, $J = 15.1, 11.2$ Hz, H-4), 6.62 (1H, dd, $J = 15.2, 11.2$ Hz, H-8), 6.74 (1H, d, $J = 15.1$ Hz, H-5) and 7.78 (1H, dd, $J = 15.1, 11.2$ Hz, H-3), two methylenes at δ_H 1.50 (1H, t, $J = 14.7$ Hz, H-12), 1.75–1.82 (1H, m, H-12) and 1.97–2.04 (2H, m, H-15). Through analysis of the ^{1}H-^{1}H COSY spectrum, it was demonstrated that four partial structures, represented by thick lines in Figure 16.10, were included in the structure of **7**. Connectivities of these partial structures were also defined on the basis of the HMBC correlations also shown in Figure 16.10.

The configuration of the double bonds and the relative stereochemistry in **7** were established by analysis of ^{1}H-^{1}H coupling constants, and NOE difference experiments. The 1,3-diaxial relationship between H-10 and H-16 was determined by NOE correlations from H-10 to H-6 and the large spin coupling constants of H-10 ($J_{10,11} = 10.7$ Hz). Furthermore, NOE correlations were observed from H-21 to H-17 and H-18, from H-17 to H-21 and from H-10 to H-8, indicating that the epoxy group at C-17 and C-18 and OH-19 were β-oriented. The geometrical configurations were determined to be 2 E, 4 E, and 8 E by the large coupling constants ($J_{2,3} = 15.1$ Hz, $J_{4,5} = 15.1$ Hz, and $J_{8,9} = 15.2$ Hz). The E stereochemistry of trisubstitute Δ^6 double bond is based on the NOE correlations from Me-23 to H-4 and H-8 and from H-7 to H-5. In addition, an NOE correlation between Me-22 and Me-25 was observed, indicating that the stereochemistry of the double bond Δ^{20} was also assigned to be E (Figure 16.10).

The molecular formula of anthracobic acid B (**8**), $C_{25}H_{34}O_4$, was determined by HRFABMS, indicating that **8** had the same molecular formula as that of **7**. The UV

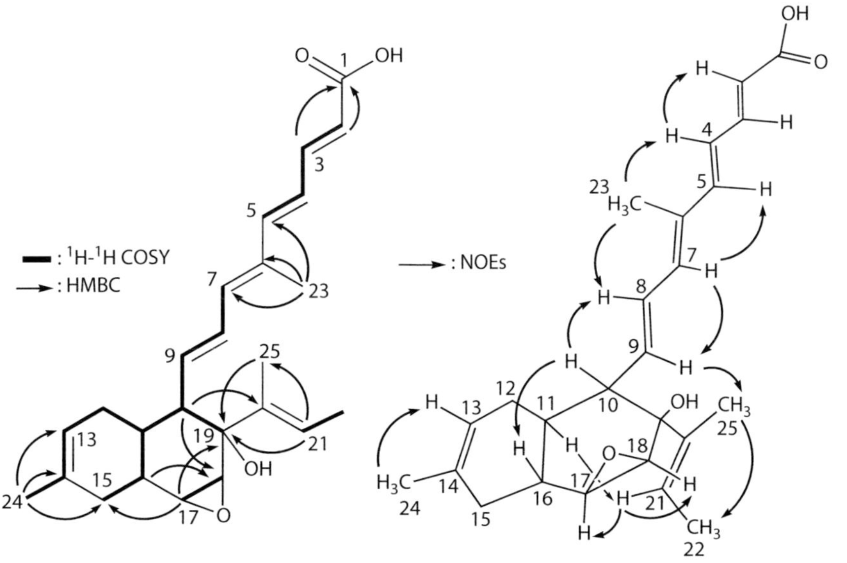

FIGURE 16.10 Important ^{1}H-^{1}H COSY, HMBC correlations, and selected NOEs observed for **7**.

and IR spectra were closely similar to that of **8**. Thus, **8** was considered to be an isomer of **7**. The ¹H-NMR spectrum of **8** was analyzed by means of a careful comparison of its ¹H-¹H COSY, HSQC, and HMBC with those of **7**. Clear differences were found in the chemical shifts around the tetraene moiety (H-1 to H-9) of **8**. In the NOESY experiment, the strong NOE interactions was observed between H-2 and H-4, between H-3 and H-5, between H-5 and H-8, between H-7 and H-9 and between Me-23 and H-8 and between H-5 and H-7, and the interaction between Me-23 and H-6 disappeared. Therefore, **8** was determined to be a configurational *cis* isomer of **7** at C-6, C-7 double bond.

16.5.2 BIOLOGICAL ACTIVITY

Anthracobic acids A (**7**) and B (**8**) were evaluated by paper-disc diffusion method against gram-positive and gram-negative bacteria, yeast, and fungus strains at the concentration of 40 µg/disc. Compounds **7** and **8** showed inhibition zones of diameter against bacteria as follow: *Staphylococcus aureus*; (**7**:18 mm, **8**:18 mm), *Bacillus subtilis*; (**7**:11 mm, **8**:18 mm), and *Escherichia coli*; (**7**:11 mm, **8**:18 mm). None of **7** and **8** showed activity against *Candida albicans* and *Mucor miehei*.

Anthracobic acids A (**7**) and B (**8**) have the same carbon skeleton as the fusarielins A (**9**) and B (**10**) (Kobayashi et al., 1995), ICM0301 A (**11**) and B (**12**) (Kumagai et al., 2004), and F2928-1 (**13**) and -2 (**14**) (Kanai et al., 2005). Fusarielins and F2928s are antibiotics isolated from the cultures of a *Fusarium* sp. and *Cladobotryum* sp., respectively (Figure 16.9).

These compounds possess a functionalized decalin moiety in their structures and in particular **9** and **11** differ from **7** and **8** in having one more epoxide at C-13, C-14, and a modification at the tetraene moiety at C-10. Due to our interest in the discovery of new natural products exhibiting Ca^{2+}-signal transduction inhibitory activity, we decided to apply **7** and **8** to the mutant yeast assay. Under the same screening system as eremoxylarins, **7** showed restored growth activity against the mutant strain in a dose-dependent manner with a clear inhibition zone (Figure 16.11). This phenotype is similar to those of eremoxylarins (Figure 16.6).

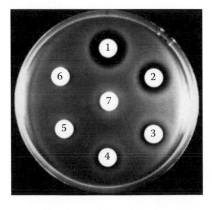

FIGURE 16.11   Growth-restored activity of **7** against the mutant yeast. No. 1–6: 40, 20, 10, 5.0, 2.5, 1.3 µg/disc, respectively; 7: FK506 0.02 µg/disc.

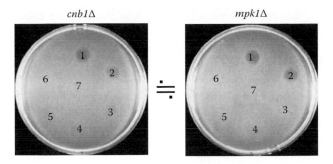

FIGURE 16.12 Presumption of the molecular target of **7** using the character of synthetic lethal effect. No. 1–6: 10, 5.0, 2.5, 1.3, 0.63, 0.31 µg/spot, respectively; 7: FK506 0.1 µg/spot.

Isogenic strains differing only in the presence or absence of functional calcineurin or MAPK were tested for sensitivity to **7** in growth inhibition assays. It showed the growth inhibition zone against both *cnb1Δ* and *mpk1Δ* strains (Figure 16.12) different from the character of eremoxylarins (Figure 16.7). We are currently analyzing the molecular target of this compound.

16.6 BENZOXEPIN METABOLITES FROM *PHOMOPSIS* SP. KS-37-2

16.6.1 ISOLATION AND STRUCTURE DETERMINATION

Two benzoxepin metabolites, **15** and **16** were obtained from the extract of a rice culture of the endophytic fungus *Phomopsis* sp. KS 37-2 isolated from the stem of a cherry tree in Yamagata, Japan, and selected through the chemical screening. One of the metabolites was identified as the known xylarinol A, on the basis of its spectral data that were indistinguishable from those of xylarinol A (**15**) (Figure 16.13).

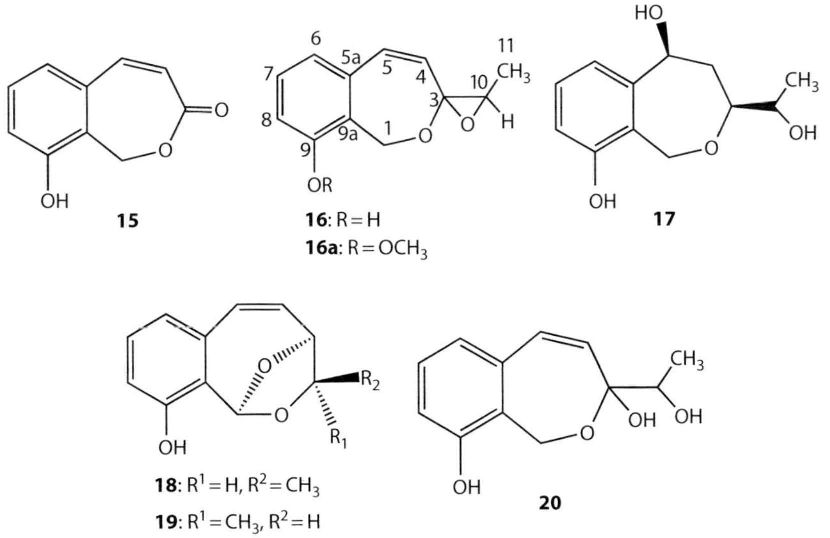

15

16: R = H
16a: R = OCH$_3$

17

18: R^1 = H, R^2 = CH$_3$
19: R^1 = CH$_3$, R^2 = H

20

FIGURE 16.13 Structure of **15–20**.

This compound has recently been isolated from the fruiting bodies of *Xylaria polymorpha* and characterized as a radical scavenger (Lee et al., 2009).

The other metabolites resulted to be a new compound, which was thus named benzophomopsin A (**16**) (Figure 16.13) (Shiono et al., 2009). The molecular formula of **16** was $C_{12}H_{12}O_3$, which required 7 degrees of unsaturation, as revealed by HRFABMS. The UV spectrum of **16** exhibited the presence of aromatic ring. The IR spectrum showed absorption bands at 3384, 1584, and 1465 cm^{-1} indicating the presence of the hydroxyl and aromatic groups. Formation of a monomethoxyl derivative **16a** [$C_{13}H_{14}O_3$ (FAB-MS: *m/z* 241 [M + Na]$^+$); δ_H 3.83 (3H, s, OMe)] after treating **16** with trimethylsilyldiazomethane confirmed the presence of a phenolic hydroxyl group.

The ^{13}C NMR (100 MHz, CDCl$_3$) spectrum showed the presence of 12 carbon signals, and analysis of the DEPT experiment revealed that the ^{13}C NMR signals consisted of six methines, four quaternary carbons, one methylene, and one methyl group. The seven unsaturation equivalents implied by the molecular formula indicated that **16** has three rings. The ^{1}H NMR (400 MHz, CDCl$_3$) spectrum of **16**, analyzed using ^{1}H-^{1}H COSY and HMQC coupling correlations, indicated the presence of a vicinal sp^2 spin network [δ_H 6.82 (1H, d, $J = 7.6$ Hz, H-8), 6.89 (1H, d, $J = 7.6$ Hz, H-6), 7.09 (1H, t, $J = 7.6$ Hz, H-7)], a two protons of a *cis* double bond [δ_H 5.95 (1H, d, $J = 12.7$ Hz, H-4) and 6.82 (1H, d, $J = 12.7$ Hz, H-5)], an isolated oxymethylene [δ_H 4.63 (1H, d, $J = 13.9$ Hz, H-1), 5.12 (1H, d, $J = 13.9$ Hz, H-1)], a doublet methyl group [δ_H 1.05 (3H, d, $J = 6.5$ Hz, Me-11)], and an oxymethine group [δ_H 4.15 (1H, q, $J = 6.5$ Hz, H-10)]. The aromatic moiety in **16** was confirmed on the basis of HMBC correlations from H-6 and H-8 to C-9a and from H-7 to C-5a and C-9 (Figure 16.14).

HMBC correlations from H-1 to C-3, C-5a, and C-9, from H-4 to C-10, and from H-5 to C-3, C-6, and C-9a give rise to presence of a 1,3-dihydro-benzo[*c*]oxepine moiety in **16**. The presence of an epoxy group at C-3 and C-10 was determined from the molecular formula, the chemical shifts of the ^{1}H and ^{13}C NMR signals at these positions, and the large $^1J_{CH}$ values for C-10 ($J_{C-10,H-10} = 153$ Hz) (Someno et al., 2004), requiring C-9 to bear free hydroxyl group. The relative stereochemistry of **16** was deduced from NOE experiments. An NOE was observed between H-4 and Me-11. Thus, the structure of benzophomopsin A was established as shown in **16** (Figure 16.13).

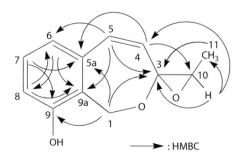

FIGURE 16.14   HMBC correlations of **16**.

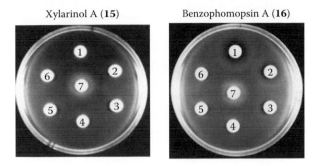

FIGURE 16.15 Growth-restored activity of **15** and **16** against the mutant yeast. No. 1–6: 40, 20, 10, 5.0, 2.5, 1.3 μg/disc, respectively; 7: FK 506 0.02 μg/disc.

16.6.2 Biological Activity

Benzophomopsin A (**16**), which possesses benzo-fused oxacycloalkene moiety of polyketide origin, was closely related to xylarinol B (**17**), cladoacetals A (**18**) and B (**19**), and heptacyclosordariolone (**20**) that were previously isolated from cultures of *Xylaria polymorpha* (Lee et al., 2009), an undetermined fungicolous hyphomycete (Höller et al., 2002), and *Sordaria macrospora* (Bouillant et al., 1989), respectively (Figure 16.13). The differences between **16** and these known precedents involved the presence of the variations in position of oxidation, and the identities of the functional group present. Since it has been reported that **18** showed the antimicrobial activity against *Staphylococcus aureus*, **15** and **16** did not show the antimicrobial activity. The Ca^{2+}-signal transduction inhibitory activity test has also been examined on **15** and **16**.

Benzophomopsin A (**16**) is characterized by possessing an epoxide ring, differently from the known compound **15**. As shown in Figure 16.15, **16** has growth restoring activity against the mutant yeast dose dependently, but **15** has no activity at the same concentration. This phenotype is a little different from those of eremoxylarins and anthracobic acid A. Benzophomopsin A (**16**) shows the growth inhibition zone against *mpk1Δ* strain, but it has salt tolerance activity on the wild type strain with 0.16 M LiCl. We think that these structure-activity relationships and the molecular target of **16** are of some interest.

16.7 PYRROSPIRONES AND PYRROCIDINES FROM *NEONECTRIA RAMULARIAE* WOLLENW KS-246

16.7.1 Isolation and Structure Determination

Chromatographic separation of the extract of unpolished rice media fermented with *Neonectria ramulariae* Wollenw KS-246 resulted in the isolation of four compounds. Two of them seemed to be new compounds, which were thus named pyrrospirones A (**21**) and B (**22**), and another two were identified as previously known compounds, pyrrocidine(s) A (**23**) and B (**24**) (Figure 16.16) (Shiono et al., 2008).

Pyrrocidines A (**23**) and B (**24**) were identified on the basis of their high resolution electron ionization mass spectra (HREIMS) and ^{1}H-, ^{13}C, and 2D NMR data, which

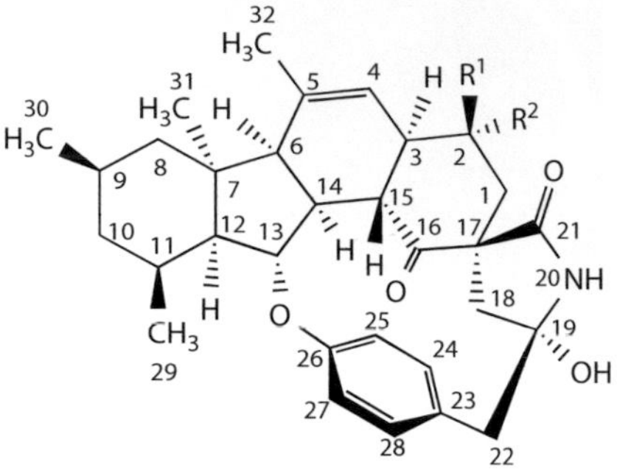

FIGURE 16.16 Structures of **21–31**.

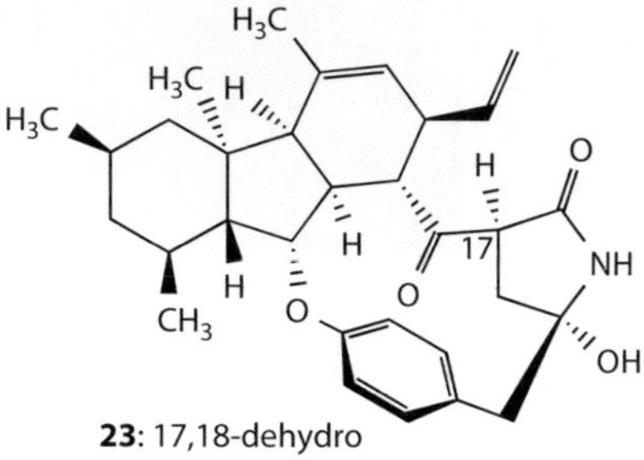

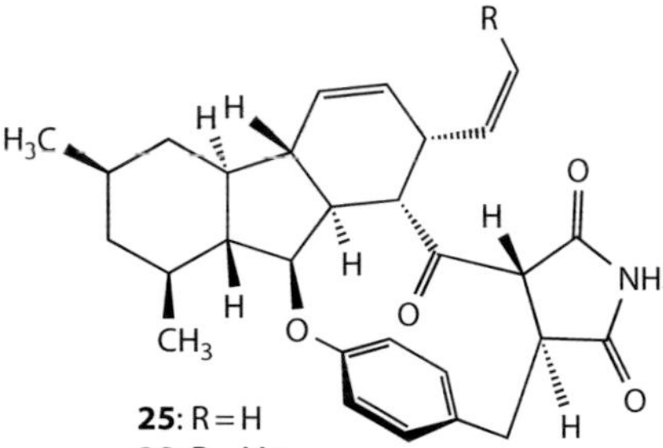

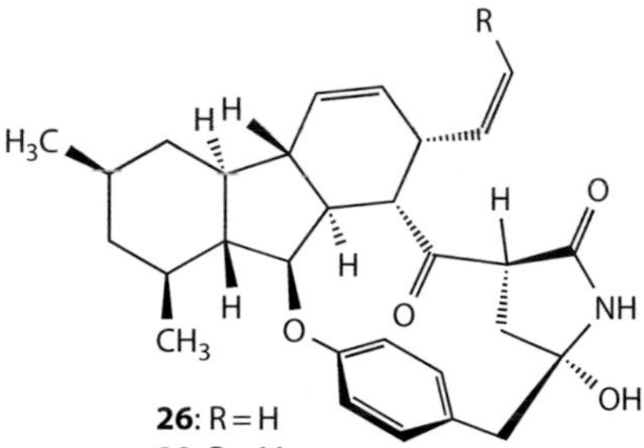

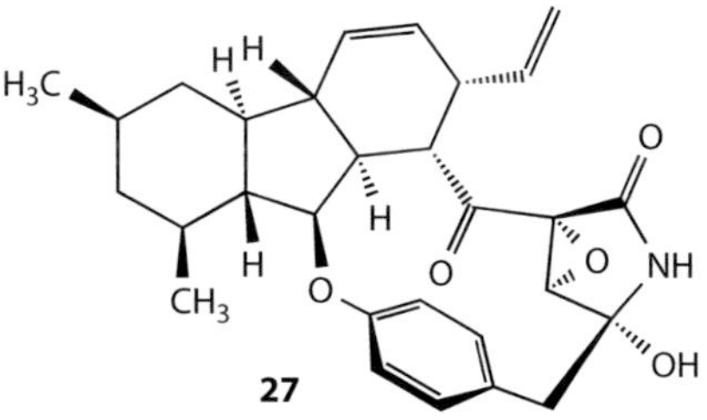

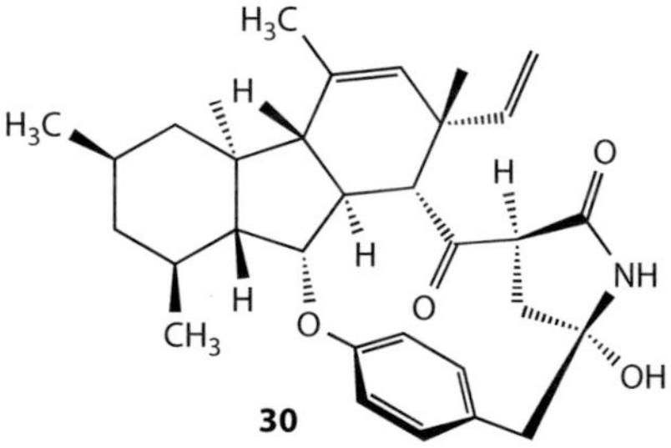

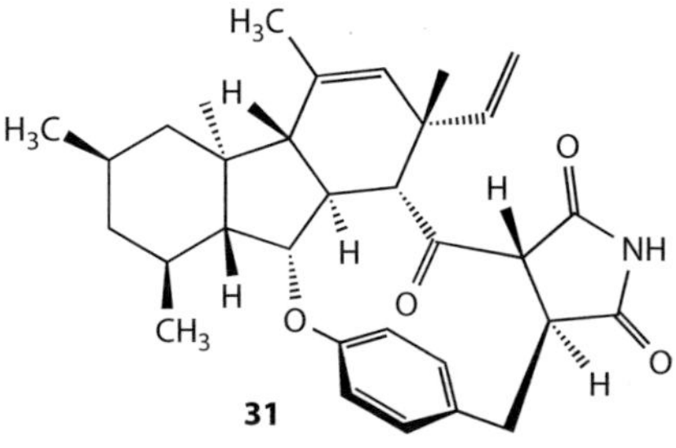

resulted identical to those previously reported in the literature (He et al., 2002). In addition, the antimicrobial activities against gram-positive bacteria of these compounds were also reported.

Pyrrospirone A (**21**) has the molecular formula $C_{31}H_{39}NO_5$ as determined by HRFABMS. The IR spectral data showed absorption bands at 1716 (CO), 1671 (NHCO), 1606, and 1508 cm^{-1} (aromatic ring). The UV spectrum exhibited an absorption maximum at 229 nm, indicating the presence of a phenyl group. The presence of the hydroxy and amide groups was confirmed by the existence of two exchangeable downfield protons that have chemical shifts at δ_H 8.02 and 9.82 in the ^{1}H-NMR (400 MHz, C_5D_5N) spectrum. The ^{13}C-NMR (100 MHz, C_5D_5N) spectrum shows 31 carbons that are classified by an analysis of the DEPT spectra into four methyls, four methylenes, fifteen methines, six quaternary carbons, and two carbonyls, suggesting a heptacyclic ring skeleton for **21**. The ^{1}H-NMR spectrum of **21** displayed a quaternary methyl [δ_H 1.33 (3H, s, Me-31)], two secondary methyls [δ_H 0.89 (3H, d, J = 6.1 Hz, Me-30), 1.24 (3H, d, J = 6.2 Hz, Me-29)], an olefinic methyl [δ_H 1.92 (3H, s, Me-32)], two oxygenated methines [δ_H 4.57 (1H, t, J = 7.9 Hz, H-13), 6.04 (1H, m, H-2)], four aromatic protons [δ_H 7.12 (2H, m, H-25, 27), 7.30 (1H, d, J = 6.4 Hz, H-24), 7.51 (1H, d, J = 6.4 Hz, H-28)], and a trisubstituted olefinic proton [(δ_H 5.95 (1H, br s, H-4)]. Detailed analyses of the ^{1}H-^{1}H COSY spectrum disclosed the presence of a partial structure shown as a bold line in Figure 16.17.

To establish the connectivity of this fragment demarcated by bold lines in the figure, HMBC experiments were carried out (Figure 16.17). The presence of spiro[cyclohexano[a]decahydrofluoren-2,3'-pyrrole]-1,2'-dione (6/5/6/6-5 rings) and 1,4-disubstituted benzyl moieties was established by ^{1}H-^{1}H COSY and HMBC correlations. Further investigation of HMBC spectrum provided the connectivity between the benzyl moiety and the 6/5/6/6-5 ring substructure through an ether linkage forming a 13-membered macrocyclic ring. These spectral data led to the planar structure of **21** shown in Figure 16.18. In addition to the ^{1}H-NMR spectrum of **21**, the characteristic proton signals of the benzene ring do not exhibit a simple A_2B_2-type coupling pattern due to the inequality between H-24 and H-28, and between H-25 and H-27. In the NOE difference experiment, NOEs from H-13 to H-15 and from H-2 to H-15 indicated that they are all positioned on the β side of the molecule. Interpretation of the NOE's data allowed us to assign the relative stereochemistry of **21**. Compound **21** shows NOEs from H-3 to H-6, from H-6 to H-12, and from H-14 to Me-31. These NOEs show that there must be a *cis*-juncture between the A and B rings and between the B and C rings; further, they indicate that a *trans*-juncture should occur between the C and D rings, with a β-orientation of H-15 and an α-orientation of H-3. Additionally, NOEs from Me-31 to H-9 and from Me-31 to H-11 suggest that the methyl groups at C-9 and C-11 are in equatorial positions. A chair-like conformation of ring A and ring D that is consistent with the results of the NOE experiments is shown in Figure 16.18. Although two aromatic methine signals (H-25 and 27) were unresolved in the ^{1}H-NMR spectrum of **21** in C_5D_5N, the spectrum of (S)-(-)-2-methoxy-2-phenyl-2-(trifluoromethyl)acetic acid (MTPA) ester **21a** in CDCl$_3$ enabled these two to be resolved. NOEs from H-22 β to NH-20 and H-28, from H-14 to H-25, and from H-24 to H-18 β and H-22 α as

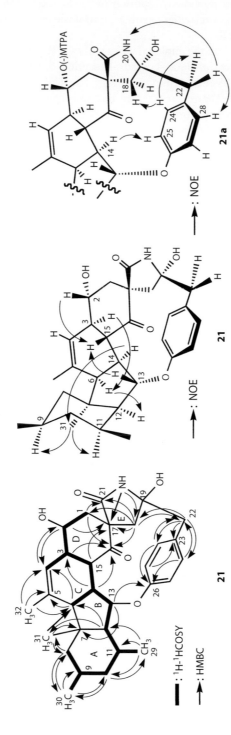

FIGURE 16.17 Important ^{1}H-^{1}H COSY, and HMBC correlations observed for **21** and selected NOEs observed for **21** and **21a**.

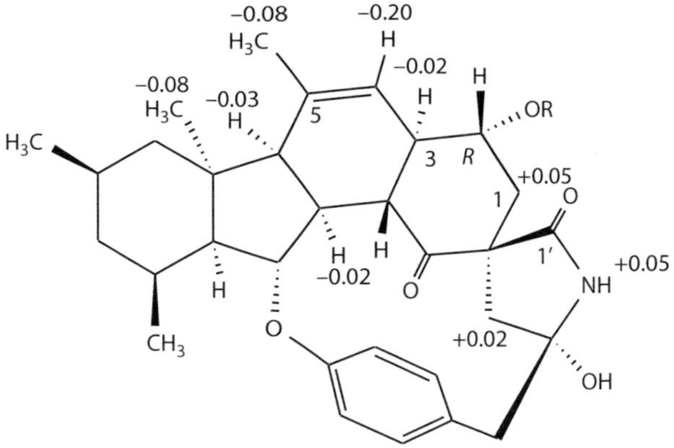

21a: R = (S)-(–)-MTPA
21b: R = (R)-(+)-MTPA

FIGURE 16.18   Δδ values [δ (–)-δ (+)] for the MTPA ester (**21a** and **21b**).

determined in the NOE difference experiment of **21a**, indicate that the relative configuration of the spiro carbon (C-17) is as shown. Thus, the relative stereochemistry of **21** was established (Figure 16.17).

To determine the absolute configuration of **21**, we applied the modified Mosher's method. The ^{1}H chemical shift differences between the (S)-(-) and (R)-(-)-MTPA esters (**21a** and **21b**) of **21** are shown in Figure 16.18. The results indicated that the R configuration is attributable to the C-2 chirality, thus establishing the absolute structure of **21**.

The molecular formula of **22** was determined to be $C_{31}H_{39}NO_5$ by HRFABMS, indicating that **22** had the same molecular formula as **21**. The IR, ^{1}H and ^{13}C-NMR spectra of **22** resembled those of **21**. It was hypothesized that **22** was a diastereomer of **21**. A detailed comparison of the chemical shifts and coupling constants between the ^{1}H-NMR spectrum of **22** and that of **21** revealed differences only in the signals of the methine proton (H-2 and H-4), while the other signals remained unchanged. This difference suggested that **22** was the C-2 epimer of **21**. This was confirmed by NOEs experiments, in which correlations between 2-OH and H-15 were observed; however, an NOE between H-2 and H-15 was not visible, indicating that the structure of **22** was the C-2 epimer of **21**. The unambiguous assignments of the signals in the ^{1}H and ^{13}C-NMR spectra were based on HMBC experiments.

16.7.2 BIOLOGICAL ACTIVITY

Hirsutellones A (**25**), B (**26**), C (**27**), D (**28**), E (**29**), GKK1032 A$_2$ (**30**), and B (**31**) are compounds structurally related to those reported above, which were isolated from *Hirsutella nivea* BCC 2564 (Isaka et al., 2005, 2006) and *Penicillium* sp. GKK1032 (Koizumi et al., 2001), respectively (Figure 16.16). These metabolites are characterized by 12- or 13-membered rings consisting of a decahydrofluorene core and phenyl ether and γ-lactam or succinimide. It has been reported that **30** is biosynthetically originated from L-tyrosine and a nonaketide chain flanked with five methyl groups being derived L-methionine, followed by unusual 13-membered macroether formation between the L-tyrosine hydroxy group and the polyketide chain (Oikawa, 2003). These compounds are becoming synthetic target molecules due to their interesting molecular skeletons. While hirsutellones are reported to have strong antituberculosis activity and cytotoxicity, GKK1032s have antimicrobial and antitumor activities.

Compounds that affect cell cycle could be good candidates for developing antitumor drugs. First we used a screening method for cell-cycle blockers similar to the case of Ca^{2+}-signal transduction inhibitor, that is, active substance could be detected as growth zone of *Saccharomyces cerevisiae cdc2-1 rad9Δ* strain (Tsuchiya et al., 2010). But pyrrocidines and pyrrospirones showed only the inhibition zone against the mutant yeast. Thus, we directly examined the cytotoxicity of pyrrospirones and pyrrocidines on human cancer cells, HL60 (human promyelocytic leukemia), K562 (human chronic myelogenous leukemia), and LNCaP (human prostate carcinoma) cell lines by using the 3-(4,5-dimethyl-2-thiazolyl)-2,5-diphenyl-2H-tetrazolium bromide (MTT) methods. All the compounds inhibited the cell growth activity against these cell lines and **23** was the most potent with IC_{50} values of $0.12\,\mu M$ in HL60 (Shiono et al., 2008) (Table 16.1). As only the known compound **23** resulted about 50 times more potent than the other related compounds, we compared the activity of **23** with other related compounds focusing on apoptosis. From the structure's difference among these compounds, the double bond moiety at the C-17 and C-18 position of **23** might have an important role to exhibit the potent activity due to the formation of adducts to some molecular target(s) via Michael-type addition. Such a proposed binding was supported by a methanol addition via Michael-type addition to the α,β-unsaturated carbonyl group of **23**, when dissolved in methanol. As shown in Figure 16.19, **23** showed the DNA ladder dose dependently at wide range of concentration, but other related compounds without the α,β-unsaturated carbonyl group showed DNA ladder at limited and higher concentration. Further work would be required to clarify whether Michael-type addition of **23** was due to activation of apoptotic pathway and we have been analyzing the mechanism of **23** against HL60 cells and molecular target.

TABLE 16.1
Cytotoxic Activities of 21–24

Compounds	IC$_{50}$, μM		
	HL60	K562	LNCaP
Pyrrospirone A (**21**)	14.9	9.96	17.97
Pyrrospirone B (**22**)	7.44	9.27	20.24
Pyrrocidine A (**23**)	0.12	0.45	0.48
Pyrrocidine B (**24**)	6.92	8.64	11.74

Pyrrocidine A (**23**)

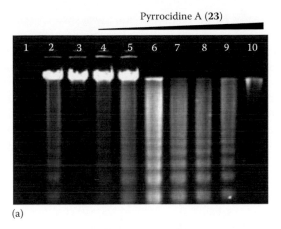

(a)

Pyrrocidine B   Pyrrospirone A   Pyrrospirone B
(**24**)   (**21**)   (**22**)

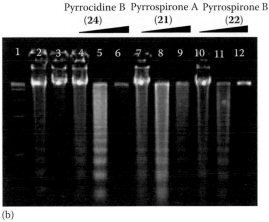

(b)

FIGURE 16.19 Compounds **21–24** induce DNA fragmentation in HL60 cells. (a) Lane 1, DNA marker; lane 2, camptothecin 0.1 µM; lane 3, control; lanes 4–10, **23**: 0.16, 0.31, 0.63, 1.25, 2.5, 5.0, 10 µM, respectively. (b) Lane 1, DNA marker; lane 2, camptothecin 0.1 µM; lane 3, control; lanes 4–6, **24**: 10, 20, 30 µM, respectively, lanes 7–9, **21**: 20, 30, 45 µM, respectively, lanes 10–12, **22**: 20, 30, 45 µM, respectively.

16.8 CONCLUSIONS

Endophytes are widespread in nature and presumably present in all plants. During 4 years from 2003 to 2006, hundreds of new biologically active substances have been obtained from endophytes (Gunatilaka, 2006). Thus, the potential of endophytes as sources of novel bioactive compounds is now widely recognized. Then, how many endophytes live inside plants? For example, if we assume that a few species of endophytes live in each of 250–350 thousands species of plant growing on the earth, then there are over 500–600 thousand species of endophytes in total, thereby forming an enormous quantity of microbial resources (Sieber, 2007). In addition, endophytes living in healthy trees constantly change their phases depending on the part of trees and the season, and their roles, being inhibitory against plant eaters and disease-causing bacteria, etc., vary greatly. However, they do not always work for the benefit of their host plants, but sometimes turn into phytopathogenic agents or saprophytes. What causes this change into harmful fungi is not known, but the balance of their interaction is important. In any event, since plants as they grow in the natural environment are constantly exposed to various external stresses and cannot grow without their symbiotic relationship with endophytes, the role that endophytes play for their host trees can never be ignored. For this reason, further research is conducted all over the world on the interaction in the symbiotic relationship between endophytes and plants (Kogel et al., 2006). Considering the various abilities that endophytes have acquired in the course of their evolution and close relationship with plants, it is understandable that many bioactive compounds have been discovered from endophytes in recent years, and many of them may become new lead compounds useful in drug discovery. In conclusion, it is presumable that endophytes will continue to attract more and more attention in the future as a source of exploration of new microorganisms and useful biologically active substances (bioprobes) with agrochemical, pharmaceutical, and industrial potential.

ACKNOWLEDGMENTS

Authors express their thanks to emeritus Prof. Tokichi Miyakawa, Hiroshima University, for providing the chance to use Ca^{2+}-signaling transduction inhibitory activity test system.

REFERENCES

Abel, U., C. Koch, M. Speitling, and F.G. Hansske. 2002. Modern methods to produce natural-product libraries. *Curr. Opin. Chem. Biol.* 6: 453–458.

Baba, Y., N. Hirukawa, N. Tanohira, and M. Sodeoka. 2003. Structure-based design of a highly selective catalytic site-directed inhibitor of Ser/Thr protein phosphatase 2B (calcineurin). *J. Am. Chem. Soc.* 125: 9740–9749.

Bouillant, M.L., J. Bernillon, J. Favre-Bonvin, and N. Salin. 1989. New hexaketides related to sordariol in *Sordaria macrospora. Z. Naturforsch.* 44c: 719–723.

Bugni, T.S. and C.M. Ireland. 2004. Marine-derived fungi: A chemically and biologically diverse group of microorganisms. *Nat. Prod. Rep.* 21: 143–163.

Chanklan, R., E. Aihara, S. Koga, H. Takahashi, M. Mizunuma, and T. Miyakawa. 2008. Inhibition of Ca^{2+}-signal-dependent growth regulation by radicicol in budding yeast. *Biosci. Biotechnol. Biochem.* 72: 132–138.

Clapham, D.E. 1995. Calcium signaling. *Cell* 80: 259–268.

De Bary, H.A. 1884. *Vergleichende Morphologie und Biologie de Plize Mycetozoen und Bacterien.* Verlag von Wilhelm Engelmann, Leipzig, Germany.

Ganesan, A. 2008. The impact of natural products upon modern drug discovery. *Curr. Opin. Chem. Biol.* 12: 306–317.

Grabley, S. and R. Thiericke. 1999. Bioactive agents from natural sources: Trends in discovery and application. *Adv. Biochem. Eng. Biotechnol.* 64: 104–154.

Gunatilaka, A.A.L. 2006. Natural products from plant-associated microorganisms: Distribution, structural diversity, bioactivity, and implications of their occurrence. *J. Nat. Prod.* 69: 509–526.

He, H., H.Y. Yang, R. Bigelis, E.H. Solum, M. Greenstein, and G.T. Carter. 2002. Pyrrocidines A and B, new antibiotics produced by a filamentous fungus. *Tetrahedron Lett.* 43: 1633–1636.

Höller, U., J.B. Gloer, and D.T. Wicklow. 2002. Biologically active polyketide metabolites from an undetermined fungicolous hyphomycete resembling *Cladosporium. J. Nat. Prod.* 65: 876–882.

Huang, X.H., H. Tomoda, H. Nishida, R. Masuma, and S. Omura. 1995. Terpendoles, novel ACAT inhibitors produced by *Albophoma yamanashiensis.* I. Production, isolation and biological properties. *J. Antibiot.* 48: 1–4.

Isaka, M., W. Prathumpai, P. Wongsa, and M. Tanticharoen. 2006. Hirsutellone F, a dimer of antitubercular alkaloids from the seed fungus *Trichoderma* species BCC 7579. *Org. Lett.* 8: 2815–2817.

Isaka, M., N. Rugseree, P. Maithip, P. Kongsaeree, S. Prabpai, and Y. Thebtaranonth. 2005. Hirsutellones A-E, antimycobacterial alkaloids from the insect pathogenic fungus *Hirsutella nivea* BCC 2594. *Tetrahedron* 61: 5577–5583.

Kanai, Y., Y. Tatsumi, T. Tokiwa, Y. Watanabe, T. Fujimaki, D. Ishiyama, and T. Okuda. 2005. F2928-1 and -2, New antifungal antibiotics from *Cladobotryum* sp. *J. Antibiot.* 58: 507–513.

Kobayashi, H., R. Sunaga, K. Furihata, N. Morisaki, and S. Iwasaki. 1995. Isolation and structures of an antifungal antibiotic, Fusarielin A, and related compounds produced by a *Fusarium* sp. *J. Antibiot.* 48: 42–52.

Kogel, K.H., P. Franken, and R. Huechelhoven. 2006. Endophyte or parasite-what decides? *Curr. Opin. Plant. Biol.* 9: 358–363.

Koizumi, F., K. Hasegawa, K. Ando, T. Ogawa, and A. Hara. 2001. Antitumor GKK1032 manufacture with *Penicillium*, Jpn. Kokai Tokkyo Koho 2001, JP 2001247574.

Kumagai, H., T. Someno, K. Dobashi, K. Isshiki, M. Ishizuka, and D. Ikeda. 2004. ICM0301s, new angiogenesis inhibitors from *Aspergillus* sp. F-1491. I. Taxonomy, fermentation, isolation and biological activities. *J. Antibiot.* 57: 97–103.

Lee, I.K., Y.W. Jang, Y.S. Kim, S.H. Yu, K.J. Lee, S.M. Park, B.T. Oh, J.C. Chae, and B.S. Yun. 2009. Xylarinols A and B, two new 2-benzoxepin derivatives from the fruiting bodies of *Xylaria polymorpha. J. Antibiot.* 62: 163–165.

Liu, J., J.D. Farmer, Jr., W.S. Lane, J. Friedman, I. Weissimann, and S.L. Schreiber. 1991. Calcineurin is a common target of cyclophilin-cyclosporin A and FKBP-FK506 complexes. *Cell* 66: 807–815.

Mann, J. 2001. Natural products as immunosuppressive agents. *Nat. Prod. Rep.* 8: 417–430.

McChesney, J.D., S.K. Venkataraman, and J.T. Henri. 2002. Plant natural products: Back to the future or into extinction? *Phytochemistry* 68: 2015–2022.

McDonald, L.A., L.R. Barbieri, V.S. Bernan, J. Janso, P. Lassota, and G.T. Carter. 2004. 07H239-A, a new cytotoxic eremophilane sesquiterpene from the marine-derived xylariaceous fungus LL-07H239. *J. Nat. Prod.* 67: 1565–1567.

Mejia, L.C., E.I. Rojas, Z. Maynard, E.A. Arnold, P. Hebbar, G.J. Samuels, N. Robbins, and A.E. Herre. 2008. Endophytic fungi as biocontrol agents of *Theobroma cacao* pathogens. *Biol. Control* 46: 4–14.

Miyakawa, T. and M. Mizunuma. 2007. Physiological roles of calcineurin in *Saccharomyces cerevisiae* with special emphasis on its roles in G2/M cell-cycle regulation. *Biosci. Biotechnol. Biochem.* 71: 633–645.

Mizunuma, M., D. Hirata, K. Miyahara, E. Tsuchiya, and T. Miyakawa. 1998. Role of calcineurin and Mpk1 in regulating the onset of mitosis in budding yeast. *Nature* 392: 303–306.

Nakamura, T., Y. Liu, D. Hirata, H. Namba, S. Harada, T. Hirokawa, and T. Miyakawa. 1993. Protein phosphatase type 2B (calcineurin)-mediated, FK506-sensitive regulation of intracellular ions in yeast is an important determinant for adaptation to high salt stress conditions. *EMBO J.* 12: 4063–4071.

Nakamura, T., T. Ohmoto, D. Hirata, E. Tsuchiya, and T. Miyakawa. 1996. Genetic evidence for the functional redundancy of the calcineurin- and Mpk1-mediated pathways in the regulation of cellular events important for growth in *Saccharomyces cerevisiae*. *Mol. Gen. Genet.* 251: 211–219.

Newman, D.J., G.M. Cragg, and K.M. Snader. 2003. Natural products as sources of new drugs over the period 1981–2002. *J. Nat. Prod.* 66: 1022–1037.

Nishikawa, K., N. Aburai, K. Yamada, H. Koshino, E. Tsuchiya, and K. Kimura. 2008. A bisabolane sesquiterpenoid endoperoxide, 3,6-epidioxy-1,10-bisaboladiene was isolated from *Cacalia delphiniifolia* as an anti-tumor substance and induced apoptosis. *Biosci. Biotechnol. Biochem.* 72: 2463–2466.

Ogasawara, Y., J. Yoshida, Y. Shiono, T. Miyakawa, and K. Kimura. 2008. New eremophilane sesquiterpenoid compounds, eremoxylarins A and B directly inhibit calcineurin in a manner independent of immunophilin. *J. Antibiot.* 61: 496–502.

Oikawa, H. 2003. Biosynthesis of structurally unique fungal metabolite GKK1032A$_2$: Indication of novel carbocyclic formation mechanism in polyketide biosynthesis. *J. Org. Chem.* 68: 3552–3557.

Osada, H. 2000. *Bioprobes*. Springer-Verlag, Tokyo, Japan.

Robinson, T., D. Singh, and P. Nigam. 2001. Solid-state fermentation: A promising microbial technology for secondary metabolite production. *Appl. Microbiol. Biotechnol.* 55: 284–289.

Rydzewki, R.M. 2008. *Real World Drug Discovery*. Elsevier, Slovenia.

Schulz, B. and C. Boyle. 2005. The endophytic continuum. *Mycol. Res.* 109: 661–686.

Schulz, B., U. Wanke, S. Draeger, and H.J. Aust. 1993. Endophytes from herbaceous plants and shrubs: Effectiveness of surface sterilization methods. *Mycol. Res.* 97: 1447–1450.

Shiono, Y. 2006. Anthracobic acids A and B, two polyketides, produced by an endophytic fungus *Anthracobia* sp. *Chem. Biodivers.* 3: 217–223.

Shiono, Y. and T. Murayama. 2005. New eremophilane-type sesquiterpenoids, eremoxylarins A and B from xylariaceous endophytic fungus YUA-026. *Z. Naturforsch.* 60b: 885–890.

Shiono, Y., A. Nitto, K. Shimanuki, T. Koseki, T. Murayama, T. Miyakawa, J. Yoshida, and K. Kimura. 2009. A new benzoxepin metabolite isolated from endophytic fungus *Phomopsis* sp. *J. Antibiot.* 62: 533–535.

Shiono, Y., K. Shimanuki, F. Hiramatsu, T. Koseki, M. Tetsuya, N. Fujisawa, and K. Kimura. 2008. Pyrrospirones A and B, apoptosis inducers in HL-60 cells, from an endophytic fungus, *Neonectria ramulariae* Wollenw KS-246. *Bioorg. Med. Chem. Lett.* 18: 6050–6053.

Shitamukai, A., M. Mizunuma, D. Hirata, H. Takahashi, and T. Miyakawa. 2000. A positive screening for drugs that specifically inhibit the Ca^{2+}-signaling activity on the basis of the growth promoting effect on a yeast mutant with a peculiar phenotype. *Biosci. Biotechnol. Biochem.* 64: 1942–1946.

Sieber, T.N. 2007. Endophytic fungi in forest trees: Are they mutualists? *Fung. Biol. Rev.* 21: 75–89.

Simon, J.A. and A. Bedalov. 2004. Yeast as a model system for anticancer drug discovery. *Nat. Rev.* 4: 481–488.

Singh, S.B., D. Zink, J. Polishook, D. Valentino, A. Shafiee, K. Silverman, P. Felock, A. Teran, D. Vilella, D.J. Hazuda, and R.B. Lingham. 1999. Structure and absolute stereochemistry of HIV-1 integrase inhibitor integric acid. A novel eremophilane sesquiterpenoid produced by a *Xylaria* sp. *Tetrahedron Lett.* 40: 8775–8779.

Singhania, R.R., A.K. Patel, R.S. Carlos, and A. Pandey. 2009. Recent advances in solid-state fermentation. *Biochem. Eng. J.* 44: 13–18.

Smith, C.J., N.R. Morin, G.F. Bills, A.W. Dombrowski, G.M. Salituro, S.K. Smith, A. Zhao, and D.J. MacNeil. 2002. Novel sesquiterpenoids from the fermentation of *Xylaria persicaria* are selective ligands for the NPY Y5 receptor. *J. Org. Chem.* 67: 5001–5004.

Someno, T., H. Kumagai, S. Ohba, M. Amemiya, H. Naganawa, M. Ishizuka, and D. Ikeda. 2004. ICM0301s, New angiogenesis inhibitors from *Aspergillus* sp. F-1491. II. Physico-chemical properties and structure elucidation. *J. Antibiot.* 57: 104–109.

Strobel, G., B. Daisy, U. Castillo, and J. Harper. 2004. Natural products from endophytic microorganisms. *J. Nat. Prod.* 67: 257–268.

Tsuchiya, E., E. Tsuchiya, M. Yukawa, M. Ueno, K. Kimura, and H. Takahashi. 2010. A novel method for screening cell' cycle blockers as candidates of anti-tumor reagent by using yeast as a screening tool. *Biosci. Biotechnol. Biochem.* 74: 411–414.

Vega, F.E., F.J. Posada, M.C. Aime, M. Pava-Ripoll, F. Infante, and S.A. Rehner. 2008. Entomopathogenic fungal endophytes. *Biol. Control* 46: 72–82.

Watanabe, K., K. Hotta, A.P. Praseuth, M. Searcey, C.C. Wang, H. Oguri, and H. Oikawa. 2009. Rationally engineered total biosynthesis of a synthetic analogue of a natural quinomycin depsipeptide in *Escherichia coli. Chembiochem* 17: 1965–1968.

Wicklow, D.T., S. Roth, S.T. Deyrup, and J.B. Gloer. 2005. A protective endophyte of maize: *Acremonium zeae* antibiotics inhibitory to *Aspergillus flavus* and *Fusarium verticillioides*. *Mycol. Res.* 109: 610–618.

Yanai, K., N. Sumida, K. Okakura, T. Moriya, M. Watanabe, and T. Murakami. 2004. Para-position derivatives of fungal anthelmintic cyclodepsipeptides engineered with *Streptomyces venezuelae* antibiotic biosynthetic genes. *Nat. Biotechnol.* 22: 848–855.

Zhang, H.W., Y.C. Song, and R.X. Tan. 2006. Biology and chemistry of endophytes. *Nat. Prod. Rep.* 23: 753–771.

17 Biologically Active Natural Products from Australian Marine Organisms

Robert J. Capon

CONTENTS

17.1 INTRODUCTION

Before embarking on a discussion of biologically active natural products from Australian marine organisms, it is useful spending a moment considering the term *biologically active*, particularly as relates to the interconnecting themes of marine chemical ecology and drug discovery. The reasons are twofold. First, some may question the premise that the chemical ecology achievements of primitive marine creatures could inspire the development of new drugs—particularly drugs applicable to human disease. In response, history has successfully and repeatedly traversed the path from terrestrial plants, animals, and microbes, through traditional medicines to modern pharmacology and therapeutics, so it should come as no surprise that we turn to the marine "road less travelled" in our search for new routes to better drugs. Second, the proposition that *biologically active* metabolites have value inevitably implies the corollary view that *biologically inactive* metabolites have no (or lesser) value. The danger of judging by these criteria is that all too often the assessment of *biologically active* versus *inactive* is based on a limited set of bioassays. Choose the wrong bioassay(s) and an exquisitely potent and selective biologically active marine metabolite can be mistakenly categorized and dismissed as inactive. Put another way, simply because we may not currently know or appreciate the *biological activity* of a particular marine metabolite, it does not mean that it is *biologically inactive* and thereby lacking in drug discovery potential. But why should we give marine metabolites the benefit of the doubt? Surely, if they do not register in the "bioassay of the day," it is reasonable to assume they are inactive, dismiss them, and move on. The answer, a resounding no, goes to the heart of what natural products are, how they came to exist, and how we might make best use of them. Conventional wisdom has marine natural products providing host organisms with a survival advantage, typically referred to as a chemical defense. The host organisms can themselves be the biosynthetic source of these metabolites, or they can acquire them from dietary sources and/or symbiotic/associated organisms (bacteria or microalgae). These chemicals can protect the host from infection (antibiotics, antiparasitics), repel or dissuade predators (antifeedants, toxins), inhibit the development and growth of competitors (selective cytotoxins and cell growth inhibitors), or even guard against UV radiation (sunscreens). They may also enhance reproductive outcomes (sperm attractants), or improve the ability to feed by rapidly immobilizing prey (venoms). This ecological significance can on occasion be tested experimentally. For example, a marine algal metabolite that elicits a feeding avoidance response when added to the food pellets of aquaria fish might have an ecological role as an antifeedant against herbivorous reef fish. Similarly, a marine tunicate metabolite that is antibacterial toward laboratory strains of human pathogenic bacteria might protect the tunicate from opportunistic pathogens present in seawater, while a marine sponge metabolite that selectively kills fast growing human cancer cells in tissue culture might inhibit growth and development of the avalanche of larval species that threaten overgrowth of filter feeding organisms such as sponges. That such insights into marine chemical ecology are achieved by proxy in the laboratory is an inevitable consequence of the challenges associated with replicating complex marine ecosystems. Although a number of noteworthy exceptions exist, for the large part we remain ignorant of, or at best hypothesize on, the ecological role played by the vast majority of known marine metabolites. This limitation notwithstanding, as our knowledge of marine metabolites has grown over the last three to four decades to include many thousands of diverse structures from numerous species, it has become clear that these chemicals are disproportionally represented in sedentary or slow moving lineages such as invertebrates and algae. As these life forms typically lack the physical attributes to flee (fins, legs), fight (claws, spines) or take cover (shells), and include some of the oldest marine life forms, it is little wonder that they have experienced both the selection pressure and time to evolve chemicals to enhance survival. Similarly with marine microbes, which have embraced chemistry to enhance survival and successfully compete for resources in unique niche ecosystems. Under this scenario, all marine metabolites should be viewed as implicitly biologically active within their ecological setting. Such metabolites represent an extraordinary preassembled pool of biologically active molecular diversity, programmed by evolutionary processes to be potent and selective modulators of key biopolymers (i.e., DNA, proteins, etc.), cells, tissues, organs, and animals. In essence, these metabolites represent Nature's intellectual property, gleaned from the evolutionary equivalent of a billion year global drug discovery program, boasting an unlimited budget and a workforce of trillions. Clearly a very, very, long-term investment, but one with an impressively successful intellectual property portfolio! The privileged biologically active chemical structures that have emerged from this (ongoing) investment can inform, guide, and inspire modern drug discovery, allowing us to repurpose ecological advantage to pharmaceutical benefit. To extend a nautical metaphor,

> ...Marine metabolites can be viewed as natural "molecular" waypoints, guiding our exploration of the infinite reaches of chemical space, serving as both inspiration and compass as we search out the very elusive islands and archipelagos of biological activity, and the even rarer hidden safe harbors of new drugs.

Given this broader appreciation of the concept *biologically active*, this chapter will review metabolites from Australian marine organisms, where the biological activity may be known or unknown, accepting the premise that the very existence of these remarkable compounds is suggestive of an intrinsic ecological purpose, and as such a pharmacological potential.

17.2   PORIFERA (SPONGES)

By far the most studied and productive source of Australian marine metabolites, sponges are not only rich in novel chemistry, but they can be collected by hand with relative ease on a scale suitable for laboratory extraction and isolation, making them an ideal resource for natural products chemists. One of the earliest reports on Australian marine sponge metabolites, by Sharma et al. from Columbia University, New York, described the antibacterial dibromophakellin (**1**) from a Great Barrier Reef *Phakellia flabellate* (Sharma and Burkholder, 1971). Building on the growing international interest in marine natural products chemistry, the Roche Research Institute for Marine Pharmacology (RRIMP), Sydney, was the first Australian facility to undertake intensive large-scale collections of geographically and taxonomically diverse marine organisms, including sponges, for the purpose of drug discovery. In operation for less than a decade (1974–1981), RRIMP achieved many firsts—discovering a wide array of novel structure classes and laying the groundwork for future generations of Australian marine natural products researchers. Two noteworthy discoveries included methylaplysinopsin (**2**) from a Great Barrier Reef specimen of *Alysinopsis reticulata*, (Taylor et al., 1981) as a serotonergic agent and potential antidepressant, and 1-methyliso-guanosine (**3**) from a New South Wales specimen of *Tedinia digitata*, (Cook et al., 1980) as an adenosine analog with potential as an orally active muscle relaxant, antihypertensive, and anti-inflammatory. Although neither **2** nor **3** progressed to a drug, they served as the inspiration for considerable future research, and provided a valuable proof-of-principle that sponges could yield metabolites with drug-like properties (chemistry and pharmacology). Another memorable contribution from RRIMP was the robustly named bastadins (i.e., **4**) from *Ianthella basta*, (Kazlauskas et al., 1981) a structure class that has since grown in diversity and attracted considerable attention for a range of biological properties. While it is difficult to do justice to the extraordinary diversity of metabolites reported from Australian sponges over the intervening decades, a selection of achievements are listed below. Early international interest in Australian sponges was further demonstrated by discovery of the vasodilatory xestospongins (i.e., **5**), isolated from *Xestospongia exigua* by researchers from the Suntory Institute for Biomedical Research, Japan (Nakagawa et al., 1984), and the ichthyotoxic 9,11-secosterol herbasterol (**6**), isolated from a Great Barrier Reef sample of *Dysidea herbacea* by researchers from Scripps Institution of Oceanography, University of California, San Diego (Capon and Faulkner, 1985). That Australian researchers were enthusiastic in taking up the challenge of exploring sponge metabolites was evidenced by a wealth of scientific publications. These included a series of studies into antibacterial and cytotoxic norterpene endoperoxides (i.e., **7**) from New South Wales and Victorian sponges of the genera *Mycale, Latrunculia*, and *Sigmosceptrella* spp., which lead to empirical rules for assigning relative configurations about this unique functionality (Capon and MacLeod, 1985), and documented the rare trunculin carbon skeleton (i.e., **8**) (Capon et al., 1987). The antibacterial trikentrins (i.e., **9**) from a Northern Territory specimen of *Trikentrion flabelliforme* displayed an unusual indole skeleton, which became a favored target for synthetic chemists (Capon et al., 1986), while a New South Wales specimen of *Clathria pyramida* yielded 5-thio-D-mannose (**10**) as the first, and still the only known naturally occurring 5-thiopyranose monosaccharide (Capon and Macleod, 1987). It is interesting to note that the related compound 5-thio-D-glucose first synthesized in 1962 as an "anti-metabolite," was found to be an inhibitor of insulin release and sperm-cell development, as well as glucose transport. The synthesis of 5-thio-D-glucose and later isolation of 5-thio-D-mannose represents a case where synthetic chemistry (almost) pre-empted Nature—or more accurately our knowledge of Nature. Other examples where chemical compounds enjoyed a prior literature existence as synthetics before being discovered as marine metabolites include the acetylcholine mimic esmodil (**11**) first synthesized in 1935 before being isolated from a southern Australian *Raspalia* sp. and resynthesized (Capon et al., 2004), the cycloxanthosine **12** first synthesized in the 1960s before being isolated from a Great Australian Bight *Erylus* sp. and resynthesized (Capon and Trotter, 2005), and the secoxanthine hymeniacidin (**13**) first synthesized in 1959 before being isolated from a southern Australian *Hymeniacidon* sp. and again resynthesized (Capon et al., 2002). Lest Australian sponges be seen as biosynthetically unimaginative, the A–Z series of luffarins (i.e., **14**) recovered from a Great Australian Bight specimen of *Luffariella geometrica* represents one of the largest sets of structurally related cometabolites to be identified from a single sponge specimen (Butler and Capon, 1992). Of even greater structural complexity were the cytostatic phorboxazoles (i.e., **15**) and salicylihalamides (i.e., **16**), from Western Australian *Phorbas* sp. (Searle and Molinski, 1995) and *Haliclona* sp. (Erickson et al., 1997), respectively. The topoisomerase I inhibitory cyclopropylidene amphimic acids (i.e., **17**) were reported from a Great Barrier Reef *Amphimedon* sp. (Nemoto et al., 1997), while a Western Australian *Cymbastela* sp. returned the insecticidal bromopyrrole agelastatins (i.e., **18**) (Hong et al., 1998). The protein phosphatase inhibitor dragmacidin E (**19**) from a southern Australian deepwater *Spongosorites* sp. was also noteworthy in being a bright yellow pigment with a tenacious ability to stain glassware (Capon et al., 1998b). Pursuing the pigment theme, a Great Barrier Reef sponge *Crella spinulata* yielded benzylthiocrellidone (**20**) as a yellow pigment with a strong UV absorption suggestive of a natural sunscreen agent, being the first recorded example of a natural product containing a dimedone moiety (Lam et al., 1999). The norsesterterpene endoperoxide hydroperoxide sigmosceptrellin D (**21**) from a Great Australian Bight *Sigmosceptrella* sp. was pivotal in inspiring a plausible biosynthetic pathway and attractive biomimetic synthetic strategy for this structure class (Ovenden and Capon, 1999). A search for nematocidal agents capable of treating gastrointestinal parasitic infections in livestock yielded the polyketide tetramic acid geodin A magnesium salt (**22**) and the unprecedented enamino lactone/lactam amphilactams (i.e., **23**), from Great Australian Bight *Geodia* (Capon et al., 1999,) and *Amphimedon* sp. (Ovenden et al., 1999), respectively. A Na,K-ATPase inhibitor, the dimeric polybrominated benzofuran iantheran A (**24**) was isolated from a Great Australian Bight *Ianthella* sp. (Okamoto et al., 1999), while the nucleoside microxine (**25**) recovered from a South Australian *Microxina* sp. was determined to be a cdk2 kinase inhibitor and possible anticancer agent (Killday et al., 2001). The lipid thiocyanatins (i.e., **26**) from a Western Australian *Oceanapia* sp. displayed promising nematocidal activity (Capon et al., 2001), while bromotyrosine alkaloids (i.e., **27**) isolated from another Australian *Oceanapia* sp. were determined to be the first known inhibitors of the mycobacterial enzyme mycothiol S-conjugate amidase, and as such have potential application in the treatment of tuberculosis (Nicholas et al., 2001). Two separate Great Barrier Reef sponges of the Family Dysideidae yielded dysinosin A (**28**), a potent inhibitor of Factor VIIa and thrombin, suggestive of a pharmacophore with possible application in the management of blood clotting (Carroll et al., 2002), and the sterol **29** as the first possible inhibitor of the MDR1-type efflux pump in azole-resistant *Candida albicans* (Jacob et al., 2003). A Great Australian Bight *Echinodictyum* sp. yielded as nematocidal agents the two methylated amino acid analogs (−)-echinobetaine A (**30**) (Capon et al., 2005a) and (+)-echinobetaine B (**31**) (Capon et al., 2005b). A northern Australian, Gulf of Carpentaria *Oceanapia* sp. yielded petrosamine B (**32**) as an inhibitor of the *Helicobacter pylori* enzyme aspartyl semialdehyde dehydrogenase (Carroll et al., 2005). Isoprenylcysteine methyltransferase (Icmt) catalyzes the carboxyl methylation of oncogenic proteins, and its inhibition represents an attractive and novel anticancer target. Two Queensland specimens of *Pseudoceratina* sp., one from the Great Barrier Reef and the other the Sunshine Coast, yielded the Icmt inhibitory bromotyrosine alkaloids spermatinamine (**33**) (Buchanan et al., 2007b) and aplysamine 6 (**34**) (Buchanan et al., 2008a), respectively. The stylissadines (i.e., **35**) from a Great Barrier Reef specimen of *Stylissa flabellate* are potent antagonists of P2X$_7$ receptor, a ligand gated cation channel and target for the treatment of inflammatory diseases, osteoarthritis, rheumatoid arthritis, and chronic obstructive pulmonary disease (COPD) (Buchanan et al., 2007a). Clavatadine A (**36**) is a bromotyrosine inhibitor of Factor XIa from the Great Barrier Reef sponge *Suberea clavata*, with antithrombotic potential (Buchanan et al., 2008b). More recently a set of Great Australian Bight sponges yielded distinct structurally unique families of cytotoxic alkaloids, with selective activity against human cancer cell lines. These included the trachycladindoles (i.e., **37**) from *Trachycladus laevispirulifer* (Capon et al., 2008), the phorbasins (i.e., **38**) from a *Phorbas* sp. (Zhang and Capon, 2008), and the bistellettazines (i.e., **39**) from a *Stelletta* sp. (El-Naggar et al., 2008) (Figure 17.1).

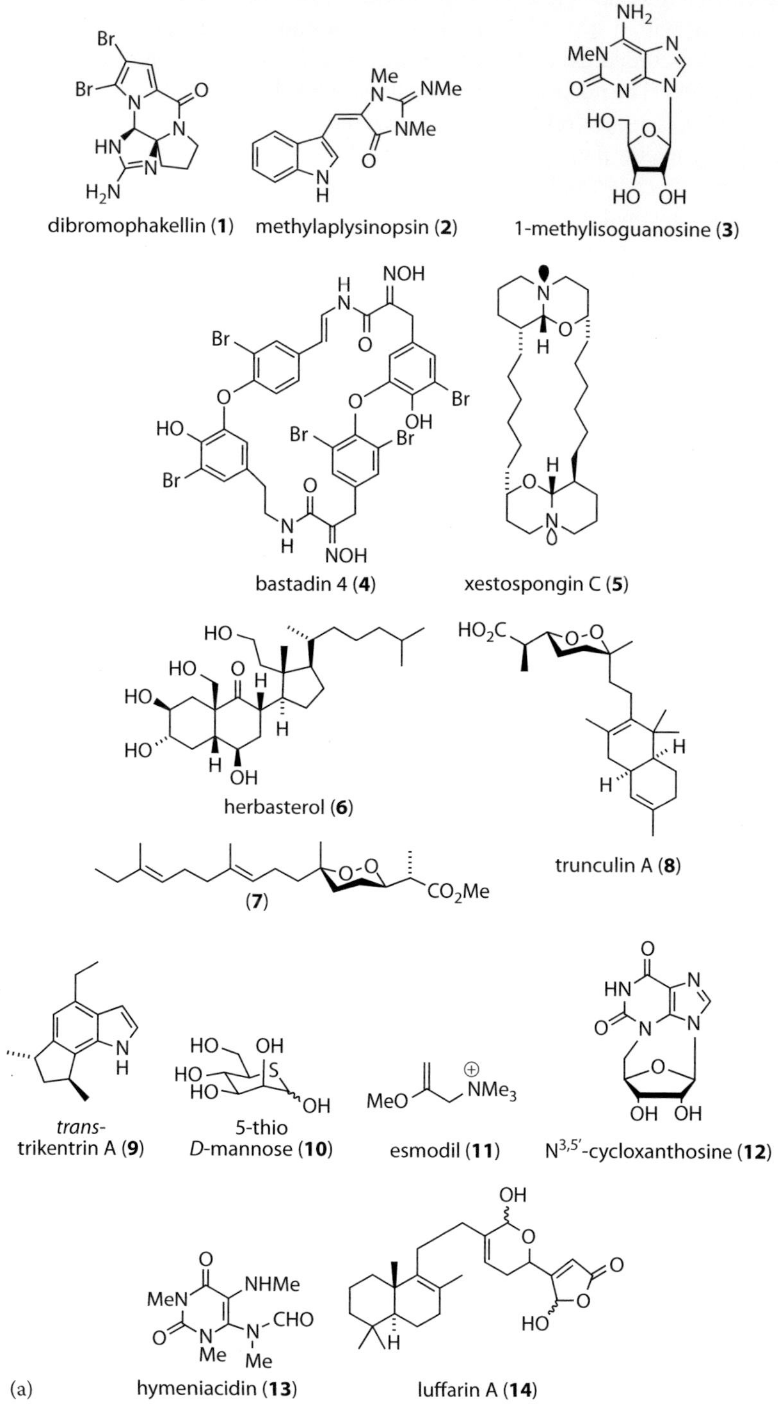

dibromophakellin (**1**)   methylaplysinopsin (**2**)   1-methylisoguanosine (**3**)

bastadin 4 (**4**)   xestospongin C (**5**)

herbasterol (**6**)

trunculin A (**8**)

(**7**)

trans-
trikentrin A (**9**)

5-thio
D-mannose (**10**)

esmodil (**11**)

N^3,5'-cycloxanthosine (**12**)

(a)   hymeniacidin (**13**)   luffarin A (**14**)

FIGURE 17.1   (a–d) Porifera (sponge) metabolites.

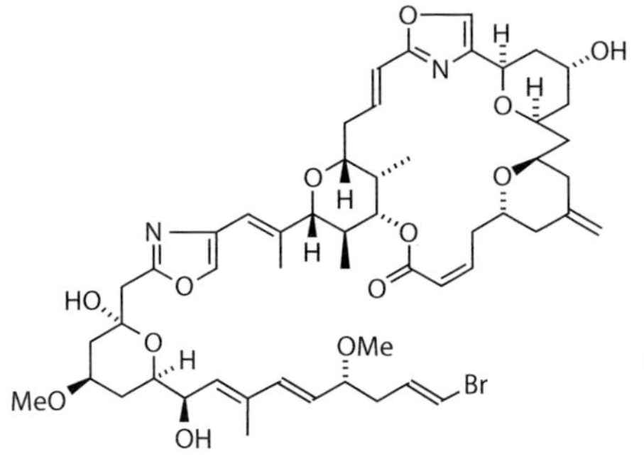

phorboxazole A (**15**)

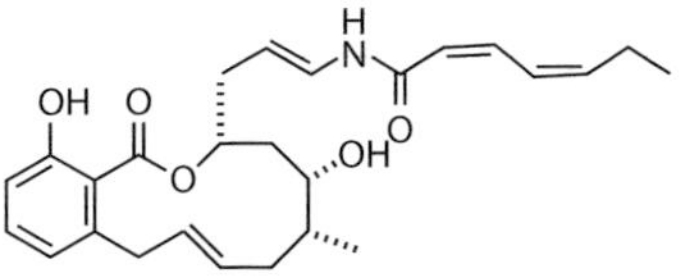

salicylihalamide A (**16**)

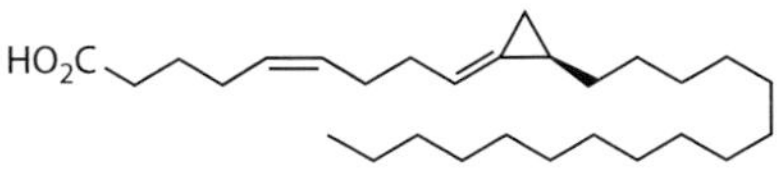

amphimic acid A (**17**)

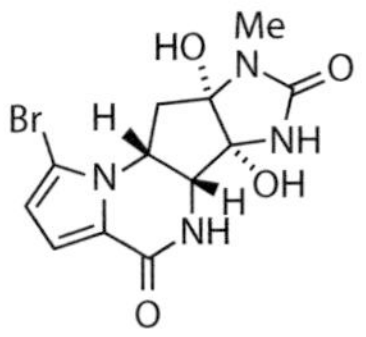

agelastatin C (**18**)

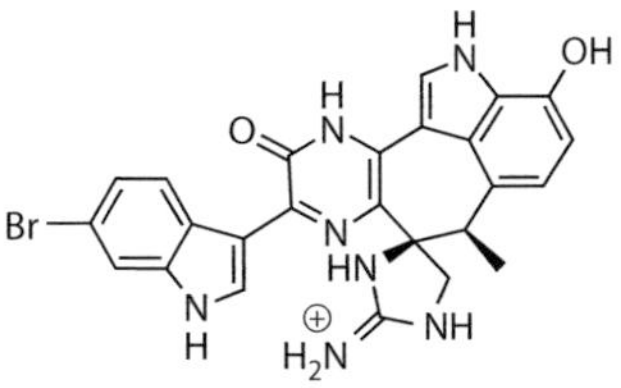

dragmacidin E (**19**)

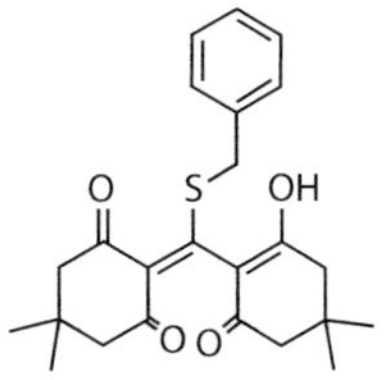

benzylthiocrellidone (**20**)

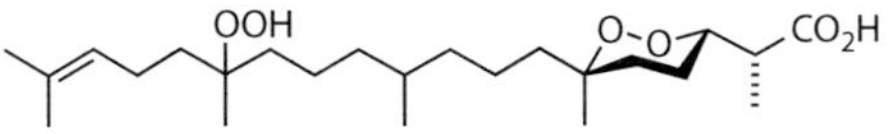

sigmosceptrellin D (**21**)

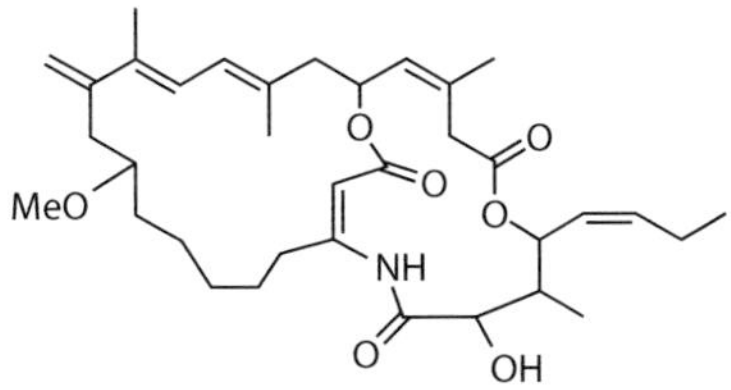

amphilactam A (**23**)

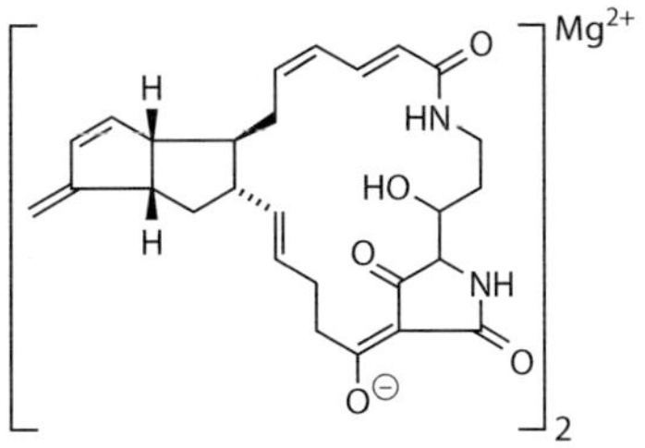

geodin A Mg salt (**22**)

(b)

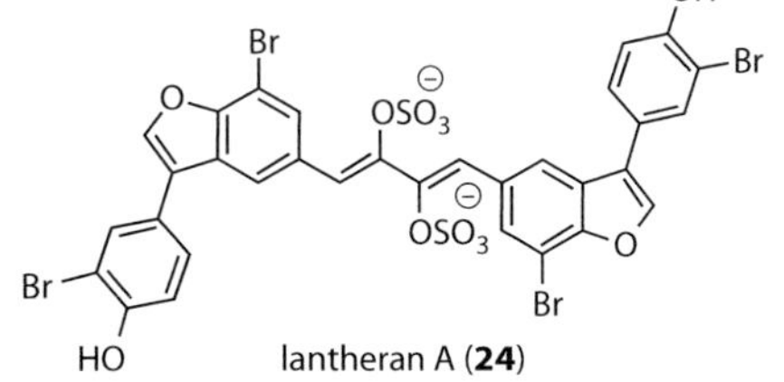

lantheran A (**24**)

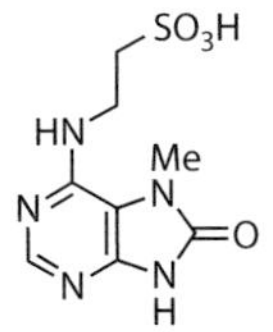

microxine (**25**)

FIGURE 17.1 (continued)

(continued)

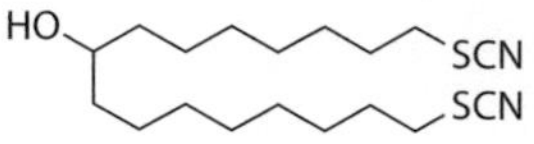

thiocyanatin A (**26**)

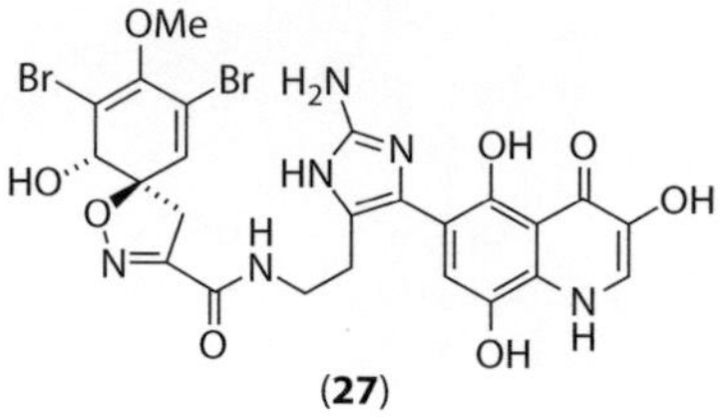

(**27**)

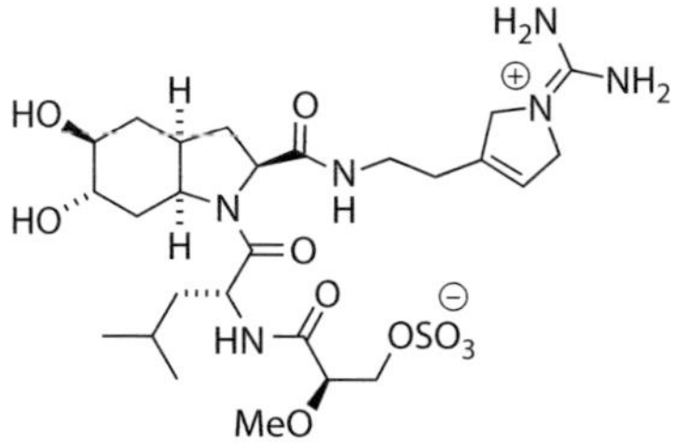

dysinosin A (**28**)

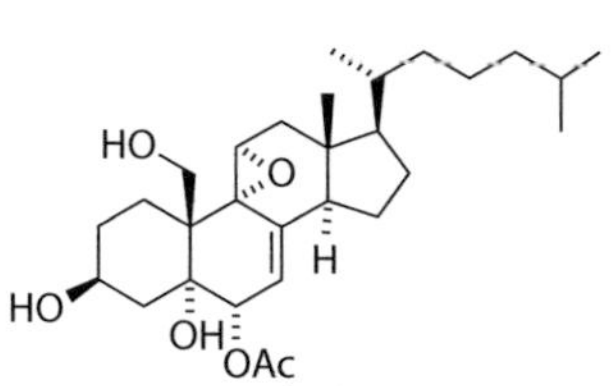

(**29**)

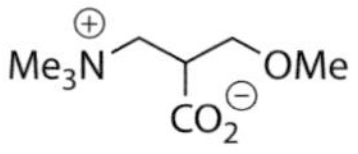

(−)-echinobetaine A (**30**) 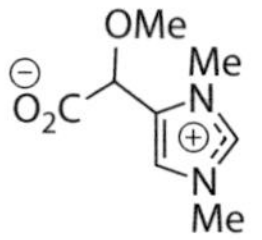   (+)-echinobetaine B (**31**)

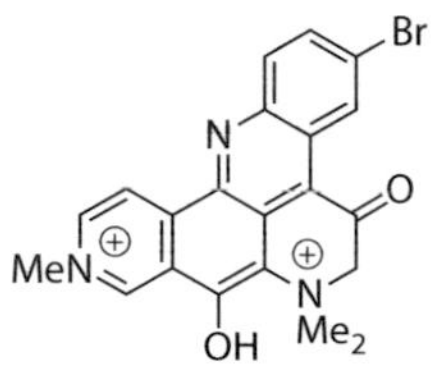

petrosamine B (**32**)

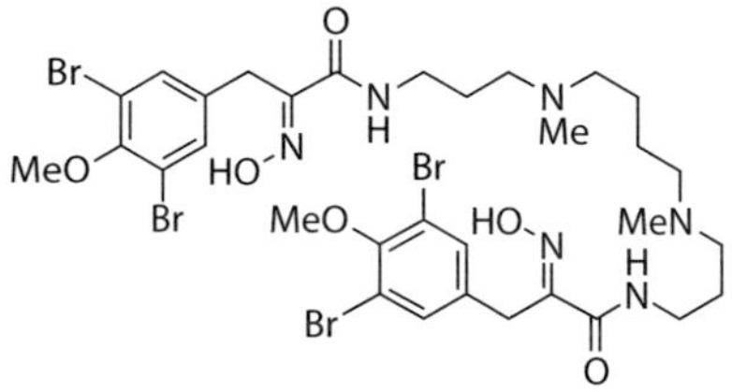

spermatinamine (**33**)

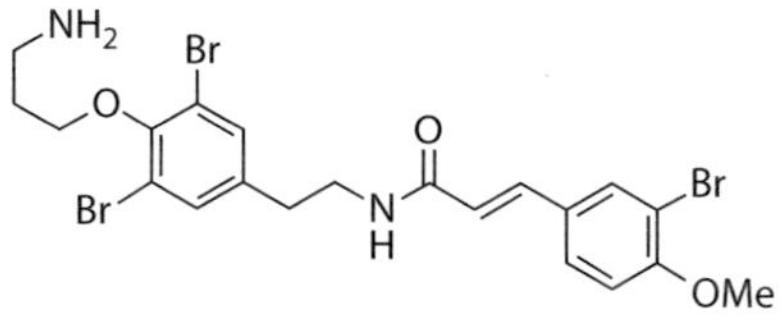

aplysamine 6 (**34**)

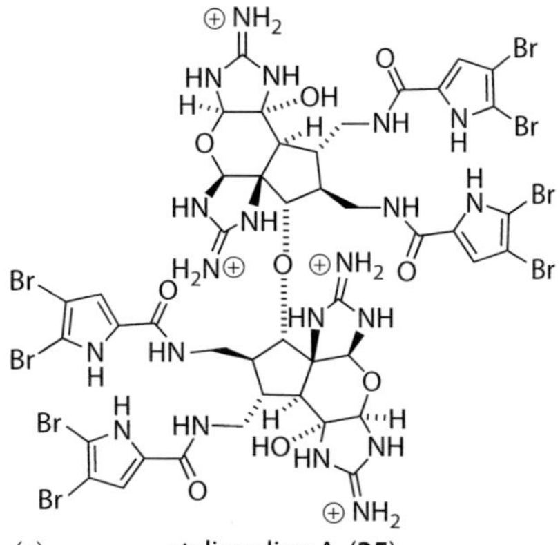

stylissadine A (**35**)

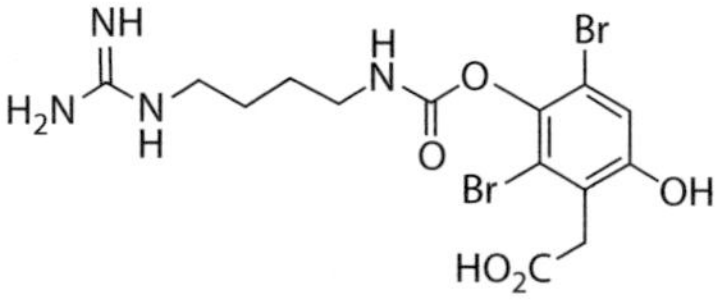

clavatadine A (**36**)

(c)

FIGURE 17.1 (continued)

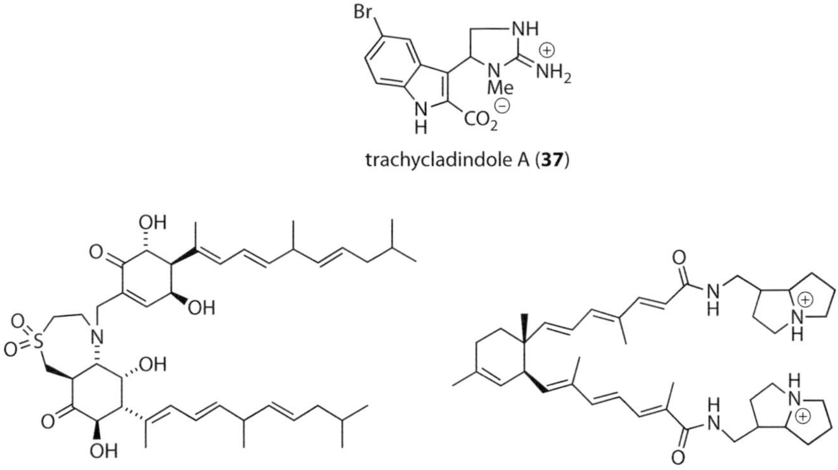

trachycladindole A (**37**)

(d)   phorbasin E (**38**)   bistellettazine A (**39**)

FIGURE 17.1 (continued)

17.3 CHORDATA (TUNICATES AND ASCIDIANS)

Slower to catch the attention of chemists than sponges, tunicates have nevertheless proven to be a remarkable source of novel alkaloids. The first metabolite reported from an Australian tunicate was the prodigiosin analog **40** from an unidentified Western Australian compound tunicate collected from the Abrolhos Islands (Kazlauskas et al., 1982). Prescient of future interest in marine microbes, at the time of its discovery, **40** was already known from a mutant strain of the bacteria *Serratia marcescens*. Over the next decade, a number of cytotoxic cyclic peptides rich in oxazole and thiazole heterocycles were reported from Australian tunicates, an indicative example being patellamide D (**41**) from a Great Barrier Reef specimen of *Lissoclinum patella* (Degnan et al., 1989). Other cytotoxic metabolites from Great Barrier Reef tunicates include the iodinated tyrosine analog **42** from an *Aplidium* sp. (Carroll et al., 1993a), and the tetrahydrocannabinol analog **43** from *Synoicum castellatum* (Carroll et al., 1993b). The antifungal lissoclinotoxin A (**44**) recovered from a Great Barrier Reef *Lissoclinum* sp. was noted for unusual stereoisomerism due to restricted inversion about the benzopentathiepin ring (Searle and Molinski, 1994), while lamellarin S (**45**) from a New South Wales *Didemnum* sp. displayed unusual optical properties attributed to slowly racemizing atropisomers (Urban and Capon, 1996). The growing realization that microbes may be the biosynthetic source of some metabolites isolated from macro marine invertebrates such as tunicates was reinforced with the discovery of the *Streptomyces* antibiotic enterocin (**46**) in a Western Australian *Didemnum* sp. (Kang et al., 1996). Other Western Australian discoveries included the cytotoxic unstable dimeric disulfide alkaloid polycarpine hydrochloride (**47**) from a specimen of *Polycarpa clavata*, (Kang and Fenical, 1996), the siderophoric ningalins (i.e., **48**) from a *Didemnum* sp. (Kang and Fenical, 1997b), and the zwitterionic hydroxyadenine aplidiamine (**49**) from an *Aplidiopsis* sp. (Kang and

Fenical, 1997a,b). More recently, Great Barrier Reef tunicate chemistry discoveries have extended to the antimalarial lepadins (i.e., **50**) from a *Didemnum* sp. (Wright et al., 2002), selective A_1 and A_3 adenosine receptor substrate methylthionucleoside diesters (i.e., **51**) from *Atriolum robustum* (Kehraus et al., 2004), and neuronal nitric oxide synthase (nNOS) inhibitory eusynstyelamides (i.e., **52**) from *Eusynstyela latericus* (Tapiolas et al., 2009) (Figure 17.2).

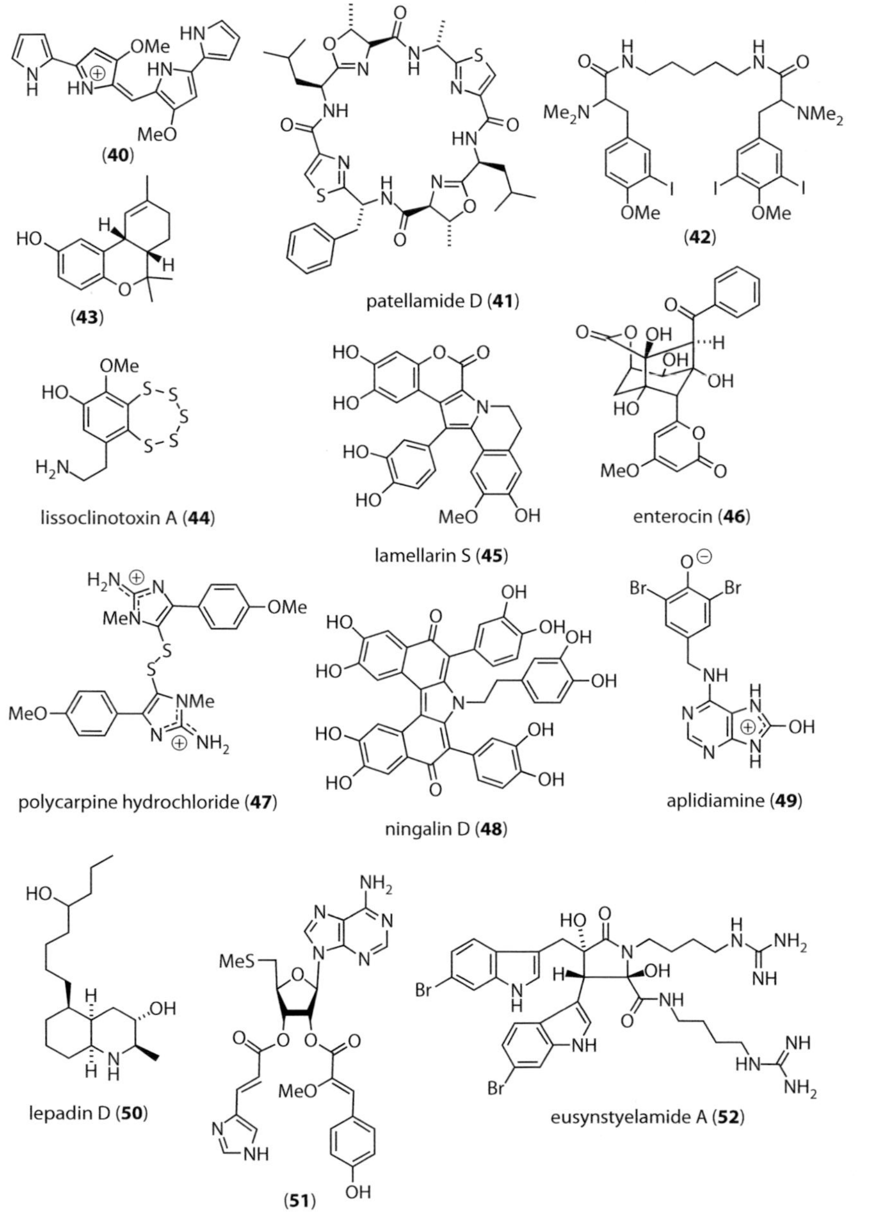

FIGURE 17.2 Chordata (tunicate and ascidian) metabolites.

17.4   CNIDARIA (SOFT AND STONY CORALS)

Current knowledge of the chemistry of Australian soft corals can be largely attributed to the efforts of Coll et al. at James Cook University of North Queensland, who lead a prolific laboratory that pioneered this field and trained generations of Australian marine natural products chemists in the process. This era of discovery commenced with the report of cembranoid diterpenes (i.e., **53**) from a Great Barrier Reef *Sarcophyton* sp. (Coll et al., 1977), and went on to achieve a systematic survey of Australian soft coral chemistry, yielding a bewildering array of new terpene skeletons rich in oxygen functionality. An early example of a biologically active soft coral chemistry was that of the antiinflammatory flexibilide (**54**), from a Great Barrier Reef sample of *Sinularia flexibilis* (Kazlauskas et al., 1978b), and cardiotonic cometabolite phenylethylamides (i.e., **55**) (Kazlauskas et al., 1980a). Although not commented on at the time, the structural similarity between the phenylethylamides and the plant defensive compound capsaicin, a potent agonist of the noxious heat and pain receptor TRPV1, is noteworthy, and may be suggestive of a common chemical ecology and pharmacology. The water soluble natural sunscreen agent mycosporine-methylamine:threonine (**56**) was described from Great Barrier Reef corals *Pocillopora damicornis* and *Stylophora pistillata* (Wu Won et al., 1995). Perhaps the most promising bioactive metabolite to be reported from an Australian soft coral was the diterpene glycoside eleutherobin (**57**), from a Western Australian *Eleutherobia* species (Lindel et al., 1997). Eleutherobin was reported to be an extremely potent cancer cell inhibitor (IC_{50} 10–15 nM) with a taxol-like mode of action, and its synthesis and biology attracted considerable attention. While the published chemistry of stony corals is less diverse than that of their soft bodied cousins, it does include the mycalolides (i.e., **58**) and montiporic acids (i.e., **59**) from Great Barrier Reef collections of *Tubastrea faulkneri* (Rashid et al., 1995) and *Montipora digitata* (Fusetani et al., 1996), respectively (Figure 17.3).

17.5   BRYOZOA (LACE CORALS)

Despite access to a rich taxonomic diversity, knowledge of the chemistry of Australian bryozoa remains limited, and for the most part can be attributed to the efforts of Blackman et al. based at the University of Tasmania. Blackman's studies commenced with a report on amathamides (i.e., **60**), and concluded with the discovery of nematocidal alkaloids such as convolutindole A (**61**), from Tasmanian collections of *Amathia wilsoni* (Blackman and Matthews, 1985) and *Amathia convoluta* (Narkowicz et al., 2002), respectively (Figure 17.4).

17.6   ECHINODERMATA (SEA AND FEATHER STARS)

The crinoids, also known as feather stars, were one of the earliest Australian marine taxa to attract the attention of natural products chemists, and our current knowledge of this field is due almost entirely to the efforts of Sutherland et al. at the University

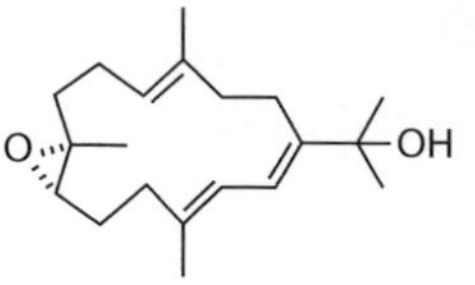

(53) 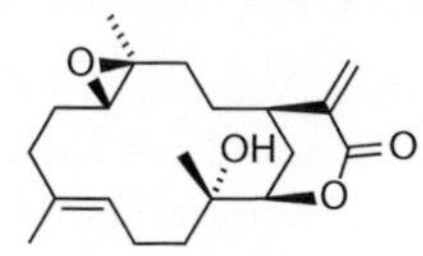   flexibilide (54) 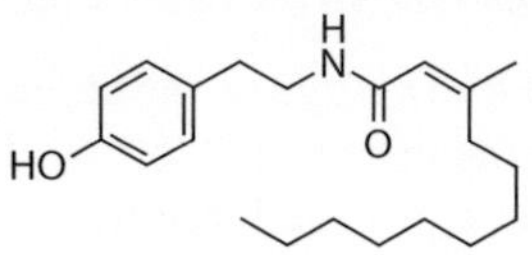   (55)

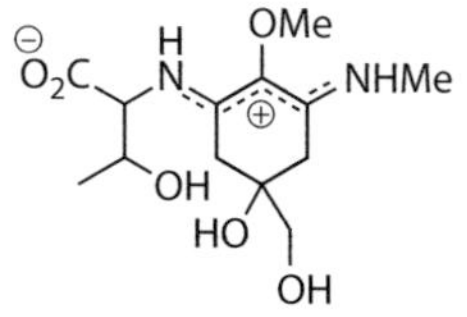

(56) 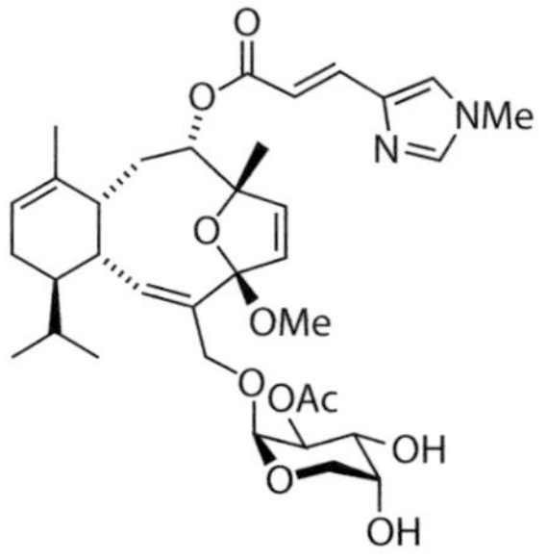   eleutherobin (57)

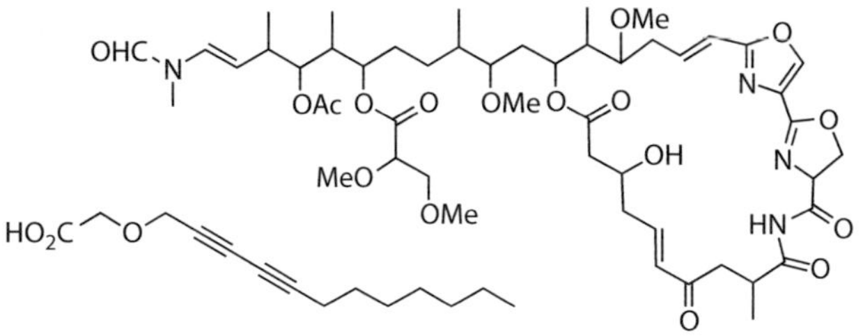

montiporic acid A (59)   mycalolide D (58)

FIGURE 17.3   Cnidaria (soft and hard coral) metabolites.

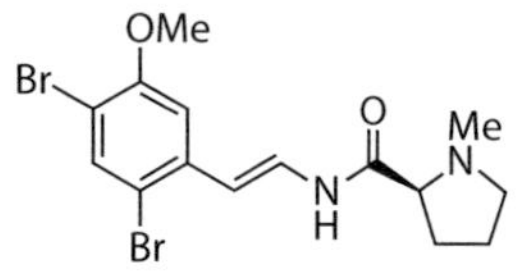

amathamide A (60) 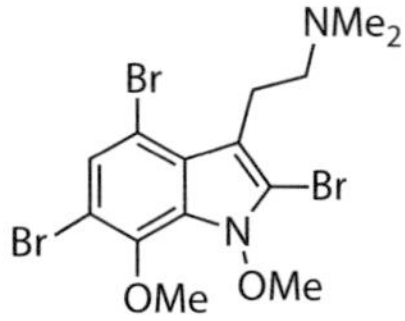   convolutindole A (61)

FIGURE 17.4   Bryozoa (lace coral) metabolites.

of Queensland, and is synonymous with the chemistry of pigments. The first contribution to Australian crinoid chemistry described hydroxyanthraquinones (i.e., **62**)
from a Moreton Bay collection of *Comatula pectinata* (Sutherland and Wells, 1959),
and the last, over a quarter century later, was also by Sutherland, on bianthrones
(i.e., **63**) from a Moreton Bay collection of *Lamprometra palmate* (Rideout and
Sutherland, 1985) (Figure 17.5).

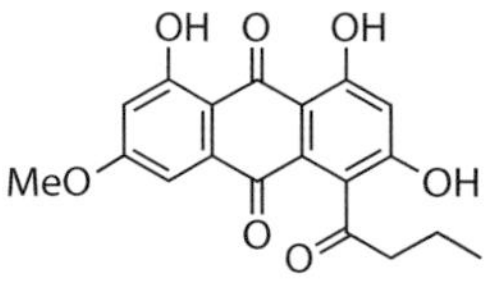

rhodocomatulin 6 methyl ether (**62**) 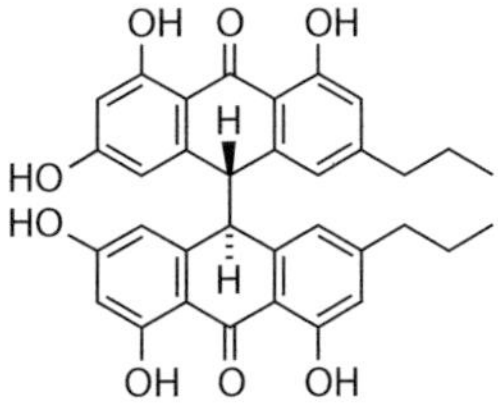   (**63**)

FIGURE 17.5   Echinodermata (sea and feather star) metabolites.

17.7   MOLLUSCA (GASTROPODS, CLAMS, SIPHONARIA, NUDIBRANCHES, OCTOPUS, CONE SHELLS)

Following the marine pigment theme, Sutherland et al. also launched Australian marine mollusc chemistry with a description of tyrindoxyl sulfate (**64**) from the gastropod *Dicathias orbita* as the *missing* biosynthetic source of tyrian purple (6,6′-dibromoindigotin) (Baker and Sutherland, 1968). It should be noted that while tyrian purple had its genesis as a Royal Dye several thousand years earlier (with historic accounts dating to 2000 BC), tyrian purple does not actually exist within the *Murex* mollusc, but is a photo and oxidative handling artifact. Sutherland's achievements were all the more impressive given that they predated many commonly used morden analytical technologies, such as high-performance liquid chromatography (HPLC) and high field Fourier transform nuclear magnetic resonance (FT NMR). An arsenic containing nucleoside (**65**) recovered from the kidney of the Western Australian giant clam *Tridacna maxima* (Edmonds et al., 1982, Francesconi et al., 1991) was implicated in the cycling of environmental arsenic. The air breathing mollusk, *Siphonaria denticulate*, collected from intertidal locations along the coast of New South Wales, yielded the polypropionate denticulatins (i.e., **66**) (Hochlowski et al., 1983), a forerunner of many new polypropionates to be isolated from Australian specimens of the subclass Pulmonata. The gastropod *Nerita albicilla* collected from One Tree Island, on the Great Barrier Reef, yielded the alkaloid isopteropodine (**67**) (Martin et al., 1986), while oxygenated diterpenes (i.e., **68**) were reported from a nudibranch *Chromodoris* sp. collected off Mooloolaba, Queensland (Yong et al., 2008). Turning to two iconic Australian venomous marine creatures, the blue-ring octopus *Hapalochlaena maculosa* was found to contain tetrodotoxin (**69**) (Sheumack et al., 1978), while many species of cone shells, *Conus* spp., have been shown to contain neurotoxic peptides (i.e., **70**) (Hu et al., 1998), some with high potency and selectivity for therapeutically important ion channels, making them excellent pharmacological scaffolds (Figure 17.6).

17.8   PHAEOPHYTA (BROWN ALGAE)

Farnesylacetone analogs (i.e., **71**) from *Cystophora moniliformis* were noted for their structural similarity to insect juvenile hormones (Kazlauskas et al., 1978a), while the antiinflammatory principle from *Caulocystis cephalornithos*, collected on the

tyrindoxyl sulfate (64) 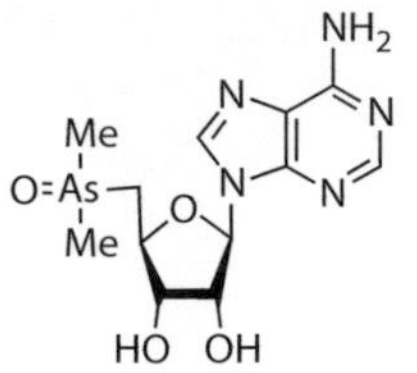   (65) 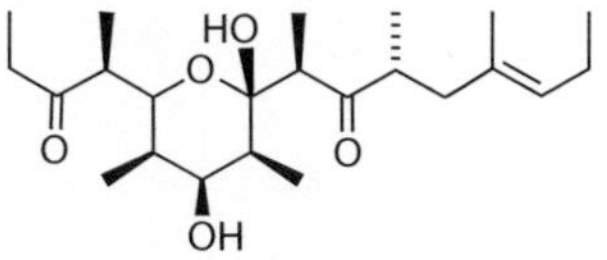   denticulatin A (66)

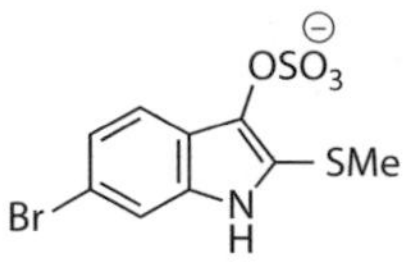

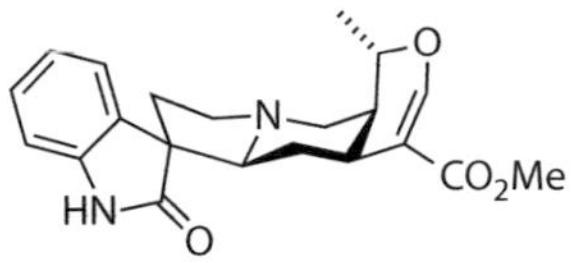

isopteropodine (67) 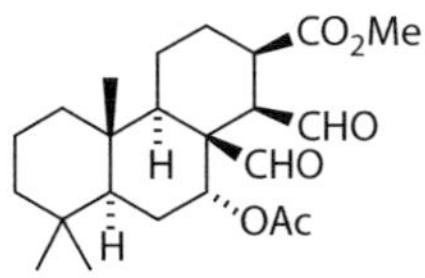   (68) 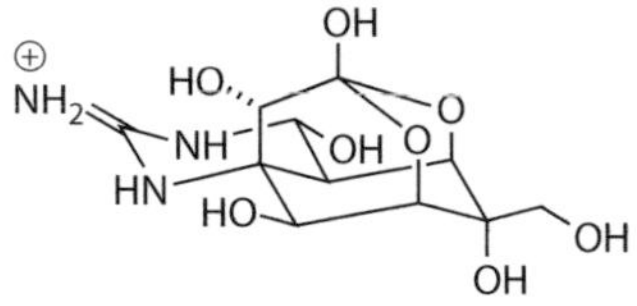   tetrodotoxin (69)

GCCSDPRCNMNNPDY(SO₃H)C-NH₂

α-conotoxin EpI (70)

FIGURE 17.6   Mollusca metabolites.

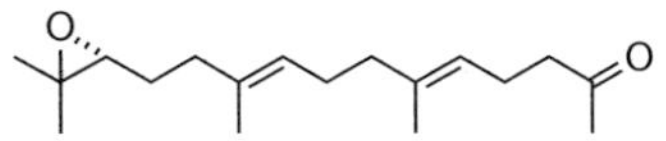

(71) 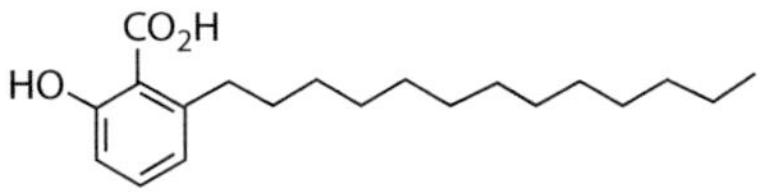   6-tridecylsalicylic acid (72)

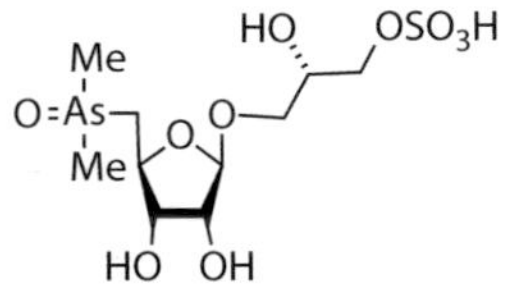

(73) 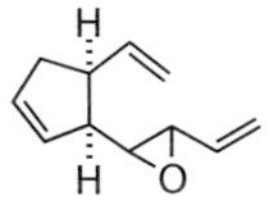   caudoxirene (74) 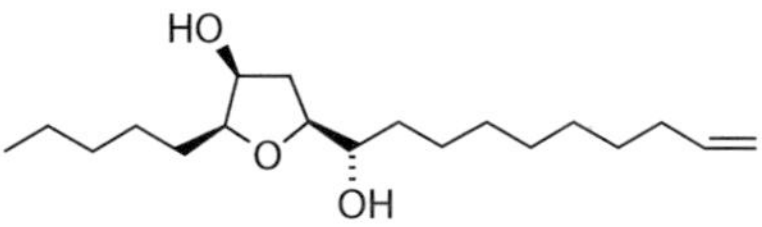   (75)

FIGURE 17.7   Phaeophyta (brown algae) metabolites.

coast of Victoria, was determined to be 6-tridecylsalicylic acid (**72**) (Kazlauskas et al., 1980b). The Western Australian kelp *Ecklonia radiata* yielded arseno-sugars (i.e., **73**) (Edmonds and Francesconi, 1981), while the spermatozoid-releasing and attracting factor in Victorian and Tasmanian collections of *Perithalia caudata* was identified as the epoxylipid caudoxirene (**74**) (Mueller et al., 1988). Epoxylipids (i.e., **75**) from Victorian collections of *Notheia anomala* were shown to possess promising nematocidal properties (Capon et al., 1998a,b) (Figure 17.7).

17.9   RHODOPHYTA (RED ALGAE)

Red algae are renowned for their ability to produce halogenated terpenes and acetogenins. The first report of a brominated diterpene from a marine red alga (and at that time only the second ever reported natural occurrence of a

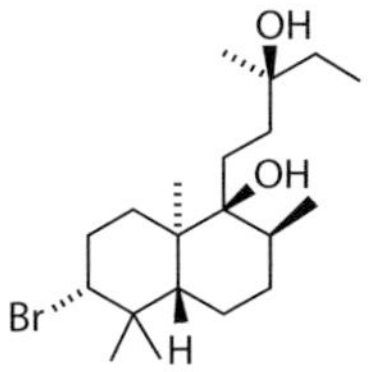

concinndiol (**76**) 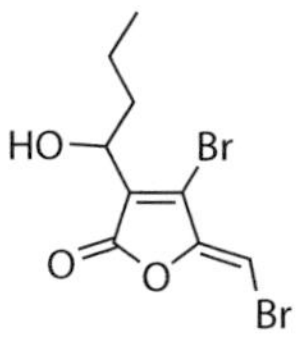   (**77**)

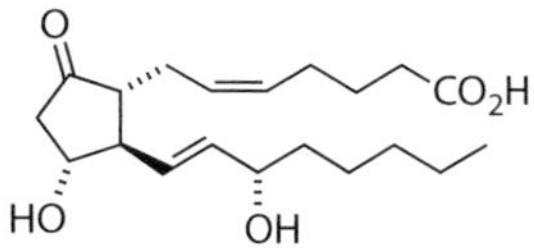

PGE$_2$ (**78**) 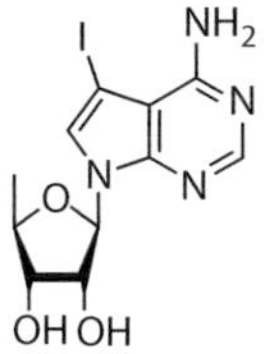   5′-deoxy-5-iodotubercidin (**79**)

FIGURE 17.8   Rhodophyta (red algae) metabolites.

brominated diterpene) was of concinndiol (**76**) from a New South Wales collection of *Laurencia concinna* (Sims et al., 1973). The bromolactone fimbrolides (i.e., **77**) reported from numerous Australian collections of *Delisea fimbriata* (Kazlauskas et al., 1977), went on to attract significant notoriety and application as *N*-acylhomoserine lactone mimics in the field of bacterial quorum sensing. Analysis of the antihypertensive extract derived from a Victorian collection of *Gracilaria lichenoides* revealed known prostaglandins, including PGE$_2$ (**78**) (Gregson et al., 1979), while an extract derived from a Western Australian collection of *Hypnea valendiae* produced pronounced muscle relaxation and hypothermia in mice, attributed to 5′-deoxy-5-iodotubercidin (**79**) (Kazlauskas et al., 1983) (Figure 17.8).

17.10   CHLOROPHYTA (GREEN ALGAE)

The chemistry of Australian green alga is largely associated with that of various species and collections of the genus *Caulerpa*. The *Caulerpa* are well known for their ability to produce 1,4-diacetoxybuta-1,3-diene, a functionality viewed as a masked 1,4-dialdehyde, which can exert biological effects by forming Schiff's bases to key amine residues. The first report of this structure class included flexilin (**80**) from a Tasmanian collection of *Caulerpa flexilis* (Blackman and Wells, 1978). Although many related terpenes have been reported from other Australian *Caulerpa* species, a selection of Western Australian *Caulerpa* lacked significant quantities of terpenes but did produce the pigment caulerpin (**81**) (Capon et al., 1983b). The novel bicyclic lipid dictyosphaerin (**82**) was isolated from southern Australian collections of *Dictyosphaeria sericea* (Rochfort et al., 1996), while vanillic acid protein tyrosine phosphatase 1B inhibitors (i.e., **83**) were isolated from a Queensland sample of *Cladophora socialis* (Feng et al., 2007) (Figure 17.9).

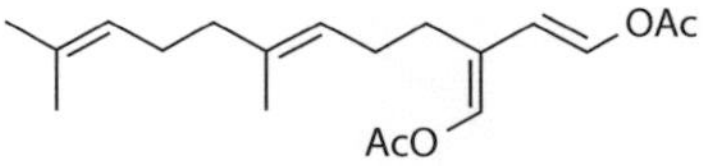

flexilin (**80**)

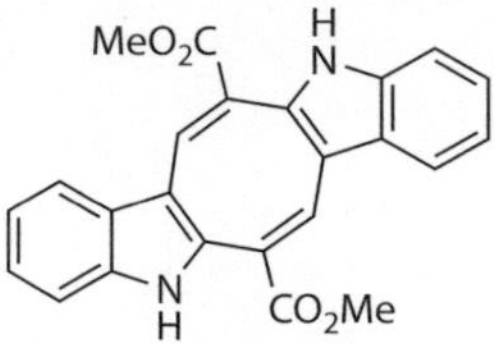

caulerpin (**81**)

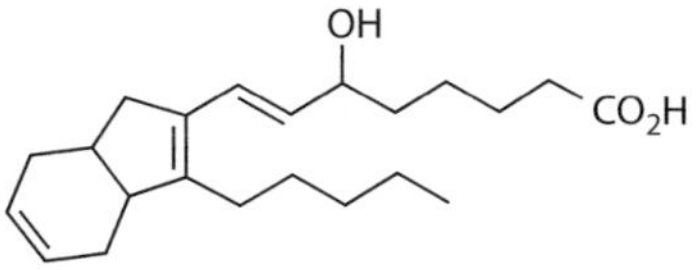

dictyosphaerin (**82**)

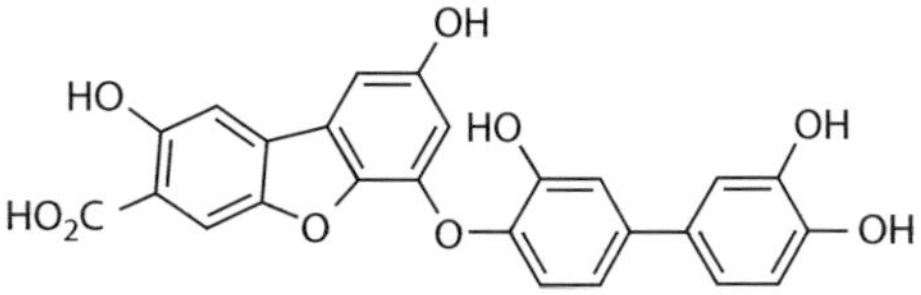

(**83**)

FIGURE 17.9 Chlorophyta (green algae) metabolites.

17.11 CYANOBACTERIA (BLUE-GREEN ALGAE)

Knowledge of the chemistry of Australian marine blue-green algae is limited but includes antiinflammatory polybrominated bisindoles (i.e., **84**) from Victorian collections of *Rivularia firma* (Norton and Wells, 1982). The characterization of poly-3-hydroxypentanoates (i.e., **85**) from Western Australian hypersaline blue-green algal mats (Capon et al., 1983a) was also noteworthy in that it predated by a decade their commercial production and use as bioplastics and biomaterials (Figure 17.10).

17.12 MARINE-DERIVED BACTERIA AND FUNGI

The first account of natural products from Australian marine-derived bacteria can be attributed to Russian researchers who reported surfactin-like cyclic depsipeptides (i.e., **86**) from an isolate of *Bacillus pumilus* recovered from a sponge

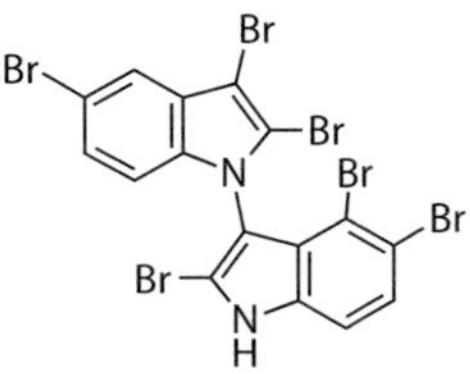

(**84**)

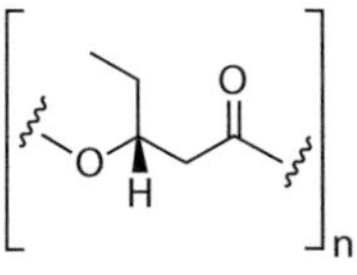

(**85**)

FIGURE 17.10 Cyanobacteria (blue-green algae) metabolites.

Ircinia sp. collected off the Great Barrier Reef (Kalinovskaya et al., 1995). Since then, U.S. researchers have described plant growth promoting diketopiperazines (i.e., **87**) from an undescribed bacterial isolate obtained from a Great Barrier Reef *Palythoa* sp. (Cronan et al., 1998), and a team of German and Russian researchers have described the cytotoxic bromoalterochromides (i.e., **88**) from an isolate of *Pseudoalteromonas maricaloris* obtained from a Great Barrier Reef specimen of the sponge *Fascaplysinopsis reticulata* (Speitling et al., 2007). The first report on natural products from Australian marine-derived fungi appeared as recently as 2003, with an account of the antiparasitic alkaloid marcfortine A (**89**) and depsipeptide aspergillicins (i.e., **90**) from a Tasmanian estuarine isolate of *Aspergillus carneus* (Capon et al., 2003). Later reports described lipodepsipeptide acremolides (i.e., **91**) and cyclopentapeptide cotteslosins (i.e., **92**) from Tasmanian and Western Australian isolates of an *Acremonium* sp. (Ratnayake et al., 2008) and *Aspergillus versicolor* (Fremlin et al., 2009), respectively (Figure 17.11).

17.13 PAST, PRESENT, AND FUTURE: A PERSONAL VIEWPOINT

The early history of Australian marine natural products chemistry was dominated by a collegiate community of academics pursuing largely curiosity driven research, targeting metabolites with novel chemical structures some of which were attributed interesting and potentially valuable biological properties. This golden age of enquiry science was pleasantly uncomplicated, and was a time when new discoveries lay scattered along the beach (literally). Many researchers were drawn into this new frontier of marine natural products chemistry, where they survived and prospered. In more recent years, as the field matured, aspirations and expectations were realigned such that biological activity and drug discovery became the prism through which much success was measured—both in terms of research funding (grants, contracts) and outcomes (papers, patents). In keeping with this shift in focus, collaborations across disciplinary boundaries (chemistry, microbiology, ecology, pharmacology, cell biology, molecular biology, bioinfomatics, etc.), awareness of and attention to market and community need (therapeutics, agrochemicals, etc.), and engagement with industry, became the norm. The resulting goal-oriented applied science of marine biodiscovery generated significant successes (as illustrated by some examples listed above), but not without challenges. In addition to the technical, intellectual, and funding imperatives that define modern science, today's marine biodiscovery researchers must contend with the dichotomy of publish (share knowledge) versus patent (protect knowledge), restrictions on the collection and transport of biodiversity, and legislative requirements for benefit sharing at local, national, and international levels. Researchers must also guard against premature disclosure and loss of intellectual property through the use of such instruments as commercial-in-confidence and material transfer agreements—introducing a level of bureaucratic oversight that can be viewed as both constructive and helpful, and also intrusive and distracting. Such are the pressures on modern marine biodiscovery that only a handful of Australian academic laboratories maintain an active presence in this field, far less than was the case only a few years earlier.

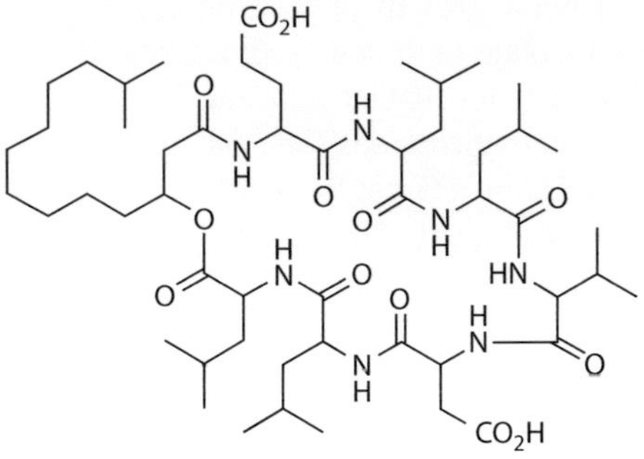

bacirine 3 (**86**)

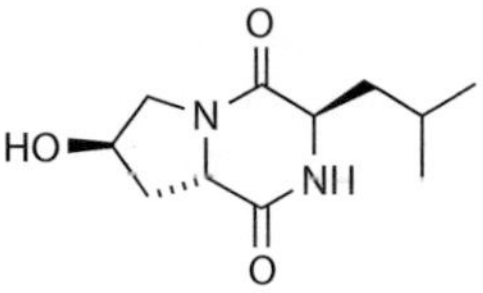

(**87**)

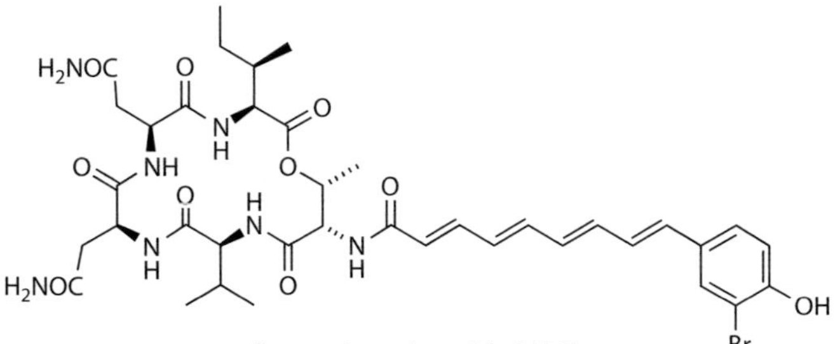

bromoalterochromide A (**88**)

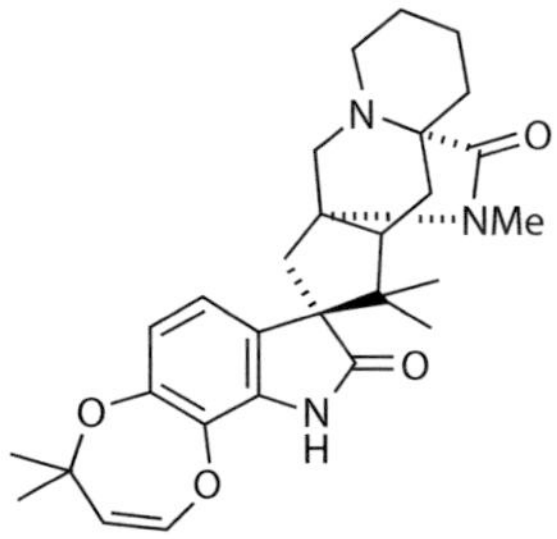

marcfortine A (**89**)

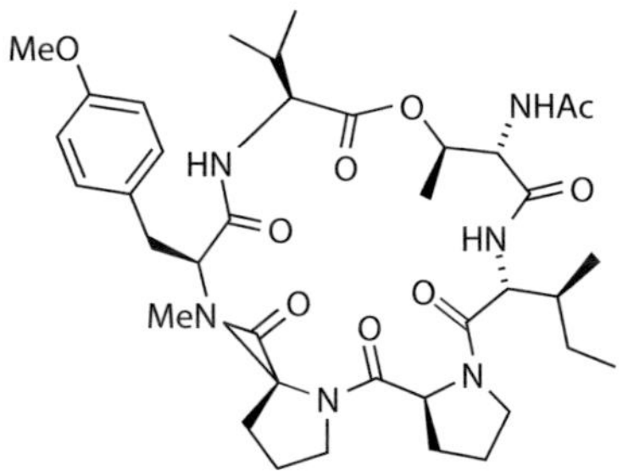

aspergillicin A (**90**)

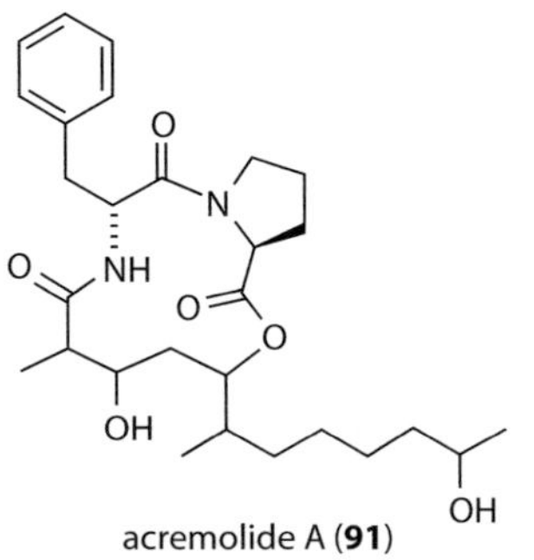

acremolide A (**91**)

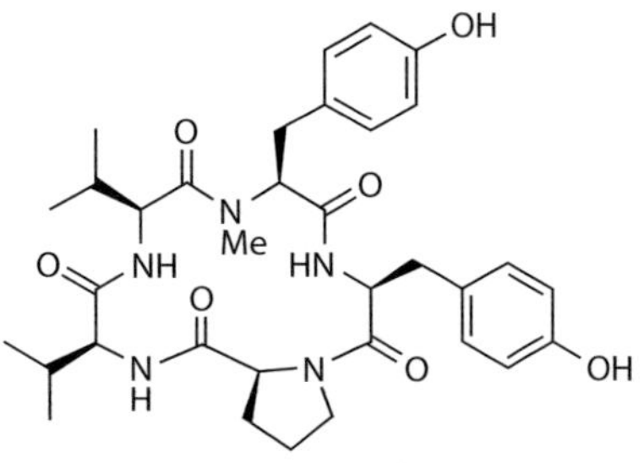

cotteslosin A (**92**)

FIGURE 17.11 Marine-derived bacteria and fungi metabolites.

The history of bioactive metabolites from Australian marine organisms has been one of great productivity and success, with significant contributions from numerous Australian and international laboratories and researchers. Building on these achievements, and with extraordinary reserves of unexplored marine biodiversity, the future for Australian marine biodiscovery could be exceptionally bright, provided talented young researchers are properly mentored and supported, and the field in general attracts appropriate levels of investment. It is the view of the author that this will be best achieved through collaboration and networking, and a modest shift in the "commercialization" pendulum back toward a more collegiate culture, such as that which served science and society so well in the past. That Australian marine biodiversity remains a valuable resource rich in biologically active metabolites, featuring new molecular motifs, scaffolds, and pharmacophores, is incontestable. How successful future researchers will be in realizing this potential remains to be seen, but we can be cautiously optimistic.

REFERENCES

Baker, J. T. and Sutherland, M. D. 1968. Pigments of marine animals. 8. Precursors of 6,6'-dibromoindigotin (tyrian purple) from mollusc *Dicathais orbita* Gmelin. *Tetrahedron Lett.*, 43–46.

Blackman, A. J. and Matthews, D. J. 1985. Amathamide alkaloids from the marine bryozoan *Amathia wilsoni* Kirkpatrick. *Heterocycles, 23*, 2829–2833.

Blackman, A. J. and Wells, R. J. 1978. Flexilin and trifarin, terpene 1,4-diacetoxybuta-1,3-dienes from two *Caulerpa* species (Chlorophyta). *Tetrahedron Lett.*, 3063–3064.

Buchanan, M. S., Carroll, A. R., Addepalli, R., Avery, V. M., Hooper, J. N. A., and Quinn, R. J. 2007a. Natural products, stylissadines A and B, specific antagonists of the P2X$_7$ receptor, an important inflammatory target. *J. Org. Chem., 72*, 2309–2317.

Buchanan, M. S., Carroll, A. R., Fechner, G. A., Boyle, A., Simpson, M. M., Addepalli, R., Avery, V. M., Hooper, J. N. A., Cheung, T., Chen, H. W., and Quinn, R. J. 2008a. Aplysamine 6, an alkaloidal inhibitor of isoprenylcysteine carboxyl methyltransferase from the sponge *Pseudoceratina* sp. *J. Nat. Prod., 71*, 1066–1067.

Buchanan, M. S., Carroll, A. R., Fechner, G. A., Boyle, A., Simpson, M. M., Addepalli, R., Avery, V. M., Hooper, J. N. A., Su, N., Chen, H. W., and Quinn, R. J. 2007b. Spermatinamine, the first natural product inhibitor of isoprenylcysteine carboxyl methyltransferase, a new cancer target. *Bioorg. Med. Chem. Lett., 17*, 6860–6863.

Buchanan, M. S., Carroll, A. R., Wessling, D., Jobling, M., Avery, V. M., Davis, R. A., Feng, Y. J., Xue, Y. F., Oster, L., Fex, T., Deinum, J., Hooper, J. N. A., and Quinn, R. J. 2008b. Clavatadine A, a natural product with selective recognition and irreversible inhibition of Factor XIa. *J. Med. Chem., 51*, 3583–3587.

Butler, M. S. and Capon, R. J. 1992. The luffarins (A–Z), novel terpenes from an Australian marine sponge, *Luffariella geometrica. Aust. J. Chem., 45*, 1705–1743.

Capon, R. J., Barrow, R. A., Rochfort, S., Jobling, M., Skene, C., Lacey, E., Gill, J. H., Friedel, T., and Wadsworth, D. 1998a. Marine nematocides: Tetrahydrofurans from a southern Australian brown alga, *Notheia anomala. Tetrahedron, 54*, 2227–2242.

Capon, R. J., Dunlop, R. W., Ghisalberti, E. L., and Jefferies, P. R. 1983a. Poly-3-hydroxyalkanoates from marine and fresh-water cyanobacteria. *Phytochemistry, 22*, 1181–1184.

Capon, R. J. and Faulkner, D. J. 1985. Herbasterol, an ichthyotoxic 9,11-secosterol from the sponge *Dysidea herbacea. J. Org. Chem., 50*, 4771–4773.

Capon, R. J., Ghisalberti, E. L., and Jefferies, P. R. 1983b. Metabolites of the green algae, *Caulerpa* species. *Phytochemistry, 22*, 1465–1467.

Capon, R. J. and Macleod, J. K. 1985. Structural and stereochemical studies on marine norterpene cyclic peroxides. *Tetrahedron, 41*, 3391–3404.

Capon, R. J. and Macleod, J. K. 1987. 5-thio-D-mannose from the marine sponge *Clathria pyramida* (Lendenfeld)—The first example of a naturally occurring 5-thiosugar. *J. Chem. Soc. Chem. Comm.*, 1200–1201.

Capon, R. J., Macleod, J. K., and Scammells, P. J. 1986. The trikentrins—Novel indoles from the sponge *Trikentrion flabelliforme. Tetrahedron, 42*, 6545–6550.

Capon, R. J., Macleod, J. K., and Willis, A. C. 1987. Trunculin A and trunculin B, norsesterterpene cyclic peroxides from a marine sponge, *Latrunculia brevis. J. Org. Chem., 52*, 339–342.

Capon, R. J., Peng, C., and Dooms, C. 2008. Trachycladindoles A–G: Cytotoxic heterocycles from an Australian marine sponge, *Trachycladus laevispirulifer. Org. Biomol. Chem., 6*, 2765–2771.

Capon, R. J., Rooney, F., Murray, L. M., Collins, E., Sim, A. T. R., Rostas, J. A. P., Butler, M. S., and Carroll, A. R. 1998b. Dragmacidins: New protein phosphatase inhibitors from a southern Australian deep-water marine sponge, *Spongosorites* sp. *J. Nat. Prod., 61*, 660–662.

Capon, R. J., Skene, C., Lacey, E., Gill, J. H., Wadsworth, D., and Friedel, T. 1999. Geodin A magnesium salt: A novel nematocide from a southern Australian marine sponge, *Geodia. J. Nat. Prod., 62*, 1256–1259.

Capon, R. J., Skene, C., Liu, E. H. T., Lacey, E., Gill, J. H., Heiland, K., and Friedel, T. 2001. The isolation and synthesis of novel nematocidal dithiocyanates from an Australian marine sponge, *Oceanapia* sp. *J. Org. Chem., 66*, 7765–7769.

Capon, R. J., Skene, C., Liu, E. H. T., Lacey, E., Gill, J. H., Heiland, K., and Friedel, T. 2004. Esmodil: An acetylcholine mimetic resurfaces in a southern Australian marine sponge *Raspailia (Raspailia)* sp. *Nat. Prod. Res., 18*, 305–309.

Capon, R. J., Skene, C., Stewart, M., Ford, J., O'Hair, R. A. J., Williams, L., Lacey, E., Gill, J. H., Heiland, K., and Friedel, T. 2003. Aspergillicins A–E: Five novel depsipeptides from the marine-derived fungus *Aspergillus carneus. Org. Biomol. Chem., 1*, 1856–1862.

Capon, R. J., Skene, C., Vuong, D., Lacey, E., Gill, J. H., Heiland, K., and Friedel, T. 2002. Equilibrating isomers: Bromoindoles and a seco-xanthine encountered during a study of nematocides from the southern Australian marine sponge *Hymeniacidon* sp. *J. Nat. Prod., 65*, 368–370.

Capon, R. J. and Trotter, N. S. 2005. N-3,5'-cycloxanthosine, the first natural occurrence of a cyclonucleoside. *J. Nat. Prod., 68*, 1689–1691.

Capon, R. J., Vuong, D., Lacey, E., and Gill, J. H. 2005a. (-)-Echinobetaine A: Isolation, structure elucidation, synthesis, and SAR studies on a new nematocide from a southern Australian marine sponge, *Echinodictyum* sp. *J. Nat. Prod., 68*, 179–182.

Capon, R. J., Vuong, D., Mcnally, M., Peterle, T., Trotter, N., Lacey, E., and Gill, J. H. 2005b. (+)-Echinobetaine B: Isolation, structure elucidation, synthesis and preliminary SAR studies on a new nematocidal betaine from a southern Australian marine sponge, *Echinodictyum* sp. *Org. Biomol. Chem., 3*, 118–122.

Carroll, A. R., Bowden, B. F., and Coll, J. C. 1993a. Studies of Australian ascidians. 2. novel cytotoxic iodotyrosine-based alkaloids from colonial ascidians, *Aplidium* sp. *Aust. J. Chem., 46*, 825–832.

Carroll, A. R., Bowden, B. F., and Coll, J. C. 1993b. Studies of Australian ascidians. 3. A new tetrahydrocannabinol derivative from the ascidian *Synoicum castellatum. Aust. J. Chem., 46*, 1079–1083.

Carroll, A. R., Ngo, A., Quinn, R. J., Redburn, J., and Hooper, J. N. A. 2005. Petrosamine B, an inhibitor of the *Helicobacter pylori* enzyme aspartyl semialdehyde dehydrogenase from the Australian sponge *Oceanapia* sp. *J. Nat. Prod., 68*, 804–806.

Carroll, A. R., Pierens, G. K., Fechner, G., De Almeida Leone, P., Ngo, A., Simpson, M., Hyde, E., Hooper, J. N. A., Bostrom, S. L., Musil, D., and Quinn, R. J. 2002. Dysinosin A: A novel inhibitor of factor VIIa and thrombin from a new genus and species of Australian sponge of the family *Dysideidae. J. Am. Chem. Soc., 124*, 13340–13341.

Coll, J. C., Hawes, G. B., Liyanage, N., Oberhansli, W., and Wells, R. J. 1977. Studies of Australian soft corals. 1. New cembrenoid diterpene from a *Sarcophyton* species. *Aust. J. Chem., 30*, 1305–1309.

Cook, A. F., Bartlett, R. T., Gregson, R. P., and Quinn, R. J. 1980. 1-methylisoguanosine, a pharmacologically active agent from a marine sponge. *J. Org. Chem., 45*, 4020–4025.

Cronan, J. M., Davidson, T. R., Singleton, F. L., Colwell, R. R., and Cardellina, J. H. 1998. Plant growth promoters isolated from a marine bacterium associated with *Palythoa* sp. *Nat. Prod. Lett., 11*, 271–278.

Degnan, B. M., Hawkins, C. J., Lavin, M. F., Mccaffrey, E. J., Parry, D. L., Vandenbrenk, A. L., and Watters, D. J. 1989. New cyclic-peptides with cytotoxic activity from the ascidian *Lissoclinum patella. J. Med. Chem., 32*, 1349–1354.

Edmonds, J. S. and Francesconi, K. A. 1981. Arseno-sugars from brown kelp (*Ecklonia radiata*) as intermediates in cycling of arsenic in a marine ecosystem. *Nature, 289*, 602–604.

Edmonds, J. S., Francesconi, K. A., Healy, P. C., and White, A. H. 1982. Isolation and crystal-structure of arsenic-containing sugar sulfate from the kidney of the giant clam, *Tridacna maxima*—X-ray crystal-structure of (2S)-3-[5-deoxy-5-(dimethylarsinoyl)-β-D-ribofuranosyloxy]-2-hydroxypropyl hydrogen sulfate. *J. Chem. Soc. Perkin Trans. 1*, 2989–2993.

El-Naggar, M., Piggott, A. M., and Capon, R. J. 2008. Bistellettazines A–C and bistellettazole A: New terpenyl-pyrrolizidine and terpenyl-imidazole alkaloids from a southern Australian marine sponge, *Stelletta* sp. *Org. Lett., 10*, 4247–4250.

Erickson, K. L., Beutler, J. A., Cardellina, J. H., and Boyd, M. R. 1997. Salicylihalamides A and B, novel cytotoxic macrolides from the marine sponge *Haliclona* sp. *J. Org. Chem., 62*, 8188–8192.

Feng, Y. J., Carroll, A. R., Addepalli, R., Fechner, G. A., Avery, V. M., and Quinn, R. J. 2007. Vanillic acid derivatives from the green algae *Cladophora socialis* as potent protein tyrosine phosphatase 1B inhibitors. *J. Nat. Prod., 70*, 1790–1792.

Francesconi, K. A., Stick, R. V., and Edmonds, J. S. 1991. An arsenic-containing nucleoside from the kidney of the giant clam, *Tridacna maxima. J. Chem. Soc. Chem. Commun.*, 928–929.

Fremlin, L. J., Piggott, A. M., Lacey, E., and Capon, R. J. 2009. Cottoquinazoline A and cotteslosins A and B, metabolites from an Australian marine-derived strain of *Aspergillus versicolor. J. Nat. Prod., 72*, 666–670.

Fusetani, N., Toyoda, T., Asai, N., Matsunaga, S., and Maruyama, T. 1996. Montiporic acids A and B, cytotoxic and antimicrobial polyacetylene carboxylic acids from eggs of the scleractinian coral *Montipora digitata. J. Nat. Prod., 59*, 796–797.

Gregson, R. P., Marwood, J. F., and Quinn, R. J. 1979. Occurrence of prostaglandins-PGE$_2$ and prostaglandins-PGF$_{2\alpha}$ in a plant—Red alga *Gracilaria lichenoides. Tetrahedron Lett., 20*, 4505–4506.

Hochlowski, J. E., Faulkner, D. J., Matsumoto, G. K., and Clardy, J. 1983. The denticulatins, two polypropionate metabolites from the pulmonate *Siphonaria denticulata. J. Am. Chem. Soc., 105*, 7413–7415.

Hong, T. W., Jimenez, D. R., and Molinski, T. F. 1998. Agelastatins C and D, new pentacyclic bromopyrroles from the sponge *Cymbastela* sp., and potent arthropod toxicity of (-)-agelastatin A. *J. Nat. Prod., 61*, 158–161.

Hu, S. H., Loughnan, M., Miller, R., Weeks, C. M., Blessing, R. H., Alewood, P. F., Lewis, R. J., and Martin, J. L. 1998. The 1.1 angstrom resolution crystal structure of [Tyr(15)]EpI, a novel alpha-conotoxin from *Conus episcopatus*, solved by direct methods. *Biochemistry, 37*, 11425–11433.

Jacob, M. R., Hossain, C. F., Mohammed, K. A., Smillie, T. J., Clark, A. M., Walker, L. A., and Nagle, D. G. 2003. Reversal of fluconazole resistance in multidrug efflux-resistant fungi by the *Dysidea arenaria* sponge sterol 9α,11α-epoxycholest-7-ene-3β,5α,6α,19-tetrol 6-acetate. *J. Nat. Prod., 66*, 1618–1622.

Kalinovskaya, N. I., Kuznetsova, T. A., Rashkes, Y. V., Milgrom, Y. M., Milgrom, E. G., Willis, R. H., Wood, A. I., Kurtz, H. A., Carabedian, C., Murphy, P., and Elyakov, G. B. 1995. Surfactin-like structures of five cyclic depsipeptides from the marine isolate of *Bacillus pumilus. Russ. Chem. Bull., 44*, 951–955.

Kang, H. and Fenical, W. 1996. Polycarpine dihydrochloride: A cytotoxic dimeric disulfide alkaloid from the Indian Ocean ascidian *Polycarpa clavata. Tetrahedron Lett., 37*, 2369–2372.

Kang, H., Jensen, P. R., and Fenical, W. 1996. Isolation of microbial antibiotics from a marine ascidian of the genus *Didemnum. J. Org. Chem., 61*, 1543–1546.

Kang, H. J. and Fenical, W. 1997a. Aplidiamine, a unique zwitterionic benzyl hydroxyadenine from the Western Australian marine ascidian *Aplidiopsis* sp. *Tetrahedron Lett., 38*, 941–944.

Kang, H. J. and Fenical, W. 1997b. Ningalins A–D: Novel aromatic alkaloids from a Western Australian ascidian of the genus *Didemnum. J. Org. Chem., 62*, 3254–3262.

Kazlauskas, R., Lidgard, R. O., Murphy, P. T., Wells, R. J., and Blount, J. F. 1981. Brominated tyrosine-derived metabolites from the sponge *Ianthella basta. Aust. J. Chem., 34*, 765–786.

Kazlauskas, R., Marwood, J. F., and Wells, R. J. 1980a. 2-Phenylethylamides of a novel lipid acid—Atrial stimulants from the soft coral *Sinularia flexibilis. Aust. J. Chem., 33*, 1799–1803.

Kazlauskas, R., Marwood, J. F., Murphy, P. T., and Wells, R. J. 1982. A blue pigment from a compound ascidian. *Aust. J. Chem., 35*, 215–217.

Kazlauskas, R., Mulder, J., Murphy, P. T., and Wells, R. J. 1980b. New metabolites from the brown alga *Caulocystis cephalornithos. Aust. J. Chem., 33*, 2097–2101.

Kazlauskas, R., Murphy, P. T., Quinn, R. J., and Wells, R. J. 1977. New class of halogenated lactones from red alga *Delisea fimbriata* (Bonnemaisoniaceae). *Tetrahedron Lett., 1*, 37–40.

Kazlauskas, R., Murphy, P. T., and Wells, R. J. 1978a. Two derivatives of farnesylacetone from brown alga *Cystophora moniliformis. Experientia, 34*, 156–157.

Kazlauskas, R., Murphy, P. T., Wells, R. J., Bairdlambert, J. A., and Jamieson, D. D. 1983. Halogenated pyrrolo[2,3-D]pyrimidine nucleosides from marine organisms. *Aust. J. Chem., 36*, 165–170.

Kazlauskas, R., Murphy, P. T., Wells, R. J., Schonholzer, P., and Coll, J. C. 1978b. Cembranoid constituents from an Australian collection of soft coral *Sinularia flexibilis. Aust. J. Chem., 31*, 1817–1824.

Kehraus, S., Gorzalka, S., Hallmen, C., Iqbal, J., Muller, C. E., Wright, A. D., Wiese, M., and Konig, G. M. 2004. Novel amino acid derived natural products from the ascidian *Atriolum robustum*: Identification and pharmacological characterization of a unique adenosine derivative. *J. Med. Chem., 47*, 2243–2255.

Killday, K. B., Yarwood, D., Sills, M. A., Murphy, P. T., Hooper, J. N. A., and Wright, A. E. 2001. Microxine, a new cdk2 kinase inhibitor from the Australian marine sponge *Microxina* species. *J. Nat. Prod., 64*, 525–526.

Lam, H. W., Cooke, P. A., Pattenden, G., Bandaranayake, W. M., and Wickramasinghe, W. A. 1999. Structure and total synthesis of benzylthiocrellidone, a novel dimedone-based vinyl sulfide from the sponge *Crella spinulata. J. Chem. Soc. Perkin Trans., 1*, 847–848.

Lindel, T., Jensen, P. R., Fenical, W., Long, B. H., Casazza, A. M., Carboni, J., and Fairchild, C. R. 1997. Eleutherobin, a new cytotoxin that mimics paclitaxel (Taxol) by stabilizing microtubules. *J. Am. Chem. Soc., 119*, 8744–8745.

Martin, G. E., Sanduja, R., and Alam, M. 1986. Isolation of isopteropodine from the marine mollusk *Nerita albicilla*—Establishment of the structure via two-dimensional NMR techniques. *J. Nat. Prod., 49*, 406–411.

Mueller, D. G., Boland, W., Becker, U., and Wahl, T. 1988. Caudoxirene the spermatozoid-releasing and attracting factor in the marine brown alga *Perithalia caudata* Phaeophyceae Sporochnales. *Biol. Chem. Hoppe-Seyler, 369*, 655–660.

Nakagawa, M., Endo, M., Tanaka, N., and Lee, G. P. 1984. Structures of xestospongin A, xestospongin B, xestospongin C and xestospongin D, novel vasodilative compounds from marine sponge, *Xestospongia exigua. Tetrahedron Lett., 25*, 3227–3230.

Narkowicz, C. K., Blackman, A. J., Lacey, E., Gill, J. H., and Heiland, K. 2002. Convolutindole A and convolutamine H, new nematocidal brominated alkaloids from the marine bryozoan *Amathia convoluta. J. Nat. Prod., 65*, 938–941.

Nemoto, T., Yoshino, G., Ojika, M., and Sakagami, Y. 1997. Amphimic acids and related long-chain fatty acids as DNA topoisomerase I inhibitors from an Australian sponge, *Amphimedon* sp.: Isolation, structure, synthesis, and biological evaluation. *Tetrahedron, 53*, 16699–16710.

Nicholas, G. M., Newton, G. L., Fahey, R. C., and Bewley, C. A. 2001. Novel bromotyrosine alkaloids: Inhibitors of mycothiol S-conjugate amidase. *Org. Lett., 33*, 1543–1545.

Norton, R. S. and Wells, R. J. 1982. A series of chiral polybrominated biindoles from the marine blue-green alga *Rivulaira firma*. Application of ^{13}C NMR spin-lattice relaxation data and ^{13}C-^{1}H coupling constants to structure elucidation. *J. Am. Chem. Soc., 104*, 3628–3635.

Okamoto, Y., Ojika, M., and Sakagami, Y. 1999. Iantheran A, a dimeric polybrominated benzofuran as a Na,K-ATPase inhibitor from a marine sponge, *Ianthella* sp. *Tetrahedron Lett., 40*, 507–510.

Ovenden, S. P. B. and Capon, R. J. 1999. Nuapapuin A and sigmosceptrellins D and E: New norterpene cyclic peroxides from a southern Australian marine sponge, *Sigmosceptrella* sp. *J. Nat. Prod., 62*, 214–218.

Ovenden, S. P. B., Capon, R. J., Lacey, E., Gill, J. H., Friedel, T., and Wadsworth, D. 1999. Amphilactams A–D: Novel nematocides from southern Australian marine sponges of the genus *Amphimedon. J. Org. Chem., 64*, 1140–1144.

Rashid, M. A., Gustafson, K. R., Cardellina, J. H., and Boyd, M. R. 1995. Mycalolide D and mycalolide E, new cytotoxic macrolides from a collection of the stony coral *Tubastrea faulkneri. J. Nat. Prod., 58*, 1120–1125.

Ratnayake, R., Fremlin, L. J., Lacey, E., Gill, J. H., and Capon, R. J. 2008. Acremolides A–D, lipodepsipeptides from an Australian marine-derived fungus, *Acremonium* sp. *J. Nat. Prod., 71*, 403–408.

Rideout, J. A. and Sutherland, M. D. 1985. Pigments of marine animals. 15. Bianthrones and related polyketides from *Lamprometra palmata gyges* and other species of crinoids. *Aust. J. Chem., 38*, 793–808.

Rochfort, S. J., Watson, R., and Capon, R. J. 1996. Dictyosphaerin: A novel bicyclic lipid from a southern Australian marine green algae, *Dictyosphaeria sericea. J. Nat. Prod., 59*, 1154–1156.

Searle, P. A. and Molinski, T. F. 1994. Five new alkaloids from the tropical ascidian, *Lissoclinum* sp—Lissoclinotoxin A is chiral. *J. Org. Chem., 59*, 6600–6605.

Searle, P. A. and Molinski, T. F. 1995. Phorboxazole A and phorboxazole B—Potent cytostatic macrolides from marine sponge *Phorbas* sp. *J. Am. Chem. Soc., 117*, 8126–8131.

Sharma, G. M. and Burkholder, P. R. 1971. Structure of di-bromo phakellin a new bromine containing alkaloid from the marine sponge *Phakellia flabellata. J. Chem. Soc. Chem. Comm.*, 151–152.

Sheumack, D. D., Howden, M. E. H., Spence, I., and Quinn, R. J. 1978. Maculotoxin—Neurotoxin from venom glands of octopus *Hapalochlaena maculosa* identified as tetrodotoxin. *Science, 199*, 188–189.

Sims, J. J., Lin, G. H. Y., Wing, R. M., and Fenical, W. 1973. Concinndiol, a bromo-diterpene alcohol from red alga, *Laurencia concinna. J. Chem. Soc. Chem. Comm.*, 470–471.

Speitling, M., Smetanina, O. E., Kuznetsova, T. A., and Laatsch, H. 2007. Marine bacteria. XXXV. Bromoalterochromides A and A′, unprecedented chromopeptides from a marine *Pseudoalteromonas maricaloris* strain KMM 636. *J. Antibiot., 60*, 36–42.

Sutherland, M. D. and Wells, J. W. 1959. Anthraquinone pigments from the crinoid *Comatula pectinata. Chem. Ind.,* 291–292.

Tapiolas, D. M., Bowden, B. F., Abou-Mansour, E., Willis, R. H., Doyle, J. R., Muirhead, A. N., Liptrot, C., Llewellyn, L. E., Wolff, C. W. W., Wright, A. D., and Motti, C. A. 2009. Eusynstyelamides A, B, and C, nNOS inhibitors, from the ascidian *Eusynstyela latericius. J. Nat. Prod., 72*, 1115–1120.

Taylor, K. M., Bairdlambert, J. A., Davis, P. A., and Spence, I. 1981. Methylaplysinopsin and other marine natural-products affecting neurotransmission. *Fed. Proc., 40*, 15–20.

Urban, S. and Capon, R. J. 1996. Lamellarin-S: A new aromatic metabolite from an Australian tunicate, *Didemnum* sp. *Aust. J. Chem., 49*, 711–713.

Wright, A. D., Goclik, E., Konig, G. M., and Kaminsky, R. 2002. Lepadins D–F: Antiplasmodial and antitrypanosomal decahydroquinoline derivatives from the tropical marine tunicate *Didemnum* sp. *J. Med. Chem., 45*, 3067–3072.

Wu Won, J. J., Rideout, J. A., and Chalker, B. E. 1995. Isolation and structure of a novel mycosporine-like amino acid from the reef-building corals *Pocillopora damicornis* and *Stylophora pistillata. Tetrahedron Lett., 36*, 5255–5256.

Yong, K. W. L., Salim, A. A., and Garson, M. J. 2008. New oxygenated diterpenes from an Australian nudibranch of the genus *Chromodoris. Tetrahedron, 64*, 6733–6738.

Zhang, H. and Capon, R. J. 2008. Phorbasins D–F: Diterpenyl-taurines from a southern Australian marine sponge, *Phorbas* sp. *Org. Lett., 10*, 1959–1962.

Natural Source Index

This index includes only the species which are sources of compounds cited in the book. The species cited for other reasons, such as microorganisms involved in metabolic engineering or bioassaying, are included in the Subject Index.

Subject Index

A

Aβ; *see* β-Amyloid
A12-2 (terpenoid antiviral agent), 174, 176
A-498 (renal tumor cells), 307
A549 (lung adenocarcinoma), 219, 321
A-10, 328
17-AAG, *see* 17-*N*-Allylamino-17-
demethoxygeldanamycin, 195
Ab initio
calculation of theoretical CD spectra, 157
in silico comparison, 392
molecular orbital (MO) calculation,
135, 139–140
theoretical methods, 321
ABCB1, 410, 413–414; *see also* P-glycoprotein
daunomycid efflux mediated by, 428
drug efflux mediated by, 415
flavonoids as modulators of, 412
inhibitors, 411, 418, 428
jathropane diterpenes as inhibitors of,
417–426
key pharmacophoric elements for the activity
of, 427
overexpression in acute myeloid
leukemia, 412
steroidal saponins as inhibitors of, 427
ABCC1, *see* MRP1
ABCG2 inhibitors, 428–433
ABC transporter, 410, 412, 417, 440–441
Abraxane (paclitaxel), 404
Absolute configuration (AC)
determination by CD methods, 138–163, 482
determination by modified Mosher's method,
468, 571
ACE (angiotensin converting enzyme), 122
Acetoacetyl-CoA:acetyl-CoA transferase, 75
Acetylbakuchiol, 247, 249
Acetylcholinesterase, 123, 498, 505–506
Acetylcholine, 499, 501, 503–507, 582
3′-*O*-Acetylhamaudol, 266–271
AchE, *see* Acetylcholinesterase
Ach, *see* Acetylcholine
Aclacinomycin A (aclarubicin), 241, 244, 463
Acridone, 434, 438–439
Actinomycin D, 192, 206
Actinophyllic acid, 154–155
Actinoplanes utahensis NRRL12052, 81
Actinorhodin, 73, 82
Actin polymerization inhibitors, 243, 245

Activity profiling, 111–112, 122
Acute myeloid (lymphoblastic) leukemia, 51
Acyclovir, 168
Acyl-CoA ligase, 79–80
Adefovir dipivoxil, 169–170
Adenylate cyclases, 454, 457, 486
ADME (adsorpion, distribution, metabolism,
excretion), 116–117, 122
Adriamycin; *see also* Doxorubicin
leukemia resistant to, 323
efflux from colon cancer cells, 412
as anti-CaM metabolite, 463
AD, *see* Alzheimer's disease
ADS, *see* Amorpha-4, 11-diene synthase
Afeletecan, 391
Aflatoxin conjugates efflux, 410
Agelastatin C, 585
Agrobacterium
mediated transformation, 87
rhizogenes, cultures transformed by, 402
tumefaciens, cultures transformed by, 87
AGS (human gastric adenocarcinoma),
217, 231, 249, 321
AIDS (acquired immunodeficiency syndrome)
CaM role in, 454
patients treated with atovaquone, 542
patients treated with valganciclovir, 170
Alanine scan, 18
Albafuran A, 225, 231
Alcaligenes sp lipase, 314
Alkaloids, 2, 8; *see also* related terms
as CaM inhibitors, 456, 467, 478–489
as HIF-1 inhibitors, 217, 244
configuration established by ECD, 152, 157
containing cyclobutane, 121
docking calculations, 122
from Australian marine organisms,
583, 587, 589
from *Cinchona* bark, 533
from engineered *Escherichia coli*, 74
from *Vinca* (*Catharanthus*) sp., 48, 218
17-*N*-Allylamino-17-demethoxygeldanamycin, 195
Allylic benzoate method, 149
Alpinumisoflavone, 237
Alvespimycin, 43
Alvocidib, 51
Alzheimer's disease, 340, 350, 363, 454, 497, 558
Amaranzole A, 150
Amathamide A, 590
9-Aminocamptothecin, 391